Characterisation of Porous Solids IV

Characterisation of Porous Solids IV

Edited by

B. McEnaney, T. J. Mays
University of Bath, UK

J. Rouquérol
CTM du CNRS, Marseille, France

F. Rodríguez-Reinoso
University of Alicante, Spain

K.S.W. Sing
Universities of Bristol and Exeter, UK

K.K. Unger
Johannes Gutenberg University, Mainz, Germany

THE ROYAL
SOCIETY OF
CHEMISTRY
Information
Services

Proceedings of the Fourth IUPAC Symposium on the Characterisation of Porous Solids, held at the University of Bath, UK, on 15–18 September 1996.

Special Publication No. 213

ISBN 0-85404-782-4

A catalogue record for this book is available from the British Library

Published by The Royal Society of Chemistry,
Thomas Graham House, Science Park, Milton Road,
Cambridge CB4 4WF, UK

Printed by MPG Books Ltd, Bodmin, Cornwall

Preface

The fourth in the series of Symposia on Characterisation of Porous Solids, COPS-IV, was held at the University of Bath, UK, on 15-18 September 1996. Previous Symposia in the series were held in Marseilles, France (1993), Alicante, Spain (1990) and Bad-Soden, Germany (1987). COPS-IV was sponsored by the International Union of Pure and Applied Chemistry and supported by the Particle Characterisation Group of the Royal Society of Chemistry and The British Carbon Group.

The regular series of COPS Symposia evolved from earlier meetings, starting with the Colston Symposium on Structure and Properties of Porous Materials held at the University of Bristol, UK, on 24-27 March 1958. The Proceedings of the Colston Symposium were edited by Professor DH Everett and Professor FS Stone and it was a great pleasure to be able to welcome both of them to COPS-IV. The Guest of Honour at the Symposium was Dr SJ Gregg of the University of Exeter, UK. Dr Gregg is the doyen of the subject and is well known as the co-author, with Professor KSW Sing, of the classic text 'Adsorption, Surface Area and Porosity'. At the Symposium dinner, Dr Gregg revealed that it was the 70th anniversary of the submission of his PhD thesis!

The local arrangements were made by Professor B McEnaney and Dr TJ Mays from the University and Mrs E Wellingham from Conference Secretariat who provided excellent administrative support. COPS-IV was the first campus-based Symposium in the COPS series and the arrangements appeared to work well, assisted by some wonderful autumn weather. In common with the earlier meetings, the Symposium was truly international in character with 186 delegates from 28 countries. The technical programme consisted of 160 papers comprising 31 oral presentations and 129 poster papers. The papers were selected by the Editors from those submitted. It is noteworthy that two oral paper sessions were devoted to molecular simulations of adsorption and mesoporous materials such as MCM-41, reflecting the considerable interest in these topics at the present time.

This book contains about 90 of the papers presented at the Symposium that have been refereed and edited by the Editors and others. The papers have been supplied by the authors as camera-ready copy. As with the previous Proceedings, the Editors hope that this book will provide a useful reference for developments in characterisation of porous solids until the next Symposium, COPS-V, which will be held in Heidelberg, Germany in 1999.

The Editors

Contents

THE KELVIN EQUATION AND ADSORPTION HYSTERESIS

A.J.Brown, C.G.V.Burgess, D.H.Everett[*] and S.Nuttall

School of Chemistry,
University of Bristol,
Bristol, BS8 1TS.

ABSTRACT

Experimental data on the adsorption of Xe by Vycor Porous Glass(PG) in the temperature range 121 to 242K; of CO_2 by PG from 173 to 273K and of Xe by active carbon from 173 to 273K are used to discuss the validity of choosing bulk values for the surface tension (σ) and the molar volume (v) in applying the Kelvin equation to derive pore size distributions. By attributing the knee of the desorption isotherm to a percolation threshold, characterised by a critical pore size, a plausible method of obtaining this size, and hence σv is presented. This product for condensed liquid is considerably lower than that for the bulk liquid. Both are linear functions of temperature and extrapolate to zero at, respectively, the hysteresis critical temperature (at which hysteresis disappears) and the bulk critical temperature. The molar volume at the knee can be calculated from the uptake and pore volume, and hence a value obtained for the surface tension, which in these instances, is lower than that for bulk liquid. This contrasts sharply with an increase predicted by various theories.

1 INTRODUCTION

The status of the Kelvin equation and the validity of the assumptions underlying its conventional application to the determination of pore size distributions[1] raise important questions to which there are as yet no satisfactory answers.[2] This paper offers some suggestions, and experimental evidence, which while not leading to definite conclusions, at least seeks to identify the nub of the problem and to indicate new theoretical and experimental approaches.

2 GENERAL STATUS OF THE KELVIN EQUATION

The form of the Kelvin equation is dictated by the fundamental assumption that the lowering of the chemical potential in a pore is directly proportional to the mean curvature (C) of the surface separating the condensed material from the vapour:

$$RT\ln(f/f^O) = KC \qquad (1)$$

where non-ideality of the vapour is taken account of by using fugacities: f and f^O are, respectively, the fugacities of condensed and bulk liquid states ($C = 2/r$ for a cylinder of radius r; $= 2/h$ for a slit of width h). K, apart from a numerical factor, must therefore have the dimensions of (molar energy x length) which may be factorised to (energy/area) x (molar volume) i.e surface tension (σ) x molar volume (v). No other factorisation can be represented by simple measurable quantities. Thus σ and v only enter into the equation as the product σv: they cannot be obtained separately from the equation. Setting the numerical factor at unity implies complete wetting of the solid by the condensed fluid (i.e. the contact angle $\theta = 0$; $\cos\theta = 1$). The equation thus takes on its conventional form

$$RT\ln(f/f^O) = C\sigma v. \qquad (2)$$

3 THE KELVIN EQUATION AND HYSTERESIS

Capillary condensation hysteresis is assumed to depend on the formation and movement of fluid/vapour interfaces within a porous medium. The thermodynamic properties of these interfaces are encapsulated in the term σv, where in the conventional application of the Kelvin equation it is assumed that σ and v have the same values as in the bulk liquid. Doubt is thrown immediately on the universality of this assumption by data on the temperature dependence of hysteresis loops. Typically the size of the loop decreases as the temperature is raised, and hysteresis disappears at a critical temperature well below that of the bulk liquid. It is then reasonable to suppose that at this temperature, just as in the case of bulk material, liquid and vapour in the pore become indistinguishable and there is no longer a liquid/vapour interface: i.e. the surface tension has gone to zero. Clearly one cannot identify the surface tension in the condensed state with that in the bulk which goes to zero at the (higher) bulk critical temperature. This behaviour can be expressed in terms of 'hysteresis phase diagrams' which reflect the influence of confinement on phase behaviour.[3] (Figure 1)

The existence, for a given porous medium, of a temperature above which a liquid/vapour interface cannot exist, is matched by a critical pore size below which condensed liquid and vapour cannot be distinguished.[4]

Hysteresis loops and the families of 'scanning curves' within them contain a great deal of information which is not always easy to extract. Mason[5] and others[6] have done much to show how such data can be analysed. However, one feature to which little attention has been directed is the sharp knee which is often observed in the desorption branch of the isotherm, especially with porous media having a narrow pore size distribution.

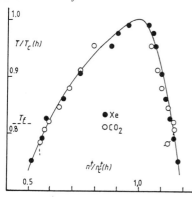

Figure 1: *Hysteresis phase diagrams for Xe/PG and CO_2/PG in reduced units. $T_c(h)$ = hysteresis critical temperature $n_c(h)$ = total adsorption at the hysteresis critical point.*

3 DESORPTION AS A PERCOLATION PHENOMENON

The sharp break in the desorption isotherm at the 'knee' is widely accepted as evidence for a percolation threshold beyond which the desorption process, which up to this point has been inhibited, suddenly becomes increasingly possible.[5] Initially desorption takes place from wider external pores which have access to the vapour. As the vapour pressure is decreased the menisci retreat into narrower pores until they encounter a pore much smaller than that corresponding to the current vapour pressure. If such small pores (windows) are present in sufficient concentration then regions of the pore structure will become completely surrounded by such pores, and pores within them will have no open routes to the vapour. Only when the pressure is reduced so that the windows open can desorption proceed. The vapour pressure at which this occurs identifies the size of the windows causing the blockage.

4 SOME TYPICAL EXAMPLES

That the co-ordinates at the knee might contain interesting information appeared from the fact that for the Xe/Vycor porous glass (PG) system $\ln(f/f^o)$ at the knee was a strictly linear function of $1/T$ from 121K to 242K,[7,8] provided that below the triple point temperature f^o was taken as the extrapolated value of the liquid fugacity. The use of the fugacity of the solid led to no such simple behaviour. It is tempting to take this as evidence that adsorbed Xe remains in a liquid-like state below the triple point. This empirical relationship implies that $RT\ln(f/f^o)$ at the knee is a linear function of T. Figure 2 illustrates this, the line extrapolating to zero at the critical temperature $[T_c(h)]$ at which hysteresis disappears. If we assume that the critical radius is a property of the pore structure and is independent of temperature, then the slope of this line provides information on the temperature variation of σv. The linear variation of σv with temperature for condensed liquid revealed by this relation, may be compared with the behaviour of the same quantity for bulk liquid. It turns out

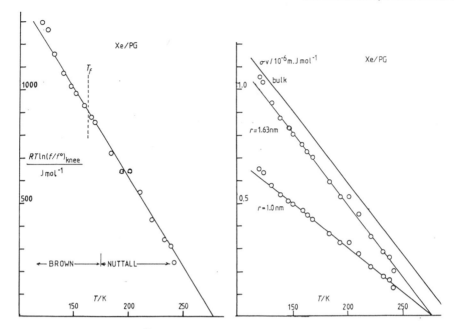

Figure 2: $RT\ln(f/f^{\circ})$ at the knee of the desorption iso-therm as function of T for the Xe/PG system.

Figure 3: σv as function of T for Xe/PG for two values of the critical window radius. Choice of $r = 1.63nm$ leads to σv parallel to the bulk value.

that, at least for simple liquids, σv is also a linear function of T, extrapolating to zero at the bulk critical temperature (Figures 3,6).

To derive values of σv from Figure 2 it is necessary to have a value for the meniscus curvature in the critical windows. The choice is essentially arbitrary. One possibility is to choose a value such that the line of σv against T is parallel to that for the bulk. The experimental data for the Xe/PG system conform to this if the window radius is chosen as 1.63nm (Figure 3). This seems to be an entirely reasonable value and can be compared with 1-2nm estimated by Mason[5] for the maximum in the window size distribution for Vycor (but based on the use of the Kelvin equation with bulk parameters).

Having obtained σv, it is in principle possible to calculate values of σ if v at the knee is known. We assume that a reasonable value can be obtained from the uptake at the knee and the pore volume. Various estimates suggest that the pore volume of Vycor is around 0.2cm^{3}g^{-1}. The experiments on Xe below 168K (Brown[7]), from 183-273K (Nuttall[8]), and for CO_2 from 173-273K (Burgess[9]) were on different samples over a twelve year period. They had been subjected to a variety of pretreatments (e.g. treatment with oxygen at 400°C to remove organic contaminants). To obtain internal consistency of the following analysis the assumed

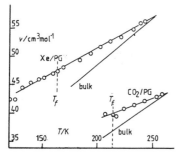

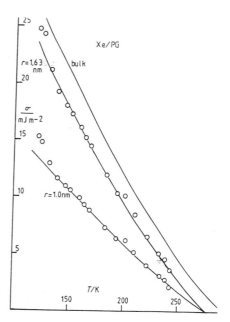

Figure 4: *Molar volumes of Xe and CO_2 at the knee as function of temperature compared with values for bulk liquids*

Figure 5: *Surface tension of condensed Xe as function of T for Xe/PG compared with the bulk, and for two assumed values of the critical window radius.*

pore volumes were taken, respectively, as 0.217, 0.200 and 0.220 $cm^3 g^{-1}$. Figure 4 shows the molar volumes of adsorbed Xe at the knee point as a function of temperature compared with those of bulk liquid. Using the above pore volumes, Brown's data and those of Nuttall fall on the same straight line over the range 130 to 240K, both above and below the triple point. This provides further evidence that the adsorbed Xe retains liquid-like properties below the triple point. It also suggests that as the temperature is raised the molar volumes of bulk and condensed Xe converge. The thermal expansion coefficient of the condensed phase is thus less than that of the material in bulk.

Accepting the calculated molar volumes, the surface tension of adsorbed Xe is obtained, and is plotted in Figure 5 and compared with the surface tension of liquid Xe. It is concluded that in small pores the surface tension of adsorbed fluid is less than that in bulk. This contrasts with several theoretical predictions, based on the effect of curvature on surface tension, that σ should be enhanced in small pores.[2,10].

A similar process applied to the CO_2/PG system leads to very similar conclusions. In Figure 6, $RTln(f/f^o)$ at the knee, and σv are plotted against temperature. The appropriate value of r needed to make the temperature coefficient of σv the same for bulk and adsorbed CO_2 was 1.46nm, reasonably close to the value derived from the Xe data. Using the molar volumes in Figure 4, the surface tensions shown in Figure 7 are obtained; again a reduction of σ in the pores is indicated.

The experimental results for the Xe/active carbon system show the same linear temperature dependence of

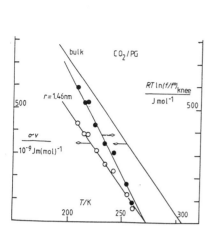

Figure 6: *RTln(f/f°) at the knee, and σv assuming a critical window radius of 1.46nm for the CO₂/PG system.*

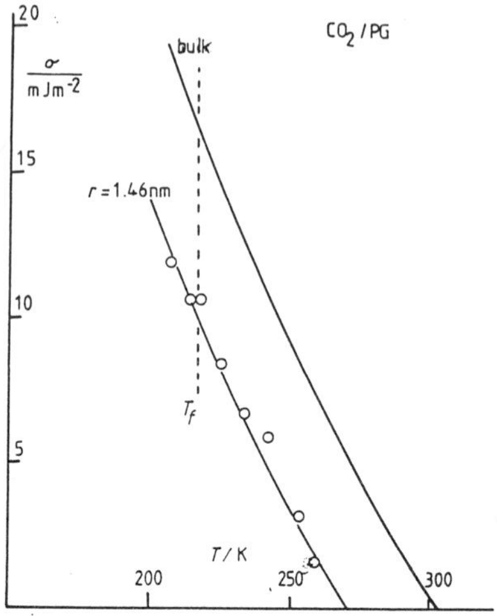

Figure 7: *Surface tension of CO₂ compared with the bulk for an assumed critical radius of 1.46nm.*

$RT\ln(f/f^{o})$ at the knee, Figure 8. Assuming that it is reasonable to apply the Kelvin equation to slit pores, a critical slit width of 0.635 nm is indicated. That this is in the micropore range might suggest that the blocking effect is caused by micropores which prevent mesopores from desorbing. This is not inconsistent with other evidence on the structure of active carbons. It is not possible in this case to obtain values for the surface tension because the uptake at the knee, needed to calculate the molar volumes, was affected by extraneous factors: other work has shown that when, at saturation pressure, the isotherm exhibits a vertical rise then the resulting desorption curve depends to

Figure 8: *σv as a function of T for the Xe/active carbon system, assuming a slit width of 0.635nm, compared with values for the bulk.*

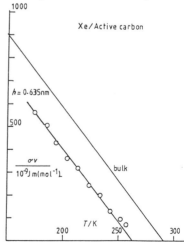

some extent on the maximum adsorption reached before desorption is commenced.

5 SUMMARY AND CONCLUSIONS

The analysis suggested in this paper rests on a number of postulates and approximations, *viz.*

1. That the hysteresis is associated with the behaviour of condensed fluid/vapour interfaces.

2. That the general form of the Kelvin equation is accepted.

3. That the knee in the desorption isotherm is a percolation threshold, characterised by a definite value of the critical window radius which is a characteristic of the adsorbent.

4. That the curves of σv against temperature for condensed fluid and the bulk liquid are parallel. Although this is a useful and plausible approximation leading to values for the critical window radius, it has, as yet, no theoretical basis.

5. That the molar volume of fluid at the knee can be calculated from the uptake at the knee and an assumed pore volume; for both Xe and CO_2 these molar volumes are larger than those for bulk liquid.

Subject to the above reservations, the present analysis indicates that in porous media the surface tension of a capillary-condensed fluid/vapour interface is reduced below that for bulk liquid. This conclusion contrasts sharply with theoretical predictions that for concave surfaces the surface tension increases with increased curvature.[2,10] However, it is difficult to see, if the surface tension of condensed fluid is greater than that in bulk, how the hysteresis critical temperature can be lower than the bulk critical temperature. This discrepancy needs to be investigated. It may be relevant to comment that the theoretical studies relate to the tension in the spherical surface of a small bubble, whereas for a liquid in a cylindrical capillary the surface is a hemisphere, and interactions with the surrounding solid may need to be taken into account.

To carry forward and to consolidate the proposed interpretation it will therefore be important to investigate theoretically the form of σ and σv as a function of temperature in narrow pores. On the experimental side, the range of adsorptives should be extended, and work done on other porous materials having both larger and smaller pores than Vycor, and exhibiting a knee on the desorption isotherm. This will enable one to assess whether the position of the knee is truly a characteristic of the pore structure and whether the reduction in surface tension depends on pore size.

REFERENCES

1. IUPAC,' Recommendations for the Characterisation of
   Porous Solids', *Pure & Applied Chem.*,1974,**66**,1739.
2. e.g. S.J.Gregg & K.S.W.Sing,'Adsorption, Surface Area and
   Porosity', Academic Press, London, 2nd.Edition, 1982,
   pp.153-154
3. C.G.V.Burgess, D.H.Everett & S.Nuttall, *Pure & Applied
   Chem.*, 1989,**61**,1845.
4. e.g. R.Evans, U.M.B.Marçoni & P.Tarazona, *Faraday
   Trans.2*, 1986,**82,**1569.
5. G.Mason, *Proc.Roy.Soc.London*, 1988,**A415**,453.
6. e.g.D.K.Efremov & V.B.Fenelonov, in ' Characterisation of
   Porous Solids II' (Eds. F.Rodriguez-Reinoso, J.Rouquerol,
   K.S.W.Sing & K.K.Unger), Elsevier (Studies in Surface
   Science and Catalysis vol.**62**), Amsterdam,1991,p.115;
   A.V.Neimark,*ibid.*,p.67; V.Mayagoitia,*ibid.*,p.51.
7. A.J.Brown, Ph.D. Thesis, Bristol, 1963.
8. S.Nuttall, Ph.D. Thesis, Bristol, 1974.
9. C.G.V.Burgess, Ph.D. Thesis, Bristol, 1971.
10.e.g.R.Defay, I.Prigogine, A.Bellemans & D.H.Everett,
   'Surface Tension and Adsorption', Longmans, London, 1966,
   p.256; E.A.Guggenheim, *Trans.Faraday Soc.*, 1940,**36**,407;
   J.C.Melrose, *Amer.Inst.Chem.Eng.J.*, 1966,**12**,986.

MICROPORE SIZE DISTRIBUTION IN CARBON MOLECULAR SIEVES BY IMMERSION CALORIMETRY

J.C. González, A. Sepúlveda-Escribano, M. Molina-Sabio and F. Rodríguez-Reinoso

Departamento de Química Inorgánica. Universidad de Alicante. Alicante. SPAIN

1 INTRODUCTION

The characterization of microporous activated carbons, including carbon molecular sieves, is commonly carried out by physical adsorption of gases and vapours[1]. Nevertheless, immersion calorimetry is emerging in the latest years as a useful technique for the characterization of the above mentioned materials[2,3], mainly because it allows a rapid determination of the pore size distribution, specially for micropores which are narrower than 0.8-1.0 nm. Thus, by using liquids of different molecular sizes and a non-porous reference material with a surface chemistry similar to that of the adsorbents to be studied, the surface area accesible to each liquid used can be obtained, assuming that in this case the immersion enthalpy is proportional to the surface covered, irrespective of the size and shape of micropores[4].

Carbon molecular sieves are characterized by a very narrow micropore size distribution, this producing a high adsorption selectivity. The methods proposed for the preparation of carbon molecular sieves are mainly based on the modification of the pores created upon the carbonization of the starting material, either by the controlled activation of chars[5] to widen the microporosity or by carbon deposition from organic vapours at the mouth of the pores[6], to reduce their size. Recently, a new method has been proposed based on the activation of a carbonized material with KOH[7,8]. Although this activation method is commonly related to the preparation of powdered carbons with very high surface areas (up to 3000 m^2g^{-1})[9], granular activated carbons with a narrow pore size distribution could be obtained by carefully controlling the preparation conditions and using lignocellulosic materials as precursors[10].

This paper reports the results obtained in the characterization of KOH activated carbons with molecular sieve properties, by immersion calorimetry and physical adsorption of nitrogen and carbon dioxide.

2 EXPERIMENTAL

Several series of activated carbons have been prepared by KOH chemical activation. The starting materials used have been olive stones (series OL), crushed and sieved to 2.8-3.5 mm and the subsequent chars obtained by carbonization in an inert atmosphere (flowing nitrogen) at 300 (series C3), 500 (series C5) and 800°C (series C8). The activation process involved a first impregnation step with an aqueous KOH solution (0.005-3 g KOH/g precursor) followed by a heat treatment for two hours under flowing nitrogen at 500, 700, 850 and 900°C. Finally, the material was washed with aqueous HCl followed by deionized water.

The carbons have been identified by a label indicating starting material-impregnation ratio-heat treatment temperature, and have been characterized by physical adsorption of CO_2 at 273 K and N_2 at 77 K and immersion calorimetry. The heats of immersion of the activated carbons and the non-porous reference carbon (V3G) into liquids with different molecular sizes (dichloromethane (DCM, 0.33 nm), benzene (Bz, 0.37 nm) and 2,2-dimethylbutane (DMB, 0.56 nm)) were obtained at 303 K with a Tian-Calvet type differential microcalorimeter (Setaram, C80D), in order to determine their molecular sieve properties and micropore size distribution. The samples (about 0.15-0.20 g of activated carbon and 0.4 g of V3G) were placed in a glass bulb with a brittle end and degassified at 523 K and 10^{-5} Torr for 4 h; then, the bulb was sealed and placed into the calorimeter cell containing 7 ml of the wetting liquid. Once the thermal equilibrium was achieved in the calorimeter block, the brittle end was broken and the liquid allowed to enter into the bulb and wet the sample, the heat flow evolution being monitored as a function of time. Thermal effects related to the breaking of the bulb and the evaporation of the liquid to fill the empty volume of the bulb with the vapor at the corresponding vapour pressure were calibrated by using empty bulbs of different volumes.

3 RESULTS AND DISCUSSION

The nitrogen adsorption isotherms at 77 K are type I for all the activated carbons, this indicating their microporous character. When very mild activation conditions are used (for example, with impregnation ratios of 0.05 and 500°C), the amount of nitrogen adsorbed is negligible, what is characteristic of carbonized olive stones in the absence of KOH. As the impregnation ratio or the heat treatment temperature are increased, the adsorption capacity increases and the isotherm's knee opens, thus indicating the development and widening of microporosity. Table 1 shows the micropore volumes of some carbons from OL series, obtained from N_2 adsorption at 77 K ($V_0(N_2)$) by application of the Dubinin-Radushkevich (DR) equation. The increase of $V_0(N_2)$ detected when the impregnation ratio and/or the temperature are increased would indicate that a reaction between the carbon and KOH is taking place upon the thermal treatment, this producing the elimination of carbon atoms from inside the particles and yielding an increase in the micropore volumes.

Table 1 *Micropore volumes (cm^3/g) obtained from N_2 at 77 K and CO_2 at 273 K (DR equation)*

Carbon	$V_0(N_2)$	$V_0(CO_2)$
OL-0.005-850	0.12	0.20
OL-0.05-850	0.18	0.27
OL-0.1-850	0.29	0.32
OL-1-850	0.73	0.50
OL-0.05-500	-	0.14
OL-0.05-700	0.10	0.22
OL-0.05-900	0.24	0.27

Table 1 also reports the micropore volumes obtained by CO_2 adsorption at 273 K $V_0(CO_2)$. For carbons prepared by using mild activation conditions, the micropore volumes obtained from CO_2 adsorption at 273 K are always higher than those obtained from N_2 adsorption at 77 K, this indicating the presence of activated diffusion of nitrogen due to the narrow microporosity. When the impregnation ratio and/or the heat treatment temperature are increased, the micropore volume also increases, and the differences between V_0 (N_2) and V_0 (CO_2) become smaller due to the dissapearance of constrictions; finally, $V_0(N_2)$ are higher than V_0 (CO_2) for high impregnation ratios.

This behavior is not restricted to carbons from series OL, but is also observed in carbons from series C3, C5 and C8 as can be seen in Figure 1, where the micropore volumes obtained from CO_2 adsorption are plotted versus those obtained from N_2 for all activated carbons obtained.

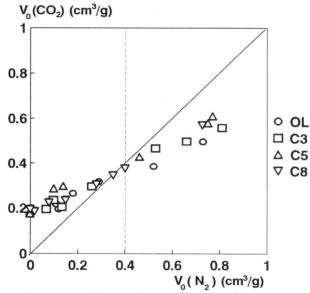

Figure 1 *Micropore volumes deduced from the adsorption of CO_2 (273 K) and N_2 (77K)*

It can be observed that the behaviour is similar for all the carbons, irrespective of the starting material used (OL, C3, C5 or C8). There is a first zone, at low impregnation ratios, in which V_0 (CO_2) > V_0 (N_2) due to limited diffusion of nitrogen through the very narrow microporosity[11]; when the intensity of the activating treatment increases a second zone appears in which V_0 (CO_2) and V_0 (N_2) are getting closer, this indicating the progressive dissapearence of constrictions; finally, there is a third zone for higher impregnation ratios, in which V_0 (CO_2) < V_0 (N_2), due to the widenning of the narrow microporosity[12,13]. The main difference among the OL, C3, C5 and C8 series is that the impregnation ratios needed to achieve a given micropore volume are lower when the starting material is OL, and progresively higher for C3, C5 and C8. In this way, for an impregnation ratio of 0.2 and an heat treatment temperature of 850°C, the micropore volume ($V_0(N_2)$) obtained when using OL is 0.33 cm^3g^{-1} , being 0.26, 0.10 and 0.08 cm^3g^{-1} when starting from C3, C5 and C8, respectively. This points to an activation mechanism in which the KOH distribution inside the particle plays a major role. When the starting material is the lignocellulosic one, OL, a better diffusion and distribution of

KOH inside the particle is obtained during the impregnation step, due to the high interaction between the impregnating solution and the vegetal structure of the precursor, which is facilitated by the basic hydrolysis produced by potassium hydroxide. When the vegetal precursor is carbonized before impregnation, potassium hydroxide would be more heterogeneously distributed through the carbon matrix, in such a way that the activation reaction between carbon and KOH would be produced to a lesser extent. It is to notice that, irrespective of the starting material, 0.4 cm^{3}g^{-1} is the maximum micropore volume that can be achieved compatible with granular activated carbon; above this volume, the carbons obtained are all of them powdered, which can be related to the activation mechanism[7].

Immersion calorimetry has been used to assess the size distribution of the narrow microporosity in these activated carbons, by using dichloromethane (0.33 nm), benzene (0.37 nm) and 2,2-dimethylbutane (0.56 nm) as probe liquids. The surface areas being accesible to each probe liquid have been determined by the method proposed by Denoyel *et al.*[4], using a non-porous carbon black (V3G) as a reference[14].

The surface areas obtained by immersion calorimetry for some of the carbons prepared from olive stones previously carbonized at 300°C (C3), 500°C (C5) and 800°C (C8) are reported in Table 2.

Table 2 *Specific Surface Areas (m^2/g) obtained from Immersion Calorimetry*

Carbon	Dichloromethane (0.33 nm)	Benzene (0.37 nm)	2,2-Dimethylbutane (0.56 nm)
C3-0.02-850	419	254	61
C3-0.24-850	739	569	165
C3-0.5-850	1516	1500	1075
C3-2-850	1818	1839	1522
C5-0.04-850	172	16	18
C5-0.18-850	451	183	93
C5-0.5-850	1350	1303	787
C5-3-850	2031	1957	1882
C8-0.21-850	362	264	37
C8-0.5-850	842	669	284
C8-3-850	1649	1687	1485

It can be seen that carbons prepared with low impregnation ratios contain a high proportion of very narrow micropores and a very low amount of wider micropores, as evidenced by the high values of surface areas accesible to dichloromethane (0.33 nm) and benzene (0.37 nm), and the low values of the surface areas accessible to 2,2-dimethylbutane (0.56 nm). The increase of the surface area accessible to 2,2-dimethylbutane, the biggest molecule used in this study, when the impregnation ratio is increased reflects the dissapearence of the molecular sieve effect in these carbons due to the widening of the microporosity as the activation proceeds. This aspect has been mentioned above when describing the evolution of the micropore volumes deduced from nitrogen and carbon dioxide adsorption measurements.

In Figure 2 the surface areas accesible to the different probe molecules for carbons of series C8 are plotted as a function of their molecular sizes; for each carbon, the micropore volume obtained from CO$_2$ adsorption (DR equation) is also included. These plots represent the distribution of micropore size expresed as cumulative surface area.

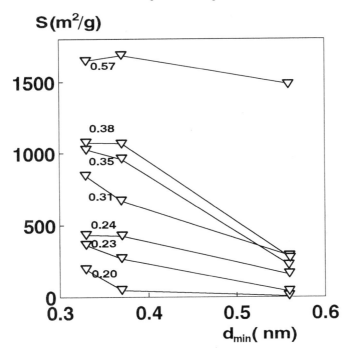

Figure 2 *Surface area accessible to liquids of different molecular dimension (series C8)*

The micropore volumes (CO_2, 273 K) and the surface areas accessible to each probe molecule used increase with the intensity of the KOH treatment. Carbons prepared with the lowest impregnation ratios show small $V_0(CO_2)$ values and a molecular sieve effect for molecules larger than 0.56 nm, as evidenced by the low surface area accessible to 2,2-dimethylbutane. The microporosity develops as the impregnation ratio is increased, this effect being more important for pores smaller than 0.37 nm. The carbon prepared with the h ghest impregnation ratio, C8-3-850 ($V_0(CO_2)$ = 0.57 cm^3g^{-1}), shows a surface area accessible to 2,2-dimethylbutane which is similar to that accessible to dichloromethane or benzene and thus, the molecular sieve effect in this range of molecular sizes is no longer present. This behaviour agrees with that shown by adsorption data. Carbons showing molecular sieve effect are placed in zones 1 and 2 in Figure 1 (V_0 (CO_2) $\geq$ V_0 (N_2)), being both of them equal or lower than 0.4 cm^3g^{-1}. For higher impregnation ratios, the whole microporosity widens, this being evidenced by an important increase of the surface area being accessible to 2,2-dimethylbutane; then, V_0 (CO_2) < V_0 (N_2) and carbons are placed in zone 3 in Figure 1.

This agreement between adsorption and immersion calorimetry results does not depend on the precursor used (OL or the carbonized products C3, C5 and C8). Figure 3 plots the micropore size distribution curves for four carbons obtained from different precursors, and includes the micropore volumes calculated from carbon dioxide adsorption. The micropore volumes are very similar for all the samples (about 0.30 cm^3g^{-1}), and the micropore size distribution is also similar for all of them. This means that the previous carbonization step has a little effect on the micropore size distribution, the impregnation conditions being the main factor which determines the final results. Immersion calorimetry allows to determine that the small difference in micropore volumes is due to pores smaller than 0.37 nm, whereas the surface area accessible to 2,2-dimethylbutane is similar for all the samples.

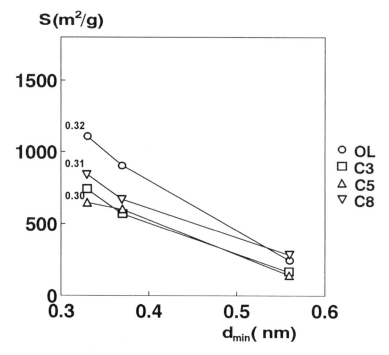

Figure 3 *Surface area accessible to liquids of different molecular dimension (all series)*

Table 1 also showed the effect of the heat treatment temperature on the micropore volume of the activated carbons. For low heat treatment temperatures $V_0(CO_2) > V_0(N_2)$, and the increase of the temperature produced an increase of the micropore volumes in such a way that both values become similar at 900°C. This behavior can be also observed by using the surface areas obtained from immersion calorimetry. Table 3 reports the surface areas accesible to dichloromethane, benzene and 2,2-dimethylbutane in carbons prepared from olive stones by activation at different temperatures and using two different impregnation ratios.

For carbons obtained at low temperatures (500°C and 700°C), the surface areas accessible to benzene and 2,2-dimethylbutane are lower than those accessible to dichloromethane, this indicating a molecular sieve effect for molecules larger than 0.37 nm; also, nitrogen adsorption at 77 K shows activated diffusion in these carbons due to their very narrow microporosity. Higher heat treatment temperatures produce an increase in the surface areas accessible to the three molecules, and the values corresponding to dichloromethane and benzene become more and more close as the temperature increases, this evidencing the widening of the microporosity. The surface area accessible to 2,2-dimethylbutane also increases with the heat treatment temperature, but it does not approach to those of the smaller molecules, this showing the molecular sieve character of these carbons for molecules of about 0.56 nm and bigger.

Table 3 *Specific Surface Areas (m²/g) from Immersion Calorimetry*

Carbon	Dichloromethane (0.33 nm)	Benzene (0.37 nm)	2,2-dimethylbutane (0.56 nm)
OL-0.05-500	524	41	17
OL-0.05-700	529	164	14
OL-0.05-850	588	484	178
OL-0.05-900	789	665	231
OL-0.1-500	715	355	35
OL-0.1-700	1097	853	89
OL-0.1-850	1108	904	247
OL-0.1-900	1100	1052	267

The above results show that all these granular carbons show molecular sieve properties for molecules in the range studied (0.33 nm - 0.56 nm). Then, they could be used for the separation of gas mixtures such as, for example, CH_4/CO_2. There are two main applications where methane-carbon dioxide separation is required: landfill gas, which contains about 50% CO_2 as an impurity, and tertiary oil recovery, where the effluent gas contains up to 20-80% CO_2 and CH_4[15]. Figure 4 shows the adsorption kinetics of methane and carbon dioxide on carbons OL-0.05-700 and OL-0.01-850, measured in a static system at room temperature and atmospheric pressure. Both of them show important differences in the ability for adsorbing methane and carbon dioxide; best results are obtained with sample OL-0.05-700 which shows a much higher difference between the adsorption ratios of both gases at small adsorption times. This behaviour can be related to data in Table 3 showing that sample OL-0.05-700 contains a lower contribution of micropores of about 0.37 nm or bigger, with a narrow microporosity centered at about 0.33 nm, as compared with sample OL-0.01-850 (S_{DCM}=492 m²g⁻¹, S_{Bz}=303 m²g⁻¹ and S_{DMB}=50 m²g⁻¹).

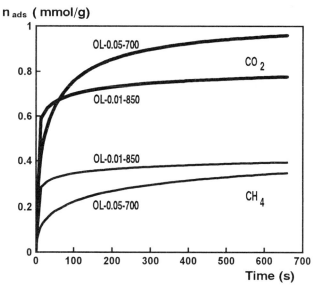

Figure 4 *Adsorption Kinetics of methane and carbon dioxide on carbons OL-0.05-700 and OL-0.01-850 (T= 298 K, P₀= 760 Torr)*

4 CONCLUSIONS

Immersion calorimetry is a valuable technique to obtain the micropore size distribution in microporous materials in a rapid way and with results in agreement with those obtained by physical adsorption of CO_2 and N_2. Furthermore, it can be used with materials containing very narrow micropores in which nitrogen adsorption at 77 K shows activated diffusion.

Activation with KOH can produce activated carbons in granular form, with molecular sieve properties, and this effect dissapears with increasing the severity of the chemical activation (higher impregnation ratio and heat treatment temperature). On the other hand, activated carbons produced from the vegetal precursor and from the carbonized materials show similar micropore development and size distribution, although the chars need higher impregnation ratios and heat treatment temperatures.

Acknowledgements
This work was supported by DGICYT (Proyect No. PB94-1500) and Fundacion CAM

References

1. R. C. Bansal, J. B. Donnet, H. F. Stoeckli, 'Active Carbon ', Marcel Dekker, New York, 1988.
2. H. Fritz Stoeckli, *Carbon*, 1990, **28**, 1.
3. F. Kraehenbuehl and H. F. Stoeckli, *Carbon*, 1986, **24**, 483.
4. R. Denoyel, J. Fernández-Colinas, Y. Grillet and J. Rouquerol, *Langmuir*, 1993, **9**, 515.
5. J. V. Company-Heras, A. Sepúlveda-Escribano, M. Molina-Sabio and F. Rodríguez-Reinoso, *Proc. European Carbon Conference*, 543, Newcastle, 1996.
6. C. Nguyen and D. D. Do, *Carbon*, 1995, **33**, 1717.
7. Z. Hu and E.F. Vansant, *Carbon*, 1995, **33**, 561.
8. Z. Hu and E.F. Vansant, *Microporous Materials*, 1995, **3,** 603.
9. H. Marsh and D. S. Yan, *Carbon*, 1984, **22**, 603.
10. J. C. González, A. Sepúlveda-Escribano, M. Molina-Sabio, F. Rodríguez-Reinoso, *Proc. European Carbon Conference*, 519, Newcastle, 1996.
11. F. Rodríguez-Reinoso, J. D. López-González and C. Berenguer, *Carbon*, 1982, **20**, 513.
12. J. Garrido, A. Linares-Solano, J. M. Martín-Martínez, M. Molina-Sabio, F. Rodríguez-Reinoso and R. Torregrosa, *Langmuir*, 1987, **3**, 76.
13. F.Rodríguez-Reinoso, J. Garrido, J. M. Martín-Martínez, M. Molina-Sabio and R. Torregrosa, *Carbon*, 1989, **27**, 23.
14. M. T. González, A. Sepúlveda-Escribano, M. Molina-Sabio and F. Rodríguez-Reinoso, *Langmuir*, 1995, **77**, 215.
15. R.T. Yang and A. Kapoor, *Chem. Eng. Sci.*, 1989, **44**, 1723.

MAGIC-ANGLE SPINNING NMR STUDIES OF ADSORPTION ONTO ACTIVATED CARBON

Robin K. Harris, Timothy V. Thompson, Jonathan Shaw and Paul R. Norman*
Department of Chemistry
University of Durham
Durham DH1 3LE, U.K.

*CBD Porton Down
Salisbury SP4 0JQ, U.K.

1 INTRODUCTION

The adsorption of materials onto porous solids, particularly activated carbon, has been the subject of much study for a considerable number of years. However, until comparatively recently (see reference 1 and references therein), it has not been possible to observe phenomena occurring inside the pore structure of these materials, and much of the existing information on the processes occurring in the pores has been derived by mathematical interpretation of bulk experimental properties such as the adsorption isotherm and heats of adsorption.

We have recently exploited advances in NMR spectroscopy, and more specifically the technique known as magic-angle spinning, to study molecules adsorbed in the pore structure of activated carbon at a molecular level in real time. The viability of the technique has been demonstrated with respect to water adsorption onto activated carbon[1], and wider application of the technique to studies at a molecular level on sorbate exchange phenomena[2] and the effects of adsorbent pore structure on the adsorption process[3] have been demonstrated.

Additionally, although this paper only discusses results obtained with ^{2}H and ^{31}P NMR, the amenability of many other nuclei to monitoring by NMR means that the technique can be applied to a very wide number of probe molecules. Indeed, we have already confirmed many of the observations made with respect to deuterium oxide adsorption by observation of ^{17}O resonances in ^{17}O enriched water adsorption.

It is the intention of this paper to expand on these observations and make some comment on the information, relevant to adsorption phenomemena, that can be derived from magic-angle spinning NMR (MAS-NMR).

2 DEUTERIUM OXIDE ADSORPTION ISOTHERM

2.1 Gravimetric and NMR Determination of the Isotherm

Initial experiments on the adsorption of deuterium oxide (used to avoid ^{1}H,^{1}H dipolar interactions) by commercial activated carbons demonstrated that an adsorption/desorption isotherm obtained by NMR methods was broadly comparable to that obtained by gravimetric means and displayed the characteristic type V shape expected for water adsorption on activated carbon (see figure 1). In addition, quantitative data obtained by MAS-NMR (using an external calibration) was in good agreement with that obtained by the classical method; this observation is important since it demonstrates that there is no apparent loss of signal intensity as a result of either quadrupole broadening or shielding effects in the graphene layers of the adsorbent. Comparable data was obtained for deuterium oxide adsorption on a range of coal-based and nutshell carbon materials.

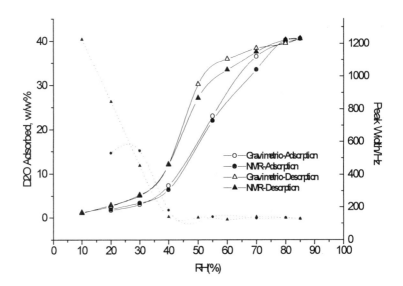

Figure 1 *Deuterium Oxide Isotherms on aged BPL activated carbon*

3 RELATIONSHIP OF PORE SIZE AND THE WIDTH OF THE NMR PEAK

3.1. Origin of the NMR Peaks

Deuterium MAS-NMR spectra of deuterium oxide on activated carbon typically show two characteristics. A comparatively sharp signal (peak width $\sim$ 10-30 Hz) at a chemical shift close to 0 ppm (the reference being an external sample of liquid ^2H_2O) accompanied by a broader peak (peak width $\sim$ 100-1000 Hz) shifted to the low frequency side of the reference peak. This shift is typically of the order of 6-9 ppm. Figure 2C shows the spectrum of deuterium oxide at a 40% w/w loading adsorbed onto a nutshell carbon with nitrogen surface area of 767 m^2 g^{-1}. The peak at $\sim$ 0 ppm has been attributed to "liquid like" deuterium oxide on the surface of the carbon or in wide pores which are open to the surface, whilst the low-frequency peaks are produced by ^2H_2O in the pore system. The chemical shift of the broad peak can be attributed to the shielding effects of the graphene layers of the activated sample. Calculations based on a simple model developed for studying proton chemical shifts in benzene[4] and studies on 1,4-polymethylene benzenes[5] would predict that the sign and magnitude of the shifts observed in these experiments described within are correct and that the shift observed with respect to the studied carbon samples could well be due to aromatic ring current effects.

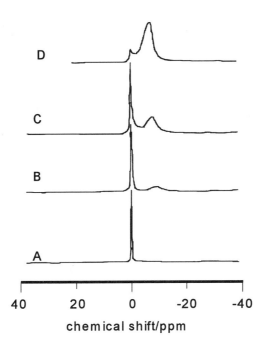

Figure 2 *30.72 MHz Deuterium MAS-NMR Spectrum of ^2H_2O adsorbed onto nutshell carbon (Pica) at a loading of 40% w/w, obtained using 64 transients with a recycle delay of 2 seconds and a pulse duration of 5.5 μs.*

Table 1   *Characteristics of Carbon Samples*

Carbon	CO_2 Surface Area^a/m^2g^{-1}	Total Pore Volume^b/cm^3g^{-1}
Nutshell A	171	0.06
Nutshell B	632	0.22
Nutshell C	767	0.27
Nutshell D	1210	0.46

a. Determined from CO_2 adsorption at 273K using the Langmuir isotherm
b. From N_2 adsorption data at 77K

It has also been demonstrated[3] that, as the pore volume of the carbon material is increased as shown in table 1, then the NMR spectrum changes as shown in figures 2A-D. The larger the pore volume, the more intense is the low frequency-peak, indicating that the low-frequency signal is indeed associated with pore filling.

3.2 Low-Frequency Peak Shape: Spin Relaxation and Restricted Motion

The origins of the broadening of the low-frequency peak are twofold. Firstly, restricted motion of the adsorbate molecules, either as a result of binding to hydrophilic surface heterogeneities or as a result of adsorption into narrow pores where motion is inhibited spatially. Similar conclusions have been drawn by other workers[6] investigating the adsorption of 2-chloroethyl phenyl sulphide on activated carbon.

A second reason for the broad nature of the low-frequency peak is that it is the sum of a series of overlapping smaller peaks. Measurement of spin relaxation parameters was seen as a method to gain more information about the elements comprising the broad peak. Drago *et al.*[7] have described a technique referred to as pore-resolved NMR porisimetry, which in part utilises changes in T_1 and T_2 relaxation data to assign different resonances in ^1H NMR spectra of adsorbates (CCl_4 and C_6H_6) to different sized pores.

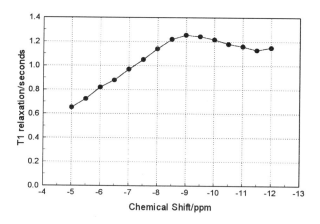

Figure 3   *Phosphorus-31 Relaxation Times of Low-Frequency Peak*

Preliminary spin lattice relaxation studies have been carried out by us in order to determine if there is a difference in relaxation times as the broad low-frequency peak is traversed. The data derived for trimethyl phosphate (TMP) on SCII carbon are shown graphically in figure 3. The variation in T_1 as the peak is crossed are marked, and the results have been shown to be repeatable. Variable temperature work is planned in order to fully identify which side of the broad peak has the least freedom of motion.

These data confirm that the low-frequency peak associated with TMP adsorption in the pore system is clearly inhomogeneous in that it comprises of many overlapping peaks. It is likely, both from this study, and Drago's work on ^{1}H relaxation times[7], that deconvolution of the complex peak could provide useful information on the pore size of the adsorbent with respect to the probe molecule since, the variations in relaxation time across the peak are ultimately associated with the local environment in which the adsorbate molecules find themselves.

4 STUDIES OF MOLECULAR EXCHANGE PROCESSES

The considerations above show that it is possible to use MAS-NMR to distinguish between deuterium oxide adsorbed at different sites in the porous adsorbate. However, many different elements and isotopes may be accessed by NMR. This offers the potential to study exchange processes occurring on and in the carbon material which otherwise would not be possible. The two experiments described below serve to illustrate this principal, but the use of carefully selected and NMR-labelled probe molecules offer huge potential for the study of dynamic processes associated with adsorption onto carbon.

4.1 Exchange between dissimilar molecules

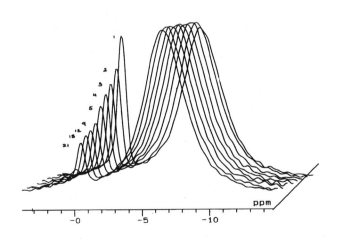

Figure 4   *Stacked ^{31}P NMR spectra obtained at 81.02 MHz for TEP adsorbed onto SCII carbon. The time in hours after addition of TEP is indicated for each spectrum.*

A sample of a commercial SCII carbon was pre-equilibrated with ^2H_2O at 80%RH, after which 35%w/w triethyl phosphate (TEP) was added. Prior to addition of TEP the

^{2}H spectrum was measured and was substantially the same as that shown in figure 2D. Immediately on addition of TEP the ^{2}H spectrum was again measured (this takes approximately 1 minute). The spectrum was significantly different; the low-frequency peak being being barely discernible on the base line with the sharp peak at *ca.* 0 ppm being the most dominant. The inference from this is that the TEP has rapidly displaced ^{2}H$_2$O from the pore system.

The corresponding ^{31}P spectrum shows a large broad signal shifted to low-frequency by *ca.* 7 ppm from an external TEP standard , indicating loosely bound TEP. Thus, not unexpectedly, there is rapid exchange of adsorbates at easily accessible surface sites. The adsorption process is however not complete. If the sample is monitored further for several hours, changes occur in both the ^{2}H and ^{31}P spectra. These are best illustrated by the data shown in figure 4 which depicts stacked ^{31}P NMR spectra . The time after addition of TEP (in hours) is indicated against each peak. Clearly, the sharp ^{31}P peak (associated with essentially free TEP) is decreasing; the broad peak increases concurrently, but this is difficult to see because of the broad nature of the low-frequency peak. The implication of this is that the TEP is moving from surface-accessible sites into the pore system. If this is true, there should be an equivalent migration of ^{2}H$_2$O (which was pre-adsorbed into the pores) out of the pore system, i.e. the rate at which the sharp ^{31}P signal disappears should be the same as which the sharp ^{2}H peak increases after the initial adsorption. This is indeed the case, and figure 5 shows the increase in the ^{2}H signal at *ca.* 0 ppm and the corresponding decrease in ^{31}P signal at *ca.* 0 ppm. Clearly, valuable information about the dynamics of adsorption and competition between adsorbates may be explored using this approach.

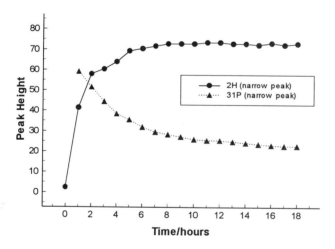

Figure 5   *Kinetic plot of TEP and ^{2}H$_2$O exchange*

4.2 Exchange between TMP molecules

Two-dimensional NMR exchange experiments result in a simple spectrum which gives information upon whether there is chemical exchange between different sites in a system. The theory underlying the technique is complex, but we have demonstrated that the 2D technique can be applied to studies on TMP adsorbed onto activated carbon. An example of the resulting spectrum is shown in figure 6. The results shown above were derived with a "mixing time" (derived from the NMR experiment, not a physical mixing process) of 10 milliseconds. The two principal peaks (the sharp ~0 ppm peak and the broad ~-6-10 ppm peaks already discussed) are shown lying on the 10,10 to -20,-20 diagonal of the plot. However, lying off this diagonal are smaller peaks which may indicate that some chemical exchange is occuring between the two types of TMP site under the experimental conditions employed on the 10 millisecond time scale. If these smaller, off-diagonal peaks were absent, it would suggest that no exchange was occurring. The current evidence is not clearcut and further experiments of this type are being undertaken.

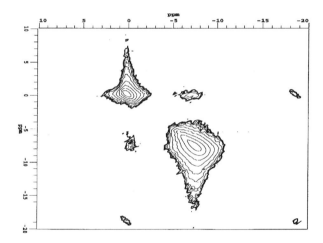

Figure 6 *Two-dimensional exchange experiment for the ^{31}P signal of TMP on SCII carbon*

5 CONCLUSIONS

The brief discussion herein demonstrates that MAS-NMR is powerful tool for the study of adsorption processes on porous materials. It provides a method of studying the adsorption of molecules, not only at the surface, which is a limiting factor of many other experimental techniques, but also within the pore system itself. Furthermore it enables dynamic measurements to be made of exchange and competition effects occurring within the pore system. Other phenomena have been noted during the course of these studies and these will be reported at a later date. These include the effects of the high spin rates used for MAS-NMR on the distribution of adsorbed molecules, and anomalous spectra which apparently result from the adsorption of chiral and some asymmetric molecules; these data may hold information about the orientation of the adsorbates on the carbon surface.

References

1.   R.K. Harris, T.V. Thompson, P.R. Norman, C. Pottage and A.N. Trethewey
        J.Chem.Soc. Faraday Trans., 1995, **91**(2), 1795

2.   R.K. Harris, T.V. Thompson, P.R. Norman and C. Pottage
        J.Chem.Soc. Faraday Trans., 1996, **92**(14), 2615

3.   R.K Harris, T.V. Thompson, P.R.Norman, C.Pottage, K.M. Thomas, N.J.
        Foley and P. Forshaw, *Carbon* accepted for publication

4.   A. Carrington and A.D. McLachlan, p60, "Introduction to Magnetic
        Resonance", Harper, New York, 1980.

5.   J.S. Waugh and R.W. Fessenden, *J.Amer.Chem.Soc.*, 1957, **79**, 846

6.   G.W. Wagner, B.K. MacIver and Y-C. Yang, *Langmuir*, 1995, **11**, 1439

7.   R.S. Drago, D.C. Ferris and D.S. Burns, *J.Amer.Chem.Soc.*, 1995, **117**, 6914

Analysis of HRADS Adsorption

E. Maglara, R. Kaminsky and W. C. Conner

Department of Chemical Engineering
University of Massachusetts
Amherst, Massachusetts 01003 USA

1 ABSTRACT

The technique of high resolution adsorption, HRADS, has been developed and applied to the adsorption in zeolites and other micoporous materials. We have further developed this technique to yield consistent equilibrium data for adsorption of N_2 at 77K or Ar at 77 or 87K*. Several approaches have been proposed to interpret such data including simulations by Horvath Kawazoe[1], Saito and Foley[2] and several others. We find that each of these approaches only seems to fit the data for a limited range of pressures and/or pore dimensions. In addition, density functional theory has been applied to analyze such data[3]. We again find that there are many inconsistencies. The purpose of these studies was to determine if a modified form of the Kelvin equation could give a more consistent interpretation of the relationship between pore dimensions and the pressure at which adsorption occurs over a broader range of pressures.

A broad series of zeolites was studied by HRADS from rho to VPI-5 for the adsorption of N_2 or Ar at 77 and/or 87K (and in a few cases at higher temperatures). We used an apparatus specifically designed for the study of HRADS[4] which provides consistent, efficient measurement of equilibrium adsorption data[5].

We used single parameters changes to the modified Kelvin equations: either as an adjustment to the pore radius or as a change in the effective surface tension as functions of radius. We first find that either of these approaches could fit the predictions of the prior simulations[2,3]. Further, we find that these relationships can fit the experimental data much more consistently than the simulations based on *first principals*..

2 INTRODUCTION

Many models have been proposed in the literature for this purpose. The most basic method for obtaining pore size distributions of porous media from adsorption isotherms relies on the well known Kelvin equation to relate the relative pressure to the dimensions of pores being filled or emptied. This approach has proven to be satisfactory for many mesoporous systems, but it breaks down for microporous systems. The basic reason for this is that the method is based on bulk liquid properties. Horvath and Kawazoe(1) proposed an intriguing approach to treat microporous media with idealized slit pores, which is currently widely used. Saito and Foley (2) and Baksh and Yang (3) extended this approach to cylindrical and spherical pores. Although the Horvath method is relatively simple to apply, its theoretical background is somewhat imprecise. Kaminsky et al. (4) have proposed a more precise model, but still many interaction parameters have to be determined. In our work, besides the models mentioned above, we studied empirical, Kelvin-like equations for the transformation of adsorption data into structural information.

3 EXPERIMENTAL

We developed a modified static technique and apparatus to employ this technique(5-7). When the adsorbate gas is dosed into the sample cell, the pressure goes up. As adsorption takes place, the pressure decreases. The system is allowed to equilibrate before the next aliquot of adsorbate is dosed. Equilibrium is reached and the pressure no longer changes with time. The time required for equilibrium to be achieved varies from more than 30 minutes for the initial doses at residual pressures below 10^{-4} torr to less than 3 minutes for pressures approaching one torr (P/Po ~10^{-3}). These experimental(5) and technique requirements(6) are summarized below:

a) The pressure of adsorption is measured by more than one pressure transducer and/or employing transducers that have more than a single maximum range. They should accurately cover the ranges of the pressure from 0.001 to 760torr.

b) Gas is dosed over the sample and the volume of the gas dosed between measurements is changed as the measurement pressure changes.

c) The time between doses varies as the time required to achieve equilibrium.

d) The tubing connecting the pumping system and the sample and the sample and the pressure transducers is of diameter greater than 0.25 inch; however, the tubing immediately connected to the sample may be smaller.

e) The pumping system employed to evacuate the closed system into which the adsorbing gas is dosed must be capable of reaching relative pressures less than 10^{-8} .

Argon and Nitrogen adsorption isotherms were determined over the following samples: silicalite; ZSM-5; ZSM-11; zeolite Y; zeolite 13X; Linde Types 5A and L; zeolite T and VPI-5. As an example, the Argon Isotherms at 77K (liquid nitrogen temperature) are shown in Figure 1.

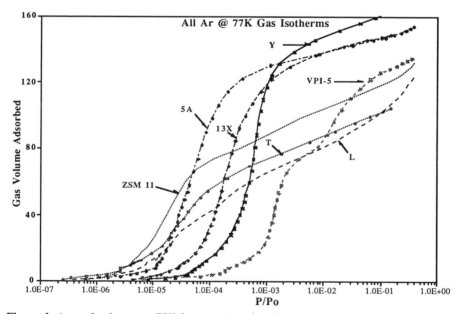

Figure 1 *Argon Isotherms at 77K for a variety of zeolites.*

4 ANALYSES

The models available can be divided into three categories. The first includes the Kelvin equation (1) where P is the pressure, P_o the saturation pressure, g the surface tension, V_L the liquid molar volume, R the gas constant, T the temperature and r_k the Kelvin radius:

$$ln\ P/P_o = -2\gamma V_L/r_k RT \tag{1}$$

This model assumes that the majority of adsorbate molecules in a pore interact with a uniform zero strength potential field. This is reasonably valid for large pores and is widely applied to pores greater than 2nm in diameter. However, this assumption is not realistic for small pores where the adsorbent-adsorbate potential is nonzero in all regions of the pore.

The second category of models are based on statistical thermodynamics of the adsorbed gas molecules on surfaces. They basically extend the Kelvin equation by modeling the adsorbed molecules as a fluid influenced by a uniform potential field. A simple mean-field description of the interaction between the pore walls and the adsorbate molecules is modelled. This includes the slit model by Horvath and Kawazoe(1) as well as two extensions of this model: the cylindrical model by Saito and Foley(2) and the spherical model by Baksh and Yang(3). It also includes a new model proposed by Kaminsky et al.(4).

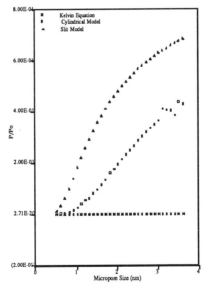

Figure 2 *Comparison of slit-like, cylindrical and Kelvin models for P/Po vs. R.*

Figure. 2 shows a comparison between the Kelvin equation, the slit model and the cylindrical model for the case of nitrogen as adsorbate and oxide ion as adsorbent. It is clear that the Kelvin equation does not apply to micropores and gives a very different results that the Horvath approach(1). However, in the following we will show that small modifications of the Kelvin equation can give exactly the same results as the Horvath method.

We could say that, besides the two categories mentioned above, there is also a third category which consists of the equations we propose for the determination of pore size distribution of zeolites. These are simple, very easy to apply Kelvin-like equations.

Besides the models mentioned above, more sophisticated statistical thermodynamics approaches exist. These theories are based on integral equation theory, density functional theory, or direct molecular simulation. Although these approaches are quite important for the fundamental understanding of the physics pertaining to adsorption phenomena, they are most of the times computationally complex and have too many experimentally unknown parameters to be practical for engineering use.

In the models proposed in the literature the estimation of the pore radius includes several assumptions. For example, Saito and Foley(2) assume that the inside wall of the cylindrical pore is made up of a single layer of spherical atoms (oxide atoms in the case of zeolites). The adsorbate atom (argon in this case) is assumed to be spherical too. But while the argon atom approximates very much a perfect sphere, we can not say the same for the nitrogen atom. Also, it is important to realize that the calculation of pore dimensions from crystallographic analysis assumes the ionic dimension for oxygen (1.32Å). It is not certain whether the covalent radius for oxygen (~0.7Å) would be more proper. That is why we thought to modify the pore radius in Kelvin equation, so that instead of r we have $(r\text{-}delta)$, where $delta$ is an empirical which can take negative or positive values. This constant corrects for the uncertainty in the evaluation of the pore radius. With this modification the Kelvin equation becomes:

$$"average" = ln\,(P/P_0) * (r\text{-}delta) \tag{2}$$

"average" and *"delta"* are constants to be determined.

For large values of r, equation (4) becomes the Kelvin equation with the 'average" given by $-2gV_L/RT$. As pores become increasingly small, the curvature of the walls changes the nature of the interface and thus the meniscus curvature. Melrose had estimated these changes in the surface tension as pores became smaller (7). We chose a reasonable fit to his estimates of this change as:

$$\gamma^* = [\gamma(r(r\text{-}delta)/r^2)] \tag{3}$$

In order to take into account this dependence of the surface tension on the pore radius, we studied the following equivalent equation:

$$"average" = ln(P/P_0) * (r^2)/(r\text{-}delta) \tag{4}$$

In this expression the constant delta modifies the surface tension γ rather than the pore radius as in equation 2. Again, equation 4 has the form of the Kelvin equation for large values of r as the surface tension of the adsorbate tends to the surface tension of the bulk fluid. For small values of r we have the modified surface tension in equation 3, γ^*.

Figure 3 shows the comparison between the cylindrical model and the equations (2) and (4). As we easily realize, one can see little difference between the curves, while it is obvious that equations (2) and (3) are much simpler than the expression of the cylindrical model. A similar fit can be made for the slit-like model. Although the Kelvin equation gives entirely different results than the Horvath method, modified Kelvin equations can give the same results with the models and their extensions. Thus, much simpler expressions than the slit and the cylindrical model can be applied for the determination of the pore size distribution, without affecting the accuracy of the results. This result becomes even more important, if we take into account that the slit model is currently the one most widely used to obtain pore size distribution from adsorption data.

Finally, we applied the models to the experimental data to investigate how well they can predict the pore size distribution of zeolites. Figure 4 shows a comparison between the cylindrical model [Saito and Foley, 1991], the spherical model [Baksh and Yang, 1991], Kaminsky et al. model (for the cases of cylindrical and spherical pore geometry) [1994], and the experimental data, for the case of argon as adsorbate at 77K, is given. For this plot, the parameters for the models were set to predict exactly the pore size of ZSM-5.

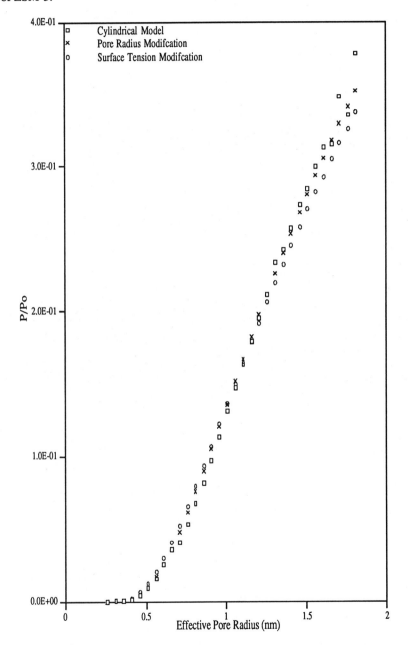

Figure 3 *Comparison between the cylindrical model and the empirical equations 2 and 4.*

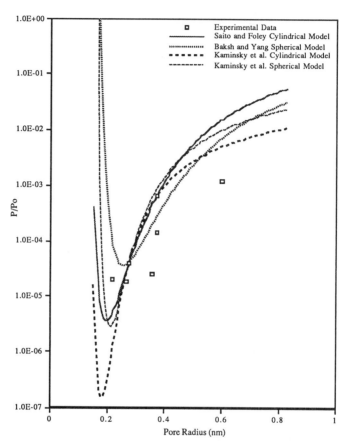

Figure 4 *Comparison between the Models and the experimental data for zeolites- Ar @ 77K.*

From these comparisons we see that the Horvath and the Kaminsky approach fail to predict the pore size of zeolites. They may predict the pore size of a particular zeolite, but then they underestimate or overestimate the size of the rest. We suggest that these single models may not predict the pore size distribution for a variety of zeolites since the pore size calculations are very sensitive to the assumed interaction parameters and especially the pore geometry. So, if the interaction parameters are set based on a spherical pore geometry, then the same model can not predict the pore size of a zeolite with slit or cylindrical pores. In the best of cases, a single model could probably predict the pore size of a certain group of zeolites. These models are more important for developing fundamental understanding of the behavior of microporous media than they are for predicting their pore size distribution. From this point of view we believe that Kaminsky model is more precise than the models proposed before in the literature, and reflects more closely the behavior of real systems.

Figure 5 compares the modified Kelvin equations 2 and 4 and the experimental data, for the argon as adsorbate at 77K. The pore radius refers to the dimensions of the opening of the pore, so the smaller dimension. We also applied the empirical equations to the experimental data for the case the dimension of the cage is taken as the pore radius and found similar results. The empirical equations give better results and in some cases can give fair to good estimates, as they should.

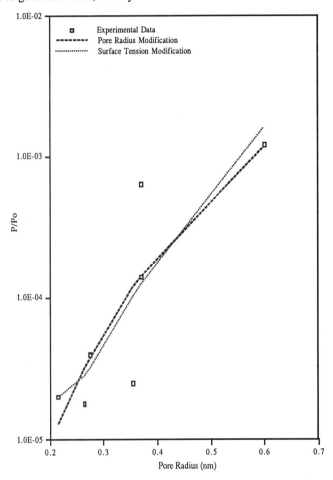

Figure 5 *Comparison between the empirical equations 2 and 4 and the experimental data (opening dimensions) - Ar @ 77K.*

These empirical equations can potentially be used for the pore size characterization of micropores, since no interaction parameters have to be determined for their application. We summarize the empirical constants that we determined from the isotherms for zeolites and also show how these empirical equations could be used to express the relationships found in the Horvath Kawazoe, and Saito Foley models for oxide or carbon surfaces in the table below. We represent these for Ar adsorbed at 77K which is recommended for the characterization of zeolites (5, 7). This is because the interaction potentials for Ar are the easiest to understand (e.g., spherical with little quadrpole interactions) and convenient with common access to liquid nitrogen. The deviations between the models rarely exceeds five percent and is often better that 2 %. Other data are available in the thesis of Maglara (8).

TABLE 1 *Comparisons between data and models for Argon Adsorption at 77K.*

Radius Modifications
Surface Tension Modifications

$$average=ln(P/Po)*(R\text{-}delta)$$
$$average=ln(P/Po)*R^2/(R\text{-}delta)$$

	delta(Å)	"average"	deviation	delta(Å)	"average"	deviation
Empirical						
Opening	-0.361	-6.488	0.056	0.109	-4.721	0.059
Cage	-0.142	-6.198	0.078	0.090	-5.897	0.077
Cylindrical (S-F)						
Oxides	0.136	-1.750	0.024	-0.278	-1.600	0.016
Carbons	0.133	-4.780	0.022	-0.242	-4.480	0.016
Slit (H-K)						
Oxides	0.288	-1.570	0.006	-0.551	-1.440	0.007
Carbons	0.278	-4.320	0.007	-0.520	-3.988	0.005

5 SUMMARY and CONCLUSIONS

For the transformation of adsorption data into structural information, we saw that the filling of the pores does not occur at a single relative pressure but over a range of relative pressures. However, all the models proposed in the literature assume a sudden filling of the pores, and therefore, they do not closely reflect the behavior of real systems.

Furthermore, comparisons between the models and the experimental data indicate that a single model can not predict the pore size of the whole range of zeolites. This is due to the fact that the pore size calculations depend upon the pore geometry assumed. However, the models can provide an insight in the behavior of fluids interacting with microporous media

The empirical equations we studied are much easier to apply, since no interaction parameters have to be determined, and they can be used for the pore size characterization of zeolites.

References

1. G. Horvath and K. Kawazoe, *J. Chem. Eng. Japan* 1983, **16(6)**, 470.

2. A. Saito and H. C. Foley, *AIChE Journal* 1991, **37**, 429.

3. M.S.A. Baksh and R. T. Yang, "Model for Spherical Cavity Radii and Potential Functions of Sorbates in Zeolites", *AIChE Journal*, 1991, **37**, 923.

4. R. Kaminsky, E. Maglara and W. C. Conner, *Langmuir*, 1994, **10**, 1556.

5. W. C. Conner, E. Maglara, A. Pullen and D. Sullivan, *Langmuir,* 1994, **10**, 1467.

6. W. C. Conner, U.S. Patent pending.

6. W. C. Conner, Proceedings of the NATO ASI ON PHYSICAL ADSORPTION, May 1996, Elsevier, pbs. in press.

7. J. C. Melrose, *AIChE Journal*, 1966, **12**, 986.

8. E. Maglara, "Use of Adsorption for the Characterization of Microporosity", MS thesis, University of Massachusetts, May 1994.

NEUTRON SCATTERING INVESTIGATIONS OF ADSORPTION IN POROUS SOLIDS

J. D. F. Ramsay* and E. Hoinkis[†]

*Laboratoire des Matériaux et Procédés Membranaires, UMR CNRS 5635, Université Montpellier II, 34095 Montpellier, France.
[†]Berlin Neutron Scattering Centre, Hahn Meitner-Institut Berlin GmbH, D-14109 Berlin, Germany.

I   INTRODUCTION

Small angle scattering measurements of both X-rays and neutrons can provide structural details of porous materials on a scale covering a range from 1nm to 1000nm. Neutron scattering arises from variations of scattering length density ρ_b, which occur over distances exceeding the normal interatomic spacings. Such variations occur when solids contain pores, and details of the porosity and surface area can be obtained from measurements of the angular distribution of the scattered intensity.

The appropriate angular range (2θ) where this information is contained is defined by the scattering vector q, and the size of the pore, d, where

$$q = 4\pi\sin\theta/\lambda \qquad (1)$$

and λ is the wavelength of radiation.

An analysis of the scattering in the range $0.1 < qd < 1$ provides details of the size and form of the scattering object (pores); information of surface properties may be obtained at larger angles ($qd \gg 1$).

Recent investigations of porous materials using small angle scattering techniques have been reviewed by several authors.[1-3] Progress in the last ten years has been made through experimental developments, and advances in theoretical analysis. An important development has been the application of contrast variation methods. Thus if we consider a two phase system, for simplicity, the intensity of scattering, I(q), is proportional to the contrast, i.e. the square of the scattering length density difference between the two phases:

$$I(q) \sim K\,(\rho_b(1) - \rho_b(2))^2 \;=\; K\,(\Delta\rho_b)^2 \qquad (2)$$

For an evacuated porous solid, where $\rho_b(1) = \rho_b(s)$ the situation is simple since $\rho_b(2) = 0$. However more detailed information may be derived if the pores are filled, or partially filled with an adsorbed vapour. Thus scattering from pores filled with a condensed liquid adsorbate may be eliminated if the scattering length, $\rho_b(1)$ is chosen to be the same as the solid, viz $(\Delta\rho_b)^2 = 0$. This feature may be used to distinguish open and closed porosity, the latter being inaccessible to the adsorbate, for example. Using the same principal

Table 1   *Scattering Length Densities, ρ_b, and Electron Densities, ρ_e, of Different Materials*

Substance	$\rho b/10^{10} cm^{-2}$	$\rho e/10^{24} cm^{-3}$
H_2O	−0.56	0.334
D_2O	6.36	
benzene-h_6	1.16	0.285
benzene-d_6	5.43	
cyclohexane-h_{12}	−0.28	0.268
cyclohexane-d_{12}	6.70	
silica	3.47	0.661
graphite	7.67	0.683
zirconia	5.25	1.568

Hoinkis and Allen[4] were able to measure the selective filling of micropores by exposing graphic carbon to C_6D_6 at $p/p_0 \approx 0.25$.

In further applications of the contrast variation method the growth of adsorbed multilayers and capillary condensation processes have been investigated by SANS[5-7]. Other important developments have been made using SAXS and SANS in particular to probe the structure of the surfaces of porous materials which have a fractal character[8-9].

The majority of these earlier studies were made with samples which were pre-exposed to a fixed adsorbate pressure before the scattering measurement. This procedure, although simple, is limited for several reasons. Firstly because of uncertainties in the equilibrium relative pressure, secondly due to irreproductibility between different solid samples, and thirdly the impossibility of making non-equilibrium and kinetic adsorption measurements. Advances have however been made recently[10] using SAXS with an in situ gravimetric system to study the isothermal adsorption of dibromomethane (CH_2Br_2) on Vycor glass, both of which have comparable scattering densities. The use of CH_2Br_2 in these investigations highlights the important limitation of SAXS compared to SANS. Thus with SANS there is a wide flexibility in the choice of adsorbate (e.g. water, hydrocarbons, alcohols etc.) due to the control in scattering length density which can be achieved by isotopic substitution of H for D in the molecule (see Table 1). SAXS measurements have however been confined to halogenated hydrocarbons since these have an electron density, ρ_e, which is sufficiently large enough to match that of porous solids, such as graphite and silica (see Table 1). This limitation precludes SAXS in studies of a wide range of simpler molecules, where evidently adsorption is not dominated by the interaction of the halogen atoms with the surface.

To exploit the advantages of SANS we have used a specially designed apparatus which allows in situ measurements on thermostatted samples under closely controlled relative vapour pressures of different adsorbates. Technical details of this system have already been given[11] and some feature are described in the next section. The advanced design and flexibility of this apparatus gives scope for equilibrium studies with mesoporous solids close to the SVP and during an adsorption/desorption cycle. Furthermore measurements at very low p/p_0, which are important in studies of microporous materials, are also feasible. Non - equilibrium and kinetic measurements, which are relevant to the rate of diffusion in the pore system are also possible. Here we illustrate the application of the contrast variation technique using this apparatus in investigations of the adsorption of benzene in a mesoporous silica gel. This gel (S4) has been prepared by a sol-gel process, and has a well-defined model pore structure which has been characterised extensively previously.[5-12]

2 EXPERIMENTAL

The silica gel (S4) was prepared by slowly evaporating a concentrated sol in air as earlier described[12]. The primary particle size of the sol particles was ~30nm. Surface and porous properties, as measured by N_2 adsorption at 77K, together with the grain size range are given in Table 2.

The gel scattering length density, ρ_b, as calculated previously was $3.47 \times 10^{10} cm^{-2}$. Benzene adsorption measurements were made at 310K, with an isotopic mixture (59% C_6D_6), corresponding to contrast matching with the gel. The SANS adsorption apparatus, had three main components:

(i) a flange, supporting a cylindrical thermostat bath, fitted with a central tube-containing the sample, and two openings for the incident and scattered neutron beams. These beams passed through sapphire windows (thickness 1mm) which were sealed onto each opening.

(ii) the vapour source, and (iii) a vacuum manifold equipped with UHV-compatible components, i.e. mainly air operated valves, pumps and pressure measuring transducers. The gel sample was contained in a quartz cell (suprasil 300), path length 1mm, and outgassed at ~373K at 3×10^{-5} torr before measurements.

SANS measurements were made with the V4 2D multidetector spectrometer, at the BER II reactor, HMI, Berlin, using a wavelength, λ of 8Å, at sample to detector distances of 1.2, 2.0, 6.0 and 12m. This corresponded to a q range from 0.003 to 0.3 $Å^{-1}$. Standard programmes for the data reduction and treatment were used as previously[5].

Table 2 *Surface and Porous Properties of Silica Gel S4*

$S_{BET}/m^2 g^{-1}$	r_p/nm	$V_p/cm^3 g^{-1}$	Grain Size/μm
128	2.9	0.20	106-212

3 RESULTS AND DISCUSSION

The effect of progressive benzene (59% C_6D_6) adsorption at 310K on the SANS of silica gel S4 is shown in Figure 1. For the outgassed sample (p/p_0=0) the scattering (full line) is characteristic of a structure formed by the packing of spherical sol particles, as has been described previously[13]. Thus the pronounced maximum at a q of 0.025 $Å^{-1}$ arises from the interference in the scattering from a partially ordered structure, where the interparticle separation is given approximately by $2\pi/q_{max}$, viz., ~24nm. The inflexion at q ~ 0.045 $Å^{-1}$ results from the form factor, P(q), of the spherical particles. Thus for monodispersed spheres of radius R, the decay of P(q) has a primary maximum at qR ~ 5.9. Although this feature is smeared here, due to particle polydispersity and slight polychromaticity of the

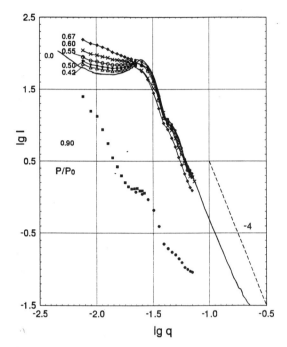

Figure 1   *Silicagel S4 exposed to increasing benzene vapour pressures*

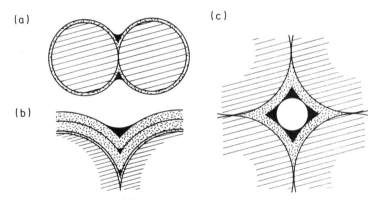

Figure 2   *Depiction of adsorption and capillary condensation in sphere packings;*
(a) adsorbed film and annular meniscus at points of sphere contacts,
(b) development at contact zone on progressive adsorption, and
(c) onset of menisci coalescence in pore throat, leading to spontaneous
filling (after ref. 14)

incident beam, the position corresponds to a particle diameter of ~26nm. This is in
reasonable aggreement with that determined previously from electron microscopy and
SANS measurements on dilute sols of this type (viz., ~ 30nm). Beyond this inflexion, the

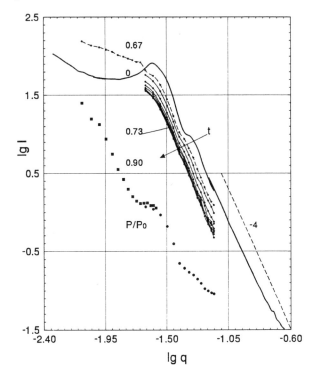

Figure 3 *Kinetics of benzene adsorption on S4 after a stepwise increase of P/P₀ 0.67 to 0.73 at 37°C. Measuring time for each curve was 305 s.*

scattering intensity, $I(q)$, decays linearly with q^{-4} in accord with the Porod law. For a system composed of two phases of different scattering length density, the scattered intensity in this region is directly proportional to the interfacial area. Here the interfacial area is the same as the specific surface area of the outgassed gel.

Changes in the SANS after equilibration with benzene having the same scattering length density as silica are shown for selected values of p/p_0. At low relative pressure ($p/p_0 \leq 0.4$) the scattering is little changed, but on further uptake progressive differences are noted: The almost total suppression of the interference peak is the most striking. Less evident are the suppression of the inflexion due to the particle form factor and the reduction in intensity in the Porod region.

The increases in intensity at low q, associated with the changes in the interference peak, can be ascribed to the growth of an adsorbed film and annular meniscus at the points of sphere contact[14] (see Figure 2). Such changes have been earlier described for the SANS during water uptake in this gel system[5]. For $p/p_0 \geq 0.67$ the changes become more marked and correspond to an enhanced uptake of benzene due to the development of capillary condensate at the sphere contact zones (Figure 2). This feature is illustrated in Figure 3, which shows the kinetics of benzene adsorption after a stepwise increase of p/p_0 from 0.67 to 0.73. Here the SANS curves correspond to sequential measurements of 305 seconds. At $p/p_0 > 0.73$ the onset of menisci coalescence in the pore throats occurs, leading to spontaneous pore filling. This results in the dramatic drop (by $\sim 10^2$) in scattered intensity

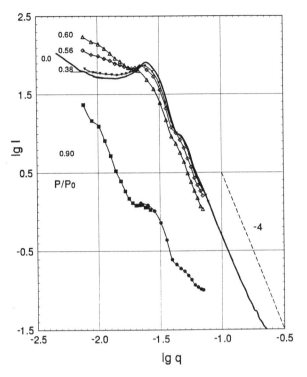

Figure 4   *Silicagel S4 exposed to decreasing benzene vapour pressures*

observed at $p/p_0 = 0.90$. The weak scattering still observed can be ascribed to some
residual contrast between the condensed benzene and the silica gel, due to a slight
mismatch between $p_b(s)$ and $p_b(l)$. The scattering at low q may furthermore arise from a
contribution due to much larger intragranular pores.

The SANS results of benzene desorption (Figure 4) are in accord with a hysteresis loop in
the adsorption/desorption isotherm. Thus on reduction of p/p_0 to 0.60 the scattering has
still not attained the form observed at 0.67 during adsorption. Nevertheless this reduction
in pressure results in a marked change in the scattering as shown by the stepwise change
from 0.90 to 0.60 in Figure 5. Here again the SANS results correspond to sequential
measurements of 305 seconds. From a more detailed analysis of these results it can be
shown that between the initial and final sequential measurements the pore filling decreases
from ≥0.9 to ~0.7 of that at saturation.

On reduction of p/p_0 to 0.56 (Figure 4) hysteresis is still apparent compared to that at 0.55
on adsorption (Figure 1), indicating non closure of the isotherm loop. However at p/p_0 of
0.38 the scattering is very similar to that noted at 0.42 on adsorption, - a feature indicating
reversibility in the cycle. Further analysis of the scattering at this pressure indicates an
adsorbed film thickness of ~0.5nm, which is in reasonable accord with that of 0.4nm based
on the Harkins and Jura relationship, as found by Dubinin[15] for benzene adsorption on
silica gel at 293K.

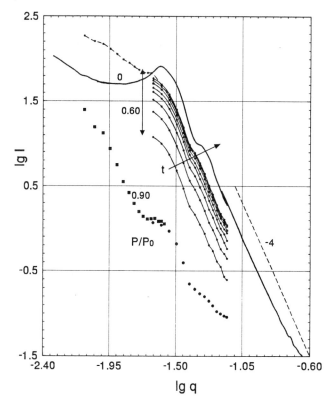

Figure 5 *Kinetics of benzene desorption after a stepwise decrease of P/P_0 from 0.90 to 0.60 at 37°C. Measuring time for each curve was 305 s.*

4 CONCLUSION

In the present study SANS has been used to obtain details of the structure of a mesoporous silica gel and the mechanisms of adsorption and condensation of benzene in the pore system. This has been achieved by following the scattering behaviour at different stages of an adsorption/desorption cycle. Such isothermal measurements have been performed using contrast matching conditions for the solid matrix and adsorbed fluid. It is shown that these conditions are readily achieved using neutron scattering for a wide range of adsorbates and solid materials. The advantages and potential scope of a specially developed sample cell and adsorption system for precise insitu measurements have been illustrated using a silica gel sample with a well-defined porous structure. Although a more detailed theoretical analysis of these measurements is precluded here, future applications of this technique can be envisaged for investigations of the mechanisms of capillary condensation and percolation in other model mesoporous structures of different geometry. There is also scope for investigations of both microporous and non-porous solids. In particular details of the confinement and packing of sorbate molecules, either in micropores or for monolayers on a non-porous surface,may in future be derived from such scattering measurements using different sorbates.

ACKNOWLEDGEMENT

Access to the neutron scattering facilities and support at the BENSC, Hahn-Meitner Institute, Berlin, Germany are gratefully acknowledged.

REFERENCES

1. J.D.F. Ramsay, 'Studies in Surface Science and Catalysis. Characterisation of Porous Solids', Eds. K.K. Unger, J. Rouquerol, K.S.W. Sing and H. Kral, Elsevier, Amsterdam, 1988, Vol. 39, p. 23.

2. P.W. Schmidt, 'Studies in Surface Science and Catalysis. Characterisation of Porous Solids', Eds. K.K. Unger, J. Rouquerol, K.S.W. Sing and H. Kral, Elsevier, Amsterdam, 1988, Vol. 39, p.35.

3. E. Hoinkis, 'Chemistry and Physics of Carbon', Ed. P.A. Thrower, Marcel Dekker, New York, Vol. 25, in press.

4. E. Hoinkis and A.J. Allen, J. Coll. Interface Sci., 1991, **145**, 540.

5. J.D.F. Ramsay and G. Wing, J. Coll. Interface Sci., 1991, **141**, 475.

6. J-C. Li, D.K. Ross and M.J. Benham, J. Appl. Cryst., 1991, **24**, 794.

7. M.Y. Lin, S.K. Sinha, J.S. Huang, B. Abeles, J.W. Johnson, J.M. Drake and G.J. Glinka, Mat. Res. Soc. Proc., 1990, **166**, 449.

8. D.W. Hua, J.V. D'Souza, P.W. Schmidt, D.M. Smith, Studies in Surface Science and Catalysis. Characterisation of Porous Solids III, Eds. J. Rouquerol, F. Rodriguez-Reinoso, K.S.W. Sing and K.K. Unger, Elsevier, Amsterdam, 1994, Vol. 87, p.

9. C.J. Ghinka, L.C. Sander, S.A. Wise and N.F. Berk, Mat. Res. Soc. Proc., 1990, **166**, 415.

10. A. Ch. Mitropoulos, J.M. Haynes, R.M. Richardson and N.K. Kanellopoulos, Phys. Rev. B, 1995, **52**, 10035.

11. E. Hoinkis, Langmuir, in press.

12. J.D.F. Ramsay and B.O. Booth, J. Chem. Soc. Faraday Trans. 1, 1983, **79**, 173.

13. J.D.F. Ramsay, Chem. Soc. Rev., 1986, **15**, 335.

14. B.G. Aristov, A.P. Karnaukhov and A.V. Kiselev, Russ, J. Phys. Chem., 1962, **36**, 1159.

15. M.M. Dubinin. 'The Modern Theory of Capillarity', Eds. F.C. Goodrich, A.I. Rusanov, Akademie-Verlag, Berlin, 1981, p. 63.

A Pore Width-Sensitive Ordered Structure of Water Molecules in a Carbon Micropore

T. Iiyama, K. Nishikawa*, T. Otowa**, T. Suzuki, and K. Kaneko
Department of Chemistry, Faculty of Science, Chiba University, 1-33 Yayoi, Inage, Chiba 263, Japan
*) Department of Phase Science, Graduate School of Natural Science, Chiba University, 1-33 Yayoi, Inage, Chiba 263, Japan
**) Research and Development Center, The Kansai Coke and Chemicals Co. 1-1, Ohama, Amagasaki, Hyogo 660, Japan

1 INTRODUCTION

Micropore filling of a noticeable adsorption even at an extremely low pressure has been one of central research subjects in physical adsorption.[1] The molecules confined in the deep potential well due to overlapping of the molecule-surface interaction potentials are interacted with each other to form the best intermolecular arrangement. Thus, molecules form an organized molecular assembly or clusters in the micropore. The formation of organized structures or clusters for NO,[2] SO_2,[3] O_2,[4] N_2,[5,6] CCl_4,[7] H_2O,[8] and benzene[9] in micropores were evidenced experimentally. Elucidation of the adsorbed states in the micropore should provide the information on the adsorbed density which is essentially important in the accurate characterization of porous solids by gas adsorption.

X-ray diffraction technique is quite effective for understanding the intermolecular structure in the micropore. Here we must choose the adsorbent which X-ray can penetrate into. The predominant micropore-walls should be neutral and nonspecific to an adsorbed molecule such as the basal plane of graphite. New activated carbon such as activated carbon fiber (ACF)[10] or superhigh surface area carbon[11] has considerably uniform slit-shaped micropores of which pore-walls are mainly composed of the basal planes of micrographites. Molecular simulation studies showed that the simple graphite slit model is quite effective for description of micropore filling of neutral molecules in micropores of activated carbon.[11-15]

In the preceding letter,[7] we reported that the electron radial distribution function of CCl_4 in micropores of activated carbon using in situ X-ray diffraction varies sensitively with the change in the pore width. As a CCl_4 molecule has a strong X-ray scattering ability, we examined the effectiveness of analysis of in situ X-ray diffraction technique for a molecular assembly system in the micropore.

The predominant intermolecular interaction of water is not governed by Lennard-Jones interaction, but by the hydrogen-bonding.[17] As the water molecule does not strongly interact with the graphitic surface, the adsorption isotherm of water on activated carbon is of Type V.   The reason why a remarkable adsorption begins at a medium relative pressure was explained by Dubinin et al[18] using the assumption of cluster formation of water molecules at a hydrophilic site without the direct evidence. The information on cluster or molecular assembly of water molecules in the micropore with the aid of in situ X-ray diffraction should be helpful to understand the water adsorption mechanism on activated carbon. Also it should be associated with water confined in mesopores[19] and the vicinal water.[20] However, the electron density difference between an adsorbed water molecule and the carbon pore-walls is not great and no negligible scattering from the interface between the adsorbed water and carbon-wall is presumed from the previous X-ray study.[19] In the preceding letter,[8] we reported the presence of the ordered structure of water molecules in the micropore. In this study, we examined more carefully the structure of water molecules adsorbed in the micropore with the in situ X-ray diffraction and determined the effect of the pore width and the measuring temperature on the water molecular structure in the carbon micropore.

2 EXPERIMENTAL

Pitch-based ACFs (P5, P10, and P20; Osaka Gas Co.) and superhigh surface area carbons (Max15 and Max30; Kansai Coke Co.) were used as the microporous carbon sample. The micropore structure was determined by N_2 adsorption at 77 K. The water adsorption isotherm was measured gravimetrically at 303 ($\pm$ 0.05) K. Gas adsorption was measured after preheating at 383 K and 1 mPa for 2 h.

The X-ray diffraction pattern of water adsorbed in micropores of activated carbon at 303 ($\pm$ 0.1) K and the saturated vapor pressure was measured by the transmission method using an angle-dispersive diffractometer (MXP3 system, MAC Science).   The diffraction measurement at 230 ($\pm$ 5) K was carried out after adsorption of water at 303 ($\pm$ 0.1) K and the saturated vapor pressure. The monochromatic X-ray from Mo Kα radiation at 50 kV and 35mA was used for the diffraction measurement and the observed range of the scattering parameter s ($= 4\pi \sin\theta/\lambda$) was from 0.7 to 12 Å^{-1}. The information on the adsorbed water structure was obtained as follows: The X-ray diffraction intensity of activated carbon with adsorbed water $I(s)$ is expressed by the additive form of the scattering by carbon wall structure $I_c(s)$, by adsorbed water $I_w(s)$, and by the adsorbed water-carbon interface structure $I_{cw}(s)$.

$$I(s) = I_c(s) + I_w(s) + I_{cw}(s) \tag{1}$$

In this analysis, we neglected $I_{cw}(s)$. Hence, we determined $I_w(s)$ by $I(s) - I_c(s)$. Therefore, we cannot discuss on the absolute $I_w(s)$, but on the relative change with the pore width or measuring temperature.

Table 1 *The micropore structure of microporous carbons by N_2 adsorption*

	Pore volume ml g^{-1}	Surface area m^{2}g^{-1}	Micropore width Å
P5	0.34	900	7.5
P10	0.62	1510	8.2
P20	0.97	1770	11.3
Max15	0.50	970	10.3
Max30	1.48	2280	13.0

at 0.4, 0.6, and 0.65 of P/P_0, respectively. The rising pressure is associated with the concentration of surface functional groups, and the lower the rising pressure, the more the surface functional groups. Consequently, the concentration of surface functional groups is greater in the order of P10, P20, and Max30.

3.2 X-Ray Diffraction

The X-ray diffraction patterns of activated carbon with adsorbed water were determined, as shown in Figure 3. Figure 3 shows the X-ray diffraction patterns of water-adsorbed Max30 [$I(s)$] and Max30 in vacuo [$I_c(s)$] at 303 K. The strong scattering below $s = 1.0$ is the small angle scattering, which can be corrected using the scattering of the carbon itself with the aid of the Porod law.[22] Their subtraction of $I(s) - I_c(s)$ leads to the approximated diffraction of water adsorbed in the micropore, which is shown by a broken line. Figure 4 shows the X-ray diffraction patterns of water adsorbed in micropores of Max15 and Max30 together with that of bulk liquid water at 303 K. The diffraction pattern of liquid water is very broad, while adsorbed water on both carbon samples has a similar broad pattern at $s = 1.95$, which is shifted toward a smaller s value from $s = 2.07$ for the liquid water. These X-ray diffraction patterns were Fourier-transformed into the electron radial distribution functions (ERDFs) to discuss the detailed difference between adsorbed water and bulk liquid water.

3.3 Effect of Pore Width on Water Molecular Assembly Structure in Micropores

The weighted structure function si(s) as defined by eq (2).

$$si(s) = s(I(s) - \Sigma f_i(s)^2) \qquad (2)$$

3 RESULTS AND DISCUSSION

3.1 Porosity and Water Adsorption Isotherms

The adsorption isotherms of N_2 on P5, P10, and Max15 were of Type I, being characteristic of uniform micropore system. The adsorption isotherms of N_2 on P20 and Max30 had a gradual linear rise above $P/P_0 = 0.1$, which is ascribed to the presence of wider micropores of more than the thickness of three to four adsorbed N_2 layers. The N_2 adsorption isotherms were analyzed by the subtracting pore effect method using high resolution α_s-plot.[10,21] The adsorption isotherm of N_2 on nonporous carbon black (Mitsubishi Chemical Co. #32) was used as the standard isotherm. Figure 1 shows the high resolution α_s-plots for Max15 and Max30. α_s-plot of Max30 has two upward swings of filling and cooperative swings below $\alpha_s = 1.0$, suggesting the presence of wider micropores around 15 Å. On the other hand, Max15 has only filling swing below $\alpha_s = 0.3$, which is characteristic of smaller micropores of less than about 10 Å. In these cases we can determine even the specific surface area from the slope of the line connecting the origin and point at $\alpha_s = 0.5$. Also the micropore volume can be determined from the α_s-plot. Then we can determine the pore width with the assumption of the pores to be nonintersecting slit pores. Table 1 summarizes these pore parameters. The average pore width varies from 7 to 13 Å.

Figure 2 shows water adsorption isotherms of P10, P20, and Max30 at 303 K. All adsorption isotherms are of Type V. Although they have a hysteresis, the hysteresis curve of only P10 is shown here. The adsorption isotherms of P10, P20, and Mx30 rise

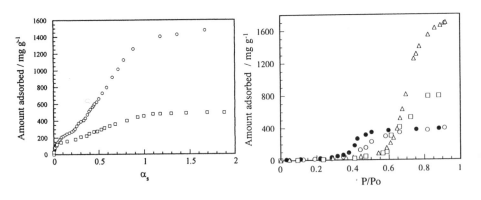

Figure 1. α_s-plots for N_2 adsorption isotherms on Max15 and Max30.

☐ : Max15,   ◯ : Max30

Figure 2. Water adsorption isotherms of P10, P20, and Max30 at 303 K

◯ : P10, Adsorption, ● : Desorption
☐ : P20,   △ : Max30

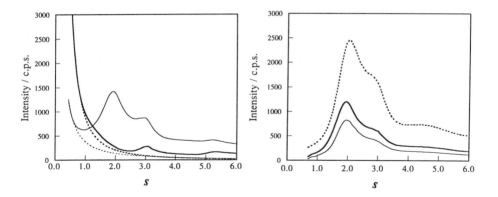

Figure 3. X-ray diffraction patterns of water-adsorbed Max-30 and Max-30 in vacuo at 303 .

Fine line: $P/P_0 =1.0$, bold line: in vacuo

Figure 4. X-ray diffraction patterns of bulk liquid water and water-adsorbed on Max15 and Max30 at 303 K.

Bold line: Max30, fine line: Max15 broken line: bulk liquid water

where $I(s)$ is the total coherent intensity per molecule and f_i is the atomic scattering factor of the i-th atom. The ERDF, $4\pi r^2(\rho(r) - \rho_0)$, is given by the Fourier transform of the weighted structure function.[21]

$$4\pi r^2(\rho(r) - \rho_0) = (2r \,/\, \pi \Sigma \; Z_i^2 \,) \; \int si(s) \sin sr \, ds \qquad (3)$$

Here, $\rho(r)$ is an approximated molecular density at r and ρ_0 is the average value of $\rho(r)$ in the system. Z_i is the number of electrons in the i-th atom. The ERDF denotes the spead of the electron cloud of each atom, which is effective for description of the multi-atom system.

The ERDFs of P5, P20, and Max30 are shown in Figure 5. The ERDF of bulk liquid water is also shown for comparison. The ERDF of liquid water has peaks at 3, 4.5, and 7 Å, which correspond to the nearest neighbor, the second nearest neighbor , and the third nearest neighbor, respectively. The ERDF is the same as the published result.[24] The first nearest neighbor peak is greater than the second nearest peak due to a fluctuated ice-like structure. The ERDF of water adsorbed on P5 has peaks at 4.5 Å and about 6 Å. In this case there is no peak at 3 Å, which is inherent to the liquid structure, indicating that water adsorbed in micropores of ACF has a different structure from liquid water. Two other ACFs have similar ERDFs to P5. The coordination sphere of the third nearest neighbor molecule has a similar size to the pore width and the molecules in the slit pore should be seriously restricted. On the other hand, the ERDF of Max30 having   the widest pores   has a shoulder at 3.5 Å and a peak at 4.3   Å in   the   shortest range, which is close to ice I. The peak at 3.5   Å is much smaller than the peak at 5 Å in case   of ice I, as shown in Figure 7. Therefore, we can estimate the ordered structure of water molecules from the relative intensities of both peaks.

Figure 6 and 7 show the changes in the maximum position and intensity of peaks at 3 - 5 Å (peak [A]) and 6.5 - 8 Å (peak [B]), respectively.   The peak position of [A] for all five samples has no definite endency,   while the peak intensity decreases with the increase of the pore width.   The peak [B] of the third nearest neighbor structure is more noticeably influenced by the difference in the pore width;the intensity decreases and the peak position shifts to a smaller value with increase of the pore width.

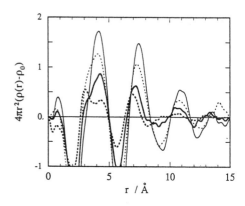

Figure 5. Electron radial   distribution functions of   bulk liquid water and adsorbed water on P5, P20, and Max30. P5 — , P20 ······ , Max30 — , water -----

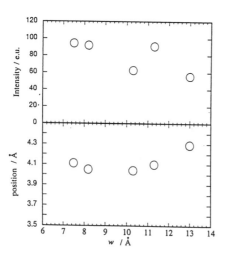

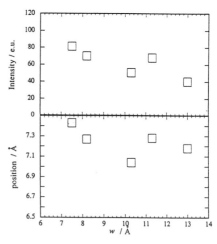

Figure 6. Changes in maximum position and intensity of the first-nearest neighbor and the second-nearest neighbor peaks with the pore width *w*.

Figure 7. Changes in maximum position and intensity of the third-nearest neighbor peak with the pore width *w*.

These results strongly indicate that water molecules form an ordered assembly structure close to ice. However, the mobility of water molecules in micropores is different from that in ice, depending on the pore width.

3.4 Comparison of Water Molecular Assembly Structure in Micropores with Ice

Figure 8 shows the ERDFs of water molecules in micropores of Max30 at 230 K and at 303 K. The ERDF of bulk ice I is also shown in Figure 8. The bulk ice has the sharp peaks at 3.5 and 4.5 Å; the intensity of the peak at 3.5 Å is much smaller than that at 4.5 Å. This intensity ratio is completely reverse for that of liquid water. The ERDF of water adsorbed in micropores of Max30 at 230 K is close to that of ice I in the peak intensity and position. Therefore, the molecular assembly in micropores of Max30 at 230 K has an ordered structure similar to ice I. The ERDF of water molecules in micropores at 303 K has a common feature to that

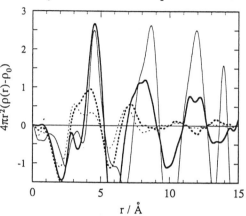

Figure 8. Electron radial distribution functions of liquid water at 303K and ice at 230 K and of adsorbed water on Max30 at 303 K and 230 K. Fine solid line: bulk at 230K, fine broken line: bulk at 303 K, bold solid line: on Max30 at 303 K, bold broken line: on Max30 at 230 K

of ice I. However, the peak intensity at 4.5 Å is much more smaller than that of bulk ice. Therefore, molecules in micropores even at 303 K have an ordered structure like ice, but the molecules should be more mobile in the ordered structure in micropores. The intensity of the [A] peak of P5 is greater than that of Max30. Although the [A] peak of P5 has no definite peak at 3.5 Å, indicating the great peak intensity near 4.5 Å. Then, water molecules adsorbed in narrowest micropores of P5 should lose their mobility to form a more definite ordered structure than that of Max30.

So far we cannot describe quantitatively the ordered state of water molecular assemblies in micropores. There is a possibility that an ordered water molecular assembly has a specific structure different from ice I, because ice has many polymorphic forms according to change in temperature and pressure.[25] Furthermore, Iijima and Nishikawa[26] revised the ice I model on liquid water by Narten and Levy[27] to give a flapping model of the six-membered oxgen-ring structure. The simulation study on water adsorption on activated carbon[28] is desirable irrespective of a difficult potential problem.

This work was supported by Grants-in-Aid for Scientific Research from the Ministry of Education, Culture, and Science of Japanese Government.

References

1. K.S.W.Sing, Carbon, 1991,**27**, 5.

2. K. Kaneko, N. Fukuzaki, and S. Ozeki, J. Chem. Phys. 1987, **87**,776.

3. Z. M. Wang and K. Kaneko, J. Phys.Chem. 1995, **99**, 16714. (1995)

4. H. Kanoh and K. Kaneko,J. Phys. Chem.1996,**100**, 755.

5. K. Kaneko, K. Shimizu, and T. Suzuki, J. Chem. Phys. 1992, **97**, 8705.

6. W.C. Conner, E.L Weist, T. Ito, and J. Fraissard, J. Phys. Chem. 1989, **93**, 4138.

7. T. Iiyama, K. Nishikawa,T. Suzuki, and K. Kaneko, Chem. Phys. Lett. in press.

8. T. Iiyama, K. Nishikawa, T. Otowa, and K. Kaneko, J. Phys.Chem.**99**,10075 (1995).

9. J. Fukasawa, K.Kaneko, C.D. Poon, and E.T. Samulski, 'Characterization of Porous Solids III,' J. Rouquerol, F. Rodriguez-Reinoso, K.S. W.Sing, and K.K. Unger eds. Elsevier, Amsterdam, 1994, p.311

10. K. Kaneko and C. Ishii, Colloid Surf. 1992,**67**,203.

11. T. Otowa,R. Tanibata, and M. Itoh, Gas Sep. Purif. **7**, 241 (1993).

12. N. A. Seaton, J.P. R. B. Walton, and N. Quirk, Carbon, 1991,**27**, 853.

13. K.R. Matranga, A.L. Myers, and E.D.Glandt, Chem. Eng.Sci. 1992,**47**, 1569.

14. R.F.Cracknell, K.E. Gubbins,M.Maddox, and D.Nicholson, D., Account Chem. Res. 1995, **28**, 281.

15. B.P.Bojan,E. Cheng, M.W.Cole, and W.A.Steele, Adsorption, 1996, **2**, 57.

16. T. Suzuki, T. Kasu,and K. Kaneko, Chem. Phys. Lett.1992, **191**, 569.

17. M. Rigby, E. B. Smith, W. A. Wakeham, and G.C. Maitland,'The Forces Between Molecules,' Clarendon Press, Oxford, 1986, Chapter 1.

18. M.M. Dubinin, E.D. Zaverina, and V.V. Serpinsky, J. Chem. Soc. 1955, 1760.

19. D.C. Steytler, J.C. Dore, and C.J. Wright, J. Phys. Chem. 1983, **87**, 2458.

20. W. Drost-Hansen and F.M. Etzler, Langmuir, 1989, **5**, 1439.

21. N. Setoyama, T. Suzuki, and K. Kaneko, Carbon, in press.

22. L.A. Feigin and D. I. Svergun, 'Sturcture Analysis by Small Angle X-Ray and Neutron Scattering,' Plenum Press, New York, 1987, p.45.

23. K. Nishikawa and M. Takematsum, Chem. Phys. Lett.1994, **226**, 359.

24. K. Nishikawa and N. Kitagawa, Bull. Chem. Soc. Jpn. 1980, **53**,2804.

25. D.Eisenberg and W. Kazmann, 'The Structure and Properties of Water,' Clarendon Press, Oxford, 1969.

26. T. Iijima and K. Nishikawa, J. Chem. Phys. 1994, **101**, 5017.

27. A.H. Narten and H.A. Levy, J. Chem. Phys. 1971, **55**, 2263.

28. E.A. Muller, L. F. Rull, L. F. Vega, and K.E. Gubbins, J. Phys. Chem. 1996,**100**, 1189.

COMPUTER SIMULATION STUDY OF SORPTION IN CYLINDRICAL PORES WITH VARYING PORE-WALL HETEROGENEITY

William Steele and Mary J. Bojan

Department of Chemistry
Penn State University
University Park, PA 16802, USA

ABSTRACT

Computer simulations of the sorption of krypton at 120 K in model pores with varying degrees of surface heterogeneity are reported. Geometric variations in heterogeneity were obtained by varying the roughness of the pore walls. Although this produced significant changes in average energy of the adsorbed atoms, the alterations observed in the capillary condensation-evaporation behavior were not large.

1 INTRODUCTION

Over the past few years, the simulation study of the phase transitions in fluids in micropores has proved to be a powerful method of characterizing these systems.[1] In particular the liquid-vapor change, which is generally described as capillary condensation, has been simulated under the Grand Canonical Monte Carlo algorithms that allow one to evaluate the isotherms of adsorption and desorption in model pores of varying size and shape. The pressures at which condensation and evaporation occur as well as the amounts of adsorbate in the pore at the beginning and end of the transitions have been evaluated for many simple pore models. It is found that the transitions are generally vertical, to within the precision of the simulations, and that the Kelvin equation which works rather well for predicting the phase changes in mesoporous systems, begins to fail rather badly as the pore size drops below roughly ten adsorbate molecular diameters. Of course, real porous materials often exhibit condensation-evaporation loops with jumps that are far from vertical.[2] It has been speculated that this is related to pore shape or, more likely, to distributions in pore size and shape in the packed powders, porous carbon blacks and similar systems often used in the laboratory.[3] Another possibility that will be explored here is that the characteristics of these phase changes depend upon the heterogeneity of the pore walls. Initially, the simulations relied upon highly simplified pore models with atomically smooth walls and simple shapes such as parallel-walled slits and straight-walled cylinders. However, as attempts were made to make the systems studied be closer to reality, the pore walls and pore shapes were taken to be more complex. A well-known example has been the modeling of zeolite sorbents, where the knowledge of their crystal structure has frequently allowed the simulator to develop models in which the atomic structure of the walls is explicitly taken into account by assuming that the total gas-pore wall interaction is taken to be a sum of two-body potentials over all the known positions of the atoms in the crystal. Indeed, one obtains reasonable potentials by including only interactions between an adsorbate molecule and the oxides in the zeolite structure. Of course, such potentials are energetically heterogeneous because the energy depends upon the position of an atom relative to the atomic structure of the wall. For example, they exhibit deeper wells when the adsorbate atom is centrally located over three or four wall atoms than when it is directly over a single atoms.[4] However, comparative simulations of sorption in zeolites are based on varying the nature of the sorbates rather than the nature of the pore walls. Here, we will take a single adsorbate (krypton) at 120 K in a pore of fixed shape which is that of a straight-walled cylinder whose radius is sufficiently large to given an initial monolayer of krypton that is, on average, 10 Å from the pore axis. The solid adsorbent is modeled by a randomly close-packed assemblage of spheres which contains a cylindrical cavity; periodic boundary conditions are applied in the direction of the pore axis, and the heterogeneity of the gas-pore wall interaction potential is varied by altering the sizes of the spheres and in two cases, their packing in the actual pore wall. One of the five pore systems obtained in this way has actually been reported upon previously,[5] but the variations in wall structure and their effect upon the properties of the sorbed fluid discussed here extends this work considerably.

We will see that the presence of vertical risers in both the adsorption and desorption branches of the isotherms is unaffected by the changes that occur as a result of the geometric structure of the pore walls. What does change are the values of the transition pressures and the structures of the fluid sorbed in these model pores.

2 PORE MODELS

We denote the diameter of a hard-sphere atom in the amorphous adsorbing solid by σ_s. As σ_s decreases in size, the atomic density of the close-packed solid increases and the roughness of its surface decreases. Of course, surface roughness is only meaningful here when measured in terms of the relative sizes of the solid and the adsorbate atom ($\epsilon/k=170$ K and $\sigma =3.60$ Å, in the usual Lennard-Jones representation of the Kr-Kr interaction), and we take $\sigma_{Kr-s}=3.45$ Å for the interaction of Kr with a solid atom - this value is assumed to be nearly independent of the spacing of the solid atoms. The three values of σ_s taken were 1.84, 2.30 (studied previously) and 3.15 Å. The diameters of the cylindrical cavities in these solids were fixed such that the average radial distance between the minimum in the Kr-pore wall potential and the pore axis, was held fixed at 10.0 ± 0.1Å. Of course, as the density of atoms in the solid is altered, the strength of the Kr-pore wall interaction, which is given by a sum over all Kr-solid atom pair-wise terms, will also vary. We wish to hold the adsorption energy of these systems constant so as to give a fair comparison of their simulated thermodynamic properties. This is done by adjusting the well depth ϵ_{Kr-s} of the pairwise interaction to give a constant value of -8.80 $\pm$ 0.1 kJ/mole for the average of the adsorption energy U_m (the minimum in the gas-pore wall potential) obtained by averaging over the surface. Both of these averages were calculated by evaluating the minimum energy and its position on a net of pore wall positions spaced 1×0.5 Å apart around the circumference and along the pore axis, respectively. (This amounts to at least 3000 values of each quantity).

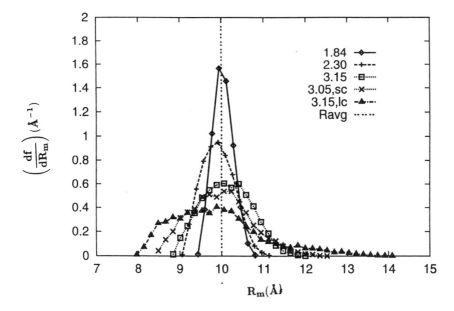

Figure 1 Distributions of R_m, the radial distance between a krypton atom at a minimum in the atom-pore wall interaction energy and the axis of the cylindrical pore, are shown here for each of the five model pores considered.

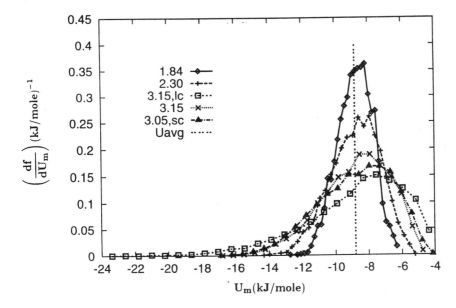

Figure 2 Distributions of U_m, the minimum Kr-pore wall interaction energy for different positions on the walls, are shown for the five model pores considered.

Unweighted numerical averages over these values were then taken to give $\overline{R_m}$ and $\overline{U_m}$. In addition, histograms were constructed that show the frequency of occurrence of given R_m and U_m. Specifically, df/dR_m shown in Figure 1 is the fraction of the total R_m values between R_m and $R_m + dR_m$, divided by dR_m. A similar plot for U_m is shown in Figure 2. The widths of these distributions are measures of the physical roughness and the energetic heterogeneity of the pore wall for a given model. As the walls approach perfect smoothness, both distributions will approach delta functions.

We will see that simulations of the thermodynamic properties of krypton sorbed in these three pores at 120 K do not show pronounced changes as the heterogeneity increases. Consequently, two other pore walls were considered, both with $\sigma_s \simeq 3.1$ Å, but with surfaces that have been roughened by randomly deleting atoms from the pore wall. One, which is denoted by "sc" ("small corrugation") has had 1/3 of the atoms originally in the exposed layer of atoms in the wall removed and the second, which is denoted by "lc" ("large corrugation") has had 1/2 the atoms deleted from the inner-most 1.5 layers of wall atoms (i.e., from the first 1.5 σ_s). The distributions for these systems given in Figs. 1 and 2 show enhanced roughness and energetic heterogeneity relative to the "uncorrugated" cases. Detailed inspection of the topography of the pore-wall indicates that "nanochasms" are formed in the lc case. These consist of areas where R_m increases rapidly, such that a nanoscale steep-walled crevise in the pore wall is created. This accounts for the small number of quite large R_m in the distribution for the lc pore in Figure 1 and also gives rise to the very large negative U_m shown in the distribution in Figure 2 for this pore - an atom in one of the nanochasms will be nearly surrounded by solid atoms and thus can have a large negative gas-solid interaction.

3 RESULTS

Table I shows some relevant parameters used as input and obtained as output from Henry's Law calculations or the GCMC computer simulations[6,7] for Kr sorbed in these five model pores. The volumes V_{hen} were obtained in the Henry's Law calculations which involve an integration of a probe atom over the pore volume. The limits of this volume were defined to be the surface where the gas-solid interaction energy becomes positive by $3 \epsilon_{Kr-s}$, which signifies a strong repulsion for positions outside that surface.

Table 1: Values of selected parameters for the Kr-pore systems

Pore structure	Z_p (nm)	ϵ_{Kr-s}/k (K)	$\sigma_{Kr-s}(\text{Å})$	$V_{hen}(\text{Å}^3)$	N_m/Z_p(atoms/nm)
σ_s=1.84Å	3.31	62.3	3.40	1.16×10^4	34
σ_s=2.30Å	4.14	101.3	3.45	1.44×10^4	33
σ_s=3.15Å	5.67	190.3	3.50	2.09×10^4	31
σ_s=3.05Å, sc	5.49	191.3	3.50	1.97×10^4	$\simeq$35
σ_s=3.15Å, lc	5.67	238.4	3.50	1.52×10^4	$\simeq$28

In the Table, N_m denotes the atoms adsorbed in the monolayers for each pore. These were obtained from plots of the radial density such as as those shown in Figures 3 and 4 - appropriate integrations over the outermost peaks in these plots yield the total

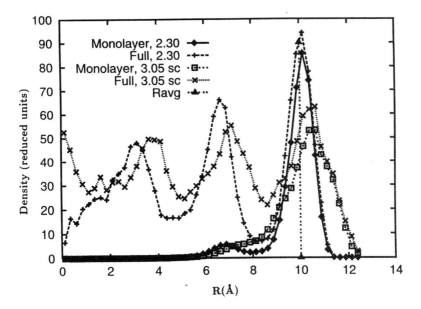

Figure 3 Krypton densities that have been averaged over the pore wall are plotted as a function of R, the distance from the pore axis for two pore loadings: one for a loading just less than the loading at which capillary condensation occurs denoted "monolayer"), and one at a loading just larger than the value at which the condensation is complete (denoted "full"). These curves are shown here for two pores, as indicated in the Figure.

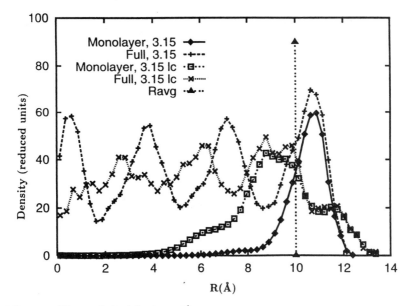

Figure 4 Same as Figure 3, but for two other pores.

atoms in the monolayers. Since the pores are of different lengths Z_p (see column 2 in the Table), the monolayer capacities per nanometer of pore length are shown. (The periodic boundary conditions applied in the axial (Z) direction mean that these pore lengths are primarily bookkeeping numbers.) Note that the monolayer becomes poorly defined in the corrugated pores where the surface roughnesses are so great that the expected peaks in the density are quite smeared out. This gives imprecise estimates of N_m relative to those for the pores with smoother surfaces. Also shown in Figures 3 and 4 are radial density curves for some nearly full pores. In this case, layering is obviously present when the pore wall is reasonably smooth, with three distinct layers formed when σ_s=2.30 Å and four layers when σ_s=3.15 Å (because the outermost layer is somewhat further from the pore axis). Moderate layering can be seen in the sc pore, but almost none in the lc case.

In addition to the Kr densities, the GCMC simulations yield two thermodynamic properties: the average potential energies of the atoms in the pore and the isotherms, which are essentially plots of the average number of atoms in the pore as a function of their chemical potential. Since chemical potential is directly related to pressure in systems such as this, the isotherms are plotted in the usual way in Figure 6. Rather than showing the average total energy of the Kr in the pore, only the part of this energy due to the gas-pore wall interaction is shown in Figure 5. Values of this energy in the limit of zero atoms in the pore were obtained from standard Henry's Law calculations[8] and agree well with extrapolations of the simulated energies at finite coverages. It is evident that the pore with σ_s=1.84 Å is almost homogeneous, with gas-solid energies that hardly vary up to monolayer completion. However, as the energy distributions shown in Figure 2 become broader, the decay in average gas-solid energy with increasing

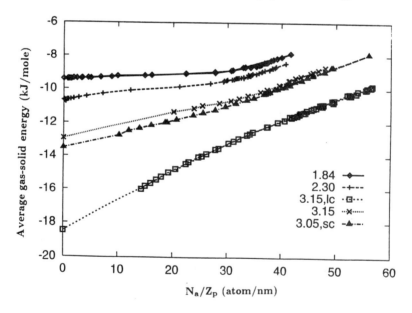

Figure 5 The average gas-solid interaction energy is plotted here for coverages up to the approximate monolayer values for the five pores considered.

coverage becomes more pronounced. It is perhaps surprising to see such a significant variation in average energy for a set of pores that have been constructed to have a fixed $\overline{U_m}$; however, one should remember that this is an unweighted average over the entire surface, whereas the statistical mechanical average is Boltzmann-weighted and gives preference to atoms with the strongest gas-solid interactions. Indeed, the Figure shows the usual situation for heterogeneous surfaces: filling of the strongest sites at low coverage, followed by gradual filling of the weaker sites as coverage increases.

Simulated isotherms are shown for these systems in Figure 6. Hysteresis loops with vertical (to within the precision of the simulations) adsorption and desorption branches were found for all five pores. For all pores, condensation begins (and evaporation ends) at coverages very slightly larger than monolayer, and condensation ends (and evaporation begins) with nearly full pores. The variations in pore-wall heterogeneity studied here primarily affect the transition pressures. The hysteresis loops are wider and the condensation pressures are generally higher for the smoother surfaces. However, these changes are not large, and lead to the conclusion that the surface heterogeneity mainly affects monolayer formation. Since evaporation-condensation is observed only in pores with completed monolayers, the heterogeneity is "shielded".

One unexpected effect is that the imposition of corrugation seems to give rise to larger capacity for pore filling per unit length than for the pores with smoother surfaces but nominally the same average radius (10.0 Å). This seems also to be at odds with

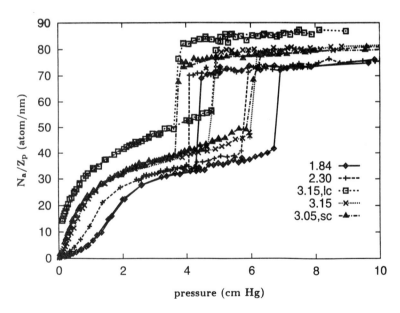

Figure 6 Sorption isotherms are shown for Kr at 120 K in the pores considered here. The amount adsorbed is given in atoms per nm of pore length. The straight-line portions of the isotherms represent capillary condensation and evaporation.

the values of V_{hen} shown in Table 1. Division of these numbers by Z_p gives effective areas of 350, 348, 369, 359, and 268 $Å^2$ for these pores (in order of the entries in Table 1). Taking 360 as a typical area, one obtains 10.7 Å for the Henry's law pore radius. This is slightly larger than $\overline{R_m}$, as expected because the Henry's law calculation is related to the locations of the points where the gas-solid interactions are positive, whereas R_m is equal to the position of a minimum in this potential. The anomaly in this calculation is for the lc pore. It appears that the surface roughness is large enough in this case to require a more sophisticated analysis than that given here, especially since the strongest interactions are found for atoms that are significantly further from the pore axis than the average because they are in nanochasms. The fluid density in these regions of strong adsorption may well be significantly larger than elsewhere in a filled pore which could cause the enhanced pore filling shown in Figure 6.

Finally, note that the strong adsorption energies found in the rough-walled pores give rise to enhanced adsorption at low coverages, as expected. This is also reflected in the Henry's Law constants for these systems. (Calculated but not listed here.)

Although the changes in geometric heterogeneity modeled here have not been sufficient to destroy the capillary condensation phenomenon in these micropores, one should remember than chemical heterogeneity effects can be very large, depending upon the type and number of the heteroatoms introduced on the solid surface. An example of this can be found in a recent simulation of water on a graphite adsorbent with associating sites.[9]

Acknowledgement This work supported by a grant from the Division of Materials Research of the NSF.

REFERENCES

1. K. E. Gubbins, M. Sliwinska-Bartkowiak and S.-H. Suh, *Mol. Sim.*, 1996, **17**, 333.
2. S. J. Gregg and K. S. W. Sing, 'Adsorption, Surface Area and Porosity, 2nd Ed', Academic Press, London, 1982.
3. 'The Structure and Properties of Porous Materials', ed. D. H. Everett and F. S. Stone, Butterworths, London, 1958.
4. W. A. Steele, 'The Interaction of Gases with Solid Surfaces', Pergamon, Oxford, 1974.
5. M. J. Bojan and W. A. Steele, in 'Fundamentals of Adsorption', Proc. 5th Conference on Fundamentals of Adsorption, ed. M. D. LeVan, Kluwer, Dordrecht, 1996.
6. M. P. Allen and D. J. Tildesley, 'Computer Simulation of Liquids', Oxford Univ. Press, Oxford, 1987.
7. D. Nicholson and N. G. Parsonage, 'Computer Simulation and the Statistical Mechanics of Adsorption', Academic Press, New York, 1982.
8. Ref. 4, chap. 3.
9. L. F. Vega, E. A. Müller, L. F. Rull and K. E. Gubbins, in 'Fundamentals of Adsorption', Proc. 5th Conference on Fundamentals of Adsorption, ed. M. D. LeVan, Kluwer, Dordrecht, 1996, p. 993.

TRANSPORT DIFFUSION OF ADSORBATES IN MICROPORES: SIMULATION IN MODEL SYSTEMS

D. Nicholson, R.W. Adams, R. F. Cracknell* and G. K. Papadopoulos

Department of Chemistry,
Imperial College,
London, SW7 2AY

1 INTRODUCTION

The rate at which a fluid can move through porous spaces is important in many applications of porous materials. Experimental studies of the dynamics of adsorption have considerable potential for characterisation, but are hindered by lack of understanding at the fundamental molecular level. Where a distribution of pore sizes exists, the smallest pore widths are likely to play the most important role in controlling transport processes. Computer studies can be a useful adjunct to the study of these processes, since it is difficult to establish well defined porous systems on which to carry out experiments. Experimental data must therefore be interpreted on the basis of some assumed model material. Simulation enables molecular processes to be analysed in some detail for a specific and well-defined model system. It can thereby provide an idealised reference system as an aid to interpretation of experiments, or as a starting point for the construction of more elaborate models. Macroscopic theories, applicable to bulk fluids, may not be valid inside very small pore spaces, partly because of constraints on packing and rotational freedom, and partly because the pore fluid is non-uniform, due to the effects of the adsorbent field. The influence of these factors can be expected to vary with adsorbent material and pore size, with temperature and with the concentration and molecular character of the adsorbate inside the pore.

Several earlier papers have reported simulation studies of flow in model micropores. Both equilibrium molecular dynamics (EMD) and non-equilibrium molecular dynamics (NEMD) have been used by Davis and co-workers to study the transport of 12-6 molecules through cylinders and slit pores[1-5]. MacElroy and co-workers[6-9] reported EMD simulation for slit and cylindrical pore models, and more recently[10] have used NEMD with 12-6 spherical molecules in random sphere packs. Fischer et al[11,12] studied flow in cylinders. Several groups have examined zeolitic systems; in many cases these studies have been restricted to the diffusion of isolated molecules, although Maginn et al[13] extended their simulations to EMD and NEMD methods covering a range of concentrations for methane in silicalite. Demi and Nicholson[14-19] used EMD to study concentration, geometry and temperature dependence of self diffusion of spherical 12-6 molecules in cylindrical pores and in a linked sphere cylinder model, self diffusion in linked cylinders has also been studied[20]. The majority of these papers confine their attention to self diffusion coefficients, or adopt the view that the flow is wholly viscous.

---

* Present address: SE7, Shell research Ltd., PO Box 1, Chester CH1 3SH, UK

More recently we have examined the transport of methane at bulk supercritical temperatures in a slit pore, as a function of concentration[20-23, 25]. The results were interpreted in terms of the theory established by Mason and co-workers[26-28] which derives from the statistical mechanical investigations of Kirkwood *et al.*[29,30]. This formulation, also employed by MacElroy[10], reveals that both the Fick's law and the viscous flow descriptions of transport are possible, but that a full description needs to incorporate both modes of transport. It also becomes clear that a full study of the phenomena is not practicable with EMD alone. Our approach is closely linked to conventional adsorption data (isotherms and heat curves)[31] and has employed GCMC, EMD and an NEMD method[23] which is still evolving.

Here we report some new results within this conceptual framework. We review briefly the background theory and its relationships to simulation techniques and demonstrate that several interesting features of transport behaviour can occur in microporous spaces.

2 SINGLE COMPONENT FLOW IN PORES

The isothermal steady state flux J of a single component in the in the r_α direction can be written in the form,

$$-J = \frac{\rho D_o}{kT} \nabla\mu + \frac{\rho B_o}{\eta} \nabla p \tag{1}$$

where ∇ here stands for $\partial/\partial r_\alpha$ and $\rho(r_\alpha)$ is the mean number density at a cross section. The first term on the right hand side of eqn. (1) is the diffusive flux, driven by the chemical potential gradient. The diffusion coefficient can be expressed as an integral over the ensemble averaged time correlation of the fluctuating streaming velocity, $u(t)$, as

$$D_o = \frac{1}{Nd} \int <u(0) \bullet u(t)> dt \tag{2}$$

where N is the total number of particles in the system, d is the dimensionality ($=2$ in a slit pore, 1 in a cylindrical pore), and the components of u can be expressed in terms of the molecular velocities by

$$u_\alpha = \sum_i v_{i\alpha}/N \tag{3}$$

In principle D_o can be obtained from EMD simulation, but in practice, long time effects make it difficult to calculate D_o accurately[13]. It is readily shown[24] that D_o can be split into two components, a cross correlation term D_ξ and a self correlation term, D_s. The latter is found from EMD, either using the molecular velocity autocorrelation function, or from the Einstein expression, using the mean square displacement,

$$D_s \atop Lt, t \to \infty = \frac{<(r_i(t) - r_i(0))^2>}{2td} \tag{4}$$

The second term on the right hand side of eqn. (1) accounts for the flux due to momentum transfer under the gradient in the transverse component of the pressure in the adsorbate fluid[25], and contains the viscosity η and the geometrical term, B_o. The latter is obtained from solution of the Navier-Stokes problem and is given by, $B_o=H^2/12$ in slits or by, $B_o=R^2/8$ in cylinders.

In spaces that are only a few molecular widths across, the continuum fluid assumptions underlying this result must be regarded as no better than a first approximation, likewise the viscosity will differ from its bulk value[2-4]. Nevertheless cooperative effects involving momentum transfer are still possible. It is easy to demonstrate within the limitations of these approximations that the second component of (1) is expected to become increasingly important as adsorbate concentration increases and as the capillary width increases[24,25].

With the aid of the Gibbs-Duhem equation and the standard relationship between chemical potential and fugacity, f ($\approx$adsorptive gas phase pressure), eqn. (1) can be put into Fickian form[24, 25], in order to relate simulation data more directly to experimental results,

$$J = -\left[D_o + \frac{kT \rho B_o}{\eta} \right] \left(\frac{\partial \ln f}{\partial \ln \rho} \right) \nabla \rho = -D \nabla \rho \tag{5}$$

It should be noted that in eqns. (1) and (5) the flux refers to the molecules in a defined "physical" volume V, (the space bounded by the centres of the graphite wall atoms in the z-direction) whereas adsorption experiments always measure excess amounts with respect to the internal or "chemical" pore volume. The term $\partial \ln f / \partial \ln \rho$ is the inverse slope of the logarithmic isotherm, also known as the Darken factor. The Darken transport diffusion coefficient is given by

$$D_{trans} = D_s \left(\frac{\partial \ln f}{\partial \ln \rho} \right) \tag{6}$$

It can be seen from eqn. (5) that this is the self diffusion contribution to the total flux. This quantity is of particular interest, since it is readily obtained from a combination of GCMC and EMD simulations, and is expected to be the dominant contribution to transport in small pores at low adsorbate concentration. Available evidence suggests that D_f is a very small fraction of D_s (~5%), even at liquid densities[32].

To gain further insight we have developed a non-equilibrium molecular dynamics method that calculates the flux directly[23]. Here we report some results for slit pore and cylindrical pore models.

3 SIMULATION METHODS

Methane was modelled as 12-6 spheres with the parameters σ=0.3812nm, ϵ/k=142.1K. The walls of the slit pores were represented as a graphitic continuum interacting with the adsorbate through a 10-4-3 potential[21-23,25] with ϵ_s/k=28K, σ_s=0.340nm, ρ_s=114nm^{-3}, Δ=0.335nm. GCMC was used to obtain isotherms and fluctuation properties[31]. EMD runs were initiated from an equilibrated GCMC configuration, temperature was maintained constant at 296.2K (T*=2.0) by Gaussian isothermal constraint. Self diffusion coefficients were found from Einstein plots (eqn. (4)) and from VACF plots. Typically 1500 timesteps (Δt*=0.005) were rejected initially and 1500 time origins used to obtain ensemble averages. The NEMD method measures the flux under a concentration gradient by direct counting at constant temperature, and has been described elsewhere[23]. Diffuse reflection boundary conditions were applied.

Slit widths in the micropore size range were chosen. The physical pore widths, H were 0.953nm, 1.144nm and 1.334nm respectively. In units of the hard sphere size of methane

(σ=0.3812nm), these widths are 2.5σ, 3.0σ and 3.5σ. The internal or chemical widths are approximately, 1.8σ, 2.3σ and 2.8σ. Thus the narrowest pore does not admit two full monolayers, the intermediate pore is large enough to do so, whilst the widest pore is just able to admit 3 layers if the central layer is close packed between the two wall layers at the Lennard-Jones minimum (1.122σ).

In order to find D_{trans} logarithmic isotherms were plotted and fitted to the function

$$\ln f = A + \ln \rho + B_1 \rho + B_3 \rho^3 + B_5 \rho^5 \qquad (7)$$

where A and B_i are parameters to be determined from the simulation data. This functional form has been substituted for that used in previous work[21,22] since it also gives excellent fits to subcritical isotherms.

4 RESULTS AND DISCUSSION

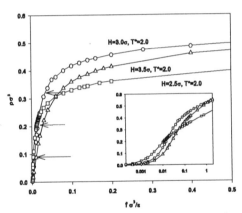

Figure 1 *Adsorption isotherms (number density versus fugacity) for methane in model graphitic slit pores at 296.2K.*

The systems are all supercritical and yield the simple type I isotherms illustrated in Figure 1. Note that in the isotherms, and in subsequent figures, the density refers to the number of adsorbate molecules inside the physical volume of the pore. The expected characteristics can be observed: (i) The initial slopes become steeper as the pore width decreases, due to the effect of the potential field overlap on the Henry law constant. (ii) The density approaches saturation at lower pressures and reaches higher values as the pore width increases and packing consequently becomes more efficient. The combined effect of these two features is to cause the isotherms to intersect each other, as can be seen from the log plots in the inset.

Although the isotherms are straightforward, and bear a close similarity to one another, the transport properties show remarkable variations with pore size and adsorbate density. Figure 2 shows reduced diffusion coefficients ($D^*=Dm^{1/2}\epsilon^{1/2}\sigma^{-1}$) over the whole range of adsorbate density. Several interesting features occur at low densities, and these have been discussed elsewhere[21,22]. The 'effective' diffusion coefficient D, defined as the proportionality between the NEMD flux and the concentration gradient (cf eqn. (5)), is shown as open triangles in Figure 2. In all cases, D increases rapidly beyond the Henry law region of adsorption. In the smallest pore, this increase takes the form of a rather steep transition. In all cases there is a strong deviation away from the Darken transport diffusion coefficient, shown as filled circles, and calculated using EMD and GCMC from eqn. (6). There is also a notable contrast in behaviour between the slits having width 2.5σ and 3.5σ and the intermediate width of 3.0σ. In the smallest and largest pores, D and D_{trans} both decrease again at very high density. This does not occur for $H=3.0\sigma$.

Thus, although the Darken transport coefficient is able to exhibit the general trends of D in these slit shaped micropores, cooperative effects are increasingly important as adsorbate density increases, and D_{trans} seriously underestimates the magnitude of D in high density regions. Thus even though the molecules can pass each other in the third dimension, the constraint on their movement imposed by the proximity of the pore walls leads to a significant difference in transport behaviour.

The adsorption isotherms in Figure 1 do not exhibit any features which correlate with the onset of cooperative effects, and it is of interest to examine fluctuation properties for indicators of these effects. Previously[25] it was shown that changes in compressibility (number fluctuations) and fluctuations in intermolecular separation correlated with the rapid increase in D at $H=2.5\sigma$. The differential heat of adsorption can be calculated from a cross fluctuation of energy and number[31].

$$Q = -\left(\partial <U_N> / \partial N\right)_{V,T}$$

$$= \frac{f(N,U_N)}{f(N,N)} \qquad (9)$$

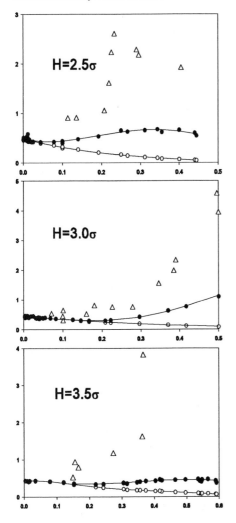

Figure 2. *Diffusion coefficients in MD units (see text) for methane at 296K in model graphitic slit pores as a function of adsorbate density, $\rho\sigma^3$. The self diffusion coefficient D_s and Darken transport coefficient, D_{trans}, are shown as open and filled circles respectively. The open triangles show the total (effective) Fickian diffusion coefficient.*

where U_N is the total potential energy and $f(X,Y)=<XY>-<X><Y>$. In simulation studies this can be separated into adsorbate-adsorbate (molecule) and adsorbate-adsorbent (wall) parts. The differential heats are shown in Figure 3. The molecule part increases in an approximately linear fashion with density until repulsions between molecules begin to make a significant contribution. There is also a decrease in the wall part at higher densities, since molecules entering (or leaving) the adsorbate are not always positioned at the potential energy minimum at higher densities. The combination of these two effects gives rise to a maximum in the overall differential heat in

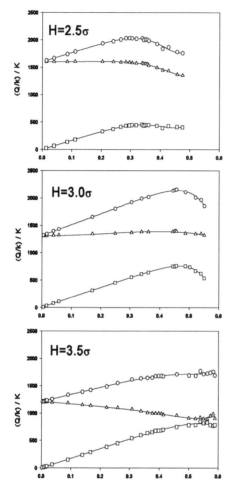

Figure 3. *Differential heats of adsorption (divided by Boltzmann constant) (see eqn. (8)) plotted against adsorbate density for methane in graphitic slit pores at 296K. The heats are separated into molecule part (squares) and wall part (triangles). The total heat is shown as circles.*

the smaller pores. The onset of these deviations from linearity occurs close to the densities where D shows a strong increase in comparison to D_{trans}, suggesting that increased momentum exchange from more repulsive collisions initially favours cooperative transport. Unfortunately these deviations are far too small to be detected experimentally and separation of experimental heats into wall and molecule parts is not possible. It should also be noted that the heats shown here are calculated with respect to the total, not the excess amount adsorbed. At high density the molecule part of the heat falls off very rapidly for $H=3.00\sigma$ in contrast to the other two pore sizes. This is because the very dense monolayers do not have available holes. In the other two pores, the adsorbate is far more disordered, so that molecular insertion is more readily accommodated.

At $H=3.5\sigma$, cooperative flow becomes significant as soon as the molecular concentration develops in the centre of the pore, enabling effective momentum exchange. The singlet distribution functions in Figure 4 illustrate the development of adsorbate concentration across the pore as the mean density increases. In the intermediate sized pore ($H=3.0\sigma$), D and D_{trans} increase continuously even at very high densities. Singlet distribution functions (Figure 4) confirm that two virtually distinct molecular layers form at this pore size with very low concentration in the central part of the slit. Transport parallel to the pore wall is therefore facilitated. In the other two slits ($H=2.5$ and $H=3.5$), there is substantial overlap between molecules in the z-direction. Motion parallel to the walls, at very high densities, is thereby frustrated by molecular repulsions, which accounts for the decline in D and D_{trans}. It is interesting to note that singlet distributions calculated in NEMD simulations are identical to those from GCMC (Figure 4) showing that transport does not result in any lateral reordering of the molecules.

The contrast in the concentration dependence of D between pores of different widths leads to interesting variations in D with pore size. In a porous material there will generally be a

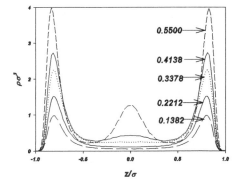 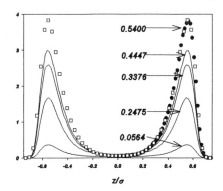

Figure 4. *Singlet distribution functions (in reduced units) for the slits of width H=3.5σ (left hand panel) and H=3.0σ (right hand panel) for the mean densities shown on the curves. The curve shown by open points is from NEMD calculations, the filled points and all the lines are from GCMC.*

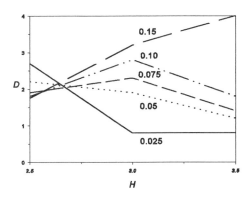

Figure 5. *Total diffusion coefficient as a function of pore width at the reduced fugacities (fσ³/ε) shown on the graphs.*

distribution of pore sizes; one way to represent this is by network modelling, and it is of interest that certain pore sizes can carry different amounts of flux, depending on adsorbate concentration. The common parameter in these circumstances will be the fugacity; this in turn determines the concentration of adsorbate in each pore through the adsorption isotherm. Figure 5 shows D as a function of pore width at a few values of fugacity for the system studied here. It may be observed that, whilst the majority of the flux will be carried by the smaller pores at low pressures, the opposite is the case at higher pressures.

CONCLUSIONS

In order to gain an understanding of transport of adsorbate in micropores, several types of simulation are needed. The results presented here, for a supercritical Lennard-Jones fluid in slit pores, show that in this system, cooperative flow becomes important at high adsorbate density. The total diffusion coefficient varies strongly with both concentration and with pore size.

Acknowledgements. We would like to thank Keith Gubbins, Neville Parsonage, Nick Quirke, Martin Schoen and Steve Tennison, for useful discussions, NATO and the CEC for support grant JOE3-CT95-0018.

References

1. J. J. Magda, M. Tirrell, and H. T. Davis, *J. Chem. Phys.*, 1985, **83**, 1888.
2. T. K. Vanderlick, and H. T. Davis, *J. Chem. Phys.*, 1987, **87**, 1791.
3. I. Bitsanis, M. Tirrell, and H. T. Davis, *J. Chem. Phys.*, 1987, **87**, 1750.
4. I. Bitsanis, S. A. Somers, H. T. Davis and M. Tirrell, *J. Chem. Phys.*, 1990, **93**, 3427
5. S. A. Somers and H. T. Davis, *J. Chem. Phys.*, 1992, **96**, 5389.
6. S-H. Suh and J. M. D. MacElroy, *Mol. Phys.*, 1986, **58** 445.
7. J. M. D. MacElroy and S.-H. Suh, *Mol. Phys.*, 1987, **60**, 475.
8. J. M. D. MacElroy and S.-H. Suh, *Mol. Simulation*, 1989, **2**, 313.
9. J. M. D. MacElroy and K. Raghavan, *J. Chem. Soc. Faraday Trans.*, 1990, **87**, 1971.
10. J. M. D. MacElroy, *J. Chem. Phys.*, 1994, **101**, 5274.
11. J. Fischer, U. Heinbuch, M. Wendland, and S. Salzmann, *Fundamentals of Adsorption III*, 1991, A. Mersmann, S. Scholl eds. p281, Engineering Foundation, New York.
12. U. Heinbuch, and J. Fischer, *Phys. Rev. A*, 1989, **40**, 1144.
13. E. J. Maginn, A. T. Bell and D. N. Theodorou, *J. Phys. Chem.*, 1993, **97**, 4173.
14. T. Demi, and D. Nicholson, *Mol. Simulation*, 1991, **5**, 381.
15. T. Demi, and D. Nicholson, *Mol. Simulation*, 1991, **7**, 121
16. T. Demi, and D. Nicholson, *Langmuir*, 1991, **7**, 2342.
17. T. Demi, and D. Nicholson, *J. Chem. Soc. Faraday Trans.* ,1991, **87**, 3791.
18. T. Demi, *J. Chem. Phys.*, 1991, **95**, 9242.
19. T. Demi, and D. Nicholson, *Fundamentals of Adsorption, IV*, 1993, ed M. Suzuki, p137, Kodanshu, Tokyo.
20. S. Sokolowski, *Mol. Phys.*, 1992, **75**, 1301.
21 R. F. Cracknell, D. Nicholson and K. E. Gubbins, *J. Chem. Soc.Faraday Trans.*, 1995, **91**, 1377.
22. R. F. Cracknell and D. Nicholson, *Fundamentals of Adsorption, V*, 1996, ed M. D. LeVan, p683, Kluwer Academic Publishers, Boston.
23. R.F. Cracknell, D. Nicholson, and N. Quirke, *Phys. Rev. Letts.*, 1995, **74**, 2463.
24. D. Nicholson, *J. Membrane Sci.*, submitted.
25. D. Nicholson, R. F. Cracknell and N. Quirke, *Langmuir*, 1996, **12**, 4050.
26 E. A. Mason and L. A. Viehland, *J. Chem. Phys.*, 1978, **68**, 3562.
27 E. A. Mason and H. K. Lonsdale, *J. Membrane Sci.*, 1990, **51**, 1.
28. E. A. Mason and L. F. del Castillo, *J. Membrane Sci.*, 1985, **23**, 199.
29 J. G. Kirkwood, *J. Chem. Phys.*, 1946, **14**, 180.
30 R. J. Bearman and J. G. Kirkwood, *J. Chem. Phys.*, 1958, **28**, 136.
31. D. Nicholson and N. G. Parsonage, Computer Simulation and the Statistical Mechanics of Adsorption, Academic Press, 1982.
32. M. Schoen, and C. Hoheisel, *Mol. Phys.*, 1984, **52**, 33.

SIMULATION STUDIES OF PORE BLOCKING PHENOMENA IN MODEL POROUS NETWORKS

M. W. Maddox[1], N. Quirke[2], and K. E. Gubbins[1]

[1]Cornell University, School of Chemical Engineering,
Olin Hall, Ithaca, New York, 14853, USA
[2]Department of Chemistry, University of Wales at Bangor,
Gwynedd, LL57 2UW, UK

1 INTRODUCTION

In recent years there have been numerous theoretical and molecular simulation studies of porous carbons based on the slit pore model. The material is modeled as having slit pores which are all of the same width, or as unconnected slit pores having a range or pore sizes.[1] Much has been learned from such studies. However, real carbons have pores of different sizes and shapes that are interconnected at various points in the matrix. The adsorption behavior is strongly influenced by such connection points, since they lead to fluid-wall intermolecular potential barriers (e.g. where a small pore joins a larger pore) and also to regions having unusually strong fluid-wall attraction (in 'corners', where the pore surface is concave as seen from the adsorbate phase). Thus, on adsorption some regions of the pore network may not fill due to the potential barriers present; while on desorption it may be difficult or impossible to empty the adsorbate trapped in corners, i.e. the adsorption may be irreversible, leading to estimates of pore volume that are too low. In this paper we consider a simple model that provides for such connectivity effects. The model consists of two narrow pores, connected into a larger pore, the connections being on opposite sides of this larger pore (Fig. 1). One end of each of the narrow pores is in contact with a bulk fluid phase. We consider the particular case where the narrower pore has a width of only two or three molecular diameters.

2 METHODS

The connected pore model used is illustrated in Figure 1. The constant pressure Grand Canonical Molecular Dynamics (GCMD) method was used for the simulations[2,3]. In the

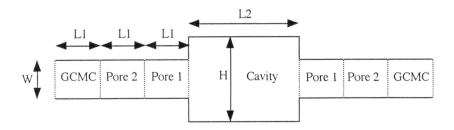

Figure 1. Schematic view of the slit pore network model.

laboratory system the regions of the narrower pore labeled GCMC would be in contact with the bulk fluid phase, and in the simulation these are equilibrated with this external phase. In this region, attempts to create and delete molecules are made in the usual GCMC fashion; 50 attempts to create, and a further 50 attempts to delete, a molecule were made between each MD step in order to equilibrate this part of the pore network with the bulk phase. In the remainder of the pore network conventional MD was used. Periodic boundary conditions (in the direction along the pore axis) and the minimum image convention were used. It is convenient to label the other sections of the pore network as shown in Figure 1. The larger pore region is termed the cavity; the pore 1 section is the part of the narrower pore in which adsorbate molecules are affected by the junction between the two pores, while in the pore 2 region the adsorbate molecules feel the same fluid-wall potential as in a pore of infinite length. The pore lengths shown in Fig. 1 were taken to be $L1=3\sigma_{ff}$ and $L2=7.2\sigma_{ff}$, where σ_{ff} is the fluid-fluid Lennard-Jones parameter. At low densities 500,000 timesteps were used for equilibration, and a further 500,000 for obtaining averages. For the higher densities somewhat shorter runs were often used, the run length being adjusted depending on the time needed to reach equilibrium.

Three network models were studied, each defined by the ratio of pore widths, H/W; pore widths were defined to be the distance separating the planes through the carbon atoms in the surface layer of each opposing wall. Since we fixed H, the values of W were restricted to H-nΔ, where $\Delta=1.79\sigma_{ff}$ is the distance between graphite layers and n is an integer. The three networks studied will be referred to as the 4/2 ($H=4\sigma_{ff}$, $W=2.21\sigma_{ff}$), 6/2 ($H=6\sigma_{ff}$, $W=2.42\sigma_{ff}$), and the 7/3 ($H=7\sigma_{ff}$, $W=3.42\sigma_{ff}$) junction. In addition, simulations of the adsorption behavior of infinitely long straight pores of the same width W as those in the network simulations were carried out, in order to enable comparisons with behavior in the small pores of the network to be made. The forces on the adsorbate molecules, as well as their intermolecular potentials, were precalculated for each point on a grid of 160x150 points, for each model pore as described previously[4].

The fluid-wall interaction potential at various points in the 6/2 network is shown in Figure 2. The potential shown is that for a single adsorbate molecule in contact with one of the pore walls, as it follows the path of minimum interaction potential along the wall. The pronounced maximum arises from the reduced interaction with the wall that

(6/2 potential)

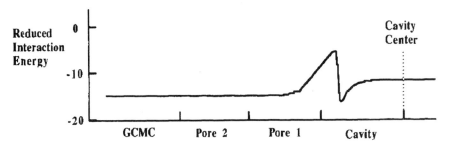

Figure 2. Fluid-wall interaction potential for a single adsorbate molecule in contact with one wall of the pore (i.e. at the potential minimum for the monolayer), as it traverses the pore network, for the 6/2 network model.

occurs at the junction of the two pores, while the deep minimum corresponds to adsorption in the corners of the cavity. Similar plots are obtained for the other two networks.

3 INTERMOLECULAR POTENTIAL MODELS

The spherical Lennard-Jones (LJ) potential was used for both the fluid-fluid and fluid-solid interactions. For the fluid the parameters were chosen to model nitrogen, σ_{ff}=0.375 nm and ε_{ff}/k=95.2 K, and the potential was cut off at a separation of $5\sigma_{ff}$, the interaction being negligible at this distance.

The pore network was made up of a slit formed by two structureless graphite slabs, on the inner surfaces of which were superimposed structured graphite layers.[4] In the cavity region the structureless walls were covered with just one such structured layer, while in the small pore region there were further structured graphite layers - a total of three such layers for the 6/2 and 7/3 networks, and two such layers for the 4/2 junction. The usual parameters for carbon atom and layer spacing for graphite were used.[7] Fluid-solid interaction parameters were obtained from the Lorentz-Berthelot rules.

4 RESULTS AND DISCUSSION

All calculations were performed at a temperature of $T^*=kT/\varepsilon_{ff}$=0.823 (T=78.3 K); at this temperature the vapor pressure of LJ nitrogen is 0.441 bar.[5] Chemical potentials were converted to pressure using standard equations of state for the LJ fluid.[5,6] Both adsorption and desorption calculations were performed. For adsorption calculations, a simulation was first performed starting with the pore network empty. The final configuration of this initial run was then used to start the next simulation at a higher bulk pressure, and this process was repeated until the relative pressure was close to unity. The

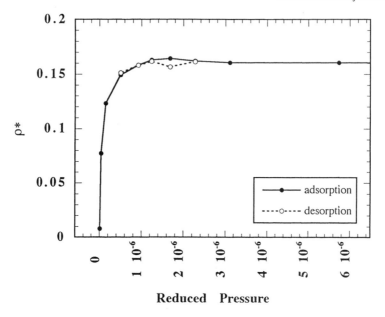

Figure 3. Adsorption and desorption isotherms for the 4/2 network model for T*=0.823, from GCMD simulations.

final configuration at the highest pressure was then used as the initial configuration for a lower pressure, and this procedure repeated to generate the desorption curve.

For the 4/2 and 6/2 networks, with small pores of widths W*=2.21 and 2.43, respectively, the nitrogen adsorbate was adsorbed readily into the small pores until they were eventually filled. But molecules did not penetrate into the inner cavity even at the highest pressures studied, due to the strong potential barrier at the junction of the two pores (see Fig. 2). A result of this behavior is that the maximum overall density in the pore network is very low, $\rho^*=\rho\sigma^3$=0.15-0.16, a direct consequence of the large amount of empty space in the cavity. The result for the 4/2 network is shown in Figure 3; for the 6/2 network the adsorption isotherm is very similar.

Another interesting feature of these networks with small pores is that a freezing transition is observed in the smaller pores, presumably as a result of the potential barrier at the pore junction. This transition occurs at $P/P^o \approx 2 \times 10^{-6}$ for the 4/2 network and at $P/P^o \approx 1 \times 10^{-4}$ for the 6/2 network. Desorption occurred along the adsorption path (see Fig. 3), except in the immediate neighborhood of the freezing transition. This freezing transition is further discussed below.

The behavior of the 7/3 network was qualitatively different from those having the narrower small pore. Although the potential barrier at the pore junction prevented adsorption into the inner cavity for pressures below $P/P^o \approx 5 \times 10^{-5}$, at or slightly above this pressure there was a steady but slow adsorption into the cavity, to form a rough monolayer. Further increase in pressure led to further adsorption, until the cavity was filled. This is shown in Figure 4. Desorption shows a hysteresis loop at high pressure (caused by the high, near-solid, density of the adsorbate in the narrow pore). There is

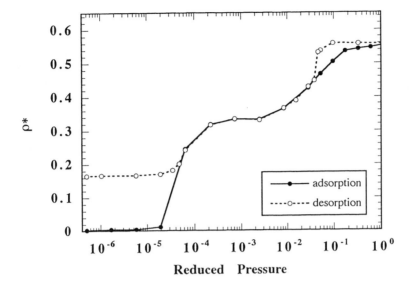

Figure 4. Adsorption and desorption isotherms for the 7/3 network model for T*=0.823, from GCMD simulations. A log scale is used for the relative pressure to show the irreversible adsorption at low pressure.

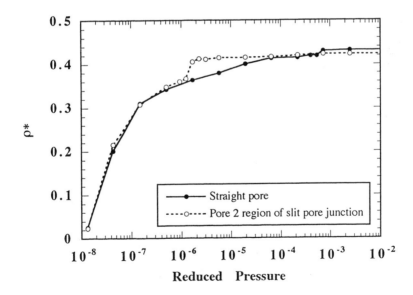

Figure 5. Comparison of the adsorption in the Pore 2 region (W*=2.21) of the 4/2 network with the adsorption in a straight pore of width W*=2.21 at T*=0.823. The jump in adsorption for the network at P/P° ≈1.3x10⁻⁶ is attributed to a freezing transition.

also residual adsorption in the cavity down to very low pressures, as seen from Fig. 4. This arises from the monolayer adsorption in the cavity; these molecules are irreversibly adsorbed as a result of the very deep potential wells in the corners of the cavity, combined with the very large potential energy barrier to adsorption that exists at the pore junction. For most adsorbate molecules in the monolayer, this latter barrier is significantly higher than the one to be surmounted to enter the cavity.

Examination of the adsorption isotherms for just the Pore 2 region of the narrower pores shows an abrupt increase in adsorption at $P/P^o \approx 2 \times 10^{-6}$ for the 4/2 network and at $P/P^o \approx 1 \times 10^{-4}$ for the 6/2 network, which we attribute to the freezing transition. No such jump in adsorption is observed in infinitely long pores of these widths (see Fig. 5).

To examine the behavior in the region of the sharp jump in adsorption for the 4/2 and 6/2 networks, we studied the fraction of molecules that remain in the network over an extended period of 200,000 MD timesteps. Molecules are created in the GCMC section, diffuse into the inner pore and cavity regions, and eventually diffuse out and are removed in the GCMC section. For a network in which the adsorbate is fluid throughout, a large fraction of molecules will cycle out of the system in the time period of 200,000 steps. However, this fraction will be drastically reduced if the adsorbate solidifies in the narrow pore region, or if significant numbers of adsorbate molecules are trapped in strongly adsorbing regions (the 'corners' of the cavity). In Figure 6 we show such results for the Pore 2 region of the 6/2 network, and compare them with results for a adsorption into a straight pore of the same width as that of the small pore in the network (W*=2.43). For the straight pore the fraction of adsorbed molecules remaining in the pore over this time period is seen to rise continuously as the pressure and adsorbate

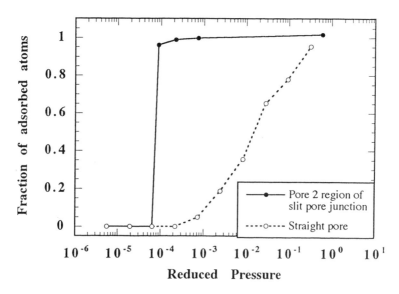

Figure 6. Fraction of adsorbate molecules remaining in the pore network after 200,000 MD steps for the pore 2 region of the 6/2 network, and for a straight pore of width W*=2.43, at T*=0.823.

density increases. For the Pore 2 region of the 6/2 network, however, there is a sharp discontinuity in the fraction of molecules remaining, and this occurs at $P/P^o \approx 8 \times 10^{-5}$, the point at which the sharp jump in adsorbate density occurs for this system. For pressures above this almost all the adsorbed molecules remain in region Pore 2 of the network. Exactly similar behavior was observed for the 4/2 network, the discontinuous jump in fraction of molecules remaining in the Pore 2 region occurring at $P/P^o \approx 1.3 \times 10^{-6}$. We attribute these jumps to the formation of a solid phase in the narrow pore (Pore 1, Pore 2 and GCMC regions), induced by the potential barrier at the point of connection of the two pores. For the 7/3 network no such sudden jump in fraction of adsorbed molecules in the Pore 2 region was seen at low pressures, although there was a sharp rise in this quantity at higher pressures of about $P/P^o \approx 9 \times 10^{-2}$.

5 CONCLUSIONS

We have found strong effects of pore connectivity that change the qualitative nature of adsorption behavior. These strong effects arise from the heterogneous nature of the fluid-wall potential as an adsorbate molecule moves through the pore network. In particular we observe two new phenomena: (a) the formation of a solid phase in the narrow part of the pore network, when this narrow pore has a width below about $W^*=3$, and (b) irreversible adsorption of adsorbate into the 'corners' of the cavity. The first effect (solidification in the narrow pores) is not observed in the corresponding straight, unconnected pore of the same material and width, and so must be due to the fluid-wall potential barrier that exists at the junction of the two pores. This suggests that similar behavior would be observed near the mouth of a straight pore with open ends. As far as we are aware this phenomenon has not been previously reported. This phenomenon is likely to occur in some real carbons. Thus, it could well lead to erroneous estimates of the pore volume, since substantial pore regions may be inaccessible to the adsorbate, even though the pores are large enough to accept the adsorbate molecules. The second effect, of irreversible adsorption, is to be expected since the fluid-wall potential in the corners of the cavity is very strongly attractive, because of the two neighbouring walls.

ACKNOWLEDGMENTS

This work was supported by the National Science Foundation under grant no. CTS-9508680, and supercomputer time was provided by NSF under a Metacenter grant (no. MCA93S011P).

REFERENCES

1.   C.M. Lastoskie, K.E. Gubbins and N. Quirke, *J. Phys. Chem.*, 1993, **97**, 4786.

2.   M. M. Cielinski, *M. S. Thesis*, Univ. of Maine, Orono, ME (1985); M. M. Cielinski and N. Quirke, unpublished (1985).

3.   R. F. Cracknell, D. Nicholson, and N. Quirke, *Phys. Rev. Lett.*, 1995, **74**, 2463.

4.   M. W. Maddox, C. M. Lastoskie, N. Quirke, and K. E. Gubbins, "Fundamentals of Adsorption", ed. by M. D. LeVan, Kluwer Academic Publishers, Boston, 1996, 571-578.

5.   A. Lotfi, J. Vrabec, and J. Fischer, *Mol. Phys.*, 1992, **76**, 1319.

6.   J. K. Johnson, J. A. Zollweg, and K. E. Gubbins, *Mol. Phys.*, 1993, **78**, 591.

7.   W. A. Steele, "The Interaction of Gases with Solid Surfaces", Pergamon Press, Oxford, 1974.

CHARACTERISATION OF MICROPOROUS CARBONS BY USING MOLECULAR SIMULATION TO ANALYSE THE ADSORPTION OF MOLECULES OF DIFFERENT SIZES

M.V. López-Ramón*, J. Jagiełło§, T.J. Bandosz† and N.A. Seaton*

*Department of Chemical Engineering, University of Cambridge, Pembroke Street, Cambridge CB2 3RA, United Kingdom.

§Department of Chemical Engineering and Materials Science, Syracuse University, 320 Hinds Hall, Syracuse, NY 13244-1190, U.S.A.

†Dept. of Chemistry, The City College of The City University of New York, Convent Avenue at 138th Street, New York, NY 10031, U.S.A.

1 INTRODUCTION

The pore size distribution (PSD) of a porous solid is closely related to the adsorption, reaction and transport properties of the solid. For many applications, the connectivity of the pore network (i.e. the way in which the pores are connected together) is also important. Connectivity is important in any process in which the presence of constrictions in the pore network inhibits the passages of one or more adsorbed species. Examples include: diffusion, particularly in poorly connected networks; catalyst fouling, where the resistance to fouling depends on the connectivity; and molecular sieving. A convenient way to quantify the connectivity is in terms of the mean coordination number of the pore network (i.e. the mean number of pores meeting at an intersection), Z.

In this paper we describe a method for obtaining both the PSD and the coordination number of the pore network for microporous solids, by using Monte Carlo simulation to analyse the adsorption of several adsorptives in the pore space of the solid. We demonstrate the method, which is applicable to microporous solids in general, by analysing the data of Jagiełło et al.[1] for adsorption of methane, carbon tetrachloride and sulphur hexaflouride on a microscopic carbon. Our approach allows a more complete picture of the PSD to be obtained than would be the case with a single adsorptive, as well as providing an estimate of Z. Because of the limited space available in these conference proceedings, this paper is very much an outline of our work; a full account will be presented elsewhere[2].

2 DETERMINATION OF THE PORE SIZE DISTRIBUTION

The PSD is related to the adsorption isotherm by:

$$N(P) = \int_0^\infty \rho(P,w)f(w)dw \tag{1}$$

Here, $N(P)$ is the experimental number of moles adsorbed, per unit mass of adsorbent, at pressure P. $\rho(P,w)$ is the "single-pore isotherm function" – the density of adsorbate in a pore of width w at pressure P – generated by the model for adsorption in individual pores. $f(w)$ is the PSD, strictly dV/dw where V is the total pore volume. The PSD is obtained by varying $f(w)$ until eq (1) is satisfied as closely as possible. The accuracy of the PSD depends on the realism of the model for adsorption in individual pores. For sufficiently large pores, classical thermodynamics provides an adequate description of adsorption at the level of individual pores. For microporous solids ($w < 2$ nm), a molecular model of adsorption is required.

Statistical mechanics provides a means of relating a molecular model for adsorption in individual pores, expressed in terms solid-fluid and fluid-fluid interactions, to the single-pore isotherm function of eq (1). The first statistical-mechanical methods for obtaining single-pore isotherms were based on density functional theories, which simplify the interactions present in the adsorbed phase and provide a computationally fast route to approximate solutions[3,4,5,6].

Another statistical-mechanical method, Monte Carlo simulation, provides a stochastic solution to the single-pore model that is, in principle, exact. That is, the accuracy of the simulation can be made arbitrarily high by carrying out a sufficiently long run. (Of course, this does not imply exact correspondence with reality, which depends also on the realism of the single-pore model.) Although Monte Carlo simulation is computationally much slower than statistical-mechanical theory, the availability of sufficiently fast computers (typically Unix workstations) has allowed Monte Carlo simulation to be used to generate the single-pore isotherms used in PSD determination. Sosin and Quinn[7] and Gusev et al.[8] have used Monte Carlo simulation to extract PSDs from data for the adsorption of methane on activated carbons.

We have used the Grand Canonical Monte Carlo (GCMC) simulation method to generate single-pore isotherms for the adsorbate species of interest - methane, carbon tetrachloride and sulphur hexaflouride - on activated carbon. We give only a brief outline of the simulations here; details of the Monte Carlo method are given elsewhere in these proceedings[9].

In a GCMC simulation of adsorption, the pore of size and shape, and the fluid-solid and fluid-fluid interactions are chosen to represent the physical system of interest. We model the pores as parallel-sided slits, bounded by infinite slabs of graphite. Other physically plausible models exist for microporous carbons. Boulton *et al.*, elsewhere in these proceedings[9], consider the effect of the choice of pore model on the characterisation results.

As the adsorbate molecules considered here are all approximately spherical, we describe them using the spherically-symmetric Lennard-Jones potential. The potential parameters for the fluid-fluid interactions are given in Table 1.

Table 1. *Lennard-Jones parameters for the fluid-fluid interactions. k_B is Boltzmann's constant.*

Species	σ (nm)	ε / k_B (K)
CH_4	0.381	148.2
CF_4	0.470	152.5
SF_6	0.551	200.9

The fluid-solid potentials are given by Steele's[10] potential, which integrates over the graphite slab the Lennard-Jones interactions between a fluid molecule and the individual atoms of the solid.

Simulated isotherms were obtained for the adsorption of CH_4, CF_4 and SF_6 in pores of various sizes, at temperatures and pressures in the range studied experimentally by Jagiełło et al.[1]: 258 K to 296 K, at pressures of up to 1 bar. A sample of the results obtained, some of the isotherms for CH_4 at 258 K, are shown in Figure 1.

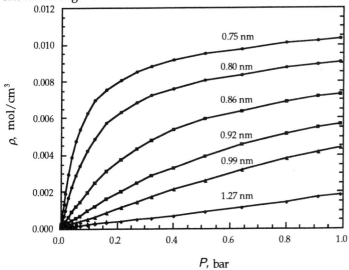

Figure 1 *Simulated single-pore isotherms for CH_4 at 258 K.*

The PSD was obtained using the SAIEUS method of Jagiełło[11]. This method represents the PSD by a linear combination of B-spline functions and solves eq 1 using a regularization method combined with non-negativity constraints. The single-pore isotherm function, $\rho(P,w)$, for each of the three adsorbate species is a correlation of the simulated isotherms, using the functional form of the Hill-de Boer isotherm. Figure 2 shows the estimates of the PSD obtained for one of the activated carbons studied by Jagiełło et al.[1], "Carbon G", using the adsorption measurements for CH_4 and CF_4 at 258 K, and SF_6 at 267 K. (The pore size is defined to be the distance between the carbon nuclei on opposing pore walls.) Figure 3 shows the corresponding fits to the experimental isotherms.

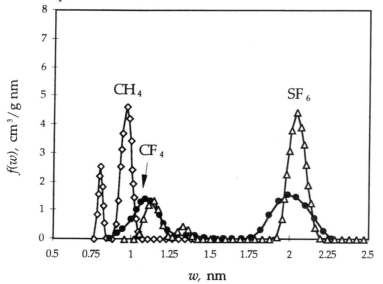

Figure 2 *Pore size distributions obtained using the three adsorbates.*

A comparison between the three PSDs serves two purposes. Firstly, as we show in the next section, the comparison between the estimated PSDs obtained with molecules of different sizes allows us to assess the pore network connectivity. Secondly, as these different PSDs are estimates of the same underlying function - the real PSD of the solid - the extent of agreement between the estimated PSDs (taking into account the different sizes of the molecules) is a measure of the consistency (and hence realism) of our PSD method. We consider the latter aspect first.

The lower limit of the PSDs is set by the smallest pore that the adsorbate molecule can penetrate to a significant degree, which we find by simulation to be: 0.75 nm for CH_4, 0.85 nm for CF_4, and 0.94 nm for SF_6. The first peak of the CH_4 PSD is below the smallest pore accessible to CF_4, the next smallest adsorbate molecule. The second CH_4 peak is within the range of the CF_4 PSD, but strikingly inconsistent with it. This discrepancy turns out not to indicate a fundamental failure of the model, but rather the existence of an effective upper limit to the PSD for a given adsorbate. Figure 1 shows that, at this temperature, and for the experimental pressure range, the CH_4 single-pore isotherms are close to linear (i.e. Henry's law) above about 0.9 nm. In the PSD analysis, it is not possible to distinguish between the sizes of pores in which Henry's-law adsorption occurs; one pore with a large Henry's constant is indistinguishable from several pores with small Henry's constants. Thus, the reliable range of the CH_4 PSD is a "window" bounded on the left by the smallest accessible pore, and on the right by the pore size at which adsorption, for the experimental temperature and pressure range, becomes substantially linear. (The precise location of the right hand boundary is arbitrary, as the sensitivity of the isotherm to pore size decreases continuously as the pore size increases.) This observation, which has also been made by Gusev et al.[8], is applicable to supercritical adsorption (i.e. in the absence of capillary condensation) in porous solids in general. The upper bound of the PSD is an increasing function of pressure[8].

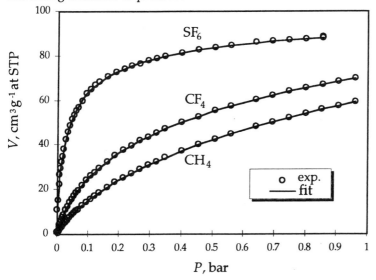

Figure 3 *Fits of simulation results to experiment at 258 K (except for SF$_6$ which is at 267 K).*

The second peak of the CH_4 PSD (counting from the left) is largely above this upper limit and is thus unreliable. The shape and location of the second peak is therefore arbitrary, except to the extent that its contribution to the total adsorption is required to fit the experimental isotherm. An almost equally good fit to the experimental data can be obtained with a CH_4 peak in any pore size range above 0.9 nm. The upper bound of the window of reliability for CF_4 is about 3 nm. The upper bound for SF$_6$ is above the largest pore size we simulated (3 nm) and, it seems, well above the size of the largest pore present in this sample.

Let us compare the first peaks of the CF_4 and SF_6 PSDs. If each species had free access to the whole pore network, these peaks should, in principle, be identical above the smallest pore size accessible to SF_6, 0.94 nm. Given the approximate nature of our single-pore model, in terms of the pore shape and the molecular interactions, one would not expect exact agreement. Nevertheless, the areas of the peaks should at least be similar. In fact, the SF_6 peak has a significantly smaller area than the CF_4 peak, indicating that some of the pores that are large enough to accommodate SF_6 are not accessible to the SF_6 molecules. The difference between these peaks is the basis of our analysis of network connectivity, which we present in the next section.

The second peaks of the CF_4 and SF_6 PSDs are broadly consistent; they have similar means and areas, although the SF_6 peak is sharper. In fact, the area of the SF_6 peak is slightly (and unphysically) larger than that of the CF_4 peak. This is not a serious discrepancy. The extent of adsorption in these pores is much less than in the first peak, so that a substantial change in the second peak has only a small effect on the total adsorption. The fitting procedure therefore tends to offset quite large changes in the second peak against small changes in the first peak, so that the uncertainty in the second peak is relatively high.

The combined use of the three adsorbates allows us to construct a detailed picture of the overall PSD. Each of the PSDs is partial[11], both in terms of its window of reliability, and because it reflects only those pores accessible to that adsorbate. Figure 4 shows the best estimate of the overall PSD, obtained by combining the first peak of the CH_4 peak and the two CF_4 peaks. Strictly speaking, this PSD is a *lower bound to the true size distribution of those pores*

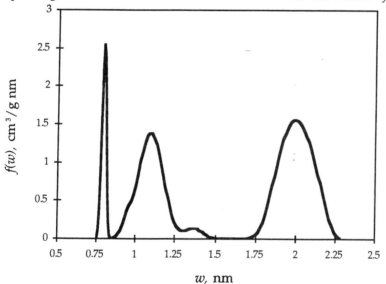

Figure 4 *Best estimate of the pore size distribution.*

that are accessible to CH_4. (It is a "lower bound" because some of the pores accessible to CH_4 above 0.9 nm are presumed to be inaccessible to CF_4, being surrounded by pores of a size that permits the passage of CH_4 but not CF_4.)

CONNECTIVITY ANALYSIS

Figure 5 shows schematically the relationship between connectivity and the PSDs measured with different adsorbates. There are four pores in this simple, model adsorbate. The smaller adsorbate species probes all the pores and its adsorption isotherm yields the complete PSD. The larger species is excluded from the smaller pores, and also from the larger pore that is "shielded" by the smaller pores. So, the PSD obtained using the larger species is (i) zero for pores that are smaller than the molecules of that species, and is (ii) smaller than the PSD for the smaller species above this pore size. This effect can be seen in the comparison of the first peaks for CF_4 and SF_6 in Figure 2. The extent to which shielding occurs in a real pore network depends on the connectivity of the network, with the shielding effect being more pronounced for less well connected networks.

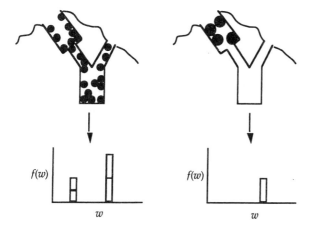

Figure 5 *Schematic illustration of the effect of connectivity on adsorption.*

Percolation theory is the natural language for describing connectivity in porous solids. The relevant percolation variables are the "bond occupation probability", X, which in adsorption terms is the fraction of pores that are that are large enough to accommodate the adsorbate of interest; and the "accessibility", A, which in adsorption terms is the fraction of pores that the adsorbate actually enters. In other words, X is the fraction of pores that the adsorbate would enter in an infinitely connected network, where each pore had direct access to the adsorbate; and the deviation of A from X is a measure of the connectivity. In our analysis, the "true" PSD, $f_t(w)$, is assumed to be given by the sum of the first CH_4 peak and the two CF_4 peaks. The PSD of the pores accessible to SF_6, our probe of connectivity, $f_s(w)$, is taken as the sum of the first SF_6 peak and the second CF_4 peak (as the second SF_6 peak is unphysically larger than the corresponding CF_4 peak). The percolation variables are calculated as follows[2]:

$$X = \frac{\int_{w_s}^{\infty} \frac{f_t(w)}{w} dw}{\int_{0}^{\infty} \frac{f_t(w)}{w} dw} \quad (2); \qquad A = \frac{\int_{w_s}^{\infty} \frac{f_s(w)}{w} dw}{\int_{0}^{\infty} \frac{f_t(w)}{w} dw} \quad (3)$$

In these equations, w_s is the size of the smallest pore accessible to SF_6. From the PSD data presented in Figure 2, we obtain $X = 0.788$ and $A = 0.648$. These values are fitted to simulation data[13] for percolation on three-dimensional lattices giving $Z = 2.3$ for this solid. Details of the connectivity analysis are given elsewhere[2], along with an investigation of the consistency of PSD and connectivity results across a range of experimental temperatures.

Acknowledgements

M.V. López-Ramón gratefully acknowledges a grant from the Spanish Ministry of Education and Science. Acknowledgement is made to the donors of The Petroleum Research Fund, administered by the ACS, for partial support of this research.

References

1. J. Jagiełło, T.J. Bandosz, K. Putyera and J. Schwarz, *J. Chem. Eng. Data*, 1995, **40**, 1288.
2. J. Jagiełło, T.J. Bandosz, M.V. López-Ramón and N.A. Seaton, in preparation, 1996.
3. N.A. Seaton, J.P.R.B. Walton and N. Quirke, *Carbon*, 1989, **27**, 853.
4. C.A. Jessop, S.M. Riddiford, N.A. Seaton, J.P.R.B. Walton, and N. Quirke, 'Characterization of Porous Solids (COPS-II). Proceedings of the IUPAC Symposium', F. Rodríguez Reinoso, J. Rouquerol and K.S.W. Sing (Eds.), Elsevier, 1991, p 123.
5. C. Lastoskie, K.E. Gubbins and N. Quirke, *J. Phys. Chem.*, 1993, **97**, 4786.
6. J.P. Olivier, W.B. Conklin and M.v. Szombathely, 'Characterization of Porous Solids III', J. Rouquerol, F. Rodríguez-Reinoso, K.S.W. Sing and K.K. Unger (Eds.), 1994, p 81, Elsevier,
7. K.A. Sosin and D.F. Quinn, *J. Porous Materials*, 1995, **1**, 111.
8. V. Yu. Gusev, J.A. O'Brien and N.A. Seaton, *Langmuir*, in press.
9. K.L. Boulton, M.V. López-Ramón, G.M. Davies and N.A. Seaton, these proceedings.
10. W.A. Steele, 'The Interaction of Gases with Solid Surfaces', Pergamon, 1974.
11. J. Jagiełło, *Langmuir*, 1994, **10**, 2778.
12. J. Jagiełło, T.J. Bandosz, K. Putyera and J.A. Schwarz, *J. Chem. Soc. Faraday Trans.*, 1995, **91**, 2929.
13. H. Liu, L. Zhang and N.A. Seaton, *Chem. Eng. Sci.*, 1992, **47**, 4393.

ORDERED MESOPOROUS MCM-41 ADSORBENTS :
NOVEL ROUTES IN SYNTHESIS, PRODUCT CHARACTERISATION AND SPECIFICATION

M. Grün [1], K.K. Unger [1], A. Matsumoto [2], K. Tsutsumi [2]

[1] Institut für Anorganische Chemie und Analytische Chemie
Johannes Gutenberg Universität, J.J. Becherweg 24, 55099 Mainz, Germany

[2] Department of Materials Science, Toyohashi University of Technology
Tempaku-cho, Toyohashi 441, Japan

1 INTRODUCTION

Since its discovery in 1992 [1,2] MCM-41 has become the most popular member of the M41S family of mesoporous silicate and aluminosilicate materials. The most prominent feature of MCM-41 is its regular pore system which consists of an hexagonal array of unidimensional, hexagonally shaped pores. The synthesis of MCM-41 can be accomplished by several routes which mainly differ in the silica source and pH during the reaction. The pore diameter of MCM-41 can be controlled by changing the carbon chain length of the surfactant. A further increase of the pore size can be achieved by addition of an auxiliary organic like toluene or mesitylene. Pore diameters of 2 to 10 nm have been reported in the literature [2]. Other physical properties of MCM-41 include a high specific surface of up to 1500 m²/g, a specific pore volume of 1.0 ml/g and a high thermal stability, which make it suitable for catalytic applications. The catalytic properties can be adjusted by incorporation of different metals eg. titanium, aluminium into the MCM-41 framework [3,4,5]. Due to its regular pore structure and pore shape MCM-41 has attracted considerable interest as a model substance the sorption of various gases [6,7]. Other metal oxides eg. zirconia and titania with an MCM-41- analogue structure have been synthesized [8,9].

MCM-41 materials with a defined morphology are promising selective adsorbents in separation techniques eg. HPLC and supercritical fluid chromatography (SFC) [10,11]. This communication introduces two novel synthesis routes and describes the aging of MCM-41 under high-pH hydrothermal conditions. All materials were characterized by X-ray diffraction, nitrogen sorption and scanning electron microscopy.

2 EXPERIMENTAL

2.1 Preparation of MCM-41

2.1.1 Synthesis of MCM-41 using ammonia as catalyst. n-dodecyltrimethyl-ammonium bromide (C12N, Aldrich), n-tetradecyltrimethyl ammonium bromide (C14N, Aldrich), n-hexadecyltrimethylammonium bromide (C16N, Aldrich), n-octadecyl-trimethylammonium bromide (C18N, Fluka) and n-eicosyltrimethylammonium bromide (C20N) were used as templating surfactants. The C12N - C18N surfactants were used without further purification. C20N was prepared according to a method given by Colichman[12]. Each surfactant is dissolved in 120 g of deionized water to give a 0.055-mol/l solution, and 9.5 g of aqueous ammonia (Merck, 25 wt %, 0.14 moles) was added to the solution. While stirring 10 g of tetraethoxysilane (TEOS, Aldrich, 98 %, 0.05 moles) was dropped slowly into the surfactant solution over a period of 15 min. Then the white product was filtered , washed with 1 l of deionized water. After drying at 363 K, the sample was heated to 823 K (rate : 1 K/min) and kept at this temperature for 5 h.

2.1.2 Aging of MCM-41 under high-pH hydrothermal conditions. MCM-41 prepared using C16N was loaded into autoclaves together with the mother liquid and aged for 10 days at 378 K and 433 K, respectively. The aged samples were filtered and washed with 1 l of deionized water. After drying at 363 K, the sample was heated to 823 K (rate : 1 K/min) and kept at this temperature for 5 h.

2.1.3 Synthesis of MCM-41 spheres. Stöber's method for the preparation of silica spheres[13] was modified for the synthesis of MCM-41 spheres. 2.5 g of C16N (0.01 moles) was dissolved in 50 g of deionized water, and 13.2 g of aqueous ammonia (Merck, 32 wt %, 0.25 moles) and 60.0 g of absolute ethanol (Merck, p.a., 1.3 moles) were added to the surfactant solution. The solution was stirred for 15 min (250 rpm) and 4.7 g of TEOS (Aldrich, 98 %, 0.022 moles, freshly distilled) was added at one time. After 2 h of stirring the white precipitate was filtered and washed with 100 ml of deionized water and 100 ml of methanol. After drying overnight at 363 K, the sample was heated to 823 K (rate : 1 K/min) and kept at that temperature for 5 h. n-hexadecyltrimethylammonium bromide was substituted by n-hexadecylpyridinium chloride (C16PYR) (99 %, Aldrich) in an additional experiment.

2.2 Characterization

Each sample was characterized by X-ray diffraction (XRD) and nitrogen sorption. XRD patterns were recorded at a interval of 0.02° theta on a Seifert TT 3000 powder diffractometer using Cu K_α radiation. Diffraction data were recorded. Adsorption and desorption isotherms for nitrogen were obtained at 77 K using a Micromeritics ASAP 2010. The samples were outgassed at 423 K and 1 mPa for 12 hours before measurements. Scanning electron micrographs were obtained on a Zeiss ZSM 962 microscope.

3 RESULTS AND DISCUSSION

3.1 Synthesis of MCM-41 using ammonia as catalyst

The adsorption and desorption isotherms of nitrogen on each sample showed the typical shape for MCM-41; the sharp step over a narrow range of relative pressure, p/p_O of 0.3-0.4. One example is shown in Fig. 2. Typical diffraction patterns usually exhibit four Bragg peaks indicating the long range order present in this material (Fig. 1). Characteristics of the MCM-41 samples prepared by use of different alkylammonium bromides are shown in Tab. 1. The specific surface area gradually decreases from 1450 m²/g (prepared by use of C12N, sample No. 1) to 980 m²/g (C20N, sample No. 5) while the values of the specific pore volume rise from 0.57 ml/g (C12N, sample No. 1) up to 1.72 ml/g (C20N, sample No. 5) with increasing carbon chain length of the alkyl group. The average pore size of materials prepared by this method is smaller than that of materials prepared by the classical hydrothermal synthesis which is also carried out in a basic medium but uses colloidal silica instead of tetraethoxysilane [2].

This novel synthesis procedure provides a convenient access to high-quality MCM-41 material in a short period (10 to 30 min) without neither additional heating nor application of external pressure. These features also make this procedure suitable for the preparation of large amounts of MCM-41.

Table 1   *Properties of MCM-41 samples prepared by use of aqueous ammonia as catalyst*

Sample No.	Number of C Atoms in Carbon Chain	Specific Surface Area [1] / $m^2\ g^{-1}$	Specific Pore Volume [2] / $ml\ g^{-1}$	Average Pore Diameter [3] / nm	Pore Diameter (XRD) [4] / nm
1	12	1450	0.57	2.1	1.8
2	14	1120	0.58	2.3	1.9
3	16	1070	0.80	2.7	2.3
4	18	1050	1.00	3.5	2.6
5	20	980	1.72	5.3	-

1 according to the BET plot, 2 according to the Gurvitch rule, 3 according to the BJH method using the desorption branch of the nitrogen isotherm, 4 using the d_{100}-value and assuming a pore wall thickness of 1.0 nm of sample No. 3.

3.2 Aging of MCM-41 under high-pH hydrothermal conditions

Figure 1 shows the change in X-ray diffractograms before and after aging under high-pH hydrothermal conditions. The Bragg peaks in the diffractogram of the sample aged at 378 K (sample No. 7, Fig. 1a) apparently become sharper and an additional fifth peak can be observed at 5.70° 2 theta indicating an improved long range order structure in the material.

On the other hand, the diffractogram of the sample aged at 433 K (sample No. 8, Fig. 1b) shows a broad Bragg peak at 1.14° 2 theta suggesting the formation of another ordered-pore structure while the Bragg peak at 2.70° 2 theta still remains indicating that the original pore arrays of 3.78 nm still remain intact. These changes in X-ray diffractograms coincide with those found in pore size distributions determined by nitrogen sorption isotherms. As shown in Fig. 2 the nitrogen isotherm of the sample aged at 378 K (sample No. 7) resembles that of the untreated one (sample No.6), as shown in Fig. 2, while the sample aged at 433 K (sample No. 8) shows a pronounced hysteresis loop. As shown in Tab. 2, specific surface area and specific pore volume of the sample without aging (sample No. 6) and that aged at 378 K (sample No. 7) are comparable while the sample aged at 433 K (sample No. 8) shows a marked decrease of both characteristics. These results also suggest the formation of wider pores.

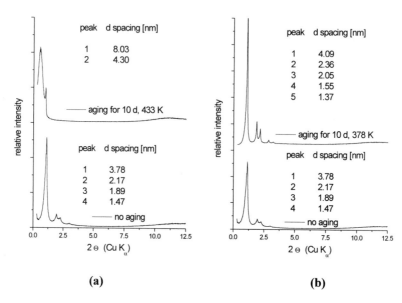

(a)   **(b)**

Figure 1   X-ray diffractograms of MCM-41 before and after hydrothermal
                    treatment at 433 K (a) and 378 K (a) for 10 days

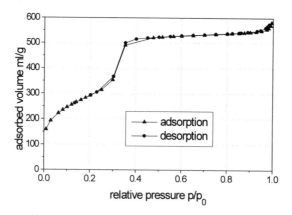

Figure 2 a   *Nitrogen isotherm of an untreated MCM-41 sample (sample No.6)*

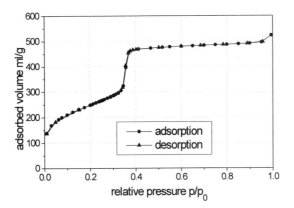

Figure 2 b   *Nitrogen sorption of an MCM-41 after high-pH hydrothermal treatment at 378 K for 10 days (sample No. 7)*

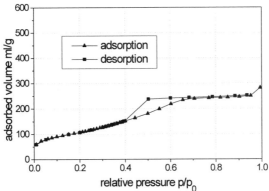

Figure 2 c   *Nitrogen sorption of an MCM-41 after high-pH hydrothermal treatment at 433 K for 10 days (sample No. 8)*

Table 2   *Properties of aged MCM-41 samples*

Sample No.	Aging Temperature / K	Aging Time / d	Specific Surface Area [1] / m² g⁻¹	Specific Pore Volume [2] / ml g⁻¹	Average Pore Diameter [3] / nm
6	293	0	1020	0.80	2.7
7	378	10	890	0.80	3.2
8	433	10	370	0.44	4.1

1 according to the BET plot, 2 according to the Gurvitch rule, 3 according to the BJH method using the desorption branch of the nitrogen isotherm.

Figure 3 a   *Scanning electron micrograph of an untreated MCM-41 sample (sample No. 6)*

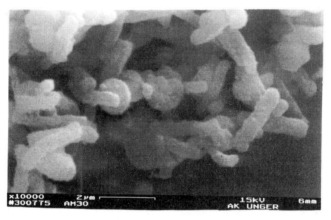

Figure 3 b   *Scanning electron micrograph of an MCM-41 sample after high-pH hydrothermal treatment at 433 K for 10 days (sample No. 8)*

Figure 3 shows the scanning electron micrograph images of the samples before (a) and after (b) aging at 433 K. The MCM-41 particles consist of agglomerated small particles before aging, therefore, that become rod-like after aging. One possible mechanism for the change of morphology is dissolution and agglomeration.

3.3 Synthesis of MCM-41 spheres

As shown in Fig. 4a, the X-ray diffractogram of the sample prepared by use of C16N (sample No. 9) exhibits three sharp Bragg peaks, which can be indexed as (100), (110) and (200). The diffractogram of the sample prepared by use of C16PYR (sample

No. 10) shows four peaks suggesting an even better long-range order present in this material (Fig. 4b). The additional fourth peak can be indexed as (210). Table 3 gives a summary of some properties of MCM-41 samples. Scanning electron micrographs of MCM-41 (sample No. 9) are shown in Fig. 5. Though the particle diameter ranges from 400 nm to 1000 nm, each particle is perfectly spherical.

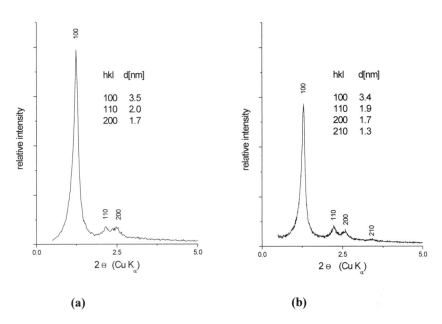

	hkl	d[nm]
	100	3.5
	110	2.0
	200	1.7

	hkl	d[nm]
	100	3.4
	110	1.9
	200	1.7
	210	1.3

(a)   (b)

Figure 4   *Diffractograms of MCM-41 spheres, surfactants used : C16N (sample No. 9, a) and C16PYR (sample No. 10, b)*

Table 3   *Properties of MCM-41 spheres*

Sample No.	Templating Surfactant	Specific Surface Area [1] / $m^2 g^{-1}$	Specific Pore Volume [2] / ml g^{-1}	Average Pore Diameter [3] / nm	Pore Diameter (XRD) [4] / nm
9	C16N	1080	0.91	2.52	3.14
10	C16PYR	1220	0.89	2.13	3.05

1 according to the BET plot, 2 according to the Gurvitch rule, 3 according to the BJH method using the desorption branch of the nitrogen isotherm, 4 using the d_{100}-value and assuming a pore wall thickness of 1.0 nm of sample No. 3.

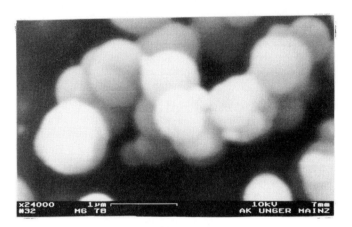

Figure 5   *Scanning electron micrograph of MCM-41 spheres, C16N was used as surfactant (sample No. 9)*

4 CONCLUSION

The use of ammonia as a catalyst provides a convenient method for the preparation of high-quality MCM-41. This method can be used both for small-scale synthesis in the laboratory and for large-scale production, which may become important if MCM-41 will be used for industrial applications.

Further research will be focussed on the behaviour of MCM-41 under hydrothermal conditions. The use of different temperatures and pH-values may lead to interesting results in terms of morphology and cristallinity. Transmission electron measurements have to be carried out to elucidate the pore structure of the aged materials.

Stöber's method for the synthesis of silica particles has been successfully modified to give spherical MCM-41 particles by the addition of a surfactant. For application as packing material in HPLC and other separation techniques larger spheres ($d_p > 3$ μm) are desirable.

Acknowledgements

This work was financially supported by the AIF (project number 10206 N). One of the authors (A.M.) is indebted to the Ministry of Education, Science, Sports and Culture of the Japanese government for financial support of his postdoctoral research in Germany.

References

1. C.T. Kresge, M.E. Leonowicz, W.J. Roth, J.C. Vartuli, J.S. Beck, *Nature*, 1992, **359**, 710.
2. J.S. Beck, J.C. Vartuli, W.J. Roth, M.E. Leonowicz, C.T. Kresge, K.D. Schmitt, C.T.-W. Chu, D.H. Olson, E.E. Sheppard, S.B. McCullen, J.B. Higgins, J.L. Schlenker, *J. Am. Chem. Soc.*, 1992, **114**, 10834.

3. A. Corma, M.T. Navarro, J.P. Pariente, *J. Chem. Soc., Chem. Commun.,* 1994, 147

4. P.T. Tanev, M. Chibwe, T.J. Pinnavaia, *Nature*, 1994 ,**368** , 321.

5. A. Corma, V. Fornes, M.T. Navarro, J. P. Pariente, *J. Catal.*, 1994, **148**, 569.

6. R. Schmidt, M. Stöcker, E. Hansen, D. Akporiaye, O.H. Ellestad, *Microporous Mater.,* 1995, **3**, 443.

7. P.J. Branton, P.G. Hall, K.S.W. Sing, *J. Chem. Soc., Chem. Commun.* 1993, 1257.

8. Q. Huo, D.I. Margolese, U. Ciesla, D.G. Demuth, P. Feng, T.E. Gier, P. Sieger, A. Firouzi, B.F. Chmelka, F. Schüth, G.D. Stucky, *Chem. Mater.*, 1994 ,**6**, 1176.

9. D.M. Antonelli, J.Y. Ying, *Angew. Chem., Int. Ed. Engl.*, 1995, **18**, 34.

10. M. Grün, A.A. Kurganov, S. Schacht, F. Schüth, K.K. Unger, *J. Chromatogr. A*, in press.

11. M. Grün, I. Lauer, K.K. Unger, *Adv. Mat.*, submitted.

12. E.L. Colichman, *J. Am. Chem. Soc.*, 1950, **72**, 1834.

13. W. Stöber, A. Fink, E. Bohn, *J. Colloid Interface Sci.*, 1968 , **26**, 62.

CHARACTERIZATION OF ZIRCONIA BASED ORDERED MESOPOROUS MATERIALS

U. Ciesla, K. Unger and F. Schüth

Institut für Anorganische Chemie
Johann Wolfgang Goethe-Universität Frankfurt
Marie Curie Straße 11
60439 Frankfurt

1 INTRODUCTION

Mesoporous silica and alumosilicates of the M41S class, which were discovered by scientists of the Mobil Oil Corp.[1], have found widespread interest as model sorbents in the range of smaller mesopores[2]. The steep capillary condensation step without hysteresis around $p/p_0 = 0.4$ for nitrogen at 77 K found for materials synthesized with cetyltrimethylammonium as surfactant, together with the fact that independent means of analysis like TEM and XRD are available to obtain independent information on the pore structure, makes these oxides ideally suited as model substances, although several problems with this class of materials still exist[3].

Similar model compounds are available in the lower micropore range, where zeolites and related materials can be used as sorption standards, because the pore system is crystallographically defined and again can be assessed with independent means. Such microporous materials are thus often used to obtain standard isotherms in order to calibrate sorption equipment.

However, the range of pore sizes between 1 and 3 nm is very difficult to analyze using gas adsorption. Various algorithms to obtain pore size distributions in this range from sorption data have been proposed. Validation of such algorithms, however, is very difficult, since materials with pore sizes in this range, which can be checked with independent methods, seem to be missing so far.

Generalization of the synthesis mechanism, which was suggested in one of the earlier publications on M41S type materials[4], could help to fill this gap, since different framework compositions could lead to different pore sizes. Following different approaches, the surfactant controlled synthesis was used to synthesize several novel mesostructured[5] and mesoporous[6] transition metal oxides. Especially promising for pore sizes in the upper micropore/lower mesopore range seemed to be ordered porous materials based on zirconia which have been developed recently[7]. In the following we show that such zirconias can be synthesized with pore sizes adjustable in the range of 1 nm to 3 nm.

2   EXPERIMENTAL

2.1   Materials

Two types of zirconium precursors were used in the preparation of the zirconium/surfactant composite, $Zr(SO_4)_2 \cdot 4 H_2O$ or $Zr(OC_3H_7)_4$. In the synthesis starting from $Zr(SO_4)_2 \cdot 4 H_2O$ typically 6.87 mmol of surfactant ($C_nH_{2n+1}N(CH_3)_3Br$, n = 14, 16, 18, 20, 22) is dissolved in 85 g of water and 12.8 mmol of $Zr(SO_4)_2 \cdot 4 H_2O$ dissolved in 15 g of water are added upon which a white precipitate forms. The mixture is stirred for 2 h at room temperature and subsequently heated in a closed polypropylene flask in an oven for 2 days at 100°C. After filtration, washing and drying at 100°C an optional treatment with 0.87 M phosporic acid for 2 h under stirring is carried out. After again drying the product at 100°C, the material is calcined at 500°C for 5 h to obtain an accessible pore system

Starting form $Zr(OC_3H_7)_4$, 6.87 mmol of surfactant is dissolved in a mixture of 115 g of water and 24.4 g of HCl (37 wt%). After the hydrolysis product has dissolved, 15.5 mmol of $(NH_4)_2SO_4$ in 23 g of water are added and stirred for 1 h. The initially clear solution is then heated to 100°C for two days after which a white precipitate forms. The material is filtered, washed, dried at 100°C and calcined at 500°C for 5h.

2.2   Equipment

The structure of the samples was analyzed using a STOE Stadi P X-ray diffractometer in Debeye-Scherer geometry equipped with a position sensitive detector covering an angle range of 7.5° (2 Θ). Analysis was performed with $Cu_{K\alpha 1}$ radiation (λ = 1.54056 nm). Typical scan time for the range to 15° (2 Θ) was 20 min. Unless otherwise stated, samples do not have reflections at angles between 10 and 60° (2 Θ). Elemental composition of the samples was determined by X-ray fluorescence on pressed pellets using a Philips PW 1400.

Sorptive analysis was carried out using a Micromeritics ASAP 2010 unit with nitrogen at 77.4 K. Samples were outgassed for 4 h at 250°C. Pore size distributions were calculated following the Horvath-Kawazoe[8] approach modified for cylindrical pores[9] or density functional theory (DFT)[10] with the routines supplied by Micromeritics.

3   RESULTS

Following both pathways, mesostructured surfactant/zirconia composites can be prepared[7]. Fig. 1 shows the X-ray diffraction pattern of a sample synthesized with $Zr(SO_4)_2 \cdot 4 H_2O$ as zirconium source and a C_{20}-surfactant. Products obtained from syntheses with C_{20}-surfactant usually showed the best resolved XRD peaks which suggests a higher degree of order of the materials as compared to materials prepared with other surfactants. Three peaks which are characteristic for the hexagonal structure of MCM-41 type materials can be distinguished. The position of the first peak corresponds to a d-spacing of 4.9 nm, which agrees well with the spacings observed for silica based material. In TEM analyses of such samples hexagonally ordered regions could be found with similar d-spacings.

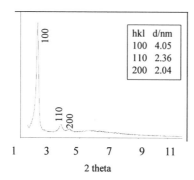

hkl	d/nm
100	4.05
110	2.36
200	2.04

Fig. 1: *XRD of a sample synthesized with $Zr(SO_4)_2 \cdot 4 H_2O$ and the C_{20}-surfactant*

However, if the material is calcined, the structure collapses. The X-ray diffraction pattern does not exhibit small angle reflections any longer, pore volume and surface area are in the range expected for calcined zirconias which have been precipitated without surfactant. The reason for this collapse is probably the high content of unreacted OH-groups and non cross linked sulfate groups in the sample (the sulfur content of such samples is typically around 2 wt%), since the pH during formation of the materials is around 1.

Since phosphate is known to react easily with zirconium, it was attempted to stabilize the composite by reacting it with phosphate to provide more extensive crosslinking. This strategy proved to be successful: After treatment with phosphoric acid, the XRD patterns usually improved, often showing a fourth peak (210). In addition, the samples can now be calcined without completely loosing the structure (fig. 2). After calcination at 500°C at least one low angle peak with high intensity is left, in some cases even higher order reflections can be distinguished. Similar to siliceous MCM-41, the position of the first XRD peak is shifted to lower angles with increasing chain length of the surfactant. However, a strong shrinking on the order between 1 and 1.5 nm occurs during calcination with the zirconia based materials. This suggests that much more condensation and reorganization proceeds in the walls of these composites as compared to the silicas. Materials after phosphatization and calcination contain only little sulfur around 0.5 %, but high amounts of phosphorus up to 10%.

The decreasing position of the XRD peak corresponds to changes in the adsorption isotherms (fig. 3). For the C_{16}-surfactant the isotherm is basically of type I, indicating a microporous material. With increasing surfactant chain length a pore filling step is observed, which, however, lies in the range below the pressures for which a real capillary condensation can be expected. Table 1 summarizes the properties of the materials synthesized. d-Spacings were calculated assuming a hexagonal array of pores according to $2d(100)/\sqrt{3}$. Pore sizes were calculated from the isotherms using DFT. Wall thickness was calculated by subtracting the DFT pore size from the lattice constant. The wall thickness is primarily controlled by the double layer potential at the interface between the surfactant and the inorganic part of the composite[4], and the shrinkage during calcination. Therefore one would not expect pronounced differences in the wall thickness if surfactants of different chain length are used, since the surfactant chain length should neither drastically influence the interface nor the degree of condensation and thus the shrinkage during calcination. The fact that the calculated wall

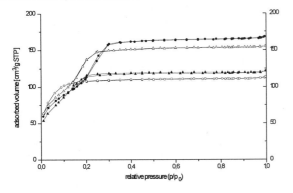

y axis: adsorbed volume [cm³/g STP]; x axis: relative pressure (p/p₀)

Fig. 3*: N₂ Isotherms corresponding to the materials given in fig. 2.* $\circ$ = C_{16}, ▲ = C_{18}, $\triangle$ = C_{20}, $\bullet$ = C_{22}

thickness is in close agreement for all samples indicates that DFT analysis gives a fairly good estimate of the pore dimensions.

If the composite is prepared using $Zr(OC_3H_7)_4$ as precursor, the X-ray diffraction pattern of the material in the as synthesized form is usually not as good as with the sulfate as precursor. However, the composite can be calcined without collapse of the structure over a relatively wide range of synthesis conditions, if the sulfate content is sufficiently high. This suggests that sulfate is a crucial component in the system, since it was so far not possible to synthesize a mesostructured material corresponding to MCM-41 from a sulfate free system on the basis of zirconium. It is well known that zirconium forms sulfate bridged oligomers in acidic solution[11]. These oligomers are a crucial component in the formation of the structured composite, and thus the sulfate content should have a strong influence on the properties of the materials.

In order to study the influence of the sulfate concentration, a series of samples was prepared from the isopropylate system using different amounts of sulfate. Fig. 4 shows the results of these experiments: Plotted are the XRD patterns of the surfactant composites. One can clearly see that the intensity of the peak and thus the quality of the material increases with increasing SO_4^{2-}/Zr-content. If the sulfate content is below a certain threshold (SO_4^{2-}/Zr < 0.6), no structured material is obtained, the highest intensity is observed for the highest ratio of 1.2. If the sulfate content is increased further, the stability of the material during calcination decreases again and has to be improved by the phosphate treatment described above. In addition to changes in peak intensities, the position of the peak is shifted to higher angles with increasing sulfate content in the synthesis mixture. Table 2 summarizes the properties of the materials synthesized with different sulfate concentrations.

Table 1: *Properties of phosphate modified zirconias synthesized with surfactants of different chain length*

Surfactant	d-spacing (100) [nm]	lattice constant [nm]	DFT pore size maximum [nm]	calc. wall thickness [nm]	pore volume [cm³/g]
C_{16}	3.02	3.49	1.6	1.9	0.17
C_{18}	3.68	4.25	2.3	2.0	0.18
C_{20}	3.88	4.48	2.6	1.9	0.24
C_{22}	4.24	4.89	2.8	2.1	0.26

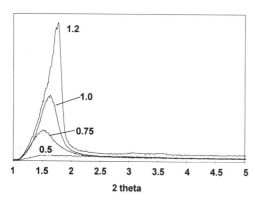

Fig. 4: *X-ray patterns of zirconias synthesized from isopropylate with different SO_4^{2-}/Zr ratios given in the figure*

The shrinkage of the walls during calcination is obviously much more pronounced for the samples synthesized with high sulfate concentration, d-spacings prior to calcination are almost identical and the density of the surfactant, which fills the pores in the as made form, can be assumed to be comparable for different sulfate concentrations as well. This suggests that the higher the sulfate concentration in the synthesis gel the higher is the amount of sulfate incorporated into the walls. Upon calcination the sulfate is removed from the walls and thus the observed shrinkage occurs. If the sulfate content of the synthesis mixture is too low, the formation of the mesostructure is hindered since insufficient condensation is present in the precursor. Increasing sulfate concentration leads to formation of oligomeric oxosulfate species which allow optimum interaction between the inorganic material and the surfactant. If the sulfate concentration is increased further, too much sulfate is incorporated in the walls. Upon calcination so many defects are created due to loss of the sulfate, that the structure collapses. There is thus an optimum sulfate concentration for the synthesis of a stable material.

The changes in the dimensions of the material are well reflected in the adsorption isotherms of the calcined materials (fig. 5, the isotherms are normalized to facilitate comparison). With increasing sulfate concentration the increasing part of the isotherm shifts to lower relative pressures, indicating smaller pores. If DFT analysis is applied, fairly sharp pore size distributions are calculated, as can be expected for this class of material (fig. 6). The increase in pore size with decreasing sulfate concentration which can

Table 2: *Properties of zirconias obtained with $Zr(OC_3H_7)_4$ as precursor and different sulfate concentrations in the synthesis mixture*

SO_4^{2-}/Zr	d-spacing (100) [nm] as made	d-spacing (100) [nm] calcined	lattice constant [nm]	DFT pore size maximum [nm]	calc. wall thickness [nm]	pore volume [cm³/g]
0.65	5.93	5.15	5.95	2.5	3.45	0.07
0.75	5.39	4.8	5.54	2.4	3.14	0.125
0.85	5.74	4.2	4.85	2.2	2.65	0.125
1.0	5.39	3.57	4.12	2.0	2.12	0.126
1.2	5.54	3.2	3.7	1.9	1.8	0.113

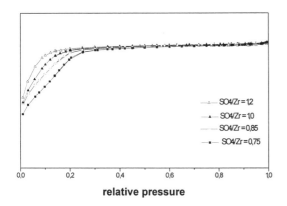

relative pressure

Fig. 5: *N₂ isotherms of samples synthesized with different sulfate content*

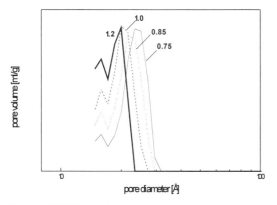

pore diameter [Å]

Fig. 6: *DFT Pore size distributions for corresponding to isotherms in fig. 5*

qualitatively be inferred from inspection of the isotherms is reflected in the calculated pore size distributions.

Even if the trend is represented nicely in DFT analysis, it is difficult to assess, how precise the absolute values for pore sizes are. The DFT algorithm used was developed for slit shaped pores, for the zirconias described here cylindrical pores are much more probable. In addition, the interaction potential between sorptive and wall is that of nitrogen on carbon. This should not influence the calculated pore sizes dramatically[12], but might affect absolute values slightly. Sufficient TEM data of the materials as independent means of calibration are not available, yet. Moreover, it is relatively difficult to obtain reliable information on wall thickness and thus pore sizes from TEM analysis, since the thickness of features strongly depends on focus conditions and careful modelling is necessary for precise analysis.

However, the pore size distributions obtained by applying DFT can at least be compared with those obtained from other methods. We thus analyzed selected zirconia samples with DFT and the Horvath-Kawazoe algorithm. It was found that the DFT method seems to give more reliable data in the mesopore range between 1 and 3 nm. Figs. 7 and 8 compare results obtained with the different methods for three selected samples (C_{18} and C_{20} each stabilized with phosphoric acid, and the microporous crystalline aluminophosphate VPI-5 with a pore size of 1.2 to 1.3 nm according to crystal structural data). As can be seen in fig. 7, VPI-5 has the steepest isotherm before and near the inflection point which suggests a narrow pore size distribution and the absence of small mesopores. The C_{20} material has a step in the isotherm at p/p_0 around 0.15 corresponding to a pore filling. The isotherm for the C_{18} material seems to lie between the one for VPI-5 and for the C_{20} sample. This conclusion, which can be inferred from inspection of the isotherm, is fairly well represented in the pore size distribution calculated from DFT (fig. 8). For VPI-5 a very narrow pore size

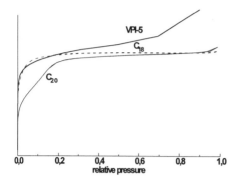

distribution with a peak at about 1 nm is obtained. This is not in such a good agreement with the crystal structural data as the H-K- value of 1.3 nm. The deviation can be explained by the use of the slit shaped pore model for the calculation of the model isotherms in the DFT program package. If the calculation following H-K is based on slit shaped pores, a pore diameter of about 0.7 nm is obtained for VPI-5.

Fig. 7: *Isotherms for zirconiumoxophosphate synthesized with C_{18} and C_{20} surfactant and VPI-5*

For the zirconia samples the H-K algorithm gives essentially identical pore sizes with sharp distributions (0.95 nm for C_{20}, 0.94 nm for C_{18}). These sharp peaks certainly do not reflect the pore structure of the materials which should have a larger pore size and a broader pore size distribution judging from preliminary TEM data, the XRDs, and the analogy to MCM-41. The sharp

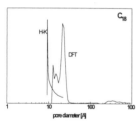
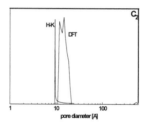
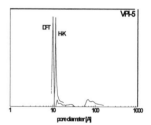

Fig. 8: *Pore size distributions calculated for the C_{18}-sample, the C_{20}-sample and VPI-5 after Horvath-Kawazoe and DFT*

peak is probably caused by the strong adsorption in the low pressure range which dominates the calculated pore size distribution. DFT also gives a peak for small pore sizes of 0.8 nm which is the first pore size used for the calculation of model isotherms from DFT. Smaller pore sizes will thus not be detected and lumped in this peak, if DFT in the form implemented in the software is used. This peak is thus omitted in the graphs given in fig. 8. In the range of larger pore sizes, the DFT pore size distributions correspond closer to expected data. The C_{18} sample has a smaller average pore size than the C_{20} sample, pore size distributions are not as sharp as calculated for VPI-5. It thus seems that for pores in the range between 1 and 3 nm which is very difficult to analyze DFT provides the most appropriate algorithm for analysis.

However, there is much room for improvement, experimentally as well as from the theoretical point of view. With respect to theory, DFT algorithms based on cylindrical pore shapes would be desirable. Such approaches exist or are being developed, but are not inplemented in commercially available software. In addition, it would be important to use interaction parameters which are sorbent specific which is again not implemented in the commercial software. As a last point, more model isotherms in the lower micropore range would probably be helpful to avoid artefacts like the strong peak at 0.8 nm. Experimentally, the samples analyzed here need to be characterized with other techniques, mainly TEM to obtain independent information on the pore sizes. In addition, for such samples with strong adsorption at very low relative pressures in the range of $p/p_0 = 10^{-5}$ more precise isotherm data are necessary. Work in these directions is being carried out in our laboratories.

Acknowledgement

This work was funded by the DFG under grant number Schu744/8-1

References

1. C.T. Kresge, M.E. Leonowicz, W.J. Roth, J.C. Vartuli, J.S. Beck, *Nature* 1992, **359**, 710.
2. P.J. Branton, P.G. Hall, K.S.W. Sing, *J.Chem.Soc.Chem.Commun.* 1993, 1257
3. U. Ciesla et al., in: T.J. Pinnavaia, M.F. Thorpe (Eds.), *Access in Nanoporous Materials,* Plenum Press, New York 1995, p. 231
4. A. Monnier et al., *Science* 1993, **261**, 1233
5. U. Ciesla et al., *J.Chem.Soc.Chem.Commun.* 1994, 1387
6. D.M. Antonelli, J.Y. Ying, *Angew.Chem.Int.Ed.Engl.* 1995, **34**, 2014
7. U.Ciesla, S. Schacht, G.D. Stucky, K. Unger, F. Schüth, *Angew.Chem.Int.Ed.Engl.* 1996, **35**, 541
8. G. Horvath, K. Kawazoe, *J.Chem.Eng.Japan,* 1983, **16**, 470
9. A. Saito, H.C. Fowley, *AIChE Journal* 1991, **37**, 429.
10. J.P. Olivier, W.B. Conklin, *International Symposium on the Effects of Surface Heterogeneity in Adsorption and Catalysis on Solids,* Kazimiersz Dony, Poland, July 1992; P. B. Balbuena, K.E. Gubbins, *Fluid Phase Equilibria* 1992, **76**, 21
11. J. Livage, M. Henry, C. Sanchez, *Progr.Solid State Chem.* 1988, **18**, 259
12. P.I. Ravikovitch, S.C. Ó Domhnaill, A.V. Neimark, F. Schüth, K.K. Unger, *Langmuir* 1995, **11**, 4765

PHYSISORPTION OF GASES BY MCM-41

Peter J. Branton, Peter G. Hall and Kenneth S. W. Sing

The Department of Chemistry, Exeter University, Exeter EX4 4QD, UK

1 INTRODUCTION

It is well known that most mesoporous adsorbents possess broad ranges of irregular pores and give rise to adsorption hysteresis. The disclosures in 1992 by Mobil scientists[1,2] of the synthesis of a new family of highly uniform mesoporous aluminosilicates has therefore attracted a considerable amount of interest. One member of this family, MCM-41 (alternatively designated M41S) has been shown to exhibit 'honeycomb' structure with hexagonal arrays of open-ended pores of diameter of *ca.* 4 nm. The physisorption results discussed here were determined on a sample of this grade of MCM-41, which had been taken from a master batch prepared by Keung[3] at Mainz University.

We have already reported the physisorption isotherms of argon, nitrogen and oxygen (at 77 K)[4,5], carbon dioxide (at 195 K)[6], sulfur dioxide (at 254 and 273 K)[6], the lower alcohols (at 290-314 K)[7] and water vapour (at 303 K)[7] - all determined on the same sample of MCM-41. The nitrogen isotherm was reversible, whereas the other isotherms exhibited hysteresis loops. The main purpose of this paper is to discuss these findings and attempt to explain the reversibility of the Type IV nitrogen isotherm.

2 EXPERIMENTAL

The MCM-41 prepared by Keung[3] and used in the present work was based on example 4 of the Mobil Oil patent[8] and used the following molar ratios: Water/silica 35, silica/alumina 29.7, hydroxide/silica 0.13, tetramethylammonium ions(TMA$^+$)/silica 0.11 and hexadecyltrimethylammonium ions (HDTMA$^+$)/silica 0.24. The material was slowly taken to a temperature of 540°C (heating rate 1°C min^{-1}) and held at this temperature for 8 h. The hydrogen form of the aluminosilicate was obtained by ion exchange with 1 mol dm^{-3} ammonium nitrate solution, vacuum filtration and air drying at room temperature and a final calcination of 550°C for 3 h. SEM showed that the product consisted of an hexagonal array of uniform mesopores with a pore diameter of 3.6 nm.

A manual volumetric technique was used to determine the adsorption isotherms of nitrogen, argon, oxygen, carbon dioxide and sulfur dioxide. A manual gravimetric technique with a quartz spring balance of the McBain-Bakr type was used to determine the adsorption isotherms of water and the alcohols. The MCM-41 was shown to be stable to heat at 350°C[4] and outgassing temperatures of 200°C were used prior to isotherm determination.

3 RESULTS

The isotherms of nitrogen, argon, oxygen, carbon dioxide, sulfur dioxide, methanol, ethanol, propan-1-ol and butan-1-ol were all of Type IV in the IUPAC classification[9], whereas the water isotherm was of Type V - indicative of the hydrophobic character of MCM-41. Isosteric enthalpies of adsorption, calculated from the sulfur dioxide isotherms, revealed strong heterogeneity in the adsorbent-adsorbate interactions. That this was not due to the presence of narrow micropores was confirmed by the appearance of the α_s-plots for nitrogen and argon[5].

Nitrogen was the only adsorptive to give no detectable hysteresis. A very narrow hysteresis loop was given by methanol, while the loops given by the other adsorptives were of the more typical Type H1 in the IUPAC classification[9].

Values of the total mesopore volume, V_p, in Table 1 have been obtained from the amounts adsorbed at $P/P_0 = 0.95$, by assuming that the pores were filled with each condensed adsorptive in its liquid state. The sharpness of the steps in the isotherms and the related α_s-plots provided clear evidence that the pore filling by each adsorptive had occurred over a narrow range of P/P_0 - as shown in Table 1. The derived values of pore diameter, d_p, have been calculated by application of the Kelvin equation (assuming the condensate meniscus to be hemispherical with zero contact angle) with allowance made for multilayer adsorption on the pore walls[10]. A surface area of 655 m^2g^{-1} was calculated from the nitrogen data using the BET analysis in the range $0.05 < P/P_0 < 0.35$ and by assuming the molecular area to be 0.162 nm^2 in the completed monolayer[10].

Table 1 *Pore characteristics derived from adsorption isotherms*

Adsorbate	T / K	V_p / cm^3g^{-1}	P/P_0	d_p / nm
Nitrogen	77	0.64	0.41-0.46	3.3-4.3
Argon	77	0.61	0.38-0.46	3.5-5.1
Oxygen	77	0.64	0.34-0.44	3.1-4.7
Sulfur Dioxide	273	0.55	0.44-0.51	3.7-5.5
Water	303	0.57	0.46-0.63	5.1-11.1
Methanol	290	0.59	0.53-0.58	4.2-6.5
Ethanol	292	0.56	0.47-0.53	4.6-6.9
Propanol	298	0.55	0.42-0.48	5.1-7.4
Butanol	314	0.54	0.35-0.42	4.9-7.2

4 DISCUSSION

Until recently, it was generally assumed that a characteristic feature of a Type IV isotherm was its hysteresis loop[9]. Thus, there appear to have been no completely reversible Type IV nitrogen isotherms reported before the first measurements[1,4] on MCM-41 in 1992 and 1993. However, it had already been established by Everett and his co-workers[11] that for a given adsorption system the size and shape of the loop is temperature-dependent. For example, over the range 208-253 K, the adsorption isotherms of CO_2 by Vycor glass exhibited hysteresis loops, which decreased in size with increase in temperature; while the CO_2 isotherms at 259 and 273 K appeared to be reversible. Murdey and Machin[12], in their recent studies of the effect of temperature on the adsorption isotherms of 2,2-dimethylpropane and other vapours by silica gels, have found a similar marked decrease in the loop size.

Many theoretical explanations have been proposed to account for adsorption hysteresis[10]. It is now widely accepted that the appearance of hysteresis loops associated with capillary condensation is governed by: (a) the limits of stability and metastability of the adsorbed multilayer[13]; and/or (b) network-percolation effects within the pore structure[14]. Unfortunately, because of the complexity of the pore structure of most mesoporous adsorbents, it is virtually impossible to separate these mechanisms and hence arrive at an unambiguous interpretation of the physisorption data. In this connection, the apparent simplicity of the pore structure of MCM-41 is potentially of great importance.

The fact that the hysteresis loops given by our sample of MCM-41 are all of Type H1 is an indication that network-percolation effects are not playing a major role in the hysteresis mechanism. This is confirmed by the linear correlation displayed in Figure 1 between the locations of the adsorption and desorption branches of the loops given by the various adsorptives. The relative pressures, P_a/P_0 and P_d/P_0, correspond respectively to the points of inflection on the adsorption and desorption branches of each hysteresis loop. The uniformity of the H1 shape illustrated in Figure 1 would not be expected if the loops were of the H2 Type, which is generally associated with the more complex percolation pore-blocking behaviour.

Figure 1   *Correlation between isotherm adsorption and desorption branch inflection points*

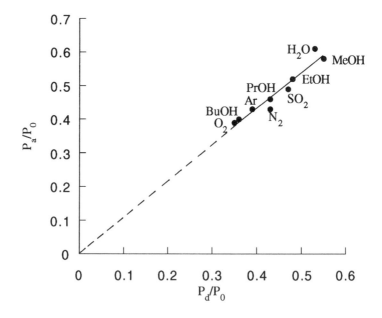

It was predicted by Evans and his co-workers[15] that for a given pore size and shape, 'the jump in adsorption associated with condensation should decrease as T is increased and should vanish at a critical temperature' and that with real porous materials this should result in the eventual disappearance of the hysteresis loop. A similar conclusion was arrived at by Findenegg and his co-workers[16], who found that the capillary critical temperature, T_{cc}, was inversely related to the pore size.

For a cylindrical pore of radius r_p,

$$1-(T_{cc}/T_c) \approx \sigma/r_p \tag{1}$$

where T_c is the bulk adsorbate critical temperature and σ is the adsorbate molecular diameter.

Among the adsorption measurements discussed here, there were considerable differences in nature of the solid-fluid and fluid-fluid interactions. However, it is of interest to ascertain whether the abnormal behaviour of nitrogen (and to some extent, also methanol) was in any respect related to its reduced temperature, T/T_c, where T was the operational temperature (77 K) and T_c its bulk critical temperature (126.2 K). Table 2 gives a comparison of the values of T/T_c for nitrogen and the other adsorptives used so far. Some results of other recent studies on MCM-41 are also included. In view of the fact that the range of T/T_c is very limited, it is not surprising to find that no firm conclusions can be drawn from this simple comparison.

When the different values of molecular diameter, σ, are also taken into account, it is possible to determine the capillary critical temperature, T_{cc}, using equation (1). For nitrogen adsorption at 77 K, application of the Kelvin equation (which assumes classical capillary condensation) gives a pore diameter range, d_p, of 3.3-4.3 nm (from Table 1). Assuming a nitrogen molecular diameter, σ, of 0.354 nm,[10] T_{cc}=99-105 K. This temperature range is significantly larger than the isotherm run temperature of 77 K and thus if classical capillary condensation were occurring we would expect to see some hysteresis at 77 K. Values of T_{cc} have been determined for the other adsorptives in the same way, assuming a pore diameter of 3.6 nm (from TEM measurements[3]), and are shown in Table 2 as T/T_{cc}. The overall trend in the T/T_{cc} values is similar to that of T/T_c, only the magnitude changes. In all cases, T is smaller than T_{cc} and hence hysteresis would be expected. It may be significant however that in the work of Ravikovitch *et al*[17] a change in temperature from 77.4 to 70.6 K was sufficient to allow hysteresis to develop.

Table 2 *Investigation of temperature dependence of hysteresis*

Adsorbate	T / K	T/T_c	T/T_{cc}	Hysteresis	
Nitrogen (d_p=3.3 nm)	77	0.61	0.68	None	
[17]Nitrogen (d_p=4.5 nm)	70.6	0.56	0.61	Yes	
	77.4	0.61	0.67	None	
	82	0.65	0.71	None	
Argon	77	0.51	0.57	Yes	
Oxygen	77	0.50	0.55	Yes	
Sulfur Dioxide	255	0.59	0.67	Yes	
	273	0.63	0.71		
[18]Carbon Monoxide (d_p=2.5 nm)	77	0.58	0.67	No	
Water	303	0.47	0.51	Yes	
Methanol	273	0.53	0.60	Small	
	290	0.57	0.63	Small	
	300	0.58	0.66	Small	Increasing
Ethanol	292	0.57	0.65	Yes	loop size
Propanol	298	0.56	0.64	Yes	with alkyl
Butanol	314	0.56	0.65	Yes	↓ chain length

Finally, it is appropriate to refer to the lower limit of the hysteresis loop[10], which in the case of nitrogen at 77 K is at $P/P_0 = ca.0.42$. It was originally believed that this was controlled by the tensile strength of the condensed liquid. The application of mean-field density functional theory by Evans[15] and others has provided a new approach to this problem, which overcomes the difficulty of relating the curvature of a liquid meniscus to the pore dimensions. Instead, we may picture a first-order transition from the adsorbed phase to the metastable condensate. The chemical potential (and therefore relative pressure) at which this phase transition occurs is dependent on temperature and the pore size and shape, but the coexistense ends at a 'capillary critical point' beyond which there is only one fluid phase (*i.e.* in the supercritical state) in the pore. The significance of this conclusion is that for a given adsorptive and temperature, a pore which is somewhat smaller than the critical dimension does not fill at a single P/P_0 and normally no hysteresis is involved. The few studies made so far on modified kanemite[19] and 2.5 nm diameter M41S[18] confirm this picture, which is also consistent with the concept of secondary (or cooperative) micropore filling[10].

References

1. C.T.Kresge, M.E.Leonowicz, W.J.Roth, J.C.Vartuli and J.S.Beck, *Nature*, 1992, **359**, 710.
2. J.S.Beck, J.C.Vartuli, W.J.Roth, M.E.Leonowicz, C.T.Kresge, K.D.Schmitt, C.T-W.Chu, D.H.Olson, E.W.Sheppard, S.B.McCullen, J.B.Higgins and J.L.Schlenker, *J.Am.Chem.Soc.*, 1992, **114**, 10834.
3. M.Pak-On Keung, Ph.D. Thesis, Brunel University, 1993.
4. P.J.Branton, P.G.Hall and K.S.W.Sing, *J.Chem.Soc., Chem.Commun.*, 1993, 1257.
5. P.J.Branton, P.G.Hall, K.S.W.Sing, H.Reichert, F.Schuth and K.K.Unger, *J.Chem.Soc., Faraday Trans.*, 1994, **90**, 2965.
6. P.J.Branton, P.G.Hall, M.Treguer and K.S.W.Sing, *J.Chem.Soc., Faraday Trans.*, 1995, **91**, 2041.
7. P.J.Branton, P.G.Hall and K.S.W.Sing, *Adsorption*, 1995, **1**, 77.
8. C.T.Kresge, M.E.Leonowicz, W.J.Roth and J.C.Vartuli, *US Pat., 5 102 643*, 1992.
9. K.S.W.Sing, D.H.Everett, R.A.W.Haul, L.Moscou, R.A.Pierotti, J.Rouquerol and T.Siemieniewska, *Pure Appl. Chem.*, 1985, **57**, 603.
10. S.J.Gregg and K.S.W.Sing, 'Adsorption, Surface Area and Porosity', Academic Press, New York, 1982, 2nd edn.
11. C.G.V.Burgess, D.H.Everett and S.Nuttall, *Pure Appl. Chem.*, 1989, **61**, 1845.
12. R.J.Murdey and W.D.Machin, *Langmuir*, 1994, **10**, 3842.
13. G.H.Findenegg, S.Gross and T.Michalski in 'Characterization of Porous Solids III', eds J.Rouquerol, F.Rodriguez-Reinoso, K.S.W.Sing and K.K.Unger, Elsevier, Amsterdam, 1994, p.71.
14. H.Liu, L.Zhang and N.A.Seaton, *Langmuir*, 1993, **9**, 2576.
15. R.Evans, U.M.B.Marconi and P.Tarazona, *J.Chem.Phys.*, 1986, **84**, 2376.
16. A.de Keizer, T.Michalski and G.H.Findenegg, *Pure Appl. Chem.*, 1991, **63**, 1495.
17. P.I.Ravikovitch, S.C.O.Domhnaill, A.V.Neimark, F.Schuth and K.K.Unger, *Langmuir*, 1995, **11**, 4765.
18. J.Rouquerol et al., *Surf.Sci.*, In Press.
19. P.J.Branton, K.Kaneko, N.Setoyama, K.S.W.Sing, S.Inagaki and Y.Fukusima, *Langmuir*, 1996, **12**, 599.

CHARACTERISATION OF POROSITY DURING THERMAL DECOMPOSITION OF NICKEL HYDROXIDE

M.M.L. Ribeiro Carrott, P.J.M. Carrott & A.J.E.G. Candeias

Departamento de Química, Universidade de Évora,
Colégio Luís António Verney, Rua Romão Romalho, 39,
7000 Évora, PORTUGAL.

Abstract

Low temperature nitrogen adsorption was used to characterise the porosity of nickel hydroxide and the products of its slow thermal decomposition under vacuum. Two different methods of preparation of the hydroxide were used and lead to samples with quite distinct adsorptive properties. The results are discussed in relation to the changes in microstructure of the samples which occur during decomposition.

1 INTRODUCTION

A convenient route for the preparation of microporous solids with slit shaped pores is by controlled thermal decomposition of an appropriate lamellar precursor. The microstructure of the porous solid is dependent, on the one hand, on the structure of the precursor and, on the other hand, on the exact conditions used to carry out the thermal decomposition. In particular, during the preparation of microporous oxides it is essential to control the pressure of water vapour in order to avoid sintering and consequent breakdown of the lamellar structure.

In previous work[1-4], some of it presented in previous COPS conferences, we studied in detail the structural transformations involved in the controlled thermal decomposition of magnesium hydroxide to give microporous magnesium oxide, with slit shaped pores of dimension between $ca.0.9$nm at low decomposition temperature and up to $ca.1.8$nm at higher temperatures, as well as the reverse transformation involving the rehydroxylation of the oxide. This work is now being extended to the system nickel hydroxide/nickel oxide.

Previous studies[5-8] have indicated that the thermal decomposition of nickel hydroxide can occur in a similar way to that of magnesium hydroxide, involving a topotactic transformation in which the major crystallographic relationship derives from the conversion of the (0001) plane of the hydroxide (CdI_2 structure) to a {111} plane of the oxide (NaCl structure). Under appropriate conditions, a uniform particle structure, consisting of oriented microcrystallites intercalated by slit shaped micropores, gradually spreads from the outside towards the centre of each crystal, actually reaching the centre at a level of decomposition of about 85-90%. At higher levels of decomposition the mean pore size increases which, at least in the case of magnesium oxide, is associated with a restructuring of the microcrystallites. The preliminary results which we will present here indicate that although the nickel system is, in certain respects, very similar to the magnesium system, there are a number of quantitative differences. Furthermore, with one of the nickel hydroxide preparations which we have studied, there are also some significant qualitative differences, notably the occurrence of interparticle, as well as intraparticle, porosity.

2 EXPERIMENTAL

Nickel hydroxide sample A was prepared by dissolution of 4g of nickel hydroxide in excess ammonia to form the hexamminonickel(II) complex, followed by removal of the ammonia by heating at *ca.* 60°C under a flow of N_2 gas. Nickel hydroxide sample B was prepared by dropwise addition of a 1.0M carbonate free solution of NaOH to a 0.5M solution of $NiCl_2$ also maintained under a N_2 atmosphere. In both cases the precipitates were separated by centrifugation and repeatedly washed until there were no traces of chloride in the washing water. The precipitates were subsequently dried at room temperature in a vacuum dessicator. Preliminary powder X-ray diffraction and electron microscopy measurements confirmed the purity of the precursor hydroxides and indicated that sample A consisted of approximately spherical aggregates of nickel hydroxide microcrystals with diameter/width ~1, while sample B consisted of smaller lamellae with a much higher diameter/width ratio.

For the adsorption experiments, the samples were decomposed *in situ* by outgassing initially at room temperature to a residual pressure of $<10^{-4}$mmHg, followed by slowly increasing the temperature while maintaining the residual vacuum $<10^{-3}$mmHg. The final outgassing temperatures are indicated in the sample designations used in the Tables and Figures. Nitrogen isotherms at 77K were determined on a conventional manual volumetric apparatus with pressure measurement was by means of strain gauge pressure transducer calibrated against a high precision Datametrics Barocell capacitance manometer.

3 RESULTS AND DISCUSSION

3.1 Evaluation of Stoichiometry

Outgassing sample A at 170°C resulted in the loss of *ca.*2-3% of the original weight of the sample at room temperature, but without alteration in the surface area, and it was therefore assumed that the outgassed weight at 170°C corresponded to the exact stoichiometry $Ni(OH)_2$. In the case of sample B, outgassing at the same temperature resulted in a much higher weight loss and it was therefore found to be necessary to take the outgassed weight at the lower temperature of 130°C as corresponding to the stoichiometric hydroxide. The corresponding weights were used to convert the weight losses at progressively higher outgassing temperatures to the percentage decompositions given in the Tables. It should be noted that, for both samples, the calculated values of %decomposition at higher temperatures are close to 100%, which appears to provide additional confirmation that sample B begins to decompose at a lower temperature than sample A. It should also be noted that all adsorption results in the Tables and Figures are expressed with respect to the equivalent mass of stoichiometric hydroxide.

3.2 Nitrogen Adsorption on Non-decomposed Samples

Nitrogen isotherms and corresponding α_s plots (standard data, silica) determined on the non-decomposed samples A and B are shown in Figure 1 and a summary of the BET and

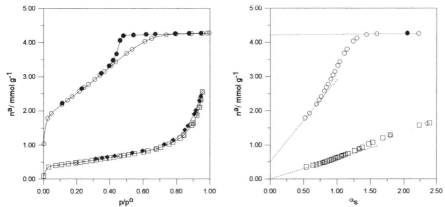

Figure 1 *Nitrogen isotherms and corresponding α_S plots on:*
$\square \, \lozenge = A/110, \, \bigcirc = B/130$

α_S analyses of the isotherm data are given in Tables 1-4. The results indicate clearly that the two different preparations used lead to samples with quite distinct textural characteristics.

Sample A gives Type II isotherms which are virtually identical at the two outgassing temperatures, 110°C and 170°C (not shown), used. The corresponding α_S plots pass through the origin indicating the absence of microporosity. However, the occurrence of a small amount of interparticle capillary condensation is evident from the upward deviations of the α_S plots in the multilayer region. If the extent of capillary condensation does not alter when the sample is decomposed at higher outgassing temperatures (and it will be seen below that this is the case) then the isotherms determined on A/110 and A/170 should be suitable for use as reference data for analysing isotherms determined on decomposed samples of sample A and we will, in fact, make use of this data in section 3.3.

Sample B is considerably more complex. The appearance of a capillary condensation hysteresis loop in the isotherm clearly indicates the presence of mesoporosity while the shape of the corresponding α_S plot, which deviates significantly from linearity at low pressures, indicates the presence of microporosity. Values of total pore volume, v_s, and external surface area, A_{ext}, can be estimated in the usual way from the intercept and slope of the linear multilayer region of the α_S plot and are included in Table 3. It can be seen that the total pore volume is comparatively high while the external surface is extremely low.

Close inspection of the sample B α_S plot indicates the presence of a fairly broad distribution of pores ranging in size from 1° micropores up to 2° micropores and small mesopores. We can make an approximate estimate of the relative contributions if we assume that 1° micropore filling is effectively complete at, for instance, p/p°<0.005 (corresponding to α_S<0.35 with the standard used) and that 2° micropore and mesopore filling involves monolayer coverage on the pore walls at lower pressures (corresponding to the linear part of the α_S plot in Figure 1 at α_S<0.7), followed by pore filling at higher pressures (corresponding to the upswing in the α_S plot at α_S>0.7). The intercept of the first linear region of the α_S plot should therefore give an approximate estimate of the 1° micropore volume. In this way we obtain v_1=0.018cm³g⁻¹ for the volume of 1° micropores and, by difference, v_2=0.128cm³g⁻¹ for the volume of the wider pores, which therefore comprise 88% of the total pore volume. An estimate of the surface area of the wider pores

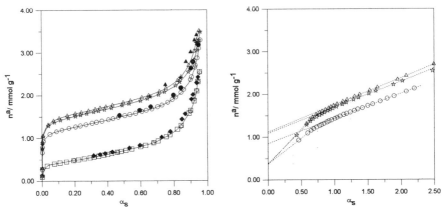

Figure 2   *Nitrogen isotherms and corresponding α_S plots on:*
$\square \lozenge = A/110$, $\bigcirc = A/220$, $\stackrel{\star}{\scriptstyle\star} = A/245$, $\triangle = A/270$

is given by the slope of the first linear region, A_s, corrected for adsorption on the external surface area, A_{ext}, the value obtained being $A_2 = 152 m^2 g^{-1}$.

As already mentioned, the combination of XRD and SEM indicated that sample B was composed of stacked thin lamellae (each composed of several HO-Ni-OH layers) and the very low external surface area is consistent with this. It seems reasonable to assume, therefore, that the microporosity occurs between adjacent lamellae and hence that the micropores are slit-shaped. With regard to the mesopores, we can't be sure about their exact shape. However, in order to obtain an estimate of the mean pore size it is convenient to assume that all the pores can be treated as slits and that a mean hydraulic pore width can be calculated from the relationship: $d_2 = 2v_2/A_2$. The value obtained is $d_2 \sim 1.7 nm$.

As sample B is microporous, even before decomposition, we can not use the isotherm as reference data to analyse isotherms determined on the decomposed samples. Fortunately, as the external surface area is very low, little error should be incurred in this case by continuing to use the silica reference data to analyse the isotherms in the next section.

3.3 Nitrogen Adsorption on Decomposed Samples

Nitrogen isotherms and corresponding α_S plots determined on decomposed samples are shown in Figures 2-5, and the results of analysis of the data by means of the α_S and BET methods are given in Tables 1-4. Once again, some quite significant differences between the two samples are found.

It can be seen from Figures 2 and 3 and Table 1 that decomposition of sample A initially results in a change in isotherm shape from Type II to mixed Type I / Type II with a corresponding upward shift of the α_S plot. The maximum micropore volume is obtained at a level of decomposition of ~89% and remains approximately constant up to ~95% decomposition. At higher decomposition levels the uptake begins to decrease and the isotherm shape alters again, returning to the Type II shape when the sample is completely decomposed. It is interesting to note from the α_S plots and Table1 that the external surface area remains practically constant, even at the highest decomposition studied. This confirms that the morphology of the particles was retained during the decomposition and indicates that during decomposition only microporosity (1° and 2°) is produced and not mesoporosity.

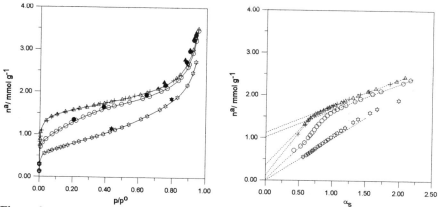

Figure 3   *Nitrogen isotherms and corresponding α_S plots on:*
$\Delta = A/270, + =A/300, \bigcirc =A/400, \Leftrightarrow =A/500$

Table 1   *BET and α_S Analysis of Nitrogen Isotherms of Sample A*

sample	% decomp.	c_{BET}	A_{BET} /m²g⁻¹	A_s /m²g⁻¹	A_{ext} /m²g⁻¹	v_s /cm³g⁻¹	v_l /cm³g⁻¹
A/110	0	220	38				
A/170	0	220	38				
A/220	76	920	103	72	36	0.030	0.013
A/245	89	770	125	94	37	0.038	0.013
A/270	91	700	130	101	39	0.039	0.013
A/300	95	430	130	122	38	0.040	0.004
A/400	97	90	104	90	39	0.035	0.001
A/500	100	90	64	60	39	0.021	0

Table 2   *Characterisation of 2° Micropores of Sample A*

sample	% decomp.	A_2 /m²g⁻¹	v_2 /cm³g⁻¹	d_2 /nm	% 2° pores
A/110	0				
A/170	0				
A/220	76	37	0.017	0.92	57
A/245	89	57	0.025	0.88	66
A/270	91	62	0.026	0.84	67
A/300	95	84	0.036	0.85	90
A/400	97	51	0.034	1.33	97
A/500	100	21	0.021	2.01	100

We can analyse the α_S plots in Figures 2 and 3 in the same way as described in Section 3.2 (for the non-decomposed sample B) in order to calculate estimates of mean hydraulic pore width for the 2° micropores. The results are included in Table 2 along with the percentage of the total micropore volume corresponding to the 2° micropores. These results show that during the initial stages of decomposition, when the micropore volume is increasing, the mean pore width is of the order of 0.9nm, about 60% of the pores being 2° micropores and the remainder 1° micropores. There is at first a small decrease in mean pore

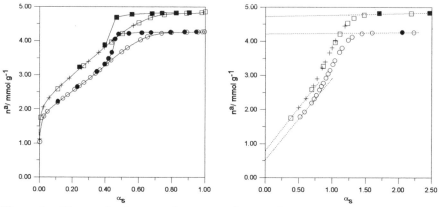

Figure 4 *Nitrogen isotherms and corresponding α_s plots on:*
$\bigcirc = B/130,$ $\square + = B/170$

Table 3 *BET and α_s Analysis of Nitrogen Isotherms of Sample B*

sample	% decomp.	c_{BET}	A_{BET} /m²g⁻¹	A_s /m²g⁻¹	A_{ext} /m²g⁻¹	v_s /cm³g⁻¹	v_l /cm³g⁻¹
B/130	0	150	201	153	1	0.146	0.018
B/170	25	170	213	160	3	0.164	0.028
B/200	62	160	246	160	3	0.168	0.031
B/250	71	120	224	148	3	0.155	0.025
B/300	83	110	94	93	3	0.109	0
B/400	87	120	41	41			
B/500	91	230	31	32			
B/600	100	370	26	27			
B/700(1)	100	330	16	17			
B/700(2)	100	270	13	12			

Table 4 *Variation in Pore Volumes of Sample B*

sample	% decomp.	Δv_s /cm³g⁻¹	Δv_l /cm³g⁻¹
B/170	25	+0.018	+0.010
B/200	62	+0.022	+0.013
B/250	71	+0.009	+0.007
B/300	83	-0.037	-0.018

width. However, when the micropore volume begins to decrease at the higher decomposition levels, the mean pore width increases very rapidly and shifts into the mesopore range at 100% decomposition. The last column of Table 2 shows that the volume of 1° micropores begins to decrease rapidly even before the total pore volume starts to decrease, and that there is in fact a shift in the pore size distribution towards wider pores at all levels of decomposition.

The results obtained with sample B, and shown in Figures 4 and 5 and Tables 3 and 4 are somewhat different to those obtained with sample A due, at least in part, to the fact that

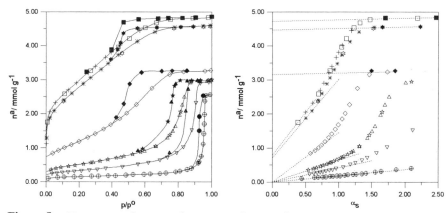

Figure 5 *Nitrogen isotherms and corresponding α_S plots on:*
$\bigcirc + =B/170$, $\ast\ \Leftrightarrow =B/250$, $\Diamond =B/300$, $\triangle\ \Leftrightarrow =B/400$, $\nabla =B/600$, $\oplus =B/700(2)$

2 separate pore systems (intra- and inter- particle) are involved. The presence of mesoporosity in the B samples, coupled with the fact that during decomposition changes may occur simultaneously in the two pore systems, makes it extremely difficult to carry out an analysis of the 2° micropore structure in the way that was presented for sample A. Hence, a Table equivalent to Table 2 is not presented for sample B. On the other hand, in order to compare the results obtained with sample B with those given by sample A it is convenient to consider the change in pore volumes relative to the original non-decomposed sample and these values are presented in Table 4.

It can be seen from Figure 4 and Table 4 that the pore volume initially increases when the sample begins to decompose, but that the maximum increase in total pore volume ($0.022 cm^3 g^{-1}$ for B/200) is significantly less than that found with sample A and is attained at a much lower decomposition level. Furthermore, the micropore volume decreases much more rapidly at higher decompositions, there already being no 1° micropores at 83% decomposition and no micropores at all at 87% decomposition. With regard to the initially present interparticle pore volume, Figure 5 indicates that, at least at decomposition levels above 62%, this pore volume decreases and the corresponding mean pore width increases significantly, having already shifted into the mesopore range at 83% decomposition. It seems likely that these changes actually begin at lower levels of decomposition thereby explaining, at least partially, the apparently low increase in micropore volume during the initial stages of decomposition.

Finally, it should be noted from Figure 5 that after complete decomposition of sample B, the sample gives a Type IV isotherm with a quite narrow Type H1 hysteresis loop, indicating that it retains a rigid, and fairly uniform, pore structure.

4 CONCLUSIONS

The results obtained with sample A are similar to those previously reported for the thermal decomposition of magnesium hydroxide[1-4]. Thus: *i)* decomposition results in the formation of a microporous product retaining the particle morphology of the precursor, *ii)* the micropore volume initially increases, reaching a more or less constant value over the approximate range of ~85-95% decomposition, and then decreases rapidly at higher levels of decomposition and *iii)* the mean micropore width initially decreases slightly, but then increases significantly when the pore volume begins to decrease at high decomposition levels.

The principal differences between the two systems are *i)* the pore volume is much less with nickel oxide, which can be attributed to a large extent, although not completely, to the difference in molar mass, *ii)* at certain levels of decomposition the isotherms obtained with magnesium oxide show evidence of stepwise character, which is almost completely absent from the nickel oxide isotherms obtained in this work and *iii)* although the estimates of mean pore width are similar, the magnesium oxide samples contained pores exclusively in the 2° micropore range, while the nickel oxide samples also contain some 1° micropores. Both ii) and iii) suggest that there is a somewhat broader pore size distribution with the nickel oxide system.

The results obtained with sample B are particularly interesting for two reasons. In the first place, they are a reminder of how the physicochemical properties of adsorbents can be significantly influenced by the state of aggregation of the solid. In the present instance, it is possible that the observed behaviour is associated with the thinness of the lamellae, although more work is still needed to check this. Secondly, the results suggest that layered nickel hydroxide prepared under suitable conditions may be a useful precursor for the preparation of model porous adsorbents with slit shaped pores in the 2° micropore or mesopore range. More work is still needed, however, to verify the pore shape of the interparticle mesopores and to improve the uniformity of the pore size.

Acknowledgements

The authors are grateful to the Junta Nacional de Investigação Científica e Tecnológica for financial support (grant n°. PBIC/C/CTM/1929/95).

References

1. M.M.L. Ribeiro Carrott, P.J.M. Carrott, M.B. Carvalho & K.S.W. Sing, 'Characterisation of Porous Solids II', ed. F. Rodriguez-Reinoso, J. Rouquerol, K.S.W. Sing & K.K. Unger, Elsevier, Amsterdam, 1991, 635.
2. M.M.L. Ribeiro Carrott & P.J.M. Carrott, 'Characterisation of Porous Solids III', ed. J. Rouquerol, F. Rodriguez-Reinoso, K.S.W. Sing & K.K. Unger, Elsevier, Amsterdam, 1994, 497.
3. M.M.L. Ribeiro Carrott, P.J.M. Carrott, M.B. Carvalho & K.S.W. Sing, *J.Chem.Soc., Faraday Trans.*, 1991, **87**, 185.
4. M.M.L. Ribeiro Carrott, P.J.M. Carrott, M.B. Carvalho & K.S.W. Sing, *J.Chem.Soc. Faraday Trans.*, 1993, **89**, 579.
5. F. Fievet and M. Figlarz, *J.Catal.*, 1975, **39**, 350.
6. R.B. Fahim and A.I. Abu-Shady, *J.Catal.*, 1970, **17**, 10.
7. C.L. Cronan, F.J. Micale, M. Topic, H. Leidheiser & A.C. Zettlemoyer, *J.Colloid Interface Sci.*, 1980, **75**, 43.
8. G.A. Nicolaon & S.J. Teichner, *J.Colloid Interface Sci.*, 1972, **38**, 172.

A THERMODYNAMIC INVESTIGATION OF PHYSISORBED PHASES WITHIN THE MODEL MESOPOROUS MATERIAL : MCM-41

P. L. Llewellyn [a], C. Sauerland [a], C. Martin [b], Y. Grillet [a], J.-P. Coulomb [b], F. Rouquérol [a] & J. Rouquérol [a]

[a] Centre de Thermodynamique et de Microcalorimétrie du C.N.R.S., 26 rue du 141° R.I.A., F-13331 Marseilles cedex 3 (France)
[b] CRMC2-Faculté des Sciences de Luminy, Université Aix-Marseille II, Case 901, F-13288 Marseilles cedex 9 (France)

ABSTRACT

The low temperature adsorption of a number of probe molecules is followed by volumetry coupled with microcalorimetry or neutron diffraction. A small increase in the adsorption enthalpies accompanies the pore filling process which may be due to an effect of the increase in close neighbours. The adsorbate phase produced seems to be amorphous in nature as observed for deuterated methane. Krypton may adsorb in a different manner to the other molecules studied as an abnormally high adsorption enthalpy is observed during the pore filling process. Finally, it seems that argon is more suitable than nitrogen for calculation of the surface area.

1 INTRODUCTION

MCM-41 [1,2] and related materials raise much interest in both fundamental and applied studies. From a fundamental point of view MCM-41 is an interesting material for adsorption studies whereas from an applied point of view, MCM-41 is seen as a potential catalyst [3] and separation material for chromatography.

MCM-41 consists of a hexagonal arrangement of pores rather like a honeycomb. In the literature pore sizes are generally obtained in the region from 2nm to 8nm although the original papers gives pore sizes of up to 20nm. In most cases a relatively narrow pore size distribution obtained. The pore walls are amorphous and generally of around 0.8nm in thickness although it is also possible to vary this [4]. These points make this material interesting for adsorption studies as material can be used of a given pore size with a narrow size distribution and of negligible external surface area.

A distinct step is observed in the isotherm for a number of probe molecules adsorbed on many MCM-41 materials [5-13]. This step is generally considered as being due to a co-operative condensation of the adsorbate. This phenomenon is rather interesting as it occurs on a "model material" (tube like) allowing a direct comparison with results obtained via a number of theoretical models.

A great number of studies have characterised these mesoporous materials via adsorption volumetry at 77K. Many other studies have used characterisation techniques such as adsorption gravimetry, X-ray diffraction, infra-red, chromatography, NMR, room temperature calorimetry as well as catalysis. These studies have essentially used highly siliceous materials of approximately 4nm in diameter.

In this study, low temperature adsorption volumetry has been coupled either with quasi-equilibrium isothermal microcalorimetry or neutron diffraction using a variety of molecules of different size and permanent moment. The aim here is to characterise from an energetic and structural standpoint the processes involved during the adsorption of simple probe molecules on MCM-41 and, in particular, during the condensation step. The influence of a small change in pore size is also examined with two samples of 2.5nm and 4.0nm in pore diameter.

Following the adsorption process using microcalorimetry allows a direct measurement of the extent of the interactions (adsorbate-adsorbent, adsorbate-adsorbate, structure changes etc...) taking place within the system. A further particularity of this technique is the method of continuous gas introduction under quasi-equilibrium conditions. This allows a particularly high resolution in both the determination of the adsorption isotherm and detection of small or sharp variations in the enthalpies of adsorption. Neutron diffraction gives an insight into the structural state of the system. The crystalline state of the adsorbate and adsorbent can be followed as a function of the quantity adsorbed and in many cases it is also possible to stop at a particular point in the isotherm to investigate the influence of temperature variations.

2 EXPERIMENTAL

The standard M41S material is synthesised according to a modified procedure to that described in reference 1. The surfactants used were dodecyl tetramethyl ammonium bromide and hexadecyl tetramethyl ammonium bromide which gave samples with pore diameters of 2.5nm and 4.0nm respectively (as calculated from the X-ray diffraction patterns and thermoporometry). The samples were calcined using Controlled Transformation Rate Thermal Analysis (CRTA) [14] to 825K to remove the surfactant species via this "soft chemistry route". CRTA is further used for outgassing as it is the only procedure which allows one to reproduce a well defined surface state of the aluminosilicate. The samples (around 50 mg) were outgassed to 473 K at a rate of 0.35 mg·h^{-1} under a residual water vapour pressure of 1.33 Pa.

The static (point by point) adsorption volumetry experiments were carried out at 77 K both on a commercial apparatus (Omnisorp 100 Analyser, Coultronics, France, S.A.) and on equipment constructed 'in house' (C.T.M., Marseille [15]). The adsorptive introduction was also performed using a quasi-equilibrium procedure (continuous technique) [16] which employs an extremely slow constant introduction of adsorptive gas, in the region of 2 cm^3·hr^{-1} for which it was verified that the quasi-equilibrium conditions were fulfilled. This latter technique coupled with isothermal adsorption microcalorimetry [17] allows direct access to a continuous measurement of the differential enthalpies of adsorption, $\Delta_{ads}\dot{h}$, during the adsorption.

The neutron diffraction experiments were carried out at the Léon Brillouin Laboratory (common laboratory CNRS-CEA, Saclay, France). Around 2 g of the M41S sample was used and the measurements carried out on the diffractometer G4-1 (wavelength $\lambda = 2.439$ Å). The results were recorded with a multi-detector and were then plotted as the peak intensity as a function of Q, the scattering vector where $Q = \{ 4\pi / \lambda \}$ sin θB (θB is the Bragg diffraction angle). The experiments were performed principally around 77 K for CD_4 and 16.4 K for D_2 with stability studies of the adsorbed phase between 3 and 77 K.

3. RESULTS & DISCUSSION

The adsorption-desorption isotherms of argon, nitrogen and carbon monoxide at 77K on the two MCM-41 samples taken for this study are shown in figure 1. These isotherms are all of type IV [18]. The distinct step, observed between $0.2 < p/p^0 < 0.5$ according to each individual system, is due to a co-operative adsorbate-adsorbate condensation within the pores. The presence or not of any hysteresis is again a function of the adsorbate-adsorbent system under question. It can however be seen that an increase in pore size leads to a greater probability of hysteresis to occur as is the case here for carbon monoxide and argon. It can also be seen that the hysteresis more often occurs when the

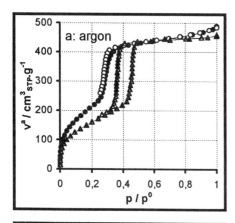

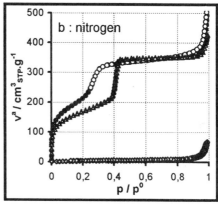

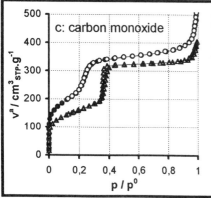

Figure 1 : *comparison of the adsorption of a : argon, b : nitrogen and c : carbon monoxide on two MCM-41 samples : diameter 2.5nm = triangles, diameter 4.0nm = circles (closed symbols = adsorption, open symbols = desorption). The third isotherm shown in figure b is that of nitrogen adsorbed on an as-synthesised sample before template removal. This allows an estimation of the extent of the external surface area which was found to be of around 2.5% of the total surface area.*

saturation vapour pressure at the measurement temperature is low. This is illustrated here for the case of the 4nm material at 77K. For nitrogen ($p^0 \approx 760$ torr) no hysteresis is observed and for carbon monoxide ($p^0 \approx 450$ torr) a small hint of hysteresis is observed, however for argon ($p^0 \approx 210$ torr) a significant hysteresis is evidenced.

The top of this step corresponds to completion of the pore filling. It is thus interesting to compare the value of the quantity adsorbed at the top of this step (V_{real}) with the value of the mesopore volume calculated using the expression $V_{calc} = dA_{BET}/4$ where A_{BET} is the surface area calculated from the BET equation and d the pore diameter. This is possible as the surface may be considered as uniquely formed of mesopores; the external surface area is less than 2.5% of the total (c.f. figure 1b).

Table 1 : *comparison of the pore volume obtained after calculation using the BET surface area (V_{calc}) with the actual volume adsorbed at the top of the step (V_{real}).*

	σ	A_{BET}	$V_{calc} = d\,A_{BET}/4$	V_{real}	$A_{recalc} = 4V_{real}/d$
N_2	$0.162nm^2$	$614m^2.g^{-1}$	$0.614cm^3.g^{-1}$	$0.510cm^3.g^{-1}$	$510\ m^2.g^{-1}$
Ar	$0.138\ nm^2$	$522m^2.g^{-1}$	$0.522cm^3.g^{-1}$	$0.536cm^3.g^{-1}$	$536\ m^2.g^{-1}$

Table 1 shows the results obtained for nitrogen and argon on an MCM-41 sample of 4.0nm in diameter. The value of the volume obtained after calculation of the nitrogen BET surface area is significantly higher than that actually observed whereas for argon, the differences are slight. This anomaly can be explained by an overestimation of the value of σ (surface area taken by one molecule) generally taken for nitrogen. This is almost certainly due to a preferential orientation of the molecule with respect to the surface. It is further interesting to recalculate the value of σ of nitrogen for this adsorption on MCM-41, a value of $0.140nm^2$ is obtained which corresponds to a position more perpendicular to the surface. Similar results have been obtained on a number of solids [19-21] which raise doubt as to whether nitrogen is the most suitable molecule for the calculation of such surface areas. It may thus be less erroneous to use argon for this calculation where possible, even at 77K where it is well known that the aron isotherm is truncated at high relative pressure.

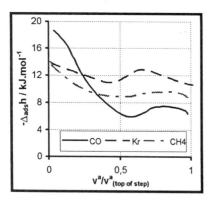

Figure 2a : *variation of the differential enthalpy of adsorption for the adsorption of Ar, CH₄ & Kr on MCM-41 (d = 2.5nm)*

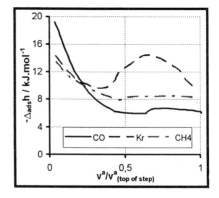

Figure 2b : *variation of the differential enthalpy of adsorption for the adsorption of Ar, CH₄ & Kr on MCM-41 (d = 4.0nm)*

Microcalorimetry results from the adsorption on the MCM-41 samples of krypton, methane and carbon monoxide are shown in figure 2. It can be seen that, as for many amorphous silica surfaces, the initial part of the curve decreases. This indicates a relatively heterogeneous distribution of adsorption sites. For all of the probe molecules studied, with the exception of krypton, a small increase in the adsorption enthalpy, above the enthalpy of liquefaction, is observed during the condensation step. Kiselev [22] observed an increase in the adsorption enthalpy just after the end of the step and explained this by a change in the meniscus shape. Franke *et al* [6], calculated the isosteric heats of adsorption for cyclopentane found an increase throughout the condensation as in this study. Surprisingly, Franke *et al* also calculated a more important increase in the isosteric heat during the plateau region of the isotherm which is almost certainly due to uncertainties in the calculation in this region. A further interesting point in this latter study is that the isosteric heats of adsorption calculated are significantly higher, throughout the adsorption, than the enthalpy of liquefaction of cyclopentane. Our values, directly observed, show regions where the adsorption enthalpy drop to those of the enthalpy of liquefaction. Maddox and Gubbins [23], using Monte Carlo calculations also show an increase in isosteric heat at capillary condensation which is encouraging however this study also reports a peak after completion of the monolayer which is not observed in the present study.

The fact that there is an increase in the adsorption enthaply during this condensation step is in contradiction with a purely capillary condensation model of the adsorbate as this process has been predicted to occur with an enthalpy of liquefaction. This phenomenon may however be logically explained if one considers the number of closest neighbours at each of the adsorption processes. During multilayer growth, the process just before condensation, the number of closest neighbours numbers 6 if the assumption of a hexagonally packed layer is taken. However during the adsorbate condensation, the number of closest neighbours increases to 12 which may thus explain this increase in adsorption enthaply at this point (i.e. an effect of adsorbate confinement).

The case of the adsorption of krypton is intriguing. The enthalpy of adsorption increases significantly during the condensation step for both adsorption on the 2.5 and 4nm material. Furthermore if one calculates the volume of liquid adsorbed at the top of the step, this is not the same as obtained with the other adsorbates which is in contradiction with the Gurvitsch rule [24]. However if one calculates the volume using the bulk solid density a much better fit results. It would thus seem that the krypton adsorbed within the mesopores has a density close to that of the bulk solid. This result has to be confirmed using a structural technique and neutron diffraction measurements have been demanded for this purpose.

Figure 3 shows the results of two different neutron diffraction experiments. Spectra obtained during the adsorption of deuterated methane (CD_4) on MCM-41 (d=2.5nm) are shown in figure 3a for three different points during the isotherm at 77K : before the step ($v^a \approx 120cm^3.g^{-1}$), during the condensation step ($v^a \approx 255cm^3.g^{-1}$) and after the step ($v^a \approx 340cm^3.g^{-1}$). A broad peak in the scattering wave vector, Q, from 1.7 to 1.9 $\mathring{A}^{-1}$ is observed in each case. This is indicative of short range order ($20 \leq L_{coher.} / \mathring{A} \leq 30$) and is also observed in the case of the adsorption of deuterium. A rather interesting result is also shown in figure 3b where the adsorption is stopped during the condensation ($v^a \approx 255cm^3.g^{-1}$) and the temperature of the system is lowered to 3.1K. This again shows the lack of any long range order with the spectrum remaining essentially unchanged. This short range order is characteristic of an amorphous phase which may be either liquid-like or solid-like.

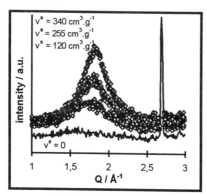
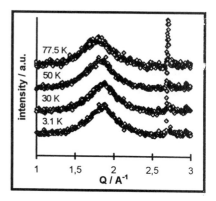

Figure 3a : *neutron diffraction spectra obtained at different points during the CD₄ isotherm (77K): before (120cm³.g⁻¹), during (255cm³.g⁻¹) and after the step (340cm³.g⁻¹).*

Figure 3b : *neutron diffraction spectra obtained during the condensation step of the CD₄ isotherm (vᵃ ≈ 255cm³.g⁻¹) at different temperatures.*

4 CONCLUSION

Variations in the adsorption behaviour of a number of simple probe molecules on two MCM-41 samples of 2.5nm and 4.0nm in diameter have been investigated by isothermal volumetry coupled with microcalorimetry and neutron diffraction. It has been shown that the condensation step leads to a slight increase in the adsorption enthalpy which may be simply due to the greater number of closest neighbours observed by the adsorbate at this point. It would seem that a phase characterised by short range order is obtained after the condensation as observed with deuterated methane. Unfortunately the experiments carried out are unable to differentiate between static disorder (amorphous solid) or dynamic disorder (fluid-like phase). Calculations of the liquid volume after pore filling would suggest that a liquid-like phase is obtained for most of the adsorbates observed. The adsorption of krypton is interesting, however; the aforementioned calculations fit best on using the bulk solid density and an unusually high value of the adsorption enthalpy is observed during the pore filling process. This may suggest the formation of a dense solid-like krypton phase at this point.

Finally it interesting to note that the BET surface area obtained for nitrogen on the MCM-41 samples gives results which are significantly higher than those obtained with argon. The subsequent calculation of the total volume, and the comparison with the actual quantity adsorbed at the top of the condensation step, suggests that it is the calculations using argon adsorption which are more more corrext. Thus, it seems that argon is more suitable than nitrogen for surface area calculations.

References

1. J. S. Beck, J. C. Vartuli, W. J. Roth, M. E. Leonowicz, C. T. Kresge, K. D. Schmitt, C. T.-W. Chu, D. H. Olson, E. W. Sheppard, S. B. McCullen, J. B. Higgens & J. L. Schlenker, *J. Am. Chem. Soc.,* 1992, **114**, 10834.

2. C. T. Kresge, M. E. Leonowicz, W. J. Roth, J. C. Vartuli & J. S. Beck, *Nature*, 1992, **359** 710.

3. Q. N. Le, R. T. Thomson & G. H. Yokomizo, U. S. Patent No. 5,134,242 (1992)

4. N. Coustel, F. Direnzo & F. Fajula, *J. Chem. Soc., Chem. Comm.,* 1994, **8** 967

5. P. J. Branton, P. G. Hall & K. S. W. Sing, *J. Chem. Soc. Chem. Comm.,* 1993, **16**, 1257.

6. O. Franke, G. Schulz-Ekloff, J. Rathousky, J. Starek & A. Zukal, *J. Chem. Soc. Chem. Comm.,* 1993, **9**, 724.

7. P. J. Branton, P. G. Hall, K. S. W. Sing, H. Reichert, F. Schüth & K. K. Unger, J. Chem. Soc. Faraday Trans., 1994, **90**, 2965.

8. P. L. Llewellyn, Y. Grillet, F. Schüth, H. Reichert & K. K. Unger, *Microporous Materials*, 1994, **3**, 345.

9. P. J. Branton, P. G. Hall, K. S. W. Sing, H. Reichert, F. Schüth & K. K. Unger, *J. Chem. Soc. Faraday Trans.*, 1994, **90**, 2965.

10. P. L. Llewellyn, F. Schüth, Y. Grillet, F. Rouquerol, J. Rouquerol & K. K. Unger, *Langmuir*, 1995, **11**, 574.

11. P. J. Branton, P. G. Hall & K. S. W. Sing, *Adsorption*, 1995, **1**, 77.

12. J. Rathousky, A. Zukal, O. Franke & G. Schulz-Ekloff, *J. Chem. Soc. Faraday Trans.*, 1995, **91**, 937.

13. R. Schmidt, M. Stöcker, E. Hansen, D. Akporiaye & O. E. Ellestad, *Microporous Materials*, 1995, **3**, 443.

14. J. Rouquerol, *Thermochimica Acta*, 1989, **144**, 209.

15. M. Boudellal, Ph.D. Thesis; Université de Provence: Marseille, France, 1979.

16. J. Rouquerol, F. Rouquerol, Y. Grillet & R. J. Ward, *in "Characterisation of Porous Solids"* (K. K. Unger, J. Rouquerol, K. S. W. Sing & R. J. Ward Eds.) Elsevier, Amsterdam, 1988; p.67.

17. J. Rouquerol, in *"Thermochimie"*, C.N.R.S., Paris, 1971, p 537.

18. K. S. W. Sing, D. H. Everett, R. A. W. Haul, L. Moscou, R. A. Pierotti, J. Rouquerol & T. Siemieniewska, *Pure & Appl. Chem.*, 1985, **57**, 603.

19. J. Rouquerol, F. Rouquerol, C. Peres, Y. Grillet & M. Boudellal, *in "Characterisation of Porous Solids"* (S. J. Gregg, K. S. W. Sing, H. F. Stoeckli, Eds.), Soc. Chem. Ind., London, 1979; p107.

20. I. M. K. , Ismaïl, *Langmuir*, 1992, **8**, 360.

21. L. Jelinek & E. S. Kováts, *Langmuir*, 1994, **10**, 4225.

22. A. V. Kiselev, in *"Solid/Gas Interface"*, Proceedings of the 2nd International Congress of Surface Activity, London, Butterworths, 1957, p.189.

23. M. W. Maddox & K. E. Gubbins, *Int. J. Thermophysics*, 1994, **15**, 1115.

24. L. Gurvitsch, *J. Phys. Chem. Soc. Russ.*, 1915, **47**, 805.

Helium Adsorption on Mesoporous Solids at 4.2K
—Density of Adsorbed Helium in Porous Systems—

N. Setoyama,[1] S. Inoue,[1] Y. Hanzawa,[1] P.J. Branton,[1] K. Kaneko[1], R.W. Pekala,[2] M.S. Dresselhaus,[3] and K.S.W. Sing.[4]

[1] Department of Chemistry, Faculty of Science, Chiba University, Chiba, JAPAN.
[2] Chemistry and Materials Science Department, Lawrence Livermore National Laboratory, Livermore, California, U.S.A.
[3] Department of Electrical Engineering and Computer Science and Department of Physics, Massachusetts Institute of Technology, Cambridge, Massachusetts, U.S.A.
[4] School of Chemistry, Bristol University, Bristol, U.K.

1 INTRODUCTION

The physical adsorption measurement of vapors is one of major methods for characterization of porous solids; it provides an important parameters such as pore volume, specific surface area, and pore width. The pore volume is a key structural parameter for porous solids to determine the adsorption capacities, using the liquid density of the adsorptive. In this case, the adsorbed state of molecules is assumed to be the same as that of the liquid. Many kinds of adsorptives support this assumption, which is called as Gurvitsch rule.

In recent years, helium adsorption measurements at liquid helium temperature have been used as a new characterization method for ultramicropore structure (micropore width <0.7nm) of activated carbons,[1-4] because nitrogen adsorption at 77K has a serious difficulty in the ultramicropore characterization; the pore blocking often occurs by adsorbed molecules at the pore entrance. The previous studies showed that the helium adsorption at 4.2K is effective for an accurate evaluation of ultramicroporosity of activated carbon fibers.[1-3] Carbon dioxide adsorption at room temperature[5] is another hopeful method for evaluation of ultramicroporosity of activated carbons, although the adsorbed density is not necessarily definite.[6] In case of helium adsorption on microporous carbons, the adsorbed density problem is serious. The pore volume from helium adsorption using 0.125g/ml of liquid density does not obey the Gurvitsch rule and it is about twice of the volume from nitrogen adsorption.

The cross sectional area of an adsorbed helium atom on the flat graphitic surface was determined with the neutron diffraction experiment by Carneiro et al.[7] The density of the first and second adsorbed layers was twice of that of the third layer. We studied about the helium adsorption on zeolites, giving 0.20-0.23g/ml of the helium adsorbed density, which is about 2 times greater than the liquid density.[8] The helium densities of different states are shown in Table 1. The adsorbed density of helium shows a medium value between liquid and solid densities, indicating an intense confinement of helium atoms in the micropores.

Although the adsorbed helium density in mesopores is not determined yet, it should be close to the liquid one. We have a fundamental question how the adsorbed density changes to the liquid one with the increase of the pore width. In recent years, new materials such as MCM-41[9] and FSM-16[10-12], which have homogeneous cylindrical

Table 1 *Helium densities of various states*

Solid at 19K *(1693atm)*	*Liquid at 4.2K*	*1st and 2nd layer on Grafoil at 4.2K by ref.7*	*Adsorbed on zeolite micropores at 4.2K by ref.8*
0.348 g/ml	0.125 g/ml	0.20-0.23 g/ml	0.23 g/ml

mesopores with the hexagonal array, have been actively studied as a model mesoporous adsorbent. Those are the fit adsorbents to elucidate the problem as mentioned above. Furthermore, mesoporous carbon aerogels developed by Pekala *et al.*,[13] which have a considerably uniform mesopore of larger size than FSM–16, can be used. In this study, we determined the density of adsorbed helium in mesoporous systems and tried to elucidate how the adsorbed density of helium changes with the pore width.

2 EXPERIMENTAL

Three types of mesoporous solids were used; two carbon aerogels (CRF 1175 and 1176) and mesoporous kanemite (FSM-16). Carbon aerogels have a graphitic surface, and FSM-16 have a siliceous one. Details of preparation are described in elsewhere.[10,11,13] Helium adsorption at 4.2K and nitrogen at 77K were determined by McBain-type gravimetric measurements. These samples were evacuated in a high vacuum prior to adsorption measurement at 383K for 2hrs for the carbon aerogels and 423K for FSM-16.

3 RESULTS AND DISCUSSION

3.1 Adsorbed Density of Helium in Mesopores

3.1.1 Carbon Aerogels. The adsorption isotherms of helium at 4.2K and nitrogen at 77K on CRF1175 and CRF1176 are shown in Figures 1 and 2, respectively. Amounts of adsorption are expressed as volume adsorbed, using the density (ρ) of 0.202g/ml and 0.125g/ml for helium adsorption isotherms according to the previous studies.[1-4] Liquid density of nitrogen (0.808g/ml) is widely used for nitrogen adsorption at 77K. The helium and nitrogen adsorption isotherms represent the type IV isotherms by IUPAC classification and show a hysteresis loop. The loop type is H3 or H4 type in helium and H1 type in nitrogen.[14]

For CRF1175 and CRF1176, the limiting adsorption amount of helium at $P/P_0 > 0.95$ does not agree with that of nitrogen when the liquid helium density (0.125g/ml) or micropore density (0.202g/ml) is used. The amounts of adsorbed helium are about 20% greater than that of nitrogen at $P/P_0 > 0.95$ when the liquid density was used. Then the adsorbed density of helium should have a medium value between liquid and micropore density.

The carbon aerogels used in this study have a little amount of micropores according to the previous study.[13] Hence, the limiting adsorption amount determined from the α_s–plot

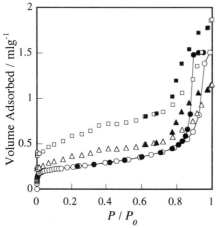

Figure 1. Adsorption (open symbols) and desorption (solid symbols) isotherms on CRF1175. Symbols definition are; nitrogen at 77K (circles), helium at 4.2K (squares; using ρ=0.125g/ml, triangles; ρ=0.202g/ml).

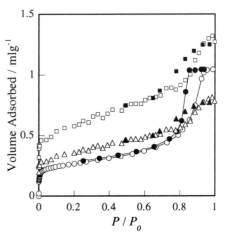

Figure 2. Adsorption and desorption isotherms on CRF1176. Definition of each symbol are same as Figure 1.

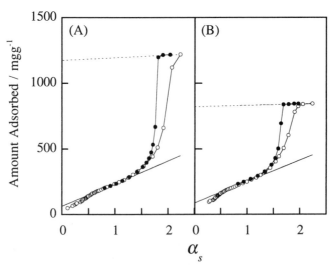

Figure 3. α_s-plots for CRF1175 (A) and CRF1176 (B).

provides the total pore volume, both of micropores and mesopores. Hence, the micropore volume must be removed from the total pore volume for the estimation of the mesopore density of helium. α_s-analysis is an effective way to separate the pore volumes as micropores and mesopores.[15,16] The α_s-plots for nitrogen adsorption isotherms on CRF1175 and CRF1176 are shown in Figure 3. The standard adsorption isotherm of nitrogen was previously determined from nonporous carbon black (Mitsubishi #32).[16] The micropore volume (W_0) is determined from the intercept of solid line in the α_s-plot. The intercept of dotted lines describes the total pore volumes (W_{TOT}). Carbon aerogels are composed by the agglomeration of spherical carbon particles with chemical bondings.[13] The micropores should be present at the contact part of primary particles.

The pore structural parameters of CRF1175 and CRF1176 from the α_s-analysis are tabulated in Table 2. The contribution of micropores to the total pore volume for each carbon aerogel was evaluated about 5% on CRF1175 and 10% on CRF1176. The amount of adsorbed helium on micropore part is subtracted from the experimental isotherms. The helium density of 0.202g/ml was used for the correction of helium adsorption amount on micropores.

The weight amount of helium adsorbed only on mesopores is calculated as 207mg/g for CRF1175. Thus the adsorbed density in the mesopores is calculated as 0.147g/ml from the mesopore volume by nitrogen adsorption (1.41ml/g). The helium amount on mesopores is 140mg/g on CRF1176, and the adsorbed density is estimated as 0.153g/ml. They are close to the density of liquid helium at 4.2K, but they are still slightly greater. A small difference of density between CRF1175 and CRF1176 is attributed to the difference of mesopore size. The greater pore size gives a smaller density, being more close to the liquid one.

3.1.2 Mesoporous Kanemite. The adsorption isotherms of helium at 4.2K and nitrogen at 77K[12] on mesoporous kanemite (FSM-16) are shown in Figure 4. The adsorbed amounts are expressed with the volumes adsorbed using the density of 0.125 and 0.202 g/ml. Both of limiting adsorption amounts of helium on FSM–16 also do not agree to that of nitrogen. A considerable vertical jump of nitrogen adsorption near $P/P_0=0.4$ is observed,[12] according to the capillary condensation into the uniform mesopores. Although the capillary condensation of helium is observed on carbon aerogels, such filling jump is not obtained on FSM–16. A broad adsorption uptake is obtained at $P/P_0=0.01–0.4$, indicating the presence of a slight mesopore size distribution which is less sensitive to the nitrogen adsorption

The nitrogen α_s-plot for FSM–16 is shown in Figure 5 using a standard nitrogen adsorption isotherm on hydroxylated nonporous silica.[17] As the straight portion below $\alpha_s \leq 0.9$ is passing through the origin, there is no micropores in FSM–16. The pore parameters by α_s-analysis are tabulated in Table 1. The pore volume (W_{TOT}) was determined from the intercept of solid line. The pore radius (R_p) of cylindrical pores can be evaluated from the geometrical calculation based on the α_s-analysis, described as follow,

$$R_p = 2000 \, W_0 / a_p \qquad \text{(in nm)} \qquad (1)$$

Table 2 *Pore parameters determined by* α_s*-analysis*

	W_0 / mlg⁻¹	W_{TOT} / mlg⁻¹	a_p / m²g⁻¹	R_p / nm
CRF1175	0.08	1.49	343	8.2
CRF1176	0.11	1.03	324	5.7
FSM–16	–	0.74	844	1.75

Here, a_p is the specific surface area of the pore wall. Then the R_p was estimated as 1.75nm, which agrees with the Kelvin pore radius.[12]

The amount of helium adsorbed at $P/P_0=0.5$ (142mg/g) is used as the limiting adsorption amount of helium in pores. W_{TOT} from α_s–analysis or amount adsorbed at $P/P_0=0.5$ ($W_{0.5}=0.78$ml/g) was used as the nitrogen pore volume. The adsorbed helium density on FSM–16 is estimated as 0.192g/ml from W_{TOT} and 0.183g/ml from $W_{0.5}$, being a intermediate density between the micropore density and liquid one. The adsorbed density on FSM–16 is larger than that on the carbon aerogels. It is evident that the adsorbed helium density is sensitively affected by the change of mesopore size, as observed on carbon aerogels.

3.2 Relationship between of Adsorbed Helium Density and Pore Size

We assume a cylindrical pore for evaluation of the adsorbed helium density as a function of the pore size. The simple pore model for calculation of adsorbed density is shown in Figure 6. Adsorbed layers grow on the pore wall toward the center of the pore. The first and second adsorbed layers (shaded zone in Figure 6) are approximated to have a high density value of 0.20–0.23g/ml and the higher layer (from third layer) to have the liquid one. This pore model is proper to FSM-16.

The adsorbed density determined in this experiment must be treated as an averaged value, which is associated with the fractions of high density layer (*i.e.* the first and second layers) and liquid-like layer (higher layer). The average density (ρ_{AV}) is geometrically given by,

$$\rho_{AV} = \rho_{ads} + (\rho_{liq} - \rho_{ads}) \, (R_p - t)^2 / R_p^2 \qquad (2)$$

where ρ_{ads} and ρ_{liq} are the adsorbed density for high density layer (0.20 or 0.23g/ml) and the liquid density, respectively. t is the thickness of the high density layer, assumed as $t = 0.6$–0.7nm which is evaluated from the packing density.[7] The relationship between pore radius (R_p) and ρ_{AV} are shown in Figure 7. A good agreement between the calculated ρ_{AV} and the experimental density is obtained when the $\rho_{ads}=0.23$g/ml is used. The calculated ρ_{AV} also agrees with the adsorbed density on micropores (ρ_{ads}), which is experimentally estimated on zeolites.[8] The difference of t is the factor to determine the limiting Rp where the ρ_{AV} shows the same value of ρ_{ads}. Hence, the adsorbed density of helium depends sensitively on the pore size even in a mesoporous range. As highly

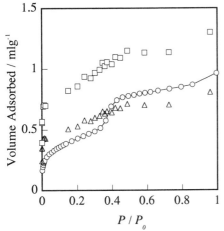

Figure 4. Adsorption and desorption isotherms on FSM–16. Definition of each symbol are same as Figure 1.

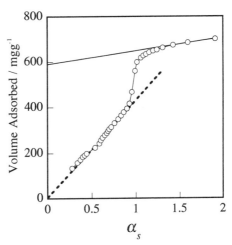

Figure 5. α_s–plots for FSM–16.

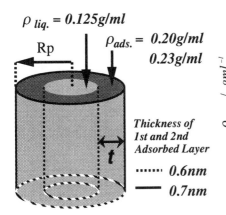

Figure 6. A schematic pore model for calculation of ρ_{AV}.

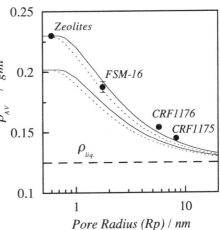

Figure 7. The relationship between ρ_{ads} and Rp. The solid symbol denotes an experimental value. The solid and broken lines are calculated curves for $t = 0.7$ and 0.6nm, respectively. The bold and fine relations are obtained from $\rho_{ads} = 0.23$ gml^{-1} and 0.2gml^{-1}, respectively.

activated carbons have micropores of more than 1.4nm in width,we must be cautious for evaluation of their pore structures.The pore width–sensitive density of helium adsorbed in pores should be taken into account on the pore characterization.

REFERENCES

[1]   H. Kuwabara, T. Suzuki, and K. Kaneko, *J. Chem. Soc., Faraday Trans.,* 1991, **87**, 1915.

[2]   N. Setoyama, M. Ruike, T. Kasu, T. Suzuki, and K. Kaneko, *Langmuir,* 1993, **9**, 2612.

[3]   K. Kaneko, N. Setoyama, and T. Suzuki, *'Characterization of Porous Solids II I* (Eds. J. Rouquerol *et al)'*, Elsevier, Amsterdam, 1994, p. 593.

[4]   N. Setoyama, K. Kaneko, and F. Roriguez–Reinozo, *J. Phys. Chem.,* 1996, **100**, 10331.

[5]   F. Rodriguez–Reinoso and A. Linares–Solano, *'Chemistry and Physics of Carbon* (Ed. P.A. Thrower)', Marcel Dekker, New York, 1989, vol. 22, pp. 1.

[6]   D. Cazorla–Amoros, J. Alcaniz–Monge, and A. Linares–Solano, *Langmuir,* 1996, **12**, 2820.

[7]   K. Carneiro, L. Passell, W. Thomlinson, and H. Taub, *Phys. Rev. B,* 1981, **24**, 1170.

[8]   N. Setoyama and K. Kaneko, *Adsorption,* 1995, **1**, 165.

[9]   P.J. Branton, P.G. Hall, and K.S.W. Sing, *J Chem. Soc., Chem. Commun.,* 1993, 1257.

[10]   T. Yanagisawa, T. Shimizu, K. Kuroda, and C. Kato, *Bull. Chem. Soc. Jpn.,* 1990, **63**, 988.

[11]   S. Inagaki, Y. Fukushima, K. Kuroda, *J Chem. Soc., Chem. Commun.,* 1993, 680.

[12]   P.J. Branton, K. Kaneko, N. Setoyama, K.S.W. Sing, S. Inagaki, and Y. Fukusima, *Langmuir,* 1996, **12**, 599.

[13]   S.S. Hulsey, C.T. Alviso, F.M. Kong, and R.W. Pekala, *Mat. Res. Soc. Symp. Proc.,* 1992, **270**, 53.

[14]   K.S.W. Sing, D.H. Evertt, R.A.W. Haul, L. Moscou, R.A. Pierotti, J. Rouquerol, and T. Siemieniewska, *Pure Appl. Chem.,* 1985, **57**, 603.

[15]   Y. Hanzawa, K. Kaneko, R.W. Pakala, and M.S. Dresselhaus, *Langmuir, in press.*

[16]   K. Kaneko, C. Ishii, M. Ruike, and H. Kuwabara, *Carbon,* 1992, **30**, 1075.

[17]   S.J. Gregg and K.S.W. Sing, *'Adsorption, Surface Area and Porosity'*, Academic Press, London, 1982, Chap. 2.

ESTIMATING THE DIMENSIONS OF ULTRAMICROPORES USING MOLECULAR PROBES

Brian McEnaney, Tim Mays, You Fa Yin, Xue Song Chen, and Francisco Rodríguez-Reinoso*

Department of Materials Science and Engineering
University of Bath
Bath BA2 7AY
UK

* Department of Inorganic Chemistry, University of Alicante, Apartado 99, Alicante
Spain

1 INTRODUCTION

Because the widths of ultramicropores are commensurate with molecular dimensions, it has long been recognised that ultramicropore sizes may also be estimated using molecular probes. In its simplest form, the molecular probe method involves estimating the capacity of the porous solid for a range of fluids with different molecular sizes. The molecular probe method can be applied using gas and vapour adsorption, or the range of molecular probe sizes may be extended by using liquids as molecular probes, either by measuring apparent densities or heats of wetting [1].

A crucial aspect of molecular probe studies is the method chosen to estimate the sizes of the probe molecules and this problem is the main focus of this paper. The dimensions of probe molecules are usually estimated from properties of the bulk fluid. These methods are reviewed and it is shown that there is a significant range of values of molecular dimensions available for a given probe fluid, depending upon the method chosen. This paper also presents a new method for estimating molecular probe sizes using molecular simulations. The new approach provides fresh insights into the molecular probe method and it also has implications for estimations using molecular simulations of densities of adsorbed phases in micropores and of pore size distributions.

2 ESTIMATING MOLECULAR SIZE FROM BULK PROPERTIES

There are numerous methods available for estimating the dimensions of molecules. These include measurement of thermodynamic and spectroscopic properties of gases, viscosity, diffusion, and solubility of gases and liquids [2], and diffraction methods applied to liquids and solids. There are significant variations in molecular dimensions calculated using different methods. These differences result from a number of factors, including the assumptions made in the calculation of molecular dimensions and the state of the substance as the relevant properties are being measured.

Noble gases have been widely used as molecular probes and, as simple spherical molecules, they provide a useful series to illustrate the calculation of molecular dimensions.

In the first method the molecular diameter, σ, is calculated from the van der Waals constant 'b' using the formula

$$\sigma = 0.926 \, b^{1/3} \quad [\text{b in dm}^3\text{mol}^{-1}] \tag{1}$$

In the second method σ is calculated from the density of the liquid, ρ, using the formula

$$\frac{4\pi}{3}\left[\frac{\sigma}{2}\right]^3 = 0.7405\left[\frac{M}{\rho N_A}\right] \tag{2}$$

where M, is the molecular weight of the liquid and N_A is Avogadro's number. A third method for estimating the molecular diameters of noble gases is from their van der Waals packing radii calculated from the separation of molecules in the solid state using diffraction methods. The fourth method estimates the kinetic diameter of noble gas molecules from gas viscosity measurements [2]. To obtain σ from viscosity measurements it is necessary to assume a function for variation of the interaction energy, u(r), between the gas molecules as a function of distance of separation of the centres of the molecules, r. The potential that is most extensively used for spherical, non-polar molecules, such as the noble gases, is the Lennard-Jones 12:6 (LJ) pair potential:

$$u(r) = 4\varepsilon\left[\left(\frac{\sigma}{r}\right)^{12} - \left(\frac{\sigma}{r}\right)^{6}\right] \tag{3}$$

where σ is the value of r at which u(r) is zero and -ε is the depth of the potential minimum. The distance parameter σ is used as the molecular diameter of the gas molecule.

Figure 1 contains examples of molecular diameters of noble gases, calculated using the four methods described above. Figure 1 shows that there are significant differences in the values of the molecular diameters calculated using the different methods. The molecular diameters calculated from Equations [1] and [2] are the smallest and largest values respectively and they differ from each other by up to ~42%. The range of values of molecular diameter in Figure 1 means that an arbitrary selection of one method to calculate the size of molecular probes may introduce a significant degree of uncertainty.

A possible approach for molecular probes is to select a molecular diameter that is appropriate to the assumed state of the adsorbate in the micropores. Thus, for liquid-like and solid-like states molecular diameters calculated from liquid densities and Van der Waals radii, respectively, could be selected, whereas, for a gas-like states values derived from Van der Waals' constants or gas viscosities could be used. A difficulty with all of these molecular diameters is that they are calculated for the bulk phases where properties are determined simply by interactions between the probe molecules. In ultramicropores, probe molecules are dispersed in aggregates with nanometric dimensions where interactions between probe molecules and the pore walls usually dominate over interactions between probe molecules. These arguments suggest that new insights into the molecular probe method for estimating pore sizes may be provided by molecular simulations in which fluid-pore wall and fluid-fluid molecular interactions are taken into account. This approach is developed in the next section.

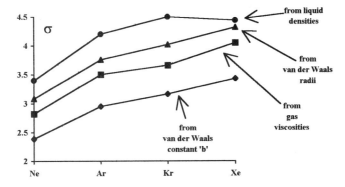

Figure 1 *Molecular diameters, σ/Å, calculated from properties of bulk fluids.*

3 MOLECULAR SIMULATIONS

The model pore used in the simulations [3], Figure 2, is a parallel-sided slit bounded by carbon walls of infinite area (in the x-y plane). It is therefore relevant to microporous carbons. The pore width, H, is the internuclear distance (along the z-axis) between carbon atoms in opposite walls. Each pore wall comprises n parallel layers of carbon atoms separated by an interlayer distance $\Delta \leq H$, and each layer contains ρ_a atoms per unit area. For perfect graphite $n = \infty$, $\Delta = 3.354$ Å and $\rho_a = 0.3818$ atoms Å^{-2}.

The model pore is considered to be immersed in a reservoir of pure adsorptive vapour or gas at constant temperature, T, and at constant pressure p. The method that was used to simulate adsorption in this system is grand canonical ensemble Monte Carlo, GCEMC, outlined below. In the GCEMC method, the pore space is divided into cells of width H in the z-direction (see Figure 2) and of equal length and breadth in the x-y plane, so that each cell has a constant volume V. An initial configuration of adsorptive molecules in one of the cells (the simulation cell) is assumed. This configuration is mirrored in all other cells in the pore (periodic

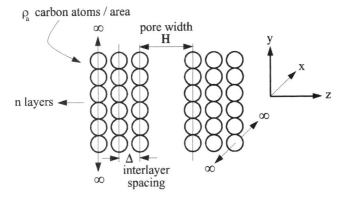

Figure 2 *The model carbon pore used in the adsorption simulations.*

boundary conditions). A measure of the total interaction energy for the configuration is the sum, U, of fluid-fluid and fluid-solid potentials for all adsorptive molecules in the simulation cell. The configuration is then changed at random by moving a molecule in, removing a molecule from or adding a molecule to the simulation cell (exactly the same change occurs in all other cells). If this change decreases U, then it is accepted unconditionally. If the change increases U, then it is accepted with a probability that depends on the Boltzmann factor for the system. For changes as a result of additions or removals of molecules to the cell, this factor involves the chemical potential, μ, of the fluid, which is uniform throughout. In this work, μ was calculated as a function of p and T from equations of state for bulk gases or vapours. A main result from the simulations is the total number, N, of molecules in the simulation cell, calculated as an average over a large number of configurations after equilibration. This result may be converted to the number density of molecules in the cell, $\rho_a{}^*$.

In common with other workers who have studied adsorption of gases and vapours on carbon, the interaction energy, u(r), between a gas molecule separated by an (effective) internuclear distance r from either another gas molecule or a C atom was assumed to be given by a Lennard-Jones (LJ) 12:6 pair potential, Equation [3]. While rather a simplification, this function is usually considered to represent the essential features of interactions in the solid-fluid system.

The simulation cell length and breadth in the x-y plane (see Figure 2) is $10\sigma_{ff}$, where the subscript ff refers to fluid-fluid interactions. In the fluid the LJ potential was assumed to be zero for internuclear separations greater than or equal to $5\sigma_{ff}$. This effectively limited the number of gas molecules in neighbouring cells that needed to be counted in determining the total fluid-fluid interaction energy in the simulation cell. The total interaction potential between a gas molecule and one wall of the model pore is given by:

$$u(z)_{wall} = 4\,\pi\,\rho_a\,\varepsilon_{sf}\sigma_{sf}^2 \left[\sum_{i=0}^{n-1} \left\{ \frac{1}{5}\left(\frac{\sigma_{sf}}{z+i\,\Delta}\right)^{10} - \frac{1}{2}\left(\frac{\sigma_{sf}}{z+i\,\Delta}\right)^4 \right\} \right], \ z > 0 \qquad [4]$$

where z is the internuclear separation between the molecule and the wall and the subscript sf refers to solid-fluid interactions. This equation (the $\Sigma10{:}4$ potential) is obtained first by integrating the LJ 12:6 solid-fluid pair potential over a single, infinite plane (assuming that carbon atoms are uniformly distributed in the plane), and then by summing this integrated potential over the n planes in the wall. The total solid-fluid interaction energy in the pore is given by:

$$u(z)_{pore} = u(z)_{wall} + u(w-z)_{wall} \ , \ 0 < z < H \qquad [5]$$

In each simulation run, of the order of 10^6 configurations were allowed for equilibration. Average values of N were subsequently determined over blocks of around 10^5 configurations.

4 COMPUTER SIMULATIONS OF MOLECULAR PROBE EXPERIMENTS

Molecular simulations of adsorption in model slit-shaped carbon pores are usually employed to generate simulated adsorption isotherms. Then, for example, comparison of simulated adsorption isotherms for pores of different widths with experimental isotherms can be used to generate pore size distributions, e.g.,[4]. Molecular simulations have also been used to estimate the maximum storage capacity of adsorbent carbons for gases such as methane, e.g.,[5]. Here, we have performed computer simulations of molecular probe experiments according to the following procedure. Simulated isotherms were generated for adsorption of each of a range of gases of different molecular size in slit-shaped carbon pores, Figure 2. The initial pore widths were chosen so that the gas molecules were adsorbed freely. A series of isotherms was then generated for each adsorbate molecule in pores of progressively decreasing width until a critical pore width, H_c, was found from which that adsorbate molecule was excluded. In the simulations pore widths were reduced in increments of 0.002Å as H_c was approached. In order to investigate the influence of pore wall structure on the molecular probe simulations, each series was repeated for model pores with one carbon layer plane in each wall, i.e., $n = 1$, and also with $n = 5$.

Figure 3 shows for both $n = 1$ and $n = 5$ that the critical pore widths, H_c, obtained from the molecular probe computer simulations increase linearly as a function of σ_{ff} for probe molecules ranging in size from Ne to Xe, where σ_{ff} is the molecular diameter for the bulk fluid obtained from Steele [6]. Linear regressions on the data yield $H_c = 3.349 + 0.768 \sigma_{ff}$ for $n = 1$ and $H_c = 3.350 + 0.746 \sigma_{ff}$ for $n = 5$.

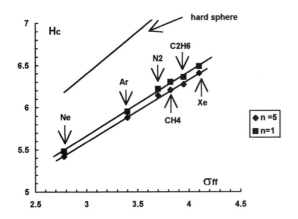

Figure 3 *Critical pore width, $H_c/Å$, as a function of molecular diameter, σ_{ff} Å of probe molecules in carbon micropores with $n = 1$ and $n = 5$. The hard sphere correlation is $H_c = 3.40 + \sigma_{ff}$.*

It is noteworthy that for both $n = 1$ and $n = 5$, the intercepts for these equations are $H_c \sim 3.35$ which is close to the interlayer spacing of graphite, 3.354Å, and the generally-accepted value [6] of σ_{cc} used in the C-C pair potential, 3.400 Å. Secondly, slopes of both lines are ~ 0.75 rather than 1.0, reflecting the response of probe molecules to the enhanced surface forces in ultramicropores as soft LJ spheres rather than hard spheres. A possible hard sphere correlation: $H_c = \sigma_{cc} + \sigma_{ff}$, i.e., $H_c = 3.400 + \sigma_{ff}$. Figure 3, predicts values of H_c that are higher than those found from the computer simulations.

5 DISCUSSION

The results of the computer simulations of molecular probes, summarised in Figure 3, suggest strongly that probe molecules are compressed by the enhanced surface forces in ultramicropores and that the critical dimensions of probe molecules are $\sim 0.75\ \sigma_{ff}$. It is also noteworthy that the slope of the correlation for $n = 5$ is smaller than that for $n = 1$. This is further evidence for the influence of surface forces on the compression of the probe molecules and it suggests that critical pore dimensions for molecular probes are influenced by the pore wall structure, although the effects are small.

The results of the simulated molecular probe experiments may also have implications for the calculation of densities of adsorbed phases using molecular simulations. In Figure 2 the pore width H is defined unambiguously as the separation between the planes of centres of carbon atoms in opposite pore walls. A definition of an effective pore width, H_{eff}, is needed to take account of the pore space that is inaccessible to the adsorbate molecules. The definition chosen for H_{eff} also has a direct bearing on the scaling of pore size distributions obtained using molecular simulations. In the model carbon pore, Figure 2, the density of the adsorbed phase, ρ_a, of molecular weight M given by

$$\rho_a = \left[\frac{H}{H_{eff}}\right]\left[\frac{10^{21} M \rho_a^*}{N_A \rho_{ff}^3}\right] \qquad [7]$$

Various definitions of H_{eff} have been proposed in the literature. For example, Everett and Powl [7] proposed that the excluded volume was simply the space occupied by the carbon atoms in the pore wall, i.e., $H_{eff} = H - \sigma_{cc}$. This definition, also used by Aukett et al.[4] who termed H_{eff} as H_{snap}, suggests that the excluded volume is independent of the nature of the adsorbate molecule, an implication that is not supported by the results in Figure 3. Also, if adsorbate molecules behave as soft spheres, it is to be expected that the excluded volume will depend upon the nature of the adsorbate molecule, the pore width, and the adsorbate pressure. adsorption Quirke and Tennison [8] in a comparative study of methane and nitrogen propose that $H_{eff} = H - 1.7\sigma_{sf}$, while Chen et al. [9] in their study of methane adsorption on carbon monoliths propose that $H_{eff} = H - \sigma_{sf}$.

The present work on molecular simulations of molecular probe experiments suggests another definition for ultramicropores: $H_{eff} = H - H_c$. Figure 4 shows the effect of pore width, H, on the relative densities, ρ', of adsorbed methane in carbon pores using different values of H_{eff}. Values of $\rho' = H/H_{eff}$ are defined relative to the density obtained using $H_{eff} = H$, see equation [7]; from Figure 3, it is also assumed that $H_c = \sigma_{cc} + 0.75\sigma_{ff}$. As expected, the differences in ρ' are smallest in the widest pores and they increase progressively as pores narrow. The choice of definition of H_{eff} is therefore critical when calculating adsorbate densities in micropores. For example, Matranga and Myers [5] have shown that the optimum value of H for storage of methane is ~ 11.2Å at 296K. From Figure 4 at this value of H, ρ' ranges from ~ 1.5-2.5, depending upon the definition chosen for H_{eff}.

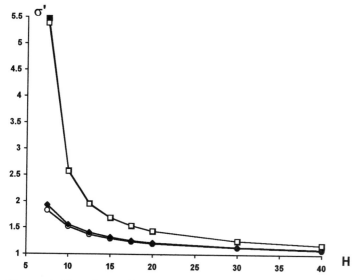

Figure 4 *The relative densities, $\sigma' = H/H_{eff}$, of methane adsorbed in model carbon pores of width, $H/\text{Å}$, using different definitions of effective pore width, H_{eff}: Open circles, $H_{eff} = H - \sigma_{cc}$; diamonds, $H_{eff} = H - \sigma_{sf}$; closed squares, $H_{eff} = H - 1.71\sigma_{sf}$; open squares, $H_{eff} = H - \sigma_{cc} - 0.75\sigma_{ff}$.*

CONCLUSIONS

The diameters of molecular probe molecules estimated from properties of the bulk fluids range widely; for noble gases the values of molecular diameter estimated in different ways vary by up to 42%. It is argued that critical pore widths estimated from properties of the bulk probe fluids are unsatisfactory, since they do not take into account interactions between pore walls and the probe molecules that dominate ultramicropore filling. A new molecular simulation method for estimating the critical dimensions, H_c, of slit-shaped model carbon ultramicropores, using simulated molecular probes, yields a linear correlation between H_c and Lennard-Jones molecular diameters of the probe molecules, σ_{ff}: $H_c \sim 3.35 + 0.75\sigma_{ff}$. The intercept of the linear correlation is close to the value of the interlayer spacing in graphite and the value of the C-C atom pair potential parameter, σ_{cc}; the slope of the linear correlation, ~ 0.75, suggests that there is compression of the probe molecules in ultramicropores. The simulation results also show that H_c depends upon pore wall structure to a small extent. The results also offer a novel way for estimating the effective width of ultramicropores that takes account the pore space excluded from the adsorptive molecules.

ACKNOWLEDGMENTS

We thank the Generalitat Valenciana, Spain for supporting a study visit by BMcE to the University of Alicante, where part of this paper was written; the British Council for supporting YFY; the Overseas Research Student Awards Scheme of the Committee of Vice Chancellors and Principals, UK, for supporting XSC.

REFERENCES

1. R.C. Bansal, J-B. Donnet, H.F. Stoeckli, 'Active Carbon', M. Dekker, New York, 1988, Chapters 3 and 4.
2. R.C. Reid, J.M. Prausnitz, and T.K. Sherwood, 'The Properties of Gases and Liquids', 3rd Ed., McGraw-Hill, New York, 1977.
3. T.J. Mays, In 'Fundamentals of Adsorption', M.D. LeVan, Ed., Kluwer, Dordrecht, 1996, pp. 603-610.
4. P.N. Aukett, N. Quirke, S. Riddiford, and S.R. Tennison, Carbon, 1992, **30**, 913.
5. K.R. Matranga, A.L. Myers, and E.D. Glandt, Chem. Eng. Sci., 1992, **47**, 1569.
6. W.A. Steele, 'The Interaction of Gases with Solid Surfaces', Pergamon, Oxford, 1974.
7. D.H. Everett and J.C. Powl, Trans. Faraday Soc. I, 1976, **72**, 619.
8. N. Quirke and S.R. Tennison, Carbon, 1996, **34**, 1281.
9. X.S. Chen, B. McEnaney, T.J. Mays, J. Alcaniz-Monge, D. Cazorla-Amoros, and A. Linares-Solano, Carbon, 1997, **35**, in press.

OFF LATTICE RECONSTRUCTED BIPHASIC MEDIA USING GAUSSIAN RANDOM FIELDS: EVALUTION FOR DIFFERENT DISORDERED POROUS SOLIDS.

P. Levitz, V. Pasquier, I. Cousin

Centre de Recherche sur la Matière Divisée
CNRS, 45071 Orléans Cedex 2, France.
Email: levitz@cnrs-orleans.fr

1 INTRODUCTION

Disordered porous solids are examples of interfacial systems where an internal surface partitions and fills the space in a complex way. These media play an important role in industrial processes related to separation science, heterogeneous catalysis, oil recovery, glass and ceramic processing. Several transport phenomena such as molecular diffusion, excitation relaxation strongly depend on the interfacial geometry of the pore network. One challenging problem deals with the ability to describe the morphology and the topology of disordered porous media. Knowledge of the 3D pore network is generally needed in order to master correctly the long range connectivity of the matrix. Serial sectioning, RMN or X ray microtomography are possible ways to reach such a goal. However for the time being, it is relatively more easy to characterise the matrix from direct inspection of 2D random cuts of the biphasic random medium. A challenging question is to find a way, if it exists, to reconstruct a realistic 3D configuration from the 2D picture, able to capture morphological and topological properties of the original matrix. From the mathematical point of view, such a problem can not have a general and exact solution thanks to the fact that interfacial Gauss curvature can only be defined on the 3D network. However, it is important to know if it exists some methods able to solve the problem with a good level of approximation and for different types of structural disorder. Joshi devised an algorithm[1] to compute an on-lattice 2D section of a porous network having a well defined bulk autocorrelation function. Based on gaussian random fields, the Yoshi 's method was extended by Quiblier[2] in order to rebuild on-lattice 3D configurations. The method was extensively used by Adler[3] to study single phase transport. Discrete reconstructions are well adapted to numerically solve partial differential equation. In any case, finite size effects, influence of the mesh spacing and control of the boundary conditions have to be carefully handled.

Discontinuous description of the pore network can raise some problems. This the case for molecular dynamics computation, simulation of brownian motion[4] or excitation relaxation[5,6] inside disordered porous medium. The aim of this short paper is to propose an off-lattice reconstruction scheme based on the Berk and Teubner 's works[7,8]. This 3D continuous reconstruction is defined by a level set (contour) of a continuous and analytical field $S(r)$. In that sense, a compact storage is needed, finite size and meshing spacing effects can be eliminated and the interface, its gradient and its curvature are well defined. In Part II, we describe the connection between bulk correlation function and $S(r)$. In Part III, we propose a critical evaluation of this 3D off-lattice recontruction for different types of structural disorder (cement, Vycor glass, artificial soil) .

2 OFF LATTICE RECONSTRUCTED BIPHASIC MEDIA

A density $\varphi(r)$, equals to 0 in the pore network and to one elsewhere else, defines completely a biphasic random medium. The autocorrelation of this field, $\varphi^2_V(r)$, can be directly estimated on 2D planar section of an isotropic disordered porous medium. This function permits also to get the porosity f and the specific surface Sv. The question is to find a way to compute an <u>off-lattice</u> 3D porous network having the same bulk autocorrelation function $\varphi^2_V(r)$. As proposed by Joshi, one introduces a correlated Gaussian random field S(r) such as $<S(r)>=0$ and $<S^2(r)>=1$. The Gaussian field correlation is noted g(r). The spectral density of the field reads:

$$P(q) = 4\,\pi\,q^2\tilde{g}(q) \qquad \text{with} \qquad \tilde{g}(q) = TF_{3D}(g(r)) \qquad Eq(1)$$

The density function $\varphi(r)$ can be computed following the relations:

$$H(\alpha - S(\vec{r})) = \varphi(\vec{r}) \quad \text{with } (1-\phi) = \frac{1}{\sqrt{2\pi}}\int_{-\infty}^{\alpha} \exp(-\frac{s^2}{2})\,ds \qquad Eq(2)$$

the next step is to find an analytical expression of S(r). To reach this goal, one first computes g(r), using a relationship between g(r) and $\varphi^2_V(r)$ derived by Berk and Teubner (7,8).

$$\varphi^2_V(\vec{r}) = (1-\phi) - \frac{1}{2\pi}\int_{Arcsin(g(r))}^{\frac{\pi}{2}} \exp\left(-\frac{\alpha^2}{1+\sin\phi}\right)\,d\phi \qquad Eq(3)$$

The integrand is a positive continuous increasing function of ϕ. In that sense, it is relatively straightforward to solve eq(3), i.e. to find g(r), knowing $\varphi^2_V(r)$.
From g(r), we can compute the spectral density P(q) which is also a probability density. Following the seminal work of Cahn (9), we write S(r) as:

$$S(\vec{r}) = \frac{\sqrt{2}}{\sqrt{N}}\sum_{i=1}^{N} \cos(\vec{q}_i.\vec{r} + \phi_i) \qquad Eq(4)$$

$\vec{q}_i$ is randomly oriented and its modulus is chosen according to the probability density P(q). f_i is a random phase. Good numerical results can be found for N ranging from 400 to 1000.
The off-lattice reconstruction of porous media having a bulk correlation $\varphi^2_V(r)$ can be used in order to generate a configuration which satisfies periodic boundary conditions. In order to build a minimal simulation cell of size L, we first choose a vector $\vec{q}_i$ as defined above. Afterwards, the nearest vector $\vec{k}_i$ defined as :

$$\vec{k}_i = \sum_{k=1}^{3} \frac{2\pi\,n_k^i}{L}\vec{j}_k \qquad n_k^i \in N \qquad (\vec{j}_l.\vec{j}_k) = \delta_{l,k} \qquad Eq(5)$$

is computed and directly used in Eq(5).

3 CRITICAL EVALUATION FOR SOME POROUS MEDIA

In this part, we propose a critical evaluation of this 3D off-lattice recontructions for different types of structural disorder. The morphology of several porous media presented in this paper was already discussed in a former paper[10].

3.1 Long range Debye randomness

A Debye random system is characterised by an exponential variation the mass and pore chord length distribution functions[10]. These properties permits the retrieval of the exponential variation of the bulk autocorrelation function and consequently the well known Debye expression for small-angle scattering[11]. Analysis of such media in term of off-lattice 3D reconstruction, permits to show that the spectral density $P(q)$ evolves as negative exponential of q. On several examples, one find a 3D reconstruction having morphological properties closely related to the starting 2D random cuts[12].

3.2 Disorder involving a length scale invariance

3.2.1 Cement paste and surface fractal. A digitised image of a thin section of hydrated cement is shown on the left part of the figure 1. This structure appears to be a surface fractal having a fractal dimension of 2.7 $\pm$ 0.05 ([10]). In such case we can prove that the spectral density evolves as $1/q^{7-2d}$ ([12]). A random cut of the 3D off-lattice reconstruction is shown on the right part of Fig (1). As shown on figure 2, the chord length distribution functions, running as $1/r^{d-1}$ in the scaling range[10], are well recovered.

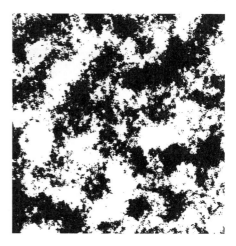

Figure 1: *On the left part, digitised picture of cement adapted from (13). On the right part, 2D random planar section of a 3D off-lattice reconstruction.*

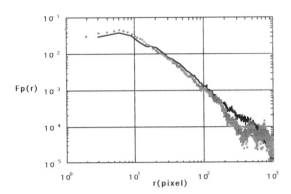

Figure 2: *Pore chord length distribution functions computed either on the digitised image of the cement paste (points) or on the 3D off-lattice reconstruction (continuous line)*

3.2.2.Mass fractal. The 3D off-lattice reconstruction uses, as an initial condition, the bulk autocorrelation function. $\varphi^2_v(r)$ can be directly estimated from a 3D Fourier transform of the small angle scattering $I(q)$. We have used this approach to build a 3D off-lattice reconstruction of a mass fractal solid. In such a case, $I(q)$ evolves as $1/q^d$ in the scaling domain. On figure 3 , we show a random planar cut off of a reconstructed 3D fractal mass solid with d=1.9. As show on figure 4, the mass chord distribution function exhibits an exponential tail at large distances. In the same range, the pore chord length distribution function evolves as $1/r^{d-1}$ as already discussed[10].

Figure 3: *Random planar cut of a reconstructed 3D fractal mass solid having a fractal dimension d=1.9.*

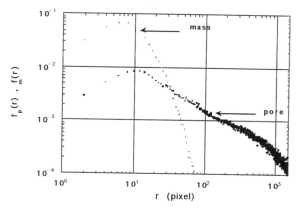

Figure 4: *Pore and mass chord length distribution functions of the mass fractal solid shown on figure 3.*

3.3 The soil as a complex medium

The natural soil is a complex medium having several levels of organisation. we will publish elsewhere a extended study of a natural soil from Beauce, France[14, 15]. In the following, we discuss the case of an artificial soil made of silica particles coated with clay mineral[16]. A random planar cut of this matrix can be observed on the left part of figure 5. 3D reconstruction is shown on the right part of this figure and evolution of the mass and pore chord distribution functions are shown on figure 6. The reconstruction is not perfect. Especially, largest particles are clearly missing. The local morphology is not exactly recovered even if angularity of the particles can be observed on the 3D reconstruction.

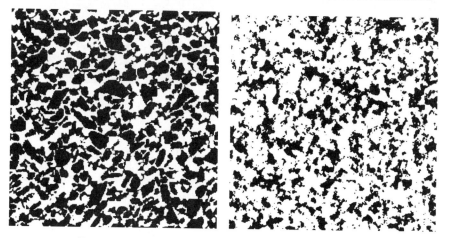

Figure 5: *Artificial soil made of silica particles (in black) coated with clay mineral (16). On the right, a random planar cut of this matrix. On the left, planar section of the 3D reconstruction.*

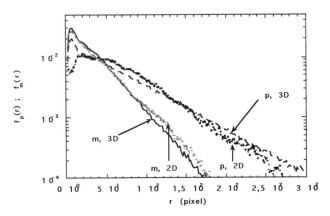

Figure 6: *Pore and mass chord length distribution functions computed either on the digitised image of the artificial soil (points) or on the 3D off-lattice reconstruction (lines) .*

3.4 Correlated disorder: The case of the Vycor glass.

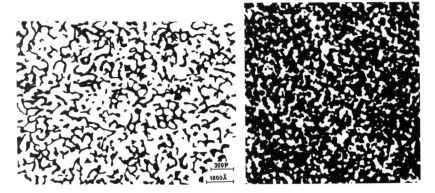

Figure 7: *Porous Vycor glass. On the left digitised pore network (in black) of a very thin section of the material. (17). On the right Planar section of the 3D reconstruction. (pore network in white)*

As shown in figures 7 and 8, a good 3D off-lattice reconstruction of a strongly correlated porous medium, mainly the Vycor glass, can be achieved. Simulations of gas kinetics inside our computed model were performed in the molecular or the Knudsen regime[6]. These studies permit to see how geometry and confinement influence these transport processes especially concerning the evolution of self-diffusion propagator[5,6]. The computed tortuosity is found around 2.6. This determination is almost independent of the mean free path. This result is close to the different experimental determinations around 3.5. The 3D reconstruction permits to construct a complete family of porous media having the same spectral density and a different porosity. In the case of the "Vycor family", the symmetrical structure at $\phi=0.5$ has a computed tortuosity around 1.55. This value is very closed to the determination[18] measured on symmetrical sponge phase (3/2).

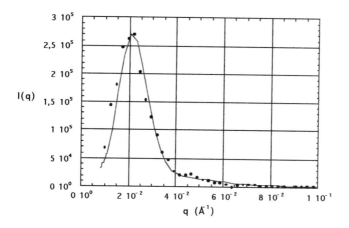

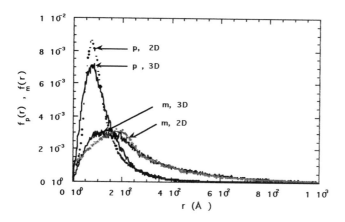

Figure 8: *On the top Experimental small angle scattering of Vycor glass (points) compared with the computed scattering curve from the 3D reconstruction (line). On the buttom Pore and mass chord length distribution functions computed either on the digitised image of the vycor glass (points) or on the 3D off-lattice reconstruction (lines)*

4 CONCLUSION

Numerical 3D reconstruction of real disordered porous media are important to understand the role of the geometrical confinement in adsorption and condensation, transport and reaction processes.

In this paper, we have described a simple way to build "off lattice" reconstructed porous media with or without periodic boundaries derived from Gaussian random fields. The possibility to have a continuous and analytical description of the internal interface can avoid finite size effect and is a prerequisite in order to study gas kinetics, fluid dynamics or adsorption process within the representative elementary volume. Critical evaluation of these computer models is performed for different types of geometrical disorder: the long range Debye randomness, the correlated disorder with a special emphasis on the structure of the Vycor glass, complex structures where length scale invariance properties (mass or surface fractals) can be observed. In the case of Vycor glass, simulation of gas kinetics inside our computer model is in good agreement with experimental determination of the tortuosity. As a possible extension of this work, we can observe that the continuous 3D reconstruction of the pore network interface is a nice template to build atomic configuration of a disordered porous medium. A step further will be to simulate for example gas adsorption/desorption processes.

Aknowlegment. Part of the numerical computations was performed on a Cray C98. YMP. We thank IDRIS, CNRS, Orsay for the grant 94281. It a pleasure also to thank Fatiha Attou and A. Bruand from SESCPF, INRA, Orléans.

References

1. M. Y. Joshi. Ph. D thesis, Univ. Of Kansas,1974.
2. J. A. Quiblier. *J. Coll. Interf. Sci.*, 1984, **98**, 84.
3. P.M. Adler, C. G. Jacquin, J.A. Quiblier. *Int. J. Multiphase Flow*. 1990, **16**, 691.
4. L. M. Schwartz, N. Martys, D. P. Bentz, E. J. Garboczi, S. Torquato, *Phys. Rev.E.*,1993, **48**, 4584.
5. Levitz., *J.Phys.Chem.* 1993, **97**, 3813.
6. V. Pasquier, P. Levitz, A. Delville. *J. Phys. Chem.* 1996, **100**,10249.
7. N. F. Berk, *Phys. Rev. A.*, 1991, **44**, 5069.
8. M. Teubner. *Europhysics letters*, 1991, **14**, 403.
9. J. W. Cahn. *J. Chem. Phys.* 1965, **42**, 93
10. P. Levitz , D. Tchoubar. *J.Phys I* , 1992, **2**, 771.
11. P. Debye, H.R. Anderson and H. Brumberger *J.Appl.Phys* .,1957, **28**, 679
12. P. Levitz., V. Pasquier, I. Cousin in preparation.
13. H. Van Damme., Ciments,Bétons,Platres,Chaux,1990, **15-17**, 782.
14. I. Cousin, P. Levitz, A. Bruand. accepted in *European J. of Soil Science* , 1996.
15. I. Cousin, A. Bruand, P. Levitz, In preparation.
16. F. Attou, Ph. D. Orléans, France. 1996.
17. P. Levitz, G. Ehret, S.K.Sinha, J.M. Drake. *J.Chem.Phys.*, 1991, **95**, 6151.
18. C. Vinches, C. Coulon, D. Roux, *J. Phys II*. 1992, **2**, 453.

MICROPORE SIZE DISTRIBUTIONS FROM EXPERIMENTAL ISOTHERM DATA AND GRAND CANONICAL MONTE CARLO SIMULATIONS

S. Samios, A.K. Stubos, N.K. Kanellopoulos
NCSR Demokritos, 15310 Ag. Paraskevi Attikis, Greece
and
R. Cracknell, G. Papadopoulos, D. Nicholson
Department of Chemistry, Imperial College of Science, Technology & Medicine, London SW7 2AY, UK

1 INTRODUCTION

Pores, and especially micropores (of size less than 2nm according to IUPAC classification), play an essential role in determining the physical and chemical properties of industrially important materials like adsorbents, catalysts, soils, biomaterials, etc. Their characterization (in terms of pore size distribution and other structural information) is indispensable for the utilization and design of improved porous systems in several applications. While for mesopores and macropores there exist a host of more or less established characterization methods, the assessment of microporosity is much less advanced despite the recent interest on microporous systems like zeolites, activated carbons and clay minerals[1]. The currently employed molecular (nitrogen) adsorption method is based on the thermodynamic approach of Dubinin. He assumed that the micropore filling process is governed by a so called adsorption potential characterizing the adsorbed molecules, and that the micropore size distribution is Gaussian. The Dubinin-Radushkevich (DR) equation relates the adsorbed amount per unit micropore volume to the temperature, relative pressure and the characteristic energy and affinity coefficients (which are in turn related to the isosteric heat of adsorption). The DR method has been subject to criticism mainly because the very mechanism of molecular adsorption in micropores is still under active debate. In fact, several studies employing both simulation and density functional theory have added to the accumulating evidence that none of the conventional adsorption methods of pore characterization is entirely satisfactory[2]. Improved approaches to the problem, based on molecular level theories, should be developed.

The Monte Carlo technique is a promising tool in the study of adsorption of pure or multicomponent gases in zeolites and other microporous solids[3-8]. In this work, the method is used in its grand ensemble variant in combination with experimental isotherm data to characterize microporous media and obtain the corresponding pore size distribution (PSD). Specifically, the mean CO_2 density inside a single slit shaped graphitic pore of given width is found on the basis of Grand Canonical Monte Carlo simulations for a predefined temperature and different relative pressures. Starting from an initial PSD guess, it is then possible to produce a computed CO_2 sorption isotherm and compare it to the measured one. After a few iterations, the procedure results in a PSD which, if desired, can be further refined at the cost of additional computational effort. The PSD of activated carbon membranes obtained by nitrogen porosimetry via the conventional (DR) approach is employed for the sake of comparison.

2 POTENTIAL ENERGY MODEL

The realistic character of the simulations and the accuracy of the results depend largely upon the potential energy model used. The carbon dioxide molecule is modeled as Lennard-Jones interaction sites on the atoms plus point charges to account for the quadrupole (three center LJ model from[9,10]). The Lennard-Jones (12-6) potential expresses interaction between sites i and j on two molecules. These interactions are cut (but not shifted) at five times the fluid molecule σ parameter. The graphitic surface is treated as stacked planes of Lennard-Jones atoms. The interaction energy between a fluid particle and a single graphitic surface is given by the 10-4-3 potential of Steele[11] where Δ, the separation between graphite layers, and ρ_s (the number of carbon atoms per unit area in the graphite layer) appear. The values used for Δ and ρ_s are 0.335 nm and 114 nm^{-2} respectively. The solid-fluid Lennard-Jones parameters are calculated by combining the graphite parameters of Table 1 with the appropriate fluid parameters according to the Lorentz-Berthelot rules. The "10-4-3" potential is obtained by the summation of the Lennard-Jones potential between a gas molecule and each carbon atom of the individual graphite planes. The external field for a single Lennard-Jones site in a slit pore of width H is the sum of the interaction with both graphitic surfaces where it should be noted that H is the C center - C center separation across the pore. The effective pore width H' (which is determined by the experiments) is in general given by H' = H - Δ. The present configuration of a slit pore bounded by stacked parallel graphite layers represents only a model of porous carbon (not necessarily the best) and no account of the surface structure of the graphite planes on the pore walls has been taken.

3 SIMULATION MODEL

The Grand Canonical Monte Carlo (GCMC) method is ideally suited to adsorption problems because the chemical potential of each adsorbed species is specified in advance[12]. At equilibrium, this chemical potential can then be related to the external (bulk) pressure making use of an equation of state. Consequently, the independent variables in the GCMC simulations are the temperature, the pressure and the micropore volume, i.e. a convenient set since temperature and pressure are the adsorption isotherm independent variables. Therefore, the adsorption isotherm for a given pore can be obtained directly from the simulation by evaluating the ensemble average of the number of adsorbate molecules whose chemical potential equals that of a bulk gas at a given temperature and pressure.

Three types of trial are used i.e. attempts to move (translate or reorient) particles, attempts to delete particles and attempts to create particles in the simulation box. A decision is made on whether to accept each trial or to return to the old configuration based on a probability which in the case of an attempted move takes the form:

$$p_{move} = \min [\exp(-\Delta U/kT); 1]$$

where $\Delta U = U_{new} - U_{old}$ is the difference in the potential energies of the new and old configurations. A detailed presentation of the method is given in refs[5,7]. Periodic boundary conditions have been applied in the directions other than the width of the slit. For a given simulation, the size of the box (i.e. the two dimensions other than H) have been selected such that ca. 200-450 molecules are present in the simulation. Statistics were not collected over the first $2x10^6$ configurations to assure adequate convergence of the simulation. The uncertainty on the final results (ensemble averages of the number of adsorbate molecules in the box and the total potential energy) is estimated to be less than 5%. Typical calculations for a single point require between 1.5 and 3.5 hours of CPU time on a Convex 3820.

4 GCMC SIMULATION RESULTS

An initial validation of the adsorbate-adsorbent potential functions can be made by comparing isosteric heats of adsorption at zero coverage with experimental data. The theoretical heat for CO_2 adsorption at 313 K is found as 11.4 kJ/mol using Monte Carlo integration with $1x10^7$ trial insertions and a large slit width. The experimental value reported in ref[13] is 12.4 kJ/mol. A more stringent test is to compare experimental and simulated isotherms on non-porous surfaces. A good level of predictive agreement has been found between measured CO_2 isotherms (at 313 K) for a well defined non-porous carbon (Vulcan 3G) and GCMC results of the code used in this study. For these comparisons (and the others to follow in this work) with experimental isotherms, the simulation results have been corrected so as to give an adsorption excess. In fact, the number of molecules which would occupy the pore space in the absence of adsorption forces has been subtracted from the average number of molecules obtained in a simulation[13].

The GCMC approach has been extensively used to simulate CO_2 sorption isotherms at 195.5 K in single graphitic pores of various sizes in the micropore range. The detailed density profiles across the pore (like the one included in fig.1) have been computed for the whole range of micropore widths H. The corresponding local orientation of the molecules is also indicated in fig.1 where the angle of the molecule axis with the vertical to the pore walls is shown across the pore width. From such information, the isotherms and the average fluid density in the micropores as a function of the pore width at various pressures are plotted in figs.2 and 3. It is observed that in all cases a sudden jump in density occurs at a pressure (fugacity) value that varies from less than 0.001 bar to almost 0.5 bar as the pore size increases. A remarkable rise in density is also identified as H increases at constant pressure. This takes place at about 1.0 nm for pressures above 0.01 bar and reflects the formation of a new layer of fluid molecules in the pore. As the pores become larger, the densities are falling to the bulk value corresponding to the prevailing pressure.

The effect of suppressing the quadrupole has also been investigated. In previous work on nitrogen adsorption in slit pores[13], the quadrupole interactions were found to have no particular significance at the high (ambient) temperatures used. In fig.4 the average fluid densities in some pores without quadrupole are given for the sake of comparison. It can be noted that densities tend to become generally lower. Differences are more pronounced at sizes larger than 1.25 nm for which the isotherms become smoother while the decline of density towards the bulk value starts at lower H at a given pressure.

5 MICROPORE SIZE DISTRIBUTIONS

A microporous carbon membrane has been tested using the Quantachrome Autosorb-1 Nitrogen porosimeter equipped with Krypton upgrade. The total pore volume has thus been found (0.54 cc/g). It compares favorably to the value estimated from the measured CO_2 isotherm (at 195.5 K). To start with, a skewed triangular pore size distribution has been postulated making sure that the total pore volume equals to the experimentally determined value. The micropore range (from 0.5 to 2.0 nm) has been subdivided in equidistant spaces with 0.1 nm width. The most probable pore size (slit width) and the standard deviation of the assumed distribution have been varied systematically. For each case, the fraction of the total pore volume associated with each class of pores (the aforementioned subdivisions of the overall pore range) has been calculated (always keeping the total pore volume of the assumed distribution equal to the measured one) and the amount of gas (CO_2) adsorbed in every class at a certain pressure has been computed using the GCMC code. In this way, a computed isotherm has been reconstructed up to pressure values of 0.9 bar. By comparing these results to the corresponding experimental isotherm counterpart, the most suitable micropore size distribution has been selected (triangular shape in fig.5 where the PSD obtained from nitrogen porosimetry at 77 K using the DA method is also shown). The most probable pore size (0.65-0.95 nm) found is in good agreement with the independent estimation of 0.7-0.9 nm coming from tests performed by BP staff[14] at Sunbury, UK. The sensitivity of the method has been demonstrated by obtaining significant deviations from the measured isotherm at certain pressures for the optimal triangular distribution and for two other cases in which the width of the distribution and the most probable pore size have been varied (again respecting the total pore volume found experimentally).

The above described trial and error procedure may become free of the need to specify a certain (readily manipulated) type of pore size distribution function. The solution (by least squares) of a minimization problem under certain constraints has been attempted numerically providing the optimal distribution that fits best the selected segment of the measured isotherm data. The resulting distribution is included in the form of histogram in fig.6. The effect of neglecting the quadrupole on the estimated pore size distribution is also presented in the same figure. It can be seen that the inclusion of quadrupole results in a wider distribution containing relatively more pores at the high end of the micropore range.

Acknowledgments
This work has been partly supported by the BRITE-EURAM Program of the European Commission (Contracts BREU-CT92-0568 and BRE2-CT94-0572).

References
1. K. Kaneko, J. Membrane Sci., 96, 59-89 (1994)
2. D. Nicholson, J. Chem. Soc. Faraday Trans., 90(1), 181-185 (1994)
3. J.L. Soto et al., J. Chem. Soc. Faraday Trans. I, 77 (1981)
4. F. Karavias and A. Myers, Mol. Sim., 8, 51, (1991)
5. D.M. Razmus and C.K. Hall, AIChE J., 37, 5, (1991)
6. R.F. Cracknell et al., Mol. Phys., 80(4), 885 (1993)
7. R.F. Cracknell et al., Mol. Sim., 13, 161 (1994)
8. R.F. Cracknell et al., J. Chem. Soc. Faraday Trans., 90(11), 1487 (1994)

9. S.F. Murthy et al., Mol. Phys., 50, 531 (1983)
10. K.D. Hammonds et al., Mol. Phys., 70, 175 (1990)
11. W.A. Steele, The Interaction of Gases with Solid Surfaces, Pergamon, Oxford (1974)
12. D. Nicholson and N.G. Parsonage, Computer Simulation and the Statistical Mechanics of Adsorption, Academic Press (1982)
13. K. Kaneko et al., Langmuir, 10(12), 4606 (1994)
14. S. Tennison, Personal Communication.

Table 1: Potential Parameters used in the Simulation

Pair	ϕ(nm)	ε/K (K)	l (nm)	Ref.
CO_2-CO_2 (3ClJ)	$\sigma_{cc} = 0.2824$ $\sigma_{oo} = 0.3026$ $\sigma_{co} = 0.2925$	$\varepsilon_{cc}/K = 26.3$ $\varepsilon_{oo}/K = 75.2$ $\varepsilon_{co}/K = 44.5$	0.2324	*
C (graphite)-C (graphite)	0.340	28.0	-	**
C (graphite) - C (CO_2) C (graphite) - O (CO_2)	0.3112 0.3213	27.1 45.9	- -	***

* C.S. Murthy, S.F. O' Shea, and I.R. MacDonald, Molec. Phys. 50, 531, 1983
  K.D. Hammonds, I.R. MacDonald, and D.J. Tildesley, Molec. Phys., 70, 175, 1990
** W.A. Steele, The Interaction of Gases with Solid Surfaces (Pergamon, Oxford), 1974.
*** by means of the Lorenz-Berthelot rules.

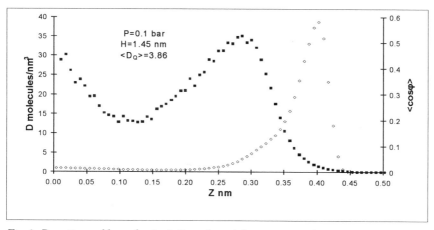

Fig.1: Density profile and orientation of particles across a micropore.

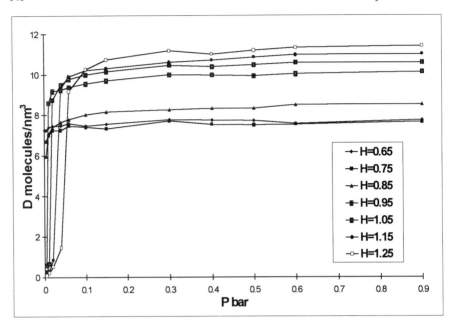

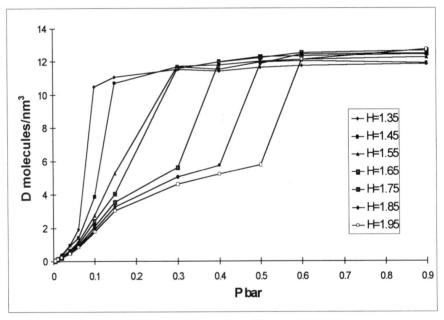

Fig.2: Computed isotherms for different pore widths.

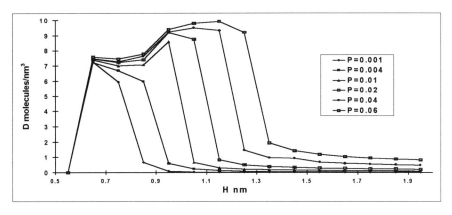

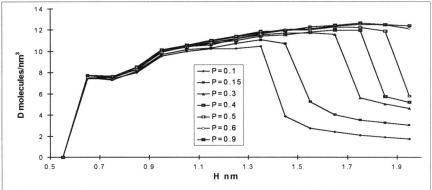

Fig.3: Density versus pore width for different pressures.

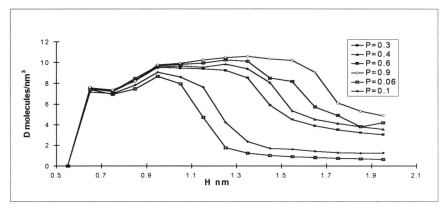

Fig.4: Density versus pore width (without quadrupole).

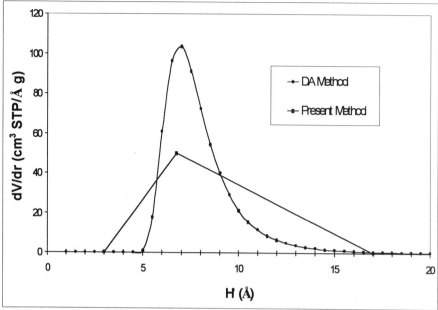

Fig.5: Pore size distributions obtained by nitrogen porosimetry (DA approach) and by the present method (assuming triangular shape).

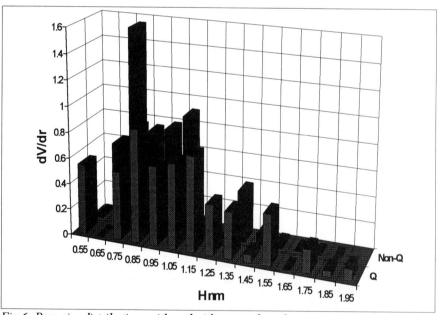

Fig.6: Pore size distributions with and without quadrupole

MEASURING AND MODELLING THE PORE-LEVEL PROPERTIES OF MESO- AND MACRO-POROUS MEDIA - AN OVERVIEW

G.P. Matthews

Department of Environmental Sciences
University of Plymouth
Plymouth PL4 8AA, U.K.

1 INTRODUCTION

The scope of this overview is limited to those pore-level properties which affect fluid migration through meso- and macro-porous media. The purpose of the overview is three-fold. The first aim is to show the range of the field of study in terms of the types of materials, the various pore-level properties, and the experimental and mathematical methods applied. The second aim is to cross-fertilise studies of different types of materials. This aim arises from the concern that studies are often inadvertently duplicated from one type of porous material to another, or limited by an inability to apply the findings from one sample type to a different one. Implicit is the gross assumption that the type of material does not matter - only its geometry. Crude though this assumption is, it does provide a starting point to compare general behaviour and the results of parallel experimental investigations on different materials. The final aim is to reveal a role for my own area of research, that of void space reconstruction.

The overview is illustrated by a selection of mostly recent publications. It is personal and eclectic rather than comprehensive, and that in itself may provide a stimulus for those interested in the properties of meso- and macro-porous media.

2 SCALES AND APPROACHES

Porous media occur over a huge range of scales. The samples may vary from a few millimetres or less, depending on how small a sample can be handled, up to the tens of kilometres of a large oil reservoir. The extremes of this scale are outside the scope of this overview. At the very lowest extreme, tiny micropores, such as those in the bentonite cladding of nuclear waste depositories, can form part of an intimate solid / fluid matrix. The properties of fluids in the bulk of such systems are difficult to study by traditional experimental methods such as surface absorption, although their surfaces are open to examination by specialist surface techniques such as field-tunnelling and environmental electron microscopy. Their theoretical interpretation lies in the realms of molecular dynamic [1] and stochastic / Monte Carlo simulations. At the other extreme, the upper end of the sample scale range lies in the realm of the oil and gas reservoir geophysicists.

Between these extremes lies much fascination and endeavour, as outlined in Figure 1. The pore sizes in these samples have been categorised [2] into micropores (< 2 nm), mesopores (2 - 50 nm) and macropores (> 50 nm). The figure shows various techniques in terms of their degree of mathematical or experimental nature, and the size of the systems which they can be used for. Some of the boxes are shaded for clarity. Their positions and boundaries are intended to give a flavour of the research area, rather than to suggest literal and unbreachable barriers.

Figure 1 *Scales and approaches*

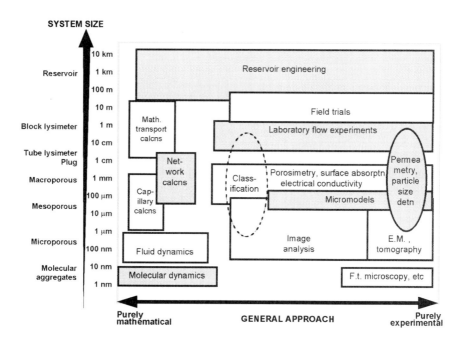

It can be seen that the purely mathematical area applied to the smallest samples in our range of interest has been labelled 'fluid dynamics'. Work in this area is based on rigorous solutions of equations such as Stokes and Reynolds. [3,4] Also in the purely mathematical category are solutions of the invasion of networks by viscous fingering, capillary fingering and stable displacement. [5-8] The category labelled 'capillary calculations' is intended to cover calculations such as the effects of hold-up and hysteresis caused by the shapes of pores. [9,10]

On a larger scale, mathematical transport calculations tend to ignore the detailed structure of the porous medium because the sample size is too big, and, in the case of soil, too complicated. Studies are therefore based on semi-empirical assumptions about the structure of the porous solid, for example that the dispersivity in solute transport increases linearly with distance, [11] or that soil water and solute transport parameters are log-normally distributed. [12]

Experiments on a plug scale tend to be based around the well established techniques of porosimetry, surface absorption, [13,14] electrical conductivity, [15] permeameter, and particle size determination. These are considered in more detail below. In principle, a perfect map of the void space of a small sample could be gained from synchrotron, [16] X-ray [17] or NMR [18] microtomographic imaging. In practice, these techniques are variously limited by their expense, ability to handle samples of sufficient size, and sufficiently accurate interpretation. Electron microscopy has been extensively used for the investigation of porous samples, coupled with the image analysis of sections. [19-21]

Solute flow experiments on large samples tend to be either at a catchment scale [22] (~ 10 km²), field scale [23] (1 - 100 m²) or laboratory scale [24] (0.1 - 1 m³), using various types of sampler array or lysimeter. The pore structure of a soil or rock sample can be deduced from its water drainage curve under increasing differential pressure. [25] An analogous method has been developed for fibrous mats and paper, in which the pore structure is deduced from the hexa-decane drainage characteristics. [26,27]

Mercury porosimetry [28] cannot be used for either soil, fibrous mats or paper, because the mercury destroys the pore structure. However, it has been widely used for rock, particularly sandstone, where both the mercury intrusion and extrusion curves yield information. [29,30] The technique has recently been extended to compressible samples such as paper coatings and ground chalk. [31] The method produced useful information, despite problems with knowing the mercury contact angle, and having to assume simplified geometries (traditionally cylinders) for the voids which it is entering. [10,32]

On a smaller scale, much work is based on electron microscopy. Micrographs of porous media, designed to indicate void space unambiguously, can be image analysed to give void sizes and shapes. [33,34] Alternatively, the micrographs can be used for serial tomography of sections, followed by a reconstruction of the void space. [21] Alternatively, the voids can be injected with Woods metal, the rock dissolved away, and the solid metal image of the pore space viewed directly. [19]

There remains a final area on the graph labelled 'Classification'. This covers work in which types of experimental behaviour, such as different shapes of intrusion curve [35,36] are linked with types of structure. In a similar vein, work is often done on model systems, such as packed spheres, [9,37,38] to correlate and classify experimental behaviour.

3 TYPES OF SAMPLE AND EXPERIMENTAL TECHNIQUE

There are a huge number of natural and synthetic meso- and macro-porous materials. Table 1 shows the more commonly studied varieties, ranging through rock, [39-41] building materials, [42,43] and soil, [44] through mats and coatings, powders, teeth, bones and cake. Many substances generated their own branch of research. Also shown in Table 1 are typical experimental techniques. The fully shaded areas of the table indicate where the technique can be successfully used on that type of sample. The cross-hatched areas show combinations where the results are dubious or inaccurate. Again, this categorisation is subjective and open to discussion.

Table 1 Types of sample and experimental technique

	Porosity	Fluid intrusion and extrusion			Gas permeametry			Particle size	Surface area	Electricl condvity	Microscopy and image analysis	NMR
		Mercury	Hexa-decane	Water	Plug flow	Spot	Gurley			Brine / cond. bridge		
Rock, e.g. sandstone and limestone	■	■			■	▨		▨	■	■	■	■
Construction materials, e.g. cements, mortars and ceramics	■	■			▨	▨		▨			▨	▨
Unconsolidated sand	■	■						■				
Soil	■			■			■	▨				
Paper coatings	■	▨	■				▨	▨			▨	
Fibrous mats (e.g. nappies, sanitary towels, catalyst supports, substrates for composites, anti-dendritic sheaths)		▨									▨	
Sintered metal powders, e.g. tantalum capacitors, silica catalysts	■	■						▨	■			
Compressed powders, e.g. medical tablets	■	■						▨				
Teeth and bone	■	■										■
Food, e.g. bread and cakes	■	▨										

4 VOID SPACE RECONSTRUCTION

Every box that was shown in Figure 1 represents an established research area. However, as indicated in the figure, there are gaps between the areas. Thus there is currently little connection between laboratory flow experiments, and the techniques of porosimetry and surface absorption. Also, there is a gulf between the various types of mathematical calculation, and the experimental methods. So can these gaps be filled? Not rigorously, certainly; the vast complexity of natural porous samples and the concomitant problems of scaling will always prevent this. However, the method of void space reconstruction can bridge the gaps in a semi-rigorous, but nevertheless useful, way.

A perfect void space reconstruction would generate an entirely accurate and representative void space map from a single experimental property, and from this map all other properties of the sample could be calculated. The limits on computing power dictate that in practice a greatly simplified network, with nominal or very simple geometry, must be used. The principle has been around for more than forty years, [45] and since then many networks and geometries have been suggested. [8,9,46-52] These have generally been based on percolation theory, but reconstructions of single pore systems have also been constructed which match the autocorrelation function of a digitised micrograph. [53]

The importance of reconstruction which then generates other properties is demonstrated by Table 1, which shows that for many types of sample, most of the investigative methods are unavailable or unsatisfactory. The problem is exacerbated by the fact that most of the experimental techniques only yield one parameter, shown fully shaded in Table 2. Some inferences of other properties may be based on these, as shown cross-hatched, but there are typically many properties left unmeasured.

Table 2 *The role of void-space reconstruction*

	Porosimetry	Image analysis	Pycnometry	Permeametry	Diffusometry	Void space reconstruction
Porosity	■	▨	■			■
Pore and throat size distribution	▨	▨				■
Connectivity		▨				■
Permeability				■		■
Tortuosity/ dispersion					■	■
Hysteresis / trapping	■					■
Particle size distribution	▨	▨				▨
Surface area	▨					▨
Correlation		■				■

Our own void-space model, Pore-Cor, generates void space maps based on non-wetting fluid intrusion, wetting fluid drainage, and/or void size distributions and correlation data from binary images of the void space. The void structure is not unique, except in terms of the model's own property-dependent fitting parameters. However, its 3-D interpretation of intrusion and drainage is undoubtedly better than the traditional 2-D interpretation for the great majority of samples, other than those such as woven carbon-fibre composite substrates. Its usefulness has also been recently proven. After a major consultancy contract concerning tantalum capacitors, we collected all the relevant production parameters and all the Pore-Cor generated parameters. A Partial Least Squares / Principal Component Analysis of all the parameters showed that the electrolytic efficiency of the capacitors was strongly correlated with the Pore-Cor simulated permeability, but not with the simulated tortuosity. There was strong correlation with one of the production parameters, but not another which had been thought important. On the basis of this information, the production process is currently being enhanced.

References

1. A. Nakano, L.S. Bi, R.K. Kalia, and P. Vashishta, *Phys. Rev. B,* 1994, **49**, 9441.

2. J. Rouquerol, D. Avnir, D.H. Everett et al. Studies in Surface Science and Catalysis - Characterization of Porous Solids III, B. Delmon and J.T. Yates, eds. , 1994.

3. V.V. Mourzenko, J.F. Thovert, and P.M. Adler, *J. Physique Ii,* 1995, **5**, 465.

4. I. Ginzbourg and P.M. Adler, *J. Physique Ii,* 1994, **4**, 191.

5. R. Lenormand, E. Touboul, and C. Zarcone, *J. Fluid. Mech.* 1988, **189**, 165.

6. O. Vizika, D.G. Avraam, and A.C. Payatakes, *J. Coll. Int. Sci.* 1994, **165**, 386.

7. K.S. Sorbie, H.R. Zhang, and N.B. Tsibuklis, *Chem. Eng. Sci.* 1995, **50**, 601.

8. F.A.L. Dullien, *Trans. Por. Med.* 1991, **6**, 581.

9. C.D. Tsakiroglou and A.C. Payatakes, *J. Coll. Int. Sci.* 1990, **137**, 315.

10. G. Mason and N.R. Morrow, *J. Coll. Int. Sci.* 1994, **168**, 130.

11. K.L. Huang, M.T. van Genuchten, and R.D. Zhang, *Appl. Math. Model.* 1996, **20**, 298.

12. W.A. Jury and G. Sposito, *Soil Sci. Soc. Am. J.* 1985, **49**, 1331.

13. W.C. Conner, S. Christensen, H. Topsoe, M. Ferrero, and A. Pullen, Studies in Surface Science and Catalysis - Characterization of Porous Solids III, B. Delmon and J.T. Yates, eds. , 1994.

14. P.J. Branton, P.G. Hall, and K.S.W. Sing, *J. Chem. Soc. ,Chem. Commun.* 1993, 1257.

15. C. Ruffet, M. Darot, and Y. Gueguen, *Surv. Geophys.* 1995, **16**, 83.

16. P. Spanne, J.F. Thovert, C.J. Jacquin, W.B. Lindquist, K.W. Jones, and P.M. Adler, *Phys. Rev. Lett.* 1994, **73**, 2001.

17. J.C. Elliot, P. Anderson, X. Gao, F.S.L. Wong, G.R. Davis, and S.E.P. Dowker, *J. X-Ray Microtom. Sci. and Tech.* 1994, **4**, 102.

18. C. Casieri, C. Deangelis, F. Deluca, G. Garreffa, and B. Maraviglia, *Magn. Res. Imag.* 1992, **10**, 837.

19. N.C. Wardlaw, Y. Li, and D. Forbes, *Trans. Por. Med.* 1987, **2**, 597.

20. D.P. Lymberopoulos and A.C. Payatakes, *J. Coll. Int. Sci.* 1992, **150**, 61.

21. M. Yanuka, F.A.L. Dullien, and D.E. Elrick, *J. Microsc.* 1984, **135**, 159.

22. C.J. Barnes and M. Bonell, *Hydrol. Process.* 1996, **10,** 793.

23. N.M. Holden, A.J. Rook, and D. Scholefield, *Geoderma,* 1996, **69,** 157.

24. D. Wildenschild, K.H. Jensen, K. Villeholth, and T.H. Illangasekar, *Ground Water,* 1994, **32,** 381.

25. D.M.W. Peat and G.P. Matthews, Characterisation of Porous Solids IV, 1996.

26. R.C. Hamlen and L.E. Scriven, Tappi Coating Conference Proceedings, 1991.

27. B. Miller and I. Tyomkin, *J. Coll. Int. Sci.* 1994, **162,** 163.

28. J. van Brakel, S. Modry, and M. Svata, *Powder Tech.* 1981, **29,** 1.

29. N.C. Wardlaw and M. McKellar, *Powder Tech.* 1981, **29,** 127.

30. G.P. Matthews, C.J. Ridgway, and M.C. Spearing, *J. Coll. Int. Sci.* 1995, **171,** 8.

31. P.A.C. Gane, J.P. Kettle, G.P. Matthews, and C.J. Ridgway, *Ind. Eng. Chem. Res.* 1996, **35,** 1753.

32. R.J. Good and R.S. Mikhail, *Powder Tech.* 1981, **29,** 53.

33. B. McEnaney and T.J. Mays, Studies in Surface Science and Catalysis - Characterization of Porous Solids III, 1993.

34. R. Ehrlich, S.J. Crabtree, O. Horkowitz, and J.P. Horkowitz, *Am. Asscn Petr. Geo. Bull.* 1991, **75,** 1547.

35. M. Day, I.B. Parker, J. Bell et al. Studies in Surface Science and Catalysis - Characterization of Porous Solids III, B. Delmon and J.T. Yates, eds. , 1994.

36. W.C. Conner, C. Blanco, K. Coyne, J. Neil, S. Mendioroz, and J. Pajares, Characterization of Porous Solids, 1988.

37. C.A. Grattoni and R.A. Dawe, *Powder Tech.* 1995, **85,** 143.

38. K. Shinohara and T. Murai, *Kagaku Kogaku Ronbunshu,* 1994, **20,** 198.

39. Y. Bernabe, D.T. Fryer, and R.M. Shively, *Geophys. J. Int.* 1994, **117,** 403.

40. J.T. Fredrich, K.H. Greaves, and J.W. Martin, *Int. J. Rock Mech. Min. Sci. & Goemech. Abstr.* 1993, **30,** 691.

41. G.P. Matthews, C.J. Ridgway, and J.S. Small, *Mar. Pet. Geo.* 1996, **13,** 581.

42. J.G. Cabrera, N. Gowripalan, and P.J. Wainwright, *Mag. Concr. Res.* 1989, **41,** 193.

43. M. Yates, M.A. Martin-Luengo, J. Cornejo, and V. Gonzalez, Studies in Surface Science and Catalysis - Characterization of Porous Solids III, 1993.

44. D. Mallants, M. Vanclooster, and J. Feyen, *Hydrol. Process.* 1996, **10,** 55.

45. I. Fatt, *Petroleum Transactions, AIME,* 1956, **207,** 144.

46. G.P. Androutsopoulos and R. Mann, *Chem. Eng. Sci.* 1979, **34,** 1203.

47. M. Blunt and P. King, *Trans. Por. Med.* 1991, **6,** 407.

48. I. Chatzis and F.A.L. Dullien, *Int. Chem. Eng.* 1985, **25,** 47.

49. E.J. Garboczi, *Powder Tech.* 1991, **67,** 121.

50. M.A. Ioannidis and I. Chatzis, *J. Coll. Int. Sci.* 1993, **161,** 278.

51. C. Ocarroll and K.S. Sorbie, *Phys. Rev. E,* 1993, **47,** 3467.

52. M. Yanuka, *Trans. Por. Med.* 1992, **7,** 265.

53. J. Sallès, J.F. Thovert, and P.M. Adler, Studies in Surface Science and Catalysis, B. Delmon and J.T. Yates, eds. , 1994.

PORE FORMATION IN CARBONS DERIVED FROM PHENOL RESIN AND THEIR GAS ADSORPTION

M. Inagaki and M. Sunahara

Faculty of Engineering, Hokkaido University,
Kita-ku, Sapporo 060, Japan

1 INTRODUCTION

Recent development in carbons with extremely high surface area has stimulated various investigations and applications, and also promoted the understanding of pore structure in these carbon materials.

We have found an irreversible adsorption of CO_2 and N_2 gases on an activated carbon prepared from phenol resin spheres; large amount of gas being trapped even after evacuation [1,2]. This irreversible adsorption of gases was studied by both volumetric and gravimetric measurements. Pyrolysis, carbonization and activation processes of phenol resin spheres under different preparation conditions were discussed through the measurements of adsorption and desorption behaviors [3].

In the present paper, the pore formation in these phenol carbons was discussed on the basis of adsorption-desorption behaviors of CO_2 and N_2.

2 EXPERIMENTAL

2.1 Sample Preparation

The precursor was phenol resin spheres. They were heat-treated at a temperature between 450 to 1200°C in a gas flow of either Ar with high purity as 99.9995%, Ar with a purity of 99.95% or CO_2 with 99.95%. Heating rate of 330 °C/h, residence time at heat treatment temperature (HTT) of 1 hour and gas flow rate of 30 cc/sec during heat treatment were employed.

The carbon samples thus prepared were found to keep spherical characteristics of the precursor. The size of the spheres was rather homogeneous in a range of 30-50 μm and no cracks were observed under scanning electron microscope.

2.2 Adsorption-Desorption Measurements

The adsorption-desorption isotherms were measured volumetrically using a swing adsorption system. Before the measurement, the samples were outgased at a temperature of 250°C for 3 hours under a vacuum of about 5 Pa. The adsorption-desorption curves of N_2 gas were measured at -196°C, and those of CO_2 gas at -72°C. When the change in gas pressure became less than 80 Pa during 2 minutes, adsorption or desorption of gas was considered to be in equilibrium, and then the gas pressure was changed to the next step.

3 RESULTS AND DISCUSSION

3.1 Pore Formation

In Fig. 1a) and b), adsorption-desorption isotherms of N_2 at -196°C were compared on the carbons heat-treated at different tempratures in high-purity Ar and CO_2, respectively. The amount of adsorbed N_2 increases and then decreases with HTT in high-purity Ar gas. In CO_2, the increase in amount of absorbed N_2 with the increase in HTT is remarkable above 900°C. Above 1000°C in CO_2 gas flow, all samples were gasified and no carbon was yielded. Most of samples heat-treated in different gas atmospheres, except the following two samples, show reversible behavior between adsorption and desorption of both N_2 and CO_2 gases. On the samples heat-treated in high-purity Ar gas above 1100°C, irreversible adsorption-desorption isotherms are observed, the desorption curve being always located above the adsorption one, as shown in Fig. 1a). This is the same phenomenon of the storage of N_2 gas in the pores of carbon as that observed on a commercial activated carbon APT which was prepared from the same phenol resin spheres. This irreversible adsorption of gases will be discussed in the next section.

In Fig. 2a), specific surface area determined from the adsorption of N_2 at the relative gas pressure up to 0.3 was plotted as a function of HTT on the carbons prepared in different gas atmospheres. Up to 800°C, specific surface

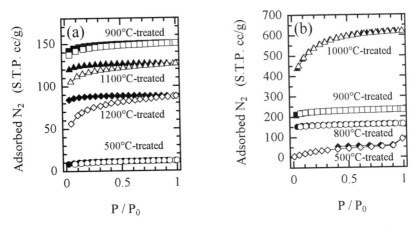

Fig. 1   Adsorption-desorption isotherms of N_2 at $-196^\circ C$ as a
         function of heat treatment temperature.
         a) Carbons heat-treated in high-purity Ar.
         b) Carbons heat-treated in CO_2.

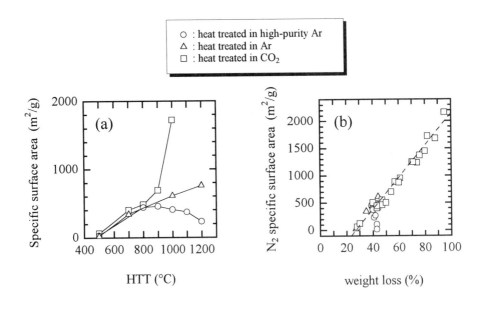

Fig. 2   Specific surface area determined from N_2 adsorption
         isotherm as a function of heating atmosphere plotted
         against heat treatment temperature HTT (a) and
         weight loss during heat treatment (b).

area increases and no difference was detected among gas atmospheres during heat treatment. Above 800°C, however, surface area increases abruptly in CO_2 atmosphere, but decreases slightly in high-purity Ar. In low-purity Ar gas atmosphere, change in surface area with HTT is intermediate between inert and oxidative gas atmospheres, showing slight increase with HTT. Similar relation was obtained on the specific surface area determined from CO_2 gas adsorption. From these relations, it can be concluded that the pore development evaluated from specific surface area up to 800°C is due to pyrolysis of the present phenol resin and that above 800°C in oxidative atmosphere is due to activation. In inert atmosphere of high-purity Ar, carbonization pro-ceeds above 800°C and causes shrinkage in pore size to result in slight decrease in surface area.

A plot of specific surface area against weight loss during heat treatment, however, gives a linear relation, independent of gas atmosphere during heat treatment, as shown in Fig. 2b). The experimental points obtained in high-purity Ar (two points of open circle) show some devia-tion from the line, which correspond to the samples showed irreversible adsorption-desorption behavior. The same linear relation was observed between the specific surface area determined from CO_2 adsorption and weight loss, where no largely deviated experimental points were found because of no irreversibility in CO_2 adsorption-desorption.

The pyrolysis of the present phenol resin results in the weight loss of about 42% and maximum specific surface area of about 400 m^2/g. The activation by CO_2 gas associat-ed with oxidation of resultant carbon can give very high surface area, near 2000 m^2/g, with very high weight loss.

3.2 Irreversible Adsorption of Gases

On the commercial activated carbon, which was said to be prepared from the same phenol resin spheres as the present ones by the heat treatment at 1000°C in a flow of CO_2 and named APT, irreversible adsorption of N_2 and CO_2 gases was observed. In Fig. 3a), adsorption-desorption isotherms of CO_2 on the sample APT at different temperatures of measurement were reproduced. The same irreversibility was observed in N_2 adsorption at -196°C. The adsorption and desorption behavior of CO_2 at 0°C was followed by gravimet-ric method and the results is shown in Fig. 3b). The ad-sorption of CO_2 gas molecules following a quick adsorption

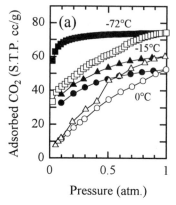

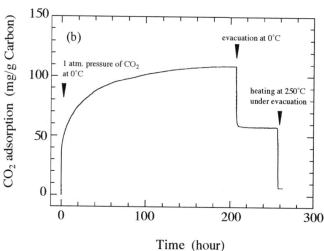

Fig. 3 Adsorption-desorption behaviors of CO_2 on the commercial sample APT prepared from the same phenol resin spheres.

a) Isotherms at different temperatures.

b) Weight change during adsorption at $0^{\circ}C$ and desorption under evacuation.

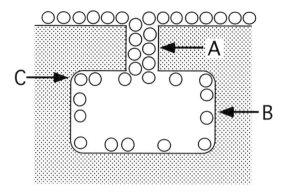

Fig. 4 A model of pore structure responsible for irreversible adsorption of gas.

is so slow that it takes more than 200 hours to reach saturation in adsorption at $0^\circ C$. By evacuation of the system, the weight decreases by approximately the same amount as quickly adsorbed one, but no decrease is observed with prolonged evacuation. Heating the adsorbed sample up to $250^\circ C$ under evacuation was effective to desorption of gas. Small but certain amount of CO_2 gas is still retained in the carbon even after heating at $250^\circ C$ for a long time.

The same irreversibility between adsorption and desorption of N_2 gas was observed, as shown in Fig. 1a), on the laboratory prepared samples in high-purity Ar at the temperature above $1100^\circ C$.

This gravimetric studies on the sample APT show clearly that the CO_2 gas molecules adsorbed can be divided into three groups, CO_2 molecules adsorbed under 1 atmospheric pressure and desorbed under evacuation quickly, of which the amount is about 40 % of the total amount, those adsorbed very slowly and desorbed only at $250^\circ C$ under evacuation, and those can not be desorbed even after heating at $250^\circ C$ of which the amount is about 5 %. Exactly the same adsorption characteristics can be expected on N_2 adsorption at $-196^\circ C$ on the sample obtained in the present work.

These three groups of CO_2 can be asigned to three groups of micropores, as discussed in our previous paper [3]. However, these are also able to be attributed to three different adsorption sites in the carbon. A study based on grand canonical Monte Carlo and molecular dynamics simulation showed that slit-shape pores, which may be characteristic for carbon materials, can cause a hysteresis loops in adsorption-desorption isotherms [4]. For a trapping of gases in the materials, it has been pointed out by various authors that pores with ink bottle shape can be responsible. A model of pore structure responsible for gas trapping is shown in Fig. 4. On the way of adsorption after covering the surface by gas molecules, the adsorption into the interior of pore (part B in Fig. 4) is reasonably assumed to be controlled by the diffusion of gas molecules at the neck part (part A), and so very slow adsorption was observed. For the desorption, the neck part A strongly disturbs the diffusion of gas molecules out of the part B. In order to desorb gases from the part B, heating up to $250^\circ C$ was necessary. Gas molecules at the part C in Fig. 4 are so strongly adsorbed that they could not be released even at $250^\circ C$.

On the sample APT, a similar irreversibility was observed on N_2 isotherm only at $-196^\circ C$, reversible adsorption and desorption at the temperature above $-72^{\circ}C$. For the samples heat-treated in high-purity Ar gas above $1100^{\circ}C$, an irreversible adsorption-desorption isotherm (Fig. 1a) was observed only on N_2 gas, not on CO_2. Therefore, the diameter of the neck part of the pores seems to be comparable with double the molecular diameter of N_2 and CO_2 (i.e. 0.66-0.70 nm) and it is very much sensitive to heat treatment conditons. However, we have also to take into account of the interaction between adsorbent molecule and pore wall, particularly at the neck part of the pores, but there has been no quantitative discussion on this interaction.

4 CONCLUSIVE REMARKS

These experimental results show that pore formation, not only its amount but also shape and size, in the carbons derived from phenol resin spheres depends strongly on the heat treatment conditions, temperature, residence time, heating rate, the content of oxidative gases in the atmosphere, etc. In order to understand the detailed mechanism of these irreversible adsorption of gases, still we need further accumulation of experimental data, determination of preparation conditions for reproducing these irreversible adsorption behaviour of various gases, pore structure analysis in the range of low partial pressure of gases and also direct observation of pores in high resolution transmission electron microscopy coupled with image analysis.

References

1. M.Inagaki and M.Nakashima, Carbon **30**, 1135(1992).
2. M.Inagaki and M.Nakashima, Tanso **1994**[No.162], 61.
3. M.Nakashima, S.Shimada, M.Inagaki and T.A.Centeno, Carbon **33**, 1301(1995).
4. M.W.Maddox, private communication.

CHARACTERIZATION OF POROUS SOLIDS: ADSORPTION POTENTIAL DISTRIBUTION AND ITS THERMODYNAMIC IMPLICATIONS

M. Jaroniec and M. Kruk
Department of Chemistry
Kent State University
Kent, OH 44242, USA

J. Choma
Institute of Chemistry
WAT
00-908 Warsaw, Poland

1 INTRODUCTION

Porous materials are widely used in many branches of modern science and technology, such as catalysis, separation of mixtures and purification of gases and liquids. A successful application of porous solids requires a thorough characterization of their surface and structural properties. Some materials, such as zeolites and M41S molecular sieves, have quite well-defined porous structures, but others (porous oxides, active carbons and carbon blacks) are usually significantly geometrically nonuniform. Moreover, many of porous solids exhibit a strong surface heterogeneity. A proper characterization of structural and surface properties of porous solids is quite involved, especially as it is often difficult to clearly distinguish between effects of surface and structural heterogeneities on properties of porous materials.

For many years, gas adsorption has been used to study properties of porous solids, [1,2,3] since it is fast, simple and informative. Many methods were developed to extract information about porosity and surface properties of materials from adsorption isotherm data. However, there is still a need for a consistent approach in adsorption characterization of porous solids. In the current work, it will be shown that such an approach can be based on the adsorption potential distribution function. [4,5,6] The differential adsorption potential distribution function X(A) provides a quantitative characterization of all changes in the Gibbs free energy for a given gas/solid sorption system. It has been demonstrated that the latter distribution function is associated via simple relationships with the differential enthalpy [7] and the differential entropy of adsorption as well as with the integral enthalpy of immersion. [8] The differential adsorption potential distribution can easily be applied to estimate mesopore [9] and micropore[10] size distributions. Moreover, the X(A) function can be used for qualitative [11] and quantitative [2,12] estimation of energetic heterogeneity of porous solids. The current study shows that the approach based on the differential adsorption potential distribution provides a convenient and comprehensive way for the thermodynamic characterization of surface and structural heterogeneities of porous materials.

2 THEORETICAL

2.1 Definition of the differential adsorption potential distribution

The adsorption potential A is defined as the change in the Gibbs free energy ΔG taken with a minus sign: [4,5]

$$A \equiv -\Delta G = RT\ln(p_0/p) \qquad (1)$$

where p and p_0 stand for the equilibrium and saturation pressures of the gaseous adsorbate, respectively, T denotes absolute temperature and R denotes the universal gas constant. Because of the fact that the adsorption potential is directly related through Eq. 1 with the equilibrium pressure of the gaseous adsorbate, the relative adsorption θ can be expressed as a function of the adsorption potential. The resulting function $\theta(A)$ is usually referred to as the characteristic adsorption curve. [4,5] It needs to be noted that the relative adsorption θ is defined as the adsorbed amount divided by the normalization factor n_0, e. g. the maximum amount adsorbed (in the case of microporous adsorbents) or the monolayer capacity (for other types of adsorbents).
    The differential adsorption potential distribution X(A) is defined as:

$$X(A) = -\frac{d\theta(A)}{dA} \qquad (2)$$

The X(A) function was shown to be a quantitative [2,12] and a qualitative[11] measure of heterogeneity of porous solids. Moreover, it was shown that the differential adsorption potential distribution can be used to calculate the differential molar enthalpy of adsorption ΔH, the differential molar entropy of adsorption ΔS and the integral enthalpy of immersion ΔH_{im}. The expressions, which provide relations between the X(A) function and thermodynamic functions for the adsorption system, were originally derived for adsorption on microporous solids [7,8], but they can be modified for other kinds of adsorbents.

2.2 Relationships of X(A) with thermodynamic functions for gas adsorption on microporous solids

The differential molar enthalpy of micropore filling ΔH can be written as a function of the adsorption potential distribution X(A) and the characteristic adsorption curve $\theta(A)$:

$$\Delta H = -A + \left(\frac{\partial A}{\partial T}\right)_\theta - \frac{\alpha T\theta(A)}{X(A)} \qquad (3)$$

The thermal coefficient α is a derivative of the logarithm of the maximum amount adsorbed in micropores n_0 with respect to temperature (taken with a minus sign): $\alpha = -d\ln n_0/dT$.
    It was shown [4,5] that in the case of adsorption on many microporous solids, the adsorption curve $\theta(A)$ is temperature-invariant: $(\partial A/\partial T)_\theta = 0$. When the temperature invariance condition is fulfilled, Eq. 3 assumes a simpler form:

$$\Delta H = -A - \frac{\alpha T\theta(A)}{X(A)} \qquad (4)$$

The expression for the differential molar entropy ΔS for gas adsorption in micropores can easily be derived from the relation $\Delta S = -(\Delta G - \Delta H) / T = (A + \Delta H) / T$ and Eq. 4:

$$\Delta S = -\frac{\alpha \theta (A)}{X(A)} \tag{5}$$

The average adsorption potential $\overline{A}$ is another important quantity suitable for characterization of energetic properties of microporous materials: [13, 14]

$$\overline{A} = \int_0^\infty A \, X(A) dA \tag{6}$$

The average adsorption potential was shown to be directly related to the integral molar enthalpy of immersion ΔH_{im}:

$$\Delta H_{im} = -(1 + \alpha T) \overline{A} \tag{7}$$

The latter quantity characterizes the immersion of a microporous solid in a pure liquid. Such a process closely resembles micropore filling at sufficiently high pressures of gaseous adsorbate.

2.3   Evaluation of micropore and mesopore size distributions

Under the assumption of the condensation approximation (CA), [15] the differential adsorption potential distribution was shown to be simply related to the pore volume distribution J(x). [9, 10] The latter function is defined as a derivative of the pore volume V with respect to the pore width x: $J(x) = dV/dx$. In terms of the CA, pores of the size x are not filled until the pressure of adsorbate reaches a condensation pressure p_c (characteristic for a given pore size x), for which they become completely filled. Therefore, for a given J(x) there exists a relation between the volume of the already filled pores and the adsorption potential A (which is directly related to the pressure through Eq. 1). When one makes an additional assumption that the density of adsorbate condensed in pores is equal to the density of the liquid adsorbate at the measurement temperature, the amount (volume) of adsorbed gas is equal to the volume of filled pores and:

$$J(x) \equiv \frac{dV}{dx} = \left(\frac{dV}{dA} \right) \left(\frac{dA}{dx} \right) = -X_n(A) \left(\frac{dA}{dx} \right) \tag{8}$$

where $X_n(A) = dV/dA = d(n_0 \theta)/dA = n_0 X(A)$ is a non-normalized differential adsorption potential distribution. The $X_n(A)$ function is directly accessible from the experiment. However, an expression for the derivative dA/dx needs to be known in order to calculate the pore volume distribution J(x). Such an expression was obtained for both microporous and mesoporous adsorption systems. For the former, the derived equations [10] are essentially equivalent to the Horvath-Kawazoe method. [16] For the latter, one can use the well-known Kelvin equation, which can also be modified in order to account for the formation of adsorbed film on pore walls. [9] The accuracy of the evaluation of the pore size distribution from Eq. 8 depends on the accuracy of the expression A(x) for the adsorption potential A as a function of the pore width and on the validity of the condensation approximation for a given adsorption system. Some of the problems related with the application of the Eq. 8 in characterization of porous solids are discussed in our previous study. [11]

3 EXPERIMENTAL

Nitrogen adsorption isotherms were measured over a wide range of pressures using an ASAP 2010 volumetric adsorption apparatus from Micromeritics (Norcross, GA, USA). Samples were outgassed for 2 h at 473 K in the degas port of the adsorption analyzer before the measurements were performed.

Sterling FT-G graphitized carbon was obtained from Laboratory of the Government Chemist (Queens Road, Teddington, Middx, UK). BP 280 carbon black (produced in the oil furnace process) was acquired from Cabot Corporation (Special Blacks Division, Billerica, MA). ROW active carbon was obtained from Norit Co. (Amersfoort, The Netherlands). The sample of a siliceous MCM-41 mesoporous molecular sieve was synthesized in the laboratory of Dr. A. Sayari. The synthetic procedure for the latter material was described elsewhere. [17]

4 RESULTS AND DISCUSSION

Shown in Figures 1 and 2 are nitrogen adsorption isotherms (in normal and logarithmic scale, respectively) for selected samples exhibiting remarkably different surface and porous properties. The differential adsorption potential distributions are shown in Figure 3. In the case of Sterling FT-G, a monolayer formation step is very well pronounced. Moreover, a distinct step arising from a formation of a second adsorbed layer can be noticed in the multilayer adsorption pressure region. An examination of the adsorption potential distribution for the sample allows for much more detailed characterization of the sample. The $X(A)$ function exhibits a peak at ca. 5.5 kJ/mol corresponding to the monolayer buildup and another one for A about 0.7 kJ/mol, resulting from the formation of the second adsorbed layer. For even lower values of the adsorption potential, the $X(A)$ curve increases steeply due to the multilayer adsorption. There is a deep minimum in the adsorption potential distribution for the adsorption potential values from ca. 1 to 5 kJ/mol, which shows that the formation of the monolayer is well-separated from the formation of further adsorbed layers. Another interesting feature of the adsorption potential distribution for Sterling FT-G is a presence of a small peak at 3 kJ/mol, which arises from disordered fluid - ordered solid phase transition of nitrogen on the surface [18] (at a relative pressure of about 0.01, see Figure 2) and indicates a high degree of surface homogeneity for the sample.

A nitrogen adsorption isotherm for the BP 280 carbon black seems to be quite similar to the isotherm for Sterling FT-G, but a monolayer formation step is significantly broader and there is no distinct step corresponding to formation of the second adsorbed layer. The similarity of these two isotherms arises from the fact that the compared samples are carbonaceous and essentially nonporous. However, the examination of the adsorption potential distributions reveals that surface properties of the materials considered are significantly different. The $X(A)$ function for the BP280 carbon black is rather smooth and featureless. Even the monolayer formation peak is poorly developed, very broad and not well-separated. Moreover, the $X(A)$ distribution for BP 280 is extended further towards higher adsorption potential values. All these findings suggest that the BP 280 carbon black exhibits a significant surface heterogeneity.

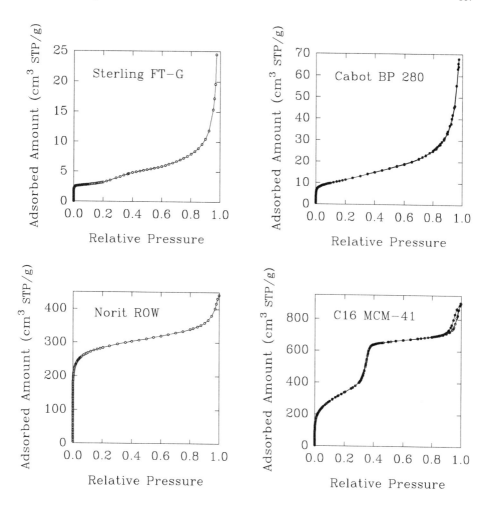

Figure 1 *Nitrogen adsorption isotherms for the samples under study.*

As can be seen in Figures 1 and 2, ROW active carbon adsorbs appreciable amount of nitrogen at very low pressures (below 10^{-4}), which indicates that the sample is considerably microporous. The distribution of adsorption potential features a sharp peak at about 7 kJ/mol, resulting from a micropore filling, and then reaches a plateau for adsorption potential values from 1 to 6 kJ/mol. Subsequently, a sharp increase of the X(A) function, which corresponds to a multilayer formation on the external surface of the adsorbent, is observed. As was shown previously, [11] the value of the adsorption potential for the micropore filling peak can be used to assess the size of micropores, especially when pertinent data obtained by such methods, as computer simulations or density functional theory (DFT) calculations, are available.

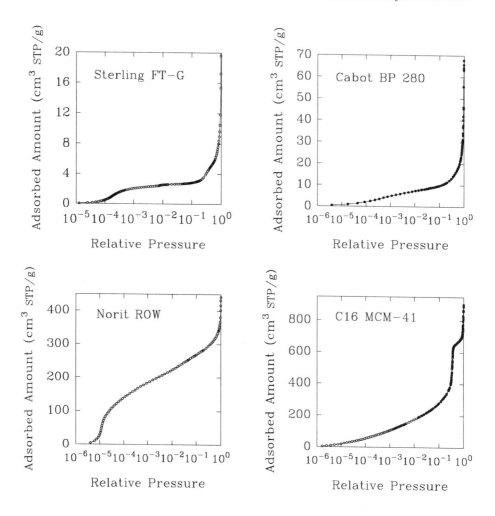

Figure 2 *Nitrogen adsorption isotherms in a logarithmic scale.*

An adsorption isotherm and an adsorption potential distribution for a typical MCM-41 mesoporous molecular sieve are shown in Figures 1-3. One can notice that the adsorbed amount increases gradually and the isotherm does not feature a distinct monolayer formation step. Moreover, even the X(A) function increases monotonously for A from 8 to 1 kJ/mol and does not show any maximum, contradictory to the behavior of the carbonaceous materials considered above. These findings show that the siliceous walls of the MCM-41 material are strongly heterogeneous, which can be expected because of the presence of different silanol and siloxane surface groups. Very similar results of low pressure nitrogen adsorption measurements were obtained for various silica gels and siliceous mesoporous molecular sieves. [17, 19]

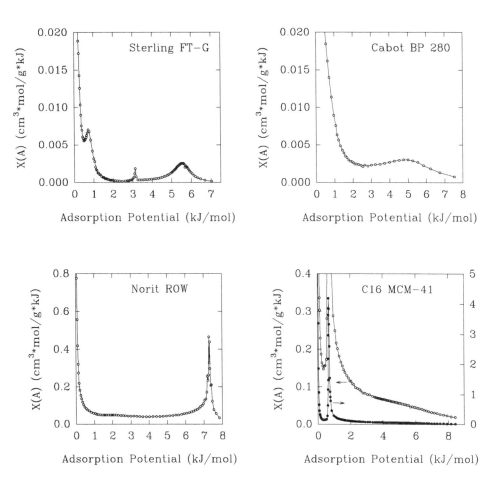

Figure 3 *Adsorption potential distributions for the samples under study.*

The isotherm for the MCM-41 material exhibits a sharp and well-pronounced step arising from the condensation of adsorbate in primary mesopores. The corresponding peak on the adsorption potential distribution can be seen for the adsorption potential value of 0.67 kJ/mol. The A value at the maximum of the peak can be used to estimate the size of the pores in the material. The relation between the pore size and the adsorption potential of the peak maximum can be found either on the basis of theoretical predictions (such as computer simulations or DFT calculations) or from experimental data. [17]

5 CONCLUSIONS

The adsorption potential distribution function can be used as a starting point of detailed characterization of thermodynamic and structural properties of adsorption systems. It is directly related with such thermodynamic functions, as the differential enthalpy of adsorption, the differential entropy of adsorption and the integral enthalpy of immersion. The adsorption potential distribution can be used to characterize surface and structural heterogeneities of porous solids. The X(A) function is suitable for an estimation of micropore and mesopore size distributions. Moreover, it provides a simple way of qualitative characterization of surface heterogeneity of nonporous and porous solids. As the adsorption potential distribution is a model independent quantity, it can be used to obtain information complementary to the adsorption energy distribution functions, which are model dependent and therefore may be somehow biased by artefacts arising from the methods of their calculation. Because of the richness of information, which can be inferred on the basis of adsorption potential distributions, further studies in that direction are needed.

6 ACKNOWLEDGMENTS

The authors acknowledge Dr. A. Sayari, Cabot Corporation (Special Blacks Division) and Norit Co. for providing the samples.

References

1.   S. J. Gregg and K. S. W. Sing, 'Adsorption, Surface Area and Porosity', Academic Press, London, 2nd ed., 1982.
2.   M. Jaroniec and R. Madey, 'Physical Adsorption on Heterogeneous Solid Surfaces', Elsevier, Amsterdam, 1988.
3.   W. Rudzinski and D. H. Everett, 'Adsorption of Gases on Heterogeneous Solid Surfaces', Academic Press, London, 1991
4.   M. M. Dubinin, *Prog. Surf. Membrane Sci.*, 1975, **9**, 1.
5.   M. M. Dubinin, *Chem. Phys. Carbon*, 1966, **2**, 55.
6.   M. Jaroniec and R. Madey, *J. Phys. Chem.*, 1989, **93**, 5225.
7.   M. Jaroniec, *Langmuir,* 1987, **3**, 795.
8.   M. Jaroniec and R. Madey, *J. Phys. Chem.*, 1988, **92**, 3986.
9.   M. Jaroniec, 'Access in Nanoporous Solids', T. J. Pinnavaia and M. F. Thorpe, Eds., Plenum Press, New York, 1995, p. 255.
10.   M. Jaroniec, K. P. Gadkaree and J. Choma, *Colloids and Surfaces*, in press.
11.   M. Kruk, M. Jaroniec and J. Choma, *Adsorption*, in press.
12.   M. Jaroniec and J. Choma, *Chem. Phys. Carbon*, 1989, **22**, 197.
13.   M. Jaroniec, R. Madey, X. Lu and J. Choma, *Langmuir*, 1988, **4**, 911.
14.   M. Jaroniec and R. Madey, *Carbon*, 1988, **26**, 107.
15.   S. Ross and J. P. Olivier, 'On Physical Adsorption', Wiley, New York, 1964.
16.   G. Horvath and K. Kawazoe, *J. Chem. Eng. Japan*, 1983, **16**, 470.
17.   M. Kruk, M. Jaroniec and A. Sayari, *J. Phys. Chem.*, submitted.
18.   T. T. Chung and J. G. Dash, *Surf. Sci.*, 1977, **66**, 559.
19.   Y. Bereznitski, M. Jaroniec, M. Kruk and B. Buszewski, *J. Liq. Chromatogr.*, in press.

Within the Hysteresis:
Insight into the Bimodal Pore-Size Distribution of Close-Packed Spheres

HERBERT GIESCHE

New York State College of Ceramics at Alfred University

2 Pine St., Alfred, NY 14802, USA

ABSTRACT

Model pore structures were prepared from dispersions of submicron monodispersed silica particles by a sedimentation process. Ordered dense sphere packing structures were observed with scanning electron microscopy. Nitrogen sorption- as well as Hg-porosimetry measurements confirmed the calculated values of the pore openings in those structures. In Hg-porosimetry measurements a two step extrusion curve was observed, when the pore system was only partially filled during the intrusion process. This two step curve was not observed in case the pore system was filled with mercury to more than 95% during the intrusion run. The mercury porosimetry results can be interpreted in terms of the coexistence of octahedral and tetrahedral voids (pores) in the examined sphere packing structure and their special arrangement within the structure (connectivity). The height of the two steps in the extrusion curve depended on the relative filling status during the preceding intrusion process, whereas the width of the hysteresis depended on the sample pre treatment. A calcination at elevated temperature changed the surface behaviour as well as the surface curvature in the pores and entrance openings. As a result the extrusion curve was shifted towards lower pressures. Nitrogen adsorption and desorption measurements on the same samples did not reveal any fine structure within the hysteresis range.

INTRODUCTION

Hysteresis phenomena in gas-sorption and mercury porosimetry have been the subject of several studies in the past.[1-10] The hysteresis between intrusion and extrusion branch in Hg-porosimetry measurements is explained by various models or mechanisms. In order to check the validity of the concepts, models and related equations, studies on porous model substances are extremely useful. Such model substances should exhibit a highly uniform pore size, well-defined pore shape and a controlled pore connectivity. A common means to achieve these defined pore geometry is to compact non-porous particles of a given shape (spheres, rods, needles, plates) and of uniform size into porous aggregates. Studies on similar compacts made of monodispersed silica spheres using nitrogen adsorption, Hg-porosimetry and other techniques have been reported by several researchers.[3-10]

EXPERIMENTAL

The silica spheres were prepared by hydrolysis of tetraethoxysilane in ethanol-water-ammonia mixtures. The model pore compacts were made by gravity settling. Detailed experimental conditions of those techniques and conditions are described elsewhere.[9-13]

On account of the narrow particle size distribution and uniform shape of the starting silica particles, dense ordered aggregates with a high coordination number (10 to 12) per sphere could be expected and were confirmed by scanning electron micrographs. Various examples were shown in previous publications.[9,12,14] Fig. 1 provides a schematic drawing of those model structures revealing the positions of octahedral and tetrahedral voids between the packed spheres.

Fig. 1: Schematic of a dense sphere packing structure. Octahedral (O) and tetrahedral (T) voids are indicated between the two layers of particles.

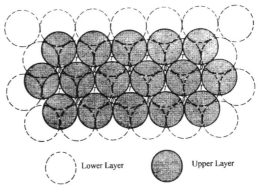

Lower Layer   Upper Layer

Mercury intrusion/extrusion curves were measured with a Micromeritics 9220 Autopore. The maximum pressure applied during those experiments was 400 MPa. The instrument determines the intruded volume of mercury in a discontinues analysis mode. The equilibration time was set to > 300 s. Individual pressure points were chosen by the operator and the instrument kept/adjusted the actual pressure to within ± 0.1% of the set value.

Nitrogen sorption measurements were performed on a Micromeritics ASAP 2010 in a similar way as the mercury porosimetry. Equilibration time was set to 60 s and the target relative pressure was kept constant to within 0.02%. Those extreme analysis conditions were necessary due to the very narrow adsorption/desorption hysteresis of the samples and the very high relative pressures where the adsorption (condensation) occurred. The equilibration between individual data points within the hysteresis range took several hours and it was necessary to check the saturation pressure (P_0) every 60 min. Further details on the kinetics of the ad-/de-sorption process within the hysteresis range were described in previous publications.[9,10]

RESULTS AND DISCUSSION

When mercury porosimetry was used for the pore characterisation, the regular analysis, using the intrusion branch up to maximum pressure and the corresponding extrusion branch back to atmospheric conditions, was in accordance with publications by other groups and agreed well with calculations for the assumed packing structure and the size of the silica particles. Further details of those measurements are described elsewhere.[9,10] The intrusion curve was controlled by the largest opening towards the pores. The experimentally observed *break through pressure* was in close agreement with models by Smith & Stermer,[18] Iczkowski[19,20] and Mayer & Stowe[21,22] derived for packed sphere structures. The extrusion branch, which is thought to be determined by the inner size of the pores, however, did not show the expected bimodal pore size distribution, which was predicted from the presence of tetrahedral and octahedral cavities (different size) in the close packed sphere structure. Yet, the extrusion curve indicated only one pronounced step at a 'critical' pressure, which indicated the sudden release of intruded mercury. The initial part of the extrusion branch (before the sudden release of Hg) corresponded well with theoretical calculations by Mayer and Stowe derived for the intrusion of mercury into the neck region between touching spheres. An assembly of spheres having a coordination number of 10 and at a mercury-solid contact angle of about 140° were assumed.[21,22] However, when the range within the hysteresis was scanned in more detail, a very distinct "*fine structure*" was revealed, which could be interpreted in terms of octahedral and tetrahedral voids. Fig. 2 shows the mercury porosimetry data for the situation when the pressure was decreased and then again increased at various points along the intrusion curve (A), as well as in-/decreasing the pressure at points along the extrusion curve (B). In both cases the intruded volume of mercury did not change much during those cycles. Combining the curves in (A) with the results in (B) would predict a monotonous transition curve between extrusion and intrusion branch of the primary curve. Yet, when the pressure was decreased over a wider range, a distinct two step extrusion behaviour was noticed (C). This behaviour was only observed when the pore system was not completely filled with mercury at the point of pressure inversion. The extrusion/intrusion was completely reversible when the measurement stayed within the "flat" parts of the curves. Yet, when the 'steep' during the extrusion was surpassed, then extrusion/intrusion cycles revealed a clear hysteresis. One experiment, showing the combination of various increasing and decreasing pressure cycles, is given in fig. 2 (D) The sequence of reaching the various points is indicated by numbers. It was obvious that starting at a specific point in the diagram did not result in the same extrusion curve. The behaviour depended very much on the history of how this point was reached. The results further indicated that there were two distinct bending points in the extrusion curves for an incomplete filled pore system (see fig 2 C). The first step occurred at a pressure of about 2300 to 2400 psi (155 MPa) and the second at about 1400 to 1500 psi (100 MPa). A preliminary comparison of those findings was consistent with the ratio of octahedral to tetrahedral voids (0.4142 particle radius, R / 0.2247 R). Thus, the observed fine structure within the hysteresis was could be attributed to the different sizes of octahedral and tetrahedral pores and the different pressures at which those pores are drained during the extrusion process. It was obvious from the measurements at different filling states that this filling state had an influence on the height of the extrusion step within the hysteresis range. The latter step was largest at a prior pore filling of about 50% during the intrusion process. With the fact that octahedral and tetrahedral voids have the same entrance pore throat, it has to be assumed that they are filled at the same pressure and with equal chances. From those facts it was obvious that the smaller tetrahedral pores can only be emptied if they were filled with mercury during the intrusion cycle. Thus, more pore volume is available for the extrusion process when it starts at a higher intrusion fillings. Second, pores can only be emptied at this earlier stage (higher pressure) when there is an empty pore adjacent. Thus, less tetrahedral void are to be measured during this first step, if too many pores are filled

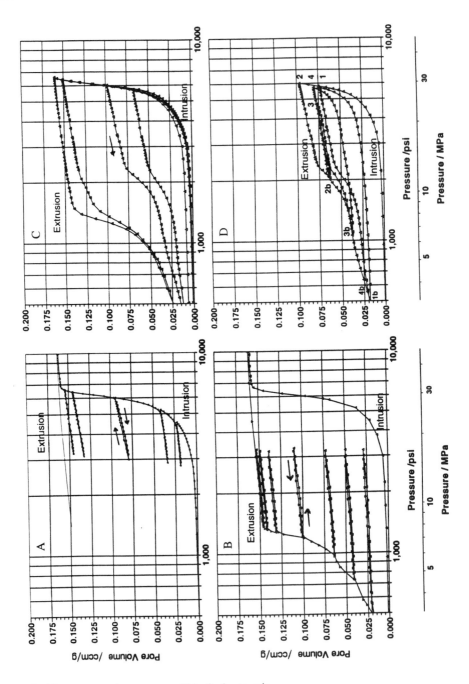

Fig. 2: Mercury porosimetry runs within the hysteresis.

during the intrusion. The latter statement is the most essential observation of this study. The "fine structure" within the hysteresis depend on the existence of those open spaces in the pore network which can act now as starting points for the extrusion process. The "Energy Barrier Model" will explain this effect in more detail.

Additional hysteresis effects were observed on another silica sample that had been calcined under vacuum at 1000°C for 10 min. and 31 hours respectively (see Fig. 3). The heat treatment under those conditions causes only a minor sintering (densification) of the compact. The pore volume decreased by about 0.01 cm^{3}g^{-1}. It was shown previously[10] that temperatures of 1100°C and longer times are needed to cause major changes in the structure. Accordingly, the intrusion curve was not at all influenced by this treatment. However, the extrusion curve as well as the behaviour within the hysteresis range showed a distinct difference. Similar observations were made in earlier studies and it could be explained in terms of the various surface treatments of those samples.[9,10] The extrusion

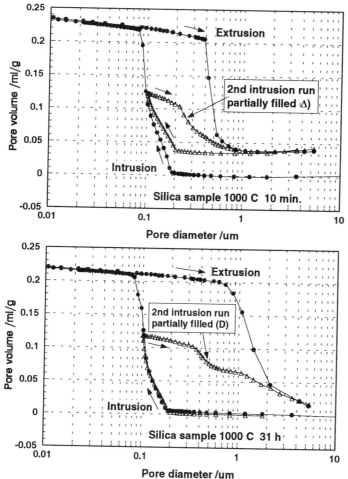

Fig. 3 Hysteresis effects of silica samples calcined a 1000°C for 10 min and 31 hours.

curve of the sample calcined for 31 hours was shifted towards lower pressure as well as the step within the hysteresis was shifted. This shift could be explained by the change in pore geometry. However, at the same time this explanation should also cause a shift of the intrusion curve. The latter, however, did not change. It is interesting to note that the factor between the first step within the hysteresis and the extrusion curve was the same in both samples. The factor of about 1.7 to 1.8 was also observed in samples which were per treated under different conditions. Further investigations are necessary to analyse and interpret those results.

COMPARISON WITH HYSTERESIS MODELS

Usually three hysteresis models are used and mentioned in the literature, the *ink-bottle theory,*[15] the *connectivity model*[6] and the *contact angle hysteresis.*[16,17] However, the existing theories do not explain why the same two-step extrusion curve is not observed when the extrusion starts from a sample completely filled with mercury. This evidently is depending on the special connectivity between all pores and an activation barrier for the extrusion process. The following description of the "*Energy Barrier Model*" explains those hysteresis effects in a better way. A "*Counter Contact Model*" was described earlier and will not be covered at this point.[14]

Table 1: Example, 1 μm cylindrical pore; $\gamma_{Hg}= 0.48$ N/m; $\Theta= 140°$

length of pore / μm	$P_{Intrusion}$ MPa	$P_{Extrusion}$ MPa	P_I / P_E
100	0.735	0.726	1.01
10	0.735	0.650	1.13
5	0.735	0.566	1.31
2	0.735	0.309	2.78
1.5	0.735	0.166	7.58
1.4	0.735	0.050	14.66

ENERGY BARRIER MODEL

The Washburn equation, which is commonly used for the determination of pore sizes, assumes cylindrical pores of infinite length. Thus, any change of liquid (mercury) / vapor interface could be neglected. However, in most real samples the ratio of pore length to pore entrance size is much smaller. For the intrusion of mercury into the pores, the latter simplification has no major effect. Yet, for the extrusion of mercury this liquid/vapor interface plays an important role. The liquid/vapor interface has to be newly created and thus needs an extra amount of energy.

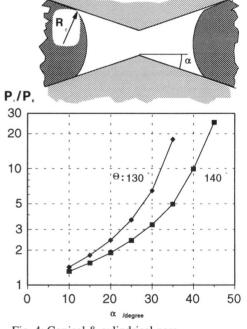

Fig. 4: Conical & cylindrical pore

In other words, the extrusion occurs at a lower pressure ($\Rightarrow$ hysteresis). Depending on the ratio of length to diameter this can be a substantial difference (see example in table 1). In a similar way the retarded extrusion was calculated for conical cylinder pores (see fig. 4). At various "slope-angles" and for the two assumed contact angles of 140 and 130°, the retardation factor, P_I / P_E, can be very large. Even so one may argue that finally the effect due to the created liquid/vapor interface becomes insignificant, the "first" disruption of the mercury network in the sample is the critical part in the extrusion curve. Thus, the here proposed mechanism could explain, why the *fine-structure* was not observed when the sample was completely filled with mercury. In the later, completely filled, pore system those breaking points have to be created and this requires an extra energy (lower pressure). Yet, when there are a sufficient number of open connections, those disrupted positions will act as starting points for the extrusion, which then in turn allows the smaller tetrahedral voids to be emptied at slightly larger pressures of about 2350 psi (160 MPa). In contrast, a completely filled pore network requires a critical extrusion pressure of about 1450 psi (100 MPa).

NITROGEN ADSORPTION

Nitrogen sorption measurements were performed in a similar way as previously described for the mercury porosimetry measurements. The hysteresis range between the adsorption and the desorption branch was scanned. However, the data did not indicate any kind of a two step adsorption or desorption curve. Fig. 5 a) shows results when the desorption was started at various points along the adsorption branch and fig. 5 b) indicates the results of increasing the relative pressure at different pressures along the initial desorption curve. Along the different desorption curves (shown in fig. 5 a)) the downward bend occurs always at approximately the same relative pressure point. This would indicate that in those cases the size distribution of the available pore openings is very similar in all occasions. Connectivity is certainly important for the hysteresis effect, but it has a different influence in this situation. In general it is accepted to apply the Cohan equation to calculate the pore size from the adsorption curve, whereas the Kelvin equation is used for the desorption branch. The two equations differ by a factor of 2 which originates in the different assumed surface curvature of empty and filled cylindrical pores. However, in a real sample one has to consider the fact that during the adsorption experiment the mechanism actually changes from empty to filled pores. Once condensation has occurred in one pore the actual surface at the entrance point to a neighbouring pore is actually described by the Kelvin equation. The capillary condensation in an open pore network starts according to the Cohan equation at the smallest pore radius. However, once the liquid meniscus has formed, it will now obey the Kelvin equation and consequently pores which are about twice as large as the starting size will be filled with condensate. Thus, the reason for the hysteresis is entirely due to the different filling and emptying mechanisms at those entrance pores. In effect, each pore has to have an open connection to the outside in order to empty at its corresponding pressure. Since octahedral as well as tetrahedral pores have the same kind of triangular pore openings (throats), both will be emptied at the same pressure. Once the entrance to a pore becomes free the whole pore now 'drains'. During the adsorption process, however, the condensation will start at the smallest pore size. In the given situation these places are the pore throats. As a liquid nitrogen meniscus is formed at the pore entrance. the condensation/pore filling process will continue until the corresponding

"Kelvin" meniscus is stable. For the given pore structure this would correspond to an automatic filling of tetrahedral voids as soon as the condensation process has started at the pore entrances to that pore. Octahedral voids are slightly larger than this critical size and they may induce a temporary interruption of that filling process. However, the interconnection between tetrahedral and octahedral pores as well as the small retardation effect due to their size makes it impossible to distinguish between those pores.

The different results of mercury porosimetry can best be explained by the idea that mercury porosimetry tries to empty the smaller pores (throats) first. Whereas the situation is inverted in the nitrogen desorption process, where larger internal pores are supposed to be emptied earlier (at higher relative pressures), however, since all entrance connections to the pore are smaller the pores have to 'wait' until the first pore entrance empties. Then it can evaporate as well. In summary: mercury porosimetry drains the pores from small to larger sizes, and nitrogen desorption would like to drain larger pores before the smaller ones, yet it needs an open connection to the pore.

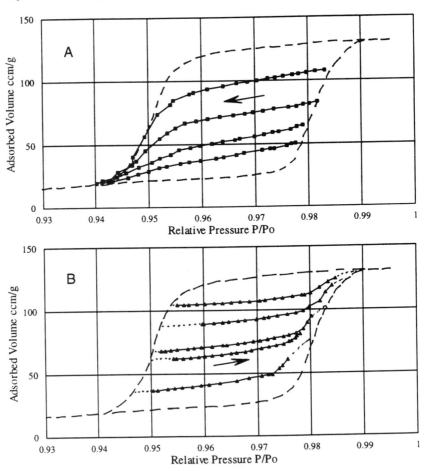

Fig. 5 Nitrogen ad-/desorption within the hysteresis

CONCLUSION

Mercury porosimetry measurements on model pore compacts made of ordered monodispersed silica spheres revealed a pronounced *fine structure* within the hysteresis, which could be attributed to octahedral and tetrahedral voids in the structure. Two *Energy Barrier Model*, has been described, which give additional information about the cause of and factors influencing the hysteresis in mercury porosimetry measurements. The model has to be further tested. Moreover precise simulations could give a better insight into the extrusion process. This may finally reveal additional information about pore size distributions and/or pore shapes in unknown samples. Nitrogen sorption measurements on the same samples did not reveal the same hysteresis fine structure, which could be explained by the fact that octahedral as well as tetrahedral pores have the same entrance size and in order to empty both kinds of pores during desorption they actually need the same relative pressure. Desorption is controlled by the entrance size to a pore. Moreover, the adsorption process is more complicated than anticipated in most models, since one has to consider the switching in mechanisms during the pore filling process.

REFERENCES

1   S. J. Gregg and K. S. W. Sing; Academic Press, London (1982)
2   D. H. Everett, in K. K. Unger, J. Rouquerol, K.S.W. Sing and H. Kral (Eds.), Vol. 39, Elsevier, Amsterdam, 1988, pP. 1-21
3   A.P. Karnaukhov and A.V. Kiselev, Russ. J. Phys. Chem., **34**, (1960) 1019
4   S. Bukowiecki, B. Straube and K. K. Unger; pp. 43-55 in 'Principles and Applications of Pore Structural Characterization', ed. J. M. Haynes; P. Rossi-Doria; Arrowsmith, Bristol (1985)
5   W. C. Conner, A. M. Lane, K. M. Ng and M. Goldblatt; J. Catal.; **83**, 336-45 (1983)
6   W. Curtis Conner and Alan M. Lane; J. Catal.; **89**, 217-25 (1984)
7   W. C. Conner, A. M. Lane and A. J. Hoffman; J. Colloid Interface Sci.; **100**[1], 185-93 (1984)
8   W. C. Conner, J. F. Cevallos-Candau, E. L. Weist, J. Pajares, S. Mendioroz and A. Cortes; Langmuir; **2**, 151-4 (1986)
9   H. Giesche, K. K. Unger, U. Müller and U. Esser; Colloids Surfaces; **37**, 93-113 (1989)
10 Herbert Giesche; Dissertation, Universität Mainz (1987)
11 Werner Stöber, Arthur Fink and Ernst Bohn; J. Colloid Interface Sci.; **26**[1], 62-69 (1968)
12 Herbert Giesche; J. Eur. Ceram. Soc.; **14** [3], 189-204 (1994)
13 Herbert Giesche; J. Eur. Ceram. Soc.; **14** [3], 205-15 (1994)
14 Herbert Giesche; pp. 505-10 in 'Advances in Porous Materials' Mater. Res. Soc. Symp. Proc. Vol. 371, eds. S. Komarneni, D. M. Smith, and J. S. Beck, Materials Research Society, Pittsburgh (1995)
15 James W. McBain; J. Am. Chem. Soc.; **57**, 699-700 (1935)
16 R.J. Good and R.S. Mikhail; Powder Technology, **29**, 53 (1981)
17 J. Kloubek; Powder Technology, **29**, 63 (1981)
18 D. M. Smith and D. L. Stermer; J. Colloid Interface Sci.; **111**[1], 160-8 (1986)
19 Raymond P. Iczkowski; Ind. Eng. Chem. Fundam.[4], 516-9 (1966)
20 Raymond P. Iczkowski; Ind. Eng. Chem. Fundam.; **6**[2], 263-5 (1967)
21 Raymond P. Mayer and Robert A. Stowe; J. Colloid Sci.; **20**, 893-911 (1965)
22 Raymond P. Mayer and Robert A. Stowe; J. Phys. Chem.; **70**[12], 3867-73 (1966)

PROPERTIES OF NANO-ENGINEERED TWO-DIMENSIONAL SILICA-BASED POROUS NETWORKS

P. A. Sermon , M. S. W. Vong and D. Grosso

Fractal Solids and Surfaces Research Group,
Department of Chemistry,
Brunel University,
Kingston Lane, Uxbridge,
Middlesex, UB8 3PH, UK

ABSTRACT

Sol-gel methods have been used to produce two-dimensional networks of silica (of controlled pore size, geometry and connectivity) which are of value as optical coatings (ion-sensors, composites and catalytic membranes)[1]. The preparation of these nanoengineered materials is illustrated by base-catalysed routes to porous optical coatings, as is their characterisation by adsorption, FTIR and TEM, etc.. FTIR suggests that these coatings are at least as responsive to humidity as bulk gels. Such silica *coatings* may require adsorption of phenol from solution to assess their total surface area and pore structure in situ.

1 INTRODUCTION

Graham's colloid science of silicic acid today finds use in paints, medicines, catalytic agents and coatings and now extends to a whole range of inorganic oxides[2]. When these colloidal particles are allowed to aggregate and dry subcritically xerogels are produced, but drying supercritrically produces aerogels. Such gels can are produced in bulk, but are also produced as two-dimensional layers by spin-coated[3], dip-coated[4] or meniscus-coated[5] to give porous coatings of a controlled thickness t. Such coatings will tend to be xerogels because their drying takes place very quickly, without good control of the temperature, although alcoholic partial pressures have on occasions been controlled[6].

Alkoxides are especially relevant[7] to silica sol-gel chemistry[8] and can be hydrolysed-condensed with the help of acid and base catalysts; the most frequent starting point is tetra-ethyl-orthosilicate $Si(OC_2H_5)_4$ (TEOS)[9]. In this case the H_2O/TEOS ratio r, the time, the viscosity and the pH[6,10] are all important in defining the sol-gel properties.

For a plane wave of wavelength λ_o is at normal incidence to a dielectric coating[11] (thickness t and refractive index n_2) which sits at the interface between air (refractive index n_1 (which equals 1.0)) and a silica-based glass (refractive index n_3 (which may equal 1.5)) then total reflectivity R[12] depends directly upon t and n_2[13] (see Figure 1) as a result of phase differences experienced (which equal $2\pi/\lambda_o tn_2$). Indeed R is zero when t is an odd integer multiple of $\lambda_o/4$ and when $n_2=(n_1.n_3)^{0.5}$. If R is to be nanoengineered to be zero using a silica-based overcoat then n_2 needs to be about 1.22 and (which is lower than that achieved

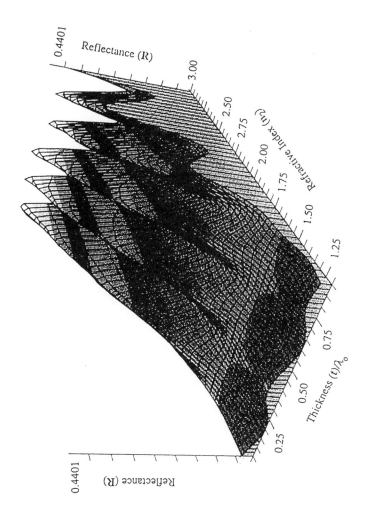

Figure 1 Variation in reflectance of porous silica coatings thickness t and refractive index n_2.

by MgF_2 coatings (i.e. 1.38)). This situation can be achieved using a silica sol-gel of about 53% porosity with an average particle size of about 20nm[14]. n_2 for such a porous silica coating can be controlled since it *decreases*[14] with increasing percentage porosity P, i.e.

$$n_2 = ((100n_B^2 - Pn_B^2 + P)^{0.5}/10, \text{ where } n_B \text{ is the bulk refractive index and}$$

decreases with decreasing average particle size d since $P \propto 1/\text{lnd}$. In addition scattering will be satisfactory if pores or inter-particle voids in the coating are smaller than λ_o, i.e.

are mesopores (2nm<d<50nm) when λ_o =351nm or 1064nm

Here then the authors have been anxious to explore the properties of sol-gel derived two-dimensional porous silica coatings[15] in-situ, realising that the deductions made may be relevant to other materials (e.g. tantala[18] or zirconia[19]).

2 EXPERIMENTAL

2.1 Materials

Base catalysed silica sols were prepared[15] containing 3w/w% silica using TEOS (98% purity; Aldrich) in pre-dried ethanol and ammonium hydroxide (M&B Pronalys; AR 33w/w%) diluted with HPLC-grade water). The substrate used for FTIR was BaF_2 (Specac).

2.2 Methods of Sol Preparation

Base catalysed silica sols were prepared[15] containing 3w/w% silica with particles of a uniform size (selected to be somewhere in the range 5nm<d<100nm) using TEOS in pre-dried ethanol. To this was added an ammonium solution of a strength and a volume to give r (the H_2O/TEOS) equal to 2.3, 4.0, 6.0 or 10.0 (although the rate of growth of SiO_2 particles of diameter d during base-catalysed TEOS hydrolysis-condensation may be independent of r^{16}) and a pH of 1.05.

2.3 Methods of Coating

Substrates were dip coated with the silica sols. t was engineered to be about 300nm and so the coating would be 15 particles thick on average if these particles were 20nm diameter on average.

2.4 Methods of Characterisation

Transmission electron microscopy (Jeol cx100) and dynamic light scattering (Malvern Zetasizer 3) at 90° were used to determine d in the silica sols. The total surface area and average pore diameter of the silica gel were estimated using BET N_2 adsorption at 77K. FTIR (Perkin Elmer 1710) was used to study water desorption at 297-677K and re-adsorption at 296K (from a N_2+H_2O (1.3kPa) stream flowing at 101kPa at 15cm³/min).

3 RESULTS

3.1 Sol Characterisation

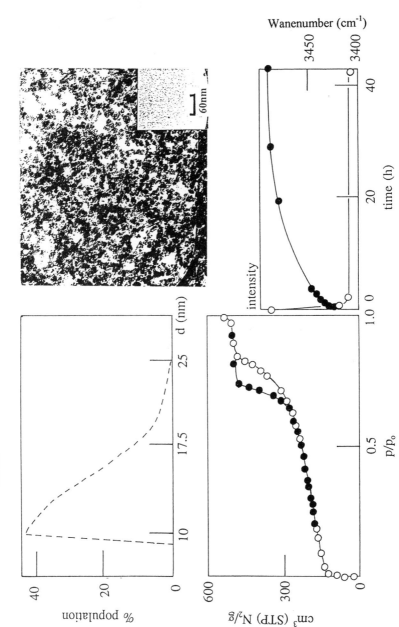

Figure 2 Particle size d in silica sol-gel seen by PCS and TEM.

Figure 3 Adsorption isotherm of N_2 at 77K on silica xerogel. Filled symbols are desorption points.

Figure 4 Variation in absorption and absorption maximum at $3400\text{-}3450\text{cm}^{-1}$ as a function of time.

Light scattering and transmission electron microscopy[15] indicated that the most frequent diameter d of the primary silica particles in the sols when r=4, pH=10.5-11.0 was 11nm after 7 days (see Figure 2) and 20nm after more than 7 days. Hence for the coating sample subjected to in-situ characterisation the average primary particle size d was ~20nm.

3.2 Xerogel Characterisation

Acid-catalysed xerogels have previously[17] been characterised for fractal dimensions using adsorption of C_{5-8} alkanes at 295K for which type I isotherms were seen. Figure 3 shows the type IV N_2 adsorption isotherm at 77K for the bulk xerogel produced after 16h at 313K when this had been outgassed at 702K for 2-5h. The isotherm corresponds to a mesoporous material with a mean pore diameter of 5nm (which is about a third of the size of the primary particles). Analysis of this isotherm suggested a total surface area S_{BET} of 700m^2/g, which is three times higher than the theoretical[14] value (237m^2/g) derived assuming the close packing of silica spheres of uniform size (i.e. 11nm (see Figure 2)) assuming that these had a bulk density of 2,300kg/m^3. This means that the particle packing is not close packing.

3.3 Coating Characterisation

The authors appreciated that the above bulk xerogel (if reproduced in a two-dimensional coating) would undoubtedly be able to release (or re-adsorb) water or ethanol vapours below say 873K, while its pores were inter-connected (i.e. while there was no significant pore closure or collapse). FTIR was used[15] to characterise such adsorption-desorption processes in the coating.

FTIR spectra were obtained[15] for the silica sol-gel derived coating, of thickness (t) ≈300nm, while this was held in flowing N_2 (20cm^3/min) at 296.5K and then as it was subject to drying at increasing temperature. These showed:

Si-OH hydroxyl bands at 950cm^{-1} and
Si-O-Si siloxane bands at about 1020cm^{-1}

There are several points that should be made about these data in the sense that as the temperature *increased*:

(i) the intensity of the Si-OH band *decreased* as dehydroxylation occurred and water was released from the coating (and this was accompanied by a simultaneous reduction in intensity of the stretching vibration of water in the 2600-3800cm^{-1} region),

(ii) the Si-O-Si bands *increased* in intensity (and new bands at 1078, 1083, 1089, 1093, 1148, 1182, 1270 and 1280cm^{-1} appeared; indicative of Si-O-Si bond formation) by further polycondensation[20], and the Si-O-Si band shifted to higher wavenumber (i.e. 1020=>1039cm^{-1}) as a result of Si-O bond shortening

during coating drying. None of this is surprising.

Now consider what happened when the coating saw humidity again in the prevailing atmosphere at 294.8K and 1.3kPa H_2O. Figure 4 shows that the Si-OH absorption maximum shifted in wavenumber and increased in intensity slowly as a function of time. In addition the shape and frequency of the Si-O-Si bands in the 1300-900cm-1 region also changed during water sorption. Thus the silica surface relaxes during water sorption. This is a process which can be described as breathing. The silica xerogel-like coatings not only retain some water and ethanolic residues, but even when these are removed (thermally or

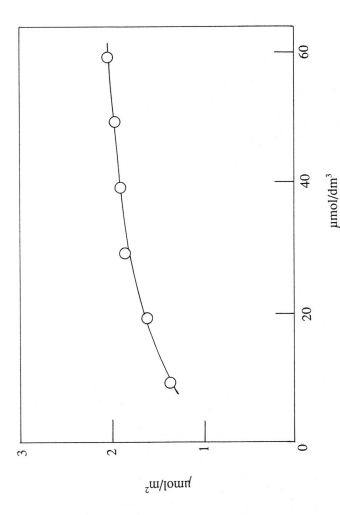

Figure 5 Adsorption isotherm for phenol from aqueous solution onto quartz.

under evacuation, then they re-accept (sorb) alternatives with relish and their lattices (networks) distend in the process. This is in keeping with measurements on bulk gels[21]. This may involve $(SiO)_3$ cyclic trisiloxane rings, which gives rise to Raman bands[22] at 606-610cm[-1]:

$H_2O_{(g)}$

Previously, symmetrical vibrationally-coupled breathing modes[21] have been seen for amorphous silica, but the present observations are novel for porous silica coatings.

4 DISCUSSION

Based catalysed sol-gel chemistry can lead to porous silica coatings of thickness up to a few µm with potentially useful optical properties. These must be characterised in terms of their adsorptive properties. The spectroscopic approach to the analysis of porous coatings is well established as shown here. These coatings are it appears genuinely response to the atmospheres and temperatures at which and in which they are held. However, such analyses need to be normalised by measurements of surface area, texture and porosity. There is in this context a problem- and possibly a solution.

It is of course possible to measure porosity and surface areas of adsorption of samples of low area or low pore volume per unit sample weight, but here for coatings this is compounded by having large integrated sample volumes (caused by the sample shape). Consider, coatings produced on a cylindrical substrate (45mm diameter x 2mm thickness). This would have a surface area of about $20cm^2$. If a silica xerogel coating 300nm thick, 53% porosity and surface area $700m^2/g$ (but particle density of 2.3 g/cm^3) was introduced on both sides of the substrate, its surface area would be 1.35 m^2. It would not be easy to assess this area by Kr adsorption for such a bulky sample.

It may be that adsorption of phenol (smaller and more suitable for interaction with oxides than p-nitrophenol) would be useful, since preliminary results on low area quartz which have produced the results in Figure 5 are relevant. These data indicate a type I adsorption isotherms. The adsorption coefficient was 18.45µmol phenol/m^2 at 303K on low surface area silica[25], and decreased by 30-40% on raising the temperature by 20K.   Such data suggest that the above substrate in contact with an aqueous phenol solution would adsorb up to only 0.006µmol of phenol at 293K while the coating described above would adsorb 24.98µmol (or 1.1mg) phenol under the same conditions. Hence the coating surface area could be assessed to be at least 3%.

Such measurements might be undertaken by immersing the sample in $25cm^3$ of an aqueous solution containing $60mg/dm^3$ of phenol at room temperature until equilibrium had been obtained as assessed by UV spectroscopy. In addition if the adsorption of the phenol were too slow then it could be accelerated by raising the temperature, although this would have the effect of decreasing the extent of adsorption as already recorded:

The proposed approach is by no means new, but it does have some advantages. Type I adsorption isotherms of p-nitrophenol from aqueous solution on activated carbon[23] *cannot* however be used for silica surfaces, and although the type I adsorption isotherms of methylene blue *can* be used to probe *mesoporous* silica gels[24], but the dye will apparently *not* penetrate pores 3.6-5.8nm in diameter in a silica (683m^2/g in surface area) because of dye micelle formation. The area of the surface occupied by a methylene blue molecule appears to be 1.20nm^2. A recent report[26] suggests that methylene blue dye adsorption from solution could be used to assess the surface area of porous silica coatings, but only if one accepts that the method can suffer from the problem of pore accessibility in xerogel coatings, when the pores are small relative to the molecular adsorbing species. Giles suggest that methylene blue adsorption is only suitable if pores are bigger than 2.7nm, but certainly not for micropores.

5 CONCLUSIONS

It may be that the adsorption of phenol in porous silica coatings is as effective at determining their surface area, as FTIR is in observing their surface properties, but this remains to be confirmed by the authors.

The methods of analysis will also in due course have to be extended to other oxide coatings and also such coatings produced by other methods and of various degrees of porosity.

Acknowledgments

The authors gratefully acknowledge the support of MSWV and DG by the Ministry of Defence, UK.

References

1. P.A. Sermon et al papers presented at PACRIM2
2. E. Matijevic Pure Appl.Chem. (1992),**64**,1703; J.W. McBain Colloid Sci. (1950),163
3. L.E. Scriven Mater.Res.Soc.Symp.Proc. (1988),**121**,717; S.D. Rege and H.S. Fogler AIChE J. (1989),**35(7)**,1177
4. Guglicimi, P. Colombs and S. Zenezini Chem.Phys. (1988),**23**,453
5. J.A. Britten Chem.Eng.Commun. (1993),**120**,59
6. B.V. Derjaguin Coll.Surfaces (1993),**79**,1
7. N. Kajiwara Porima Daijesuto (1989),**41**,2; O. Kuramitsu and S.Noguchi Jap.pat. (1989),0119305
8. I.M. Thomas Appl.Optics (1986),**28**,4013
9. S. Rojas, P. Serra, W.S. Wu, F. Sautarelli, G.C. Sarti and F. Minni J.Phys.Colloq. (1989),C5-C90; Y. Takahshi, K. Tsuda, K. Sugiyama, H. Minoura, D. Makino and M. Tsuiki J.Chem.Soc.Far.Trans. I (1981),**77**,1051
10. J.J. Brinker et al. J.Non.Crystal. Solids (1990),**121**,294; M.Guglielmi and S. Zenezini J.Non.Crystal Solids (1990),**121**,303; D. Meyerhofer J.Appl.Phys. (1978),**49**,3993
11. M. Born and E. Wolf 'Principles of Optics' (1980) Pergamon 6th.edn.; H. Alius and R. Schmidt Rev. Sci. Instrum. (1990), **61**,1220
12. J.W. Strutt Brit.Assoc.Report (1887),585; Edin. Trans. (1886),**XXXIII**,157;

Philo.Mag. (1871),**XLII**,81

13. R. Messner Z.Nachr. (1943),**4**,253; K. Hammer Z.Tech.Phys. (1943),24,169; B.E.Yoldas and T.W. O'Keeffe Appl.Optics (1979),**18**,3133
14. B.E. Yoldas Appl.Optics (1980),**19**,1425
15. M.S.V. Vong and P.A. Sermon Thin Solid Films (in press)
16. K. Konno, H.Inomata, T. Matsunaga and S. Saito J.Chem.Soc.Jap. (1994),**27**,134
17. P.A. Sermon, Y. Wang and M.S.W. Vong J.Coll.Interf.Sci. (1994),**168**,327
18. Y. Sun, P.A. Sermon and M.S.W. Vong Thin Solid Films (1996),**278**,135
19. M.S.W. Vong, P.A. Sermon and Y. Sun Proc. 1st Annual Int. Conf. Solid State Lasers for Application to Inertial Confinement Fusion, Monterey SPIE (1995), **2633**,446
20. A. Bertoluzza, C. Fagnano, M.A. Morelli, V. Gottardi and M. Guglielmi J.Non.Crystl.Solids (1982),**48**,117
21. B. Humbert, A. Burneau, J.P. Gallas and J.C. Lavalley J.Non.Crystl. Solids (1992),**143**,75; A. Burneau, B. Humbert, O. Barres, J.P. Gallas and J.C.Lavalley Adv.Chem.Ser. (1994),**234**,199
22. B. Humbert J.Non.Crystal. Solids (1995),**191**,29
23. M.M. Lynam, J.E. Kilduff and W.J.Weber J.Chem.Educa. (1995),**72**,80; K. Kamegawa and H Yoshida Bull.Chem.Soc.Jap. (1990), **63**,3683
24. C.H. Giles, D.C. Havard, W. McMillan, T. Smith and R. Wilson ..267
25. F.Mahmood, K.Holland and P.A. Sermon ..(unpublished data)
26. T.M.Harris and E.T.Knobbe J.Mater.Sci.Lett. (1996),**15**,153

THE PRESENCE OF "SHADOWED" MACROPORES IN THE POROUS NETWORK OF AN ACTIVATED ALUMINA

M.S. Goldstein and J.D. Carruthers

CYTEC Industries Inc.
Stamford Research Laboratories,
1937 W. Main St., Stamford,
CT 06904-0060
U.S.A.

1 INTRODUCTION

In the design of a catalyst, much thought is given to the type of porous structure the catalyst must possess for maximum performance in a given application. The catalyst must display a sufficient surface area for the dispersion of the active phase and that area is often contributed by a highly porous support. Catalysts used in Petroleum and Petrochemical applications undergo some level of deactivation usually by carbonaceous by-products building up in the porous structure generating a condition referred to as pore pluggage. The network of pores that provide active surface for the various chemical reactions eventually becomes blocked and the catalyst must be regenerated or replaced(1).

The importance of pore structure increases as the molecular size of the reactants reach close to the size of the pores. Diffusion limitations then outweigh reaction kinetics. This is especially true for catalysts processing heavy oils(2,3). In this instance, the diffusion limitation is related to the pore "throats"(4) in the structure. Coke deactivation in the catalyst is more related to the pores behind the "throats"(5).

Mercury porosimetry and nitrogen adsorption/desorption porosimetry are often used to investigate the porous structure of a catalyst. Mercury porosimetry, in particular appears to provide a quick understanding of the "throat" structure. However, as pointed out by Conner et al.(6), since mercury invades the particle from the outside in, not all of the pore "throats" of a particular size range in the catalyst can be accessed by the mercury at any time. Some are "shadowed" because they are not accessible until a pore to which the throat is connected has been filled. Even large throats in the interior of the particle are not intruded by mercury until the adjoining pores have been penetrated by the mercury.

This paper describes the characterization of two alumina substrates which were produced from the same alumina powder, but with differing subsequent processing to bring about changes in the pore structure.

2 EXPERIMENTAL

Nitrogen porosimetry analyses were obtained using a Quantachrome Autosorb-6 Automated Nitrogen Porosimeter. Porosity calculations involved a BJH (7) analysis using t-data of de Boer (8) from adsorption and desorption isotherms at 77.4 K. Samples were outgassed in vacuum (1.33 Pa) at 200 deg. C for 1 hour.

Mercury porosimetry analyses were generated using a Quantachrome Autoscan-60 mercury porosimeter using approximately 1.0 g. of alumina support. Samples were calcined in a muffle furnace in air at 300 deg.C followed by outgassing in vacuum (1.33 Pa) for 1 hr. at room temperature. An arbitrary mercury contact angle of 140 degrees and a surface tension of 0.48 N/m(480 dynes/cm) were used to calculate pore size distribution data from the mercury intrusion-extrusion curves.

Helium densities were measured using a Quantachrome Stereopycnometer using approximately 1 g. of sample.

Scanning Electron Microscopy images were obtained using a Zeiss DSM 9820 Gemini FEG-Scanning Electron Microscope. Samples were prepared by cross-sectioning the alumina support and mounting on carbon tape followed by carbon coating.

3 RESULTS

Intrusion and extrusion mercury porosity curves are shown in Figure 1 for the two alumina substrates. Both substrates exhibit very little pore filling in pores larger than about 30 nm diameter. All of the porosity resides in pores of 8 - 10 nm diameter, with sample IN-697 showing almost 0.3 cc/g pore volume greater than that for sample IN-729. Both extrusion curves show hysteresis and both show mercury entrapment (the lack of closure of the intrusion-extrusion curves). Once again, the entrapment for IN-697 is almost double that of IN-729.

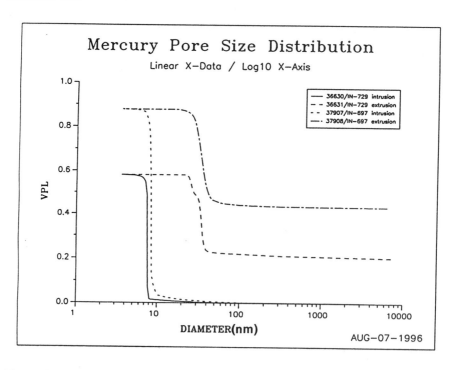

Figure 1 *Mercury Pore Size Distribution Analyses for IN-729 and IN-697 aluminas.*

Nitrogen isotherms and Pore size distributions are shown in Figures 2 and 3 and the data compared in Table 1.

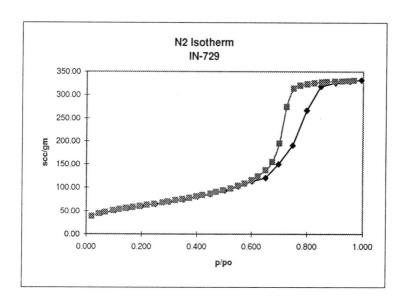

Figure 2a *Nitrogen Adsorption and Desorption isotherm for IN-729 Alumina.*

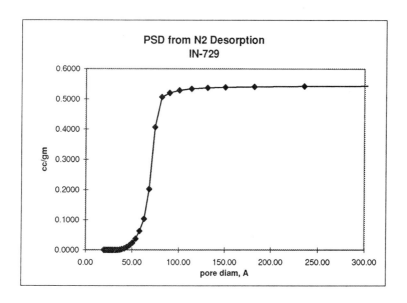

Figure 2b *Pore Size Distribution from Nitrogen Desorption Isotherm : IN-729.*

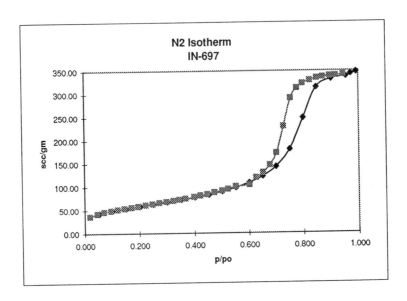

Figure 3a *Nitrogen Adsorption & Desorption Isotherm for Alumina IN-697*

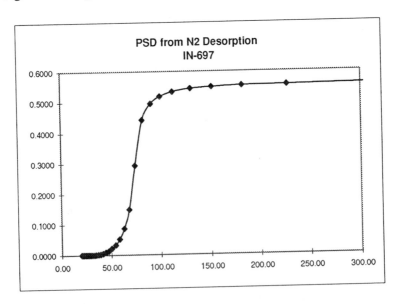

Figure 3b *Pore Size Distribution from the Nitrogen Desorption Isotherm : IN-697*

Table 1 *Porosity Characteristics for Alumina Substrates IN-729 & IN-697*

Property	IN-729	IN-729	IN-729	IN-697	IN-697	IN-697
SA(N$_2$) m^2/g	224	222	219	228	223	214
PV(Hg)60K cc/g	0.57	0.58	0.58	0.85	0.83	0.88
PV(N$_2$) cc/g	0.51	0.51	0.51	0.54	0.52	0.54
PV(H$_2$O) cc/g		0.57			0.78	
CBD g/cc		0.71			0.53	
Helium Density g/cc			3.53			3.46

(Three porosimetry analyses are reported for three samples of each alumina)

It is immediately apparent that the nitrogen data reveal almost identical pore volumes for the two aluminas, at about 0.51 - 0.54 cc/g, comparable with the mercury pore volume for IN-729. Water pore volume (PV(H$_2$O)) measurements indicate pore volumes comparable with the mercury pore volume measurements for both samples and similarly the compacted bulk densities (CBD) are appropriate for the water pore volumes reported but not the nitrogen pore volumes. Helium densities are almost identical indicating, as expected, that the aluminas do not exhibit sealed pores within the structure; helium is able to access essentially all of the pore volume on both samples.

4 DISCUSSION

A significant difference in nitrogen and mercury pore volumes in a sample can usually be explained by the presence of very large pores (larger than 100 nm) which are only filled, with difficulty, at pressures close to saturation in the nitrogen adsorption experiment. When successfully detected, the adsorption curve sweeps up near saturation and desorption proceeds by rapid nitrogen evolution and an almost vertical drop in the desorption curve. However, when samples exhibit such phenomena, they usually reveal a similar pattern in the mercury intrusion curve. Very large pores begin filling at extremely low pressures and the intrusion curve moves away from the abscissa at calculated pore diameters which are quite large.

There is absolutely no evidence for the presence of such large pores in the mercury porosity intrusion curves for either of the two samples analyzed here. Rather, the intrusion curves do not begin to move from the x-axis until pressures are reached comensurate with pores sized near 30 nm diameter.

An explanation of these measurements can be suggested only through recourse to 3-dimensional pore network theory(9,10,11). One clue is provided by examination of the mercury extrusion (retraction) curves for the two samples. A volume of mercury is entraped in sample IN-697 following mercury extrusion which is of similar magnitude to the additional mesopore volume of IN-697 over sample IN-729. Such a result could be explained by the presence of large pores within the particles which are completely inaccessible to mercury except via smaller, mesopore 'throats'. The large macropores are 'shadowed' and can only be detected indirectly.

Fortunately, in this case, the shadowed macropores were of sufficient size as to be readily detected by Scanning Electron Microscopy. In Figures 4 and 5 Scanning Electron Micrographs are shown for the two samples. A high concentration of large, spherical voids of approximately 1000 - 10000 nm can be seen in the micrograph of IN-697 but are in much lower concentrations in the micrograph of IN-729. These are the pores which can

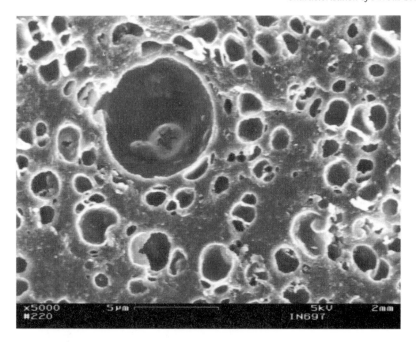

Figure 4 *Electron Micrograph of IN-697 (x5,000) showing a high concentration of large voids in the alumina structure*

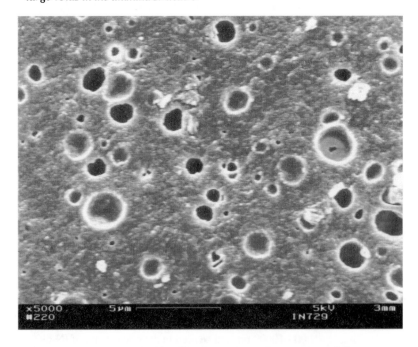

Figure 5 *Electron Micrograph of IN-729 (x5000) showing a much lower concentration of voids in the alumina structure compared to IN-697*

only be reached by mercury from the outside of the particle via the 8 - 9 nm diameter mesopores. Their size is so large that they are extremely difficult to detect by nitrogen porosimetry and are only indirectly measured via the degree of mercury entrapment indicated by the mercury extrusion curve.

5 CONCLUSIONS

Large macropores in an alumina-based catalyst support were not accessed by mercury during a mercury porosimetry experiment except via connecting mesopores. The resulting intrusion curve gave a result which implied a much larger volume of mesopores in the structure than were actually present. Nitrogen porosimetry and surface area measurements helped correct this illusion. Mercury extrusion curves and ultimately Scanning Electron Microscopy revealed the true nature of the material.

6 ACKNOWLEDGEMENTS

The authors wish to acknowledge the contributions of Dr. Eduardo Kamenetzky for the Electron micrographs, the porosity analyst, Mark Ryan and to Cytec Industries Inc. and Criterion Catalyst Company L.P., for permission to publish this work.

References

1. T.H. Fleisch, B.L. Meyers, J.B. Hall, G.L. Ott, J. Catal., 1984, **86**,147.

2. B.J.Smith, J. Wei, J. Catal.,1991, **132**, 41.

3. D.B.Dadyburjor, A.P. Raje, J. Catal., 1994, **145**, 16.

4. W.C. Conner, A.M. Lane, K.M. Ng, M. Goldblatt, J. Catal., 1983, **83**, 336.

5. J. Baumgart, Y. Wang, W.R. Ernst, J.D. Carruthers, J. Catal., 1990, **126**, 477.

6. W.C. Conner, A.M. Lane, J. Catal., 1984, **89**, 217.

7. E.P. Barrett, L.G. Joyner, P.H. Halenda, J. Amer. Chem. Soc., 1951, **73**, 373.

8. J.H. de Boer, ' The Structure and Properties of Porous Materials', 1958, Butterworths, London, p.68.

9. S. Reyes, K.F. Jensen, Chem. Eng. Sci., 1985, **40**, 9, 1723.

10. P.N. Sharratt, R. Mann, Chem. Eng. Sci., 1987, **42**, 1565.

11. R. Mann, P.N. Sharratt, Chem. Eng. Sci., 1988, **43**, 1875.

12. M. Day, I.B. Parker, J. Bell, R. Fletcher, J. Duffie, K.S.W. Sing, D. Nicholson, Studies Surface Science & Catalysis, 1994, **87**, 225.

NUCLEAR MAGNETIC RELAXATION STUDY OF WATER ADSORBED IN MESOPOROUS ACTIVATED CARBONS

S. V. Mikhalovsky[*], V. A. Trikhleb[#] and V. V. Strashko[#]

* Department of Pharmacy, University of Brighton,
Brighton BN2 4GJ, U.K.

[#] Institute for Sorption and Problems of Endoecology,
32-34 Prospect Palladina, Kiev-142, Ukraine 252142

1 INTRODUCTION

It is a truism to say that activated carbons (AC) belong to the most difficult subjects of study. Obtaining data on the fate of species adsorbed by AC requires strenuous efforts and it is scarcely available. High resolution NMR technique is, perhaps, a unique method providing direct information about the mobility and environmental state of molecules in the adsorbed layer[1-3], porous structure of adsorbents[4,5] and structure of adsorption sites[6]. Particularly advantageous is the ability of NMR to study adsorbed molecules in wet samples and at the solid-liquid interface in partially or fully saturated porous media[7].

Despite a variety of nuclei which can be studied with NMR, proton magnetic resonance of adsorbed water molecules remains the most important object of investigation. Nuclear relaxation times, both longitudinal, T_1, and transverse, T_2, are very sensitive to changes in the molecular environment and mobility and thus affected by the presence of surface[4,8,9]. Unlike the rate of the longitudinal relaxation, the rate of the transverse relaxation, $R_2 = 1/T_2$, is very sensitive to the local gradients of the magnetic field[10], hence, T_2 reflects the relaxation processes in the nearest vicinity of the surface and provides information on the structure of the adsorbed layer and the nature of adsorbate-adsorbent interaction.

An interesting and important phenomenon is adsorption of proteins from aqueous solutions onto the solid surface. This process is arguably responsible for biocompatibility of foreign materials such as artificial organs and its understanding is far from comprehensive[11]. The role of water is usually neglected in the study of adsorption from the aqueous solutions because it is difficult to evaluate. NMR is probably the only method allowing to do this. Dynamic properties of water in protein solutions were studied using NMR relaxation technique[12,13]. Using the water molecule as a probe it appeared possible to make conclusions about the structure of proteins in solution. At least two different states of water were distinguished, that is protein bound water with lower mobility and bulk water. The same concept of two types of water is also used for explaining of enhanced proton relaxation rates in porous systems although the real mechanism of relaxation may be quite different from protein solutions[14].

---

* To whom correspondence should be addressed.

Study of the behaviour of bound water in the system 'protein solution - solid surface' could provide valuable information about the nature of protein - surface interaction in the adsorbed layer. The aim of this work was to determine the transverse relaxation time of water protons in the presence of polymer-pyrolysed activated carbons used for hemoperfusion (adsorptive blood purification)[15]. Relaxation effects were studied for pure water as well as for aqueous protein solutions from which proteins were either physically adsorbed or covalently bound to the carbon surface. Covalent binding of proteins allows synthesis of carbon-based immunoadsorbents with a potential for biomedical applications[16].

2 MATERIALS AND METHODS

2.1 Activated Carbons

Polymer-pyrolysed activated carbons were produced by pyrolysis of synthetic polymers and resins. If the polymer-precursor has a fixed mesoporous structure, the final AC can retain a very similar structure. Thus AC with an unusual pore size distribution were synthesised[17,18].

Commercially available mesoporous styrene-divinylbenzene copolymer containing and vinylpyridine-divinylbenzene copolymer were used for the synthesis of AC which were given names SCS and SCN, respectively[17,18]. Each copolymer contained 10% of a cross-linking reagent, divinylbenzene. Both SCN and SCS carbons were produced by step pyrolysis of the polymer-precursor at 350-900°C and 400-1100°C, respectively, and further steam activation at 800-950°C for 0.5-4 hours. Initial copolymers had spherical granules and the carbons obtained also had regular spherical granules with the main size of 0.5-1.6 mm (95% of particles). Initial polymers had well developed mesopores, and in the course of activation carbons retained these mesopores and also developed micropores. Polymer-pyrolysed carbons thus obtained are of high chemical purity which is controlled by the chemical quality of the polymers-precursors. Micro- and mesoporous structure of AC was determined from adsorption-desorption isotherms of benzene at $25 \pm 0.5°C$ (Table 1). Prior to measurements all samples were heated in vacuum for 24 hours at 400°C. Macro- and mesoporous structure of AC was studied by a mercury porosimetry technique using Pore Sizer 9300 (Micromeritics, USA). In the calculations of pore size distribution the pores were considered as cylindrical. True and bulk density of the samples were determined by picnometric method using mercury and benzene, respectively. The carbons studied did not contain macropores and had a distinctive peak in the mesopore region[17,18]. For SCN carbon the peak maximum corresponded to the radius of 35 nm, and for the SCS series it was at 12 nm.

2.2 Protein Adsorption

Bovine serum albumin, BSA (molecular weight 68 kD), Cohn fraction V, A4503 from Sigma, USA, was further recrystallised to obtain essentially globulin-free protein. HPLC-purified mouse γ-globulin, IgG (158 kD), was obtained from Palladin Institute of Biochemistry, Kiev, Ukraine. A water-soluble carbodiimide, CMC, was purchased from Sigma. In the experiments with physically adsorbed proteins 0.2% solution of protein was shaking with AC for 1 hour at room temperature before NMR measurements.

Table 1 *Some physical characteristics of polymer-pyrolysed AC**

Carbon Parameter	SCS1	SCS2	SCS3	SCS4	SCS5	SCS6	SCN
Burn-off, %	0	3.0	13	27	60	75	85
ρ, g/cm^3	0.54	0.57	0.53	0.49	0.41	0.36	0.20
d, g/cm^3	1.49	1.53	1.59	1.68	1.87	1.91	2.1
V_Σ, cm^3/g	0.38	0.60	0.73	0.87	1.3	1.5	1.9
V_{ma}, cm^3/g	0.10	0.09	0.10	0.11	0.11	0.12	0.12
V_{me}, cm^3/g	0.18	0.27	0.32	0.41	0.64	0.76	1.2
V_{mi}, cm^3/g	0.10	0.23	0.29	0.35	0.50	0.65	0.64
S_{me}, m^2/g	84	110	170	230	360	530	300
W_0, cm^3/g	0.091	0.21	0.27	0.35	0.41	0.39	0.72
X_0, nm	0.742	0.716	0.729	0.743	0.851	0.866	0.97
Σ_{mi}, m^2/g	124	290	370	470	480	460	660
r, nm	0.66	0.62	0.58	0.84	1.0	1.2	1.35

* ρ - bulk density, d - true density (from benzene adsorption), V_Σ - total pore volume, V_{ma} - macropore volume, V_{me} - mesopore volume, V_{mi} - micropore volume, S_{me} - mesopore surface area (from benzene adsorption).
Parameters of the micropore structure according to the Dubinin's micropore zone model: X_0 - half-width of the micropore slit, Σ_{mi} - geometric surface of micropores, r - radius of the circular micropore entry.

Covalent binding of proteins to the surface was carried out using a carbodiimide-coupling technique as described elsewhere[16].

2.3 Nuclear Magnetic Relaxation Measurements

N.m. relaxation measurements with SCS carbons were carried out on "Minispec P20" spectrometer (Bruker, Germany) at 30°C and 19.8 MHz. The ratio C_{H2O}/C_{AC} was varied by adding of AC granules to the same amount of water (0.2 ml) in the n.m.r. tube. To determine the spin-spin relaxation rate $R = T_{2\tau}^{-1}$, decay of the magnetisation amplitude A of the NMR signal was measured using standard sequence of 90°-τ-180° pulses. For good reproducibility of the results it is very important to keep the amount of water, C_{H2O}, constant which gives the same initial amplitude (A_0 = const. at $2\tau = 0$) in all experiments. For each experimental point 3 to 5 measurements were done and standard deviation was less than 5%.

N.m. relaxation of water with SCN was studied using a high resolution NMR spectrometer WP-100SY (Bruker, Germany) at 100 MHz with bandwidth 50 kHz. Relaxation of the transverse magnetisation versus time was measured by a Carr-Purcell (spin echo) pulse sequence[19]. Temperature in the sample was kept with the accuracy of ± 0.5°C, error in measuring the signal intensity was less than 10%. To decrease the contribution of bulk water in the transverse relaxation, a 'freezing-out' technique was used[6,20]. For this purpose relaxation measurements were carried out at temperatures below 273 K, the freezing point of water. The contribution of bulk ice to the NMR signal can be neglected; thus only the signal from unfrozen water in the adsorbed layer or in any other state affected by proximity to the surface is measured.

3 RESULTS AND DISCUSSION

3.1 Nuclear Magnetic Relaxation of Water Adsorbed on SCS Active Carbons[21]

As $A_{2\tau} = A_0 \exp(-2\tau R)$, then:

$$\log (A^a_{2\tau}/A^a_{2\tau}) - \log (A^f_{2\tau}/A^a_{2\tau})_0 = 0.434 \, \Delta R_{2i} \, 2\tau \qquad (1)$$

where indexes (f) and (a) refer to the protons of water without adsorbent and in the presence of adsorbent, respectively, and $\Delta R_{2i} = R^a_{2\tau} - R^f_{2\tau}$. According to the experimental conditions $\log (A^f_{2\tau}/A^a_{2\tau})_0 = 1$, hence,

$$\log (A^f_{2\tau}/A^a_{2\tau}) = 0.434 \, \Delta R_{2i} \, 2\tau \qquad (2)$$

and ΔR_{2i} can be determined from the slope of the straight line plotted in the co-ordinates: $\log (A^f_{2\tau}/A^a_{2\tau})$ vs. 2τ. There are four regions with different ΔR_{2i} (Fig.1). Taking $0.48 \, s^{-1}$ as $R^f_{2\tau}$ for pure water, transverse relaxation rates for the water in the presence of AC were calculated (Table 2 and Fig. 2). For one sample, SCS5, only three regions I, II and IV were found (Table 2). Taking into account that the transverse relaxation rate increases at shorter distances between nuclear spins and at closer distances to other paramagnetic centres such as paramagnetic surface of AC, one could come to a conclusion that different values of R_{2i} reflect different nature of water binding sites on the surface of AC. Thus region II with the shortest T_2 could be attributed to the water molecules inside narrow micropores where the protons are most close to the paramagnetic surface. The transverse relaxation time is also shorter for an ensemble consisting of smaller number of molecules rather than a big molecular cluster[8]. Both these factors (surface proximity and small molecular ensemble) are consistent with regard to porous adsorbents. Regions I and III could represent longer bridge structures of water adsorbed in mesopores and wide micropores, and region IV may arise from molecules adsorbed in macropores and on outer surface. In the latter case H_2O molecules are easier available for exchange with bulk water which results in increased T_2. To obtain more detailed information about the binding efficiency of AC surface sites, we used Hill's method. It was originally suggested for calculation of binding constants of co-operative interaction between small molecules (ions) and macromolecules[22]. To validate this approach, the following conditions were accepted. 1. For the 'AC surface - water' interaction the adsorbent can be considered as a single large molecular matrix. AC surface is isotropic and continuous. 2. This interaction can be both local and co-operative. 3. For water adsorption disturbances caused by possible phase transition are negligible compared to the changes in proton spin exchange rate in different steric and chemical environment. 4. Molecular diffusion in bulk water is almost the same as in the n-th layer of adsorbed water, i.e. rotational correlation time is approximately constant. An effective binding constant K_n could be formally expressed as:

$$K_n = [E] [H_2O]^n / [E(H_2O)_n] \qquad (3)$$

where [E] is a formal concentration of the isotropic continuous matrix of adsorbent, n is a number of molecules bound by a co-operative site. Defining the fraction of surface occupied by water θ as:

$$\theta = [E(H_2O)_n] / \{[E(H_2O)_n] + [E]\} \qquad (4)$$

n and K_n were determined according to the Hill's equation (Fig. 2, Table 3):

$$\log\{\theta/(1 - \theta)\} = n\log C_{H2O} - \log K_n \qquad (5)$$

θ is determined from the water concentration dependence of ΔR_{2i} (Fig. 3).

Assuming that n should be an integer, the region I probably corresponds to the bridge structure of the adsorbed trimeric H_2O molecules with a significant contribution from polar interactions strengthened by inter- and intramolecular hydrogen bonds (3-4 molecules per site). In the region II adsorbed H_2O molecules may form cluster structures consisting of 4-5 molecules. pK_n values are similar to those in the region I. Binding of H_2O to the surface is weaker than in the region I but hydrogen bonds play a more important role within the clusters. These parameters can be attributed to the adsorption in micropore volume. The region III with n ≅ 1.0, pK_n ≈ 0.95 and lower relaxation rate could be ascribed to the H_2O molecules adsorbed in mesopores. In such an environment adsorbed water molecules are sheltered from the bulk and can form a monolayer. Respectively, the region IV can be interpreted as adsorption in macropores and on the outer surface of AC, as, although n ≅ 1.0, pK_n is lower (0.5÷0.6), reflecting diminishing interaction of H_2O with matrix and more intensive interaction with water molecules in the bulk.

It is interesting that SCS5 and to a lesser extent, SCS6 differ from the rest of SCS samples by their parameters of water binding. This behaviour has not been completely understood as yet. It is probably related to the fact that in SCS carbons different processes dominate at lower degree burn-off (up to 40%) and higher degree of burn-off. Activation at low burn-off developed initially microporosity which was accompanied also by a mesopore volume and surface increase. From the analysis of the microporous structure using the Dubinin's micropore zone model[23] it follows that activation at 0 - 40% burn-off led to the narrowing of the micropore slit X_0. Further activation caused rapid widening of the micropore width. Micropores partly transformed into supermicropores and mesopores. Similar changes happened with the micropore volume and the radius of the circular micropore entry. Changes in the geometric surface also

Table 2 *The transverse relaxation rates of water protons, R_{2i} s^{-1}, in the presence of SCS carbons at C_{H2O}/C_{AC} = 0.15 mol/g*

Region	SCS1	SCS2	SCS3	SCS4	SCS5	SCS6
I	1.73	1.76	1.73	1.86	1.91	1.83
II	2.11	2.11	2.10	2.21	2.21	2.11
III	1.51	1.51	1.53	1.53	-	1.53
IV	1.06	1.01	1.02	1.03	1.02	1.03

Table 3 *Parameters of water binding obtained from nuclear magnetic relaxation measurements for systems: 'SCS - water'*

Sample Region	SCS1		SCS2		SCS3		SCS4		SCS5		SCS6	
	n	pK_n	n	pK_n	n	pK_n	n	pK_n	n	pK_n	n	pK_n
I	0.77	0.61	0.74	0.64	0.76	0.68	0.76	0.69	1.04	0.84	0.81	0.74
II	1.96	1.01	1.97	1.02	1.99	1.04	2.01	1.08	2.48	1.31	2.09	1.19
III	0.94	0.91	0.95	0.92	0.94	0.94	0.97	0.95	-	-	1.03	0.97
IV	1.07	0.54	1.07	0.60	1.09	0.61	1.08	0.62	1.11	0.61	1.08	0.64

reflected the microstructural changes in course of activation: micropore narrowing was accompanied by a sharp increase of the surface area at low burn-off. At higher burn-off this parameter showed a tendency for stabilisation (Table 1). With an increase of burn-off the carbon lattice becomes more ordered. The evidence for this process was given by an increase of the picnometric density. The macropore volume altered insignificantly in the whole range of burn-off studied.

3.2 Nuclear Magnetic Relaxation of Water Adsorbed on SCN from Protein Solutions

T_2 for water adsorbed on SCN carbon is relatively low in the studied temperature range 200 - 250 K. This transverse relaxation time arises only from non-frozen molecules. They do not exchange with ice in the bulk and in the pores and slowly exchange between themselves. Existence of non-frozen water at sub-zero (Celsius) temperatures is due to the steric reasons[10]. Spatial distribution of water molecules inside narrow pores does not correspond to the ice lattice and they exist in a pseudoliquid state. The molecules inside pores form clusters. Exchange of molecules between clusters occurs via diffusion mechanism[6,10]. Assuming an Arrhenius type of T_2 dependence on temperature and plotting it in the co-ordinates: ln T_2 vs. (1/T), several parts can be clearly distinguished in Fig. 4 for the systems studied. There is not enough data yet to identify the structure of water and its environment in each case but different exchange behaviour indicates presence of several types of water clusters in the porous matrix similarly to the results discussed above. Introduction of a water-soluble carbodiimide CMC brought significant changes in the behaviour of both 'SCN-water-BSA' and 'SCN-water-IgG' systems. Interestingly, ln T_2 vs. (1/T) is linear without breaking points in the whole temperature range for the systems 'SCN-water-BSA' and 'SCN-water-CMC-IgG'. It possibly means that in these systems protein molecules adsorbed (BSA) or covalently attached (IgG) to the surface displace water molecules. BSA is a molecule with a flexible structure. Physical adsorption of albumin on porous AC is highly irreversible because of its multi-site binding[24]. Contrary to BSA, IgG has a rigid structure which is not affected significantly by the surface. Hence, the latter cannot substitute for pre-adsorbed water in micro- and narrow mesopores. Amino acid residues of physically adsorbed BSA penetrate pores which are much smaller than the whole protein molecule (ellipsoid 12 x 4 nm) irreversibly fixing it on the surface. Added CMC binds protein molecules via amino groups to the activated carboxylic surface functional groups. On the other hand, carbodiimide cross-links protein molecules via their own amino and carboxylic groups. Due to the covalent binding and cross-linking BSA loses its high flexibility. In the case

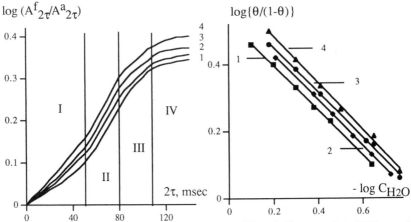

Fig. 1 *The differential NMR relaxation curves in "SCS-water" system at $C_{H2O}/C_{AC} = 0.075$ mol/g. Number corresponds to the SCS sample*

Fig. 2 *Characteristic Hill's graphs in "SCS-water" system. Region I. Number corresponds to the SCS sample*

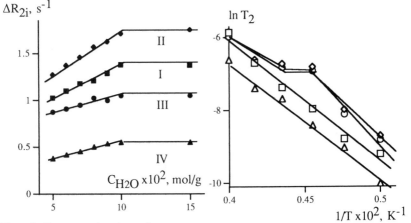

Fig. 3 *Concentration dependence of the relaxation rate in the system 'SCS4-water'*

Fig. 4 *Temperature dependence of ln T_2 for: □ - SCN-BSA, ◇ - SCN-IgG, ○ - SCN-CMC-BSA, △ - SCN-CMC-IgG*

of IgG a multi-site attachment of protein to the surface via CMC-activated surface groups is more significant than cross-linking in the solution and it leads to the substitution for the adsorbed water molecules. Adding of CMC to the system increases the temperature of freezing water in the pores by 10 - 15 K compared to the pure 'SCN-water-protein' systems. More accurate picture of the process would be possible to draw after further investigation which has been started recently.

Acknowledgements

This work was done with the support of International Science Foundation Grant U4Q200 and INTAS Grant 94-3033.

References

1. E. Almagor and G. Belfort, *J. Colloid Interface Science*, 1978, **66**, 146.
2. J. A. Glasel and K. H. Lee, *J. Am. Chem. Soc.*, 1974, **96**, 970.
3. R. M. Pearson, *J. Catalysis*, 1971, **23**, 388.
4. D. P. Gallegos, K. Munn, D. M. Smith and D. L. Stermer, *J. Colloid Interface Science*, **1986**, 119, 127.
5. R. S. Drago, D. C. Ferris and D. S. Burns, *J. Am. Chem. Soc.*, 1995, **117**, 6914.
6. V. V. Turov, R. Leboda, V. I. Bogillo and J. Skubiszewska-Zieba, *Langmuir*, 1995, **11**, 931.
7. J.-P. Korb and F. Devreux in: 'Proton conductors' (Series 'Chemistry of Solid State Materials'), P. Colomban, Ed., Cambridge University Press, 1992, Chapter 28, p. 432.
8. D. J. E. Ingram, 'Spectroscopy at Radio and Microwave Frequencies', 2nd Ed., Butterworths, London, 1967.
9. K. Overloop and L. Van Gerven, *J. Magn. Res., Ser. A*, 1993, **101**, 147.
10. V. V. Mank and N. I. Lebovka, 'Nuclear Magnetic Resonance Spectroscopy in Heterogeneous Systems' (Russ.), Naukova Dumka Publ., Kiev, Ukraine, 1988.
11. J. D. Andrade and V. Hlady, in: 'Blood in Contact with Natural and Artificial Surfaces', E. F. Leonard, V. T. Turitto and L. Vroman, Eds., Annals of the NY Academy of Sciences, Vol. 516, New York, 1987, p. 158.
12. R. Bruschweller and P. E. Wright, *Chem. Phys. Letters*, 1994, **229**, 75.
13. M. Fukuzaki, N. Miura, N. Shinyashiki, D. Kurita, S. Shioya, M. Haida and S. Mashimo, *J. Phys. Chem.*, 1995, **99**, 431.
14. P. Gillis, S. Peto and R. N. Muller, *Magn. Res. Imaging*, 1991, **9**, 703.
15. V. V. Strelko and S. V. Mikhalovsky, in: Xth World Congress of the International Society for Artificial Organs. Abstract Volume, Taipei, Taiwan, 1995, p. 77.
16. S. V. Mikhalovsky, V. V. Strelko, T. A. Alexeyeva and S. V. Komissarenko, *Biomat., Art. Cells, Art. Org.*, 1990, **18**, 671.
17. N. T. Kartel, A. M. Puziy and V. V. Strelko, in: 'Characterization of Porous Solids II'. F. Rodriguez-Reinoso, et al., Eds., Elsevier, Amsterdam, 1991, p. 439.
18. A. Bagreev, V. Strelko, J. Dentzer and J. Lahaye, in: 22nd Biennial Conference on Carbon, Extended Abstracts, San Diego, CA, 1995, p. 442.
19. H. Y. Carr and E. M. Purcell, *Phys. Rev.*, 1954, **94**, 630.
20. G. Belfort, J. Scherfig and D. O. Seevers, *J. Colloid Interface Sci.*, 1974, **47**, 106.
21. V. V. Strashko, V. A. Trikhleb and S. V. Mikhalovsky, in: Carbon'94. Extended Abstracts. Granada, Spain, 1994, p. 394.
22. G. M. Barrow, 'Physical Chemistry for the Life Sciences', McGraw-Hill, New York, 1974.
23. S. J. Gregg and K. S. W. Sing, 'Adsorption, Surface Area and Porosity', 2nd Ed., Academic Press, London - New York, 1982.
24. S. Mikhalovsky, M. Levchenko, I. Larionova and V. Strelko, in: Proceedings of 'Carbon'92' International Conference on Carbon, Essen, Germany, 1992, p. 268.

ANALYSIS OF THE TOPOLOGICAL AND GEOMETRICAL CHARACTERISTICS OF THE PORE SPACE OF PERMEABLE SOLIDS USING SERIAL TOMOGRAPHY, MERCURY POROSIMETRY, AND THEORETICAL SIMULATION

by

Christos D. Tsakiroglou and Alkiviades C. Payatakes
Institute of Chemical Engineering and High Temperature
Chemical Processes, ICE/HT-FORTH
and
Department of Chemical Engineering, University of Patras,
GR-26500 Patras, Greece

ABSTRACT

A method is developed for the determination of the topological and geometrical parameters of the pore space of porous solids, in terms of chamber-and-throat pore networks. Several parameters, such as the chamber-diameter distribution and the mean coordination number of the pore network, are obtained by applying serial sectioning analysis to double porecasts of the porous solid. This information is used in the computer-aided construction of a chamber-and-throat network which is to be used for further analysis. Mercury porosimetry curves can be fitted to either 2-parameter or 5-parameter non-linear analytic functions. A simulator of mercury intrusion/retraction (incorporating the results of serial tomography) in conjuction with the experimental mercury porosimetry curves of the porous solid are used to estimate the throat-diameter distribution, the correlation coefficients of pore-sizes and the parameters characterizing the pore-wall roughness. Estimation of the parameter values is performed by fitting the simulated mercury porosimetry curves to the experimental ones in terms of the macroscopic parameters of the analytic functions. The validity of the pore space characterization can be evaluated through the correct prediction of macroscopic properties such as absolute permeability, formation factor, effective diffusivity etc. A demonstration of the method in the case of a typical sandstone is given.

1   INTRODUCTION

The analysis of transport phenomena in porous media (e.g. one-phase and two-phase flow in oil-bearing porous rocks, gas diffusion in catalysts, soil drainage, capillary rise etc) requires detailed information about the pore structure. Every method of characterization of the pore structure of porous materials comprises three elements: (a) one or more experimental techniques, (b) a parametric geometrical model that represents the pore structure of the material and (c) a theoretical procedure for the determination of the values of parameters of the model which give the best fit to the experimental data. The applicability of the various techniques depends mainly on the range of the pore sizes. The choice is usually made based on experience gathered from the analysis of materials with similar pore structure and size ranges.

The analysis of the pore structure of oil-bearing rocks of sedimentary origin (sandstones, limestones, dolomites etc) is of prime importance in reservoir engineering. From studies of porecasts[1] it has been found that the majority of sedimentary rocks has a pore structure which can be considered as network of large pores ("chambers") which are interconnected through narrow pores ("throats"). Depositional and diagenetic processes taking place during the formation of these rocks alter geometrical properties (e.g. pore shapes and sizes) significantly. They also alter, to some extent, topological properties (e.g. the genus of the pore network) as well as pore scale heterogeneities and pore-size correlations. The microstructural characteristics of sedimentary rocks strongly affect their petrophysical properties and the efficiency of multiphase transport processes[2] such as secondary and tertiary oil recovery floods, etc. Thus, there is a strong need for the development of reliable methods of analysis for the characterization of their pore structure. A detailed and sophisticated analysis of the pore structure of rocks can be attained with the combination of serial tomography of double porecasts and mercury porosimetry, using appropriate theoretical algorithms for the interpretation of the experimental data.

In the present work, a systematic procedure for the characterization of the pore structure of permeable porous media such as sandstones, by deconvolving the mercury intrusion / retraction curves is developed. The procedure of combining the analysis of data from serial tomography with the analysis of mercury porosimetry data using a recently developed mercury intrusion/retraction simulator is demonstrated in the case of a Grey Vosgues sandstone. The predictability of the absolute permeability of the porous sample, which is under investigation, is used for the evaluation of the validity of parametervalues.

2   SERIAL SECTIONING ANALYSIS OF PORECASTS

A   Experimental technique

The experimental procedure of serial tomography as applied on a porous sample consists of the following stages[3,5]. (a) construction of a double porecast, (b) sequential grinding and polishing the surface of the porecast, (c) acquisition of the image of each polished cross section under the microscope and (d) digitization of the image and storage of the data in a computer.

(a) The sectioning and the microscopical investigation of porous rocks becomes flexible and accurate with the construction of porecasts[1]. First, the porous sample is evacuated and then the pore space is impregnated with heated liquid Wood's metal (T=85-90°C), by applying a sufficiently high pressure (P=100-150 atm) for a period of 2-4 hrs. Then, the system is cooled at room temperature so that Wood's metal is solidified. The host rock is dissolved with hydrofluoric acid and the pore cast is impregnated with a suitable epoxy-resin (ERL-4206) coloured with Oracet Red. The whole system is heated to 60-70°C and is kept at this temperature for 8-12 hrs, until polymerization of the resin is completed and a double porecast is obtained.

(b) A small piece of the porecast is sliced and is properly embedded in araldite. The plastic material with the sample near its centre is glued on a microscope glass slide. The production of polished successive sections is carried out using a Minimet grinding/polishing machine (BUEHLER). The sequential removal of varying amounts of material is attained using 200-600 grit Carbimet paper and polishing cloth. The speed, the load and the operating time of the equipment are properly regulated. The mean variation of the sample thickness is measured with a micrometer after several (4-6) serial sections have been removed.

(c) Each new polished surface of the sample is placed under an optical microscope (ZEISS), its image is taken with a videocamera (SONY), and is sent to an image analysis card (Imaging Technology) which is installed in the host computer. Alignment of the successive sections is attained by mounting the sample on a plastic holder which is permanently glued on the base of the microscope that is allowed to move only upwards and downwards.

(d) The digitization of images is carried out automaticaly through a software package which uses as parameters the length of resolution of the picture (number of pixels / area) and the breakthrough grey-level between Wood's metal (which appears white) and resin (which iappears red). The digitized images are stored for further analysis.

The technique was applied to a sample GV-1 of a Grey Vosgues sandstone and data of 44 serial sections were obtained. The active surface area of the porecast available to microscopical analysis was 2.04mmX2.04mm. The mean distance between two successive sections was 7.3μm and the resolution of digitization was 482X500 pixels per image. Two successive sections of the sandstone studied are shown schematically in Fig.1.

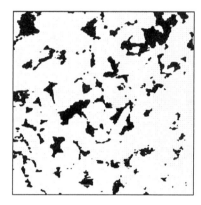

Figure 1 Two successive sections of a porecast of Grey Vosgues sandstone (2mmX2mm). The distance separating the two sections was about 7μm.

B. Pore-chart analysis of serial section data

For the determination of the topological and geometrical properties of porous media (e.g. connectivity) an algorithm has recently been developed[4] which reconstructs the 3-D pore network from data of 2-D pore features of serial sections (x- and y- coordinates of two-dimensional pore sectionals). Primarily, the algorithm is used for the determination of genus which is a direct measure of the connectivity of a multiply connected closed surface. The genus G is defined[3] as the number of (non-self-intersecting) cuts that may be made upon the surface without separating it into two disconnected parts and is equal to the first Betti number p_1 of its deformation retract, which is a bond-to-node network topologically equivalent to the studied surface. In general, if a network is composed of a number p_0 of disconnected subnetworks (zeroth Betti number) then

$$p_1 = b - n + p_0 \qquad\qquad (1)$$

where b and n are the number of bonds and nodes respectively. In addition to the genus, the algorithm[4] reconstructs a chamber-and-throat network that has the same topology with the actual pore network and approximates its geometry. Pore channels presenting small changes of the cross section along their length are identified as "throats" whereas all other features are identified as "chambers". From the exact calculation of the volume and the length of each throat as well as of the volume of each chamber, the throat-"equivalent cylindrical diameter" and the chamber-"equivalent spherical diameter" distributions are detrmined. Some other topological (distribution of the coordination number) and statistical (pore size correlations) parameters are also obtained with the same algorithm with an accuracy that depends on the degree of resolution of the serial sectioning experimental technique[4]. For our purposes we will pay attention mainly to the specific genus (genus/volume) and the chamber-size distribution as they are obtained by reconstructing the 3-D pore space.

The results of the application of the algorithm to data of 44 serial sections of GV-1 are given in Fig.2a,b.Using the Marquardt parameter estimation method[5], the CSD was fitted to a bimodal distribution composed of two lognormal density functions (c_f is the contribution fraction of the distribution of larger sizes).

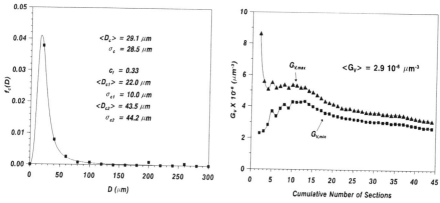

Figure 2. Results of serial sectioning analysis of a porecast of Grey Vosgues
(a) Chamber-diameter distribution   (b) Specific genus (genus per volume)

3   ANALYTIC DESCRIPTION OF MERCURY POROSIMETRY CURVES

In the simplest case, mercury porosimetry curves can be fitted with 2-parameter non-linear functions of the following form

(a)   Mercury intrusion curves

$$S = \frac{\left(1+a_I\right)e^{-b_I/P}}{1+a_I\,e^{-b_I/P}} \; , \; b_I > 0 \qquad\qquad S = \frac{\left(1+a_I\right)}{1+a_I\,e^{-b_I/P}} \; , \; b_I < 0 \qquad\qquad (2)$$

(b)   Mercury retraction curves (S_0 is the residual mercury saturation and is provided by the experiment)

$$S = \frac{S_0 +\left(1+a_R -S_0\right)e^{-b_R/P}}{1+a_R\,e^{-b_R/P}} \; , \; b_R > 0 \qquad\qquad S = \frac{\left(1+a_R -S_0\right)+S_0 e^{-b_{RI}/P}}{a_R +e^{-b_R/P}} \; , \; b_R < 0 \quad (3)$$

The median pressures P_I , P_R and the mean slopes s_I , s_R of the curves are defined at saturation values $S=1/2$ for mercury intrusion and $S=(1+S_0)/2$ for mercury retraction respectively and they are given below

(a)   Characteristic values of mercury intrusion curves

$$P_I = \frac{b_I}{\ln\left(a_I +2\right)} \qquad s_I = \frac{\left(a_I +2\right)\ln^2\left(a_I +2\right)}{4b_I\left(a_I +1\right)} \qquad , \qquad b_I > 0 \qquad (4a)$$

$$P_I = \frac{b_I}{\ln\left(\dfrac{a_I}{2a_I +1}\right)} \qquad s_I = -\frac{\left(2a_I +1\right)\ln^2\left(\dfrac{a_I}{2a_I +1}\right)}{4b_I\left(a_I +1\right)} \qquad , \qquad b_I < 0 \qquad (4b)$$

b.   Characteristic values of mercury retraction curves

$$P_R = \frac{b_R}{\ln\left(a_R +2\right)} \qquad s_R = \frac{\left(1-S_0\right)\left(a_R +2\right)\ln^2\left(a_R +2\right)}{4b_R\left(a_R +1\right)} \qquad , \qquad b_R > 0 \qquad (5a)$$

$$P_R = -\frac{b_I}{\ln\left(a_I +2\right)} \qquad s_R = -\frac{\left(1-S_0\right)\left(a_R +2\right)\ln^2\left(a_R +2\right)}{4b_R\left(a_R +1\right)} \qquad , \qquad b_R < 0 \qquad (5b)$$

The parameter values (a_I, b_I) and (a_R, b_R) of intrusion and retraction curves respectively are estimated seperately for each sample using a properly adjusted unconstrained Marquardt method of non-linear least squares[5]. The experimental curves of a limestone and a sandstone are compared to the corresponding ones given by the analytic functions in Fig.3a,b. Although the 2-parameter functions fit satisfactorily to curves of narrow pressure range (limestone C-1, Fig.4a) they are unable to fit curves of wide pressure range (sandstone S-1, Fig.3b). In order to improve the fitting to mercury porosimetry curves 5-parameter functions, resulting from the linear combination of two 2-parameter functions, are used below

(a)   Intrusion curve

$$S = c_1 S_1\left(P; a_{11}, b_{11}\right) + (1 - c_1) S_2\left(P; a_{12}, b_{12}\right) \tag{6a}$$

$$S_j\left(P; a_{1j}, b_{1j}\right) = \frac{\left(1 + a_{1j}\right) e^{-b_{j1}/P}}{1 + a_{1j}\, e^{-b_{1j}/P}} \qquad j = 1, 2 \tag{6b}$$

(b)   Retraction curve

$$S = c_R S_1\left(P; S_0, a_{R1}, b_{R1}\right) + (1 - c_R) S_2\left(P; S_0, a_{R2}, b_{R2}\right) \tag{7a}$$

$$S_j\left(P; S_0, a_{Rj}, b_{Rj}\right) = \frac{S_0 + \left(1 + a_{Rj}\right) e^{-b_{Rj}/P}}{1 + a_{Rj}\, e^{-b_{Rj}/P}} \qquad j = 1, 2 \tag{7b}$$

The characteristic values of median pressure and mean slope of each component curve are given by Eqs.4,5. The fitting to the curves of sandstone S-1 is improved substantially with the 5-parameter function as shown in Fig.3b. The description of mercury porosimetry curves through the parameters of specific analytic functions makes easier and more efficient any comparison between experimental curves of different porous rocks, or between experimental and theoretical ones provided by simulation. Besides, qualitative information about the range of pore sizes can be obtained directly with reference to the median pressures and the mean slopes of analytic functions whereas the coefficients c_I, c_R are roughly related to the relative contributions of macropores (interconnected pore network) and micropores (fractal roughness features) on the total pore volume.

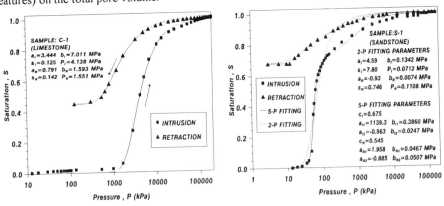

Figure 3 Fitting of mercury porosimetry curves to 2- and 5-parameter functions
(a) Fitting to curves of limestone   (b) Fitting to curves of sandstone

4   SIMULATION OF MERCURY INTRUSION / RETRACTION IN
     CHAMBER-AND-THROAT NETWORK MODELS

A detailed description of the theoretical simulator of mercury intrusion and retraction that is used in the present work is reported elsewhere[6-8]. In principle, the porous medium is modelled as a 3-D primary network of spherical pores (chambers) which are interconnected through long cylindrical pores (throats), having as parameters CSD (Chamber-diameter distribution), TSD (Throat-diameter distribution), the primary porosity ε_p, and $<c_n>$ (mean coordination number) or G_V (mean specific genus). The assignment of sizes to chambers and throats can be done in such a way that an uncorrelated, a c-t correlated or a c-c & c-t correlated pore network is obtained[7]. Pore-wall roughness features are added to the pore network using an approach based on ideas of fractal geometry[8]. The roughness features are right triangular prisms for the throats and right circular cones for the chambers of the primary network. With regard to the roughness model[8], some modifications have been made here: (a) the number of the layers of roughness features extends to infinity (fractal structure of infinite specific surface area), (b) the parameters required for the determination of the roughness of such a network are the characteristic number of features per linear dimension of the previous layer n_r and the surface fractal dimension D_s. The common angle of sharpness is different for throats w_t and chambers w_c and is related to the other parameters so that the surface fractal dimension of the pore network D_s is equal to the self-similar surface fractal dimensions of the roughness of throats D_{st} and chambers D_{sc}, which in turn are given by

$$D_{st} = \frac{\ln\left(n_t{}^2 \sin w_t\right)}{\ln\left(n_t \sin w_t\right)} \qquad D_{sc} = \frac{2\ln\left(n_c\right)}{\ln\left(2n_c \sqrt{\dfrac{\tan w_c}{\pi}}\right)} \qquad (8)$$

Stepwise mercury porosimetry is modelled as a sequence of flow events that are caused in the throats and chambers of the pore network by each (positive or negative) increment of the externally applied pressure. These flow events proceed until capillary equilibrium is attained at the new pressure value. The algorithm used to simulate intrusion is different from that used to simulate retraction because of the different mechanisms involved. The motion of menisci within the primary pore network is decided in accordance with the capillary resistance encountered at entrances of pores which is a known function of the local pore geometry and dimensions. The mechanism of snap-off in throats and the permanent entrapment of mercury in the form of disconnected blobs (ganglia) during mercury retraction are taken into account. The filling (intrusion) and the emptying (retraction) of the fractal roughness with mercury is dependent not only on the angular geometry but also on the accessibility properties of roughness features and such events are reflected on the shape of mercury porosimetry curves over the high pressure region. (The parameter values γ=480 dyn/cm, θ_I=40°, θ_R=60° are used in simulations).

The absolute permeability of the pore network is calculated by analogy to the solution of electric circuits[9]. The fractal roughness features of throats are included in the calculation of hydraulic conductances as cylindrical tubes of equivalent cross-section area, whereas the roughness features of chambers are considered as dead-end pores and are ignored.

5   METHODOLOGY FOR THE ESTIMATION OF PARAMETERS OF
     PERMEABLE POROUS SOLIDS

The form of mercury porosimetry curves of a porous material is the cooperative result of several parameters of the pore space[6-8,10] and any information obtained by other techniques is incorporated in the algorithm intended to characterize the pore structure in terms of a chamber-and-throat network.. The mean specific genus $<G_V>$ and the chamber-diameter distributSD (Fig.2) are provided by serial tomography, whereas the total porosity of the porous sample is obtained with the imbibition method[11]. These parameters are further incorporated in the construction of the theoretical pore networks.

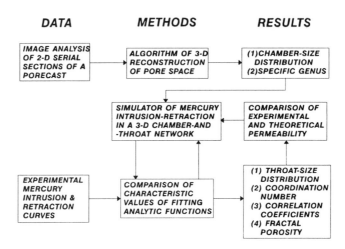

Figure 4   Characterization of the pore structure of permeable solids with network models

First, a network of given CSD and assumed TSD and ε_p with the skeleton of cubic lattice is computer-aided constructed. The initial estimation of TSD is based on the differentiation of the first part of experimental mercury intrusion curve taking into account overall network effects which have been clarified by theoretical simulation[6-8]. Then the mean coordination number is reduced so that the specific genus of the network matches that of the real pore structure. Afterwards, the parameters of fractal roughness features (D_s, n_r) are chosen in such a way that the total porosity ε_t becomes equal to the experimental one. The simulation of mercury intrusion in and retraction from the chamber-and-throat network gives the theoretical curves which in turn are fitted with the 5-parameter analytic functions. Comparison between the characteristic values of the functions, which fit the experimental and theoretical curves, as well as between the theoretical and the experimental value of absolute permeability leads to a new estimation of TSD, ε_p and D_s, n_r. The procedure can be continued until both the mercury porosimetry curves and the permeability of the porous solid are predicted (Fig.4).

6   APPLICATION TO A GREY VOSGUES SANDSTONE

(a) The CSD and $<G_V>$ are given in Fig.2a,b.

(b) The experimental mercury porosimetry curves are fitted with the 5-parameter analytic functions and the optimum values of parameters are determined (Fig.5). Using this procedure the mercury intrusion curve is decomposed to two curves extending over the low and high pressure range (Fig.5). The value of coefficient c_l is roughly comparable to the ratio of primary porosity to the total one. Mercury penetration in the interconnected pore network is reflected in the curve of low pressure range which in turn is differentiated according to the conventional method of analysis and a lognormal TSD with $<D_t>=12.2\mu m$ and $\sigma_t=2.45\mu m$ is obtained. Theoretical simulation have shown[6] that the real TSD is wider and is affected by network connectivity.. So as a first estimation a lognormal TSD with parameters $<D_t>=12.5\mu m$ and $\sigma_t=4.0\mu m$ is chosen whereas a value $\varepsilon_p=0.11$ ($\varepsilon_p/\varepsilon_t=0.6$) is assumed.

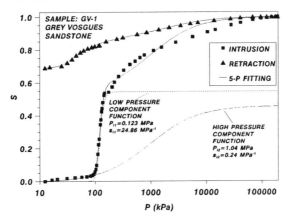

Figure 5. Mercury porosimetry curves of Grey Vosgues (ε_t=0.18 , k=110 md)

(c) After the cubic network has been constructed with the above values of parameters, an iterative procedure of sequential removal of throats is carried out so that both the specific genus of the pore network becomes comparable to the experimental value and the connectivity of each pore to the external boundary is preserved. This stage reduces the mean coordination number to $<c_n>$=5.85 whereas any changes of TSD and ε_p are negligible.

(d) The variation of the total porosity of network ε_t as a function of D_s and n_r is calculated. The choice of these parameters D_s=2.45 and n_r=12 (ε_t=0.182) is made so that the simulated curves are sufficiently wide over the high pressure range (namely s_{l2} is low enough).

(e) Simulated mercury intrusion/retraction curves are obtained and they are fitted with the 5-parameter functions (Table 1). The absolute permeability of the pore network is also determined with network analysis (Table 2).

(f) The parameters of the theoretical pore network are updated so that both the mercury porosimetry curves (in terms of the characteristic values of fitting functions) and the absolute permeability of the porous sample are predicted satisfactorily.

The results of some sample calculations are given in Tables 1 and 2 whereas one pair of simulated curves is compared with experimental ones of VG-1 in Fig.6

Table 1   Optimum parameters of fitting analytic functions

Sample	c_I	P_{I1} (MPa)	s_{I1} (MPa^{-1})	P_{I2} (MPa)	s_{I2} (MPa^{-1})	c_R	P_{R1} (MPa)	s_{R1} (MPa^{-1})	P_{R2} (MPa)	s_{R2} (MPa^{-1})
GV-1	0.545	0.1233	24.86	1.04	0.243	0.59	0.057	2.65	2.11	0.037
Simul-1	0.775	0.1134	20.67	1.16	0.215	0.64	0.043	16.60	1.56	0.084
Simul-2	0.825	0.1174	18.56	1.17	0.214	0.68	0.051	13.90	1.94	0.062
Simul-3	0.790	0.130	13.68	1.01	0.218	0.74	0.053	8.68	2.86	0.043

Table 2   Parameters of theoretical networks (20X20X20) used in simulations

Simulation	$<D_t>$,μm	σ_t, μm	$\kappa=\varepsilon_p/\varepsilon_t$	$<c_n>$	D_s	n_r	ε_t	k , md
1	12.5	4.0	0.60	5.85	2.45	12	0.182	180.0
2	12.5	4.0	0.66	4.95	2.40	14	0.180	140.0
3	11.5	4.0	0.65	4.3	2.45	16	0.184	95.0

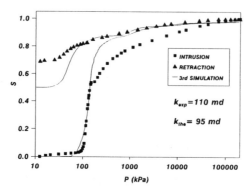

Figure 6 Comparison of experimental to simulated mercury intrusion / retraction curves

7   CONCLUSIONS

- The determination of geometrical and topological parametrs of a permeable porous solid in terms of chamber-and-throat networks requires the combination of some experimental techniques with theoretical simulators.
- Serial sectioning analysis is used to obtain mainly the chamber-diameter distribution and the specific genus of the pore network whereas the estimation of some other important parametrs of the pore structure (throat-diameter distribution, fractal roughness porosity) is carried out through the comparison of experimental and simulated results of mercury porosimetry (intrusion / retraction curves) and one-phase flow (absolute permeability).
- Non-linear 2-P and 5-P analytic functions can be used to describe any mercury intrusion and retraction curve and comparison between theoretical and experimental mercury intrusion / retraction curves is made easier and more flexible in terms of the characteristic parameters of these fitting functions.
- The simultaneous prediction of experimental mercury porosimetry curves and absolute permeability of a porous medium through pore network simulations is a laborious procedure and some further improvements of the code of analysis are required so that the computational effort is minimized.
- In order to evaluate the present method, further applications to the characterization of the structure of porous solids of varying origin are required.

REFERENCES

1. N. C. Wardlaw and J.P. Cassan, *Bull. Canad. Pet. Geol.*, 1978, **26**, 572.
2. M. Sahimi, *Rev. Mod. Phys.*, 1993, **65**, 1393.
3. R. T. DeHoff, E. H. Aigeltinger and K. R. Craig, *J. Microsc.*, 1972, **95**, 69.
4. D. P. Lymberopoulos and A. C. Payatakes, *J. Colloid Interface Sci.*, 1992, **150**, 61.
5. Y. Bard, 'Nonlinear Parameter Estimation', Academic Press, 1974, Chapter V, p.94.
6. C. D. Tsakiroglou and A. C. Payatakes, *J. Colloid Interface Sci.*, 1990, **137**, 315.
7. C. D. Tsakiroglou and A. C. Payatakes, *J. Colloid Interface Sci.*, 1991, **146**, 479.
8. C. D. Tsakiroglou and A. C. Payatakes, *J. Colloid Interface Sci.*, 1993, **159**, 287.
9. G. N. Constantinides and A. C. Payatakes, *Chem. Eng. Commun.*, 1989, **81**, 55.
10.M. A. Ioannides and I. Chatzis, *Chem. Eng. Sci.*, 1993, **48**, 951.
11.F. A. L. Dullien, 'Porous Media: Fluid Transport and Pore Structure', Academic Press, 1979, Chapter 3, p.77.

MORPHOLOGY OF BIPHASIC RANDOM MEDIA: CHORD DISTRIBUTIONS AND 2 POINT CORRELATION FUNCTIONS.

P. Levitz

Centre de Recherche sur la Matière Divisée
CNRS, 45071 Orléans Cedex 2, France.
Email: levitz@cnrs-orleans.fr

1 INTRODUCTION

A disordered porous medium can be considered as a complex interface filling the 3D space. Describing the interfacial geometry of this interfacial system is a challenging problem. Almost three levels of description are encountered in the literature:

(i) Determination of some scalar quantities which give either the overall amount of interface per unit of volume (Sv) or the fraction of void space (porosity: ϕ).

(ii) Information's concerning the morphology of the pore network such as the "average pore shape", the mean curvature (<H>) , the surface roughness, different 2 point correlation functions (bulk, surface and surface-pore)....

(iii) Analysis of the topology of the matrix which is closely related to the long range connectivity of pore network (Gauss curvature of the interface <K>, deformation retract, genus of the interface).

Some questions are however ill defined or ambiguous. As example, this is the case for the definition and the determination of the average pore shape of a disordered pore network. The aim of this short paper is to show how some functions such as chord length distribution functions provide a synthetic stereological analysis of the medium and give some inside in describing the pore network morphology (levels 1 and 2). These morphological descriptors are known for a long time[1-9]. In part 2, we present different definitions of theses functions and discuss their main analytical properties in 2D or in 3D. In part 3, we show how these functions can be used to get good approximations of 2 point correlation functions such as bulk, surface autocorrelation or surface-pore correlation functions. Application of these tools to other processes such as energy transfer, excitation and magnetic relaxations or gas diffusion are rapidly mentioned in the conclusion. Extended proofs of several relations reported in this communication will be presented elsewhere[10].

2 MORPHOLOGICAL PROPERTIES OF CHORD LENGTH DISTRIBUTION FUNCTIONS

A chord is a segment belonging either to the pore network (p) or to the solid matrix (m) and having its two extremities on the interface (i). There are several ways to define chord length distribution functions depending of the chosen angular average. In the following, we will use the classification discussed by R Coleman[11].

We first discuss different properties of the μ-chords. In this case, determination of the chord statistics starts with the definition of a random direction inside the matrix. A first point belonging to the interface is located along this direction and the associated pore or mass chord is computed. The probability density of these chords can be written as:

$$f_p^\mu(r) = \frac{1}{S} \int_S ds \int_{2\pi} d\widehat{\omega} \frac{\widehat{(\overrightarrow{\sigma_r} \cdot \overrightarrow{\omega})} \, \delta(R_\omega - r)}{2\pi} \qquad (Eq\ 1)$$

where notations are explained on figure 1. These μ-chords are in fact the usual chords used either in stereology[7,8] or in different works where possible connection between chord length distributions functions and bulk autocorrelation[1-6, 9] is attempted. For a smooth and curved interface having a set of sharp angularities (such as wedge), small r expansions of μ-chord distributions function read (10) :

$$\lim_{r \to 0} f_{p,m}^\mu(r) = A_{p,m} + C_{p,m}\, r \quad ; \quad C_{p,m} = \frac{1}{2} < k_1 k_2 >_{p,m} + \frac{3}{16}\left((k_1 - k_2)^2\right)_{p,m} \qquad Eq(2)$$

In this expression, the curvature of the interface is involved in a complex way. There is no simple method to extract from Eq(2) either the mean ($<k_1 + k_2>/2$) or the Gauss curvature $<k_1 k_2>$.

In order to check these different theoretical results, we have performed direct simulations on a 3D numerical porous medium. A random planar (2D) cut of this matrix is shown in Fig 2. The pore network was build according to an algorithm proposed by Chen et al[12].

Evolution of the pore and the mass μ-chords distribution functions are shown in Figure 3. At small distances, both functions start from the origin and are proportional to r. This result is in good agreement with the properties of the smooth and curved interface of the numerical pore network.

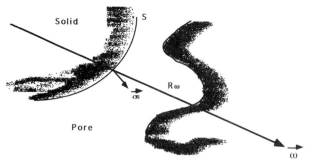

Figure 1 : *Notations used in the different definitions of chord length distributions functions (see Eq 1 and Eq 3)*

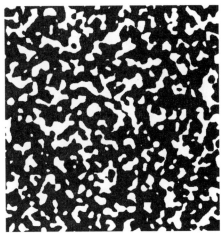

Figure 2: *Random planar section of a 3D numerical porous medium built according to ref (12)*

We will now introduce a second type of probability densities associated with the so called s-chords. As shown below, these functions are interesting morphological descriptors. However, to the best of our acknowledge, their 3D properties were scarcely discussed in the literature. Determination of the s-chord statistics begins with the definition of a starting point belonging to the interface (Oi). The pore or mass s-chord is then computed along a random direction starting from Oi. The probability density of these pore s-chords can be written as :

$$f_p^s(r) = \frac{1}{S} \int_S ds \int_{2\pi} d\hat{\omega} \frac{\delta(R_\omega - r)}{2\pi} \qquad Eq(3)$$

Similar expression can be used for the mass s-chord length distribution function.
For smooth and curved interface, it is relatively straightforward to show that we can directly measure the mean curvature of the porous interface.

$$\lim_{r \to 0} f_{p,m}^s(r) = \frac{1}{4}(k_1 + k_2)_{p,m} \qquad Eq(4)$$

Moreover, we have

$$\lim_{r \to \infty} f_{p,m}^s(r) = f_{p,m}^\mu(r) \qquad Eq(5)$$

As shown in figure 3, predictions of Eqs (4) and (5) are numerically verified.
These different probability density functions defined by Eq(1) or Eq(3) permit an integrated stereological analysis of a disordered and isotropic porous medium. First, these functions can be estimated either on 3D sample or on 2D random planar cut. As shown in Fig 4, this point was numerically verified on different reconstructed porous media. Three Minkowski functionals can be determined using either μ-chord (porosity and specific surface) or s-chord(mean curvature) length distributions functions. The last Minkowski functional, i.e. the average Gauss curvature, can only be determined on a 3D system and is directly related to the connectivity of the pore network. Finally, analytic form of chord length distribution function is

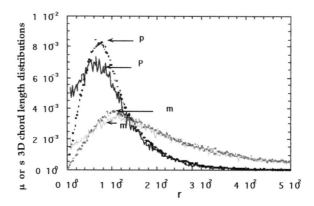

Figure 3: *Comparison of μ-chord (points) and s-chord (lines) length distribution functions computed on the 3D numerical porous medium shown in figure 2*

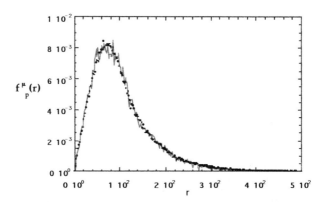

Figure 4: *Comparison of pore μ–chord length distribution functions computed either inside the 3D numerical porous medium (points) or on a random planar section (line)*

sensitive to the type of structural disorder. This point is discussed in a former paper[9] and permits to consider s or μ chords distribution function as fingerprints of the local and semi-local morphology of the pore network.

3 CORRELATIONS AND CHORD LENGTH DISTRIBUTION FUNCTIONS

Two point correlation functions such as the bulk and the surface autocorrelation or the pore-surface correlation functions are of basic importance for a morphologic characterisation of porous medium. they plays a central role in different processes involving energy, excitation or molecular transport. Moreover, as shown in a seminal work of Doi[13-15], these three correlation functions are directly involved in an upper bound limit of the permeability. Clearly a statistical geometrical analysis of these two point correlations functions is suitable. One of the first contribution in this direction was performed by Debye, Anderson and Brunberger[16]. As shown in the following, these function can be approximated using chord length distributions functions.

3.1 Bulk autocorrelation and small angle scattering

Small angle scattering is an experimental way to probe the bulk autocorrelation function via a 3D Fourier transform. This last function reads:

$$\varphi_v^2(r) = \frac{1}{4\pi V}\int_{V_m} d\vec{r_1} \int_{V_m} d\vec{r_2} \int_{4\pi} d\vec{\omega}\, \delta(\vec{r_1}+r\vec{\omega}-\vec{r_2}) \qquad Eq(6)$$

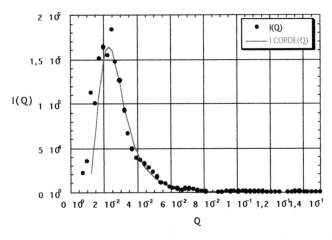

Figure 5: *Small angle scattering of the porous matrix shown on figure1. Comparison between direct 3D computation and Eq 7-8 using chord distributions functions.*

Properties of the second derivative of the bulk autocorrelation function quantitatively define the level of connection between small angle scattering and the statistical properties of the pore network interface. The general expression of this second derivative involves a "delicate" surface integral. A tractable expression, involving the "pore" and "mass" chord distributions functions has been proposed under some general hypothesises[4,9]: statistical isotropy of the matrix and uncorrelation between adjacent chords along a random line. An explicit relation between bulk autocorrelation function, small angle scattering and chord length distribution functions can be written as followed:

$$\widetilde{\varphi_v^2}(q)= \int_{-\infty}^{+\infty} \varphi_v^2(r)\, \exp(iqr)\, dr = \frac{S_v}{2\,q^2}\mathrm{Real}(\,1-\widetilde{G}(q)) \qquad \widetilde{G(q)} =\frac{\widetilde{f_m^\mu}(q) +\widetilde{f_p^\mu}(q)-2\widetilde{f_m^\mu}(q)\widetilde{f_p^\mu}(q)}{1-\widetilde{f_m^\mu}(q)\widetilde{f_p^\mu}(q)}$$

$$Eq(7)$$

and

$$I(q)= \frac{2\pi}{q}\frac{d}{dq}\Big[\mathrm{Real}(\widetilde{\varphi_v^2}(q))\Big] \qquad Eq(8)$$

As shown in figure 5, computation of the small angle scattering from Eqs 7-8 is in good agreement with a direct 3D estimation. Interesting enough[10], Eqs 7-8 and low r expansion of chord distributions can be used to retrieve several high q evolutions of I(q) directly related to the curvature (Kirste-Porod correction), the angularity or the roughness of the interface.

3.2 Surface autocorrelation

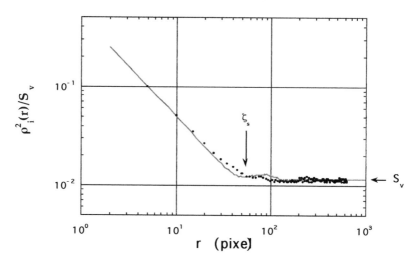

Figure 6: *Surface autocorrelation function of the porous matrix shown on figure 2. Comparison between 3D computation (points) and Eq 10. (line)*

Surface autocorrelation function reads

$$\varphi_I^2(r) = \frac{1}{4\pi V} \int_{4\pi} d\hat{\omega} \int_S dS_i \int_S dS_j \, \delta(\vec{r_{ij}} - r\vec{\omega}) \qquad Eq(9)$$

Evolution of the surface autocorrelation function can be subdivided in two parts:(i) below persistence length ξ_S, a correlated angular regime takes place where local curvature and/or flatness of the interface play a central role; (ii) above ξ_S, an density distribution regime has to be considered where angular correlations are small. In this regime, a analytic development of eq(9) is possible when one assumes a statistical uncorrelation between adjacent chords along a random line. A good approximation is given by:

$$\frac{\varphi_I^2(r)}{S_V} \approx \frac{1}{2r} \exp(-(r/\xi_S)^2) + f_m^\mu(r) + f_p^\mu(r) + \sum_{j=1}^{\infty} (2\delta(r) + f_m^\mu(r) + f_p^\mu(r)) * (f_p^\mu(r) * f_m^\mu(r))^{j*}$$

$$Eq(10)$$

$$\xi_S \approx (2 S_V)^{-1} \qquad\qquad Lim_{r\to\infty} \frac{\varphi_I^2(r)}{S_V} = S_V$$

As shown on figure 6, this equation works relatively well.

3.3 surface-pore correlation

This last two point correlation function reads:

$$\varphi_{ip}(r) = \frac{1}{4\pi\,V} \int_{V_{pore}} d\vec{r_1} \int_S ds \int_{4\pi} d\vec{\omega}\; \delta(\vec{r_1}+r\vec{\omega}-\vec{r_s}) \qquad Eq(11)$$

Its first derivative can be approximated as :

$$\frac{d(\varphi_{i,p}(r))}{dr} \approx \frac{S_v}{2}\left[(f_m^s(r)-f_p^s(r)) + \sum_{j=1}^{\infty} (\,f_m^u(r)-f_p^u(r))*(f_m^u(r)*f_p^u(r))^{j}\,^{*} \right] \qquad Eq(12)$$

and a simple integration permits to get the function

$$\varphi_{i,p}(r) \approx \frac{S_v}{2}\left[1+ \int_0^r \frac{d(\varphi_{i,p}(r))}{dr}\,dr \right]$$ with following properties:

$$Lim_{r\to 0}\frac{d(\varphi_{i,p}(r))}{dr} = \frac{S_v}{8}(k_1+k_2) \qquad Lim_{r\to\infty}\,\varphi_{i,p}(r)= \phi\,S_v$$

As shown in figure 7, computation of the pore-surface correlation function using Eq 12 is in good agreement with a direct 3D estimation.

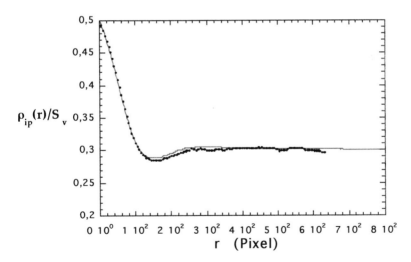

Figure 7: *pore-surface correlation function of the porous matrix shown on figure 2. Comparison between direct 3D (points) computation and Eq 12 (line)*

4 CONCLUSION

Chord length distribution functions are interesting tools that contribute to characterise the morphology of porous systems. In the case of isotropic media, these functions can be estimated either on a random 2D cut or inside the 3D matrix. In that sense, they permit an integrated stereological analysis and a direct determination of three Minkowski functionals (porosity, specific surface, mean curvature). The last one, i.e. the average Gauss curvature, can only be determined on a 3D system and is directly related to the connectivity of the pore network. Chord length distribution functions are sensitive to the type of structural disorder and can be used as fingerprints of the local and semi-local morphology of the pore network.

Under some general hypothesises, 2 point correlation functions can be estimated using chord length distribution functions. A good knowledge of these correlations functions is important for a more complete morphological characterisation (small angle scattering and the bulk autocorrelation function). However, they also play a central role in different processes, such as energy and excitation transfer[8,17], gas diffusion[18,19] and magnetic relaxation[19], taking place inside a disordered porous medium. Finally they can be used in order to evaluate rigorous upper bound for the permeability of the porous matrix[13-15].

Acknowledgment. Part of the numerical computations was performed on a Cray C98. YMP. We thank IDRIS, CNRS, Orsay for the grant 94281.

References

1. A. Guinier A, G. Fournet., " Small angle scattering of X rays",John Wiley & son eds., 1955, p 12-13
2. G. Porod, in "Small angle X rays scattering" Syracuse 1965,H.Brumberger Ed., Gordon and Breach Science Publ. 1967,p 1-15
3. M. Coster and J.L. Chermant, "Précis d'analyse d'image", eds CNRS, 1985, 521 p.
4. J. Mering.and D. Tchoubar., *J.Appl.Cryst.* 1968,**1**, 153.
5. H. Wu andP. W. Schmidt, *J.Appl.Cryst.* 1971, **4**, 171.
6. H. Wu andP. W. Schmidt., *J.Appl.Cryst.* 1974,**7**, 131.
7. J. Serra., " Image analysis and mathematical morphology", Academic Press (1982), p 329.
8. P. Levitz, G. Ehret, S.K.Sinha, J.M. Drake. *J.Chem.Phys.*, 1991, **95**, 6151.
9. P. Levitz , D. Tchoubar. *J.Phys. I* , 1992, **2**, 771.
10. P. Levitz, submitted to *Phys Rev. E*
11. M. Coleman, *J. Appl. Prob.*, 1965, **2**, 169.
12. S. H. Chen, S.L. Chang, R. Strey, *J. Appl. Cryst.* 1991,**24**, 721.
13. M. Doi, *J. Phys.Soc. Jpn.*1976,**40**, 567
14. H. Reiss, *J. Phys. Chem.* 1992, **96**, 4736
15. S. Torquato. *Appl. Mech. Rev.* 1991, **44**, 37
16. P. Debye, H.R. Anderson and H. Brumberger *J.Appl.Phys.* ,1957, **28**, 679
17. P.Levitz, Material Society Symposium Proceedings, published by J.M.Drake,D.D.Awschalom, J.Klafter, R. Kopelman, 1993, **290**, 19718
18.Levitz, *J.Phys. Chem.* 1993, **97**, 3813.
19. V. Pasquier, P. Levitz, A. Delville. *J. Phys. Chem.* 1996, **100**, 10249.

INTERPARTICLE CAPILLARY CONDENSATION AND THE PORE SIZE DISTRIBUTION OF SEVERAL ADSORBENTS

W.D. Machin and R.J. Murdey

Department of Chemistry
Memorial University of Newfoundland
St. John's, Newfoundland, Canada
A1B 3X7

Pore size and shape are often used to characterize porous adsorbents. Various methods have been devised to measure these quantities and those based on the phenomenon of adsorption-desorption hysteresis are widely used[1,2,3,4] for adsorbents that yield Type IV isotherms with H1 or H2 hysteresis loops.[5] Determination of the pore size distribution (PSD) from such isotherms can be effected with appropriate assumptions about the adsorbent, capillary condensate and the pore shape. Pore models such as slit or cylinder are often selected, but more complex models based on the packing of spheres or a general model where cavities are connected through smaller "windows" are sometimes used.[6,7,8]

Capillary condensation and evaporation is governed by the Kelvin equation[1], *i.e.*

$$RT\ln P/P^\circ = -\gamma V \left[1/r_1 + 1/r_2 \right] \cos\theta \tag{1}$$

where γ, V are the surface tension and molar volume of the capillary liquid at temperature T; θ is its contact angle; r_1 and r_2 are the principal radii of curvature of the liquid-vapour interface and P/P° the relative pressure at which a pore fills or empties. Since r_1 and r_2, differ for adsorption and desorption a given pore will fill and empty at different relative pressures giving rise to the phenomenon of hysteresis. For a given pore shape there is an explicit relationship between $(P/P^\circ)_a$ and $(P/P^\circ)_d$, where the subscripts a, d denote adsorption and desorption respectively. This relationship is illustrated with the following model pores:

1. Cylinder: As adsorption proceeds a liquid-like multilayer film forms on the walls of the pore. The principal radii of curvature are $r_1 = r_p - t$ and $r_2 = \infty$, where r_p is the pore radius and t is the thickness of the adsorbed film, hence from eq 1

$$\ln (P/P^\circ)_a = -(\gamma V/RT) \left[1/(r_p - t_a) \right] \tag{2}$$

assuming that the liquid wets the pore wall ($\cos\theta = 1$). All pores are filled at saturation pressure and desorption proceeds by evaporation across the hemispherical interface capping the pore. In this case, $r_1 = r_2 = r_p - t_d$ and

$$\ln (P/P^\circ)_d = -(2\gamma V/RT) \left[1/(r_p - t_a) \right] \tag{3}$$

We define the quantity K (Kelvin radius) as

$$K = -(2\gamma V/RT) \left[1/\ln(P/P^\circ) \right] \tag{4}$$

and since $t_d < t_a$, then from eq 2 and 3,

$$K_a/K_d \;=\; 2(r_p - t_a)/(r_p - t_d) \le 2.0 \tag{5}$$

This value of K_a/K_d is characteristic of cylindrical pores irrespective of their radius.

2. Slit: Adsorption on the flat walls of the pore occurs with $r_1 = r_2 = \infty$, hence $K_a = \infty$. Desorption from the filled pore proceeds across a hemicylindrical liquid-vapour interface where $r_1 = r_p - t_d$, where r_p is one-half the slit width, and $r_2 = \infty$. Hence $K_d = 2(r_p - t_d)$ and $K_a/K_d >> 2$.

3. Pores formed by the packing of spherical particles. The pore structure depends on the packing arrangement but it can be regarded as a series of cavities connected through smaller windows. The absolute size of the cavities and windows depends on the size of the spheres, but their relative size depends only on the packing. The cavity radius, r_c, is defined as the radius of the largest sphere that can be inscribed within a given cavity and r_w is the radius of the largest circle that can be inscribed within the largest window leading into the cavity. Avery and Ramsay[9] have derived r_c and r_w for a number of regular packings. Their results show that while both r_c and r_w increase with increasing porosity, the ratio, r_c/r_w, is relatively constant at *ca.* 1.3-1.5. Cavities fill when $K_a = r_c - t_a$, empty when $K_d = r_w - t_d$ and $K_a/K_d \approx 1.3$-1.5.

4. General pore structure: Spherical cavities connected by short cylindrical throats. Cavities will fill when $K_a = r_c - t_a$ and empty when $K_d = r_w - t_d$. Since $r_c > r_w$, then $K_a/K_d > 1$.

The ratio K_a/K_d for a given adsorbent may be used to select a suitable model pore structure. It is also evident that this ratio is a property of the adsorbent and should be independent of the adsorptive. The ratio is most conveniently determined from the position of the upper closure point of the hysteresis loop (K_a) and the position of the shoulder on the desorption branch (K_d). The values of K_a/K_d in Table 1 have been so estimated.

We will attempt to define the conditions required to obtain reliable PSD, for mesoporous adsorbents, from adsorption isotherms. To this end, we summarize the assumptions pertaining to the properties of the adsorbent and the capillary condensed liquid.

A.   Adsorbent
1. The adsorbent is inert, rigid and non-swelling, *i.e.* the pore volume is constant.
2. All pores have the same shape, *i.e.* K_a/K_d is independent of pore size.
3. Pores fill and empty independently of other pores.

B.   Capillary liquid
1. The surface tension, γ, and molar volume, V, are the same as bulk liquid.
2. Capillary condensate wets the adsorbent, *i.e.* the contact angle is equal to zero.
3. As relative pressure increases, the growth of liquid-like multilayers on the pore walls is followed by capillary condensation when eq 1 is satisfied.
4. As relative pressure decreases, hemispherical menisci form at pore entrances and capillary condensate evaporates when eq 1 is satisfied leaving a multilayer film. In the case of slit-shaped pores hemicylindrical menisci form at pore entrances.

If this model for adsorbent structure and capillary condensate is valid then the following criteria should be satisfied.

C.   Criteria
1. PSD derived from the adsorption branch of the hysteresis loop should be the same

as that derived from the desorption branch.

2. PSD should be independent of the adsorptive and temperature.
3. Cumulative surface area from the PSD should agree with the nitrogen BET area.
4. Cumulative pore volume should agree with the total pore volume.
5. PSD should agree with other, independent, methods if such are available.

These are applied to isotherms for adsorptives and adsorbents in Table 1.

Sterling FT-G (2700); SCI-IUPAC-NPL surface area standard, graphitized carbon black, 11.1 ± 0.8 m²g⁻¹. Designated as FTG.

Vulcan 3-G (2700); SCI-IUPAC-NPL surface area standard, graphitized carbon black, 71.3 ± 2.7 m²g⁻¹. Designated as VUL. Both of these adsorbents have been described elsewhere.[10] The primary adsorbent particles are non-porous, quasi-spherical polyhedrons which readily agglomerate to form porous networks.[11]

Controlled pore glass; CPG/100 from Pierce Chemical Co., a fine porous powder, particle size 37-74 microns with normal pore diameter *ca.* 100Å. Designated as CPG.

Silica gel SG3; a mesoporous silica gel derived from Davison Silica Gel grade 03. The properties of this adsorbent are described elsewhere.[12] Designated as SG3.

Vycor porous glass; Corning 7930 porous glass. The properties of this adsorbent are described elsewhere.[13,14] Designated as VPG.

Surface areas for all adsorbents have been determined from nitrogen adsorption at *ca.* 77 K; the densities of the two carbon blacks were provided by the National Physical Laboratory (U.K.) and the densities of the three silica-based adsorbents were assumed to be equal to the density of a standard reference silica, also provided by the National Physical Laboratory (U.K.). Purity of the various adsorptives and examples of several isotherms have been published.[12,15,16,17,18] The following abbreviations are used for three adsorptives; BUT, n-butane; CFC, trichlorofluoromethane; DMP, 2,2-dimethylpropane. Properties for each adsorptive are given in Table 1.

Some example isotherms are shown in figure 1 plotted as liquid volume adsorbed against the Kelvin radius, K. The parameters for each isotherm are summarized in Table 1. Monolayer volume is the volume of liquid required to cover the surface with a monolayer having same molecular arrangement as bulk liquid.[1] Pore volume is the liquid volume adsorbed at saturation pressure and the porosity, ϵ, has been calculated from this quantity and the adsorbent density. The relative amount adsorbed (A/A°) at the lower closure point indicates the fraction of pore volume that is filled prior to the onset of capillary condensation. The number of multilayers present at this point is also given. The ratio K_a/K_d has been determined directly from isotherms such as those in figure 1.

Pore size distributions, shown in figure 2 have been calculated using the method of Roberts.[1,19] Multilayer thicknesses have been estimated from the Frenkel-Halsey-Hill (FHH) equation fitted to the isotherm for coverages below the onset of hysteresis but above the monolayer. A full description of the method is presented elsewhere.[20] Parameters for the various PSD are summarized in Table 1. Surface areas and pore volumes are cumulative values for each PSD, the minimum pore radius corresponds to K, the Kelvin radius (eq 4), at the lower closure point of the hysteresis loop. The actual radius will be larger by an amount corresponding to the multilayer thickness. The Evans pore radius, r_E, is based on the concept of a capillary critical radius as expressed by the equation[21] $r_E = d/(1 - T/T_c)$. An adsorptive having molecular diameter d and critical temperature T_c will not exhibit irreversible capillary condensation (hysteresis) at temperature T in pores with radii less than r_E. We have used this concept to develop an

Table 1   Adsorbent Properties, Isotherm Parameters and PSD Parameters

	FTG					VUL				
ADSORBENT	11.1					71.3				
Surface Area/m² g⁻¹										
Density/g cm⁻³	2.14					2.0				
ADSORBATE	N₂	CFC		DMP		N₂	CFC		DMP	
Temperature/K	77.8	283.35	290.05	273.15	279.85	77.8	283.35	290.05	273.15	279.85
Surface Tens./m J m⁻²	8.81	20.36	19.41	14.40	13.66	8.81	20.36	19.41	14.40	13.66
Molar Volume/cm³ mol⁻¹	34.67	0.92	91.87	117.70	119.06	34.67	90.92	91.87	117.70	119.06
Molec. Diam./nm	0.383	0.581	0.581	0.632	0.632	0.383	0.581	0.581	0.632	0.632
ISOTHERM PARAMETERS										
Monolayer Vol./cm³ g⁻¹	0.0039	0.0054	0.0055	0.0059	0.0059	0.025	0.035	0.035	0.038	0.038
Pore Vol./cm³ g⁻¹	0.283	0.269	0.265	0.264	0.260	0.856	0.832	0.829	0.810	0.798
Porosity	0.377	0.365	0.362	0.361	0.357	0.631	0.625	0.624	0.618	0.615
Lower Closure Point A/A°	0.13	0.09	0.09	0.10	0.10	0.14	0.15	0.13	0.19	0.18
Multilayers	9.4	4.1	4.1	4.1	4.2	4.8	3.4	2.9	3.9	3.7
K_a/K_d	-	1.2	1.2	1.4	1.3	1.2	1.3	1.4	1.4	1.4
PSD PARAMETERS										
ADSORPTION BRANCH										
Surface Area/m² g⁻¹	9.8	11.4	10.8	11.1	11.2	69	68	65	70	68
Pore Vol./cm³ g⁻¹	0.276	0.258	0.251	0.252	0.258	0.830	0.777	0.780	0.771	0.771
Mean Pore Radius/nm	68	58	58	58	59	25	26	27	25	26
DESORPTION BRANCH										
Surface Area/m² g⁻¹	10.7	11.0	10.7	11.2	11.0	69	67	67	70	68
Pore Vol./cm³ g⁻¹	0.282	0.266	0.263	0.263	0.263	0.847	0.790	0.793	0.784	0.784
Mean Pore Radius/nm	68	63	63	65	61	26	28	28	27	27
Minimium Pore Radius/nm	15.9	18.8	22.0	20.6	16.8	8.2	13.5	11.6	11.7	10.9
Evans Radius/nm	1.0	1.5	1.5	1.7	1.8	1.0	1.5	1.5	1.7	1.8

Table 1 (cont'd)   Adsorbent Properties, Isotherm Parameters and PSD Parameters

ADSORBENT	CPG				SG3				VPG	
Surface Area/m^2 g^{-1}	134.3				232				184	
Density/g cm^{-3}	2.2				2.2				2.2	
ADSORBATE	N$_2$	N$_2$	Xe	Xe	N$_2$	Xe	BUT	Kr	Xe	Xe
Temperature/K	91.0	100.0	239.0	250.0	77.8	168.0	273.35	117.41	163.32	168.25
Surface Tens./m J m^{-2}	5.93	4.12	5.80	4.23	8.81	17.89	14.85	16.05	18.78	17.84
Molar Volume/cm^3 mol^{-1}	37.82	40.68	55.72	58.63	34.67	44.74	96.87	34.51	44.24	44.77
Molec. Diam./nm	0.383	0.383	0.433	0.433	0.383	0.433	0.586	0.387	0.433	0.433
ISOTHERM PARAMETERS										
Monolayer Vol./cm^3 g^{-1}	0.049	0.050	0.056	0.057	0.082	0.090	0.116	0.065	0.070	0.071
Pore Vol./cm^3 g^{-1}	0.667	0.667	0.667	0.667	0.331	0.268	0.284	0.189	0.180	0.180
Porosity	0.595	0.595	0.595	0.595	0.421	0.371	0.385	0.294	0.284	0.284
Lower Closure Point A/A°	0.27	0.32	0.25	0.30	0.47	0.28	0.38	0.35	0.34	0.35
Multilayers	3.5	4.2	3.0	3.5	1.8	0.8	0.9	1.0	0.9	0.9
K$_a$/K$_d$	1.5	1.5	1.5	1.4	2.0	2.2	2.3	2.1	2.1	1.9
PSD PARAMETERS										
ADSORPTION BRANCH										
Surface Area/m^2 g^{-1}	105	110	106	100	170	145	144	110	114	112
Pore Vol./cm^3 g^{-1}	0.539	0.547	0.511	0.459	0.260	0.211	0.218	0.143	0.140	0.139
Mean Pore Radius/nm	8.2	8.2	8.2	8.2	1.7	1.9	1.9	1.5	1.6	1.6
DESORPTION BRANCH										
Surface Area/m^2 g^{-1}	82	86	77	72	162	155	150	114	110	109
Pore Vol./cm^3 g^{-1}	0.560	0.579	0.534	0.504	0.281	0.242	0.251	0.159	0.156	0.154
Mean Pore Radius/nm	8.2	8.2	8.2	8.2	2.5	2.5	2.6	2.1	2.2	2.2
Minimium Pore Radius/nm	5.6	3.8	5.4	5.2	1.2	1.3	1.6	1.1	1.3	1.4
Evans Radius/nm	1.4	1.8	2.5	3.2	1.0	1.0	1.6	0.9	1.0	1.0

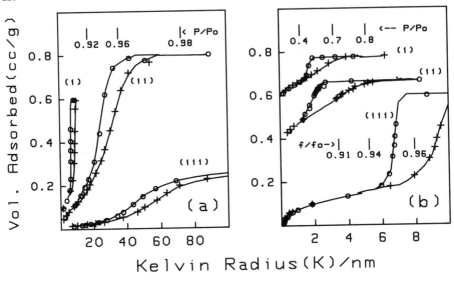

Figure 1 *Adsorption isotherms; plotted as liquid volume against Kelvin radius (eq 4)*
a. (i) xenon on CPG at 239 K; (ii) CFC on VUL at 290 K; (iii) CFC on FTG at 290 K.
b. (i) xenon on VPG at 168 K, offset 0.6 cm³g⁻¹; (ii) xenon on SG3 at 168 K, offset
0.4 cm³g⁻¹; (iii) xenon on CPG at 239 K.

alternative method of PSD analysis[17,18] based on the temperature dependence of hysteresis.
In contrast to traditional PSD analyses[1,19], the method does not depend on the shape of
either branch of the hysteresis loop, nor does it require knowledge of multilayer thickness.
PSD determined using this method for CPG, SG3 and VPG[17,18,22] are shown in figure 2.

   FTG: The pore structure of this adsorbent has been discussed in detail elsewhere[20].
Briefly, it has been shown that the pore structure results from the packing of spherical
particles and that the PSD derived from the particle size distribution is in good agreement
with that derived from the various isotherms. All criteria, C1-C5, are well satisfied. Note
that thick multilayers are present at the onset of hysteresis and that r_E is much smaller than
the minimum pore radius. This indicates that all of the pores are active with respect to
capillary condensation. Note also that only a small fraction of the total pore volume is
filled at the onset of hysteresis (A/A° ≅ 0.1).

   VUL: While no independent PSD is available for this adsorbent the ratio
K_a/K_d≅1.4 and porosity, ϵ ≅ 0.63, suggest a relatively open packing of spherical particles.
It is clear from Table 1 and figure 2 that criteria C1-C4 are well satisfied. We conclude
that the PSD is valid. As with FTG, thick multilayers are present at the onset of
hysteresis, r_E is much less than the minimum pore radius and only a small fraction of the
pore volume is filled at the lower closure point of the hysteresis loop.

   CPG: As with VUL, the ratio K_a/K_d≅1.5 and porosity, ϵ = 0.595, suggest that
the adsorbent structure may be described by a relatively open packing of spheres.
However it is clear from Table 1 and figure 2 that few of the criteria, C1-C5, are satisfied.
Results from the temperature dependence of hysteresis indicate a narrow PSD and this
agrees reasonably well with the PSD derived from the desorption branch of the hysteresis
loop, but the adsorption branch yields a broader PSD. We conclude, in this case, that only

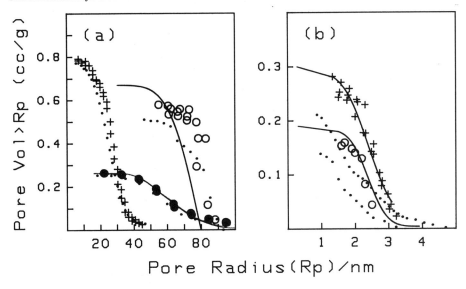

Figure 2 *Cumulative pore size distributions*
a. (+), VUL; (●) FTG; (○) CPG (divide radius by 10).
b. (+), SG3; (○) VPG. Solid lines in (a) and (b) present independent PSD. Smaller dots in (a) and (b) show PSD determined from the adsorption branches of the isotherms.

the desorption branch of the isotherms yield satisfactory PSD. Cumulative surface areas and pore volumes are smaller than the corresponding BET (N_2) area and total pore volume. A relatively large fraction of the pore volume is filled at the lower closure point ($A/A° \cong 0.3$) and r_E is only slightly smaller than the minimum pore radius. Therefore, some pores may be inactive with respect to capillary condensation but adsorb by a reversible pore filling mechanism. Hence we would expect PSD cumulative surface areas and pore volumes to be reduced.

SG3: The ratio $K_a/K_d \cong 2.2$ which suggests that the pores can be regarded as cylinders. It is also clear from Table 1 and figure 2 that criteria C1-C5 are not satisfied. However, the PSD derived from the desorption branch of the various isotherms show good agreement with that derived from the temperature dependence of hysteresis[17], but PSD derived from the adsorption branch do not. Cumulative surface areas and pore volumes are smaller than corresponding BET (N_2) areas and total pore volume. A large fraction (up to $A/A° \cong 0.47$) of the pore space is filled at the lower closure point even though only a thin adsorbed film is present. The smallest pores that exhibit capillary condensation are only slightly larger than r_E indicating that a significant fraction of the pore volume is found in pores that fill and empty reversibly.

VPG: As with SG3, the ratio $K_a/K_d \cong 2.0$ suggests that the pores are cylinders. Comments noted above with respect to SG3 are applicable to VPG. Note that PSD derived from the desorption branch of the various isotherms are in agreement with that derived from the temperature dependence of hysteresis.[17]

Taken together, these five adsorbents cover the entire mesopore range. Only FTG and VUL yield isotherms that satisfy all the criteria applicable to the determination of PSD from adsorption hysteresis; all pores are active with respect to capillary condensation and thick adsorbed films are present during capillary condensation and evaporation. With the other adsorbents, CPG, SG3 and VPG, thick multilayers are not present so that surface tension and molar volume of the adsorbed film are unlikely to be the same as bulk liquid. Hence the model for capillary condensation described in B1-B4 is not valid and the adsorption branch of the hysteresis loop cannot yield realistic PSD. These difficulties are less likely to affect the desorption process since the state of the multilayer film is less important. Reasonable PSD should be obtained if the contact angle is zero and if the properties of the core capillary liquid are similar to bulk properties. PSD derived from the desorption branch for isotherms on CPG, SG3 and VPG are all in good agreement with PSD derived from the temperature dependence of hysteresis.

We conclude, therefore, that when seeking to determine the PSD of mesoporous adsorbents that (i) more than one adsorptive should be used and (ii) if PSD derived from the two branches of the hysteresis loop do not agree then the one derived from the desorption branch is to be preferred.

The authors thank Professor P.D. Golding for useful comment and discussion. This work was supported in part by the Natural Sciences and Engineering Research Council of Canada.

References

1. S.J. Gregg and K.S.W. Sing, "Adsorption, Surface Area and Porosity", Academic Press, New York, 2nd edition, 1982, chapter 3.

2. "Characterization of Porous Solids - COPS I", editors K.K. Unger, J. Rouquerol, K.S.W. Sing, H. Kral, Elsevier, Amsterdam 1988.

3. "Characterization of Porous Solids - COPS II", editors, F. Rodriguez-Reinoso, J. Rouquerol, K.S.W. Sing, K.K. Unger, Elsevier, Amsterdam 1991.

4. "Characterization of Porous Solids - COPS III", editors, J. Rouquerol, F. Rodriguez-Reinoso, K.S.W. Sing, K.K. Unger, Elsevier, Amsterdam 1994.

5. K.S.W. Sing, D.H. Everett, R.A.W. Haul, L. Moscou, R.A. Pierotti, J. Rouquerol, T. Siemieniewska, *Pure Appl. Chem.*, 1985, **57**, 603.

6. J.M. Haynes, *Colloid Sci.*, 1975, **2**, 101.

7. G. Mason, *Proc. Roy. Soc., London; Ser. A.*, 1983, **390** (1798), 47.

8. H. Liu, L. Zhang and N.A. Seaton, *J. Colloid Interface Sci.*, 1993, **156**, 285.

9. R.G. Avery and J.D.F. Ramsay, *J. Colloid Interface Sci.*, 1973, **42**, 597.

10. D.H. Everett, G.D. Parfitt, K.S.W. Sing and R. Wilson, *J. Appl. Chem. Biotechnol.*, 1974, **24**, 199.

11. R. Wilson, Division of Chemical Standards, National Physical Laboratory U.K., private communication.

12. W.D. Machin and P.D. Golding, *Langmuir*, 1989, **5**, 608.

13. C.G.V. Burgess, Ph.D. Thesis, University of Bristol, 1971.

14. S. Nuttal, Ph. D. Thesis, University of Bristol, 1974.

15. W.D. Machin, *J. Chem. Soc., Faraday Trans. 1*, 1982, **78**, 1591.

16. W.D. Machin and P.D. Golding, *J. Chem. Soc., Faraday Trans.*, 1990, **86**, 171.

17. W.D. Machin, *Langmuir*, 1994, **10**, 1235.

18. R.J. Murdey and W.D. Machin, *Langmuir*, 1994, **10**, 3892.

19. B.F. Roberts, *J. Colloid Interface Sci.*, 1967, **23**, 266.

20. W.D. Machin and R.J. Murdey, *Langmuir*, accepted for publication, 1996.

21. R. Evans, U.M.B. Marconi, P. Tarazona, *J. Chem. Soc., Faraday Trans. 2*, 1986, **82**, 1763.

DUAL SITE-BOND MODELS IN THE PRESENCE OF CORRELATIONS

A. Adrover[a], and M. Giona[b]

[a]Dipartimento di Ingegneria Chimica
Università di Roma "La Sapienza"
Via Eudossiana 18, 00184, Roma, Italy
[b]Dipartimento di Ingegneria Chimica,
Università di Cagliari
Piazza d'Armi 09123, Cagliari, Italy

1 INTRODUCTION

The analysis of adsorption phenomena on heterogeneous surfaces requires a detailed description of the statistical and correlation properties of the energy distribution, since the spatial structure of the energy field plays a fundamental role as regards the behaviour of the resulting macroscopic quantities (e.g. adsorption equilibria), especially in those cases in which the correlation properties of the energy field influence surface-hopping phenomena between nearest neighbouring sites (surface diffusion).

We focus on the Dual Site-Bond (DSB) model proposed by Mayagoitia, Zgrablich and coworkers[1-4], which intrinsically contains a description both of the energy-site distribution (associated with the adsorbing centres) and of the saddle-point energies between two nearest neighbouring sites (minima between two local maxima of the adsorption energy) which control particle hopping between nearest neighbouring sites.

The generation of a DSB model with prescribed distribution functions $F_S(E)$ and $F_B(E)$, respectively for the site and bond energies E_S and E_B, is subject to two constitutive laws expressing globally and locally the fact that the bond energies are the local minima between two energy maxima. These law are: (I) $F_B(E) - F_S(E) \geq 0$, (II) $p_{SB}(E_S, E_B) = 0$ for $E_S < E_B$, where $p_{SB}(E_S, E_B)$ is the joint probability density function of nearest neighbouring site-bond energies.

There are two ways of introducing correlations in the DSB model, either directly in the assigment process or at a global level. In the first case, energy correlations are introduced at a local level, by assuming a given correlation function $\phi(E_S, E_B)$ in the joint pdf $p_{SB}(E_S, E_B) = p_S(E_S) p_B(E_B) \phi(E_S, E_B)$ of nearest neighbouring site and bond energies[5]. The function $\phi(E_S, E_B)$, may (as for the so-called Self-Consistent Case) or may not be a functional of the site and bond energy distribution functions.

The local correlation function $\phi(E_S, E_B)$ and the resulting Markovian generation process do not make it possible to control long-distance correlation properties of the energy landscape.

In the latter case, by considering the site and bond energy fields $E_S(\mathbf{x})$ and $E_B(\mathbf{x})$ as a dual manifestation of the same entity $E(\mathbf{x})$, it is possible to define a single spatial correlation function $C_{2E}(\mathbf{x})$ for the whole energy landscape. For example, for a one-dimensional lattice model, $E(x) = E_S(x)$ for even values of the lattice coordinate x

and $E(x) = E_B(x)$ for odd values of x, and the spatial correlation function $C_{2E}(\mathbf{x})$ can be defined as

$$C_{2E}(2x) = \frac{1}{2}\left[C_{2E_S}(x) + C_{2E_B}(x)\right], \qquad C_{2E}(2x-1) = C_{E_SE_B}(x), \qquad (1)$$

where the correlation functions $C_{2A}(x)$ $(A = E_S, E_B)$ and $C_{E_SE_B}(x)$ are given by

$$C_{2A}(x) = \frac{< (A(x+y)- < A >)(A(y)- < A >) >}{< A^2 > - < A >^2} \qquad (2)$$

$$C_{E_SE_B}(x) = \frac{< (E_S(x+y)- < E_S >)(E_B(y)- < E_B >) >}{\sqrt{< E_S^2 > - < E_S >^2}\sqrt{< E_B^2 > - < E_B >^2}}. \qquad (3)$$

Consequently, the aim of the generation process is to obtain a DSB energy landscape with a given correlation function $C_{2E}(x)$ defined by eq. (1). This problem is similar to the inverse problem in the reconstruction of porous media, and indeed the same formal mathematical apparatus can be applied[6-7].

No matter what kind of assigment process[a] and what kind of method for inducing correlation (local or global) are chosen, the fulfilment of the local exclusion principle (law II of the DSB model) induces a distortion in the shape of the local distribution functions, and the effective site and bond energy distribution functions $F_S^*(E)$ and $F_B^*(E)$, obtained from the generation of the lattice by implementing the assigment rules, differ from $F_B(E)$ and $F_S(E)$.

The aim of this article is to analyze the influence of correlations on the effective site and bond energy distribution functions, and to obtain a closed-form expression for the functional relation relating $F_B(E)$ and $F_S(E)$ to $F_S^*(E)$ and $F_B^*(E)$.

Since this article is methodological, we analyze those assigment processes which highlight in the clearest and simplest fashion the distortion induced in the effective distribution function upon the assignment of given correlation properties.

The remainder of the article is organized as follows. In the next section we obtain an analytic expression for the effective site and bond energy distribution functions for a given nearest neighbouring correlation function $\phi(E_S, E_B)$. We then present a general method to generate site DSB models with prescribed site and bond energy distribution functions and given correlation function $C_{2E}(x)$ of the whole energy field $E(\mathbf{x})$.

2 EFFECTIVE DISTRIBUTIONS IMPLEMENTING LOCAL CORRELATIONS

For the sake of simplicity, we shall consider one-dimensional models. In this case, the SC assignment coincides with the alternating assignment of site and bond energies

---

[a] Three different generation methods are usually considered, referred to Case A, Case B and the Self-Consistent Case. In Case A, all the site energies are assigned first according to the given local distribution function $F_S(E)$. Subsequently, energy values are assigned to all the bonds according to the local distribution function $F_B(E)$ and to the local exclusion principle.

In Case B, the assignment of bond and site energies proceeds through an iterative process. In the one-dimensional case, the energy of the first site of the linear chain is assigned first, then the energy of the adjacent bond, in accordance with the local exclusion principle, and so on with the next site and the next bond. The Self-Consistent Case is similar to Case B, but nearest neighbouring site and bond energies are assigned according to the joint pdf $p_{SB}(E_S, E_B) = p_S(E_S)p_B(E_B)\phi(E_S, E_B)$.

$E_S(x)$, $E_B(x+1)$ characterizing Case B procedure, but the joint probability distribution $p_{SB}(E_S, E_B)$ for nearest neighbouring sites and bonds is given by $p_{SB}(E_S, E_B) = p_S(E_S)p_B(E_B)\phi(E_S, E_B)$, and the local conditional probabilities for bond and site energies are $F_B(E; E_S) = \int_{B_1}^{E} p_B(\epsilon)\phi(E_S, \epsilon)d\epsilon$, $F_S(E; E_B) = \int_{S_1}^{E} p_S(\epsilon)\phi(\epsilon, E_B)d\epsilon$.

For each bond, its distribution function in the assigment $F_B^*(E; E_{S1})$ depends on the energy value of the preceding site E_{S1}, and the distribution function from which the energy of each site is generated depends on the energy value of the preceding bond E_{B1}.

In the presence of a correlation function $\phi(E_S, E_B)$, by taking into account the local exclusion principle, the effective conditional probabilities are given by

$$F_B^*(E; E_{S1}) = \begin{cases} F_B(E; E_{S1}) & \text{if } E_{S1} \geq B_2 \\ F_B(E; E_{S1})/F_B(E_{S1}; E_{S1}) & \text{for } E \leq E_{S1} \text{ if } E_{S1} \leq B_2 \\ 1 & \text{for } E \geq E_{S1} \text{ if } E_{S1} \leq B_2 , \end{cases} \quad (4)$$

and

$$F_S^*(E; E_{B1}) = \begin{cases} F_S(E; E_{B1}) & \text{if } E_{B1} \leq S_1 \\ 0 & \text{for } E \leq E_{B1} \text{ if } E_{B1} \geq S_1 \\ \frac{F_S(E; E_{B1}) - F_S(E_{B1}; E_{B1})}{F_S(S_2; E_{B1}) - F_S(E_{B1}; E_{B1})} & \text{for } E \geq E_{B1} \text{ if } E_{B1} \geq S_1 \end{cases} \quad (5)$$

The effective distributions $F_B^*(E)$ and $F_S^*(E)$ of bond and site energies are the average of eqs. (4)-(5) respectively over $F_S^*(E)$ and $F_B^*(E)$, i.e.

$$F_B^*(E) = \int_{S_1}^{S_2} F_B^*(E; E_{S1})p_S^*(E_{S1})dE_{S1} , \quad F_S^*(E) = \int_{B_1}^{B_2} F_S^*(E; E_{B1})p_B^*(E_{B1})dE_{B1} .$$

This leads to a pair of coupled nonlinear integral equations for $F_B^*(E)$ and $F_S^*(E)$:

$$F_B^*(E) = F_S^*(E) + \int_E^{B_2} \frac{F_B(E; \epsilon)p_S^*(\epsilon)}{F_B(\epsilon; \epsilon)}d\epsilon + \int_{B_2}^{S_2} F_B(E; \epsilon)p_S^*(\epsilon)d\epsilon , \quad (6)$$

$$F_S^*(E) = \int_{B_1}^{S_1} F_S(E; \epsilon)p_B^*(\epsilon)d\epsilon + \int_{S_1}^{E} \frac{F_S(E; \epsilon) - F_S(\epsilon; \epsilon)}{F_S(S_2; \epsilon) - F_S(\epsilon; \epsilon)}p_B^*(\epsilon)d\epsilon . \quad (7)$$

By differentiating with respect to E, an explicit relation between $p_B^*(E)$, $p_S^*(E)$, $p_B(E)$, $p_S(E)$ and $\phi(E_S, E_B)$ is obtained:

$$p_B^*(E) = p_B(E)\left[\int_E^{B_2} p_S^*(\epsilon)\phi(\epsilon, E)\left(\int_{B_1}^{\epsilon} p_B(\epsilon')\phi(\epsilon', \epsilon)d\epsilon'\right)^{-1}d\epsilon + \int_{B_2}^{S_2} p_S^*(\epsilon)\phi(\epsilon, E)d\epsilon\right],$$

$$p_S^*(E) = p_S(E)\left[\int_{B_1}^{S_1} p_B^*(\epsilon)\phi(E, \epsilon)d\epsilon + \int_{S_1}^{E} p_B^*(\epsilon)\left(\int_{\epsilon}^{S_2} p_S(\epsilon')\phi(\epsilon', \epsilon)d\epsilon'\right)^{-1}d\epsilon\right], \quad (8)$$

which expresses the influence of the correlation function ϕ on the effective distribution functions. A more detailed discussion of the relationship between local and effective distribution functions is developed by Giona[8].

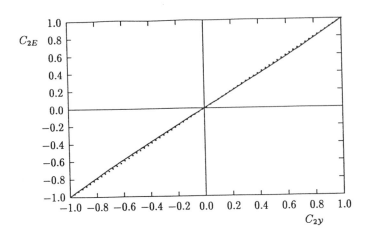

Figure 1: C_{2E} vs C_{2y} for uniform effective bond and site energy distributions. The dotted line is the curve C_{2E} vs C_{2y}.

3 GLOBAL CORRELATION IN DSB MODELS

In the approach discussed in the last section, based on the local correlation function $\phi(E_S, E_B)$, it is very difficult to control the global correlation properties of the energy landscape. Such properties, expressed by the correlation function $C_{2E}(x)$, play a significant role in surface phenomena (especially if long-range correlations exist, $C_{2E}(x) \sim x^{-\alpha}$). For these reasons, it is interesting to develop a method for generating DSB energy landscapes with prescribed global correlation properties.

In this section we outline the mathematical formulation of a generation method for globally correlated DSB surfaces, focusing for the sake of simplicity on Case A assigment on the one-dimensional line. This method can be extended with no major difficulty to problems of higher dimensions and to other kinds of assigment processes.

The basic idea is that site and bond energy fields $E_S(x)$ and $E_B(x)$ can be generated from a single Gaussian and correlated stochastic process $\mathcal{Y}(x)$ by applying the nonlinear filters $E_S(x) = \mathcal{G}_S^*(\mathcal{Y}(x))$ (for even values of the lattice coordinate x) or $E_B(x) = \mathcal{G}_B^*(\mathcal{Y}(x))$ depending on the Gaussian distribution F_y of $\mathcal{Y}$ and on the effective site and bond distribution functions $F_S^*(E)$ and $F_B^*(E)$.

For each point x, the corresponding value of the site or bond energy field $E_A(x)$ ($A = E_S, E_B$) can be obtained by solving the probability balance equation

$$F_y(\mathcal{Y}(x)) = \frac{1}{\sqrt{2\pi}} \int_{-\infty}^{y} \exp\left(-y^2/2\right) dy = F_A^*(E(x)) , \tag{9}$$

so that the nonlinear filter $\mathcal{G}_A^*$ is therefore given by $\mathcal{G}_A^*(\mathcal{Y}) = F_A^{*-1}(F_y(\mathcal{Y}(x)))$.

Eq. (9) ensures statistically that the $E_A(x)$ admits the effective distribution function $F_S^*(E)$. Since the stochastic Gaussian process $\mathcal{Y}$ is correlated with correlation function $C_{2y}(x)$, the resulting energy fields $E_S(x)$ and $E_B(x)$ also exhibit correlation properties.

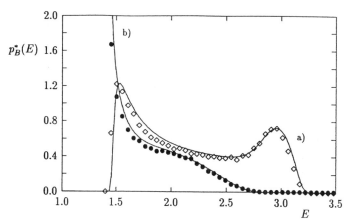

Figure 2: $p_B^*(E)$ vs E $(1.4 \leq E_B \leq 3.4$ a.u.$)$ for uniform site energy distribution $(2 \leq E_S \leq 4$ a.u.$)$ and for the nonlinear filter H_B given by eq. (17). The dots are the numerical results and the lines the theoretical predictions, eq. (22). a) $\mu = 1$; b) $\mu = 4$. $\lambda = 0.05$ and the number of sites in the simulation is $N = 2 \cdot 10^5$.

The relation between $C_{2E}(x)$, eq.(1), and $C_{2y}(x)$ is given for even x by

$$C_{2E} = \frac{1}{2} \left\{ \int_{-\infty}^{\infty} dy_1 \int_{-\infty}^{\infty} dy_2 \left[\frac{(\mathcal{G}_S^*(y_1) - <E_S>)(\mathcal{G}_S^*(y_2) - <E_S>)}{<E_S^2> - <E_S>^2} \right] p(y_1, y_2, C_{2y}) \right.$$

$$\left. + \int_{-\infty}^{\infty} dy_1 \int_{-\infty}^{\infty} dy_2 \left[\frac{(\mathcal{G}_B^*(y_1) - <E_B>)(\mathcal{G}_B^*(y_2) - <E_B>)}{<E_B^2> - <E_B>^2} \right] p(y_1, y_2, C_{2y}) \right\},$$

$$(10)$$

and for odd x by

$$C_{2E} = \int_{-\infty}^{\infty} dy_1 \int_{-\infty}^{\infty} dy_2 \left[\frac{(\mathcal{G}_S^*(y_1) - <E_S>)(\mathcal{G}_B^*(y_2) - <E_B>)}{\sqrt{(<E_S^2> - <E_S>^2)(<E_B^2> - <E_B>^2)}} \right] p(y_1, y_2, C_{2y}),$$

$$(11)$$

where $p(y_1, y_2, C_{2y})$ is a bivariate Gaussian density function with correlation constant $C_{2y}(x)$.

Figure 1 shows the behaviour of $C_{2E}(x)$ vs $C_{2y}(x)$ in the case of uniform effective pdfs $p_S^*(E)$, $p_B^*(E)$. As can be seen, C_{2E} is very close to C_{2y}. This situation also occurs for exponentially distributed site and bond energies, and it is reasonable to assume that

$$C_{2E} \simeq C_{2y}. \tag{12}$$

Eq. (12) simplifies the generation of correlated DSB energy landscapes.

Let us now consider the generation of DSB one-dimensional energy landscapes on the assumption of Case A assignment. The local exclusion principle implies that the bond energies E_B of a generic bond are assigned according to a nonlinear relation $E_B = f(\mathcal{Y}, E_{S1}, E_{S2})$, depending on the energies E_{S1}, E_{S2} of the two sites connected by the bond. Let $K_B(\mathcal{Y})$ be a generic continuous monotonically increasing function mapping $(-\infty, \infty)$ onto $[0,1]$, then

$$f(\mathcal{Y}, E_{S1}, E_{S2}) = \begin{cases} B_1 + (B_2 - B_1)K_B(\mathcal{Y}) & \text{if } E_{S1}, E_{S2} \geq B_2 \\ B_1 + (E_{S1} - B_1)K_B(\mathcal{Y}) & \text{if } E_{S1} \leq B_2, E_{S2} \geq E_{S1} \\ B_1 + (E_{S2} - B_1)K_B(\mathcal{Y}) & \text{if } E_{S2} \leq B_2, E_{S1} \geq E_{S2} \end{cases} \quad (13)$$

Eq. (13) can be rewritten in compact form as

$$f(\mathcal{Y}, E_{S1}, E_{S2}) = B_1 + (\min\{B_2, E_{S1}, E_{S2}\} - B_1)K_B(\mathcal{Y}) . \quad (14)$$

Let $p(E_{S1}, y, E_{S2})$ denote the probability density function for the three-dimensional random variable (E_{S1}, y, E_{S2}), where $E_{S1} = \mathcal{G}_S^*(y_i)$, $y = y_{i+1}$, $E_{S2} = \mathcal{G}_S^*(y_{i+2})$ and $y_i = \mathcal{Y}(i)$.

By definition, the effective bond energy distribution function $F_B^*(E_B)$ is given by

$$F_B^*(E_B) = \int\int\int_{f(y,E_{S1}E_{S2})\leq E_B} p(E_{S1}, y, E_{S2})dE_{S1}dE_{S2}dy . \quad (15)$$

After some steps, we obtain

$$\begin{aligned} F_B^*(E_B) &= \int_{B_2}^{S_2} dE_{S1} \int_{B_2}^{S_2} dE_{S2} \int_{-\infty}^{H_B(E_B,B_2)} p(E_{S1}, y, E_{S2})dy + \\ &+ 2\int_{E_B}^{S_2} dE_{S1} \int_{S_1}^{E_B} \bar{p}(E_{S1}, E_{S2})dE_{S2} + \int_{S_1}^{E_B} dE_{S1} \int_{S_1}^{E_{S1}} \bar{p}(E_{S1}, E_{S2})dE_{S2} + \\ &+ 2\int_{E_B}^{B_2} dE_{S1} \int_{E_{S1}}^{S_2} dE_{S2} \int_{-\infty}^{H_B(E_B,E_{S1})} p(E_{S1}, y, E_{S2})dy , \end{aligned} \quad (16)$$

where $\bar{p}(E_{S1}, E_{S2}) = \int_{-\infty}^{\infty} p(E_{S1}, y, E_{S2})dy$ and $H_B(E_B, \epsilon)$ is given by

$$H_B(E_B, \epsilon) = K_B^{-1}\left(\frac{E_B - B_1}{\epsilon - B_1}\right) . \quad (17)$$

Note that in the two triple integrals on the right-hand side of eq. (16) containing $H_B(E_B, \epsilon)$, the value of ϵ ($\epsilon = B_2$ in the first, $\epsilon = E_{S1}$ in the second) is always greater than or equal to B_2. By taking the derivative with respect to E_B, it follows that

$$\begin{aligned} p_B^*(E_B) &= \int_{B_2}^{S_2} dE_{S1} \int_{B_2}^{S_2} dE_{S2}\, p(E_{S1}, H_B(E_B, B_2), E_{S2})\frac{dH_B(E_B, B_2)}{dE_B} \\ &+ 2\int_{E_B}^{B_2} dE_{S1} \int_{E_{S1}}^{S_2} dE_{S2}\, p(E_{S1}, H_B(E_B, E_{S1}), E_{S2})\frac{dH_B(E_B, E_{S1})}{dE_B} . \end{aligned} \quad (18)$$

The pdf $p(E_{S1}, y, E_{S2})$ can be written in terms of the pdf $p_Y(y_i, y_{i+1}, y_{i+2})$ for the three-dimensional Gaussian random variable (y_i, y_{i+1}, y_{i+2}) as

$$p(E_{S1}, y, E_{S2}) = p_Y(\mathcal{G}_S^{*-1}(E_{S1}), y, \mathcal{G}_S^{*-1}(E_{S2}))\left|\frac{d\mathcal{G}_S^{*-1}(E_{S1})}{dE_{S1}}\frac{d\mathcal{G}_S^{*-1}(E_{S2})}{dE_{S2}}\right| , \quad (19)$$

where the pdf $p_Y(y_i, y_{i+1}, y_{i+2}) = p_Y(\mathbf{y})$ is given by the usual expression for a multi-variate Gaussian random variable

$$p_Y(\mathbf{y}) = \frac{1}{(2\pi)^{3/2}\det(\mathbf{C})} \exp\left[-\frac{1}{2}\mathbf{y}^t\mathbf{C}^{-1}\mathbf{y}\right] , \quad (20)$$

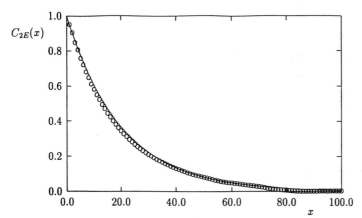

Figure 3: $C_{2E}(x)$ vs x in the case $\mu = 1$, $\lambda = 0.05$. The dots are the numerical results and the line is the theoretical energy correlation function $C_{2E} = \exp(-\lambda x)$.

and the covariance matrix $\mathbf{C}$ for the three-dimensional normalized Gaussian random variable $\mathbf{y}$ attains the form

$$
\mathbf{C} = \begin{pmatrix} 1 & c_1 & c_2 \\ c_1 & 1 & c_1 \\ c_2 & c_1 & 1 \end{pmatrix}
\tag{21}
$$

where $c_1 = C_{2y}(1)$, $c_2 = C_{2y}(2)$, since y_i, y_{i+1} and y_{i+1}, y_{i+2} are associated with nearest neighbouring sites and y_i, y_{i+2} with next nearest neighbouring sites, and the correlation function $C_{2y}(x)$ is expressed in lattice units. The determinant of the covariance matrix is $\det(\mathbf{C}) = 1 - 2c_1^2 - c_2^2 + 2c_2 c_1^2$.

By performing a coordinate transformation, eq. (18) can be rewritten as

$$
\begin{aligned}
p_B^*(E_B) &= \int_{y(B2)}^{\infty} dy_i \int_{y(B2)}^{\infty} dy_{i+2}\, p_Y(y_i, H_B(E_B, B_2), y_{i+2}) \frac{dH_B(E_B, B_2)}{dE_B} + \\
&\quad + 2 \int_{y(E_B)}^{y(B_2)} dy_i \int_{y_i}^{\infty} dy_{i+2}\, p_Y(y_i, H_B(E_B, \mathcal{G}_S^*(y_i)), y_{i+2}) \frac{dH_B(E_B, \mathcal{G}_S^*(y_i))}{dE_B},
\end{aligned}
\tag{22}
$$

where $y(E_B) = \mathcal{G}_S^{*-1}(E_B)$ and $y(B_2) = \mathcal{G}_S^{*-1}(B_2)$.

As an example, let us take the case of a nonlinear filter $H_B(E_B, \epsilon)$ given by

$$
H_B(E_B, \epsilon) = \tan \left[\pi \left(\frac{E_B - B_1}{\epsilon - B_1} \right)^{1/\mu} - \frac{\pi}{2} \right],
\tag{23}
$$

where μ is a positive constant. Correspondingly,

$$
K_B(y) = \left(\frac{\arctan(y) + \pi/2}{\pi} \right)^{\mu}.
\tag{24}
$$

Let us further consider an exponentially correlated energy landscape $C_{2E}(x) = \exp(-\lambda x)$, where λ is related to the correlation length $L_c = 1/\lambda$ (expressed in lattice units).

Figure 2 shows the behaviour of the effective bond energy pdf $p_B^*(E)$ obtained in the case of a uniform site energy pdf ($S_1 = 2$, $S_2 = 4$ a.u.) for $\lambda = 0.05$ and for two different values of μ obtained from the numerical simulation of the DSB model compared with the theoretical prediction eq. (22). Figure 3 shows the comparison of the prescribed correlation energy functions and those obtained from the simulations.

Eq. (22) can of course be applied to solve the inverse problem, i.e. to find the function H_B which furnishes a given effective bond energy pdf with prescribed correlation properties.

4 CONCLUDING REMARKS

This article presents a method to generate globally correlated DSB energy landscapes starting from a prescribed correlation function $C_{2E}(x)$. This method can be viewed as an extension of the inverse reconstruction method of porous media developed by Adler et al.[9] and by Giona and Adrover[6]. Although the analysis has been restricted to one-dimensional structures, extension to higher dimensions is straightforward. The physical implications of long-range correlation in surface phenomena will be discussed elsewhere.

We have also obtained a closed-form expression for the nonlinear functional equations relating the effective and the local distribution functions in DSB model implementing local and global correlations. These equations can be also applied in an inverse way, i.e. to determine $F_B(E_B)$ and $F_S(E_S)$ starting from given expressions of $F_B^*(E_B)$ and $F_S^*(E_S)$.

References

1. J. L. Riccardo , V. Pereyra, J. L. Rezzano, D.A. Rodriguez Saá and G. Zgrablich, *Surf. Sci.*, 1988, **204**, 289.
2. V. Mayagoitia, F. Rojas F, J.L. Riccardo, V. D. Pereyra and G. Zgrablich, *Phys. Rev. B*, 1990, **41**, 7150.
3. J.L. Riccardo, V. Pereyra and G. Zgrablich, *Langmuir*, 1993, **9**, 2730.
4. M. J. Cruz, V. Mayagoitia and F. Rojas *J. Chem Soc. Faraday Trans. 1*, 1989, **85**, 2079.
5. J.L. Riccardo, W. A. Steele, A. J. Ramirez Cuesta and G. Zgrablich, *Langmuir*, 1995, (accepted for publication).
6. M. Giona and A. Adrover, *AIChE J.*, 1996 **42**, 1407.
7. A. Adrover and M. Giona, 1996, (submitted to COPS IV).
8. M. Giona, *Langmuir*, 1996, (submitted).
9. P.M. Adler, C.G. Jacquin C.G. and J.A. Quiblier, *Int. J. Multiphase Flow*, 1990, **16**, 691.

DETERMINATION OF THE MICROPORE SIZE IN CARBONS BY MEANS OF EXPERIMENTAL AND CALCULATED ISOTHERM DERIVATIVES

F. Ehrburger-Dolle[1], M. T. Gonzalez[2], M. Molina-Sabio[2] and F. Rodriguez-Reinoso[2]

[1] Institut de Chimie des Surfaces et Interfaces, CNRS, F-68057 Mulhouse (France)
[2] Departamento de Quimica Inorganica, Universidad de Alicante, E-03080 Alicante (Spain)

1 INTRODUCTION

Adsorption of gases remains the most widely used experimental method for the characterisation of microporous solids because molecules can be considered as rulers of different sizes for the measurement of geometrical features at nanometric scales. The interpretation of the experimental data in terms of adsorbent geometry, was made possible by means of classical theories, like the Polanyi volumetric micropore filling model, leading to the Dubinin-Radushkevich (DR) equation. This approach and its modifications were reviewed in several books[1-3]. Since a few years, there is a renewed interest in the adsorption data analysis, mainly because the classical potential model gives poor results for large micropores. New approaches, based on molecular models of adsorption involving molecule-solid and molecule-molecule interactions were presented and described among others, in the Proceedings of the two former COPS Symposia[4,5]. In the present paper, one will consider the adsorption isotherms predicted by the density functional theory (DFT) using the nonlocal approximation (NLA) calculated by Lastoskie et al.[6] yet refined versions[7] using a smoothed density approximation (SDA), yielded improvement for the analysis of experimental isotherms, particularly in the case of non microporous solids.

It was shown recently[8,9] that the derivative curves of adsorption isotherms can be fitted to power laws. The physical meaning of the exponents, was inferred[9] in terms of fractal geometry. The aim of the present paper is twofold..Firstly, it is to show that the micropore size of activated carbons can be determined by comparing the exponents measured on experimental curves to the ones characterising the curves calculated for adsorption of nitrogen at 77 K in slit pores of different widths. Secondly, it is to show that the existence of a characteristic line, involved in Dubinin's model of micropore filling, is a natural consequence of the similarity of the exponents.

2 ADSORPTION ISOTHERMS

The first step of the present study will be devoted to the comparison between the derivative of experimental and model[6] isotherms of adsorption of nitrogen at 77 K on a microporous carbon. The adsorbent is an activated carbon, D52, obtained by carbonisation of olive stones followed by carbon dioxide oxidation at 1098 K up to 52 % burn-off.

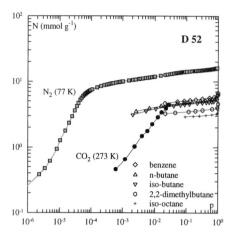

Figure 1 *Isotherms of adsorption of different molecules on D52[10-13]. Adsorption temperatures are 273 K for n- and iso-butane and 298 K for benzene, iso-octane and 2,2-dimethylbutane*

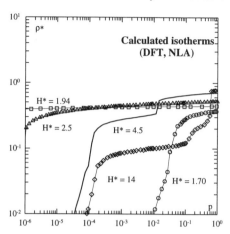

Figure 2 *Model isotherms for nitrogen adsorption in carbon slit pores redrawn from publication of Lastoskie et al.[6]; p = P/P_o is the relative pressure*

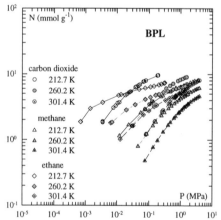

Figure 3 *Isotherms of adsorption on BPL, drawn from the data published by Reich et al.[14]*

Details concerning the preparation of the sample and the measurement of the adsorption isotherms (Figure 1) were described in previous papers[10-13]. The BET surface area of D52 is close to 1200 $m^2 g^{-1}$.

The model isotherms (Figure 2) were calculated by Lastoskie *et al.*[6]. The pore width H and the mean fluid density ρ are scaled with respect to the fluid-fluid molecular diameter σ_{ff}. Thus, for nitrogen at 77 K, one has: $H^* = H/\sigma_{ff}$ with $\sigma_{ff} = 0.357$ nm and $\rho^* = \rho/\sigma_{ff}^3$ with $\rho = 17.4$ molecule nm^{-3}. The data points shown in Figure 2, were obtained by digitising the curves obtained for different pore widths, as follows:

- pores able to accommodate two or less than two nitrogen molecules in width: H^*=1.70, 1.94 and 2.5 corresponding respectively to H=0.61, 0.69 and 0.89 nm

- pores able to accommodate more than two N_2 molecules in width: H^*=4.5 and 14 corresponding respectively to H=1.61 and 5 nm.

BPL[14] is a commercially available activated carbon, characterised by a BET surface area close to 990 $m^2 g^{-1}$. Details on the determination of the adsorption isotherms plotted in Figure 3, are indicated in the paper by Reich *et al.*[14]. Adsorption isotherms were measured at pressures larger than the atmospheric pressure.

3 ANALYSIS OF THE ISOTHERM DERIVATIVES

The derivative of an adsorption isotherm is determined from the experimental data by calculating the differences N_i-N_{i-1} and the corresponding p_i-p_{i-1} between two successive values. As a first approximation, the difference p_i-p_{i-1} can be considered as small enough for the ratio $(N_i-N_{i-1})/(p_i-p_{i-1}) = (\Delta N/\Delta p)_i$ to be assimilated to $(dN/dp)_i$ which will be plotted as a function of $p = (p_{mean})_i = (p_i+p_{i-1})/2$, in logarithmic co-ordinates.

3.1 Adsorption of nitrogen at 77 K

3.1.1 Experimental data. Figure 4 shows the derivative of the N_2 adsorption isotherm on the activated carbon D 52. In order to compare eventually this curve with the calculated ones, the data are expressed as n σ^3 where n is expressed as molecules and $\sigma = \sigma_{ff} = 0.357$ nm for N_2[6].

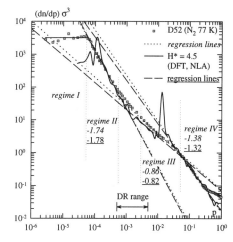
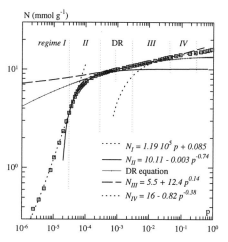

Figure 4 *Derived curves of experimental isotherm and of fitting model isotherm*

Figure 5 *Experimental isotherm fitting curves*

It was already shown[9], that the derived curves of isotherms of adsorption of nitrogen on microporous carbons exhibit two or four different regimes characterised by different power laws. The exponents corresponding to the different regimes are indicated, *in italics*, on Figure 4. The equations of the fitting curves, determined by the same method as the one used for graphitized Vulcan[9], are indicated in Figure 5.

In regime I, the derivative can be considered as constant, as a first approximation, and corresponds to $1.19 \; 10^5$ mmol g^{-1}. In the case of nonmicroporous carbons in the low pressure range (regime I), the integration constant was negligible[9], yielding the Henry law. However, in the present case, the constant cannot be omitted. It indicates the existence of highly energetic sites which are filled at very low relative pressures.

Regime II was attributed[9] to the formation of the first adsorbed layer. In this case, the constant of integration $N^*_{II} = 10.11$ mmol g^{-1}, is the number of molecules in a monolayer, yielding the surface area $S_{II} = 986$ m^2 g^{-1}. This value is smaller than S_{BET} (=1200 m^2 g^{-1}) determined between $p = 0.05$ and $p = 0.12$ for which one obtains a positive C value (C =

350). Meanwhile, as also observed for Grafoil, the BET plot exhibits a change of slope near p = 0.1. By considering the second line, between p = 0.12 and 0.2 (leading to a negative C value), one obtains $S_{BET} = 999$ m^2 g^{-1}, which is very close to the value obtained by using N*$_{II}$.

Regime III, characterised by a Freundlich type equation, was attributed[9] to simultaneous monolayer and multilayer adsorption.

Regime IV, which is not observed for nonmicroporous, carbons is characteristic of the secondary micropore filling. The fitting equation yields N*$_{IV}$ = 15.2 mmol g^{-1}, at p = 1. One may assume that, at this point, the fluid density inside the micropores is equal to the liquid density ($\rho = 0.808$ g cm^3) and deduce the following value of the micropore volume $W_o = 0.53$ cm^3 g^{-1}.

3.1.2 Fit to the calculated curve. From the comparison between the derivatives of the series of calculated curves and that of the experimental one (Figure 4), it appears that the curve obtained for H* = 4.5, i.e. H = 1.6 nm, is a reasonable fit. This result agrees fairly well with the maximum of the distribution determined by using Monte Carlo simulations[15] However, there are some differences between the two curves:
- the first one appears in the low pressure range, corresponding to regime I
- the second one is a cross-over region between regime II and III, which cannot be fitted to a power law; this region corresponds to the one fitting to a DR equation
- the peak located between regime III and IV is not observed on the experimental curve. However, it was shown[7] that the use of more refined approximations yields a smoothening of the transitions corresponding to this peak.
- in the high pressure range (p > 0.2), the experimental curve exhibits a behaviour somewhat similar to that of regime III, with an exponent, if any, slightly negative. This behaviour could be attributed to the onset of a third layer in pores of sizes H* ranging between $\approx$ 6 (H $\approx$ 2 nm) and $\approx$ 8 (H $\approx$ 3 nm), but probably not to adsorption on a free surface (external surface) or in mesopores for which the derivative increases at p > 0.6.

3.1.3 Origin of regime I. As in the previous cases[9], the discrepancies between the experimental data and the model isotherm, calculated by assuming infinite, perfect graphitic layers, can be attributed to structural defects such as poorly organised and/or finite size graphitic layers and edge carbon atoms. When regime II begins, the number of adsorbed molecules is equal to 3.6 mmol g^{-1}. If one assumes that it corresponds to the coverage of all these sites by one nitrogen molecule, one may determine the corresponding specific surface area, $S_I = 350$ m^2 g^{-1}.

3.1.4 Cross-over between regime II and regime III and DR equation. Interestingly enough, the region of pressures ranging between 2.5 10^{-4} and 2.5 10^{-3}, in which the derivative does not follow a power law (Figure 4), corresponds to the domain of linearity of the DR plot. The corresponding DR equation, plotted on Figure 5, writes:

$$\ln (N_{DR}) = \ln (N^*_{DR}) - [RT/\beta E_o \ln (1/p)]^2 \tag{1}$$

with $\beta = 0.33$ and $E_o = 23.7$ kJ mol^{-1} and N*$_{DR}$ = 13.8 mmol g^{-1}.

N*$_{DR}$ is larger than N*$_{II}$ (= 10.11 mmol g^{-1}), corresponding to the monolayer capacity, and smaller than N*$_{IV}$ (= 15.2 mmol g^{-1}), at the saturation pressure, assumed to be the maximum capacity of the micropores.

Although further refinements have been proposed[3,16,17], the pore width H is generally determined from the DR fitting equation by means of the following relation[18]:

$$H = 26/E_o \qquad (2)$$

From Eqs. (1) and (2), one obtains $H = 1.1$ nm, which is smaller than $H = 1.6$ nm. Interestingly enough, the same result[9] is obtained from the DR plot determined for the model isotherm ($H^* = 4.5$), in the same range of pressure, whereas Eq. (2) yields the expected value $H = 1.6$ nm, in the second linear region of the DR plot, corresponding to $2 \cdot 10^{-2} < p < 0.1$, i.e. to the beginning of regime IV.

Finally, Figure 5 shows that the experimental data are also fairly well fitted to the equation describing regime II, up to $p \approx 10^{-3}$ and to the equation describing regime III above this limit.

3.1.5 Micropore shape. The above discussion suggests the existence of two types of carbon surfaces: surfaces consisting of poorly organised carbon atoms and graphitic layers. If one assumes slit shaped pores between two graphitic walls, the relation between the surface of the walls S_g, the pore volume W_o and the width H writes:

$$S_g = 2000 \ W_o/H \qquad (3)$$

By using the values $W_o = 0.53$ cm^3 g^{-1} and $H = 1.6$ nm determined above, one obtains $S_g = 662$ m^2 g^{-1}. By taking into account the area displayed by non graphitic carbon atoms, $S_I = 350$ m^2 g^{-1}, one obtains a total surface area $S_t = 1012$ m^2 g^{-1}. Because this value differs from the surface area S_{II} (= 986 m^2 g^{-1}) by only 2.6 %, one may conclude that the whole analysis is consistant. These results suggest that the micropore network, evidenced by SAXS[19] may consist of more or less rectangular channels between two graphitic layers, distant by H, and two walls made of less organised carbon atoms.

3.2 Adsorption of other molecules on D52

Figure 6 shows the derivatives of isotherms of adsorption of molecules of different sizes. In the case of CO_2 (273 K) the derivative remains constant for $p < 5 \cdot 10^{-3}$, involving, as for N_2 (77 K) during regime I, a Henry type equation for the integral curve. However, this constant is much smaller than in the case of N_2, as a result of the higher adsorption temperature. Above this pressure, the experimental data overlap that of nitrogen, in the regime III domain. The exponent $v_{III} = -0.88$ is close to that obtained for N_2 (= -0.86).

It is worth to indicate that regime I corresponds, in the DR plot[11], to the well known low pressure deviations which is positive in the case of CO_2 adsorption at 273 K and negative, in the case of N_2 adsorption at 77 K.

On the opposite, the data obtained for all other molecules are located below the N_2 curve. The size of N_2 and CO_2 molecules is quite similar ($\sigma_{ff} = 0.357$ nm for the former and $\sigma = 0.34$ nm for the latter) whereas all other molecules are significantly larger. If one assumes that the pore width is equal to 1.6 nm, the width H^* expressed in molecular size units $\sigma = [V_M/6.02 \ 10^{23}]^{1/3}$), is close to $H^* = 3$ for benzene ($\sigma = 0.53$ nm), for n-butane ($\sigma = 0.54$ nm) and for iso-butane ($\sigma = 0.56$ nm) and close to $H^* = 2.5$ for iso-octane ($\sigma =$

0.65 nm) and 2,2-dimethylbutane ($\sigma = 0.60$ nm). In other words, for all these molecules, there is no possibility of secondary micropore filling.

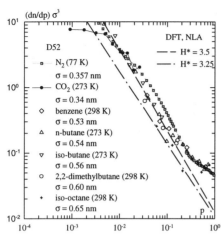

Figure 6 *Derivative curves of the isotherms of adsorption of different molecules on D52*

As a first approximation, the data corresponding to the first series of adsorbates can be fitted to the curve calculated for H* = 3.5 (instead of H* = 3 expected) and the second ones, to the curve calculated for H* = 3.25 (instead of H* = 2.5), which is acceptable owing to the uncertainty on the σ values, which are not the true σ_{ff} values, and the limited number of experimental data available for each adsorbate.

It follows that the curves calculated for the adsorption of N_2 at 77 K also describes the pore filling mechanism for any molecule, at any temperature, provided the interaction ε_{sf} with the surface, is similar. Obviously, these curves could not be used in the case of water. However, it is likely that reducing ε_{sf} would have the same effect than reducing H* below a critical value.

3.3 Adsorption of methane, ethane and carbon dioxide on BPL

The derivatives of the adsorption isotherms obtained for CO_2 are shown in Figure 7a. It appears that all curves exhibit two regions:
- above a given pressure P*(T) which increases with temperature, all experimental data are located on the same line with a slope equal to $v_{CO_2} = -1.04$. It follows that the mechanism of pore filling, corresponding to this regime, is independent on the temperature of adsorption.
- below this pressure, the data are located below this line. Unfortunately, the range of pressure does not extend down to very low pressure. Anyhow, if a power law would be evidenced, the absolute value of its exponent would be smaller than 1.

The same features are observed for adsorption of methane at supercritical temperatures. For P > P*, the data are located on the same line as the CO_2 ones, thus: $v_{CH_4} = -1.04$. The corresponding fitting line is plotted in Figure 7b.

Below P*, the behaviour of ethane (Figure 7b) is somewhat different. Above P*, all data also belong to the same line, but the corresponding exponent is slightly different ($v_{C_2H_6} = -1.10$) from the one obtained for carbon dioxide and methane.

As before, one will search the value of H* leading to a derivative curve with the same exponent. One finds: H* = 1.75 ($v = -1.12$), H* = 1.94 ($v = -1.06$) and H* = 2.25 ($v = -1.04$). It follows that, for methane and CO_2, the pore size H* is close to 2.25. This corresponds respectively to H = 0.86 nm and H = 0.77 nm. Owing to the uncertainty on the molecular size σ, H = 0.81 ± 0.05 nm is inferred. Thus, for ethane ($\sigma_{C_2H_6} = 0.425$ nm)

H* should be close to 1.9 as expected for $v \approx -1.08$. It follows that BPL exhibits pores able to accommodate less than two layers of ethane molecules. By a different approach, involving adsorbate-adsorbent interactions, Rudzinski *et al.*[20] have shown that adsorption of ethane cannot be fitted to the same equation as adsorption of methane. Our results suggest that differences in the isotherm equation can also be explained by differences in the molecule size or bulkiness.

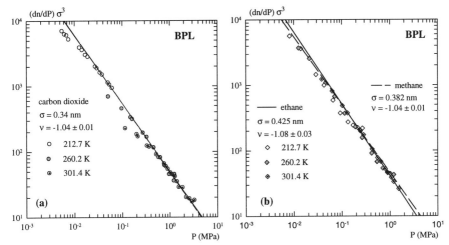

Figure 7 *Derivatives of the isotherms of adsorption of (a) carbon dioxide and (b) ethane on BPL*

It is worth mentioning, on the one hand, that the pore width (H = 0.8 nm), corresponding to the maximum of the distribution determined by means of the DFT Plus Micromeritics Software[21], from an argon adsorption isotherm (87.3 K), reported by Olivier[21], agrees with the result obtained by the method presented here. On the other hand, the DR equation, plotted for CO_2 (with $P_o = 3.44$ MPa at 260 K) yields $W_o = 0.39$ cm^3 g^{-1}. By using Eq.(3), with $S_{BET} = 990$ m^2 g^{-1}, one obtains H = 0.79 nm, which is agrees with the present result. On the opposite, the pore width (H = 1.18 nm) deduced from $E_o = 22$ kJ/mol (with $\beta = 0.35$) by means of Eq. (2), is significantly larger.

4 CONCLUSION

From the work described in the present paper, the following conclusions can be inferred:

- Within a given range of pressure, the exponent v is a fingerprint of the process of adsorption occurring within a given range of pressure. As we have shown previously[9], v can be related to a fractal dimension.

- Integration yields classical isotherm equations, fitting the experimental data within the corresponding pressure range, from which the surface area of the micropore walls and the micropore volume, are obtained.

- For nitrogen adsorption at 77 K, the micropore width is deduced from that of the model isotherm, calculated for a given H* value, characterised by the same exponent(s).

- Providing similar molecule-carbon interactions, the v value characteristic of micropore filling, depends only on the pore width (expressed as molecular size units), H*, but not on adsorption temperature (on the contrary to surface coverage); it follows that the derivative curves of model isotherms can be used to fit data obtained for molecules other than nitrogen. This conclusion is the same as the one involved in Dubinin's approach of primary micropore filling.

Acknowledgements

The authors wish to thank Wladyslaw Rudzinski (University Maria Curie-Sklodowska, Lublin, Poland), Peter Pfeifer and Haskell Taub (University of Missouri, USA) for stimulating discussions.

References

1. S. J. Gregg and K. S. W. Sing, "Adsorption, Surface Area and Porosity", Academic Press, London, 1982, Chapter 4, p. 195.
2. K. K. Unger, J. Rouquérol, K. S. W. Sing and H. Kral, "Characterization of Porous Solids", Elsevier, Amsterdam, 1988.
3. W. Rudzinski and D. H. Everett, "Adsorption of Gases on Heterogeneous Surfaces", Academic Press, London, 1992.
4. F. Rodriguez-Reinoso, J. Rouquérol, K. S. W. Sing and K. K. Unger, "Characterization of Porous Solids II", Elsevier, Amsterdam, 1991.
5. J. Rouquérol, F. Rodriguez-Reinoso, K. S. W. Sing and K. K. Unger, "Characterization of Porous Solids III", Elsevier, Amsterdam, 1994.
6. C. Lastoskie, K. E. Gubbins and N. Quirke, *J. Phys. Chem.*, 1993, **97**, 4786.
7. J. P. Olivier, W. B. Conklin and M. v. Szombathely, ref.[5], p. 81.
8. F. Ehrburger-Dolle, *Langmuir*, 1994, **10**, 2052.
9. F. Ehrburger-Dolle, *Langmuir*, 1996, to appear.
10. F. Rodriguez-Reinoso, J. M. Martin-Martinez, M. Molina-Sabio, R. Torregrosa and J. Garrido-Segovia, *J. Colloid Interface Sci.*, 1985, **106**, 315.
11. J. Garrido, A. Linares-Solano, J. M. Martin-Martinez, M. Molina-Sabio, F. Rodriguez-Reinoso and R. Torregrosa, *Langmuir* 1987, **3**, 76.
12. J. Garrido-Segovia, A. Linares-Solano, J. M. Martin-Martinez, M. Molina-Sabio, F. Rodriguez-Reinoso and R. Torregrosa, *J. Chem. Soc., Faraday Trans. 1*, 1989, **83**, 1081.
13. F. Rodriguez-Reinoso, J. Garrido, J. M. Martin-Martinez, M. Molina-Sabio and R. Torregrosa, *Carbon* 1989, **27**, 23.
14. R. Reich, W. T. Ziegler and K. A. Rogers, *Ind. Eng. Chem. Process Des. Dev.*, 1980, **19**, 336.
15. M. Molina-Sabio, F. Rodriguez-Reinoso, D. Valladares and G. Zgrablich, ref.[5], p.573.
16. McEnaney, B., *Carbon* 1987, **25**, 69.
17. F. Stoeckli, P. Rebstein, and L. Ballerini, *Carbon* 1990, **28**, 907.
18. M. M. Dubinin, and F. Stoeckli, *J. Colloid Interface Sci.* 1980, **75**, 34.
19. F. Ehrburger-Dolle, P. Pfeifer, P. W. Schmidt, T. Rieker, F. Rodriguez-Reinoso, M. Molina-Sabio and M. T. Gonzalez, *MRS 1995 Fall Meeting*, (MRS, Pittsburgh, 1995), p. 410.
20. W. Rudzinski, K. Nieszporek, J. M. Cases, L. I. Michot and F. Villeras, *Langmuir* 1996, **12**, 170.
21. *Micromeritics*, Part No 201-42812-01 (1996), p. 5-24.

STUDY OF ACTIVATED PHYSICAL ADSORPTION BY ESR-METHOD

V.V.STRELKO

Institute for Sorption and Problems of Endoecology,
National Academy of Sciences of Ukraine,
32/34, Palladin Ave., 252180, Kiev, Ukraine

1 INTRODUCTION

There are still M.M.Dubinin's and coauthors' works [1] in which they mentioned that there could be three principally different cases in the processing of adsorption on micro porous adsorbents. The critical size of the adsorbed molecules could be less than them effective diameter of pore inlets ($d_{cr.} < d_{eff.}$), more than them ($d_{cr.} > d_{eff.}$) and, lastly is very close to them ($d_{cr.} \sim d_{eff.}$).

In the first case pores are easily available for the adsorbed molecules and they can fill the whole volume of micro pores by theory [1].

The second case is important for the separation of mixtures. This is the case when we use adsorbents (zeolites and carbons) as molecular sieves.

And lastly the third case could not be frequently found and is not actually studied well. The specific feature here is that in the certain range of temperatures micro pores are not available for the adsorbed molecules to enter them. It appears that the value of adsorption rises as the temperature goes up only to the certain point, reaching some maximum and, then one observes its reduction according to the laws of physical adsorption. The typical dependence of adsorption of perfluorocyclohexane on micro porous carbon AC-3 at the same pressure ($p/p_s = 0.102$) in equilibrium of the process is shown in Fig.3 [1]. As it follows from Fig.1 the micro pores volume is filled completely at 16 °C.

The atypical behavior noted in a number of cases (Fig.1) could be for the molecules which are very close in their size to the effective diameter of the entrance into pores. This is the case of activated adsorption. In other words in order to penetrate into pores molecules have to be activated and they need a considerable energy to be given to them. That is why the value of adsorption rises now as the temperature goes up. As a result of such a activated adsorption there could be a molecular-sieved effect in the certain range of temperature and the essential reduction of dynamic activity of adsorbent [1,2].

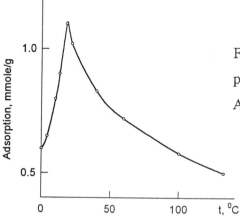

Fig. 1. Dependence of adsorption of perfluorocyclohexane on carbon AC-3 upon temperature

In this research we attempted to use ESR-method to study processes of the activated physical adsorption in order to estimate the mobility of molecules in zeolites which have commensurable pores [3,4]. While studying the activation of a number of comparatively large organic molecules on zeolites NaX and CaX we had shown [2] that the process had to be activated only when $d_{cr.} \sim d_{eff.}$. That is in order to overcome repulsion between the walls in pores and molecules, when they try to penetrate inside the matrix of adsorbent, these molecules need to be activated and adsorption takes place at high temperature.

If one stops the process of activation at any stage by decreasing the temperature for example, one can actually have some molecules fixed in the channels of zeolite. Exposing the "host-guest" systems to γ-rays fixed in the channels of zeolite according to the scheme:

$$\overset{\text{γ-irr.}}{R - H \;\rightarrow\; \cdot R + \cdot H} \tag{1}$$

a hydrocarbon   free
fixed in the pore   radicals fixed in the pore

Free radicals produced by γ-rays from hydrocarbons are studied well by ESR. The peculiar feature of these radicals is that they can exist in the hydrocarbon matrix only at -196 °C. While unfreezing such systems the temperature rises and free radicals become more mobile. Their recombination reactions don't need any activation and as a result free radicals are annihilated by means of recombination:

$$\cdot R_1 + \cdot R_2 \rightarrow R_1 - R_2 \tag{2}$$

Free radicals, produced at γ-radiolysis of hydrocarbons, have specific ESR-spectra, which can be easily recorded.

In other words the presence of molecules or free radicals fixed in the channels of zeolites can be described by their spectra taken for example at the temperature of their fixation in the pores of zeolites. On the other hand, if the temperature rises then at certain point the activation of adsorption begins. In accordance with scheme (2) free radicals are annihilated in this process which can be visually observed while studying the intensity of the ESR spectra. To solve this problem we studied the adsorbance of triethylbenzene (TEB), diethyl ester of dimethyl malonic acid (DEA) and a tertiary butyl benzene (TBB) on zeolites CaX and NaX (see Table) as they did it in [2]. Critical sizes of these molecules are very close to the diameter of inlets into the channels of zeolites.

Table. Characteristics of adsorbents and adsorbed molecules.
Activation energy values in the process of annihilation of free radicals being trapped in the channels of zeolites.

System studied	Pore diameter in zeolites, d_{eff}, Å	Critical diameter (d_{cr}) of adsorbed molecules, Å	Activation energy, kCal/mole (Arrhenius)
TEB-CaX	8	9	14.4 ± 1.3
TEB-NaX	9	9	6.5 ± 0.1
DEA-CaX	8	$7 \div 8$	-
DEA-NaX	9	$7 \div 8$	-
TBB-CaX	8	6	-
TBB-NaX	9	6	-

As it follows from the above Table TEB and DEA molecules have a size which is close to the diameter of pores in zeolites CaX and NaX, whereas the critical diameter of TBB is less than the effective diameter of entrances into the channels in these adsorbents.

2 EXPERIMENTAL PART

First the samples (m ~ 0.05 g) were heated at high temperature in open air to burn off organic from zeolites and then they were evacuated in the closed vacuum system at 350 °C within 10 hours.

Adsorption of TEB, DEA and TBB on these species was studied in the same system at 150-200 °C at pressure of their saturation during 6-8 hours. It was printed out in [2] that these temperatures are favorable for the activation of adsorption. Later the samples were cooled till ~ 20 °C. The excess of adsorbates was removed by evacuation at 20 °C. Then the tubes were open in vacuum and exposed to X-rays of ^{60}Co at -196 °C. As a rule a total radiation dose was about 5-15 Mrad.

Stabilised free radicals were registered by ESR radiospectrometer RE-1301 at -196°C and at room temperature.

For recombination of free radicals the samples were heated at 30-250°C in the oven placed into a resonator at 4 sec. variation of the magnetic field.

3 DISCUSSION

As it follows from the ESR spectra recorded for the activated adsorption of TEB, DEA and TBB on zeolites NaX and CaX followed by γ–irradiation at -196°C (Fig. 2a, b, c), only free radicals exist in these conditions. There are no peaks corresponding to the defects in the structure of zeolites. The ESR spectra of TEB and DEA on zeolites after being exposed to γ–rays at -196°C have five main peaks. In the case of TBB we noted a poorly resolved multicomponent signal. After being heated to the room temperature the samples changed their spectra (Fig. 2a', b'). We recorded a good resolution of SFS lines (five quadruplets with resolution of 19.00±1.27 and 4.7±1.2 e). In the case of DEA-CaX and DEA-NaX spectra remained practically unchanged (quintet with resolution of 21.3±1.2 e). There were no peaks in the spectra of TBB-CaX and TBB-NaX at 20°C which was probably due to the annihilation of radicals because of their recombination (see scheme). This is probably due to the fact that $d_{cr.}$ for TBB molecules (6 Å) is less than $d_{eff.}$ of the inlets into micropores in zeolites CaX and NaX (9 Å and 8 Å). That is why free radicals produced in pores as a result of γ–irradiation at -196°C can freely move inside the pores leaving them on heating to 20°C.

Analysis of the ESR spectra of systems TEB-CaX and TEB-NaX, DEA-CaX and DEA-NaX exposed to γ–rays proves that they belong to the free radicals represented by Scheme I and Scheme II [3,4].

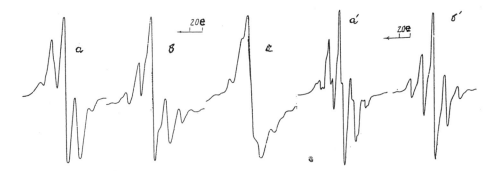

Fig. 2. ESR spectra of free radicals formed after γ-irradiation of the systems: TEB-CaX (a), DEA-CaX (b) and TBB-CaX (c) at - 196°C and systems TEB-CaX (a') and DEA-CaX (b') after further heating to 20°C.

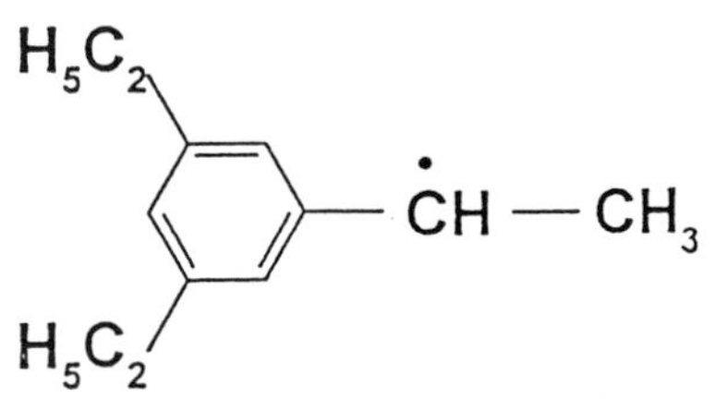

$$H_5C_2 \diagdown$$

Scheme I

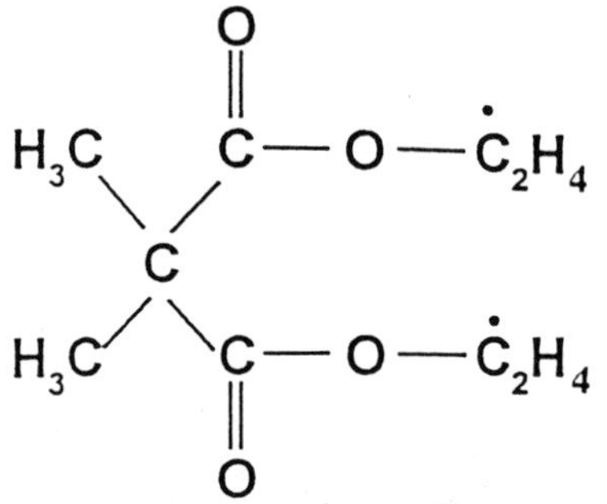

Scheme II

The aim of this research was to study the thermal stability of radicals fixed in the pores of zeolites due to activated adsorption of some molecules followed by their γ-ray exposure.

Kinetic curves of annihilation of these radicals in the range of temperatures 30-85 ⁰C are shown in Fig.3. These curves become linear in $1/C_i$-t coordinates what proves that annihilation process follows reactions of the order of two.

On the basis of dependence of the effective constant of reaction rate upon the value of inverse temperature one can calculate the activation energy (see Table) which describes the process of activation of adsorption, that is the energy which is necessary to overcome the repulsion between walls in pores and molecules or radicals when they try to enter the pores of zeolites.

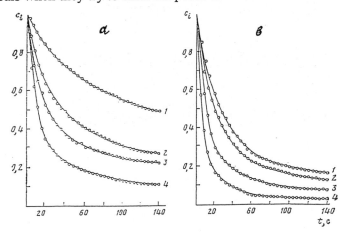

Fig. 3 Kinetic curves of annealing of Free radicals in γ-exposed systems TEB-CaX (a) and TEB-NaX (b). a: 1 - 30 ⁰C; 2 - 40 ⁰C; 3 - 50 ⁰C; 4 - 60 ⁰C. b: 1 - 30 ⁰C; 2 - 35 ⁰C; 3 - 60 ⁰C; 4 - 85 ⁰C.

On one hand the data obtained proves a good stability of free radicals which are normally very reactive due to their fixation in the commensurable with their size pores. On the other hand they enable us to study the intricate mechanisms of movement of molecules and radicals in pores provided that their diameters are very close. This method is obviously good for the examination of size of ultra pores particularly when their amount is very small in adsorbents or inorganic membranes. This method is very sensitive and makes possible to use molecules (or radicals) with continuously changing size.

That is why we studied the activated adsorption of methane followed by its γ-ray exposure. The ESR-method was used to determine a thermal stability of free radicals produced in pores. It makes possible to examine cavities (pores) in the matrix of silicas (aerosil, silica gel).

4 EVALUATION OF METHOD

There are data available that such compact silicas as tridymite and cristobalite can dissolve fairly well hydrogen, helium, neon and water as well. There are also some references on the formation of inclusion compounds (host-guest compounds) for I_2 - silica gel. In this process iodine molecules can penetrate in the matrix consisting from elementary globules in silica gel lattice being heated together in closed tubes.

These facts can be explained by the formation of cavities (channels) in the lattice of silica gel due to the connection of six-eight SiO_4 tetrahedrons. These channels appear to be within easy reach for small atoms and molecules.

Fig. 4 shows the model of amorphous silica lattice with cavities present there. The evaluation of inlets in these cavities proves that a silicon-oxygen ring Si_6O_{18} has a ~2.6 Å inlet in the cavity. That is the critical diameter of water molecules exactly corresponds to this cavity ($r_{cr.}(H_2O) = 1.38$ Å).

In order to obtain the independent data concerning ultra pores (channels) in the lattice of silica globules we studied the disappearance (annihilation) of free radicals, produced due to γ-irradiation of methane and ammonia, being adsorbed by aerosil and silica gel.

Producing of radicals:

$$CH_{4ads.} \xrightarrow{\gamma\text{-irr.}} \cdot CH_3 + \cdot H \qquad\qquad NH_{3ads.} \xrightarrow{\gamma\text{-irr.}} \cdot NH_2 + \cdot H$$

Annihilation of radicals:

$$\cdot CH_3 + \cdot CH_3 \rightarrow C_2H_6 \qquad\qquad \cdot NH_2 + \cdot NH_2 \rightarrow N_2H_4$$

Procedure of experiments

Aerosil and silica gel were evacuated at 500 °C within 5 hours to remove water and gases previously trapped by their lattice. Then they were placed in contact with methane and ammonia during long time at room temperature. Such conditions were considered to be good for their molecules to penetrate into ultra pores. After being exposed to γ-rays at -196°C we recorded ESR spectra at -196°C and room temperature. Fig.5 shows the ESR spectrum for ·CH, radicals being extremely reactive can acquire stability in a hydrocarbon matrix only at -196°C.

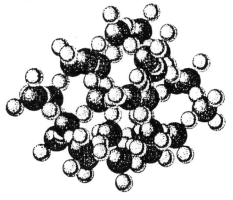

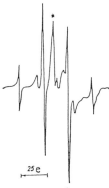

Fig. 4. Covalent model of amorphous silica.   Fig. 5. ESR spectra of ·CH₃
    Si atoms - black circles,   radicals stabilised in
    O atoms - white circles.   aerosil globules at 20°C.

In our experiments these radicals remain trapped in pores even when the temperature rises up to 150°C. The amount of such radicals fixed in pores increases linearly as the contact time between aerosil and methane longer before evacuation of samples at room temperature (removing of methane that did not enter pores) and their further exposure to γ-rays. This fact probably proves that methane molecules are firmly fixed in ultra pores of aerosil silica globules as far as methane can not be adsorbed on the outer surface of silica particles because of its poor affinity to this surface.

The thermal stability of methyl radicals is described by kinetic curves of annealing of samples like of the case of the zeolites.

Results of could be represented in Arrenius's coordinates and adapted using the least squares method. The mean value of activation energy in the process of annihilation of radicals was found to be 12.2 kCal/mole. This fact most likely proves that annihilation of radicals is slow due to their poor diffusion in pores as far as they have to overcome rather strong forces of adsorption.

We studied the adsorption of ammonia in ultra pores of silica followed by γ-irradiation. The ·NH₂ free radicals appeared to be also stable in the pores commensurable with their size which was proved by ESR.

Finally it has to be mentioned that the ESR is very sensitive method. It can give reliable data and enabled us to measure about 10^{14}-10^{15} free radicals in 1 g of a sample, that is about 10^{-8}-10^{-9} mole/g.

In other words if a sample (inorganic adsorbent or membrane) has even a small portion of ultra pores they could be examined by a proposed procedure.

It is worth to be mentioned that this method have some limits. If the material to be studied as a semi-conductor with a narrow width of the forbidden band then its electrons can recombine with free radicals. In this case free radicals are not recorded by ESR. This problem is discussed in our work [6], in which we studied the stability of $\cdot CH_3$ radicals in the lattice of SiO_2, Al_2O_3, TiO_2 and ZrO_2 which are dispersed oxides and insulators (SiO_2, Al_2O_3 and ZrO_2) whereas TiO_2 is a semi-conductor.

CONCLUSION

1. The activated physical adsorption takes place when the critical diameter of adsorbed molecules $d_{cr.}$ is very close to the diameter of pore entrances. There is the growth of adsorbed substance for the certain range of temperature in this case.
2. To study these processes we first used the ESR-method combined with γ-irradiation of molecules (radicals) being trapped in micro pores during activated adsorption.
3. On the basis of direct measurements of annihilation of the trapped free radicals by ESR we found the activation energies of their penetration into micro pores.
4. The ESR-method combined with γ-irradiation was shown to be very sensitive and can be used to examine the sizes of ultra pores in inorganic materials (adsorbents, membranes) even if their portion in adsorbent is extremely tiny.

REFERENCES

1. M.M.Dubinin, K.M.Nikolaev, N.C.Polyakov, 'Kinetics and Dynamics of Physical Adsorption', Nauka, Moscow, 1973, p.26.
2. K.M.Nikolaev, N.C.Polyakov, 'Adsorption and Porosity', Nauka, Moscow, 1976, pp.236, 275.
3. N.T.Kartel, N.N.Tsiba, V.V.Strelko a.o., *Izvestia Acad. of Sci. of USSR*, 1978, **10**, 2243.
4. V.V.Strelko, D.I.Shvets, N.T.Kartel, 'Radiation-Chemical Process in geterogeneous Systems Based on Dispersed Oxides' by V.V.Strelko and A.M. Kabakchi, Energoizdat, Moscow, 1981.
5. V.V.Strelko, N.T.Kartel, T.N. Burushkina, *Dokl. USSR Acad. of Sci.*, 1974, **216**, No 2, 360.
6. N.T.Kartel, A.M. Kabakchi, V.V.Strelko, *J. of Applied Spectroscopy*, 1977, **26**, No 6, 1078.

Synthesis and characterization of a zeolite A with detergent characteristics

M.J.Pérez Zurita, M. Matjushi, G. Giannetto*, C.Urbina, L. Garcia*, C. Bolivar and C. Scott
Universidad Central de Venezuela, Facultad de Ciencias, Escuela de Química, Centro de Catálisis, Petróleo y Petroquímica, Apartado Postal 47102, Caracas, Venezuela
* Facultad de Ingenieria

P. Bosch and V. H. Lara
Universidad Autónoma Metropolitana - Iztapalapa, Departamento de Química
Apartado Postal 55- 534, México D. F., México

1 INTRODUCTION

Synthetic zeolites are used in processes as diverse as ionic exchange[1], catalysis[2] or detergents[3]. The most requested is the A zeolite (related to faujasite) as it seems to be an adequate substitute of tripolyphosphates. Zeolite A can be synthesised from gels that, after an "induction time", generate the zeolite. The mechanism for crystallisation of these systems is strongly directed by the components making up the reaction system and the conditions of the synthesis. The gel dissolves continuously and the dissolved species from the gel are transported to the nuclei crystals in the solution phase[4].

If zeolite A is prepared to be used in detergents, the gel and the crystallisation have to be controlled in such a way that not only a highly crystalline zeolite is obtained: The crystals have to be fairly small (circa 1 micron) and the usual cubic shape has to be eroded as the material can not be abrasive. The way to generate such materials is only found in the patent literature and characterization techniques have to be chosen carefully in order to fully understand the process and control the characteristics of the final solid.

In this work we present the details of this kind of synthesis and we characterize each step of the preparation by infrared spectroscopy (FTIR), scanning electron microscopy (SEM) and X-ray diffraction (XRD). Crystallization kinetics are presented and the optimal conditions to obtain detergent type zeolite A are defined. Finally the structural modifications of the initial mixture used to crystallise zeolite A are studied.

2 EXPERIMENTAL

2.1 Synthesis

The synthesis of the solids was carried out by studying systematically the effect of varying the Na_2O / SiO_2 , SiO_2 / Al_2O_3 and H_2O / SiO_2 molar ratios. Table 1 shows the studied ranges. The procedure of synthesis was as follows :

The required amount of the sodium hydroxide was dissolved in distilled water and this solution was added to the sodium silicate under shaking (SOLUTION 1). A SOLUTION 2 was prepared by completely dissolving the sodium aluminate in distilled water. Both solutions 1 and 2 were heated to 80 °C and then solution 1 was added to solution 2 under vigorous shaking until complete homogeneity was reached, leaving under shaking for approximately 45 min at minimum temperature. The final pH of this gel was 11.5.

Table 1. *Parameters of synthesis studied*

SERIES	Na_2O/SiO_2,	SiO_2/Al_2O_3	H_2O/SiO_2
1	0.8	2	72
2	1.0	2	72
3	1.2	2	72
4	1.4	2	72
5	1.6	2	72
6	1.4	1	72
7	1.4	3	72
8	1.4	2	108
9	1.4	2	144

Subsequently, the mixture was divided into 60 ml portions, sealed in 120 ml poly-propylene flask and placed in a paraffin bath at 100 °C and autogen pressure.

Samples were taken out from the bath at the desired time intervals and were filtered to separate the solution and the solid phase. The solid phase was then thoroughly washed with distilled water and dried for 12 h at 110 °C.

The most crystalline zeolite from each series was named using the following general formula ZA N^o series. In this way, the most crystalline zeolite of series 3 was named ZA3. For comparative purposes a commercial zeolite was obtained.

In order to determine the structural differences until the sample was fully crystalline three samples were taken at 1, 1.5 and 2 hours during crystallization of sample ZA4. A crystalline zeolite sample to be used as reference was also studied. These sample were named ZA4 (time of crystallization). Therefore they are ZA4 (1), ZA4 (1.5) and ZA4 (2).

2.2 Characterization

All the obtained solids were characterized by X-ray diffraction (XRD) and Fourier transform infrared spectroscopy (FTIR). The most crystalline zeolite of each series was further characterized by scanning electron microscopy (SEM), BET adsorption and chemical analyses.

The IR measurements were carried out on a IR-435 V-04 from Shimadsu Corporation and the SEM micrographies were obtained with a Hitachi S-500 Scanning Electron Microscope. Ca, Na, Al and Si analyses were made on a VARIAN Techtron AA6 and surface area measurements were conducted on a multipoint surface area Flowsorb II 2300 from MICROMERITICS.

The diffraction analyses were conducted either on Phillips PW 1730-10 X-ray powder diffractometer using $CuK\alpha$ radiation or on a D-500 Siemens diffractometer also coupled to a copper anode tube. The $CuK\alpha$ radiation was selected with a diffracted beam monochromator. A graphite internal standard was mixed with the aluminosilicates to measure the cell parameter of the zeolites. Percentage of crystallinity was obtained considering that all peaks appearing at angles lower than $2\theta = 15°$ are affected by the degree of hydration of the solids. Therefore, the total area of the peaks between $2\theta = 15°$ and $40°$ was calculated for each sample. The obtained values represent the overall crystallinity.

Hence, the crystallinity percentage was defined as the ratio of the total peak area of the same sample to the reference sample that had the largest value in each series.

To calculate the radial distribution functions the diffractograms were obtained with a molybdenum X-ray tube (MoKα). The intensities, read as intervals $\Delta 2\theta = 0.08°$ from $2\theta = 2°$ to $2\theta = 120°$ (where 2θ is the diffraction angle), were the input data for the Magini and Cabrini program[5]. The radial distribution function was then obtained and interpreted in terms of the reported interatomic distances.

3 RESULTS AND DISCUSSION

3.1 Chemical Analyses

Table 2 presents the chemical analyses and surface areas obtained for the most crystalline zeolite of each series.

Table 2 *Chemical and surface analyses of the most crystalline zeolite of each series.*

SOLID	%Na	Si/Al	Surface Area (m^2/g)
ZA1	16.02	1.0	615
ZA2	16.97	1.1	530
ZA3	16.88	1.0	570
ZA4	15.97	1.0	560
ZA5	15.66	1.0	570
ZA6	15.56	1.0	550
ZA7	16.06	1.0	580
ZA8	16.50	1.0	570
ZA9	16.32	1.0	560
Commercial	16.16	1.0	600

As expected, Si/Al ratios were 1 in all cases. Surface area values demonstrated the high degree of crystallinity of the solids and therefore their purity.

3.2 Kinetic Studies

Kinetics studies of crystallisation were conducted while studying all synthesis parameters. As an example, Figures 1 show the dependence of crystallinity versus reaction time when studying the effect of the Na_2O/SiO_2 ratio. Crystallisation time increases with increasing Na_2O/SiO_2 ratios (Figure 2). The opposite effect is observed for H_2O/SiO_2 ratio where the time of crystallisation increase with H_2O concentration. In contrast, SiO_2/Al_2O_3 (Figure 3) ratio did not exert important changes on the rate of crystallisation, however, when the SiO_2/Al_2O_3 was 3, the presence of another crystalline phase was evidenced.

3.3 Infrared Spectroscopy

These kinetics studies were followed as well by IRFT spectroscopy by recording the progress of the relative intensity with 560 cm^{-1} and 980 cm^{-1} signals (I560/I980).

The results obtained were comparable with the XRD results as far as the time of maximal crystallisation is concerned. Figure 4 shows an example of one of the kinetics experiments followed by IRFT.

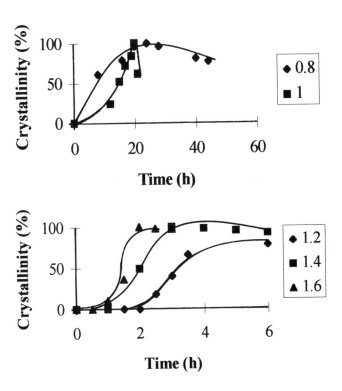

Figure 1 *Dependence of crystallinity with reaction time. Effect of Na_2O/SiO_2 ratio*

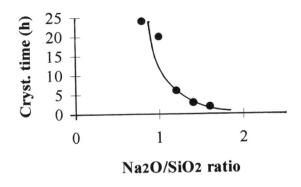

Figure 2 *Effect of the Na_2O/SiO_2 ration on the crystallization time*

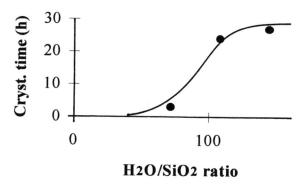

Figure 3 *Effect of the H_2O/SiO_2 ration on the crystallisation time*

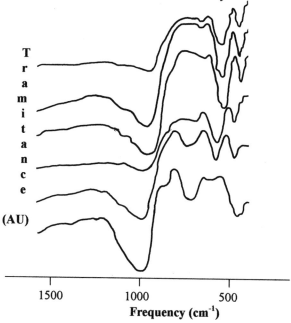

Figure 4 *Kinetic study follow by IRFT Spectroscopy*

3.4 Scanning Electron Microscopy

The effect of the different parameters studied on the crystal size and morphology of the prepared zeolites was assessed. Table 3 and Figure 5 show that the shape and size of the crystals are greatly influenced by the parameters studied.

Only synthesis with Na_2O/SiO_2 ratio higher than 1.2 produce cubic crystals with bevelled or round edges or spherical crystals whose crystallite size decreases as Na_2O/SiO_2 ratio increases. If SiO_2/Al_2O_3 ratio is modified (keeping constant H_2O/SiO_2 ratio = 72 and Na_2O/SiO_2 = 1.4) the crystal size is small for low or high values but increases if

$SiO_2/Al_2O_3 = 2$ which is, indeed, the stoichiometric value; as expected (Na_2O/SiO_2 ratio higher than 1.2) the crystallites had bevelled edges. Finally, the crystallite size was constant if H_2O/SiO_2 was less than 100 but increased abruptly for higher values. Therefore, the crystallite shape seems to be strongly dependent of the Na_2O/SiO_2 ratio but the crystallite size turn to be mainly determined by the SiO_2/Al_2O_3 as well as the H_2O/SiO_2 ratios. Hence, if small crystals with bevelled or rounded edges have to be obtained Na_2O/SiO_2 ratio has to be higher than 1.2 and, if small crystals are required, high Na_2O/SiO_2, high SiO_2/Al_2O_3 and lower than 100 H_2O/SiO_2 values are recommended.

Table 3 *Effect of parameters of synthesis on the morphology and size of the crystals*

SOLID	Na_2O/SiO_2*	SiO_2/Al_2O_3*	H_2O/SiO_2*	Crystal size (μ)	Crystal shape
ZA1	0.8	2	72	2.7	Cubic
ZA2	1.0	2	72	10.5	Cubic
ZA3	1.2	2	72	6.2	Cubic
ZA4	1.4	2	72	5.5	Cubic with bevelled edges
ZA5	1.6	2	72	4.0	Cubic with rounded edges
ZA6	1.4	1	72	1.2	Cubic with bevelled edges
ZA7	1.4	3	72	0.3	Spherical
ZA8	1.4	2	108	5.8	Cubic
ZA9	1.4	2	144	9.9	Cubic with bevelled edges
Commercial				1.5	Cubic with bevelled edges

(*) nominal values

3.5 X-ray Diffraction

A typical x-ray difracction pattern of the obtained crystalline A zeolites shows its 12 characteristics peaks. X-ray difractograms of samples ZA4(1), ZA4(1.5), ZA4(2) show a non crystalline phase attributed to the crystallising gel and a crystalline phase whose interplanar distances can be interpreted as NaA-zeolite. Although peak intensity increases with crystallisation time, no clear peak broadening was observed, showing that crystallite sizes are beyond 3 μ. The measured cell parameters reproduce the values reported by von Ballmoos and Higgins[7] for Na-A zeolite within the error range: $a_o = 24.58$ Å.

From the radial functions obtained for the samples ZA4(1), ZA4(1.5) and ZA4(2) it was evidenced that, as crystallinity increases, i.e. as long range order is better defined, the peaks become sharper and more intense; instead, in the less crystalline material, peaks found at

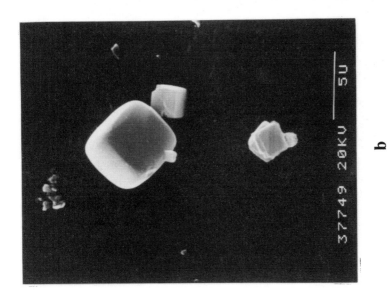

b

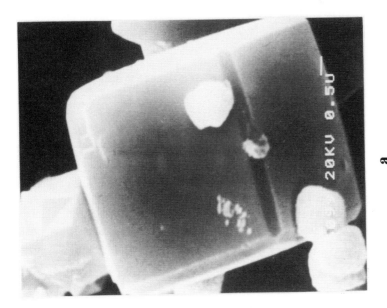

a

Figure 5 *Scanning Electron Micrographs of the most crystalline zeolite of **a**) series 4 and **b**) series 5*

radial distances higher than 0.07 μ are small and undefined, probably indicating that small crystalline nuclei are 0.14 μ diameter. The peak found at 0.016-0.0167 μ can be attributed to a (Si,Al)-O distance in the (silicon, aluminium) tetrahedron, the peaks at 0.032-0.033 μ are the (Al,Si)-(Si,Al) distances. These distances determine the angle between the (Si,Al) tetrahedra which is $155^0 \pm 5$, this value is the average between the Si-O-Si angle present in the zeolite as well as the Si-O-Si angles found in the non crystalline aluminosilicate. In the crystalline zeolite A the value was 144^0. It was observed that a peak at 0.034-0.035 μ becomes more defined as the crystallinity increases. Also the unresolved peaks found at 0.065 to 0.068 μ become a clear triplet in the ZA4(2) sample.

4 CONCLUSION

The conditions to obtain a zeolite A appropriate to be used in detergent formulation have been defined. A correlation between SiO_2/Al_2O_3, Na_2O/SiO_2 and H_2O/SiO_2 and the crystal morphology was found. Na_2O/SiO_2 and H_2O/SiO_2 have to be higher than 1.2 and lower than 100 respectively if bevelled or rounded edges have to be obtained. If small crystallites are required these conditions have to be restricted and the ratio H_2O/SiO_2 has to be reduced as Na_2O/SiO_2 and SiO_2/Al_2O_3 are increased.

The crystallisation time is favoured if low (less than 100) H_2O/SiO_2 and high (more than 1.2) Na_2O/SiO_2 ratios are used. These conditions, as stated previously, favour small bevelled edges crystallites. The estimated crystallising time was around 3 hours for a zeolite with detergent characteristics.

The radial distribution functions obtained from the partially crystallised zeolite A to be used in detergents showed how (Si,Al) tetrahedra orient until the crystalline structure is formed.

5 ACKNOWLEDGEMENTS

This work was performed in the frame of the CYTED-D program, the financial assistance is gratefully acknowledge to Conycet and Conacyt. V.H. Lara and P. Bosch thanks the Instituto Mexicano del Petróleo for the use of the X-ray diffraction instrumentation.

References

1. M.T. Olguin, P. Bosch, J.M. Dominguez and S. Bulbulian, *Zeolites*, 1993, **13**, 493
2. E. Pietri de Garcia, M.R. de Goldwasser, C. Franco Parra and O. Leal, *J. Appl. Catal.* 1989, **50**, 55
3. Y. Yamane and T. Nakazawa, 'New developments in zeolite science and technology', Proceedings of the 7th International Zeolites Conference. Y. Murakami, A. Iijima and J.W.W. Kodansh Eds., Elsevier, Tokyo, Japan, 1986, p. 991
4. R. Szostak, 'Molecular Sieves, principles of synthesis and identification', Van Nostrand Reinhold, New York, U.S.A., 1989, p. 191
5. M. Magini and A. Cabrini, *J. Appl. Crystallogr.*, 1972, **5**, 14
6. R. von Ballmoos and J.B. Higgins, 'Collection of simulated XRD powder patterns for Zeolites', 2nd revised edition, Butterworth-Heinemann, Stoneham, U.S.A., 1990, p. 428S

GRAVIMETRIC AND VOLUMETRIC MEASUREMENTS OF HELIUM ADSORPTION EQUILIBRIA ON DIFFERENT POROUS SOLIDS

R. STAUDT, S. BOHN, F. DREISBACH, J.U. KELLER

Inst. of Fluid- and Thermodynamics
University of Siegen, D-57068 Siegen, Germany

ABSTRACT

Adsorption equilibria of Helium on several types of porous solids like activated carbon, activated carbon fibres, molecular sieve, and Trockenperlen (mixture of activated carbon and zeolite) have been measured at different temperatures (278 K - 323 K) in the pressure range 0,1 MPa - 15 MPa using a microbalance. The dynamic behavior of this systems reaching an adsorption equilibrium has been examined. The resulting adsorption isotherms and uptake rates will be shown and discussed.

1   INTRODUCTION

To investigate adsorption equilibria of gases on porous solids gravimetric or volumetric methods are used [1, 2]. These measurements only allow to determine the reduced mass $\Omega = m - \rho^f V^{as}$. This quantity is defined as the difference of the mass m adsorbed on a porous solid and the product of the buoyancy related volume V^{as} of the adsorbent / adsorbate system and the density of the adsorptive ρ^f. In both experimental methods the volume of the porous solid must be known to calculate the buoyancy correction in gravimetric measurements or the free space in the volumetric techniques.

To calculate the adsorbed mass m from the measurable quantities Ω and ρ^f, it is necessary to measure the volume of the adsorbent or to introduce a model or an assumption on V^{as} [1, 2]. Often V^{as} is identified with the so-called He-volume V_{He} and is considered to be constant (i.e. independent of T, p). It can be measured in gravimetric or volumetric experiments by a helium adsorption isotherm with the assumption that helium will not be adsorbed at all ($m_{He} = 0$). This assumption seems to be acceptable for some porous solids at high temperatures and low pressures. But there is experimental evidence, that it does not hold for ultramicroporous substances like activated carbon fibres. Also it can be questioned in view of the characterization of ultramicropores with Helium at 4.2 K [3]. To proove this assumption at higher temperatures we investigated adsorption equilibria of Helium on different porous solids at ambient temperatures.

2   EXPERIMENTAL

The experimental setup for gravimetric measurements of adsorption equilibria is sketched in Figure 1 [2, 4]. It mainly consists of a microbalance (Sartorius 4104S, 0...15 MPa), a thermostat (Haake F3-C/01), and a gas supply system (DruVa, SMD 617-84). The tubes are made of stainless steel. Valves have been chosen from DruVa (BVR 11, BVR 12). Pressures are measured with a high precision pressure receiver (VDO, BR 3201) with measuring range 0...16 MPa. The temperature is measured by a resistance thermometer Pt100 in the thermostat and by thermocouples in the adsorption chamber. The thermal conditions within the vessels could be maintained within a range of temperature $|\Delta T| \leq 0.005$ K. To measure uptake rates we used another installation including a microbalance (Mettler AT201, Giessen, Germany) and a magnetic suspension (Rubotherm, Bochum, Germany). To determine the density of porous solids additional volumetric measurements were carried out using a Micromeritics AccuPyc 1330.

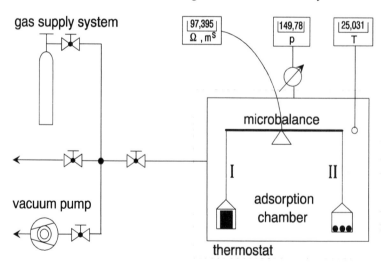

Figure 1 Experimental setup for gravimetric measurements of gas adsorption equilibria.

3   RESULTS AND DISCUSSION

Adsorption equilibria of Helium on several types of porous solids like activated carbon (AC), activated carbon fibres (ACF), molecular sieve (MS), and Trockenperlen have been measured at ambient temperature (278 K - 323 K) in the pressure range 0,1 MPa - 15 MPa. The Trockenperlen (Engelhard, Hannover, Germany) are mixtures of activated carbon and zeolites to combine the advantages of both porous solids for adsorption processes. The TP BR 4916 consists of 40% AC and 60% zeolite.

Figure 2 shows the adsorption isotherms of Helium on the activated carbon Norit R1 Extra (Norit Deutschland GmbH, Düsseldorf, Germany) at the temperatures 298 K and 323 K. The waiting time between the pressure step and the „adsorption equilibrium" was 15 min. The isotherms increase steeply at low pressures (p < 1 bar) and than reach a plateau with a saturation load of 1.9 mg/g at 298 K and 0.8 mg/g at 323 K. Activated carbon SC 40 Extra (Silcarbon, Kirchhundem, Germany) and activated carbon fibres (Carbo Tech, Essen, Germany) show the same behavior as the Norit R1 Extra. As we expect the amount of Helium being adsorbed nearly to be constant or at least to increase monotonously with increasing pressure, the data (Figure 2) indicate adsorption of Helium in parts of the experimental installation, most probably in the resin used for coating electric wires and/or the pressure receiver/transducer system.

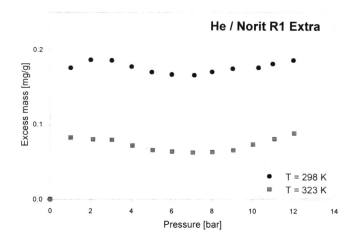

Figure 2 Adsorption isotherm of He on AC Norit R1 Extra, T = 298 K and T = 323K.

The isotherms of Helium on the molecular sieve 5A (Union Carbide) were measured under the same conditions as the He/Norit isotherm. After a waiting time of 15 min from the pressure step to the equilibrium, we observe at higher temperatures a higher load of Helium (Figure 3). We assume that this behavior is due to faster diffusion at higher temperatures.

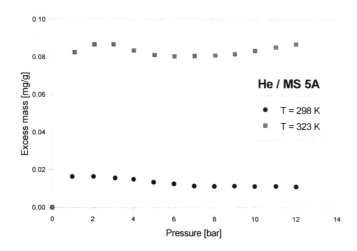

Figure 3 Adsorption isotherm of He on molecular sieve 5A, T = 298 K and 323 K.

Figure 4 shows the reduced mass $\Omega = m - \rho^f V^{as}$ (microbalance signal) for the adsorption of Helium on Trockenperlen TP BR 4916 for different temperatures. From the slope of this curves, it is possible to calculate the so-called Helium volume (true density) of the porous solid for different temperatures [1, 2, 5]. For a constant pressure, we observe increasing of the reduced mass with increasing temperature. That means that the excess mass adsorbed increases with a decrease of the temperature. Other Trockenperlen (TP BR 4942, TP BR 4942/1) or activated carbons (Norit R1 Extra, SC 40 Extra) and activated carbon fibres show the same temperature dependence in Helium adsorption.

Uptake rates of Helium on AC Norit R1 Extra and molecular sieve 5A are measured at T = 298 K, after the „adsorption equilibrium" (15 min, cp. Figure 2 and 3) at p = 0.1 MPa was reached using a magnetic suspension microbalance. The result for the system He/Norit R1 Extra is shown in Figure 5. During a measuring time of 60 h, the mass of the activated carbon increases due to slow but continuing adsorption of Helium. The pressure and the temperature were nearly constant, $p = 0.105$ MPa ± 0.001 MPa and $T = 297.94$ K ± 0.02 K. We assume that this slow increase of the sample mass is an effect of Helium absorption in the crystal lattice of the porous solid. The system He/molecular sieve 5A shows the same behavior, but the increase of the sample mass due to ad- or absorption is much smaller.

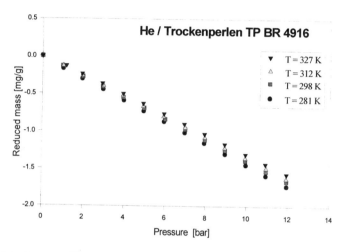

Figure 4 Reduced mass of He on Trockenperlen TP BR 4916 at different temperatures.

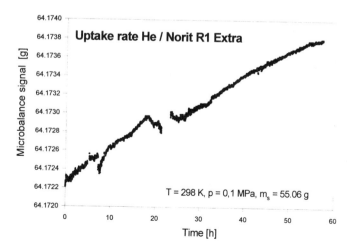

Figure 5 Uptake rate of He on Norit Re Extra, $p = 0.1$ MPa, $T = 298$ K, sample mass $m_s = 55.06$ g.

4 CONCLUSION

Our investigations of the Helium adsorption show that Helium will be adsorbed in different porous solids at ambient temperatures only in the low pressure region ($p < 0.1$ MPa). The adsorption occurs very fast followed by a slow ad- or absorption process.

The determination of the volume of porous solids by volumetric or gravimetric methods can be incorrect by doing these experiments with Helium neglecting adsorption. Measurements of Helium adsorption isotherms should be performed at the same conditions as the volume determination experiments. Using these two measurements the effect of Helium adsorption occurring at low pressures can be nearly eliminated and the „true volume" of the adsorbent can be calculated.

References:

[1] S.J. Gregg and K.S.W. Sing: *Adsorption Surface Area and Porosity*; 2nd Ed. Academic Press, London, 1982.

[2] R. Staudt, G. Saller, M. Tomalla, J.U. Keller: *A Note on Gravimetric Measurements of Gas-Adsorption Equilibria*; Ber.Bunsenges.Phys.Chem. 97, (1993), p. 98-105.

[3] K. Kaneko, N. Setoyama, T.Suzuki: *Ultramicropore caracterization by He adsorption*; Proceedings of Caracterization of Porous Solids 3; Marseille May 1993, J. Rouquerol, Ed., Elsevier, Amsterdam, 1994, p. 593-602.

[4] R.Staudt, F.Dreisbach, J.U.Keller: *Buoyancy Corrections of Gravimetric Gas Adsorption Equilibria Measurements,* Proceedings of the 25th Int. Conf. on Vacuum Mikrobalance Techniques; April 1995, Marrakech, Marokko, in print.

[5] E. Robens, J.U.Keller, C.H. Massen, R. Staudt: *Sources of error on Sorption and Density Measurements*; Symposium on Characterization of Porous Solids 4; Bath, UK.

MULTISCALING TEXTURE CHARACTERISATION OF FREEZE-DRIED RESORCINOL-FORMALDEHYDE AEROGELS

B. Mathieu, S. Blacher, F. Brouers, J.P. Pirard
Liège University, Laboratoire de Génie Chimique, Bât. B6, B-4000 Liège, Belgium

B. Diez, R. Sobry, G. Van den Bosshe
Liège University, Laboratory of Experimental Physics, Bât. B5, B-4000 Liège, Belgium

ABSTRACT

A new drying method to prepare resorcinol-formaldehyde aerogels was studied: freeze-drying. The obtained material has a low density, presents large specific surface area S_{BET} (up to 450 m^2/g) and an important porous volume V_{Hg} (up to 2.5 cm^3/g). The molar ratio Resorcinol/Catalyst was modified to obtain a complex porous structure in a very large range from the nanometer to the micrometer length-scale. Several characterization methods where used in order to get a picture of the aerogel at different scales: nitrogen adsorption-desorption (including the classical and the fractal interpretations), mercury porosimetry, Transmission Electron Microscopy (TEM) and Small Angle X-ray Scattering (SAXS).

1 INTRODUCTION

Recently, Pekala et al.[1-6] developed organic resorcinol-formaldehyde aerogels. In the route proposed by these authors, resorcinol-formaldehyde gels are dried by CO_2 supercritical drying, after exchange of the water with an organic solvent compatible with liquified carbon dioxyde. The material obtained in that way is an excellent thermal insulator and once pyrolyzed in an inert atmosphere, it forms high surface area carbon aerogels which can be used for various electrical applications. The purpose of this work is to study the texture of resorcinol-formaldehyde aerogels dried by freeze-drying. The aerogels obtained with this drying method present a more complex structure than classical organic aerogels. The pores extend in a very large range: from nanometer to micrometer length-scale. In the perspective of a catalytic application, this very opened structure deserves to be studied in detail. Moreover, it must be pointed out that freeze-drying is an industrial-scale method, easier to use and cheaper than supercritical drying. The final texture of freeze-dried aerogels depends on several independant parameters. Taking a standard freezing rate, the influence of the molar ratio Resorcinol/Catalyst (R/C) has been studied in detail. Several characterization methods provide us with a representation of the morphology from the nanometer to micrometer scale: nitrogen adsorption-desorption isotherms (classical and fractal interpretations), mercury porosimetry, Transmission Electron Microscopy (TEM) and Small Angle X-ray Scattering (SAXS).

2 EXPERIMENTAL

2.1 Aerogel Preparation

The first stage of the gel preparation consists in mixing under stirring the four following reagents: i) a determined weight of resorcinol (Aldrich), ii) a determined volume of a 37% weight formaldehyde solution, keeping the molar ratio Resorcinol/Formaldehyde equal to its stoichiometric value, 0.5, iii) a determined weight of sodium carbonate Na_2CO_3 (Aldrich), in accordance with the imposed molar ratio R/C, iiii) 25 ml deionized water. The resulting solution is poured into a 250 ml erlenmeyer. After 24 hours at 70°C, the obtained dark red resorcinol-formaldehyde gel is left at room temperature during 24 hours.

Freeze-drying consists in eliminating the solidified solvent by low-pressure sublimation.[8] To produce sublimation at the interface, the pressure in the freeze-dryer must be lower than the vapor pressure of the solidified solvent at the considered temperature. These conditions are easily obtained in the case of resorcinol-formaldehyde aerogels synthesized using water as solvent. The resulting aerogels is a brown material with a fibrous appearence. This aspect has its origins in the growth of the ice crystals during the freezing step before the drying process. The growth step may result in different shapes of the ice crystals and thus in different textures of the dried material.[9] In our case, the samples were frozen in liquid nitrogen (77 K). The samples characterized by molar ratios R/C of 50, 100, 150, 200, and 250 are respectively designed by RC50, RC100, RC150, RC200 and RC250.

2.2 Methods

The used freeze-dryer was a Heto CT110 with a pump Balzers F5, characterized by a condenser temperature of 163 K. The nitrogen adsorption-desorption isotherms were determined at liquid-nitrogen boiling temperature (77 K) by the classical volumetric method with Sorptomatic Carlo Erba Series 1900 apparatus. The analysis of the isotherms was performed in accordance with the BET theory and extensions.[10,11] The mercury porosimetry measurements were made using a porosimeter Carlo Erba 2000 between 0.1 and 200 MPa. To study the macroporosity of the samples, we also worked under atmospheric pressure using a manual porosimeter. According to Washburn's law, this allows to study the pores of diameter d_p between 7.5 nm and 150 μm. The SAXS spectra were obtained with a Kratky compact camera, using a step scanning procedure. The Kratky X-ray tube was supplied with a power of 1.5 kW and Cu-Kα radiation was used. The explored domain was in the wave-vector range $0.003 \le q \le 0.6$ Å^{-1}. The scattering of the device without sample was systematically measured and subtracted from the intensities recorded with the sample. The collimated effects were mathematically eliminated ("desmearing") using an original method.[12] Aerogels were observed by transmission electron microscopy (TEM) using a Philips CM100 electron microscope with a resolution of 100 kV.

3 RESULTS

Figure 1 shows the nitrogen adsorption-desorption isotherms of the studied samples. As the molar ratio R/C increases, one can observe a transition from an isotherm of type I of the de Boer classification[11] for the sample RC50 to an isotherm of type IV for RC250. For increasing R/C ratios, the hysteresis and the adsorbed volume increases indicating that the microporous solids become more and more mesoporous. Table 1 presents the BET specific surface area (S_{BET}), the cumulative specific area (S_{cum}^{BdB}) of the mesopores calculated by the Broekhoff-De Boer method, the porous volume (V_p) calculated from the adsorbed volume at saturation and the total microporous volume (V_{DR}) calculated by the

Dubinin-Radushkevich equation[11]. When R/C increases, S_{BET} and V_{DR} do not significantly change. Simultaneously we can observe an important increase of the cumulative specific area S_{cum}^{BdB} of the mesopores and of the porous volume V_p, which indicates a development of a mesoporous and macroporous texture in the solid.

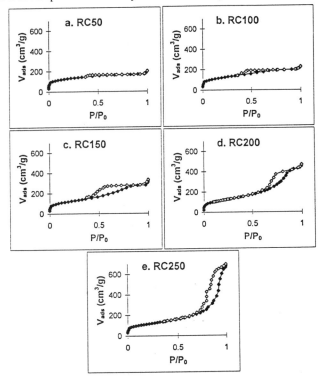

Figure 1 *Nitrogen adsorption-desorption isotherms of the various samples.*

Mercury porosimetry measurements show that V_{Hg} does not significantly change with the R/C ratio. As it is shown in Table 1 the same porous volume corresponding to different pore size distributions depending on R/C is obtained.

The classical FHH equation[13], $\ln(P_0/P) = \alpha/z^m$, describes the continuous growth of an adsorbate film with thickness z on a flat surface when P/P_0 tends to 1, where P_0 is the saturated vapour pressure of the adsorbate and α and m depend on the solid-adsorbate interaction. The value of m is determined experimentally plotting the experimental isotherm $N/N_m = (\ln(P_0/P))^{-1/m}$ where N/N_m is the number of layers adsorbed on the surface at the relative pressure P/P_0 and in general $2 \leq m \leq 3$.

Fractal FHH type equations were proposed by Avnir et al.[14], Pfeifer et al.[15,16] and Yin[17]. In spite of the different adsorption mechanism for multilayer formation considered by Avnir et al. and Yin, in the case of micropores and the mesopores size distributions they obtained the same isotherm $N/N_m = (\ln(P_0/P))^{-(3-D_s)}$ where $2 \leq D_s \leq 3$. Table 1 shows the values of m and the relative pressure range over which this parameter was calculated. Figure 2 shows the log-log plot of N/Nm vs ln (P_0/P) for the studied samples. For R/C $\leq$ 100, the linear variation is too narrow to yield an accurate value of m. However, m seems slightly greater than 3. For greater R/C, the self-similarity

Table 1 *Main results of the sample characterization*

Samples	Nitrogen adsorption-desorption							Mercury porosimetry		Megalopores	SAXS
	Classical interpretation				Fractal interpretation		Total porous volume	Meso +macropores			Giration radius
	S_{BET}	S^{BdB}_{cum}	V_P	V_{DR}	m	$\Delta P/P_0$	V_{Hg}	V_{Hg} (d<10μm)		V_{Hg} (d>10μm)	R_g
	(m^2/g)	(m^2/g)	(cm^3/g)	(cm^3/g)			(cm^3/g)	(cm^3/g)		(cm^3/g)	(A)
RC50	464	165	0.310	0.188	-	-	2.105	0.108		1.997	25.58
RC100	420	214	0.350	0.197	-	-	2.074	0.132		1.942	34.62
RC150	423	267	0.517	0.167	2.50	0.01 -0.90	2.010	0.439		1.571	38.92
RC200	419	320	0.722	0.180	2.23	0.01 -0.90	2.101	0.601		1.500	44.11
RC250	409	341	1.073	0.165	2.45	0.01 -0.94	2.508	1.222		1.286	42.54

extends over a large P/P_0 length-scale and we obtain $m \cong 2.5$. The value of m is determined by the regime of physical adsorption[15]: $2 < m < 3$ for the capillary condensation regime, in which the dominating surface-adsorbate force is the surface tension and $m > 3$ for the FHH regime, in which the dominating surface-adsorbate force is the substrate potential. For $R/C \leq 100$, the inaccurate value of m leads to $D_s \cong 3$ which, taking into account the great value of S_{BET}, would characterize a very rough surface. In the capillary condensation regime, for $R/C > 100$, $D_s \cong 2.6$ characterize the structure of the pore network surface.

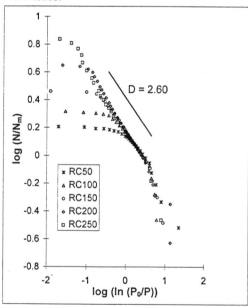

Figure 2 *Adsorption isotherms in the coordinate system defined by the FHH equation*

In the small-angle scattering measurements, the scattered intensity, I, is measured over a range of scattering vectors q defined as $q = 4\pi\lambda^{-1} \sin(\theta/2)$ where λ is the wavelength of the incident photons and θ is the total scattering angle. For fractal structures, the intensity has a power law dependence on the scattering vector such that $I(q) \sim q^{-P = -D_{m/p}}$ with $1 < D_{m/p} < 3$ for mass/pores fractals and $I(q) \sim q^{-P = -(6-D_s)}$ with $3 < 6 - D_s < 3$ for surface fractals[19]. Figure 3 shows the log-log plots of the SAXS intensity curves I versus q for the studied samples. Table 1 shows the values of the large particulate radious of giration R_g. For R/C = 50, the low-q regime is almost flat and we found a very narrow linear region with $P \cong 2.64$. For greater R/C, the linear regime extends over a length-scale which broadens with increasing R/C values and R_g increases from $\cong 26$ to 45 Å. In this case, we obtain $D_{m/p} \cong 2.80$ which indicates that the fractal pore size distribution follows the same scaling law for all samples with R/C > 50. The deviation observed for R/C = 50 is probably due to a different pore structure. At small values of q, deviations from the scaling behavior are observed.

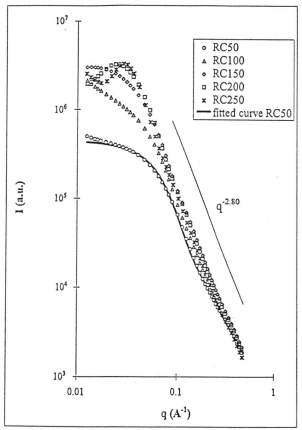

Figure 3. *Small Angle X-Ray Scattering: R/C dependence of scattering profiles*

As R/C increases, a peak begins to develop. For R/C = 250, a pronounced maxima at about $2\pi/q \cong 20$ nm suggest that there are some correlations in the structure in the aerogel.

The observed self-similar pore distribution is related with the important microporosity observed by adsorption-desorption measurements. The almost constant value obtained of S_{BET} for all R/C values could be explained if we consider that all the samples present the same pore size distribution at small scale. TEM observations show that the studied aerogels are composed of interconnected particles with sizes of about 10 to 20 nm for all the R/C ratios. Then, most of the surface area results from the pore fractal nature of the constitutive particles.

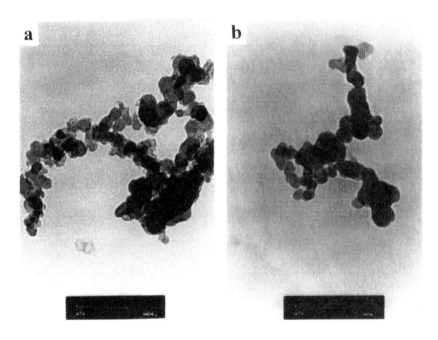

Figure 4 *TEM photograph (1.5cm=50nm) of the samples RC50(a) and RC250(b).*

4 DISCUSSION

The use of several characterization methods give us a picture of the texture of freeze-dried resorcinol-formaldehyde aerogels and of the influence of the R/C molar ratio on this texture. An important and almost constant surface area S_{BET} (up to 464 m²/g) and a high microporous volume V_{DR} (about 0.180 cm³/g) are observed in all samples. TEM micrographs show us that the size of the constituting particles is about 10 to 20 nm and does not significantly change. Taking into account an average particle size of 15 nm and that the apparent density ρ of samples is almost constant ($\rho = 1.6$ g/cm³), the estimated value of the geometric specific surface area S_G is 250 m²/g. From these considerations and from the pore distribution pointed out by SAXS measurements, we conclude that the microporosity is more of inner than inter-particle nature. The meso- and macroporous character of the aerogel develops as R/C increases. This development is pointed out by the important increase of S_{cum}^{BdB} (from 165 to 341 m²/g) and of V_p (from 0.310 to 1.073 cm³/g). It is also detected by the transition from the FHH regime to the capillary condensation regime as R/C increases. It is finally confirmed by the evolution of V_{Hg} for $d_p < 10$ μm, which increases from 0.108 to 1.222 cm3/g. SAXS measurements show that the porous distribution has a self-similar character which extends over a range which increases when R/C increases. For the highest R/C ratios, we obtain $D_p \cong 2.80$ in a range from 1 to 20 nm. The fractal FHH theory points out that the parameter R/C controls not only the width of the pore distribution but also their fractal roughness. For the highest R/C ratios, we obtain $D_s \cong 2.60$ in a range of $0.01 < P/P_0 < 0.95$. We point out that the length scale over which fractal dimension was calculated by SAXS and FHH methods is different. For SAXS measurements, it extends from 1 to 20 nm describing a self-similar distribution of micropores in the interior of the constitutive particles of aerogels. FHH fractal theory is applied on a range from 3.5 Å (N_2 monolayer) to more or less 20 Å describing the self-similarity of the accessible surface. This last scale is not large enough to allow us describe the global inner structure of particles; it provides only information on the rugosity of the system at the molecular length scale. Finally, for R/C = 250, the aerogel scattering curve presents a pronounced maxima at about $2\pi/q \cong 20$ nm suggesting the existence of some correlations in the structure. The studied aerogels exhibit an important megaloporous character which can be explained by crystals growth during the freezing step of the drying: V_{Hg} for $d_p > 10$ μm varies from 1.997 to 1.286 cm3/g. For the highest value of R/C (R/C=250), a very opened structure presenting a microporous, a mesoporous, a macroporous and a megaloporous character is obtained.

The very clear textural evolution described by the different techniques is the result of two distinct influences of the R/C ratio during the aerogel synthesis. The amount of catalyst plays indeed a major role in the formation mechanism of the precursor gel but it has also a great influence on the freezing step of the freeze-drying by changing the nucleation and crystals growth mechanisms[9]. Finally, we underline that the studied aerogels have a structure which is quite different of CO_2-supercritical dried resorcinol-formaldehyde aerogels .[2-7] It must be reminded that the two synthesis methods are not exactly the same and that the drying technique has certainly a major influence.

5 CONCLUSION

We synthetized organic resorcinol-formaldehyde aerogels using a new drying method: freeze-drying. By this simple method, we showed that we could produce a material with a very large and complex porous distribution. The combination of independent characterization techniques probing different scales of observation provides an improved and coherent picture of the structure of these complex materials. Using a multidisciplinary approach, it is thus possible to obtain a very detailed description of the influence of the molar ratio R/C, which is one of the parameter which can help us to tailor the aerogel texture in order to develop catalyst supports with controlled porosity or performant absorbants.

ACNOWLEDGEMENT

The authors are very much indebted to the "Service de la Recherche Scientifique du Ministère de l'Education, de la Recherche et de la Formation de la Communauté Francaise de Belgique" and to the "Ministère de la Région Wallonne, Direction générale des Technologies et de la Recherche" for financial support.

REFERENCES

1. R.W. Pekala, "Low Density, Resorcinol-Formaldehyde Aerogels", issued March 5, 1991, US Patent # 4.997.804 .
2. J.D. LeMay, R.W. Hopper, L.W. Hrubesh and R.W. Pekala, *Mater. Res. Soc. Bull.*, 1990,**15**, 19.
3. R. Pahl, U. Bonse, R.W. Pekala and J.H. Kinney, *J. Appl. Crystallogr.*,1991, **24**, 771.
4. G.C. Ruben, R.W. Pekala, T.M. Tillotson and L.R. Hrubesh, *J. Mater. Sci.*,1992, **27**, 4341.
5. R.W. Pekala and D.W. Schaefer, *Macromolecules*, 1993, **26**, 5487.
6. D.W. Schaefer, R.W. Pekala and G. Beaucage, *J. Non-Cryst. Solids*, 1995, **186**, 159.
7. C.J. Brinker and G.W. Scherer, in "Sol-Gel Science: The Physics and Chemistry of Sol-Gel Processing" , Academic Press, London, 1989.
8. J.D. Mellor, in "Fundamentals of Freeze Drying" , Academic Press, New-York, 1978.
9. A.P. MacKenzie, International Symposium on freeze-drying of biological products, Develop. biol. Standard, 1977, Vol. 36 .
10. S. Brunauer, P.H. Emmet and E. Teller, *J. Am. Chem. Soc.*, 1938, **60**, 309.
11. A.J. Lecloux, in "Catalysis, Science and Technology", J.R. Andersen, P. Boudard (Eds), Springer-Verlag, Berlin, 1981, Vol. 2, p.171
12. R. Sobry, Y. Rassel, F. Fontaine, J. Ledent and J.M. Liégeois, *J. Appl. Cryst.*, 1991, **24** , 692.
13. S.J. Gregg and K.S.W. Sing, in "Adsorption, Surface Area and Porosity", 2nd ed. Academic Press, London, 1982, 89.
14. D. Avnir and M. Jaroniec, *Langmuir*, 1989, **5**, 1431.
15. P. Pfeifer and M.W. Cole, *New J. Chem.*, 1990, **14**, 221.
16. P. Pfeifer, M. Obert and M.W. Cole, *Philos. Trans. R. Soc. Lond. A.*, 1989, **169**, 423.
17. Y. Yin, *Langmuir*, 1991, **7**, 216.
18. J.E. Martin and A.J. Hurd, *J. Appl. Cryst.*, 1987, **20** , 61.
19. D.W. Schaefer, in: "Better Ceramics through Chemistry II", Mater. Res. Soc. Symp. Proc., C.J. Brinker, D.E. Clark and D.R. Ulrich (eds), Materials Research Society, Pittsburgh, PA, 1986, .Vol. 121

COMPARISON OF t-PLOTS FOR CARBONS, SILICAS AND ALUMINAS

Diane R. Milburn and <u>Burtron H. Davis</u>
University of Kentucky Center for Applied Energy Research, 3572 Iron Works Pike, Lexington, KY 40511

1 ABSTRACT

Multilayer thickness (t) values calculated from nitrogen adsorption are compared using the methods of Lippens, Linsen and DeBoer,[1] Payne and Sing,[2] and Lecloux and Pirard[3] for nonporous carbons, aluminas and silicas. The "universal t-curve" proposed by DeBoer, et al.[1,4-6] based on aluminum oxides and hydroxides is in good agreement with the t-values calculated for an aluminum oxide which has been wetted, dried and has developed some mesoporosity. The t-curve proposed by Payne and Sing is in superior agreement for a nonwetted aluminum oxide (type II isotherm), suggesting that some mesoporosity may have been present in the aluminum oxides utilized for the DeBoer studies. Additionally, several spherical silicas and nonporous carbon blacks were compared using the t-values calculated from the DeBoer universal t-curve, the values proposed by Lecloux and Pirard, which vary according to BET c-values calculated for these materials and a t-curve for Degussa aluminoxide-C. Generally the data suggest that samples containing mesopores show greater agreement with the DeBoer type t-curve and the dry alumina. The nonporous materials are more accurately compared with the standards proposed by Lecloux and Pirard.

2 INTRODUCTION

Nitrogen physisorption is possibly the most common analytical technique utilized to describe porosity in solids. Since the concept of a "universal t-curve" was introduced by de Boer, *et al.* to describe nitrogen adsorption on various oxides and hydroxides,[1] numerous research groups have examined the applicability and limitations of the method (eg. refs 2-4,7,8). Sing introduced the alpha (α_s) plot to compare porosities of materials by normalizing the adsorbed volume to the volume of nitrogen adsorbed on a nonporous reference material at a relative pressure equal to 0.4.[9] The curves produced by this method are essentially the same as the t-plots, although on a different scale. Lecloux and Pirard have presented evidence indicating the importance of the C_{BET} constant, which reflects the heat of interaction between the adsorbate and adsorbent.[3] According to their work, the C_{BET} of the reference material should be matched as closely as possible to the sample if the isotherm comparison is to be valid. They further proposed a set of "standard" isotherms corresponding to specific ranges of C_{BET} for use in calculating t or α_s.

Current work involves applying these models to a variety of carbons, aluminum oxide and silicon oxides.

3 EXPERIMENTAL

Nitrogen sorption measurements were made using a Quantachrome Autosorb 6 instrument. Outgassing was performed prior to analysis at 200°C and <10^{-3} torr for a minimum of 12 hours.

4 RESULTS AND DISCUSSION

Table 1 lists the BET surface areas, BET C constant (C_{BET}) values, isotherm and hysteresis classifications of the samples examined in the present study. The Degussa Aluminoxide-C is a low density nonporous material which, when wetted with water to the incipient wetness point and dried (100°C, ambient pressure), develops some mesoporosity as the result of particle agglomeration (Figure 1). The nitrogen

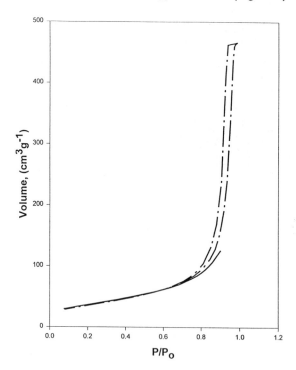

Figure 1 Nitrogen isotherms for (——) wetted and dried Degussa Al$_2$O$_3$-C and (—·—) the aluminum oxide standard from ref. 1.

isotherm measured by Lippens, *et al.* for a non-porous aluminum oxide and proposed for their universal t-curve is shown in Figure 1 along with the wetted and dried Degussa Aluminoxide-C. The similarity suggests the presence of some mesoporosity in the "universal" sample, possibly due to exposure to moisture. This is further supported by the t-plots, shown in Figure 2. Excellent agreement is observed for the Degussa Al_2O_3 reported by Payne and Sing (filled circles) and for the non-wetted Degussa Al_2O_3 from this study (filled squares), while the wetted sample (open circles) is in better agreement with the line representing the Lippens *et al.* data, represented by the line.

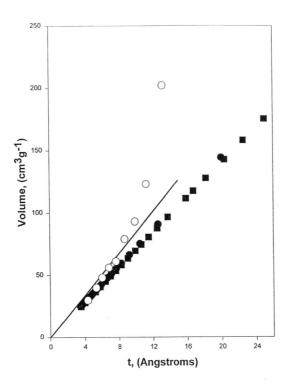

Figure 2 t-plots comparing (●) aluminum oxide from ref. 2, (■) Degussa Aluminoxide-C and (○) wetted and dried Degussa Aluminoxide-C with the aluminum oxide standard reported by Lippens, Linsen and DeBoer (——).[1]

The two aluminas appear to compare equally well with the deBoer isotherm and the Lecloux isotherm (Figure 3). Following the argument of Lecloux and Pirard, these would be expected to agree well due, not only to the chemical similarity, but also to the agreement in C_{BET} values (Table 1).

Table 1 *Results of nitrogen sorption analyses*

Sample	BET Surface Area (m^2g^{-1})	C_{BET}	Isotherm Type[10]	Hysteresis Type[10]
Degussa Al$_2$O$_3$-C	98	120	II	-
Wetted Al$_2$O$_3$-C	110	131	IV	H1
Shell Silica S980 B1.5	270	85	IV	H1
Shell Silica S980 G2.3	88	137	IV	H1
Shell Silica S980 H1.5	100	114	IV	H1
Graphite	6	22	II	-
Graphon Carbon	89	97	IV	H1
Sterling SO1 Carbon	37	43	II	-
Vulcan 3 Carbon	72	91	IV	H1
Vulcan 6H Carbon	101	64	IV	H1

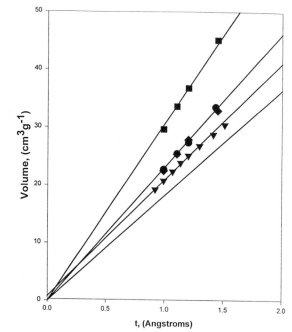

Figure 3 t-plots calculated from Lecloux and Pirard standards[3] for (●) Degussa Aluminoxide-C, (■) wetted and dried Degussa Aluminoxide-C, (◆) Shell S980 H1.5 SiO$_2$, (▼) Shell S980 G2.3 SiO$_2$ and (——) Shell S980 B1.5 SiO$_2$.

The silica materials B1.5, G2.3 and H1.5 possess approximately 1.0-1.5 cm^3g^{-1} total pore volume, all in the mesopore range.[11] The deBoer t-plots (Figure 4) reveal small positive y-intercepts for the two higher surface area materials and positive deviations at higher P/P_o, suggesting the presence of both micropores and mesopores. While the isotherms support the mesoporosity, they do not indicate an appreciable number of micropores. The lower surface area silica (S980 B1.5) has a zero y-intercept and does not show a significant deviation from linearity. Lower agreement in the deBoer t-plot and greater agreement with the Lecloux standard isotherm would be predicted for these samples, due to the porosity and/or chemical differences inherent between silica and the nonporous aluminas reported in the deBoer study.

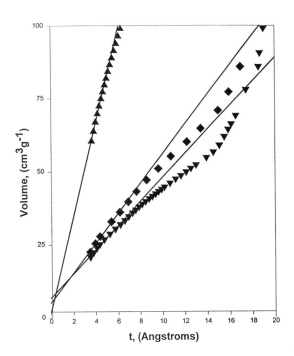

Figure 4 t-plots comparing (▲) Shell S980 B1.5 SiO_2, (▼) Shell S980 G2.3 SiO_2 and (◆) Shell S980 H1.5 using the aluminum oxide standard reported by Lippens, Linsen and DeBoer.[1]

The silicas are shown compared to the Lecloux *et al.* standard in Figure 3. While a reasonable agreement is seen with this model, limited data is available for comparison. In Figure 5 the linear fit with the silica data compared with the wetted and dried Degussa Al_2O_3 are seen to extend over a wider range of t (and P/P_o) than with the Lecloux *et al.* model and with y-intercepts closer to zero than when compared with the deBoer standard.

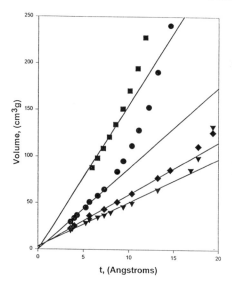

Figure 5 t-plots comparing (■) Shell S980 B1.5 SiO$_2$, (◆) Shell S980 H1.5 (▼) Shell S980 G2.3 SiO$_2$ and (●) wetted and dried Degussa Al$_2$O$_3$-C using dry Degussa Al$_2$O$_3$-C as the standard isotherm.

Type IV isotherms with very narrow hystereses were observed for three of the carbon samples, with only Graphon having greater than 1cm^{3}g^{-1} total pore volume, all in the mesopore range.[12] The deBoer t-plots, presented in Figure 6, show a relatively wide range of linearity with nearly zero y-intercepts for all carbon materials.

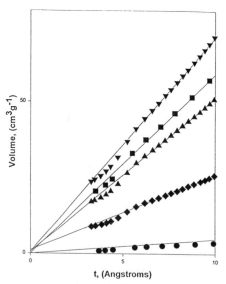

Figure 6 t-plots comparing (▼) Vulcan 6H carbon, (■) Graphon carbon, (▲) Vulcan 3 carbon, (◆) Sterling SO1 carbon and (●) graphite using the aluminum oxide standard reported by Lippens, Linsen and DeBoer.[1]

Of the five carbon samples in this study, three fall in the $C_{BET} < 100$ category and the remaining two in the $100 < C_{BET} < 300$ region. The graphite and Graphon carbons show superior agreement with the Lecloux isotherm (Figure 7), and the remaining three are in superior agreement with the deBoer isotherm. Positive deviation with increasing t is seen in the comparison with the wetted Al_2O_3, shown in Figure 8, also reflected as mesoporosity in the isotherms.

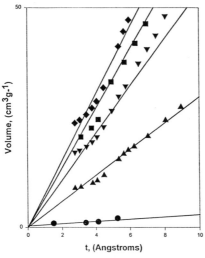

Figure 7 t-plots calculated from Lecloux and Pirard standards[3] for (●) graphite, (■) Graphon carbon, (◆) Sterling SO1 carbon, (▲) Vulcan 3 carbon and (▼) Vulcan 6H carbon.

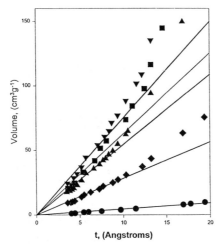

Figure 8 t-plots comparing (▼) Vulcan 6H carbon, (■) Graphon carbon, (▲) Vulcan 3 carbon, (◆) Sterling SO1 carbon and (●) graphite using dry Degussa Al_2O_3-C as the standard isotherm.

5 CONCLUSION

The agreement between the wetted and dried Degussa Al_2O_3 and the deBoer standard isotherm (Figure 1) as well as the superior agreement for samples in this study which produce type IV isotherms suggest   that the deBoer "non-porous aluminum oxide" may actually have been a mesoporous material. Thus, superior agreement with silicas, aluminas and carbons exhibiting a small volume of mesopores indicates that an important factor in determining the isotherm for comparison is more dependent on degree of porosity. For nonporous materials, the Lecloux and Pirard model, involving closely matching the C_{BET} value appears to be more accurate in this study.

References

1. B. C. Lippens, B. G. Linsen and J. H. DeBoer, *J. Catal.*, 1964, **3**, 32.
2. D. A. Payne and K. S. W. Sing, *Chem and Ind.*, (5 July, 1969), 918.
3. A. Lecloux and J. P. Pirard, *J. Colloid and Interface Sci.*, 1979, **70**, 265.
4. B. C. Lippens and J. H. DeBoer, *J. Catal.*, 1965, **4**, 319.
5. J. H. DeBoer, B. G. Linsen and TH. J. Osinga, *J. Catal.*, 1965, **4**, 643.
6. J. H. DeBoer, B. G. Linsen, Th. Van der Plas and G. J. Zondervan, *J. Catal.*, **4**, 649.
7. P. Schneider, *Appl. Catal.*, 1995, **129**, 157.
8. J. Adolps and M. J. Setzer, *J. Colloid and Interface Sci.*, 1996, **180**, 70.
9. K. S. W. Sing, *Chem. & Ind.*, 1968, 1520.
10. J. Rouquerol, D. Avnir, C. W. Fairbridge, D. H. Everett, J. H. Haynes, N. Pernicone, J. D. F. Ramsay, K. S. W. Sing and K. K. Unger, *Pure & Appl. Chem.*, 1994, **66**, 1739.
11. D. R. Milburn, B. D. Adkins and B. H. Davis, 'Characterization of Porous Solids II,' F. Rodríguez, *et al.* (eds.), 1991, p. 543.
12. D. R. Milburn, B. D. Adkins and B. H. Davis, 'Characterization of Porous Solids,' K.K. Unger, *et al.* (eds.), 1988, p. 501.

FRACTAL ANALYSIS OF AEROGELS AND CARBON BLACKS USING NITROGEN ADSORPTION DATA: COMPARATIVE STUDY OF TWO METHODS

B. Sahouli, S. Blacher and F. Brouers
Physique des Matériaux et Génie Chimique
Université de Liège, B5
Liège 4000, Belgium

ABSTRACT

In this work we discuss two methods to determine the surface fractal dimension (D) from nitrogen adsorption isotherms data. We consider: a) the fractal version of the FHH (Frenkel-Halsey-Hill) equation[1] and b) the thermodynamic method proposed by Neimark et al.[2] Experimental nitrogen adsorption of a very different porous systems, such as the aerogels (inorganic and organic) prepared in different conditions and carbon blacks (commercial and from vacuum pyrolysis of used tyres), are analysed. Recently, Jaroniec[3] has shown theoretically the equivalence between the fractal FHH type equation for capillary condensation and the thermodynamic method in some conditions. Here, we report an experimental investigation of this equivalence and we discuss the case when both methods do not give the same results. Our results show that the equivalence between these methods depends of the nature of the porosity.

1 INTRODUCTION

We consider the problem of the determination of fractal dimension from the results of an adsorption isotherm of one adsorbate (here the nitrogen) in the multilayer adsorption and capillary condensation. The most used methods are the fractal version of the BET theory, the fractal version of the FHH theory[1] and the the thermodynamic method[2]. Recently, it was shown theoretically by Jaroniec[3] and experimentally by us[4] for a serie of commercial carbon blacks samples that the thermodynamic method developped by Neimark[2] and the fractal FHH type equation are equivalent in some conditions.The main of this work is to test the validity of these two methods and their equivalence for some representative samples very different in their chemical nature and morphology. The FHH type equation has been used for the determination of the fractal dimension either for microporous solids[5] for which it has been derived initially by Avnir and Jaroniec[1b] or for mesoporous solids where the capillary condensation (CC) regime is predominant[6]. by contrast, Neimark's method is only applicable for mesoporous solids as confirmed by the use of Kiselev and Kelvin equations, i.e., when the main adsorption process is CC. However, the generality of Neimark's method lies in fact that it does not depend on any porous structure in particular on the shape of pores.

The objective of this work is to test the validity of these both methods and their equivalence for some representative samples very different in their chemical nature and morphology. The obtained results show that the equivalence between these methods depends of the nature of the porosity and it occurs in the range of applicability of the Neimark's method.

2 EXPERIMENTAL

2.1 Materials

The commercial carbon black samples were provided by by Cabot, Sarnia, Canada. The pyrolytic carbon blacks (CB_p) samples were obtained by vacuum pyrolysis of the side wall of used tyres in a laboratory scale batch reactor at temperatures and total pressures ranging from 420 to 700°C and from 0.3 to 20.0 kPa, respectively. The detailed studies of these carbon blacks have been published elsewhere[7]. In this investigation, only two representative samples (commercial sample N115 and one pyrolytic carbon black sample CBp) were considered.

The mixed SiO_2-ZrO_2 aerogels were synthesized following the way detailed elsewhere[8]. Two different samples were considered: one was prepared under acidic catalyst conditions (MIX-A) and the other one under under basic conditions (MIX-B) .

The organic resorcinol-formaldehyde (RF) aerogels were synthesized following a method described elsewhere[9-10]. A serie of the RF samples have been prepared and characterized as a function of the molar Resorcinol/Catalyst (RC)[10]. In this study, we consider two samples RC50 and RC250 corresponding to the molar ratios of 50 and 250 respectively.

2.2 Adsorption Measurements

Nitrogen adsorption experiments were performed on a SORPROMATIC CARLO-ERBA SERIES 1800 appartus controlled by by an IBM Personal Computer. Prior to the nitrogen adsorption, all samples were outgassed in vacuum at 300 °C

3 METHODS

3.1 Fractal FHH Equation

The classical Frenkel-Halsey-Hill (FHH) theory[11] describes the continuous growth of an adsorbate film according to the following assumptions: 1) the surface-adsorbate potential controls all layers, i.e., the source of potential is all the atoms in the solid which is considered as semi-infinite, 2) the multilayer formation is driven by the energy, i.e., the theory is based on the minimisation of the energy resulting from a competition between solid-adsorbate and adsorbate-adsorbate interactions forces exists. This leads to a smooth film-vapour interface. In this case, the classical FHH equation on a flat surface reads :

$$V_{ads} \propto \left[\ln\left(\frac{P_0}{P}\right) \right]^{-\frac{1}{s}}$$

(1)

V_{ads} is the volume of gas (here nitrogen) adsorbed at equilibrium pressure P on the sample. P_0 is the saturation pressure of nitrogen at the sample temperature (77 K). The exponent s describes, according to Halsey[12], how fast the solid-adsorbate interactions decrease with increasing distance from the surface.

A fractal version of the FHH equation was developed[11]:

$$V_{ads} \propto \left[\ln\left(\frac{P_0}{P}\right) \right]^{-\frac{1}{m}} \tag{2}$$

Where the exponent is a function of the fractal dimension D .This theory allows only the determination of surface fractal dimension D when the analytic form of the FHH-exponent (m) is known. Ideally, for pure Van der Waals forces, s has the value 3 and in this case Pfeifer et al.[1c] obtained :

$$\frac{1}{m} = \frac{3-D}{3} \text{ for the film thickness below a critical value } z_c \tag{4}$$

$$\frac{1}{m} = 3-D \text{ for the film thickness above a critical value } zc \tag{5}$$

According to Pfeifer et al.[1c], Eq. (2) with the exponent given in Eq. (4) characterises the polymolecular adsorption or FHH regime in which the dominating force is the substrate potential whereas Eq.(2) with the exponent given in Eq. (5) is valid for the capillary condensation (CC) regime where the dominating force is the surface tension. The crossover between these two regimes is given by the critical film thickness z_c which depends of the ratio between the surface tension and the solid-adsorbate interactions.

Thus, one obtains the practical expression of the capillary condensation prediction on a fractal surface (in log-log form):

$$\ln(V_{ads}) = \text{Constant} + (D-3)\ln(-\ln X) \tag{6}$$

where $X = P/P_0$ is the relative pressure.

3.2 Thermodynamic Method

The thermodynamic method proposed and developed by Neimark[2] is presented as an independent method. In this model, the fractal surface area

$$S(r) \propto r^{2-D} \tag{7}$$

of the condensed adsorbate-gas interface is calculated, at a given pressure, from the Kiselev equation[13]:

$$S = \frac{RT}{\sigma} \int_{N_l}^{N_h} (-\ln X)\, dV_{ads} \tag{8}$$

The yardstick r is measured in terms of the average radius of curvature of the meniscus at the interface between condensed adsorbate and gas by the Kelvin equation:

$$r = -\frac{2\sigma V_{mol}}{RT \ln X} \tag{9}$$

R, T, V_{mol} and σ are the gas constant, the temperature, the molar liquid nitrogen volume and the surface tension respectively. It must be pointed out that the Kiselev integral (Eq. 9) was initially proposed as an alternative method for the determination of the surface areas of porous and especially mesoporous solids. The upper limit N_h is the coverage adsorption when the relative pressure tends to unity while the lower limit N_l corresponds usually to the beginning of the hysteresis loop.

Taking equations 7, 8 and 9 into account, the fractal surface dimension D is derived from the slope of the linear part of the following equation :

$$\ln(S) = \text{Constant} + (D\text{-}2)\ln(\text{-}\ln X) \qquad (10)$$

In this method, the only adsorption process taken into account is the capillary condensation phenomenon which can be described by the Kelvin equation.

4 RESULTS AND DISCUSSION

Figure 1a,b and c show the nitrogen adsorption isotherms for the considered samples. It must be point out that except the RC50 sample, all the other samples (N115, CBp, RC250, MIX-A and MIX-B) present the hysteresis loop (not shown in the figure). For the carbon blacks, the isoterms are intermediate between type II and type VI according to the IUPAC classification[14]. The samples of MIX-A and MIX-B mixed SiO_2-ZrO_2 aerogels exhibit adsorption isotherms of type IV. For the organic resorcinol-formaldehyde (RF) aerogels, the adsorption isotherms depend strongly[10] on the molar ratio R/C since we observe in Fig.1b a clear transition from an isotherm of type I for RC50 to an isotherm of type IV for the RC250 sample.

The fractal analysis using the method based on the FHH equation (eq.6) and the thermodynamic method (eq.10) is illustrated in the figure 2 of the studied samples. The obtained results are reported in the table . This table includes the BET specific surface area (S_{BET}), the FHH-exponent (m), the fractal dimension D and the relative pressure range over which these D-values have been calculated.

The results (Table) can be summarized as follows: (a) The S_{BET} of the sudied samples range from 50 to 650 m^2/g depending on the porosity of each sample, (b) For both methods (FHH and Neimark methods), the range of relative pressure corresponding to the linear part of the curves (Figure 2) is not the same for all samples, (c) At first approximation, the m values seems to be depending on two factors: (i) the chemical nature of the adsorbent and (ii) the nature of porosity within a same class of solids. This trend agrees with the fact that the mesoporosity decreases the FHH exponent while the microporosity increases it, as noted previously.[15] Hence, only the RC50 sample has a dominant microporous character. In other words, and according to Pfeifer et al.[1c] the FHH exponent is related to the physical adsorption regime, (d) The fractal dimension D obtained by these methods, range from 2.4 to approximately 3. The D value for RC50 sample given by the thermodynamic method is greater than 3 and it is nonphysical from a geometric point of view.

Table *Main results of the studied samples.*

Samples	$S_{BET}(m^2/g)$	P/P0 (FHH)	m	FHH (Eq.6)	Neimark (Eq.10)	P/P0 (Neimark)
N 115	145	0.17 - 0.91	2.43	2.59	2.41	0.35-0.94
CBp	49	0.49 - 0.98	2.46	2.60	2.59	0.52-0.97
RC50	464	0.44 - 0.98	50	2.98	'3.87'	0.01-0.83
RC250	409	0.01 - 0.92	2.35	2.58	2.62	0.01-0.90
Mix-A	474	0.14 - 0.93	2.50	2.60	2.90	0.06-0.88
Mix-B	643	0.43 - 0.99	4.48	2.77	2.92	0.24-0.99

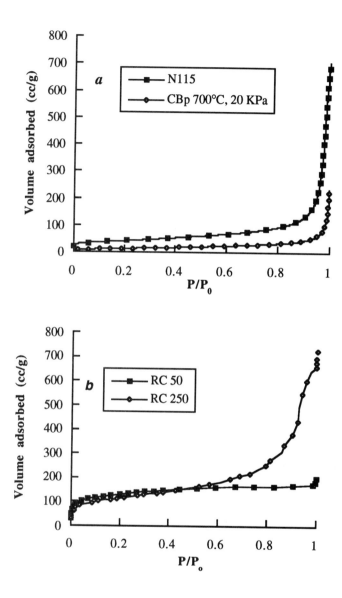

Figure 1 *(continued on p. 288)*

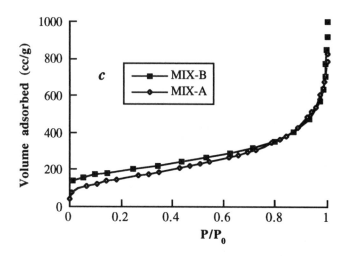

Figure 1 *Nitrogen adsortion isotherms of (a) N115 and CBp carbon blacks samples, (b) RC50 and RC250 organic resorcinol formaldehyde aerogels and (c) MIX-A and MIX-B mixed SiO₂-ZrO₂ aerogels.*

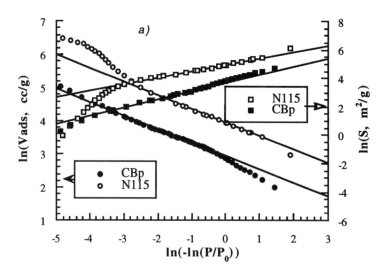

Figure 2 *(continued on p.289)*

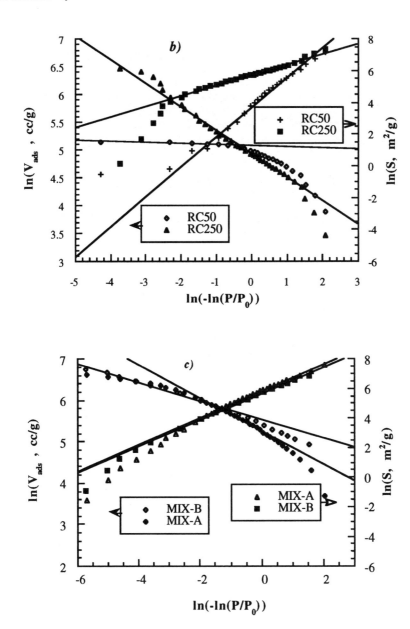

Figure 2 *FHH and Neimark's-Plots of (a) carbon blacks samples, (b) organic resorcinol frormaldehyde aerogels and (c) mixed SiO₂-ZrO₂ aerogels.*

By comparing the fractal dimension determined by the FHH equation and the thermodynamic method proposed by Neimark, one can observe that for the CBp and RC250, there is a total equivalence in the range of experimental errors. For the considerably microporous sample RC50, the FHH equation gives a D value close to 3 and the Neimark's method cannot be applied to this type of porous systems. For some samples such N115, MIX-A and MIX-B, the D values obtained by both method differ slightly. This can be understood as follows: a) If neither the microporosity nor the mesoporosity cannot be neglected, then the obtained D-value depends on the sensitivity of these methods to the nature of the porosity. In the words, if the the adsorption process is a mixture of polymolecular and capillary condensation. b) In contrast to the Neimark's method, the FHH type equation is also sensitive to the microporous structures and this most probably explains the disagreement in the case of these samples.

CONCLUSION

This study shows how experimental arguments lead to the conclusion that the equivalence between a FHH type equation and the thermodynamic method holds only in the case where the adsorption process is dominated by capillary condensation forces and that it depends on the nature of the porosity. Also, the relation of the FHH exponent (m) to the fractal dimension (D) needs further examination.

REFERENCES

1. (a) P.G. de Gennes, *in Physics of Disordered Materials*, edited by D. Adler, H. Fritzsche and S.R. Ovhinsky (Plenum, New York 1985),.(b) D. Avnir and M. Jaroniec, *Langmuir* , 1989, **5**, 1431,.(c) P. Pfeifer and M.W. Cole, *New J.Chem.* 1990, **14**, 221 .(d) Y. Yin, *Langmuir* , 1991, **7**, 216.
2. (a) A.V. Neimark, *JETP Lett.,* 1990 , **51**, 607, (b) A.V. Neimark, M. Hanson, K.K. Unger, *J. Phys. Chem.* 1993, **97**, 6011, (c) A.V. Neimark, K.K. Unger, *J. colloid Interface Sci.,* 1993,**158**,412, (d) A.V. Neimark, *Ads.Sci.Tech.,* 1990, **7**, 210.
3. M. Jaroniec, *Langmuir* , 1995, **11**, 2316.
4. B. Sahouli, S. Blacher and F. Brouers, *Langmuir* , 1996, **12**, 2872
5. K. Kaneko, M. Sato, T. Suzuki, Y. Fujiwara, K. Nishikawa and M. Jaroniec, *J. Chem. Soc. Faraday Trans.,* 1991, **87**, 179.
6. I.M.K. Ismail and P. Pfeifer, *Langmuir* , 1994, **10**, 1532.
7. (a) H. Darmstadt, C. Roy, S. Kaliaguine, B. Sahouli, S. Blacher, R. Pirard and F. Brouers, *Rubber Chem. Technol..*, 1995, **68**, 330 (b) B. Sahouli, S. Blacher, F. Brouers, R. Sobry, G. Van den Bossche, B. Diez ; H. Darmstadt, C. Roy and S. Kaliaguine, *Carbon,* 1996, **34**, 633.
8. S. Blacher, R. Pirard, J.P Pirard, B. Sahouli , F. Brouers, *Langmuir*, 1996, in press.
9. R.W. Pekala and D.W. Schaefer, *Macromolecules*, 1993, **26**, 5487.
10. B. Mathieu, S. Blacher, F. Brouers and J.P Pirard, to be presented in this conference.
11. S.J. Gregg and K.S.W. Sing, *'Adsorption, Surface Area and Porosity '*, 2nd Edition Academic, London, 1982.
12. G.D. Halsey, *J. Chem. Phys* ., 1948, **16**, 931.
13. Reference 11, pp. 171.
14. K.S.W. Sing, D.H. Everett, R.A.W. Haul. L. Moscou, R.A. Pierotti, J. Rouquerol and T. Siemieniewska, *Pure Appl. Chem.,* 1985, **57**, 603.
15. P.J.M. Carrot, K.S.W. Sing, *Pure Appl. Chem.,* 1989, **61**,1835.

CERTIFICATION OF REFERENCE MATERIALS FOR PORE ANALYSIS BY THE GAS ADSORPTION METHOD

B. Röhl-Kuhn[a], K. Meyer[a], P. Klobes[a], K.-F. Krebs[b], and Th. Fritz[a]

[a]Bundesanstalt für Materialforschung und -prüfung, Berlin, Germany
[b]Merck KGaA, Darmstadt, Germany

1 INTRODUCTION

By definition, a reference material (RM) is a material or a substance one or more of whose property values are sufficiently homogeneous and well established to be used for the calibration of an apparatus, the assessment of a measurement method, or for assigning values to materials.[1,2]

A certified reference material is accompanied by a certificate in which one or more property values are certified by a procedure, together with the uncertainty and the corresponding level of confidence. The procedure allows the traceability of the values to an accurate realization of the unit or to a normal value whose metrological properties (value and uncertainty) are generally known and accepted data.[1,2,3,4]

With more than 200 suppliers world-wide, it is often difficult for the user to find the reference material best suited for his application. In order to facilitate the search for the required reference materials, the database COMAR[5] has been developed. The database contains information on more than 8853 certified reference materials from France, Germany, Great Britain, the USA, Japan, the Commonwealth of Independent States, China, and other countries.

Certified reference materials are necessary to assure reliability and trueness of analytical results (Figure 1). Their use encompasses development and validation of procedures as well as establishment of guidelines and standards. They are used for calibration of measurement instruments and for quality management. Certification of element contents of reference materials achieved by interlaboratory comparison involves different independent procedures to determine the conventionally true value. This does not hold true for the determination of some other characteristic values such as the thickness of a layer or the density of a solid, for instance. In these cases the results obtained by different methods are not necessarily comparable.

Although CRMs are important for calibrating measurement instruments, for checking analyses and hence for the accuracy of analytical results and for quality management, the present situation regarding certified reference data for porous and disperse materials is unsatisfactory. Reference materials with certified data regarding the specific surface area are available for particular measurement ranges[5,6], but their quality standard is relatively low for various reasons, e.g. they are often not traceable to interlaboratory test comparisons, the results are unsatisfactory and data on their stability and durability are missing. Reference materials with certified data on their specific pore volume, pore size distribution and mean pore radius are missing altogether. For these reasons, manufacturers of products and instruments often use internal standards, which means, however, that the results are not comparable.[7] Furthermore, very few reference materials with large specific surface area are known due to the low long-term stability caused by the thermodynamic properties of the material.

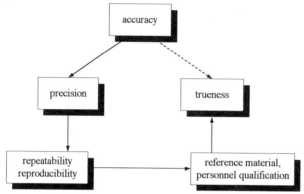

Figure 1   *Relation between precision and trueness of analytical measurements*

2 GENERAL PROCEDURE

The production of certified reference materials is based on ISO Guides 30-35,[1,8,9,10,11,12] the BCR-Guidelines[13] as well as the Standard DIN/ISO 5725[14]. REMCO[15] is ISO's (International Organization for Standardization) committee on reference materials, reporting to the ISO Technical Management Board.

According to the BCR Guidelines, the general procedure for the production of a certified reference material consists of the following steps:

− Demand analysis
− Making the raw materials
− Performance tests:
    homogeneity
    stability (thermal stability, long-term stability)
− Method development *(validation of the method or DIN/ISO/CEN standards)*
− Preparation of samples for interlaboratory test
− Certification analysis *(participation of more than 10 laboratories)*
− Specifications for participating laboratories:
    description of the method
    calibration
    number of tests
− Evaluation of certification analysis
    method of evaluation (DIN/ISO 5725 or BCR program)
− Issue of the certificate, report

3 SPECIAL PROCEDURE

When developing certified reference materials for pore analysis by the gas adsorption method, two important restrictive conditions must be taken into account:

i. A closed theory of physisorption for the entire p/p_0 range does not exist. Furthermore, standardized evaluation methods are available only for certain isotherm types of the IUPAC classification.

ii. Disperse or porous solids show fractal behavior with respect to important pore properties.[16]

It follows from i. that the material to be certified must strictly be selected according to a "suitable" isotherm type of the IUPAC classification.

From ii. one concludes for example that with porous or highly disperse solids the experimentally determined value of the specific surface area depends on the size of the probe molecule, i.e. on the area effectively occupied by the probe molecule in the monolayer, and that a "true" or absolute value of the specific surface area does not exist for fundamental reasons. The area occupied by the probe molecules thus becomes the decisive quantity for the traceability of the calculated specific surface area. Therefore the traceability required for certified reference materials can only be assured by applying strict standardized measurement and evaluation methods (Figure 2). The flux diagram of the certification procedure of a CRM for the physical characteristic values specific surface area, specific pore volume, and mean pore radius is shown in Figure 3.

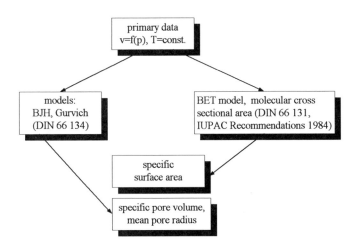

Figure 2 *Adherence to standardized procedures for assuring traceability*

For the area effectively occupied by a nitrogen molecule in a monolayer physisorbed at 77 K (molecular cross sectional area), the 1984 IUPAC Recommendations[17] give a value of 0.162 nm^2; this value has also been definitely fixed in the Standard DIN 66 131.[18]

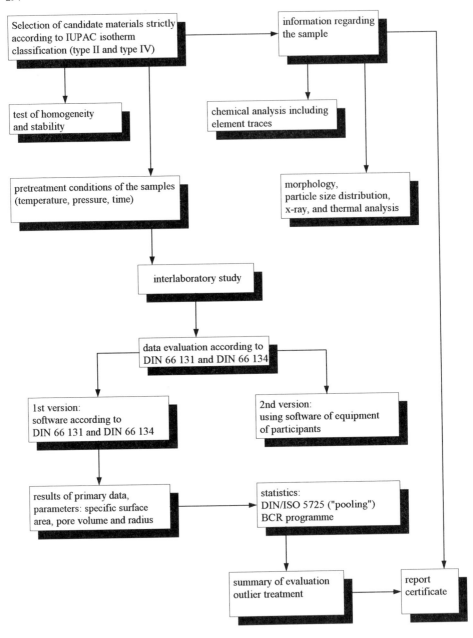

Figure 3 *Certification procedure of a CRM for the physical characteristic values specific surface area, specific pore volume, and mean pore radius*

4 EXPERIMENTAL

Four materials, silicon dioxide, α-alumina, alumina type 60, and alumina type 150, were certified with respect to the above-mentioned characteristic values. For silicon dioxide, the adsorptive used was krypton, for the other three materials nitrogen was used. For silicon dioxide and α-alumina, the specific surface area was certified; for alumina type 60 and alumina type 150 also the specific pore volume and the pore radius were certified.

The interlaboratory tests were carried out with 38 participating laboratories (44 various apparatus). Table 1 gives a list of the measurement apparatus employed.

Table 1 *The measurement apparatus employed in the interlaboratory tests*

Manufacturer	Instruments
FISONS/Carlo Erba Instruments, Milano, Italy	Sorptomatic 1800/1900/1990
Micromeritics Instrument Corp., Atlanta, USA	ASAP 2000/2010/2400 Gemini III and 2360, Flowsorb Digisorb 2400/2600
Porous Materials Inc., Ithaca, New York, USA	Automated BET Sorptometer
Quantachrome Corporation, New York, USA	Autosorb
Ströhlein West GmbH & Co., Kaarst, Germany	DEN AR MAT II

The specific surface area was determined according to the Standard DIN 66 131 (BET method),[18] while the pore distribution was determined according to the BJH method, as described in DIN 66 134.[19] Differences in the results are produced, when the different laboratories evaluate their primary data with the software of their own equipment or when the results of the different laboratories are calculated using the same software as the intercomparison laboratory, responsible for the evaluation. It is therefore important that the primary data (isotherm data) be evaluated using the same software (called version 1 in Figure 3). It has been shown in calculations of the above-mentioned characteristic values that an evaluation according to version 2 (instrument software of interlaboratory test participants) yields an enhanced number of outliers and stragglers, i.e. a generally enhanced dispersion, which is a consequence of the different implementations of the evaluation models in the software of the instrument manufacturers.

The statistical evaluation of the quantities determined from the isotherms can be carried out according to DIN/ISO 5725 or according to the BCR Guidelines. Evaluation according to DIN/ISO 5725 via the total mean value ("pooling") presupposes, however, that the number of measured values of the individual laboratories is equal and that the variances lie within a fixed range, i.e. that they are homogeneous. If this is not the case, pooling is discarded, i.e. the evaluation is not performed via the total mean value. In the present case the variances of the laboratories are heterogeneous due to the use of different measurement apparatus.

Therefore, we proceeded as follows:

- Outlier test for the mean values of the laboratories after Dixon (1%, two-sided iteration), elimination of the outliers

- Because of the mentioned heterogeneity, the laboratory dispersions were not subjected to an outlier test
- Kolmogorov-Smirnov-Lilliefors test for normal distribution of the mean values of the laboratories
- Determination of the certified value as an estimate x of the unknown value μ
- Determination of the uncertainty of the certified value as 95 % confidence interval for μ

In the case of "no pooling", the BCR Guide[13] uses a very simplified model for the calculation of the certified value. This means that only laboratory mean values are modeled and not the individual values. All data were evaluated via "pooling", according to DIN/ISO 5725 as well as via "no pooling" (mean of means); the statistician decided to choose the model "no pooling".

As an example, we summarize the statistical results of the certification of the reference material CRM BAM-PM-103. This substance with the manufacturer's designation alumina type 60 (mixture of the forms η-Al_2O_3 and γ-Al_2O_3) possesses an N_2 isotherm (Figure 4) of type IV according to the IUPAC classification. Figure 5 shows the pore size distribution as calculated by the BJH method. Figure 6 represents an example for the statistical evaluation of the certification interlaboratory test. Finally, the results of these evaluations are shown in Table 2.

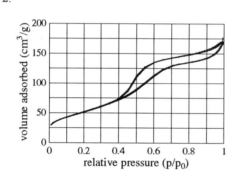

Figure 4   *Adsorption isotherm for nitrogen at 77 K on alumina type 60 (BAM CRM-PM-103, 9 measurements)*

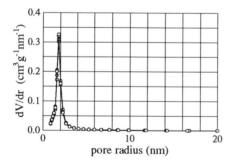

Figure 5   *Pore size distribution of alumina type 60 (BAM CRM-PM-103, 9 measurements)*

Specific surface area (m^2/g)

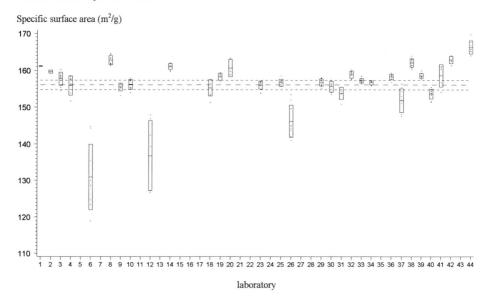

laboratory

Figure 6 *Interlaboratory study: Specific surface area S_{BET} of CRM BAM-PM-103 (alumina type 60), the wide-dashed line corresponds to the mean of means with standard deviation (narrow-dashed lines); laboratory mean values with standard deviation and individual values are also given*

Table 2 Results for CRM BAM-PM-103

Quantity	Method	Mean of means	Uncertainty	
			Standard deviation	95%-confidence interval
specific surface area	BET (DIN 66 131)	156.0 m^2/g	1.3 m^2/g	2.7 m^2/g
specific pore volume	Ads., Gurvich (DIN 66 134)	0.250 cm^3/g	0.002 cm^3/g	0.004 cm^3/g
mean pore radius	2V/S$_{BET}$ (DIN 66 134)	3.18 nm	0.02 nm	0.03 nm
most frequent pore radius	BJH (DIN 66 134)	1.93 nm	0.04 nm	0.07 nm

5 CONCLUDING REMARKS

The development of CRMs with the certified physical characteristic values specific surface area, specific pore volume, and mean pore radius was described. As an example, we presented the CRM BAM-PM-103 (alumina type 60) developed on the basis of BCR Guidelines and subjected to interlaboratory tests with 38 participating laboratories, as well as the underlying statistical model.

It is expected that further development of porous and disperse CRMs for a broad pore spectrum will allow a more objective and secure evaluation of pore analysis in the future, but much more work, also concerning standardization of pore analysis, is needed to realize this goal.

References

1. ISO Guide 30, 'Terms & definitions used in connection with reference materials', 1992
2. DIN 32811, 'Grundsätze für die Bezugnahme auf Referenzmaterialien in Normen', 1979
3. W. Hässelbarth, *GIZ Fachzeitschrift für das Laboratirium* 1995, **39**, 273
4. P. D. Bièvre, R. Kaarls, S. D. Rasberry and W. P. Reed, 'Lecture, NCSL (National Conference of Standards Laboratories) Symposium and Workshop', 1995, Gaithersburg, USA
5. COMAR, 'Database for certified reference materials', Bundesanstalt für Materialforschung und -prüfung, D-12489 Berlin
6. E. Robens and K.-F. Krebs, in: F. Rodríguez-Reinoso et al. (Editors) 'Characterization of Porous Solids II', Elsevier Science Publishers B.V., Amsterdam, 1991, 133
7. K. Meyer, P. Lorenz, B. Röhl-Kuhn and P. Klobes, *Cryst. Res. Technol.* 1994, **29**, 903
8. ISO Guide 31, 'Contents of certificates of reference materials', 1981
9. Draft ISO Guide 32, 'Calibration of chemical analysis and use of certified reference materials', 1994
10. ISO Guide 33, 'Uses of certified reference materials', 1989
11. ISO Guide 34, 'Quality system guidelines for the production of reference materials', 1994
12. ISO Guide 35, 'Certification of reference materials-general and statistical principles', 1989
13. BCR/48/93, 'Guidelines for the production and certification of BCR reference materials', 1994
14. DIN/ISO 5725, 'Präzision von Meßverfahren; Ermittlung der Wiederhol- und Vergleichspräzision von festgelegten Meßverfahren durch Ringversuche', 1988
15. ISO REMCO, 'Committee on reference materials, International Organization for Standardization', CH-1211 Genève 20
16. D. Avnir, 'The Fractal Approach to Heterogenous Chemistry', Wiley and Sons, Chichester, 1989
17. K. S. W. Sing, D. H. Everett, R. A. W. Haul, L. Moscou, R. A. Pierotti, J. Roquerol and T. Siemieniewsky, *Pure & Appl. Chem.* 1985, **57**, 603 (IUPAC Recommendations 1984)
18. DIN 66 131, 'Bestimmung der spezifischen Oberfläche von Feststoffen durch Gasadsorption nach Brunauer, Emmett und Teller (BET)', 1993
19. Draft DIN 66 134, 'Bestimmung der Porengrößenverteilung aus der Stickstoff-Sorptionsisotherme nach Barrett, Joyner und Halenda', 1995

ADSORPTION AND DISPLACEMENT OF FORMIC ACID AND ACETIC ACID ON MICROPOROUS PHENOLIC RESIN CARBONS

W. H. Lawnik,[1] O. Dietsch,[1] A. Fritz,[1] G. H. Findenegg,[1] and S. R. Tennison[2]

[1] Iwan-N. Stranski-Institut für Physikalische und Theoretische Chemie,
 Technische Universität Berlin, Strasse des 17. Juni 112, D-10623 Berlin, Germany
[2] MAST International Ltd, 62 Farleigh Road, Addlestone, Surrey KT15 3HR, England

ABSTRACT

Phenolic Resin Carbon (PRC), a new synthetic microporous adsorbent, was tested for the recovery of aqueous acids from process effluent. Chromatographic frontal analysis and batch measurements were used to study the acid adsorption step and the subsequent immiscible solvent displacement step using di-isopropyl ether (DIPE) as a displacer. The performance of the new materials was compared with that of a conventional coconut shell carbon (SSC 607C).

The analysis of the adsorption isotherms in terms of the Generalised Freundlich Equation, $q = q^{\infty} \cdot [w/(K+w)]^{m}$, reveals marked differences in the adsorption behaviour of PRC samples compared with SSC: The adsorption constant K is significantly lower than for SSC implying improved regeneration characteristics of PRC. On the other hand, K is not influenced by any parameter in the PRC production. Independent on the activation mode, acetic acid is adsorbed to a higher degree in terms of mass uptake than formic acid. The saturation uptake q^{∞} depends on the available surface area and micropore volume and is therefore controlled by the extent of activation. Chromatographic displacement studies confirm the finding of improved regenerability of PRC compared with SSC. Generally, the efficiency in displacing formic acid by DIPE is found to be lower than for acetic acid. Whereas the adsorption process is primarily controlled by bulk diffusion effects, the DIPE displacement stage is dominated by micropore diffusion which implies a dependence on the extent of activation rather than macropore volume characteristics: Neither the activation agent nor the particle size do influence the displacement kinetics. From these results it is concluded that higher levels of adsorbent activation lead to better desorption characteristics and to higher adsorption capacities of the Phenolic Resin Carbons.

1 INTRODUCTION

The adsorption process on activated carbons is widely applied for the removal of organic impurities from gas streams and from industrial or municipal effluents.[1-6] In case of wastewater technology, the spent carbon is normally regenerated by means of an ex-situ high temperature steam/air treatment resulting in a partial destruction of the adsorbed contaminants. The recovery and recycle of water soluble organics such as formic and acetic acids from process effluent exhibit the following key problems with at present no economically viable solution: (i) Compared to water-insoluble organics, these hydrophilic organics exhibit low adsorption affinity and high concentrations in the effluent. This leads to short adsorption cycles and requires an in-situ regeneration process. (ii) A conventional steam regeneration may result in an aqueous solution of similar concentration as the original process effluent. Therefore a novel three-stage process was evaluated: After the breakthrough of the acids in the adsorption step, the adsorbed acids are displaced by a water-immiscible solvent that is subsequently removed by steam regeneration.[7]

The present study is concentrated on the adsorption and displacement of formic and acetic acid on commercial and synthetic carbons. We discuss the improved adsorption behaviour of a new synthetic microporous adsorbent (Novolak Phenolic Resin derived Carbon PRC) in comparison with a commercial carbon (Coconut Shell 607C) and the influence of several parameters in the PRC production (agent and degree of activation, particle size) on the adsorption process.

2 EXPERIMENTAL SECTION

2.1 Experimental Procedure

Equilibrium adsorption isotherms of aqueous acid solutions were recorded by the conventional batch method: Weighed amounts of dried carbon (m_S) and acid solution (m_L, acid weight fraction w_0) were mixed together in small bottles. After allowing about 20 hours for equilibration in a temperature controlled environment, the equilibrium weight fractions w of the filtered solutions were determined from their refractive indices using a Brice-Phoenix differential refractometer. The specific adsorbed mass uptake q was obtained from the relation

$$q = (w_0 - w) \cdot m_L/m_S \tag{1}$$

The kinetics of acid adsorption and displacement was studied by means of liquid chromatography. The solutions were degassed by a 4-channel on-line HPLC degasser (Knauer, model A1050) before entering the HPLC pump (Waters, model 501). The pump transported the solvent to an electric 6-port valve for the selection of the flow direction through the MPLC column (Omnifit; 10 mm diameter, 250 mm length): During the adsorption stage, the aqueous solution was transported through the column from below. During the displacement stage, a downflow mode was chosen in order to produce a sharp water/DIPE interface with no fingering of the lighter organic. The acid concentration was detected on-line by means of an UV spectral photometer (Knauer, model A0293) and a differential refractometer (Knauer, model A0298). In addition, the constancy in flow was controlled by a liquid flowmeter. Flow rates of 2 ml/min (6 bed volumes/hour) and 1 ml/min were used in the adsorption and displacement process, respectively. Prior to each adsorption measurement, the column was equilibrated with pure water until stable detector signals were obtained. The displacement was recorded immediately after the adsorption run. In all kinetic studies the adsorbent was regenerated on-line using methanol and water before starting the next adsorption and displacement cycle.

2.2 Materials

Since 1970 there has been a surge in testing synthetic resins as tailor-made adsorbents in both fundamental studies and industrial applications.[5, 8-11] After controlled curing and forming, the resin precursor is carbonised and activated to the desired extent by means of

Table 1 *Physical Properties of the Carbons*

Adsorbent (activation agent)	Weight Loss by Activation	N_2 Surface Area $[m^2/g]$	Micropore Volume $[cm^3/g]$	Macropore Volume $[cm^3/g]$
SSC 607C (steam)	50 %	1245	0.55	0.14
PRC RV02 (steam)	22.2 %	800	0.37	0.15
70 µm Resin:				
PRC 46 (steam)	70.8 %	1584	0.71	0.40
PRC 47 (CO_2)	70.7 %	1427	0.63	0.39
30 µm Resin:				
PRC 55 (steam)	68.4 %	1573	0.69	0.36
PRC 49 (steam)	69.3 %	1541	0.67	0.39
PRC 50 (air)	69.3 %	1544	0.68	0.35
PRC 51 (CO_2)	71.6 %	1647	0.70	0.41

steam, CO_2 or air. The main advantages of synthetic carbons are their bimodal pore size distribution and a significant reduction in surface heterogeneity in marked contrast to commercial carbons like Coconut Shell SSC 607C. The PRC samples used in this study exhibit the following structure: Their micropores are slit-shaped with a slit width of ca. 0.8 nm which remains almost unaltered during the activation process. Their macrostructure derives from the voids between resin particles during the grinding and sintering process. Therefore the mean macropore size is directly related to the mean particle size of the resin powder: 70 µm resin samples exhibit mainly 20 µm pores; 30 µm samples mainly 12 µm pores. The physical properties are summarised in Table 1. The micropore and macropore volumes as well as the specific surface area increase with increasing extent of the PRC activation (i.e. increasing carbon weight loss). The nearly constant macropore volume of the high-activated samples corresponds to a mixture of cubic and hexagonal close packed pseudo spherical particles which indicates well cured materials with little particle coalescence. The commercial coconut shell carbon SSC 607C is characterised by a high surface area but by low values for the micropore and macropore volume.

The organic solvents, formic acid (GC purity 98%, Fluka), acetic acid (GC purity > 99%, Fluka), di-isopropyl ether (GC purity > 99%, Merck), and methanol (HPLC purity > 99.8%, Merck), were used without further purification. De-ionised water was taken from a Milli-Q 50 reagent grade water purification system (amount of total organic carbon < 200 ppb, electrical resistance ρ > 18 MΩ·cm).

3 RESULTS AND DISCUSSION

3.1 Equilibrium Adsorption Studies

Adsorption isotherms of formic acid and acetic acid on low and high-activated carbon samples are illustrated in Figures 1a+b. They cannot be represented by a two-parameter equation based on the Langmuir or the Freundlich model (dashed and dotted curves), but a combination of these relations, the Generalised Freundlich (GF) Equation,[12] fits the data (uptake q as a function of the acid concentration w in wt%) within experimental accuracy:

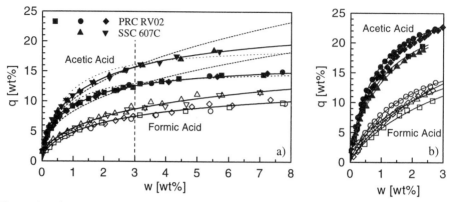

Figure 1 *Adsorption Isotherms of Dilute Acids on Two Low-activated Carbons (1a) and on Four High-activated 30 μm PRC Samples (1b) (see Table 1)*

$$q = q^{\infty} \cdot \left[\frac{w}{K+w} \right]^{m} \tag{2}$$

In the above equation q^{∞} reflects the limiting adsorbate uptake, K is inversely proportional to the adsorption strength, and the adsorbent specific exponent m is related to the heterogeneity in adsorption energy.

The results of the analysis in terms of the GF equation are listed in Table 2. As to be expected, the adsorption amount mainly depends on the available surface area. Different activation methods don't influence significantly the relative adsorption of the two species in the low concentration range (Figure 1b): For a given adsorbent, acetic acid is adsorbed to a higher degree (approximately by a factor of 2 in weight uptake) corresponding to higher saturation uptakes q_A^{∞} and lower adsorption constants K_A. On the other hand, K exhibits no significant dependence on any parameter in PRC production but a pronounced deviation from the relatively high values for Coconut Shell. The adsorbent-specific exponent m increases with the decrease in adsorbent particle size (from 70 μm to 30 μm) indicating a more uniform surface.

Table 2 *GF Parameters (Subscripts: A = Acetic Acid; F = Formic Acid)*

Adsorbent	Exponent m	K_A [wt%]	q_A^{∞} [wt%]	K_F [wt%]	q_F^{∞} [wt%]
SSC 607C	0.48 ± 0.02	4.0 ± 1.2	23.5 ± 1.8	14 ± 4	20 ± 3
PRC RV02	0.53 ± 0.02	2.1 ± 0.4	16.6 ± 0.5	5.1 ± 0.6	12.6 ± 0.6
70 μm Resin:					
PRC 46	0.60 ± 0.07	2.1 ± 1.0	22.6 ± 2.3	1.9 ± 0.2	15.1 ± 0.7
PRC 47	0.56 ± 0.02	1.8 ± 0.2	25.2 ± 0.4	4.5 ± 0.4	24.2 ± 1.0
30 μm Resin:					
PRC 55	0.83 ± 0.03	1.4 ± 0.1	30.9 ± 0.6	6.5 ± 0.6	34.6 ± 2.0
PRC 49	0.79 ± 0.04	1.4 ± 0.2	28.1 ± 0.8	6.4 ± 1.3	27.6 ± 3.3
PRC 50	0.71 ± 0.02	2.1 ± 0.2	29.5 ± 0.6	4.7 ± 0.4	24.2 ± 1.0
PRC 51	0.83 ± 0.02	1.0 ± 0.1	29.0 ± 0.3	2.6 ± 0.1	22.9 ± 0.4

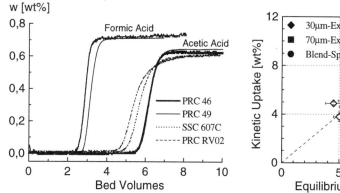

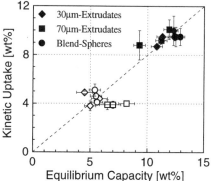

Figure 2 *Acid Breakthrough Curves*   **Figure 3** *Carbon Capacity at w = 0.6 wt%*

3.2 Breakthrough Curves

The kinetics of acid adsorption was measured by chromatographic frontal analysis at an upflow rate of 6 bed volumes/hour (2 ml/min). The acid breakthrough on high-activated and low-activated carbon samples is shown in Figure 2. For both acids, sharp breakthrough curves without any deterioration between subsequent adsorption/regeneration cycles (reproducibility within ±5-10%) are observed: The adsorption within the micropores is extremely rapid compared to the mass transfer in their macropores. The steepness in acid breakthrough, i.e. the mass transfer rate, increases with macropore volume and therefore with the extent of activation. Independent on both the activation mode and the particle size of the resin powder, the adsorption capacity of acetic acid is a factor 2 greater than the adsorption capacity of formic acid. In confirmation with the previous results of the equilibrium studies, the breakthrough uptake depends on the available surface area and the bulk density of the PRC carbons in the column packing.

As a favourable adsorption isotherm implies a constant-pattern mass transfer front,[4] the chromatographic finding is directly related to the adsorption characteristics in larger scale adsorption columns. Therefore a comparison between the kinetic breakthrough uptake and the corresponding equilibrium value serves as a guideline for the design of a technical process. In addition, it represents the PRC efficiency in adsorption kinetics at a technical relevant flow. The experimental breakthrough uptakes for all high activated carbon samples as a function of the corresponding equilibrium capacities are given in Figure 3. The data for acetic acid (full symbols) exhibit no dependence on the shape of the carbon, nor on the particle size of the resin or the activation agent: For all carbon samples the kinetic capacity is approximately 80 % of the corresponding equilibrium capacity. The analysis of the formic acid data (open symbols) reveals a similar ratio for the spheronised blends and the 30 μm extruded materials but a pronounced deviation for the 70 μm extruded samples (0.55 ± 0.11) indicating worse mass transfer characteristics with regard to their possible adsorption capacity.

3.3 Displacement Curves

The acids were displaced by di-isopropyl ether (DIPE) in the downflow mode at the half rate (3 bed volumes/hour). Typical on-line-recorded curves of the displacement stage from low and high-activated carbons are shown in Figures 4 and 5: Indicated by a vertical

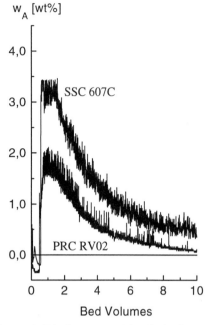

Figure 4 *Displacement of Acetic Acid*
            from Low-activated Carbons

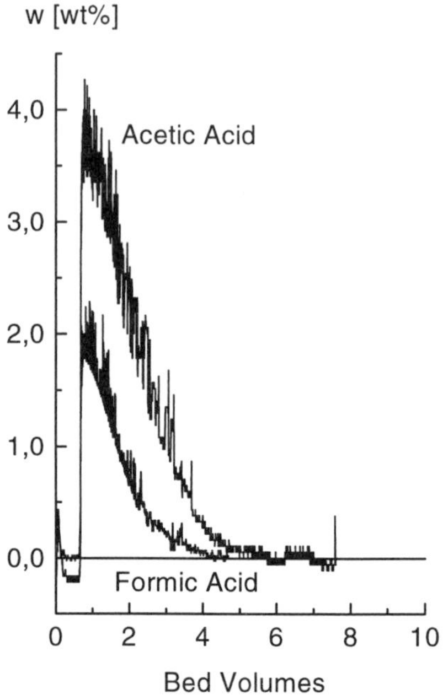

Figure 5 *Displacement of Acids from*
            High-activated PRC 46

step at about 0.6 bed volumes, the organic phase pushes the forward aqueous phase out of the column without any mixing effect. The following acid displacement shows the expected asymptotic approach to pure DIPE. In marked contrast to smooth breakthrough curves in the adsorption stage, these graphs exhibit a considerable noise caused by water microdroplets carried with the organic solvent. Because of the vertical DIPE breakthrough this finding cannot be attributed to macropore transport limitations, but to micropore diffusion. This suggestion is supported by the large decay period of the regeneration curve compared with the relatively small width of the breakthrough region (1 - 2 bed volumes).

Figure 4 illustrates the poor regenerability of the low-activated PRC sample RV02 and the commercial carbon SSC 607C: The displacement of acetic acid is still incomplete even after 10 bed volumes. The main aim of the carbon optimisation was thus (i) to increase the rate of the displacement to both reduce the required amount of DIPE and to increase the acid concentration in the DIPE and (ii) to reach a complete regeneration by the water-immiscible displacer.

Figure 5 demonstrates the improved regeneration characteristics of all high-activated PRC samples: Significant higher levels of acetic acid are observed at the beginning of the DIPE breakthrough. In addition, the displacement is already completed after 4 - 5 bed volumes. Within the experimental error (± 5-10%), none of the three series of a given precursor material exhibits any dependence of the regeneration characteristics on the activation method: Neither the relative adsorption nor the relative displacement of the two acid species is influenced by the activation agent or by the particle size of the resin powder. Although the adsorption capacity of formic acid is much lower compared with acetic acid, nearly the same retention volume (resp. bed volume) is required to regenerate a given carbon sample. The different efficiency in displacing formic and acetic acid is supported by the finding of significant differences in the relevant strengths of adsorption in the equilibrium studies.

4 CONCLUSIONS

For all PRC samples tested neither the adsorption stage nor the displacement stage exhibit any deterioration between subsequent cycles. Higher levels of adsorbent activation lead to higher adsorption capacities and to better desorption characteristics. Whereas the adsorption stage is primarily controlled by bulk diffusion effects (macropore and inter-particle diffusion), the DIPE displacement stage is dominated by micropore diffusion: Penetration of DIPE into the micropores to displace the adsorbed acid is the rate limiting process. Therefore the same retention volumes of the two acids are observed for a given carbon sample. The activation removes pore blocking decomposition products and thus leads to increasing micropore volumes and an improved regenerability of high-activated materials. The incomplete regenerability of the commercial coconut shell carbon SSC 607C is attributed to its narrow micropores from which the adsorbed acids cannot be removed easily by the displacer.

ACKNOWLEDGEMENT

This work has been funded by the CEC Environment Programme, Project No. EV5V-CT94-0557. We wish to thank our partners D. Watson, J. Carey, J. Jones, M. Roberts (BP Chemicals Ltd, Hull), S. Ragan and H. Sharrock (Sutcliffe Speakman Carbons Ltd) for enlightening discussions during our collaboration.

REFERENCES

1. P. N. Cheremisinoff and F. Ellerbusch (eds.), 'Carbon Adsorption Handbook', Ann Arbor Science, Ann Arbor, 1978.
2. I. H. Suffet and M. J. Mc Guire (eds.), 'Activated Carbon Adsorption of Organics from the Aqueous Phase', Ann Arbor Science, Ann Arbor, 1980.
3. J. R. Perrich (ed.), 'Activated Carbon Adsorption for Wastewater Treatment', CRC Press, Boca Raton, 1981.
4. D. M. Ruthven, 'Principles of Adsorption and Adsorption Processes', Wiley and Sons, New York, 1984.
5. S. D. Faust and O. M. Aly, 'Adsorption Processes for Water Treatment', Butterworths, Boston, 1987.
6. K. E. Noll, V. Gounaris, and W. Hou, 'Adsorption Technology for Air and Water Pollution Control', Lewis Publishers, Chelsea, 1992.
7. S. R. Tennison and D. J. Watson, European Patent 0 635 294 A1, 1994.
8. N. T. Kartel, A. M. Puziy, and V. V. Strelko, in 'Characterisation of Porous Solids II', Elsevier, Amsterdam, 1991, p. 439.
9. P. Fott, F. Kolar, and Z. Weishauptova, *Collect. Czech. Chem. Commun.*, 1995, **60**, 172.
10. A. Gierak, *Mat. Chem. Phys.*, 1995, **41**, 28.
11. J. Simitzis, J. Sfyrakis, and A. Faliagas, *Mat. Chem. Phys.*, 1995, **41**, 245.
12. A. Derylo-Marczewska, M. Jaroniec, *Chemica Scripta*, 1984, **24**, 239.

Characterization of Sulphated Oxides of Titanium: Infrared, Adsorption and Related studies

Christine A. Philip and Suzy A. Selim
Department of Chemistry, Ain Shams University
Nahed Yacoub and <u>Jehane Ragai</u>
Department of Science, The American University in Cairo

ABSTRACT

A series of sulphated oxides of titanium, TS, TS_1, TS_2 and TS_3, with different sulphate ion contents, respectively <0.3 %, 16.89%, 33.41% and 40.81%, was prepared, using a solution of titanium oxysulfate-sulfuric acid complex hydrate ($TiOSO_4.12H_2SO_4.12H_2O$) as starting material and ammonia solution as precipitating agent. The oxides were then heated at different increasing temperatures deemed significant according to their thermal behaviour. Nitrogen adsorption studies in conjunction with infrared and thermal studies were carried out on all samples. The surface acidity was also determined by the method described by Boehm (1). Infrared results confirm the presence of the surface sulphate ions and suggest the gradual lowering, through heat-treatment, of the symmetry of the sulphate ion by complex formation. The coordination of the sulphate ions,through heat-treatment at 335°C and 425°C of TS_1, is suggested to change, from free sulphate ions to bridging bidentate and possibly monodentate sulphate ligands. In the case of TS_2 and TS_3, depending on the thermal treatment, bridging and chelating bidentate sulphate ligands are identified at 425°C and 580°C. Sample TS does not indicate the presence of any sulphate ions. Heat-treatment as well as the presence of the sulphate ions affect the textural characteristics of the samples and drastically reduce the extent of the surface area. It is suggested that the presence of the sulphate ions allows for the appearance of strong acidic sites on the surface of the sulphated oxides.

1 INTRODUCTION

The so-called superacids have triggered a great deal of attention in recent years in view of their widespread use as catalysts (2-9). Indeed, such superacids have been shown to be active in a wide spectrum of reactions ranging from esterification, to alkene polymerization, to acylation and isomerization of hydrocarbons (10). Together with Fe_2O_3 and ZrO_2 ,TiO_2 is known to be a superacid after sulphatizing (10) and is generally prepared by heating the corresponding hydroxide impregnated with sulphuric acid or ammonium sulphate (3-6). The calcination temperature of the oxide has also been found to be of importance for its catalytic properties (8). The present work aims at complementing such previous studies by elucidating the effect of the *sulphate content* on the surface acidity and textural characteristics of titanium oxides prepared from ($TiOSO_4.12H_2SO_4.12H_2O$) . An attempt is also made to determine the state of ligation of the generated surface sulphates.

2 EXPERIMENTAL

2.1 Materials

Sulphated hydrous titanium oxide was precipitated by the addition of a 33% w/w ammonia solution to a constantly stirred 0.1 M solution of ($TiOSO_4.12H_2SO_4.12H_2O$) prepared in 1M H_2SO_4 and heated to 77°C. Precipitation was carried out until the pH of the suspension reached 10.8. The precipitate was digested at that temperature for 2 hours allowed to cool and kept in contact with the mother liquor for 2 days. It was then filtered and dried at room temperature. The dried product was divided into 4 equal parts, three of which were washed with 300 mls, 150 mls and 100 mls of distilled water to give respectively samples TS_1, TS_2 and TS_3 . The fourth portion was washed with distilled water until almost free from the sulfate ion (sample TS). All samples were dried to constant weight at 52 °C.
Elemental tests indicated respectively a sulphate content of <0.3%, 16.89%. 33.41% and 40.81% for samples TS,TS_1,TS_2 and TS_3. These samples were then heated for two hours in air to different increasing temperatures of 185°C, 335°C, 425°C, 580°C, and 780°C which were chosen according to the thermal behaviour of the oxides.

2.2 Techniques

The infrared studies were carried out using a 337 Perkin- Elmer double beam grating spectrophotometer. Solid samples were prepared in the form of a KBr pellet. Two milligrams of the compound were mixed with approximately 200mg of KBr (spectroscopic grade). The mixture was then subjected to a pressure of about 20 lb in^{-2} in a hydraulic press. A disc of pure KBr was used in the reference cell .
The thermal studies were carried out on a Stanton Redcroft STA-780 simultaneous thermal analyser series designed to give simultaneous thermogravimetric (TG), differential thermal analysis (DTA) and differential thermogravimetric records (DTG).
Adsorption measurements were carried out at 77K using a conventional volumetric technique. The sample was outgassed overnight at room temperature to residual pressures of ~25x10^{-4} torr. The gas pressures were measured on a mercury manometer with the aid of a cathetometer. The time required for each point of the adsorption or desorption isotherm to attain equilibrium was between 15-20 minutes.
X-ray diffraction patterns were determined using a Philips X-ray diffraction analyser system, PW1840, using Nickel filtered CuKa radiation.
Adsorption isotherms obtained by the adsorption of sodium hydroxide or phosphoric acid from solution were determined by shaking overnight about 0.2 g of the solid material in 10 ml of the base or acid of known normality. A series of different concentrations, ranging between 0.2N and 0.005N, was used. A blank titration was carried out first on the base or acid alone. The difference between the blank titration and that with the sample gave the amount adsorbed. In the case of the adsorption of H_3PO_4 , the PO_4^{3-} content

was estimated photometrically.

3 RESULTS AND DISCUSSION

3.1 Infrared studies

Infrared studies carried out on the original unheated TS, TS_1, TS_2 and TS_3 samples reveal in all of the cases two peaks at $1400cm^{-1}$ and $1630cm^{-1}$. The first peak is characteristic of NH_4^+ ions (11,12) presumably originating from the ammonia used as precipitating agent whereas the $1630cm^{-1}$ is related to the bending mode of molecular water (13). Additional bands at $1110cm^{-1}$, $975cm^{-1}$ and $615cm^{-1}$ are only observed with the sulphated samples TS_1,TS_2 and TS_3.These lines are respectively attributed to the $\upsilon3$, $\upsilon1$ and $\upsilon4$ vibrational frequencies of the free sulphate SO_4^{2-} ion in T_d symmetry (14).The X-ray diffraction results support such an inference by indicating the presence of the mascagnite $(NH_4)_2SO_4$ phase in all of the original sulphated oxides.

Heat-treatment of the oxides respectively to 185°C, 335°C, 425°C, 580°C and 780°C leads in all of the samples to a gradual decrease of the 1400 cm^{-1} and $1630cm^{-1}$ bands until their eventual disappearance at 780°C. This corresponds to the respective elimination of the NH_4^+ ions in the form of ammonia and of water from the oxides. In case of the TS_1, TS_2 and TS_3 samples, heat-treatment at 335°C and 425°C leads to the appearance of new bands at $1230cm^{-1}$,$1130cm^{-1}$, $1035cm^{-1}$ and $980cm^{-1}$ which persist at 425°C, diminish at 580°C and disappear at 780°C. These lines are attributed to a bridging bidentate sulphate ion and correspond to the gradual lowering, through heat-treatment, of the T_d symmetry of the sulphate ion to C_{2v} by complex formation (14). Representative infrared spectra for sample TS_2 and its thermal decomposition products are shown in Figure 1.

It is relevant to mention in this connection the findings of Saur et al(15). These authors have postulated that for sulphated oxides of titanium, in the absence of OH groups or water, a structure resembling $(Ti_3O_3)S=O$ prevails, whereas in the presence of H_2O or excess surface OH groups this is converted, as will be clarified later, to $(Ti_2O_2)S\underset{O}{\overset{OH}{\lessgtr}}$ type groups. This interpretation tallies with the present results and accounts, in view of the presence of molecular water in the present systems, for the appearance of bridging bidentate sulphate ligands in TS_1,TS_2 and TS_3.

The coexistance of monodentate sulphate ions, at this stage, with the bridging bidentate ligands is however not to be precluded as the infrared bands of the latter overlap with some of the bidentate bridging ligands (14). Samples TS_2 and TS_3 indicate that at 425°C and 580°C some of the sulphate ions apart from being bridging bidentate seem also to become chelated to the titanium metals. This is indicated by the appearance at 425°C and 580°C of additional bands at ~1275 cm^{-1}, 1215 cm^{-1}, 1155 cm^{-1} and 1085 cm^{-1} which are attributed to a chelating bidentate sulphate group (14). For all of the sulphated oxides two distinctive peaks are observed in the fundamental functional group region viz, at 3470 cm^{-1} and 3150cm^{-1}. By reference to the work of Parkyns(13) these bands may be respectively attributed to the combined symmetric and asymmetric streching modes of chemisorbed water and ammonia on titania.

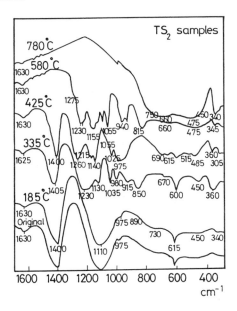

Fig 1. Representative infrared spectra for sample TS$_2$ and its thermal decomposition products.

3.2 Thermal Studies

All the original oxides of titanium gave thermograms with an initial broad endothermic peak, the maximum of which came around 120 °C. Such an endotherm may be associated with the loss of free or instertitial water coupled wih the probable evolution of ammonia(16). Themogravimetric studies indicate a decrease in weight of ~12% corresponding to such an endotherm. A marked difference is noted in the thermal behaviour of the sulphated samples. Samples TS$_1$,TS$_2$ and TS$_3$ display additional endotherms with maxima at ~320°C, 380°C, 450°C and 525°C. Evolved gas analysis studies still need to be carried out to confirm the origin of such endotherms, but evidence obtained so far by the infrared studies seems to suggest that the 320°C, 450°C and 525°C endotherms may somehow be related to the evolution of SO$_2$ stemming from ionic and differently bound sulphato ligands. In view of the different modes of attachment of these ligands it is understandable that their desorption temperatures would be different.The 380°C endotherm has been previously associated, in the case of titanium oxides, with the expulsion of water rigidly bound in the form of hydroxyl ions (16). In the case of TS$_2$ and TS$_3$ an additional endotherm is observed at 600°C Figure 2. Here also, subject to confirmation, infrared studies suggest that such an endotherm is due to the decomposition of additional sulphato ligands possibly in a chelated form. TG studies indicate respectively, for TS₁,TS₂, and TS₃ total weight losses of 32% , 55.5% and 60.5%. It should be born in mind however, in all of the present thermal studies, that DTA is a dynamic method in which equilibrium conditions are not attained so that the temperature of changes do not correspond to thermodynamic equilibrium temperatures. Representative DTA, DTG and TG curves determined on samples TS,TS₁ and TS₂ are shown in Figure 2.

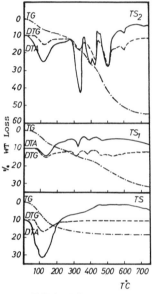

Figure 2. Simultaneous DTA, TG and DTG curves determined on samples TS,TS₁and TS₂.

3.3 Adsorption studies

Representative nitrogen adsorption isotherms for sample TS and its heat-treated products as well as for all of the original titanium oxides TS, TS₁ and TS₂ are shown in Fig. 3 (a and b). The αs plots are also shown in the same figure (c and d). In view of the very close adsorption behaviour displayed by samples TS₂ and TS₃, the adsorption isotherms and αs plots of the latter are not, for the sake of clarity, represented. In the construction of each αs plot the volume of nitrogen is plotted against the reduced adsorption, αs, as measured on a nonporous reference TiO₂ (17) (S_{BET}=34.5m²/g). As explained elsewhere (18), αs is defined as (amount adsorbed / amount adsorbed at p/po =0.4). Consideration of the main features of the isotherms and αs plots allows us to make the following conclusions: Sample TS and its heat-treated products have mainly a mesoporous texture. Heat-treatment leads to a gradual decrease in the extent of surface area and at 780°C to the development of some microporosity. Increase in the sulphate content as in the case of oxides

TS$_1$,TS$_2$ and TS$_3$ leads, together with mesoporosity, to the development of a microporous texture and to a decrease in the surface area. As shown in Figure 3d, the downward deviation from linearity in the αs plots of the original oxides TS$_1$ and TS$_2$ is the result of micropore filling followed by multilayer adsorption on a relatively small external surface. The microporous texture is preserved in all of the heat-treated products of the sulphated oxides TS$_2$ and TS$_3$.

Table I summarizes all of the adsorption results, obtained in this study, as well as the calculated surface areas and pore volumes. The values of the B.E.T. surface areas are based on the assumption that each nitrogen molecule occupies 0.162 nm^2. S$_S$ values have been calculated from the slope of the original part of the αs plots. The values in parentheses refer to microporous oxides and are therefore regarded as being apparent rather than real areas. Through the examination of Table I it is noteworthy that sulphation and heat-treatment of all the oxides generally leads to a decrease in the specific surface areas and pore volumes . The most pronounced decrease is observed in the case of the sulphated oxides TS$_2$ and TS$_3$ at 425°C and is concomitant with the development of a predominantly microporous texture. For further clarification of the mechanisms involved adsorption measurements, such as water vapor adsorption, still need to be made on such systems using smaller molecules than the nitrogen molecule. Indeed these systems may exhibit molecular sieve properties with a large number of the micropores having pore widths that are less than that of the nitrogen molecule.

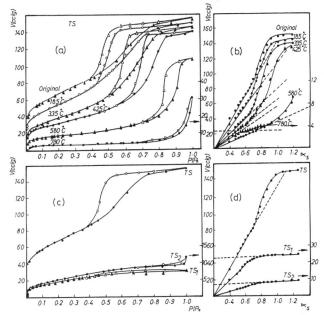

Fig. 3. Representative adsorption isotherms and αs plots for the TS sample and its thermal decomposition products curves (a) and (b) and for samples TS, TS$_1$ and TS$_2$ curves (c) and (b).

Table I. *Analysis of the nitrogen adsorption data and of the acidity and basity of the surface.*

Sample	Tempera-ture in °C	Porosity	S_{BET} m^2g^{-1}	S_s m^2g^{-1}	Pore volume cm^3g^{-1}	Amount of acidic sites in $\mu mol\,g^{-1}$	Amount of basic sites in $\mu mol\,g^{-1}$
TS	original	Meso	234	227	0.242	1330	0.0532
TS1	original	Meso+Micro	72	63	0.048	3310	0.0475
TS2	original	Meso+Micro	18	18	0.018	4690	0.0437
TS3	original	Meso+Micro	21	18	0.018	5190	0.0369
TS	185	Meso	197	179	0.234	932	0.0324
TS1	185	Meso+Micro	56	53	0.115	3320	0.0342
TS2	185	Meso+Micro	13	11	0.095	4190	0.0329
TS3	185	Meso+Micro	15	15	0.011	4500	0.0347
TS	335	Meso	144	128	0.227	588	0.0359
TS1	335	Micro+Meso	8.6	8	0.003	3790	0.0348
TS2	335	Micro+Meso	3.9	1.8	0.001	7130	0.0304
TS3	335	Micro+Meso	9.5	10	0.004	7850	0.0434
TS	425	Meso	115	104	0.218	533	-------
TS1	425	Micro+Meso	4.9	5.3	0.005	3860	-------
TS2	425	Micro	1.5	1.8	0.001	7680	-------
TS3	425	Micro	1.5	1.7	0.001	8710	-------
TS	580	Meso	62	56	0.169	389	-------
TS1	580	Meso	45	48	0.133	585	-------
TS2	580	Meso+Micro	23	20	0.089	3040	-------
TS3	580	Meso+Micro	19	17	0.077	3420	-------
TS	780	Meso	12	12	0.048	47	-------
TS1	780	Meso	19	18	0.043	73	-------
TS2	780	Meso+Micro	16	16	0.035	26	0.0018
TS3	780	Meso+Micro	20	17	0.037	27	0.0021

Samples TS, TS1, TS2 and TS3 have different sulphate ion contents, respectively <0.3%, 16.9%, 33.4% and 40.8%.

3.4 Surface Acidity

The total number of acidic sites was determined from the saturation values of the neutralization-adsorption isotherms of NaOH. The latter were Langmuir-like in shape. The values expressed in $\mu mol/g$ are indicated in Table I . The total number of basic groups were determined by phosphoric acid adsorption from solution. The uptake in $\mu mol/g$ is given in Table I .

It is interesting to note that the number of basic sites, as indicated by the phosphoric acid adsorption, is negligible as opposed to a very large amount of total acidic sites being detected. The very large number of acidic sites obtained in the case of TS2 and TS3 heat-treated at 335°C and 425°C, together with a drastic decrease in surface areas and pore volumes, is noteworthy.

Saur (15) has postulated, as mentioned earlier, that sulphated TiO_2 has the structure $(Ti_3O_3)S=O$ which upon the addition of H_2O leads to the following scheme:

In the present study such a scheme would account for the increase in acidity with sulphate content in the titanium oxide systems, and for the appearance of the largest number of acidic sites in the case of samples TS2 and TS3 heat-treated at 335°C and 425°C. In the latter case infrared results have indeed indicated a predominance of bidentate bridging sulphate ligands.

4 CONCLUSION

Sulphated oxides of titanium exhibit strong surface acidity, the effect being more pronounced with increasing sulphate content. It is suggested that the generation of surface acidity is favored through a bridging bidentate state of ligation of the sulphate ions. The latter surface complex leads in the sulphated oxides of titanium to a pronounced decrease in surface area and pore volume and to the generation of a predominantly microporous texture. Molecular sieve characteristics still need however to be investigated in the latter systems!

REFERENCES

1. H.P.Boehm, *Discussions of the Faraday Society* , 1971,**52**,264.
2. K.Tanabe ,*Mater. Chem. Phys.*,1987,**17**,1.
3. T.Yamaguchi,T.Jin,T.Ishida,K.Tanabe,*Mater.Chem.Phys.*, 1987,**17**,3.
4. K.Tanabe ,*Mater.Chem. Phys.*,1985,**13**,347.
5. N. Kitajima, Y.Ono. ,*Mater.Chem.Phys.*,1987, **7**,31.
6. K. Tanabe, T. Yamaguchi, K. Akiyama,A. Miton, K. Iwabushi,K. Isogai , *Proc. VIII Intern. Confr. Catal.*, Berlin, 1984.
7. K. Hino,M.,K.Arata, *J.Chem.Soc.,Chem.Commun.*,1980,851.
8. F. Al Mashta,N.Sheppard ,C.U.Davanzo,*Mater.Chem.Phys.* 1985 13,315.
9. T.Jin.,M.Mashida,T.Yamaguchi,K.Tanabe,*Inorg.Chem.*,1984,**23**,4396.
10. K.Hadjivanov, A. Davydov, D. Klissurki *Communications of The Department of Chemistry* , Institute of General and Inorganic Chemistry, Bulgarian Academy of Sciences, 1988, 21, 498-505.
11. M.L.Hair, *J.Phys. Chem.*, 1970, 74,1290-1292.
12. J. Ragai, *J.Chem.Tech.Biotechnol*, 1987,40,143-150.
13. N.D. Parkyns, '*Chemisorption and Catalysis*, ', Hepple Publications, London, 1970, p.150-172.
14. K. Nakamoto, 'Infrared and Raman Spectra of Inorganic and Coordination Compounds ' Wiley, New York, 1978, 3rd edition, Part III, p.239.
15. O. Saur,M. Bensitel,A.B. Mohammed, J.C. Lavalley,C.P. Tripp,B.A. Morrow, *J.Catal.*,1986,**99**,104.
16. J. Ragai, *J.Chem.Tech.Biotech.*,1987,40, 75-83.
17. J.Ragai and K.S.W. Sing , *J. Colloid Interface Sci.* ,1984,101,369
18- S.J. Gregg and K.S.W. Sing, 'Adsorption, Surface Area and Porosity', Academic Press, London, 1982, p. 98.

ANALYSIS OF PORE PROPERTIES FROM LOW PRESSURE HYSTERESIS IN NITROGEN ADSORPTION ISOTHERMS

R.R. Mather

The Scottish College of Textiles
Heriot-Watt University
Netherdale
Galashiels
Scotland TD1 3HF

1 INTRODUCTION

The occurrence of low pressure hysteresis at low relative pressures, as well as at high pressures, in the adsorption of a gas or vapour on a porous solid has been very clearly demonstrated. Low pressure hysteresis has been observed in adsorption by a wide variety of adsorbents, including carbons[1-3], clays[4], porous glass[5] and zeolites[6]. Several phenomena may give rise to low pressure hysteresis, and these have recently been clearly set out by Christensen *et al.*[7]:

(i)   diffusion of adsorbate through microporous constrictions[8];

(ii)   inelastic distortion of the adsorbent[9], such as swelling of its constituent particles[10], as the pressure of adsorbate is raised during the determination of the isotherm;

(iii)   strong bonding of the adsorbate to the adsorbent surface, for example through chemical interaction.

Whilst low pressure hysteresis has been clearly observed in the adsorption of water vapour and many organic compounds, its occurrence in the adsorption of nitrogen appears to be less well appreciated. Nevertheless, the literature reveals that a wide range of commercially important materials exhibit low pressure hysteresis in their nitrogen adsorption isotherms. These materials include organic pigments[11-13], disperse dyes[14-16], catalyst samples[7] and a sample of calcium carbonate[17]. It seems appropriate, therefore, to consider the phenomenon of low pressure hysteresis in nitrogen isotherms in some detail.

In a previous publication[12], the adsorption of nitrogen by samples of the blue organic pigment, copper phthalocyanine, was considered. In those cases where low pressure hysteresis was observed, two broad categories were distinguished: one in which the hysteresis loop broadens at relative pressures above ca. 0.4 (illustrated schematically in Figure 1 as isotherm A), and the other in which the loop broadens towards lower relative pressures (isotherm D). It was concluded that the former category of hysteresis is associated with pigment crystals aggregated into closely packed structures, and that the latter category indicates more loosely packed crystal

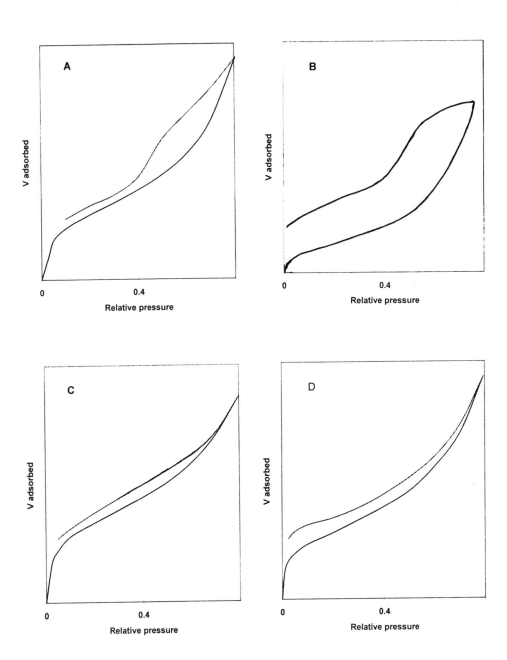

Figure 1: *Categories of low pressure hysteresis in nitrogen adsorption isotherms*

aggregate structures. However, further scrutiny of the literature has revealed that other categories of low pressure hysteresis can also be identified in nitrogen isotherms. Some of these categories are included in the present paper.

2 CATEGORIES OF LOW PRESSURE HYSTERESIS

Figure 1 illustrates schematically a number of the categories of low pressure hysteresis which have been encountered in nitrogen adsorption isotherms. In isotherm A, there is a distinct transition between classical high pressure hysteresis and low pressure hysteresis. The relative pressure at which the transition occurs is ca. 0.4, the pressure where the closure of hysteresis loops is very often observed in nitrogen isotherms[18]. Thus, at relative pressures above ca. 0.4, the isotherm appears to possess standard Type IV character, which would be associated with a rigid mesoporous structure. In isotherm D, the hysteresis loop is of a different nature. There is now no distinct transition between high and low pressure hysteresis, and indeed the loop broadens towards lower relative pressures.

It has been argued[12] that copper phthalocyanine pigments giving isotherms of the type depicted by isotherm A contain aggregates in which the constituent pigment crystals are more firmly held together. By contrast, the nature of the hysteresis loop in isotherm D has been explained by penetration of adsorbed nitrogen molecules into microporous junctions between aggregated crystals. The aggregates in these pigments are more loosely held together and hence more flexible.

Nitrogen isotherms of the type depicted by isotherm D have also been observed for samples of two types of commercial disperse dye, C.I. Disperse Blue 79 and C.I. Disperse Red 54[14-16]. Disperse dyes are unusual compared with other classes of dye in that they have very low solubility in aqueous media, and are instead dispersed in the dyebath. For both types of disperse dye, those samples whose isotherms resembled isotherm D were shown to consist of the loosest aggregate structures amongst the range of samples studied.

Other examples of low pressure hysteresis observed in nitrogen isotherms are represented schematically by isotherms B and C In isotherm C, hysteresis is most pronounced at intermediate relative pressures. Such isotherms have been observed for some organic pigments[19] and for two samples of C.I. Disperse Red 54[16]. The latter formed part of a study of a wider range of samples of this dye. From this study, differences in the nature of the nitrogen isotherms could be correlated with differences in dispersion behaviour. It was concluded that the dye samples giving isotherms resembling isotherm C consisted of aggregates which were more rigid than those dye samples whose nitrogen isotherms resemble isotherm D.

Isotherm B illustrates another pattern of low pressure hysteresis. Here, there is also evidence of a transition at a relative pressure of ca. 0.4 from low pressure hysteresis to classical Type IV high pressure hysteresis. However, in comparison to isotherm A, the gap at low pressures between the adsorption and desorption branches of the isotherm is much larger. Hysteresis at low pressures is much more pronounced. This type of isotherm plot has been observed in some organic pigments[19] and recently by Christensen *et al.* in whole extrudates of a deactivated commercial hydroprocessing catalyst[7]. These workers concluded that the low pressure hysteresis they observed arises from restricted diffusion of nitrogen molecules through microporous constrictions in the catalyst extrudate. They demonstrated that the isotherms they determined do not truly represent conditions of adsorption equilibrium. However, although this explanation accounts for the character of the isotherm at low relative pressures, the broadening of the hysteresis loop at high relative pressures above 0.4 also indicates a more rigid mesoporous structure. Christensen *et al.* found too that for crushed samples of catalyst extrudate no low pressure hysteresis was observed, and the isotherms displayed standard Type IV behaviour.

3 DISCUSSION

The different categories of low pressure hysteresis depicted in Figure 1 can be attibuted to differing flexibilities in the porous structures of the adsorbents. Isotherm A displays significant hysteresis to low pressures, but pronounced hysteresis at relative pressures above ca. 0.4. Whilst the low pressure hysteresis observed in the isotherm indicates an inelastic distortion of the pore structure as adsorption proceeds, the broader loop at higher pressures indicates extensive rigidity in the pore structure. This rigidity of the pores is manifest too in isotherm B. Low pressure hysteresis, however, is far more marked than in isotherm A and can be attributed to restricted diffusion of adsorbate through microporous constrictions[7].

Isotherm D, characterised by broadening of the hysteresis loop to lower pressures, appears to accord with a looser porous structure[12]. Where, as with copper phthalocyanine pigments, the porous structure arises through aggregation of pigment crystals, the nature of the isotherm has been explained by penetration of adsorbed nitrogen into junctions between the aggregated crystals. The aggregates expand with consequent weakening of the contacts between crystals.

Isotherm C is intriguing. At high relative pressures it resembles isotherm D, but at lower pressures it resembles isotherm A. From the study of the samples of C.I. Disperse Dye 54[16], there is some evidence to suggest that isotherm C corresponds to solids containing more rigid porous structures than does isotherm D. On the other hand, the absence of pronounced hysteresis at high relative pressures indicates greater flexibility of the porous structure than is possessed by solids represented by isotherms A and B.

It seems clear then that isotherm A corresponds to more rigid porous structures than does isotherm D. It is tentatively suggested that isotherm C may correspond to porous solids of an intermediate rigidity Isotherm B appears to represent solids containing a mixture of rigid and more flexible pores.

References

1. H.L. McDermott and J.C. Arnell, Can. J. Chem., 1955, **33**, 913.
2. H.L. McDermott and B.E. Lawton, Can. J. Chem., 1956, **34**, 769.
3. B. McEnaney, J. Chem. Soc., Faraday Trans. 1, 1974, **70**, 84.

4. R.M. Barrer and D.M. MacLeod, Trans. Faraday Soc., 1954, **50**, 980.
5. P.H. Emmett and Th. DeWitt, J. Am. Chem. Soc., 1943, **65**, 1253.
6. P.J.M. Carrott and K.S.W. Sing, Chem. Ind. (London), 1986, 786.
7. S.V. Christensen, J. Bartholdy, P.L. Hansen, W.C. Conner, J. Fraissard, J.L. Bonnardet and M. Ferrero, in 'Characterization of Porous Solids - III', J. Rouquerol, F. Rodriguez-Reinoso, K.S.W. Sing and K.K. Unger (eds), Elsevier, Amsterdam, pp165-172.
8. S.J. Gregg and K.S.W. Sing, in 'Surface and Colloid Science', E. Matijevic (ed), Vol. 9, Wiley, New York, 1976, pp232-359.
9. A. Bailey, D.A. Cadenhead, D.A. Davies, D.H. Everett and A.J. Miles, Trans. Faraday Soc., 1971, **67**, 231.
10. J.C. Arnell and H.L. McDermott, Proc. 2nd Int. Congr. Surface Activity, II, Butterworths, London, 1957, pp113-121.
11. R.R. Mather and K.S.W. Sing, J. Colloid Interface Sci., 1977, **60**, 60.
12. R.R. Mather, Colloids Surfaces, 1991, **58**, 401.
13. R.B. McKay, FATIPEC (Fed. Assoc. Tech. Ind. Peint. Vernis, Emaux Encres Impr. Eur. Cont.) Congr. XVIII, 1986, **2/B**, 405.
14. R.R. Mather, in 'Characterization of Porous Solids', K.K. Unger, J. Rouquerol, K.S.W. Sing and H. Kral (eds), Elsevier, Amsterdam, 1988, pp263-271.
15. R.R. Mather, Colloids Surfaces, 1989, **37**, 131.
16. R.R. Mather and R.F. Orr, in 'Characterization of Porous Solids - III', J. Rouquerol, F. Rodriguez-Reinoso, K.S.W. Sing and K.K. Unger (eds), Elsevier, Amsterdam, pp745-751.
17. T. Ahsan, Colloids Surfaces, 1992, **64**, 167.
18. M.R. Harris, Chem. Ind. (London), 1965, 269.
19. R.R. Mather, unpublished results.

Lloyd Abrams

DuPont Company
Central Research and Development
Experimental Station, E-228
Wilmington, DE 19880-0228

1 INTRODUCTION

ZSM-5 is a zeolite which has shown important catalytic properties including high selectivity and low tendency to coke. X-Ray and neutron diffraction studies show that ZSM-5 contains two intersecting channel systems composed of 10-membered rings formed of SiO_2 (or AlO_2) tetrahedra. The framework and schematic of the channel system is shown as Figure 1. One channel is nearly straight while the other is sinusoidal in connectivity, intersects the straight system, and is perpendicular to it.

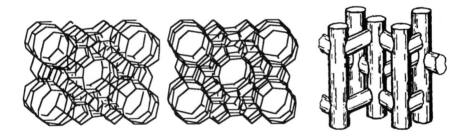

Figure 1 *Stereoscopic Depiction of the ZSM-5 Framework (left) and Schematic of the Channel Structure (right).*

Further, ZSM-5 can be synthesized using a variety of template molecules as well as wide range of Si/Al ratios, from ~10 to infinity for the pure silica polymorph, silicalite. Because of these differences, as well as the normal difficulties associated with synthesizing zeolites, characterization of ZSM-5 by adsorption and catalysis has provided a large body of literature which is not entirely consistent. For this study, over 200 samples[1] were examined for catalytic activity and were characterized by x-ray diffraction and adsorption. Independent of most of the preparation variables, several equilibrium adsorption values were obtained that are specific to the framework geometry of ZSM-5 while others seem to be kinetic in origin. Vital to this publication is the ability to correct the data for adsorption on the external surfaces of the particles versus adsorption into the ZSM-5 framework.

2 EXPERIMENTAL DETAILS

For the present study, ZSM-5 was prepared using tetrapropylammonium, TPA, and tetraethylammonium, TEA, cations. Within its detection limits, X-ray diffraction was routinely used to monitor the preparations to insure that other zeolitic phases were not present. Some samples contained amorphous debris but, as noted below, a method was developed to account for adsorption on external surfaces.

Gravimetric measurements[2] were used to follow the sorption of a vapor by a sample, generally at 20-25°C. Prior to the adsorption experiments, the samples were heated to 425°C for 16 to 24 hours at pressures less than 0.01 Pa. Typically, weight measurements were taken at 3, 20, and 44 hours of exposure. A method[3] was devised to correct for adsorption on the external surface of the particles; all of the values reported in this paper have been corrected. For some small particle preparations as well as for samples containing amorphous or non-zeolitic debris, the external surface accounted for almost 30% of the total adsorbed amount. The term 'adsorption' is used to denote sorption on the external surfaces of particles while 'absorption' is used for sorption into the framework. All of the absorbed amounts are reported as g/100G: the number of grams of a specific solvent absorbed per 100 grams of dried sample.

Probe molecules (Table 1) were selected from the alkanes and aromatics to provide information concerning the internal framework volume accessible for absorption, cross-sectional dimensions, and migration limiting features. The interaction of these molecules with the ZSM-5 framework (at room temperature) is typical of a physical adsorption process with heats of sorption < 50 kJ/mole. These solvents readily desorbed from a sample under mild heating conditions under vacuum and permitted the same sample to be used for the sorption of a variety of solvents.

3 RESULTS AND DISCUSSION

The absorption of hexane into H-ZSM-5 is fairly rapid and reaches an 'end point' or equilibrium value within 3 hours exposure. The 20 hour sorption readings are the same within experimental error. A mean value of 13.07 g/100 g (std dev 0.27) was determined for 29 highly crystalline samples prepared using the TPA ion. This value corresponds to a framework volume of 0.198 (std dev 0.004) cc/g using a value of 0.659 g/cc as the density for hexane. Using the methodology provided by Breck[4], a framework volume of 0.197 cc/g is calculated from unit cell dimensions assuming that the framework TO_2 groups occupy the same volume as they do in quartz. Within experimental error, the volumes agree suggesting that hexane fills the H-ZSM-5 framework in a 'liquid-like' manner. The samples of H-ZSM-5 that were used for this measurement had Si/Al ratios ranging from 11 to >1000; therefore, within experimental error, the amount of aluminum in the framework did not appear to affect the value of hexane absorbed.

Table 1 *Dimensions[5] of Probe Molecules and Abbreviations Used in the Text.*

Molecule	Dimensions Å	Molecule	Dimensions Å
p-xylene, pXyl	3.6x6.5x8.9	carbon tetrachloride, CCl4	5.4
o-xylene, oXyl	3.6x7.1x8.1	n-hexane, nHex	3.9x4.3x9.7
m-xylene, mXyl	3.6x7.2x8.6	3-methylpentane, 3MeP	5.1x5.9x8.0
toluene, tol	3.6x6.5x8.1	cyclohexane, 0Hex	5.1x6.0x6.5
benzene, bz	2.5x6.5x7.2	2,2-dimethylpropane neopentane	5.5x5.8x6.2

3MeP is absorbed almost as rapidly as hexane into the H-ZSM-5 framework. Its liquid density is sufficiently close to that of hexane such that the same sorption weight gain is expected. But, H-ZSM-5 samples absorb a significantly smaller amount of 3MeP compared to hexane. However, the sorption data (Figure 2) for 3MeP strongly correlates

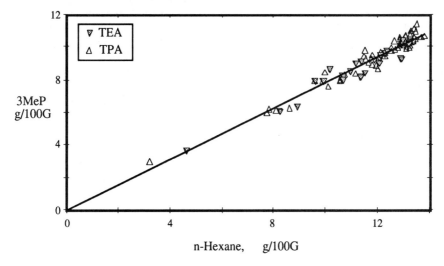

Figure 2 *Absorption Data for Hexane and 3MeP into Samples of H-ZSM-5.*

with that of hexane for a variety of H-ZSM-5 preparations independent of template. The ratio of absorbed amounts, 3MeP/hexane, is also independent of the amount of hexane absorbed with a value of 0.789 (std dev 0.023 for 39 samples). This 'packing ratio', 3MeP/Hex, is deemed to be characteristic of the H-ZSM-5 framework. Similarly, the absorption of *p*-xylene is directly related to the hexane absorption amount (Figure 3) with

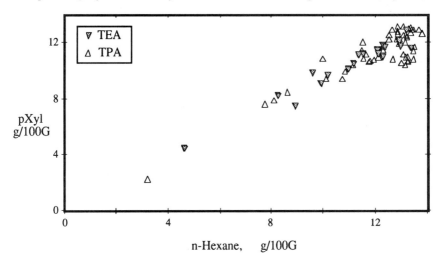

Figure 3 *Absorption Data for Hexane and p-Xylene into Samples of H-ZSM-5.*

a packing ratio of 1.218 (std dev 0.045 for 30 samples). Again, the amount of aluminum in the framework or the template molecule did not seem to affect the observed 'packing 'ratio. The crystallinity of H-ZSM-5 samples displaying these characteristic ratios, 3MeP/Hex and pXyl/Hex, is determined by dividing the hexane value sorption by 13.07.

The equilibrium sorption data for H-ZSM-5 TPA preparations, at 20 hours, for a variety of solvents are listed (Table 2). Smaller molecules, i.e., butane and pentane, pack according to their liquid densities giving the same framework volume as hexane. If the 'packing ratio' is equal to the 'density ratio' (density of the solvent/density of hexane), then the solvent would be 'liquid-like' in the absorbed state. Otherwise, it could be considered to be constrained in some manner so that it cannot fill the framework according to its liquid density. Via this definition, most of the molecules larger than hexane listed in Table 2 are constrained in some fashion.

The equilibrium sorption data for H-ZSM-5 TEA preparations are also listed (Table 3). As noted above (Figures 2 and 3), absorption of 3MeP, pXyl, hexane (and smaller) molecules cannot distinguish the TEA from TPA preparations. With larger solvent molecules, distinctions become apparent between the different templated preparations. The adsorption of cyclohexane (Figure 4) for example clearly shows that the TEA preparations absorb less than their TPA counterparts. In fact, the wide scatter in data (Table 3) for cyclohexane may be due to a kinetic effect. By using an even larger probe molecule, o-xylene, the kinetic effect on absorption is clearly demonstrated (Figure 5).

Table 2   *Characteristic Absorption Values for H-ZSM-5: TPA template, 58 samples.*

Solvent	g/ 100G	Std Dev	No. #	Packing Ratio[†]	Density Ratio[∂]
n-Hexane	13.07	0.27	29	1.000	1.00
p-Xylene	15.78	0.43	22	1.218	1.31
3-Methylpentane	10.36	0.32	27	0.789	1.01
Toluene	12.91	1.03	14	0.974	1.31
Benzene	11.29	0.53	20	0.888	1.33
CCl4	15.23	0.64	20	1.177	2.42
Cyclohexane	7.38	0.98	29	0.638	1.18

[#]   The number of samples examined to obtained the reported data
[†]   The ratio of the amount absorbed to the amount of hexane absorbed
[∂]   The ratio of the solvent density to the density of hexane

Table 3   *Characteristic Absorption Values for H-ZSM-5: TEA template, 24 samples.*

Solvent	g/ 100G	Std Dev	No. #	Packing Ratio[†]	Density Ratio[∂]
n-Hexane	12.56	0.49	10	1.000	1.00
p-Xylene	14.54	0.64	11	1.191	1.31
3-Methylpentane	9.68	0.47	10	0.777	1.01
Toluene	9.09	1.17	8	0.74	1.31
Benzene	8.09	0.73	13	0.66	1.33
CCl4	11.58	1.07	9	0.94	2.42
Cyclohexane?	4.43	1.30	13	0.34	1.18

[?]   not sure that this is an equilibrium value

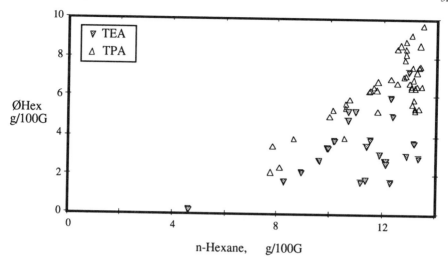

Figure 4 *Absorption Data for Hexane and Cyclohexane into Samples of H-ZSM-5.*

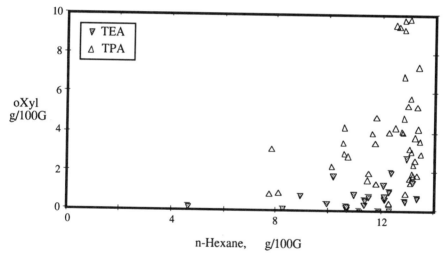

Figure 5 *Absorption Data for Hexane and o-Xylene into Samples of H-ZSM-5.*

At room temperature, the sorption of molecules, such as *o*- and *m*-xylenes, 2,2 dimethylbutane, and neopentane, did not equilibrate at 20 or 48 hours; in fact, extending the measurements to one week still did not provide equilibrium data. The absorption rate of the large molecules into H-ZSM-5 samples varies from sample to sample and suggests that their pore openings on the external surface area are different. The different rates of sorption of *o*-xylene into ZSM-5 samples can be used to distinguish the various samples from each other even though they have similar hexane sorptions (similar crystallinities); *o*-xylene absorption values at 20 hours ranging from 0.37 to 9.7 g/100 G were obtained for 55 samples of the TPA preparation while the TEA preparations had a much lower range of values (Figure 5).

As noted above, for molecules the size of hexane and smaller, the absorption data, within experimental error, is the same for the TPA and TEA template preparations. Further, the X-ray diffraction patterns are very similar, if not identical, for the two different templated materials. However, as the size of the probe molecule increases, the TEA framework absorbs (under equilibrium conditions) less than the TPA framework. This observation suggests that the TEA preparations have slightly smaller channel dimensions than the TPA preparations and this difference is magnified as the absorbing molecule gets larger. By using these kinetic (non-equilibrium) absorption measurements, the TEA preparations can be clearly distinguished from the TPA preparations (Figures 6 and 7). As the molecular probes become larger, the locus of TEA data points moves

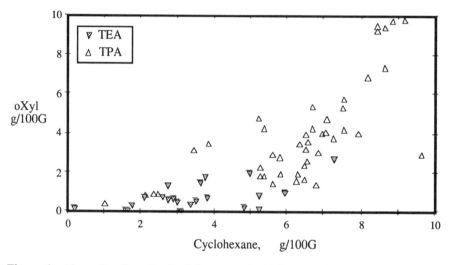

Figure 6   *Absorption Data for Cyclohexane and o-Xylene into Samples of H-ZSM-5.*

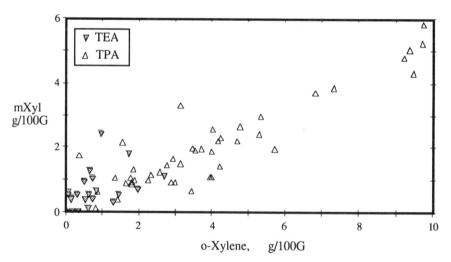

Figure 7   *Absorption Data for o-Xylene and m-Xylene into Samples of H-ZSM-5.*

closer to the origin.

As noted above (Figure 1), ZSM-5 crystals have two slightly different channel systems capable of absorbing molecules. Consider that the template ions may control the morphology of the zeolite crystals but not their crystal structure. When tetraethylammonium ion, TEA, was used as the template the resulting crystals are stubby, elliptical cylinders with flat ends. These particles are significantly different from the ZSM-5 crystal configuration of right-angle, prismatic, penetration twins obtained when TPA is used as the templating ion. Conditions during the synthesis of a sample cause one channel system to dominate over the other, thereby providing a spectrum of sorption rates for large molecules. Using TEA as the template molecule seems to alter the crystal morphology and let the smaller channels dominate. The TPA preparations, however, absorb the larger molecules faster and to a larger amount which indicates that a higher percentage of the exposed channels are larger than the TEA preps. The difference in channel dimension and relative channel population imposes greater constraint on the packing of probe molecules as they increase in size which is consistent with the data presented. Thus, the choice of template molecule seems to control the external morphology of the ZSM-5 crystals, which channel system dominates, and the rates of absorption and amount absorbed into the framework.

4 SUMMARY AND CONCLUSIONS

Absorption of probe molecules of varying geometries and sizes is used to characterize the framework dimensions and topography of H-ZSM-5. The volumes of the absorbed probe molecules, using their liquid densities, are then compared to the calculated pore volume. The constraint on the packing of the absorbed molecules is quantified by comparing their packing density to their density in the liquid state. Further, the packing of probe molecules into the same pore volume is compared via a ratio technique called the 'packing ratio'. The effect of the lattice geometry and framework dimensions on the 'packing ratios' provides a set of characteristic values for a given molecular sieve. The absorption and packing ratios provide sensitive indicators of very small changes in framework dimensions. The combination of X-ray diffraction and absorption measurements permits the crystallinity of a molecular sieve to be quantified. The absorption values and packing ratios for molecular probes into ZSM-5 are presented as expectation values for other scientists to use as bases of comparison.

Acknowledgments

The information presented took over a decade to compile with the help of numerous colleagues at DuPont, in particular, Gunter Teufer for discussions and X-ray diffraction analyses, Ed Moran for his preparations, Dave Corbin for his discussions and assistance, Brian Arters and Bill Stephens for their patience with the adsorption measurements, and Nick Turro, Columbia University, for his unbounded enthusiasm and his desire to understand the molecular basis for sieving of photolytically produced radicals. DuPont Contribution No. 7482

References

1. Samples varied in particle and crystallite size, were made using a variety of templates, had a large variance in Si/Al ratio, were synthesized in DuPont as well as at other laboratories, and sometimes had different particle morphologies.
2. L. Abrams and D. R. Corbin, *J. Catal.*, 1991, **127**, 9.

3. L. Abrams, M. Keane, and G. C. Sonnichsen, *J. Catal.*, 1989, **115**, 361.
4. D. W. Breck, 'Zeolite Molecular Sieves', Wiley and Sons, New York, 1974.
5. Fisher Scientific Co., Cat. No. 12-824, *U.S. Patent* 2,308,402 (1940).

SURFACE FUNCTIONAL GROUPS AND TEXTURE CHARACTERISATION OF CATALYTIC SUPPORTS BASED ON ACTIVATED CARBON

G. de la Puente, A. Centeno, A. Gil, P. Grange, B. Delmon

Unité de Catalyse et Chimie des Matériaux Divisés, Université Catholique de Louvain, Place Croix du Sud 2/17, Louvain-la-Neuve, Belgium. Fax : 32 10 473649.

1 INTRODUCTION

The functionalisation of supports is a powerful method to get better catalysts. This work is part of a programme aiming at the preparation of activated carbon based catalysts for upgrading biomass[1, 2]. The performances of the activated carbons as catalyst supports are determined both by their texture and surface chemistry[3]. The initial metal dispersion, its resistance to agglomeration and the catalytic behaviour of the system are all sensitive to the state of the carbon surface. The manufacturing process is not always adequate to obtain the desirable surface properties on the activated carbons for this application. Post-treatments are needed to widen the applicability of these materials.

In high surface area supports, anchorage sites for the metal precursor and active centres with acid-base or red-ox properties can be obtained with the partial oxidation of activated carbons. Selective analyses have revealed the presence of different functional groups at the surface of activated carbon such as carboxyl, phenol, quinone, lactone and carbonyl groups[4]. The nature and the amount of these oxygen-bearing functions depend upon the nature of the activated carbon and oxidation method.

In order to introduce oxygen functionality while producing only minimum change to the textural properties, we studied the influence of the conditions of oxidation on the surface properties of an active carbon. For this purpose, a single parent activated carbon was selected and oxidised using different reagents (H_2O_2 of various concentrations at 363 K and concentrated HNO_3 at various temperatures).

2 EXPERIMENTAL

2.1 Oxidation treatments

An activated carbon supplied by Merck (18-35 mesh ASTM, Merck-9631) (sample K0) was used in this study. Samples of the carbon were subjected to a number of oxidative treatments with H_2O_2 of various concentrations at 363 K and concentrated HNO_3 at various temperatures in order to introduce oxigenated surface complexes. The procedure was as follows: 200 cm^3 of concentrated nitric acid (Janssen Chimica) were heated in a glass flask equipped with a stirrer and a thermometer. After reaching the desired temperature (298, 333, and 363 K), 20 grams of activated carbon were added and the reactions were stopped after 3 hours. Following each treatment, the solutions were filtered and the carbons were repeatedly washed with distilled water until neutral pH indication of the effluent water and dried overnight under nitrogen atmosphere at 283 K. These samples will be referred to as N298, N333 and N363, respectively. In the case of the sample oxidised at reflux (NR) the reaction was conducted in a three necked flask equipped with a stirrer, a condenser and a thermometer. The H_2O_2 treatment was carried out at 363 K during 5 hours using solutions 1, 5 and 10 M (samples H1, H5 and H10, respectively).

2.2 Textural characterisation

Characterisation of the porous texture of the carbon supports was carried out by nitrogen adsorption at 77 K in a volumetric apparatus Micromeritics ASAP 2000. In order to achieve an exhaustive micropore characterisation, adsorption isotherms were obtained beginning at very low relative pressures ($\sim 10^{-6} < P/P° < 0.99$). To obtain an adequate micropore size distribution, sufficient data points at low pressures are needed, and this requires the addition of small nitrogen volumes. In this work, nitrogen adsorption data were obtained using 0.1 g of sample and successive doses of nitrogen of 0.5 ml/g until $P/P° = 0.04$ was reached. Subsequently, further nitrogen was added and the volumes required to achieve a fixed set of $P/P°$ were measured. Previous to analysis, samples were degassed at 200°C during 10 hours ($P < 10^{-3}$ mmHg).

In microporous carbons, the micropore volume is a well defined quantity, whereas the specific surface area is not. There is not ambiguity in the micropore volume except for insufficient information on the density of adsorbed molecular aggregates in the micropore. The Dubinin-Raduschkevich (DR)[5] and α_s[6] plots can provide a definite value of micropore volume. On the contrary, there are stimulating discussions on the concept of the surface area in the microporous system. If we apply the BET analysis to the adsorption data in the $P/P°$ range of 0.1-0.3, the BET area is significantly overestimated because of quasi-capillary condensation effect. Hence, the analysis of adsorption data in the relative pressure range of 0.01 to 0.05 should be recommended for microporous carbons[7]. This relative pressure range coincides with the linear region of the α_s plot between the filling and condensation swings, also corresponding to the monolayer process. Specific total surface areas (A_{Lang}) were calculated using the Langmuir equation ($0.01 \leq P/P° \leq 0.05$ interval) for monolayers, which is more suitable for microporous solids than the BET equation for multilayers[8,9]. Specific external surface areas (A_{ext}) were obtained from the t-method[10] and the specific total pore volumes (V_p) were estimated from nitrogen uptake at $P/P° = 0.99$. The Horvath-Kawazoe method[11] was used to describe the micropore size distributions. The Dubinin-Radushkevich equation was used in order to calculate the micropore volume ($V_{\mu p}$) and the characteristic energy (E).

2.3 Characterisation of surface oxygen groups

In order to determine the nature of the surface functional groups, quantitative XPS analysis, FTIR spectra and ammonia chemisorption experiments were carried out.

For the XPS analysis, the spectrometer was a Surface Science Instruments SSX-100, Model 206 with a monochromatic AlKα source (1486.6 eV), operating a 10 kV and 15 mA. The spectrometer was interfaced to a Hewlett-Packard 9000/310 computer for data acquisition and treatment. The powdered samples were pressed on stainless steel capsules which were mounted on top of the specimen holder. The samples were degassed under a minimum vacuum of $5 \cdot 10^{-7}$ mmHg before being introduced in the analysis chamber. During the analysis the pressure did not exceed $5 \cdot 10^{-8}$ mmHg. The lines C1s, O1s, and Mo3d were investigated. In all cases, the binding energy of the C1s line was found to be 284.6 ± 0.2 eV; we checked this line at the beginning and after the analysis of each catalyst and could not find any modification in its position. A non-linear, Shirley type[12] baseline and an iterative least-squares fitting algorithm were used to deconvolute the peaks, the curves been taken as 85% Gaussian and 15% Lorenzian. Surface atomic concentration ratios were calculated as the ratio of the corresponding peak areas, corrected with theoretical sensitivity factors based on Scofield's photoionisation cross sections[13].

FTIR spectra were recorded using a Brücker IFT 88 spectrometer in the 4000-400 cm^{-1} range with 2 cm^{-1} resolution. The samples were diluted in KBr to a 0.1% carbon / (carbon + KBr) content, ground and the powders pressed into discs (125 mg, 13 mm in diameter). Corrections were made with respect to the mass of carbonaceous material in the pellets. The spectra of the diluent itself was obtained and subtracted from those of the mixtures.

The acidity properties were characterised by ammonia adsorption experiments at various temperatures (323, 373 and 473 K) using a static volumetric apparatus (Micromeritics ASAP 2010C adsorption analyser) under the pressure ranging from 5 to 700 mmHg. The samples were previously degassed at 673 K for 3 hours (P<10^{-3} mmHg). The quantity of ammonia chemisorbed was obtained using the dual technique of Yates and Sinfelt, consisting in the determination of the first isotherm of NH$_3$ (ammonia physisorption and chemisorption), evacuation in helium flow (the physisorbed species are evacuated and only chemisorbed species remain on the surface), and measure of the second isotherm; the difference between both isotherms gave the quantity of chemisorbed ammonia on the activated carbons. The samples were degassed at analyses temperatures for 0.5 hour between both analyses.

3 RESULTS

3.1 Characterisation of textural properties

The nitrogen isotherms over the whole relative pressure range of the obtained activated carbons are shown in Figure 1. The starting activated carbon has been also included for comparison. The adsorption isotherms are of type I in the Brunauer, Deming and Teller (BDDT) classification[14]. From these isotherms, it can be observed that the oxidation with H$_2$O$_2$ does not produce important changes in the textural properties of the support. The treatment with 1 M (H1) produces a loss of 15% of the specific surface area and 13% of the specific total pore volume. These losses are similar for the other samples (H5 and H10). However, the treatment with HNO$_3$ produces a continuous loss of the textural properties of the activated carbon. This is more explicitly shown in Table 1, where the specific surface areas and the specific pore volumes of all the samples are summarised. The Langmuir C values, characteristic of the intensity of the adsorbate-adsorbent interactions, are also reported. The treatment at 363 K (N363) produces a lost of 32% of the specific surface area and 30% of the specific total pore volume. These losses are increased up to 88% and 86% when the activated carbon is treated in conditions of reflux (sample NR).

The Dubinin-Radushkevich plots of the samples are the characteristic curves describing the micropore filling[15]. The linear portions of these plots ($4 \leq \log^2(P°/P) \leq 10$) have been used to obtain the micropore volume (V$_{\mu p}$) and characteristic energy (E). The description of the Dubinin-Radushkevich plots and their relation with the porous accessibility are described in the literature[16-18]. The very low pressure range of the Dubinin-Radushkevich plots shows a modified adsorption ability for N363 sample as compared with the starting activated carbon. The total destruction of the textural properties is observed for sample NR. These differences can also be observed in the adsorption isotherms for P/P° values lower than 10^{-5} (see Figure 1). As can be seen in Table 1, the oxidation with HNO$_3$ produces a continuous decrease of the specific microporous volume (V$_{\mu p}$). The treatment at 298 K (N298) produces a lost of 9% of V$_{\mu p}$ and this lost is increased up to 89% when the activated carbon is treated in reflux conditions (NR). The treatment with H$_2$O$_2$ does not produce important modifications of V$_{\mu p}$, a decrease of 17% is observed when the activated carbon is treated with H$_2$O$_2$ 10 M.

The characteristic energy (E) is greater for the oxidation samples. It seems likely that this behaviour could be related with a change in the size of micropores. According to the inverse relationship between both parameters[19, 20], the increase of E can be explained by a decrease of the mean micropore diameter. In order to check this possibility, the micropore size distributions of the samples from the Horvath-Kawazoe formalism were obtained. The distributions are centred below 8 Å and a displacement of the mean micropore diameter towards the ultramicropore region, below 7 Å, can be observed for the oxidised samples with respect to the starting activated carbon, which is in agreement with the adsorption isotherms and the Dubinin-Radushkevich plots at very low pressures. This behaviour may be due to the fact that the oxygen is fixed on the walls of the micropores and, having into account their size, the creation of oxygen functional groups produces a slight decrease of micropore size and volume. On the other hand, the treatment with HNO_3 at elevated temperatures provokes the destruction of a great part of the micropores as a consequence of the lost of the walls between neighbouring pores. The attack by this strong acid was too severe and resulted in the collapse of the pore structure. It is clear that although the NR sample has the higher oxygen content, the aim to produce different carbon supports with approximately the same textural properties is obviously not met by this sample.

In relation to their porous texture, it can be considered that, despite the above mentioned differences among them, the samples treated with H_2O_2 at 363 K and with concentrated HNO_3 at room temperature and 333 K show quite similar textural properties. Since the porosity is almost constant in these supports, one can analyse the effect of oxygen surface groups separately.

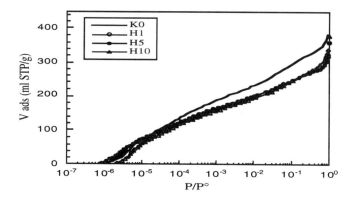

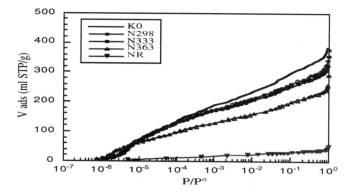

Figure 1 *Nitrogen adsorption isotherms obtained for activated carbons treated with hydrogen peroxide (top) and nitric acid (bottom).*

Table 1 *Textural characterisation of the activated carbons.*

Sample	A_{Lang} (m²/g)	A_{ext} (m²/g)	V_p (ml/g)	$V_{\mu p}$ (ml/g)	E (kJ/mol)
K0	1242 (C=333)	70	0.645	0.413	25.9
H1	1055 (C=373)	63	0.558	0.355	26.6
H5	1052 (C=414)	60	0.553	0.353	27.5
H10	1035 (C=350)	73	0.586	0.344	26.4
N25	1116 (C=365)	67	0.582	0.374	27.2
N60	1075 (C=391)	59	0.551	0.358	27.4
N90	842 (C=358)	44	0.449	0.278	27.5
NR	129 (C=269)	12	0.081	0.042	22.0

3.2 Identification of surface oxygen groups

The quantitative XPS analysis indicates that all oxidative treatments increase the oxygen content of the samples (4.3% for the parent activated carbon and 22.1% for the activated carbon treated with HNO_3 at 383 K). The surface compositions obtained by XPS analysis for the various activated carbons are displayed in Table 2. The small amount of sulphur present in the original active carbon is progressively eliminated by the nitric acid treatment. Even when samples were repeatedly washed, some nitrogen remains chemisorbed on the surface of active carbons.

The C1s signal of all samples consists in a major graphitic peak at 284.5 eV. More information can be gained from satellites of the C1s peak at higher binding energies[21, 22]. The shifts from the main peak range from 1.7 eV for carbon atoms singly bonded to oxygen[23] to 4 eV for carbon atoms in carboxyl groups or esters. The presence of a satellite of the C1s peak at 288.4 eV in the oxidised activated carbons is observed, being specially important in those treated with HNO_3 at 363 K and 383 K, where this satellite appears as a clear shoulder. This signal can be related with the presence of C=O double bonds, confirmed by FTIR analysis (appearance of the band at 1720 cm⁻¹).

The creation and increase of the surface oxygen groups were checked by FTIR spectroscopy. In addition to well-known experimental difficulties involved in obtaining carbon spectra, there is also a common consensus on the fact that the interpretation of the surface groups on carbon is a difficult task since it is unrealistic to consider these groups as isolated functions as in classic organic chemistry[24].

Figure 3 includes the FTIR spectra for the original carbon support and treated with the reagents, that must be considered only qualitatively. The oxidised carbon samples have bands at 1720 and 1580 cm⁻¹ and a broad envelope from 900 to 1300 cm⁻¹. Over the years, there have frequently been disagreements regarding band assignments, and the fact that these bands can be quite broad has not helped this situation. Nevertheless, a consensus in the assignment of band frequencies to different functional groups is possible[25]. Upon oxidation with both H_2O_2 and HNO_3, an increase of the band between 1000 and 1300 cm⁻¹ is first observed, which is attributed to C-O single bonds, such as those in ethers, phenols and hydroxyl groups. These results indicate that these groups are created and an oxidised layer is formed by ether-like bonds crosslinking the aromatic substrate.

In the more oxidised samples a new peak centred at about 1720 cm⁻¹ appears, assigned to stretching vibration of carboxyl groups on the edges of the layer planes[26, 27] or to conjugated carbonyl groups (C=O double bonds in carboxylic acid and lactone groups). This band begins to develop in the spectrum of the samples N333 and H5, and its intensity is stronger in the spectra of the samples treated at higher temperatures or higher concentrations. As can be observed in Figure 2, the development of the IR band at 1720 cm^{-1} is much more important in the HNO$_3$ oxidation series, indicating that nitric acid treatment gives rise to a greater increase in the C=O double bonds in carboxylic acid and lactone groups.

Upon the formation of the 1720 cm^{-1} band, an increase in the absorption at 1580 cm^{-1} is found. Apparently, the presence of the oxygen double bond, conjugated with the carbon basal planes, results in a stronger adsorption at 1580 cm^{-1}. This phenomenon was also described by other authors[28]. We do not observe any band in the range 1740-1880 cm^{-1}, indicating that anhydrides are not formed (on in a very low quantity) during neither HNO$_3$ nor H$_2$O$_2$ oxidation of this active carbon.

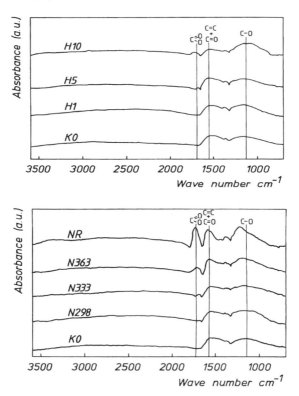

Figure 2 *FTIR spectra of the active carbons oxidised with HNO$_3$ at different temperatures (top) and H$_2$O$_2$ of various concentrations (bottom).*

Chemisorption of ammonia is a way to evaluate the surface acidity of a material[29]. The ammonia chemisorption at 323, 373 and 473 K increased with the oxidation degree. The results obtained at 323 K are summarised in Table 2. The oxidation treatments gave rise to an increase of NH$_3$ chemisorption, indicating that the treatments caused the formation of chemical groups with an acidic character. Among the treated carbons tested, those treated with an HNO$_3$ solution showed the largest amount of NH$_3$ chemisorbed.

Nitric acid treatment produced the highest amount of surface groups, especially the strongly acidic carboxylic acids. However, hydrogen peroxide as milder oxidising agent introduced phenols more than carboxylic acids or lactones. These results are in agreement with those of other authors[30].

Table 2 *Surface composition and acidic properties for the untreated and oxidised activated carbons.*

Sample	XPS				NH$_3$
	C (% at)	O (% at)	S (% at)	N (% at)	(μmol/g)
K0	95.4	4.3	0.3	-	119
H1	92.9	6.8	0.3	-	125
H5	91.4	8.3	0.3	-	249
H10	87.5	11.4	0.3	-	241
N298	90.3	8.4	0.2	1.1	134
N333	88,1	10.6	0.2	1.1	480
N363	80.4	18.1	0.1	1.4	569
NR	76.0	22.1	0.1	1.7	638

4 CONCLUSIONS

Oxidation with H$_2$O$_2$ did not produce important changes in the textural properties of the support. However, the treatment with HNO$_3$ produced a continuous loss of the textural properties of the activated carbon. Oxidation provoked a decrease of the mean micropore diameter, this behaviour might be due to the fact that oxygen was fixed on the walls of the micropores and, taking into account their size, the creation of oxygen functional groups produced a slight decrease of micropore size and volume. The treatment with HNO$_3$ at elevated temperatures provoked the destruction of a great part of the micropores as a consequence of the loss of the walls between neighbouring pores.

Quantitative XPS analysis indicated that all oxidative treatments increase the oxygen content of the samples, that will be anchorage sites for the metal active phase in the preparation of activated carbon supported catalysts. The presence of a satellite of the C1s peak at 288.4 eV in the oxidised activated carbons was observed, being especially important in those treated with HNO$_3$ at 363 K and 383 K, where this satellite appeared as a clear shoulder. This signal can be related with the presence of C=O double bonds, confirmed by FTIR analysis (1720 cm^{-1}).

FTIR spectra confirmed the formation of carbonyl, lactone, ether and hydroxyl groups during the oxidation process. Nitric acid treatment gave rise to a greater increase in the C=O double bonds in carboxylic acid and lactone groups.

The oxidation treatments enhanced NH$_3$ chemisorption, indicating that upon these treatments acidic oxygen functional groups are introduced to the activated carbon surface. Among the treated carbons tested, those treated with HNO$_3$ solution showed the largest amount of NH$_3$ chemisorbed. Nitric acid treatment produced the highest amount of surface groups, especially the strongly acidic carboxylic acids. However, hydrogen peroxide as milder oxidising agent introduced phenols more than carboxylic acids or lactones.

ACKNOWLEDGEMENT

G. de la Puente thanks the Fundación para el Fomento en Asturias de la Investigación Científica Aplicada y la Tecnología (FICYT) for her postdoctoral fellowship.

5 REFERENCES

1. R. Maggi and B. Delmon, *International Conference on Advances in Thermochemical Biomass Conversion 1992*, Bridgwater, A. V. (ed), Chapman & Hall, 1993, 1086.
2. A. Centeno, E. Laurent, and B. Delmon, *J. Catal.*, 1995, **154**, 288.
3. C. Prado-Burguete, A. Linares-Solano, F. Rodríguez-Reinoso, C. Salinas-Martínez de Lecea, *J. Catal*, 1991, **128**, 397.
4. R. C. Bansal, J. B. Donnet, F. Stoeckly, 'Active Carbon', Marcel Dekker, New York, 1988.
5. M.M. Dubinin, 'Progress and Membrane Science', D.A. Cadenhead, J.F. Danielli and M.D. Rosemberg, Ac. Press, New York, 1975, Vol. 9, 1.
6. K. S.W. Sing, *Carbon*, 1989, **27**, 5.
7. K. Kaneko, C. Ishii, M. Ruike and H. Kuwara, *Carbon*, 1992, **30**, 127.
8. M.L. Occelli and D.H.J. Finseth, *J. Catal.*, 1986, **99**, 316.
9. F. Figueras, *Catal. Rev.- Sci. Eng.*, 1988, **30**, 457.
10. B.C. Lippens and J.H. de Boer, *J. Catal.*, 1965, **4**, 319.
11. G. Horvath and K. Kawazoe, *J. Chem. Eng.* Jpn, 1983, **16**, 470.
12. D.A. Shirley, *Phys. Rev.*, 1972, **35**, 4709.
13. J. H. Scofield, *Electron. Spectrosc. Relat. Phenom.*, 1976, **8**, 129.
14. S.J. Gregg and K.S.W. Sing, 'Adsorpyion, Surface Area and Porosity', Ac. Press, London, 1991, Chapter 1.
15. M.M. Dubinin, *Carbon*, 1989, **27**, 457.
16. K.J. Masters and B. McEnaney, *J. Coll. Intf. Sci., 1983*, **95**, 340.
17. K. Kakei, S. Ozeki, T. Suzuki and K. Kaneko, *J. Chem. Soc., Faraday Trans.*, 1990, **86**, 373.
18. A. Gil and M. Montes, *Langmuir*, 1994, **10**, 291.
19. M.M. Dubinin, *Carbon*, 1985, **23**, 273.
20. J.B. Parra, J.C. de Sousa, J.J. Pis, J.A. Pajares and R.C. Bansal, *Carbon*, 1995, **33**, 801.
21. C. Kozlowski and P.M.A. Sherwood, *J. Chem. Soc., Faraday Trans.*, 1984, **80**, 2099; 1985, **81**, 2745.
22. J.B. Donnet and G. Guilpain, *Carbon*, 1989, **27**, 749
23. T.T Cheung, *J. Appl. Phys.*, 1992, **53**, 6857
24. E. Papirer and E. Guyon, *Carbon*, 1978, **16**, 127.
25. P.E. Fanning and A. Vannice,*Carbon* , 1993, **31**, 721.
26. A. Clauss, R. Plass, H.P. Bohem and U. Hofmann, *Z. Anorg. Allg. Chem.*, 1957, **291**, 14.
27. W. Scholz and H.P. Bohem, *Z. Anorg. Allg. Chem* 3, 1969, **69**, 327.
28. J. Zawadzki, *Carbon* , 1978, **16**, 491.
29. P. Carniti, A. Gervasini, A. Auroux, *J. Catal.*, 1994, **150**, 274.
30. D.J. Suh and T.-J. Park, *Carbon*, 1992, **31**, 427.

MODELLING STRATIFIED POROUS MEDIA

T. J. Mathews and G. P. Matthews

Department of Environmental Science
University of Plymouth
Drake Circus
Plymouth
PL2 8AA

1 INTRODUCTION

The aim of this ongoing study is to characterise stratified porous materials to enable further advancement of Pore-Cor, a computer model of the pore level properties of porous media[1]. Pore-Cor has already been successfully employed in simulations of a variety of porous substances, including various sandstones, paper coatings, tantalum capacitors and fibrous mats[2,3]. However, to date the process of modelling has had to assume spatially homogeneous distributions of pore size within samples, and clearly this is a limiting factor as far as more complex materials is concerned.

The modelling theory behind Pore-Cor has always been strongly supported by experimental work, utilising techniques such as permeametry, mercury porosimetry and electron microscopy to verify modelled behaviour. As such the model has managed to be successful without recourse to any arbitrary fitting parameters unlike many other more theoretical systems. Therefore this examination of stratified materials also seeks to have its basis in experimental work; current research is being carried out on two reasonably different scales, that characterising pore size correlations at the microscopic level and the transport behaviour of laboratory sand columns.

2 PORE SIZE CORRELATION

The majority of the initial development of Pore-Cor represented the pore structure of simulated samples as being randomly distributed. It quickly became apparent that many of the problems encountered by Pore-Cor in modelling the permeabilities of otherwise well simulated porous materials may have arisen from correlations in the pore structure of samples. For example, although Pore-Cor has successfully modelled the pore structure of various sandstones, difficulty was experienced simulating the permeability of those where some stratification was visually obvious. It is therefore necessary to develop some method of characterising the spatial distribution of pores. Current work is investigating the possibility of gaining correlation information using a combination of microscopy, image analysis and statistical analysis.

2.1 Method

Some visual representation of the pore space within the sample to be modelled is required. To date backscattered scanning electron micrographs of resin impregnated samples have been used, providing an unambiguous representation of the pore space. However, other methods of image acquisition could also be used, for instance optical microscopy with ultraviolet-dye impregnated resin for more macroporous materials. Obviously magnification should be kept comparatively low to include as many pores in the image as is possible given the resolution.

Once a satisfactory image has been acquired and stored in some format, image analysis can be carried out. Initially the image analyser must identify the features of interest first converting the image into binary format, with the pores represented as black. This is fairly straightforward as the high contrast backscattered electron micrograph is almost a binary image itself. All pores are then 'eroded' down by three layers of pixels and 'rebuilt' to their original sizes. Some pores become split into two or more and any 'new' features are retained after rebuilding via the creation of one pixel width boundaries. Finally, any pores overlapping the image analysers guard frame (the edge of the images in Figure 3) are omitted as incomplete.

Various measurements can then be carried out and the results stored, those of most interest here being size measurements such as length, breadth and area of pores. It is perhaps worth defining the first two terms mentioned. Length refers to the maximum distance between a user specified number of parallel tangent pairs, called ferets, around a feature and breadth is the maximum feret diameter perpendicular to this measurement.

The resultant data set is then processed using geostatistical software, GESS[4], to produce a variogram of the pore size distribution. Variograms are widely used in the statistical analysis of spatial variance and exhibit various characteristic shapes according to the distribution of features observed. A brief explanation of their calculation follows.

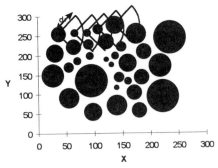

Figure 1 *Imaginary pore distribution.*

Each circle on Figure 1 represents a pore with corresponding area and position. The maximum distance separating two pores is d and this is divided into a number, n, of equal lag distances, d_1, d_2, $d_3...d_n$. For each lag GESS constructs arcs centred on each feature with radius equal to that lag (shown here for d_1), facing in a user specified direction measured from East (0° here) with a user specified angle (here 90°). It then calculates the difference in magnitude of the feature of interest, for example pore area, for every single pair found within this lag; the semi-variance for each lag distance is half the average of all these differences. The resultant variogram is a plot therefore of semi-variance versus lag distance.

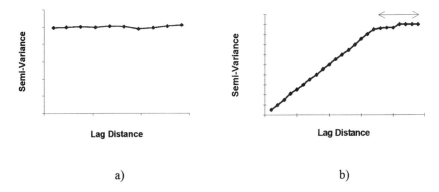

a)   b)

Figure 2 *Variograms for a sample with a) random spatial distribution and b) non-random spatial distribution.*

Figure 2a shows what might be expected were the sample in Figure 1 to be processed using GESS. The constantly high semi-variance indicates that at no distance of observation, are similar features likely to be found. Figure 2b is typical of variograms observed in ore prospecting. It suggests that within a region where soil samples were taken an area of similar ore concentrations exists. However, as the lag distance increases to values greater than the maximum diameter of this area the semi-variance increases as concentration becomes more randomly distributed. Eventually semi-variance reaches some asymptotic upper limit signifying that it is no longer a function of direction and distance.

2.1.1 Sample. For the purpose of evaluating the procedure described above an artificially stratified sample was constructed by combining two resin-impregnated sintered glass discs of markedly different pore size ranges. Using the backscattered electron detector, resin appears much darker than glass particles, due to great difference in molecular weights, giving an unambiguous representation of pore space. The images of the sample used in the study are shown in Figure 3.

2.2 Results

Initial results from this investigation are encouraging. The variogram in Figure 4 is from the electron micrographs in Figure 3. It clearly shows spatial variability in the distribution of pore size, the semi-variance remaining low over small lag distances then increasing rapidly at distances of observation greater than 1mm.

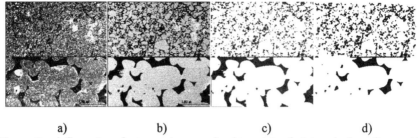

a)   b)   c)   d)

Figure 3 *a) Scanning electron micrograph of two joined sintered glass discs, with finer disc uppermost. b) backscattered electron micrograph. c) image analysed segmented image. d) segmented image minus edge features. The sample is approximately 3.4mm x 3.4mm, the boundary occurs roughly 1.4mm from the top (the top right is considered to be the origin).*

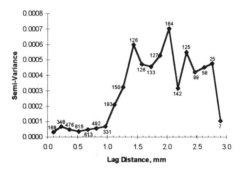

Figure 4 *Variogram of the sintered glass disc sample shown in Figure 3.*

This can readily be explained in that it is known that the width of the narrower layer was approximately 1.4mm. Therefore at lags considerably smaller than this there will be large numbers of pore pairs comprising only similarly sized pores from within the same layer and very few combinations pores from each layer ie. pores situated almost on the boundary. However, as the lag distance increases more and more pairs comprise one pore from each of the two layers with consequently greater semi-variances resulting, until at some point *most* pore pairs will consist of such a combination.

Thus at distances greater than 1mm semi-variance rapidly increases to a maximum which occurs at a distance corresponding approximately with the width of the two layers. Beyond this distance the relationship between distance and semi-variance breaks down into random 'noise'. It should be noted that the figures by the data points give the number of pore pairs used to calculate each value of semi-variance. In more recent work a cubic spline algorithm fits a curve using this statistical weighting information.

2.3 Conclusion

Thus far the results of this work allow only a qualitative appraisal of the degree of pore size correlation within materials simulated by Pore-Cor. However, this in itself is exceedingly useful information, allowing as it does the differentiation between completely homogeneous material and that exhibiting some level of correlation. Subsequent development has enabled the imposition of varying levels of pore space correlation within Pore-Cor and improvement in the resulting simulations. Further work will concentrate on quantifying the effects observed here and using this information to produce yet more accurate models of porous media.

3 TRANSPORT IN POROUS MEDIA

Although Pore-Cor can successfully model a range of features for a wide variety of porous materials, no transport processes have yet been simulated. Obviously transport is of great concern to a wide range of interests. The movement of agrochemicals and pollutants through soils has been the subject of much work as natural resources are put under increasing pressure by industry and agriculture. Also colloid transport has become a source of great interest as certain colloids come to be considered pollutants or agents of mobilisation for other harmful material, such as heavy metals or even radionuclides within radioactive waste repositories.

Therefore, as Pore-Cor has become increasingly sophisticated, and particularly as soils begin to be modelled, it is appropriate that some transport simulation is incorporated into the existing procedures. The aim of this study is to examine experimentally transport, initially of soluble material though latterly to include more complex substances, through laboratory sand columns. Consequently some adaptation of current modelling theory will be developed and built into Pore-Cor. More advanced aims include to make this model operate in more than one-dimension (in which the majority of solute transport modelling is done) and to look at transport behaviour through heterogeneous porous material in conjunction with the work mentioned previously on this page.

3.1 Apparatus

The experimental set up described here allows study of lateral, as well as vertical, dispersive transport and also the development of preferential flow paths through homogeneous material, and transport behaviour of layered samples.

The apparatus, shown in Figure 5, is adapted from a variety of sources[5,6] and allows a range of different studies to be carried out. The sample is contained in the middle of the rig in a large, cubic perspex container, open at the top and bottom, of side length 0.5 metres. Time domain reflectometry (TDR) probes installed through the sides of the container allow the sample volumetric water content to be measured throughout. The base of the sample sits on a large aluminium base plate, or grid lysimeter, described below.

Simulated rainfall and solute application are administered by the rainfall simulator, a large plastic reservoir with a 12 x 12 array of syringe needles protruding from the underside through which 'raindrops' are formed. An adjustable height overflow in the reservoir allows the head of water over the needles to be controlled, and in combination with needle size allows a broad range of application rates to be achieved. An electric motor and cam mechanism gives the rainfall simulator a degree of horizontal motion, which in conjunction with a stainless steel mesh over the sample container homogenises water/solute application to the sample surface. If lateral transport is to be studied evidently it is vital that water and solute are applied as evenly as possible to the sample; this relatively simple apparatus achieves a uniformity of application (relative standard deviation 8.8%) comparing favourably with far more complex arrangements[6,7].

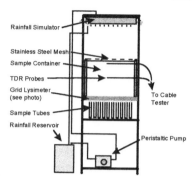

Rainfall Simulator

Stainless Steel Mesh

Sample Container

TDR Probes

Grid Lysimeter
(see photo)

Sample Tubes

Rainfall Reservoir

To Cable
Tester

Peristaltic Pump

Figure 5 *Experimental apparatus for
studying transport through porous media.*

The grid lysimeter shown in Figure 6 is based on a Canadian design[5] and allows detailed study of lateral transport and flow behaviour of a sample. The 10 x 10 array of square funnels machined into the aluminium plate have the same spatial distribution as the inner 100 needles of the rainfall simulator, while the large channels around the edges allow the bulk collection of edge flow, which is largely ignored due to possible interference from the container walls.

Figure 6 *The grid lysimeter.*

Flow through the sample drains via the grid lysimeter into either the racks of 100 sample tubes or, if more temporally detailed data is of interest, via a funnelling collector through a timed 3 way valve to an autosampler. Edge flow is collected to monitor conservation of solute mass.

3.2 Method

Sand is packed into the sample container trying to ensure that packing is as homogeneous as possible. This is actually reasonably difficult, the sand having inevitably undergone some settling/sorting during transit and packing methods are under constant re-evaluation. Because of the hydrophobicity of the dry sand column, if the rainfall simulator was started at this point water repellence would result in majority of the flow instantly being conducted into a few major flow channels. For the purposes of examining lateral transport variations this is extremely counter-productive, therefore it is necessary to first completely saturate the sand. This is achieved by stopping all the outputs from the grid lysimeter bar one, and pumping water in from the bottom as slowly as possible. For a 120mm deep sand column this typically can be carried out over 24 hours. Providing the porosity of the sample is known the volumetric water content as measured by the time domain reflectometry (TDR) probes can monitor whether saturation is achieved.

Once the sample is saturated the rainfall simulator is started and the stoppers in the grid lysimeter can be removed. Prior to the commencement of the solute transport study the system must be allowed to reach some sort of equilibrium allowing volumetric water content to decrease to some 'field capacity', this term being used to describe the level of saturation maintained by the water application rate employed.

Once this state has been achieved a number of studies can be conducted. The simplest is to monitor the spatial variation of pore water velocity output via the grid lysimeter. However, more important to the ongoing development of Pore-Cor is the investigation of transport behaviour. This is examined by applying a pulse of potassium bromide solution of known concentration via the rainfall simulator and monitoring the bromide concentration of the subsequent output.

3.3 Results

Figure 7 is a 'map' of average pore water velocity (cm hour^{-1}) through an apparently homogeneous sand column of depth 120mm. Velocities were measured hourly for each of the 100 outputs, corresponding to the grid intersections in the figure, in the underside of the grid lysimeter over seven hours prior to a solute transport run. It should be noted that the data sets contributing varied negligibly, in fact by Pearson's correlation coefficient the output of the funnels was consistently about 95% correlated from one hour to the next.

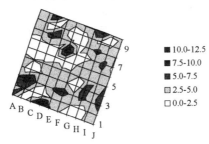

Figure 7 *Spatial distribution of pore water velocity (cm h^{-1}) through an homogeneous sand column.*

Given the apparently homogeneity of the sample it is perhaps surprising that such a heterogeneous flow pattern should be established. This phenomenon is frequently observed however, and a number of possible explanations given. Certainly it would seem that in homogeneous samples increasing depth seems to result in larger flows being conducted by less of the medium. One of the continuing aims of this work will be to look in particular at the relationship between sample depth and the development of preferential flow channels.

Figure 8 is the breakthrough curve (BTC) for a bromide transport run conducted on a 500 x 500 x 120 mm sample of Redhill 30 sand. The outputs of all 100 funnels of the grid lysimeter were combined to give the overall BTC for the whole sample, although unfortunately at this point sampling was manual hence the incomplete tail of the curve.

The aim of this experimental work is to get one BTC for the entire sample, such as the one shown here, followed by BTC's obtained from less frequent sampling of the individual funnel outputs. Thus using this detailed vertical and lateral transport information an intricate picture of the transport properties of the sample under consideration is built up. Ultimately some adaptation of solute transport modelling theory that best represents this experimental data will be incorporated into Pore-Cor.

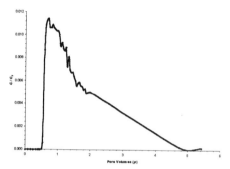

Figure 8 *Bromide breakthrough curve for Redhill 30 sand.*

3.4 Conclusion

Although the results detailed above are of a preliminary nature, they suggest behaviour likely to be observed as the study progresses and indicate the direction future development of Pore-Cor will take.

1. G. P. Matthews, C. J. Ridgway and M. C. Spearing, *J. Coll. Int. Sci.*, 1995, **171**, 8
2. T. Mathews, G. P. Matthews, A. K. Moss and G. Powell, *Journal of the Society of Petroleum Engineers: Formation Evaluation pre-prints*, 1994
3. P. A. C. Gane, J. P. Kettle, G. P. Matthews and C. J. Ridgway, *Ind. Chem. Eng. Res.*, 1996, **35**, 1753
4. R. E. Knighton and R. J. Wagenet, *Agronomy Mimeo*, 1988, 23
5. B. T. Bowman, R. R. Brunke, W. D. Reynolds and G. J. Wall, *J. .Env. Qual.*, 1994, **23**, 815
6. M. J. M. Romkens, L. F. Glenn, D. W. Nelson and C. B. Roth, *Soil Sci. Soc. .Am. J.*, 1975, **39**, 158
7. R. E. Phillips, V. L. Quisenberry and J. M. Zeleznik, *Soil Sci. Soc. .Am. J.*, 1995, **59**, 707

MODELLING OF PORE-LEVEL PROPERTIES

Cathy J. Ridgway and G.Peter Matthews

Department of Environmental Sciences
University of Plymouth
Plymouth
Devon PL4 8AA

1 INTRODUCTION

Pore-Cor is a 1.3 Mbyte Fortran 77 software package, which uses information from mercury porosimetry curves, supported by image-analysed electron micrographs, to generate a three-dimensional model of the pore-space within a porous material. The mercury intrusion curves for the simulated structures converge automatically onto the experimental curves producing a pore and throat size distribution which is used in the construction of the unit cell. This same unit cell can then be used to model other parameters such as porosity, permeability and tortuosity. The model has been applied to a number of different porous mediums including outcrop and reservoir sandstones, building materials, paper coatings, tantalum slugs, tablets, fibrous mats and soils.

2 DESCRIPTION OF THE PORE-COR MODEL

The Pore-Cor model has three main characteristics: (i) it has a physically realisable geometry; (ii) all fitting parameters are real properties related to the geometry of the unit cell; and (iii) the same network with precisely the same geometry can be used to model a wide range of properties.

    The void space within a porous solid can be regarded as a network of void volumes (pores) connected by a network of smaller void channels (throats). The Pore-Cor network comprises a three-dimensional cubic unit cell which repeats infinitely in each direction, each cell containing 1000 nodes on a regular 10 x 10 x 10 matrix. The nodes are positioned using Cartesian co-ordinates x, y and z, although, since the sample is isotropic, the allocation of these axes is arbitrary. The origin of the axes is at the corner of the unit cell adjacent to the first node. The void volume in the unit cell comprises up to 1000 cubic pores centred on the nodes. Connected to each pore are up to six cylindrical throats along the axial directions in the positive and negative x, y and z directions. The number of throats connected to a particular pore is the *pore co-ordination number*, and the arithmetic mean of this quantity over the whole unit cell is the *connectivity*. Individual pore co-ordination numbers may range from 0 to 6, while the connectivity of sandstone is around 3.[1]

    The void-space model assumes that the throat diameter distribution is log-linear, i.e. the diameters are equally spaced over a logarithmic axis. A log-normal distribution can also be used, but usually gives mercury intrusion curves which are too steep at the point of inflection. The parameters that can be varied are the range of throat diameters, the connectivity, the skew of the throat diameter-distribution, the relationship between adjacent throat and pore sizes, and the spacing between the rows of pores. The gradient of the mercury intrusion curve at its point of inflection is governed by the connectivity (Matthews et al., 1995b). The term 'throat skew' is the percentage of throats of the smallest size in a log-linear distribution. A throat-size distribution with a throat skew of 0.1%, for example, has 0.1% of the smallest throat size, 1% of the size in the middle of the logarithmic distribution, and 1.9% of the largest size. It is possible for the throat skew to be negative, which is a situation that is described in a previous publication.[2] The pore size : throat size correlation in the model is a 1:1 ratio, which is typical of sandstones but not limestones.[3,4] The row spacing of the matrix is adjusted so that the porosity of the simulated network is equal to that of the experimental sample.

In the present mercury intrusion simulation, the mercury is injected normal to the xy plane at $z=l_{cell}$, the unit cell size, in the $-z$ direction. The injection corresponds to intrusion downwards from the top surface of the unit cell. The unit cell repeats infinitely in each direction, and therefore the simulation corresponds to intrusion into an infinite sheet of thickness l_{cell}.

3 MODELLING THE MERCURY INTRUSION CURVES

The simulated mercury intrusion curve can be made to converge onto the experimental curve automatically.[5] Four criteria for closeness of fit have been described in the previous publications. They are: (i) that the curves should cross at 50% pore volume ('50% fit'), (ii) that the rms deviation of simulated from experimental throat diameters should be a minimum ('linear fit'), (iii) the rms deviation of the logarithms of simulated from experimental throat diameters should be a minimum ('logarithmic fit'), and (iv) that the rms deviation of simulated from experimental throat diameters should be a minimum for comparison points above the median point in the experimental curve ('linear top fit'). The deviations in (ii), (iii) and (iv) are measured at experimental and interpolated points.

Using criterion (i), only one point on the experimental and simulated is compared, and thus much valuable information is ignored. Criteria (ii) and (iii), based on fits to the entire intrusion curve, can be unsuitable for small samples due to edge effects. These comparatively large effects arise from large cavities on the cut surfaces of the sample. One approach to this problem is to carry out a straightening and truncation of the experimental intrusion curve below the point of inflection, but the changing of the experimental curve involves a considerable degree of subjectivity and cannot account for subtle internal void-space effects in the presence of large edge effects. The other approach is to use criterion (iv), the 'linear top fit'.

Figure 1 shows the logarithmic distribution of throat diameters and the experimental and simulated mercury porosimetry curves for a sample of Clashach outcrop sandstone. The curve has been fitted using the logarithmic fit.

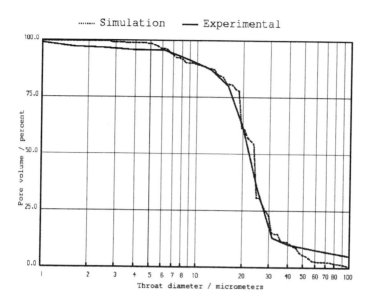

Figure 1 *Experimental and simulated mercury intrusion curves for Clashach sandstone*

4 NON-HOMOGENEOUS STRUCTURES

Non-homogeneous structures have been modelled, in the simulation of layered samples, clay included sandstones, vuggy soil samples, high porosity tantalum slugs and extremely open high-voidage fibrous mats. These structures diverge from the completely random arrangement of the pores and throats, and include banded structures, clustered arrangements where the large pores are either at the centre or at the periphery of the clusters, structures where the pore size distribution is altered to give a larger number of the larger pore sizes and structures where all the pores are of the same large size. Levels of correlation can also be introduced into these structures.[6]

5 RANGE OF SAMPLES AND CORRESPONDING UNIT CELLS

The following figures show the types of samples that have been studied. The unit cells illustrate the range of different void space structures, both homogeneous and non-homogeneous.

5.1 Random Structure

A pressed powder compact made from pharmaceutical grade crystalline lactose and an anti-inflammatory compound has been modelled, the unit cell is shown in Figure 2. A range of compaction pressures have been investigated and the results showed expected

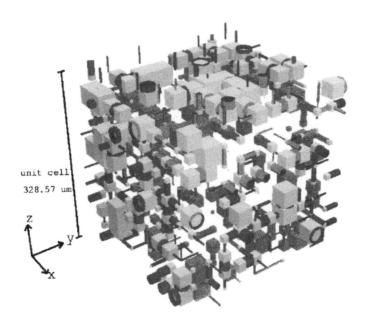

unit cell
328.57 um

z
y
x

Figure 2 *Random unit cell representing a pressed powder compact*

trends of reduced porosity and reduced permeability.[7]

5.2 High Porosity Structure

Tantalum slugs are made with the aim of having a high porosity, whilst keeping the actual physical size of the capacitor as small as possible. A parameter has been introduced into Pore-Cor whereby the pore size distribution is multiplied by this parameter; any pore whose size becomes larger than the previous maximum is allocated that maximum size. This results in a peak at the maximum size of the pore size distribution and allows much higher porosity samples to be modelled. Figure 3 shows a unit cell which has a pore skew of 1.9, it is clear to see the increased number of large pores, and that a distribution of pore sizes still exists.

5.3 Banded Structure

Sandstones are particularly prone to having layered structures as a result of the way they are formed. The Pore-Cor program models these layers as vertical bands to correspond with the required permeability calculation. It is usual for permeability measurements to be taken in the plane of maximum permeability; having vertical bands allows Pore-Cor to simulate this method. The banded structure in Figure 4 is a partially correlated structure, with a correlation level of 0.8.

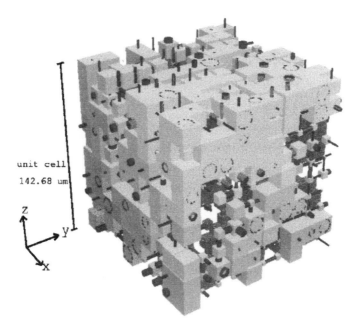

Figure 3 *High porosity unit cell representing a tantalum slug*

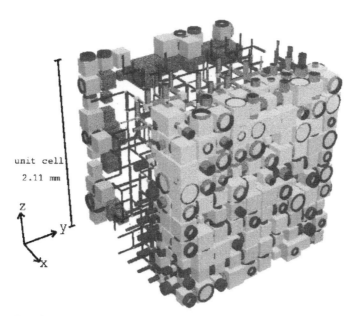

Figure 4 *Banded unit cell representing a layered sandstone*

5.4 Structure With Small Pores And Throats In The Centre

A Fontainebleau sandstone sample had small amounts of illite and kaolinite artificially deposited within it. The clay deposits coat the inside of the structure of the sandstone resulting in areas of smaller pores and throats. Pore-Cor has simulated the differences between a clean sample and a clay precipitated sample in terms of connectivity, throat skew, tortuosity and permeability.[2] A further illustration of the possible reduced area resulting in smaller pores and throats is shown in the unit cell in Figure 5.

5.5 Structure With Large Pores And Throats In The Centre

Soil can have a very open structure with clods of mud, stone and organic matter incorporated. Water drainage curves are measured and then converted into air intrusion curves for the Pore-Cor program to simulate. Figure 6 shows a structure with the large pores and throats clustered together, giving an area of open voidage as in the soil sample.

5.6 Fibrous Mat Structure

Fibrous mat materials have been modelled by Pore-Cor. The data needed is converted from hexa-decane drainage to air intrusion. The Pore-Cor structure shown in Figure 7 has all the pores the size of the largest throat diameter. The fibres of the sample lie between these large pores and where there is a pore missing, it can be interpreted as a knot of fibres.

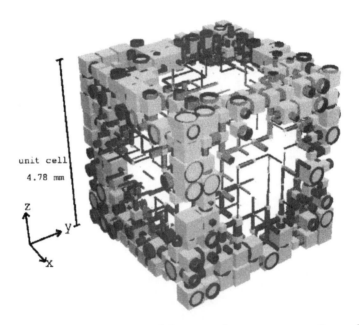

Figure 5 *Cluster of small pores and throats in the centre representing a clay included sandstone*

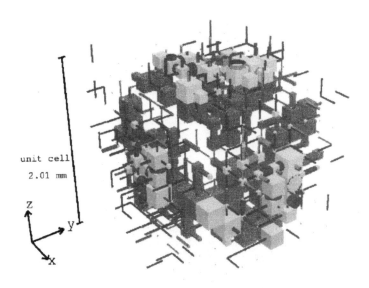

Figure 6 *Cluster of large pores and throats in the centre representing a soil structure*

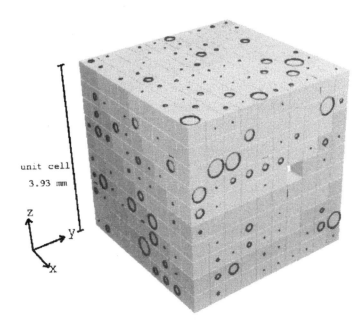

Figure 7 *Fibrous mat structure representing a nappy filling*

6 SUMMARY

Pore-Cor has been shown to be very versatile in the range of samples that can be modelled. The void space structures have covered a wide range of scales from micrometers for the tantalum slugs to millimetres for the fibrous mat. Pore-Cor has been developed to simulate correlated structures but an actual experimental measurement of the level of correlation of a particular sample is still being developed.[8] The modelling has produced useful predictions of permeability which have been used in the optimisation of production processes.

References

1 M. Yanuka, F. Dullien and D. Elrick, *J. Colloid Interface Sci.*, 1986, **112**, 24.

2 G.P. Matthews, C.J. Ridgway and J.S. Small, *Marine and Petrol. Geol.*, 1996, **13**(5), 581.

3 N. Wardlaw, Y. Li and D. Forbes, *Transport in Porous Media*,1987, **2**, 597.

4 M. Spearing and G.P. Matthews, *Transport in Porous Media*,1991, **6**, 71.

5 G.P. Matthews, C.J. Ridgway and M. Spearing, *J. Colloid Interface Sci.*,1995, **171**, 8.

6 G.P. Matthews, A.K. Moss and C.J. Ridgway, *Powder Technol.*, 1995, **83**, 61.

7 G.P. Matthews, C.J. Ridgway and K. Ridgway, *J. Pharm. and Pharmacol.*, 1996, to be submitted.

8 T.J. Mathews, G.P. Matthews, C.J. Ridgway and A.K. Moss, Measurement of Void Size Correlation in Sandstones and other Porous Media. *Transport in Porous Media*, 1995, submitted.

SORPTIVE PROPERTIES OF CLOVERITE

W. Schmidt and F. Schüth

Institut für Anorganische Chemie
Johann Wolfgang Goethe - Universität Frankfurt
Marie Curie-Str. 11, D-60439 Frankfurt, Germany

1 INTRODUCTION

The gallophosphate cloverite is one of the latest successes in the preparation of crystalline microporous materials. The crystal structure of this material shows some remarkable features. The micropores of cloverite are formed by two independent channel systems, one of them consists of large cages with approximately 30Å in diameter connected via 20 T-atom windows, the other one is built by α-cages and rpa-cages connected by 8 T-atom windows [1]. The framework density of 11.1 T-atoms per $1000Å^3$ is remarkably low. Usually framework densities of 12.5 to 20.5 $T/1000Å^3$ are observed for zeolitic structures [2]. Thus, cloverite appears as a very promising adsorbent. Unfortunately, the framework of cloverite usually collapses during calcination which is neccessary to remove water molecules and the organic template occluded inside the channel system during the synthesis.

Following the calcination process with *in situ* techniques we were able to develop a suitable procedure which allows the removal of the occluded molecules while the cloverite structure remains intact. Thus, adsorption experiments could be performed on the novel gallophosphate cloverite. Here we present first results on the adsorptive properties of cloverite.

2 EXPERIMENTAL

Synthesis

The cloverite materials used for this work were crystallized in teflon lined autoclaves at 150°C for 24 hours. The batch composition of the reaction gels can be formulated as:
$Ga_2O_3 : P_2O_5 : 6\ Q : 0.75\ HF : 71.3\ H_2O$ with (Q = quinuclidine, *Fluka*)
The amount of water in the reaction gels was determined by the water content of the gallium sulphate solution (8 wt.% Ga, *VAW AG*), the phosphoric acid (85 wt.%, *Merck*), and the hydrofluoric acid (40 wt.%, *Merck*). For the preparation of the reaction mixture the phosphoric acid was slowly poured into the gallium sulphate solution. Then the

quinuclidine and the hydrofluoric acid were added to the gel. The reaction mixture was stirred during the preparation. Finally, the homogeneous reaction mixtures were filled in teflon lined autoclaves which then were placed in a preheated oven. The crystalline product was separated by filtration, washed with demineralized water and dried at room temperature.

Characterization

In situ XRD and *in situ* IR techniques as well as TG/DTA measurements were employed in order to develop a calcination procedure which allows the removal of the template without destroying the crystal structure.

For in situ XRD measurements $Cu_{K\alpha}$ radiation and a *Paar* HTK 10 heating chamber were used. The powder samples were prepared on a platinum sample holder serving as a resistance heater. Samples were heated with a rate of 5°C/min for in situ XRD measurements and data sets were recorded every 50°C up to 500°C and down to room temperature again. One measurement took about 20 minutes.

TG/DTA experiments were performed with a *Setaram* TG-DTA 92-16 thermobalance in an air flow of 10 l/h with a heating rate of 10°C/min.

Nitrogen and argon adsorption isotherms were recorded at 77 K on a *Micromeritics* ASAP 2010 adsorption system. The samples were outgassed at 350°C under vacuum for 14 hours. For measurements in the low pressure region of the isotherms the seal frits were removed from the sample tubes during the measurements in order to avoid diffusion limitation caused by the small pore diameters of the frits. In the low pressure range up to p/p_0 of 0.15-0.2 the adsorbates were dosed onto the samples as discrete gas volumes (incremental step mode). No thermal transpiration corrections were made.

3 RESULTS AND DISCUSSION

Calcination of cloverite

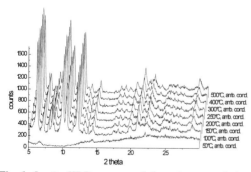

***Fig. 1**: In situ XRD pattern of cloverite recorded after heating to 500°C (top) followed by cooling down to 50°C in ambient air.*

The crystalline cloverite material consists of cube and cube octahedrally shaped single crystallites typical for cloverite with sizes of approximately 1-10 µm. Most of the measurements described in this work were performed with cloverite from the same batch (C5) for comparison of the results. Additionally the results were reproduced with samples from other synthesis batches.

Heating in ambient air also does not affect the crystallinity and the structure of cloverite as shown by XRD in fig. 1. The crystals can withstand temperatures up to 500°C without any structural damages. While no structural changes are detected down to 100°C a rapid amorphization of the crystalline cloverite is observed within a few minutes when the temperature gets below 100°C.

The collapse of the structure is related most probable to the adsorption of water from the ambient air[3]. Therefore, it is obvious that calcined samples have to be kept at temperatures above 100°C in ambient air or moisture has to be excluded, e.g. by vacuum, in order to use them for further experiments. Thus, the samples were heated to 600°C with a heating rate of 10°C/min in ambient air and kept at this temperature for four hours. TG/DTA measurements proved that no more template or coke are present in a sample calcined by this procedure.

For adsorption measurements the calcined samples were poured directly from the furnace at 600°C into a pre-heated sample tube at approximately 400°C. Since the sample had still a temperature far above 100°C, no water was adsorbed. The sample tube was sealed and then evacuated for another 14 hours at 350°C prior to the adsorption measurement.

Adsorption isotherms of cloverite

Adsorption data were processed using the report options and the DFT software provided by Micromeritics[®]. The adsorption isotherms of cloverite are of type I according to the BET classification[4] which is typical for microporous materials. Fig. 2 shows isotherms observed with nitrogen as adsorbate on different cloverite samples. For sample C5 two isotherms are shown to illustrate the variation between two independent measurements on the same sample. The resulting micropore volumes calculated from the nitrogen isotherms by the Horvath-Kawazoe method lie in the range of 0.19 - 0.21 cm^3/g.

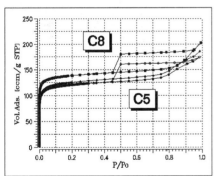

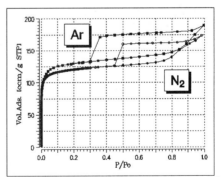

Fig. 2: Nitrogen isotherms of two cloverite samples at 77K.
■-*sample C8,* ◆-*sample C5,* ●-*sample C5;* *solid symbols: adssorption, open symbols: desorption*

Fig. 3: Nitrogen and argon isotherms of the same cloverite sample (C5) at 77K.
●-*nitrogen,* ■-*argon,*

The micropore volumes depend much more on the quality of the sample than on the experimental error. Three independent measurements on cloverite samples from the same batch, all of them calcined at 600°C, resulted in 0.191, 0.190, and 0.191 cm^3/g. Bearing in mind that the amount of the sample varied from 160.2 to 62.9 mg for these measurements, the results are in remarkably good agreement, proving the reproducibility of the calcination procedure as well as that of the adsorption experiments.

The desorption branches of the isotherms in fig. 2 do not match with the adsorption branches. Hystereses are observed, most probably due to mesopores caused by cracks inside the crystallites which are formed during the calcination. Cloverite samples activated at 200°C do not exhibit that hysteresis loop. One might suspect that the 30Å supercages of the cloverite structure could be somehow related to the hysteresis phenomenon, since the shape of the hysteresis indicates ink bottle pores which, in general, describe the supercages with its restricted windows quite well. However, according to the relative pressure where the steep desorption branch occurs one should have limiting pore openings in the order of about 40Å. The windows of the micropores of cloverite are definetely smaller and lie in the range of 3.8Å for the 8 T-atom windows and 3.5-13.2Å for the 20 T-atom windows (see fig. 4). Surprisingly the steep desorption step was observed at almost the same pressure for most of the samples, indicating very similar crack sizes. Harris and Whitaker observed similar desorption steps in the relative pressure range of 0.5 - 0.42 for nitrogen isotherms for about half the adsorbents they investigated (more than 100). They suggested that desorption from pores finer than 40Å proceeds by a mechanism independent of the pore size of the adsorbent, and only dependent on the adsorbate and the temperature, since it would be very unlikely if all the materials have the same pore sizes[4,5].

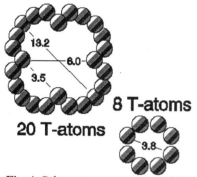

Fig. 4: Schematic presentation of the pore openings of cloverite[2] (diameters in Å)

Argon isotherms were recorded only on one sample. The calculation of the micropore volumes from the argon isotherms on that sample resulted in values of 0.165 and 0.167 cm^3/g. Again, the reproducibility of the measurements proves to be satisfactory. The micropore volumes calculated from the argon isotherms are somewhat smaller than those obtained from the nitrogen adsorption experiments. The steep drop in the desorption branch is observed with argon as adsorbate as well, but shiftet to lower relative pressure (see fig.3), which would agree with the suggestion of Harris and Whitaker.

In general the shape of the argon isotherm is similar to the nitrogen isotherm.

Low pressure range of the isotherms

Due to the different pore openings of the 8 T-atom windows and the 20 T-atom windows we expected a more or less structured isotherm in the low pressure range. Using the usual glass sample tubes, a somewhat unexpected behavior of the isotherms was observed in the range of $p/p_0 < 10^{-4}$. A sigmoidal loop occurs in the nitrogen as well as in the argon isotherms. The sigmoidal shape is less pronounced for argon isotherms. After the first doses the relative pressure decreases below the initial pressures, as shown in fig. 5. Further dosing then leads to an increase of the relative pressure again. In the low pressure range one would rather expect a linear increase of the volume adsorbed with increasing relative pressure (Henrys law[4]). Figure 5a shows two isotherms of the same sample measured with different equilibration delays for each data point in the low pressure region. For longer equilibration times the relative pressure shifts to lower values indicating that the system is not in equilibrium. The sample tubes were usually sealed with seal frits. Removal of this frit and warming up the sample to room temperature after the free space determination with helium led to isotherms shown in fig. 5b. Here the sigmoidal branch is reduced to a steep increase of the isotherm at the beginning. At higher relative pressures it meets the isotherms, measured with the frit.

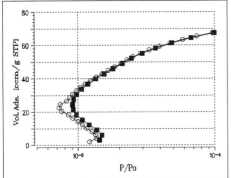 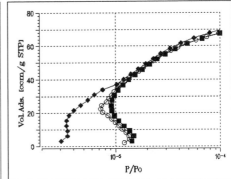

Fig. 5a: *Low pressure range of nitrogen isotherms of the same sample. Equilibration time per data point: left 30 min, right 60 min*

Fig.5b: *Same as fig. 5a. Additionally an isotherm, measured without seal frit and warmed up after free space measurement*

The seal frit seems to restict the diffusion of the adsorbate molecules from the manifold onto the sample. The diffusion through a frit has to be regarded as Knudsen diffusion and, thus, very long equilibration times have to be expected. For this reason the tubes with the cloverite samples were backfilled with dry nitrogen gas after the activation and then placed under the analysis port. Fast removal of the seal frit, followed by evacuation at the analysis port keeps the structure of the material intact.

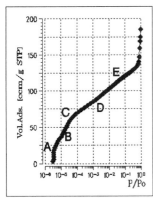

Fig. 6: *Complete nitrogen adsorption isotherm of cloverite*

After the free space measurement the sample was warmed up to room temperature to desorb helium which might still be incorporated in the molecular sieve and would also change the isotherm in the low pressure range.

Following this procedure adsorption isotherms as shown in fig. 6 can be measured in reasonable times even in the low pressure range. Inspection of the obtained data reveals no plateaus in the low pressure range but two distinct knees (A and C) and at least two points of inflection (B and D) are visible. Point E marks an additional knee, and can be assigned to the point where the micropore filling is completed. The points A,B,C, and D are reflected in the determination of micropore sizes. Calculating pore size distributions either by the Horvath-Kawazoe or the DFT method leads to pore size distributions with two distinct maxima in the micropore range.

Dependency of the pore volume on the calcination conditions

Cloverite samples have to be calcined at least at 500°C, better at 600°C, in air or in oxygen in order to obtain isotherms as shown in fig. 6. Calcination at lower temperatures leads to an incomplete removal of the organic template. The quinuclidine is removed from the pores in distinct exothermic steps within the temperature range from 300-620°C. Cloverite samples usually contain up to 14.5 wt% of quinuclidine and about 7 wt% of water which desorbs at temperatures up to 200-250°C. In the range of 800-870°C additional exothermic signals are observed which are caused by the conversion of cloverite into GaPO$_4$-tridymite, as shown by Bedard et al.[6].

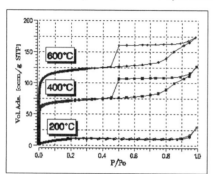

Fig. 7: Nitrogen isotherms measured on samples calcined at different temperatures

Isotherms of incompletely calcined cloverite are shown in fig. 7. In the sample calcined at 200°C only a little proportion of micropores are found. This sample was activated at 200°C prior to adsorption. Thus, only water desorbed from that sample. It still contained its whole amount of template inside the micropores as can be seen by TG/DTA experiments. Cooling this sample to room temperature in ambient air does not lead to a collapse of the structure, as observed for samples calcined at higher temperatures. The sample calcined at 400°C still contains about 7 wt % of coke which can be shown by TG/DTA.

This amount of residuals in the pore system is reflected by its nitrogen uptake. It lies in-between that of the sample calcined at 200°C and one calcined properly at 600°C. A hysteresis loop comparable to that of a sample calcined at 600°C is also observed in this sample, which is not found for the sample calcined at 200°C. Interparticular gaps in aggregates of cloverite crystallites should also be present for the sample calcined at 200°C and, thus, a hysteresis should be visible for that sample as well. Cracks, due to the combustion of the template and to thermal stress, are more likely to be responsible for the hysteresis.

Comparison of different models for the calculation of pore size distributions

Two models were used to calculate the pore size distributions of the cloverite samples: the Horvath-Kawazoe method and the density functional theory (DFT).

In principle none of the models describes the interaction of the gallophosphate cloverite with the adsorbates nitrogen or argon readily. The original Horvath-Kawazoe model is only valid for slit shaped pores[7]. For pores of the kind found in the cloverite structure cylinder pore geometry (Saito/Foley[8]) or even sphere pore geometry (Cheng/Yang[9]), describing the almost spherical cages, seem to be more realistic models. Adsorbate/adsorbent interaction parameters do not affect the pore size distribution as strong as the choice of the pore model but they have to be chosen with care. The density functional theory package provided by Micromerics® uses only two sets of model isotherms; nitrogen on a carbon substrate at 77.35 K and argon on a carbon substrate at 87.29 K. Furthermore, slit shaped pores are assumed which definetely are not present in cloverite crystals. However, since these two models are the only more or less reliable approaches at the moment, they were used to calculate the pore size distributions of the cloverite samples.

The pore size distributions according to Horvath-Kawazoe in fig. 8 were calculated from the same isotherms, only parameters like the pore shape and the kind of adsorbent were varied. It is obvious that, due to the shape of the adsorption isotherm, two distinct maxima are observed in the pore size distribution (according to points A - E in fig.6).

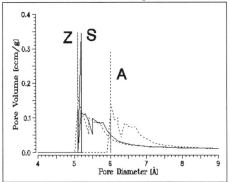

Slit Pores (original Horvath-Kawazoe)

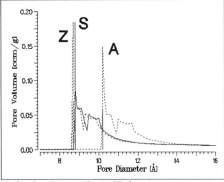

Cylinder Pores (Saito/Foley)

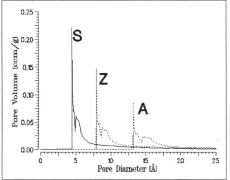

Sphere Pore Geometry (Cheng/Yang)

Fig. 8: *Pore size distributions obtained from the same sample by variation of the model for the pore geometry and the kind of adsorbent.*

S : *standard set for interaction parameters*
Z : *parameters calculated for zeolites*
A : *parameters calculated for AlPO$_4$s*

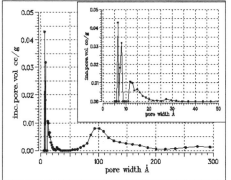

Fig. 9: *Pore size distribution calculated with DFT*

It can be inferred that the use of different parameter sets enables one to compute almost any desired pore size distribution. Which data can we rely on? For the gallophosphates no appropriate data set for the interaction parameters are provided and for the calculation a parameter set from a material has to be chosen as close as possible related to gallophosphates. Here data sets from zeolites and from alumophosphates were used. For slit pore and cylinder pore geometry fixed parameter sets were used. The maxima of the distributions vary in the range of 1-1.5 Å for identical pore types and depend basically on the diameters of the adsorbate atoms/molecules and that of the surface atoms of the adsorbent. For spherical pore geometry the parameter sets had to be calculated for each adsorbent, which leads to quite different parameter sets. Thus, the pore size distributions vary much more for the sphere pore geometry.

The pore geometry itself has an even more pronounced influence on the position of the maxima of the pore size distributions. The shape of the distributions is invariant, since it depends on the shape of the isotherm. Cylinder shaped pores describe the channel system of cloverite probably best, but nevertheless the agreement with crystallographic data for the pore openings is very poor (see fig. 4) if one suggests that the first maximum is related to the 8 T-atom windows and the second one to the 20 T-atom windows.

Pore size distributions calculated with the density functinal theory from the adsorption branch of the same isotherm as used for the calculations of fig. 8 exhibit two adsorption steps in the pore size range below 10 Å, which is in agreement with the results from the Horvath-Kawazoe calculations. In general the distribution exhibits three maxima at 6.4, 7.6, and 11.8 Å and an additional one in the mesopore range at about 95 Å. The reliability of that method in the micropore range is also limited. Especially for the steep adsorption step at low relative pressures some more model isotherms would be desirable for better fitting of the adsorption data.

CONCLUSIONS

Using appropriate calcination procedures it is possible to measure adsorption isotherms on cloverite. The micropore volumes, measured with nitrogen, vary slightly for different batches and lie within the range of 0.19-0.21 cm^3/g. With argon as adsorbate smaller pore volumes are found. The adsorption of nitrogen and of argon on cloverite takes place in at least two distinct steps, most probable due to two different openings of the micropore system. The hysteresis loop at higher relative pressures might be caused by cracks formed during the calcination of the crystals. Special care has to be taken if precise measurements in the low pressure range are desired. Appropriate diameters for the gas inlet have to be warranted to avoid restricted gas diffusion. Care should also be taken in the choice of the calculation model to determine the pore size distributions in the micropore range.

References

[1] M. Estermann, L.B. McCusker, C. Baerlocher, A. Merrouche, H. Kessler, *Nature*, 1991, **352**, 320

[2] W.M. Meier, D.H. Olson, Ch. Baerlocher, 'Atlas of Zeolite Structure Types', Elsevier, 1996

[3] W. Schmidt, F. Schüth, S. Kallus, 'Proceedings of the 11th IZC', Seoul, Korea, 1996, Elsevier, in print

[4] S.J. Gregg, K.S.W. Sing, 'Adsorption, Surface Area and Porosity', Academic Press, 1967

[5] M.R. Harris, G. Whitaker, *J. Appl. Chem.*, 1963, **13**, 348

[6] R.L. Bedard, et al., *J. Am. Chem. Soc.*, 1993, **115**, 2300

[7] G. Horvath, K. Kawazoe, *J. Chem. Eng. Japan*, 1983, **16(6)**, 470

[8] A. Saito, H.C. Foley, *AlChE Journal*, 1991, **37(3)**, 429

[9] L.S. Cheng, R.T. Yang, *Chem. Eng. Sci.*, 1994, **49(16)**, 2599

COMPARISON OF THE TRANSLATIONAL MOBILITY D_t OF HYDROGEN METHANE AND NEOPENTANE MOLECULES SORBED ON AlPO4-5.

C. Martin, J.P. Coulomb, Y. Grillet* and R. Kahn**.

C.R.M.C² - CNRS, Campus de Luminy, Case 901, 13288 Marseille Cedex 9 - France.
* C.T.M.- C.N.R.S., 26 rue du 141° R.I.A., 13003 Marseille - France.
** Laboratoire Léon Brillouin, C.E.N.- Saclay, 91191 Gif-sur-Yvette Cedex - France.

ABSTRACT

We have measured by incoherent quasi-elastic neutron scattering the translational molecular mobility D_t of hydrogen, methane and neopentane adsorbed molecules on the one-dimensional micropore network of AlPO4-5 . Due to the relative size between the $C(CH_3)_4$ molecules ($\emptyset = 6.2$ Å) and the micropore diameter ($\emptyset = 7.3$ Å), the neopentane sorbed phase on AlPO4-5 is an illustrative example of a single-file diffusion system . In such a case, D_t is expected to be strongly reduced. Surprisingly, it is the neopentane molecules which present the highest mobility in comparison with hydrogen and methane molecules.

1 INTRODUCTION

Recently a phase transition has been observed during the sorption of methane on AlPO4-5, in the high loading regime [1-3]. Such zeolitic material is very interesting owing to the one dimensional character of its micropore network and the simplicity of their inner surface composed by only one type of hexagonal sorption sites. In the low and medium loading regime the CH_4 sorbed phase is a " Brownian " fluid [4]. To investigate, the molecular size influence on the translational mobility D_t , we have measured D_t for hydrogen and neopentane sorbed molecules on AlPO4-5 . The relative size of the three sorbates species in comparison to the AlPO4-5 micropore dimensions is represented schematically on figure 1 .

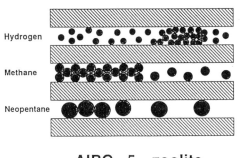

AlPO₄ -5 zeolite

Figure 1

2 EXPERIMENTAL SECTION

The $AlPO_4$-5 sample used in our experiment (m = 2.416 g) was synthesized at the laboratory of the Prof. H. Kessler (Mulhouse-France). The calibration isotherms were measured at T = 17 K, T = 96.5 K and T = 250 K respectively for hydrogen, methane and neopentane sorbed on $AlPO_4$-5 (the maximum sorption capacities are $Q_{ads.}[H_2]$ =10.0 mol./u.c.; $Q_{ads.}[CH_4]$ = 5.7 mol./u.c. and $Q_{ads.}$ [$C(CH_3)_4$] = 1.2 mol./u.c.). The incoherent quasi-elastic neutron scattering experiments have been accomplished at the Leon Brillouin Laboratory (CEA - Saclay, France) on the Mibemol spectrometer (the wavelenght λ and resolution $\Delta E_{resol.}$ were equal to 8 Å and 40 μev respectively).

3 DESCRIPTION OF THE MODEL

Incoherent quasi-elastic neutron scattering IQNS is the more direct and powerful experimental technique to determine the molecular translational mobility D_t . The scattering law $S_{inc.}(\vec{Q}, E)$ is the Fourier transform in space and time of the autocorrelation function $P(\vec{r}, t)$ which expresses the probability to find the molecule at the position $\vec{r}$ at time t (the same molecule being at the origin $\vec{r} = 0$ at time t = 0). The common assumption is that the translational and the rotational molecular motions are uncoupled. As a consequence the total scattering law is the convolution of the translational scattering function $S^{trans.}(\vec{Q}, E)$ and of the rotational scattering $S^{rot.}(\vec{Q}, E)$ function . By introducing the Debye - Waller factor , exp (- $Q^2 <u^2>$), in the quasi-elastic region the total scattering function is equal to :

$$S_{inc.}(\vec{Q}, E) = [S^{trans.}(\vec{Q}, E) \otimes S^{rot.}(\vec{Q}, E)] \exp(-Q^2 <u^2>)$$

To describe the translational diffusive motion we have used the simplest model which is the Brownian type diffusion model . In that case, the scattering function $S^{trans.}(\vec{Q}, E)$ for bulk (3d) fluid has a Lorentzian shape [5] :

$$S^{trans.}(\vec{Q}, E) = \frac{D_t Q^2}{\pi [E^2 + (D_t Q^2)^2]}$$

D_t and Q are the translational diffusion coefficient and the modulus of the scattering wave vector

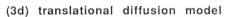

(3d) translational diffusion model

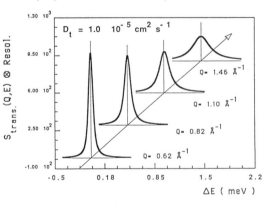

Figure 2

respectively. The Lorentzian width ΔE_{FWHM} (full width at half maximum) is directly proportional to the diffusion coefficient D_t ; $\Delta E_{FWHM} = 2\,D_t \cdot Q^2$. As represented on <u>figure 2</u> , the " signature " of the translational diffusive motion is expressed by the broadening of the IQNS peak. For confined sorbed phase in one-dimensional channels $S^{trans.}(\vec{Q}, E)$ presents a slight modification (the component of the scattering wave vector $\vec{Q}$ along the AlPO$_4$-5 micropore axis $\vec{Z}$ has to be considered) [6].

$$S^{trans.}(\vec{Q}, E) = (1/\pi)\, D_t\, Q^2 \cos^2\theta \,/\, \left[E^2 + (\,D_t\, Q^2 \cos^2\theta\,)^2 \right]$$

(θ is the angle between the scatteing wave vector $\vec{Q}$ and the zeolite micropore axis $\vec{Z}$). In addition, due to the powder character of our sample, we have to calculate the average of $S^{trans.}(\vec{Q}, E)$ over θ .

$$< S^{trans.}(\vec{Q}, E) >_{powder} = 1/2\pi \int_{0}^{\pi} S^{trans.}(Q, E) \cdot \sin\theta \cdot d\theta$$

We want to remark that the power averaging and the fact that the molecular translational mobility is confined in one-dimensional micropores, the broadening of the IQNS peak decreases when the fluid dimensionality decreases . The IQNS spectra calculated for the fluid(3d), fluid(2d) and fluid(1d) (with the same value of the diffusion coefficient $D_t = 1.0$ 10^{-5} cm^{2}s^{-1}) are represented on <u>figure 3</u> .

Quasi-elastic peak broadenings

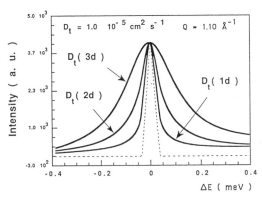

<u>Figure 3</u>

To represent the molecular rotational motion, we have used the isotropic rotational diffusion model . The powder average of $S^{rot.}(\vec{Q}, E)$ is not necessary due to the isotropic character of the rotations .

$$S^{rot.}(\vec{Q}, E) = j_0^2 (Q\rho)\, \delta(E) + (1/\pi) \sum_{l=1}^{\infty} (2l+1)\, j_l^2 (Q\rho)\, \frac{D_r\, l\,(l+1)}{(D_r\, l\,(l+1)\,)^2 + E^2}$$

$j_0 (x) \ldots j_l (x)$ are the spherical Bessel functions . D_r , ρ and $\delta(x)$, are the rotational diffusion coefficient, the radius of gyration of the molecule and the Dirac function respectively .

As represented on the figure 4 , the scattering function of the isotropic rotational motion $S^{rot.}(\vec{Q}, E)$ is characterized by both an elastic component and a quasi-elastic component. Note that the intensity of the rotational quasi-elastic signal, which is proportional to $j_1^2(Q\rho)$, decreases when Q goes to zero. On the other hand the quasi-elastic peak width ΔE_{FWHM} does not depend on the Q modulus. It is directly proportional to the rotational diffusion coefficient D_r :

$$\Delta E_{FWHM} = 2 D_r$$

Finally, in order to compare the calculated total scattering function with the IQNS measured spectra, $S_{inc.}(\vec{Q}, E)$ has to be convoluted with the instrumental resolution ΔE_{resol}. The rotational diffusion coefficient D_r of globular molecules as methane and neopentane are quite large in the bulk fluid phase $D_r \simeq 4 \ 10^{11} \ s^{-1}$. As a consequence, in the low and medium Q range the two types of broadening are different $\Delta E_{FWHM}(trans.) = 2 D_t Q^2 << \Delta E_{FWHM}(rot.) = 2 D_r$.

(3d) rotational diffusion model

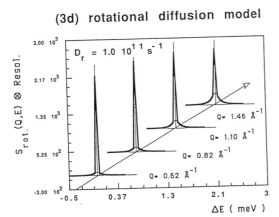

Figure 4

4 RESULTS AND DISCUSSION

The incoherent quasi-elastic neutron scattering (IQNS) spectra measured for the neopentane sorbed phase at T = 250 K on AlPO4-5 are represented on the figure 5. The loading is equal to 42 % . From the elastic peak broadening we can deduce that the translational mobility D_t of the neopentane molecules is quite large $D_t = 2.2 \pm 0.3 \ 10^{-5} \ cm^2 \ s^{-1}$. We have measured the IQNS spectra at different neopentane loading . In the loadings range 20 - 60 % the translational mobility D_t is constant. At higher loading D_t decreases strongly. We have also measured the IQNS spectra for methane and

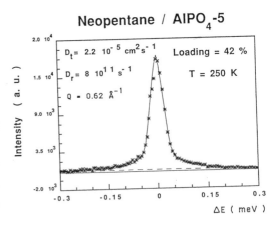

Figure 5

hydrogen molecules sorbed on AlPO$_4$-5 at respectively T = 96.5 K for a loading equal to 40 % and T = 17 K for a loading equal to 27 %. Some results obtained in the analysis of the CH$_4$ / AlPO$_4$-5 system have been already published [4]. The translational mobility D$_t$ of the methane molecules is rather large at T = 96.5 K : D$_t$ = 1.0 ± 0.02 10^{-5} cm^2 s^{-1}. An IQNS spectrum measured at the Q = 0.62 Å^{-1} is represented on figure 6. We have observed that the translational mobility of the CH$_4$ molecules decreases when increasing the methane loading [4].

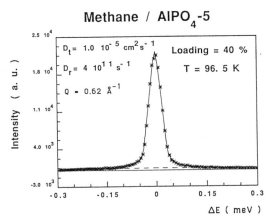

Methane / AlPO$_4$-5

D$_t$ = 1.0 10^{-5} cm^2 s^{-1}

D$_r$ = 4 10^{11} s^{-1}

Q = 0.62 Å^{-1}

Loading = 40 %

T = 96. 5 K

Figure 6

The diffusion coefficient measured at T = 17 K for the hydrogen (HD) molecules is rather small D$_t$ = 0.3 ± 0.1 10^{-5} cm^2 s^{-1}. Measured IQNS spectra in the Q range: 0.62 ≤ Q ≤ 1.45 Å^{-1} are shown on the figure 7. The considered loading is 27 %. Recently we have made a more detailed analysis of the hydrogen / AlPO$_4$-5 system at the Institut Laue Langevin on the IN-5 spectrometer with a better resolution (ΔE$_{resol.}$ = 27 μev). The translational mobility of the hydrogen molecules decreases when the sorbed gas quantities increases [7].

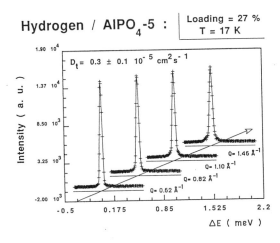

Hydrogen / AlPO$_4$-5 : Loading = 27 % T = 17 K

D$_t$ = 0.3 ± 0.1 10^{-5} cm^2 s^{-1}

Q = 1.45 Å^{-1}
Q = 1.10 Å^{-1}
Q = 0.82 Å^{-1}
Q = 0.62 Å^{-1}

Figure 7

We have to note that both for neopentane and methane sorbed phases on AlPO$_4$-5 the molecular rotational motions are characterized by high frequency values (4 ≤ D$_r$ ≤ 8 10^{11} s^{-1}). Results which are in agreement with the globular morphology of such molecules.

The IQNS spectra measured for hydrogen (HD) , methane (CH_4) and neopentane ($C(CH_3)_4$) sorbed phase on $AlPO_4$-5 for comparable relative loadings and reduced temperatures T / T_c (T_c is the bulk critical temperature of the each considered gas) are represented   figure 8 . Our results show that the largest translational mobility is observed for the neopentane molecules. Such observation is rather surprising . Indeed the neopentane sorbed phase on $AlPO_4$-5 is an illustrative example of single-file system (the $C(CH_3)_4$ molecules cannot pass each other in the $AlPO_4$-5 micropores). In a single-file diffusion system, a displaced molecule is more likely to return to its original position than to proceed further; as a consequence, D_t is expected to be strongly reduced [8,9].

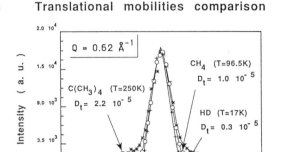

Figure 8

The smallest translational mobility has been measured for the hydrogen molecules. Our results seem to indicate that the diffusion coefficient D_t decreases concomitantly with the size of the molecular sorbed species on $AlPO_4$-5 . The corrugation adsorption energy of the $AlPO_4$-5 inner surface should have a pronounced influence on molecules of small sizes.

5 CONCLUSION

The comparison of the translational mobilities D_t of the considered three sorbed species on $AlPO_4$-5 is surprising in term of single file diffusion behaviour. Indeed in appearance D_t increases when increasing the molecular size . Several factors can influence the translational mobility in microporous materials. Among them, the energy corrugation of the inner surface is important (the energy corrugation $\Delta U_\perp$ is the variation of the potential of interaction between the sorbed molecule and the zeolite when the molecule is displaced on the inner surface). Large $\Delta U_\perp$ induces stabilization effect which restrains the molecular translational mobility . Such a phenomenon has been recently observed for the methane / VPI-5 system [10] .

ACKNOWLEDGMENTS

The authors would like to thank kindly J. Patarin and H. Kessler (Laboratoire des Matériaux Minéraux, Université de Haute - Alsace, Mulhouse, France) for AlPO$_4$-5 zeolite samples supply .

REFERENCES

1. Y. Grillet, P. Llewellyn, N. Tosi-Pellenq and J. Rouquerol, Proceedings of the 4th Int. Conference on " Fundamentals of Adsorption ", Kyoto - Japan (1992) p. 235.
2. N. Tosi-Pellenq, Y. Grillet, J. Rouquerol and P. Llewellyn, Thermochimica Acta 204 (1992) 79.
3. J.P. Coulomb, C. Martin, Y. Grillet and N. Tosi-Pellenq, Proceedings of the 10th Int. Zeolite Conference, Garmisch - Partenkirchen (1994) p. 445.
4. C.Martin, J.P. Coulomb, Y. Grillet and R. Kahn, in " Fundamentals of Adsorption ", M.D. LeVan (ed.), Kluwer Academic Publishers Boston, Massachusetts (1996) p. 587.
5. F. Volino, in " Microscopic Structure and Dynamics of Liquid ", edit. J. Dupuy and A.J. Dianoux, Plenum Press (1978) p. 221 .
6. H. Jobic, M. Bee and A. Renouprez, Surface Sci. 140 (1984) 307.
7. C. Martin, J.P. Coulomb and M. Ferrand (submitted to Europhysics Letters).
8. J. Kärger, Physical Review A 45 (1992) 4173.
9. J. Kärger, M. Petzold, H. Pfeifer, S. Ernst and J. Weitkamp, J. Catal. 136 (1992) 283.
10. J. Kärger, W. Keller, H. Pfeifer, S. Ernst and J. Weitkamp, Microporous Mat. 3 (1995) 401.

CHARACTERIZATION OF THE SOLID TEXTURE OF POROUS PARTICLES BY APPLICATION OF A NON-CATALYTIC GAS-SOLID-REACTION IN A THERMOBALANCE

G. Aichinger and G. Staudinger

Department of Apparatus Design and Particle Technology
Graz University of Technology
A-8010 Graz

1 INTRODUCTION

Shape and size of the non-porous constituents of a porous particle are summarized with the term "solid texture". The question is, how can a texture, such as the graphite texture from **Figure 1**, be characterized?

Methods based on N_2-adsorption or mercury intrusion measurements[1] are useful for the characterization of the void space texture rather than the solid texture. Image analyses of electron microscopic photographs give access to details of a texture. Fractal dimensions[2] do not characterize a solid texture in a manner chemical engineers need.

The texture of solid reactants has been characterized by the grain model[3], a modification of it[4,5], or by random pore models[6,7,8]. In all of these models an idealized texture is presumed: In grain models the solid is considered either plate shaped, cylindrical or spherical; in random pore models the pores are considered cylindrical with solid matter around them.

In this paper a simple method is presented that allows the characterization of solid textures relevant for chemical gas-solid reactions precisely and that is not based on assumed idealized textures. Prerequisites are the conduction of a gas-solid reaction under chemical control, e.g. in a thermobalance, and knowledge of the initial surface area. To ensure chemical control a reaction should be selected whose solid product has a lower molar volume than the solid reactant. The plot of mass against time can be transformed into a function that characterizes the texture of the solid reactant: Relative surface area, $a_{A0\ (s)}$ against front path, s.

This function can be visualized in thought by repeatedly peeling off a layer from the solid surface and measuring the surface area. The procedure is derived in general and demonstrated on known solid textures. Finally, the texture of graphite is evaluated.

2 EXPERIMENTAL

Between 0,5 and 1 mg of pulverized graphite from Sigma Aldrich (99.95% C) was oxidized in air (1 Nl/min) at temperatures from 500 to 750°C, using a thermobalance supplied by DMT (Deutsche Montan Technologie), Germany. A specially designed sample holder - the HITHERM-bag whose details and merits are yet to be published - was used to avoid transport limitations influencing the reaction.

The BET-surface area was gained from N_2-adsorption measurements using an ASAP unit from Micromeritics, USA.

3 MODEL

3.1 Front velocity, k_s

The chemical reaction rate of a gas-solid reaction,

$$a A_{(s)} + b B_{(g)} \rightarrow c C_{(s)} + d D_{(g)}$$
Scheme 1

referred to the solid reactant, A, is

$$\frac{dN_A}{dt} = - k_{g(T)} \cdot S_{A (t)} \cdot f(c_B^0{}_{(T)} - c_B) \qquad [mol/s] \qquad (1)$$

where
dN_A/dt .... is the amount of solid reactant oxidized per unit time $[mol/s]$,
$k_{g(T)}$ .... is the temperature-dependent reaction coefficient,
$S_{A (t)}$ .... is the absolute reaction surface area of the sample, which changes with reaction time $[m^2]$,
$f(c_B^0{}_{(T)} - c_B)$ ... is a function of the distance of the actual conditions from thermodynamic equilibrium related to the gaseous key reaction component, e.g. reactant $B_{(g)}$.

Since the reaction conditions - temperature, pressure and gas concentrations - stay constant during a thermobalance experiment, the term, $k_{g(T)} \cdot f(c_B^0{}_{(T)} - c_B)$, also remains constant over the course of reaction. It can be summarized as, k_s $[m/s]$, the front velocity, the velocity at which the reaction front progresses into the solid reactant.

With m_E, the mass at 100% conversion,

$$m_E = m_0 \cdot \left((1 - w_{inert}) \cdot \frac{c \cdot C}{a \cdot A} + w_{inert} \right) \qquad [mg] \qquad (2)$$

where
m_0 .... is the initial sample weight $[mg]$,
w_{inert} .... is the weight fraction of inert matter in the unreacted sample $[-]$,
c, a .... are the stoichiometric constants of the solid reaction components $[-]$,
C, A .... are the molar weights of the solid reaction components $[g/mol]$,

and with the definition of conversion, $\zeta_{(t)}$,

$$\zeta_{(t)} = \frac{m_0 - m_{(t)}}{m_0 - m_E} \qquad [-] \qquad (3)$$

the front velocity, k_s, can be derived from the reaction start, see equation (1)

$$k_s = \frac{1}{a_{V0}} \cdot \frac{d\zeta}{dt}\Big|_0 \qquad [m/s] \qquad (4)$$

where
a_{V0} .... is the initial specific surface area with respect to volume $[m^2/m^3]$.

3.2 Front path, X

Since the front velocity, k_s, stays constant during a chemically controlled gas-solid reaction, the path covered by the reaction front, s, at any reaction time, t, is

$$s = k_s \cdot t \qquad [m] \qquad (5)$$

3.3 Relative reaction surface area, $a_{A0\,(s)}$

The reaction surface area usually changes as the reaction front progresses into the solid reactant. It continually decreases if the texture is convex shaped, e.g. spherical, or if the textures consists of plates of different thickness. It increases at first if the texture is concave shaped, e.g. the solid is located around cylindrical pores, and later decreases because of pore overlap. It only stays the same if the texture consists of plates of a single thickness.

Whatever the case, the reaction surface area with respect to the initial surface area - the relative reaction surface area, $a_{A0\,(s)}$ - is

$$a_{A0\,(s)} = \frac{d\zeta}{dt}\Big|_t \Big/ \frac{d\zeta}{dt}\Big|_0 \qquad [-] \qquad (6)$$

which is gained by putting equation (4) into equation (1).

The function, $a_{A0\,(s)}$, is a universal function characterizing the solid texture. It is easily determined from the course of a chemically controlled gas-solid reaction.

4 DEMONSTRATION

The use of the relative reaction surface area, $a_{A0\,(s)}$, against front path, s, is demonstrated by
- textures made of equal spheres, long cylinders or flat plates, see **Figure 2**.
- textures made of cubes with rectangular recesses in each edge with different initial pore volumes, see **Figure 3**.

The front path, s, and the development of the reaction surface is also indicated.

A sample of each of these textures with $a_{M0} = 5$ m²/g or $a_{V0} = 10^7$ m²/m³ respectively ($\rho_0 = 2000$ kg/m³), undergoes a gas-solid reaction with a front velocity, $k_s = 3 \cdot 10^{-10}$ m/s.

The conversion-time relationship for grains is[3]

$$\zeta_{(t)} = 1 - \left(1 - \frac{a_{V0}}{F_p} k_s t\right)^{F_p} \qquad [-] \qquad (7)$$

where
F_p.................. is the shape factor: 1 for flat plates, 2 for long cylinders, 3 for spheres.

The record of a thermobalance experiment as it results with textures made of plates, cylinders and spheres of a single thickness is shown in **Figure 4**.

For the texture from **Figure 3**, the conversion-time relationship can be derived as[9]

$$\zeta_{(t)} = 1 - \frac{w^3 - 3\,w \cdot (q + k_s \cdot t)^2 + 2\,(q + k_s \cdot t)^3}{w^3 - 3\,w \cdot q^2 + 2\,q^3} \qquad (8)$$

where
w..................... is the half cube length *[m]*,
q .... is the pore depth into the cube *[m]*, see **Figure 3**,

which both can be determined from the given specific surface area and the pore volume[9].

The record of a thermobalance experiment under chemical control for such textures with different initial pore volumes, $v_{p,M0} = 5, 50$ *and* 250 mm^3/g, is indicated in **Figure 5**. From the comparison of **Figure 4** and **Figure 5**, it can be seen that for a particular surface area the reaction time can vary by one order of magnitude.

Application of equations (4) - (6) yields the relative reaction surface area, $a_{A0\,(s)}$, against front path, *s*. The results for the known textures are illustrated in **Figure 6** and **Figure 7**. This shows the suitability of the function, $a_{A0\,(s)}$, in describing both textures with initial enlargement and textures with initial decrease of the surface area.

5 TEXTURE OF GRAPHITE

Results from graphite oxidation experiments in a thermobalance in synthetic air between 500°C and 750°C are shown in **Figure 8**. The function, $a_{A0\,(s)}$, can be determined for each experiment. As long as the reaction proceeds under chemical control, the curve of $a_{A0\,(s)}$ should stay the same, since it is characteristic for the texture. This does in fact occur, as shown in **Figure 9**, except for a slight deviation in the experiment at 650°C.

The texture of graphite is known to be plate shaped. The model assumption, however, that the graphite consists of plates of a single thickness would be wrong, compare **Figure 6** and **Figure 9**. The decrease of surface area from **Figure 9** originates from a plate thickness distribution, which can also be determined[9].

Acknowledgements

The authors would like to thank the Bundesministerium für Wissenschaft, Verkehr und Kunst, the Austrian Science Foundation (FWF), project number S06803, and the electron microscopy center at the Graz University of Technology for supporting this project.

References

1. Gregg S.J., Sing K.S.W.; 'Adsorption, Surface Area and Porosity', Academic Press, London 1982.
2. Kaye B.H. et al.; 'Strategies for Evaluating Boundary Fractal Dimensions by Computer Aided Image Analysis', *Part. Part. Syst. Charact.*, 1994, **11**, 411
3. Szekely J. et al.; 'Gas-Solid Reactions', Academic Press, London 1976.
4 Efthimiadis E.A., Sotirchos S.V.; 'A Partially Overlapping Grain Model for Gas-Solid Reactions', *Chem. Eng. Sci.*, 1993, **48**, Nr.7, 1201.
5. Lindner B., Simonsson D.; 'Comparison of Structural Models for Gas-Solid Reactions in Porous Solids Undergoing Structural Changes', *Chem. Eng. Sci.*, 1981, **36**, Nr.9, 1519.
6. Petersen E.E.; 'Reaction of Porous Solids', *AIChE J.*, 1957, **3**, 442.
7. Bhatia S.K., Perlmutter D.D., 'A Random Pore Model for Fluid-Solid Reactions: Isothermal, Kinetic Control', *AIChE J.*, 1980, **26**, Nr.3, 379
8. Gavalas G.R.; 'A Random Capillary Model with Application to Char Gasification at Chemically Controlled Rates', *AIChE J.*, 1980, **26**, Nr.4, 577.
9. Aichinger G., "Bestimmung der Feststofftextur aus chemisch kontrollierten Gas-Feststoff-Reaktionen"; Dissertation, Appendix B, Graz University of Technology, Austria, 1996,

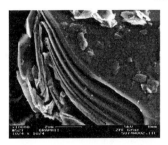

Figure 1 SEM-image depicting the texture of pulverized graphite (Sigma Aldrich, 99,95% C) using a Zeiss DSM 982 Gemini from the electron microscopy center at the Graz University of Technology. Enlargement: 10 000. The graphite has a BET surface area, a_{M0}, of 8.73 m²/g and a true density, ρ_0, of 2230 kg/m³.

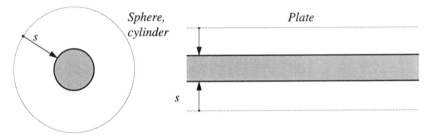

Figure 2 Front path, s, over the course of a gas-solid-reaction progressing into a single spherical particle, a long cylinder and into a single flat plate respectively. The reaction surface area continually declines (sphere, cylinder) or stays the same (single plate).

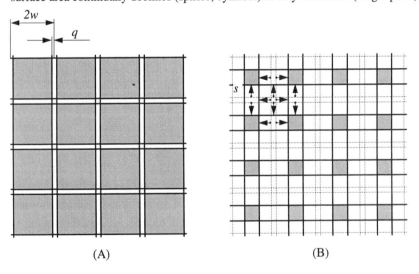

(A)   (B)

Figure 3 Illustration of a porous texture made of cubes of edge length, $2\,w$, in which square pores of half width, q, were cut at the edges (A). The reaction surface area initially increases with the front path, s, and declines again towards the end of the reaction (B).

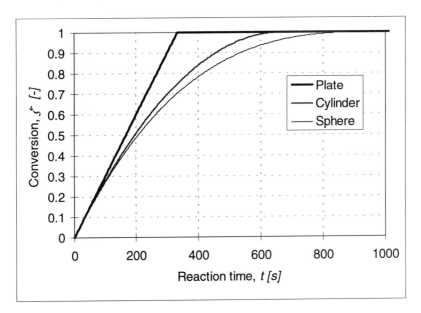

Figure 4 Course of conversion of spheres, cylinders and plates, see Figure 2, each of the same size with a specific surface area of $a_{M0} = 5\ m^2/g$ or $a_{V0} = 10^7\ m^2/m^3$ respectively ($\rho_0 = 2000\ kg/m^3$), undergoing a gas-solid reaction with $k_s = 3 \cdot 10^{-10}\ m/s$.

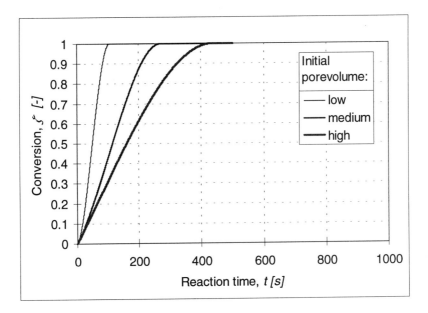

Figure 5 Conversion of the texture of Figure 3 with a specific surface area of $a_{M0} = 5\ m^2/g$ or $a_{V0} = 10^7\ m^2/m^3$ respectively ($\rho_0 = 2000\ kg/m^3$), and with different pore volumes, $v_{p,M0} = 5,\ 50$ and $250\ mm^3/g$, undergoing a gas-solid reaction with $k_s = 3 \cdot 10^{-10}\ m/s$.

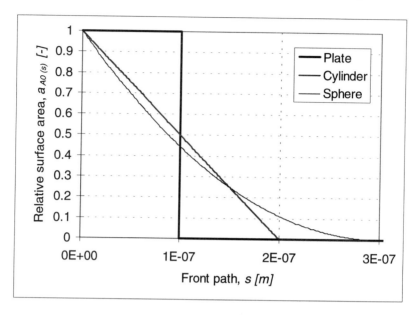

Figure 6   Characterization of a flat plate, a long cylinder and a sphere, see Figure 2 and Figure 4, in terms of relative surface area, $a_{A0\,(s)}$, against front path, s.

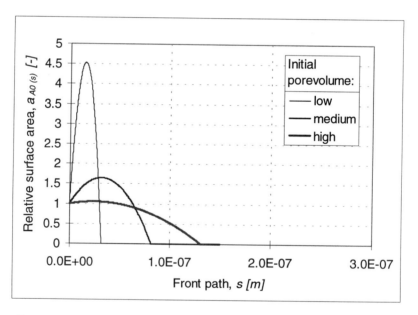

Figure 7   Characterization of a texture with surface area enlargement, see Figure 3 and Figure 5, with different pore volumes in terms of relative surface area, $a_{A0\,(s)}$, against front path, s.

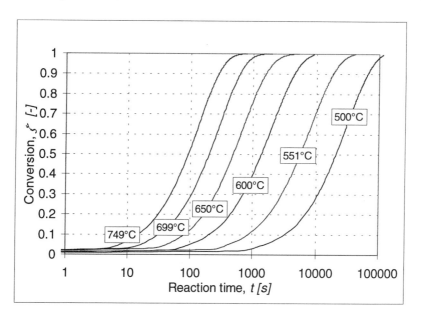

Figure 8 Oxidation of graphite, see Figure 1, in synthetic air between 500°C and 750°C in a thermobalance.

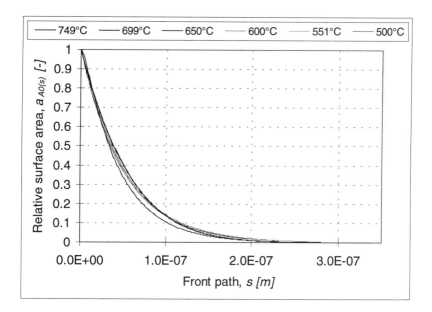

Figure 9 The texture of graphite in terms of relative surface area, $a_{A0\,(s)}$, against front path, s, see Figure 1 and Figure 8. The decrease of surface area originates from a plate size distribution.

RETENTION MECHANISMS OF PHENOL ON ACTIVATED CARBON IN A CEMENT ENVIRONMENT STUDIED BY HIGH RESOLUTION ARGON ADSORPTION AND CONTROLLED TRANSFORMATION RATE THERMAL ANALYSIS.

L. J. Michot, F. Didier, F. Villiéras, J. M. Cases, Y. Grillet[1], A. Bouchelaghem[2]

Laboratoire Environnement et Minéralurgie. INPL-ENSG URA 235 du CNRS BP 40, 54501 VANDOEUVRE CEDEX, FRANCE.
[1] Centre de Thermochimie et de Microcalorimétrie, 26, rue du 141ème R.I.A. 13003 MARSEILLE, FRANCE.
[2] INERTEC S.A. 6, rue de Watford, 92000 NANTERRE, FRANCE.

ABSTRACT

In order to improve the efficiency of artificial geochemical barriers used for the containment of landfills, a material based on a mixture of cement, smectite clays and activated carbon was designed. The retention of phenol by this material was studied. Activated carbon appears to be the only adsorbing component of the system. The performance of the material can then be enhanced by studying the adsorption mechanisms of various pollutants on activated carbons in a cement-like environment (high pH and ionic strength). For this purpose, an original methodology based on Controlled Transformation Rate Thermal Analysis (CTRTA) and high resolution argon adsorption at 77K treated by the Derivative Isotherm Summation (DIS) was developed. The combination of these two techniques appears as a very powerful tool as it allows to identify the most energetic adsorption sites for phenol molecules on activated carbons. These sites are micropores which are filled by argon between $Ln(P/Ps) = -12$ to -7. Smaller and larger pores appear less energetic. The proposed methodology could certainly be applied to numerous adsorption systems at the solid/liquid interface.

1 INTRODUCTION

In many polluted sites such as ancient landfills or ancient industrial sites, it is very often necessary, before decontamination, to first confine the pollution to a restricted area to prevent further pollution of the surrounding soil system and water table. Confinement walls made of cement-clay composites can be built around the polluted area as they exhibit a low permeability and retention capacities towards inorganic contaminants such as heavy metals. However, in many actual situations, the contamination in organic compounds is important and classical clay-cement walls are rather inefficient for such situations. For this reason, Soletanche SA, a french civil engineering company, developped clay-cement composites in which activated carbon is included as an adsorbing additive[1,2]. The retention potential of activated carbon towards organic molecules is well documented as for instance, vapor activated carbons have been used in water purification units in France since the 1970's[3].

The aim of the present study is first to assess the actual adsorbing potential of such composites towards a representant of one of the most frequently encountered classes of pollutants, phenol. The retention mechanisms of phenol can then be elucidated for further enhancement of the performances of the geocomposites. It will be shown that some techniques developed at the solid/gas interface, such as high resolution argon adsorption and controlled transformation rate thermal analysis, can be used for an indirect study of the adsorption mechanisms at the solid/liquid interface.

2 MATERIALS AND METHODS

2.1 Materials

The starting geocomposites were provided by Soletanche SA. Their loading in activated carbon varied between 3.15 and 10.5 weight %.

The powdered activated carbon (PAC) used in this study (and added in the cement-clay composite) was provided by CECA SA. Its commercial name is Acticarbone TK. It is derived from pine bark and was vapor activated.

The phenol used for adsorption studies was supplied by Merck. Its purity is > 99.5%.

2.2 Experimental

2.2.1 Phenol adsorption isotherms. Phenol adsorption isotherms were carried out in batch in glass tubes. 500 mgs of adsorbent were first preequilibrated with 45 mls of water. 45 mls of a known phenol concentration were then added. The glass tubes were agitated at 30°C during 24 hours. After solid/liquid separation the solution was analyzed by UV spectrometry (Shimadzu UV 2100) and the phenol concentration was then determined from the intensity of the absorption band at 254 nm. The adsorbed quantity could then be calculated. After adsorption, the solid was air dried and stored for further experiments.

In one experiment, in order to match the physicochemical conditions encountered in a cement environment, the activated carbon was first pre-equilibrated with 45 mls of water, which was previously equilibrated with cement. This water was at high pH, around 11, and was rich in ions mainly calcium.

2.2.2 Controlled Transformation Rate Thermal Analysis (CTRTA). CTRTA analyses were carried out on a lab-built setup. In this procedure[4-6] the rate of sample outgassing is kept constant over the entire temperature range by means of an appropriate heating loop which results in an a-priori unknown temperature program. The flow of gas evolved from the sample submitted to a dynamic vacuum is used to control the heating of the furnace and is directly analyzed by mass spectrometry (Balzers QMG 420 C mass spectrometer). The experimental conditions were : a sample mass of ≈ 0.11 g and a residual pressure of 1.2 Pa. The mass spectra corresponding to pure phenol were obtained by placing ≈ 0.2 g of phenol in the set-up using the same residual pressure as mentioned above.

2.2.3. High Resolution Argon adsorption. Low-pressure argon adsorption isotherms were recorded on a lab-built automatic quasi-equilibrium volumetric set-up[7,8]. In this method a slow, constant and continuous flow of adsorbate is introduced into the adsorption cell. From the recording of quasi-equilibrium pressures vs time, the adsorption isotherms are derived. The experimental conditions were a sample mass of ≈ 0.1g, outgassing at 0.001 Pa at room temperature or 350°C. The data were then treated using the improved Derivative Isotherm Summation (DIS) procedure designed by Villiéras et al.[8,9] to examine the surface heterogeneity of the samples. Due to the large number of experimental data points acquired by the quasi-equilibrium procedure, the experimental derivative of the adsorbed quantity as a function of the logarithm of relative pressure can be calculated accurately. The experimental derivative adsorption isotherm is simulated by the sum of local theoretical derivative adsorption isotherms on heterogeneous surfaces with random heterogeneity.

$$\theta_t = \sum_i X_i \theta_{it} = X_i \int_\Omega \theta_i(\varepsilon) \cdot \chi_i(\varepsilon) \cdot d\varepsilon \qquad (1)$$

The theoretical isotherms used[9] are obtained by first approximating the energy distribution by the condensation approximation (CA). For describing adsorption into micropores, the generalization of the Dubinin-Asthakov isotherm (DA) was proposed:

$$\theta_{it}(P/P_0) = e^{-\left[\frac{kT}{E_i}\ln\left(\frac{P_i^0}{P}\right)\right]^{r_i}} \tag{2}$$

where E_i is the variance of $\chi_i(e)$ and r_i a parameter governing the shape of the distribution function. It is a gaussian like function widened on the low energy side for $r_i < 3$ and widened on the high energy side for $r_i > 3$. Equation (2) can be extended to take into account the effect of lateral interactions by replacing $\frac{kT}{E}\ln\left(\frac{p^0}{p}\right)$ by $\Delta = \frac{kT}{E}\ln\left(\frac{p^0}{p}\right) - \frac{\omega\theta}{E}$.

A multilayer extension of this equation (MDA) has also been proposed in order to simulate peaks corresponding to multilayer adsorption, i.e. on external surfaces[9]. In this case it allows to extract the contribution of the first adsorbed layer. The first and second derivatives of all the used equations can be calculated either analytically or numerically.

In the DIS method, identification of the different parameters is obtained from the coordinates of the maximum and from the width of the derivative. The mathematical relations between E_i, r_i and P_i^0 can be easily derived from the expression of the first and second derivatives when the second derivative equals zero. Then, for given values of r and p^0, the amount adsorbed on the domain i, Xi, can be obtained from the comparison between the height of the experimental and calculated maxima. All the parameters for each domain can then be obtained. Further combination of the obtained results yields the condensation curves which can be corrected to energy distribution for each adsorption domain by applying the Rudzinski-Jagiello correction[10].

$$\theta_{1t} = \int_{\Omega} \theta_1(\varepsilon)\chi(\varepsilon)d\varepsilon = -X(\varepsilon_c) - \frac{\Pi^2}{6}(kT)^2\left(\frac{\partial\chi}{\partial\varepsilon}\right)_{\varepsilon=\varepsilon_c}$$

3 RESULTS AND DISCUSSION

3.1 Phenol adsorption

The adsorption isotherms of phenol on the various geocomposites are plotted in Figure 1A. The sample with no PAC does not exhibit any significant retention capacity whereas the adsorbed quantity increases with PAC content in the other composites suggesting that activated carbon is the sole adsorbent of the system. Figure 1B presents the same curves normalized by the weight % of PAC, together with those obtained for pure PAC in water and in water in equilibrium with cement.

It appears that for loadings $\geq 5.76\%$, the adsorption isotherms of geocomposites are superimposed on that of pure PAC in a cement environment. The retention mechanisms of phenol on the geocomposites can then be studied by simply working on the adsorption isotherm of phenol on PAC in water in equilibrium with cement. It must be noted that the same behaviour was observed for other non volatile or volatile organic compounds[11].

The shape of the adsorption isotherm of phenol on PAC reveals an important heterogeneity. Attempts to determine the state of the adsorbed molecules by infrared spectroscopy were unsuccessful as no signal corresponding to phenol molecules could be observed. For this reason an original methodology based on techniques developed for studying the gas/solid interface was implemented.

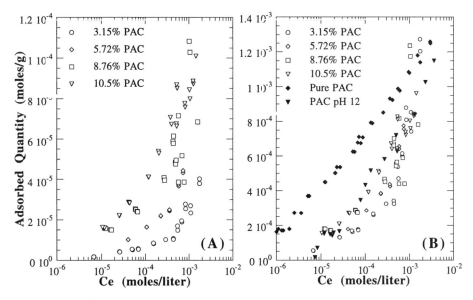

Figure 1 *Adsorption isotherms of phenol on various geocomposites and PAC. (A) Raw Data. (B) : Adsorbed quantities normalized by the weight % of PAC.*

3.2 Controlled Transformation Rate Thermal Analysis

Bare activated carbon and three samples with increasing adsorbed quantity of phenol (1.56×10^{-4}, 5.81×10^{-4} and 1.15×10^{-3} Mole.g^{-1}) were analyzed by CTRTA coupled with mass spectrometry. TPD-MS experiments have recently been used for following the desorption of organic molecules from activated carbons[12,13]. It reveals physisorbed and chemisorbed species. Due to the use of quasi-equilibrium conditions, CTRTA is even more suitable for such type of study.

The evolution of temperature with time for the four samples reveals that, in the case of bare PAC, the temperature ramp is linear. The pressure constraint was then never reached which indicates that only very few species were evolved from the surface. In the case of the sample loaded with 1.56×10^{-4} Mole/g of phenol, the curve deviates from linearity around 140°C. For the two other samples, the linearity is lost around 90°C.

By comparing the complete mass spectra obtained on these samples with the pattern obtained after using pure phenol it was shown that phenol is not cracked upon outgassing and that the major mass m/z = 94 can be used to describe accurately and quantitatively the phenol evolved from the surface[14].

Figure 2 presents the evolution of the derivative of the integrated intensity of the mass 94 as a function of temperature. The first graph (Figure 2A) corresponds to the actual values whereas in the second one (Figure 2B), the intensities were normalized with regard to the total quantity of phenol evolved from the surface. It appears that phenol is desorbed from the surface up to temperatures of around 600°C. For decreasing adsorbed amounts, the curves are shifted towards higher temperatures i.e. towards higher adsorption energies. Therefore, in the case of phenol adsorption on activated carbons, the adsorption energy distribution observed by CTRTA is principally similar to the energetic distribution at the solid/liquid interface. At least three desorption peaks are observed on these curves which suggests that more than three adsorption domains could be defined for phenol molecules.

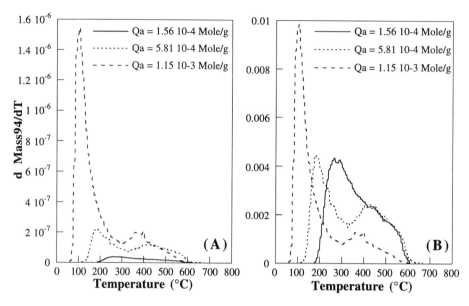

Figure 2   *Evolution with temperature of the derivative of the integrated intensity of the mass 94 recorded during controlled transformation rate thermal analysis of bare PAC and of PAC loaded with increasing phenol amounts . A. Actual Values. B: Normalized Values.*

3.3 High Resolution Argon Adsorption

The nature of phenol adsorption sites was further investigated by analyzing the four samples previously described by high resolution argon adsorption after outgassing at room temperature and 350°C. As shown by CTRTA experiments, upon outgassing at ambient temperature, no phenol is desorbed from the surface. The argon adsorption isotherm and their decomposition using the DIS method on the samples loaded with phenol can then be compared with the argon adsorption isotherms on bare activated carbon outgassed in the same conditions to observe changes in adsorption energy distributions. After outgassing at 350°C, CTRTA reveals that some phenol is still present in the activated carbon. The use of argon for such a study is appropriate as it will be mainly sensitive to the geometric distribution of adsorption sites as no specific interaction between argon molecules and surface chemical groups should be expected.

Table 1 presents the parameters obtained from the BET treatment of the argon adsorption isotherms obtained on the bare activated carbons and on three different points of the phenol adsorption curve after outgassing at 25°C and 350°C. It reveals that phenol adsorbed at the surface of activated carbon blocks some adsorption sites for argon as the amount of adsorbed argon decreases with increasing phenol adsorbed quantity. The BET C constants decrease concomitantly, which shows a less and less microporous character of the samples. After outgassing at 350°C, some sites are still blocked as the BET parameters exhibit the same trends as after outgassing at 25°C.

Table 1   *Parameters Derived from the BET Treatment of the Argon Adsorption Isotherms on PAC and PAC Loaded with Increasing Amounts of Phenol.*

Phenol Adsorbed Mole/g	Outgassing 25°C			Outgassing 350°C		
	C_{BET}	V_m $cm^3.g^{-1}$	S_s $m2.g^{-1}$	C_{BET}	V_m $cm^3.g^{-1}$	S_s $m2.g^{-1}$
0	689	114.5	425	699	150.4	558
1.56 10^{-4}	397	89.2	331	337	134.7	500
5.81 10^{-4}	172	66.8	248	227	95.1	353
1.15 10^{-3}	89	35.6	132	156	65.2	242

Figure 3 presents the derivative adsorption isotherms and their decomposition using the improved DIS method on the bare activated carbon and on the three samples loaded with different amounts of phenol after outgassing at 25°C.

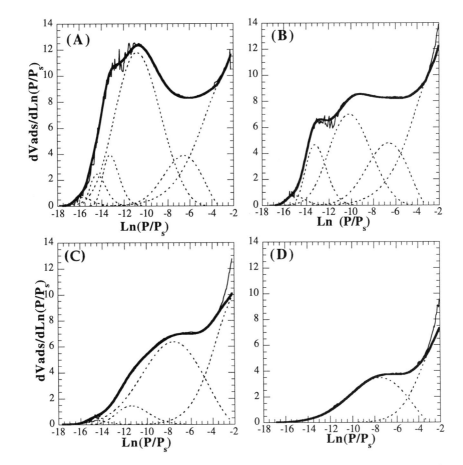

Figure 3   *Derivative Argon Adsorption Isotherms and Decomposition using the Improved DIS method on PAC Loaded with (A) 0 (B) 1.56 10^{-4} Moles/g (C) 5.81 10^{-4} Moles/g (D) 1.15 10^{-3} Moles/g of phenol.—— Experimental Curve.* —— *Simulated Curve. Outgassing 25°C.*

Table 2 *Argon Adsorption at 77K on Bare PAC and on PAC Loaded with Increasing Phenol Amounts. Set of Parameters Obtained from the DIS Method.*

Phenol adsorbed Mole.g-1	Domain	Ln(P/P$_s$)	Model	r	Ln (P^0/P$_s$)	ω/kT	E/kT	Vm cm^3.g^{-1}
	1	-15.7	DA	8	-10	0	5.8	1.3
	2	-14.4	DA	7	-10	0	4.5	4.3
0	3	-13.2	DA	6.5	-9	0	4.4	7.1
	4	-10.8	DA	3.8	-4	0	7.4	59.6
	5	-6.5	DA	2.4	-3.3	0	4	16.0
	6	-2.98	MDA	1.8	0	0.9	4.3	35.8
	1	-15	DA	7	-10	0	5.2	1.8
	2	-13.2	DA	5.5	-8.7	0	4.7	12.8
1.56 10-4	3	-10.2	DA	3.8	-4	0	6.7	38.9
	4	-6.6	DA	2.4	-3.2	0	4.3	25.3
	5	-2.75	MDA	1.8	0	0.9	4.1	39.9
	1	-14.5	DA	8	-8	0	6.6	0.7
5.81 10-4	2	-11.4	DA	5	-5	0	6.7	5.1
	3	-7.6	DA	2.6	1-.8	0	6.9	42.4
	4	-2.5	MDA	1.9	0	0.7	3.6	24.2
	1	-7.5	DA	2.3	-3	0	5.8	21.0
1.15 10-3	2	-2.5	MDA	2	0	0.5	3.4	16.3

Table 2 presents the parameters of the different local domains used to model the whole adsorption isotherm. In the case of bare PAC, the isotherm was decomposed in 6 domains. The five high energy domains, modelled using Dubinin-Astakhov isotherms likely correspond to various micropores in the sample whereas the sixth domain at lower energy takes into account the formation of multilayers and can then be assigned to the external and mesoporous surface of the carbon.

The adsorption of increasing amounts of phenol blocks the adsorption sites according to their energy. The low energy domain corresponding to the adsorption of argon on the external surfaces is affected only for the highest phenol loading. The domains corresponding to the adsorption of argon in the microporous structure disappear upon phenol adsorption. The sample with the highest adsorbed amount exhibit almost no micropores. This shows, as already suggested by CTRTA experiments, that the texture of the activated carbon controls the adsorption of phenol. This behaviour was also suggested by calorimetric experiments[15] which showed that the surface chemical groups played only a minimal role in the adsorption of phenol at the solid/liquid interface.

Using the decomposition presented in Table 2, it is possible to go further into the assignment of phenol adsorption sites. Indeed, the sites that are affected by phenol adsorption of phenol at low equilibrium concentration (adsorbed amount, 1.56 10-4 Mole.g-1) are the two sites located around Ln(P/P$_s$) = -10 and -6.5. The decrease in the amount of gaseous argon adsorbed in these two sites is equal to 11.4 cm^3.g^{-1} which corresponds to a liquid volume of 0.0142 cm^3. Taking into account a phenol density of 1.05[16], 1.56 10-4 Mole.g-1 correspond to a volume of 0.0140 cm^3. For higher phenol equilibrium concentration (adsorbed amount, 5.81 10-4 Mole.g-1) all the high energy sites for argon adsorption are modified. The loss in argon adsorption corresponds to a volume of 0.052 cm^3 whereas the adsorbed amount of phenol corresponds to a volume of 0.048 cm^3. For the highest phenol equilibrium concentration (adsorbed amount, 1.15 10-3 Mole.g-1), all the sites including the domain where multilayer formation occurs are affected. The total loss in argon corresponds to a volume of 0.108 cm^3 whereas the amount of phenol adsorbed is equivalent to a volume of 0.103 cm^3. The adsorption of phenol on activated carbon seems then to occur first in micropores which fill for P/Ps between -12 and -7. It then fills all the other micropores. Finally it adsorbs either on the external surface or in small mesopores which cannot be distinguished using the improved DIS procedure. Results obtained after outgassing at 350°C confirm this assignement[14].

4 CONCLUSIONS

The combination of controlled transformation rate thermal analysis and high resolution argon adsorption appears as a powerful tool for studying indirectly the adsorption of organic molecules at the solid-liquid interface. This is particularly useful in the case of the adsorption on activated carbon which, due to their complex nature and porous structure, are difficult to study by classical spectroscopic techniques.

The adsorption of phenol on activated carbon is mainly controlled by the porous structure of the adsorbent. CTRTA reveals that phenol is mainly physisorbed in the structure as no cracking is observed. The energy distributions obtained by argon adsorption allow to define how this phenol physisorption occurs. Micropores filled between Ln (P/Ps) = -10 and -7 are the most energetic sites.

The proposed experimental approach should be developped on other systems using different adsorbent-liquid adsorbates pairs. In particular, using various activated carbons with different micropore size distributions, should allow to confirm the adsorption mechanism suggested by the preliminary results obtained in the present paper.

Acknowledgements

This research was supported by the Program Saut Technologique "Géomatériaux et Nouveaux Adsorbants" (n° 91 F 0131) of the French MESR.

References

1. A. Bouchelaghem, D. Gouvenot, G. Raillard in Proceedings of the Second International Congress on Environmental Geotechnics Osaka, Japan.1996 *In press*
2. D. Gouvenot, A. Bouchelaghem in *"Geology and Confinement of Toxic Wastes "* Arnould, M, Barrès, M., Côme, B. Eds, Balkema, Rotterdam, 1993. Vol 1. pp 2807.
3. Degrémont, *Memento Technique de l'Eau,* 8 ed.; Technique et Doc.:Paris, 1978.
4. J. Rouquerol, *J. Thermal Anal.,* 1970, **2**, 123.
5. J. Rouquerol, *Thermochim. Acta.,* 1989, **144**, 209.
6. J. Rouquerol, S. Bordère, F. Rouquerol, *Thermochim. Acta.,* 1992, 203, 193.
7. L. Michot, M. François, J.M. Cases, *Langmuir,* 1990, **6**, 677.
8. F. Villiéras, J.M. Cases, M. François, L.J. Michot, F. Thomas, *Langmuir,* 1992, **8**, 1789
9. F. Villiéras, L.J. Michot, J.M. Cases, M. François, W. Rudzinski, *Langmuir,* 1996, In Press.
10 W. Rudzinski and D.H. Everett (eds.) 'The Adsorption of Gases on Heterogeneous Surfaces' Academic Press, 1992.
11. F. Hui, P. Sassiat, Z. Sahraoui, R. Rosset, *Analusis* 1995, **23**, 268.
12. J. Rivera-Utrilla, M.A. Ferro-Garcia, C. Moreno-Castilla, I. Bautista-Toledo, J.P. Joly,*J. Chem. Soc. Faraday Trans.* 1995, **91 (18)**, 3213.
13. M.A. Ferro-Garcia, J.P. Joly, J. Rivera-Utrilla, C. Moreno-Castilla, *Langmuir,* 1995, **11**, 2648.
14. L.J Michot, F. Didier, F. Villiéras, J.M. Cases, 1996*Submitted to Polish J. Chem.*
15. A.J. Groszek, S. Partyka, In "Proceedings of the Second ISSHAC Symposium" A. Brunovska, W. Rudzinski, and B. Wojciechowski, Eds 1995, 206208.
16. Anonymous. 'Handbook of Chemistry and Physics' 74th edition Lide D.R Ed. CRC Press 1994.

A MODEL OF MERCURY INTRUSION IN AN ELLIPTICAL OR RECTANGULAR PARALLEL CRACK AND ITS INFLUENCE ON POROSITY HISTOGRAMS OF GRANITE

P. Couchot, C. Dubois, E. Boeglin and A. Chambaudet

Laboratoire de Microanalyses Nucléaires - E.A. 473
Université de Franche-Comté
25030 Besançon Cedex - France

1 INTRODUCTION

In mercury porosimetry, porosity histograms are often plotted according to diameter classes, calculated with the Washburn's equation relating to the Circular cross-Section pore model[1] (C.S. model). But certain materials, such as granite, have a network of inter-granular and intra-granular microcracks much smaller than their lengths. In order to analyse the intrusion of mercury, under P pressure, in a granite microcrack network, we propose an intrusion model in a parallel microcrack with a Rectangular cross-Section (R.S. model). For purposes of comparison, these analyses are also carried out with the help of the C.S. model[1], as well as that of Elliptical cross-Section pores[2] (E.S. model).

2 DEVELOPMENT OF THE R.S. MODEL - COMPARISON TO THE E.S. MODEL

As in the C.S. and E.S. models, the R.S. model is founded on Dubinin's equation[3] which expresses the work supplied to mercury to penetrate a pore volume dV, under pressure P:

$$P \, dV = - \gamma \cos\theta \, dA \quad (1)$$

γ : interfacial tension between mercury and the material $\approx$ surface tension of mercury

θ : angle of contact between mercury and the material

dA : surface of the pore covered by mercury

For a random cross-section pore (Fig. 1a), (1) becomes (2):

$$P \, S = - \gamma \, C \cos\theta \quad (2)$$

S : area of the cross-section

C : perimeter of the cross-section

For a R.S. crack with a width of 2l and a length of 2L (Fig. 1b), (2) becomes (3):

$$P \, l \, L = - \gamma \, (l + L) \cos\theta \quad (3)$$

When the shape factor $\rho = L/l$ is introduced, (3) becomes (4):

$$P1 = -\gamma \frac{(\rho + 1)}{\rho} \cos\theta = -2\gamma \, K_r \cos\theta \quad (4)$$

$K_r = (\rho + 1)/2\rho$ = constant of rectangular modelisation.

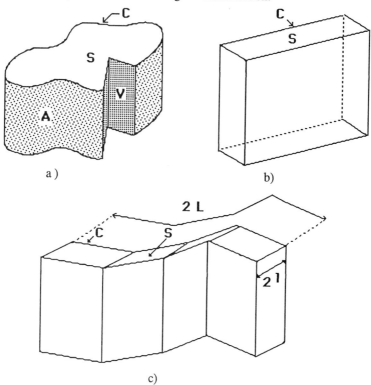

Figures 1 *Model cracks (a: random cross-section; b: rectangular cross-section; c: zig-zag-like cross-section with a constant width)*

In granites, the porosity is essentially made up of a crack network. Each crack exhibits changes in direction, while its width varies little[4]. A crack is represented in figure 1c., and the equation (4) applies to this crack morphology.

Equation (4) applies also to the E.S. et C.S. models given that the specific morphological modelisation constant is introduced as follows:

E.S. Model $K_e = [(\rho^2 + 1)/2\rho^2]^{1/2}$ with: $\rho = L/l$ (L : 1/2 big ellipsis axis)

C.S. Model $K_c = 1$ ($\rho = 1$) l = (radius of the cylindrical pore)

The variation of the modelisation constants, as a function of the shape factor ρ, is plotted in figure 2.

The limit values of the modelisation constant are obtained for the following values of ρ (within a range of 1% greater than the limit value).

R.S Model $\rho \geq 100$ $K_{r(lim.)} = 0.500$

E.S Model $\rho \geq 7$ $K_{e(lim.)} = 0.707$

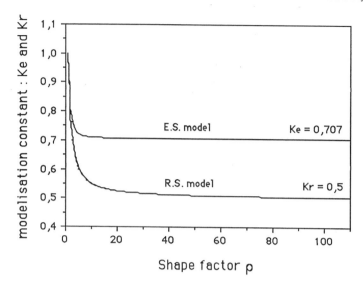

Figure 2 *Variation of the K_e and K_r modelisation constants, as a fonction of the shape factor ρ*

Beyond these $\rho_{lim.}$ values, equation (4) becomes:

$$\text{R.S. model:} \quad 1 = -\frac{\gamma}{P}\cos\theta \quad (5)$$

$$\text{E.S. model:} \quad 1 = -\frac{\gamma\sqrt{2}}{P}\cos\theta \quad (5')$$

Consequently, for sufficiently large shape factors, the mercury porosimetry only takes into account the crack's small dimension 2l. Thus, the role of the large dimension 2L becomes neglected.

3 UTILIZATION OF THE MODELS IN THE STUDY OF A GRANITE

For granite, the objective is to obtain histograms of the cracks' volumes as a function of the small dimension 2l, whatever their large dimension 2L happens to be. To achieve this objective, beforehand, the distribution of the porosity of the granite's crack system, as a function of the shape factor ρ, must be indirectly determined using image analysis.

A study was conducted on granite from Auriat, France[5] using thin sections of which cracks were impregnated with an orasol-blue epoxy resin. The S surface and the C perimeter of 7,000 cracks were recorded semi-automatically by optical microscope, with a resolution of 0.5 μm.

The shape factor ρ is expressed as a function of S and C, in the following way:

R.S. model

$$\rho = \frac{16\,S}{\left(C - \sqrt{C^2 - S}\right)^2} \quad (6)$$

E.S. model

$$\rho = \frac{4\Pi S}{C^2 - \sqrt{C^4 - 16\Pi^2 S^2}} \quad (6')$$

The distributions are different: according to the R.S. model, 41% of the porosity corresponds to cracks for which the shape factor is greater than or equal to 100 ($K_{r(lim.)} = 0.500$). According to the E.S. model, 90% of the porosity corresponds to the cracks for which the shape factor is greater than or equal to 7 ($K_{e(lim.)} = 0.707$).

For each modelisation, the 2D fissural space is divided into N classes. Each class is characterized by a value of the shape factor ρ_i, an average value Ki of the modelisation constant and a porosity rate T_i.

For the N classes: $\sum_{i=1}^{N} T_i = 1$

The volume porosity $\Phi(P)$ of granite is equal to:

$$\phi(P) = \frac{\text{Impregnated porous volume under P pressure}}{\text{Volume of the sample}} \quad (8)$$

The volume porosity was measured using a mercury porosimeter MICROMERITICS 9310 under a pressure P varying from 7 to 2000 bars[6]. The cracks penetrated by mercury, under a pressure P, have small dimensions 21, calculated according to the C.S. model, which vary from 1 µm to 6 nm.

The domains of study of the porosity by image analysis and by mercury porosimetry have a very small overlap. But the distribution of the porosity percentages, as a function of the shape factor's classes of the entire granite crack network, is considered identical to the one recorded by image analysis.

The experimental curve $\Phi(P)$ is numerized by a polynomial regression of degree 6:

$$\Phi(P) = \sum_{j=0}^{j=6} a_j \, P^j \quad (8)$$

Within a class of shape factors and for each value of the small dimension, pressure P of penetration of the mercury is given by equation (9), obtained from equation (4):

$$P = \frac{-2\gamma \cos\theta}{l} K_i \quad (9)$$

The porosity $\Phi_i(P)$, corresponding to a class characterized by ρ_i, K_i and T_i, is equal to:

$$\Phi_i(P) = T_i \, \Phi(P) \quad (10)$$

Combining the equations (8), (9) and (10) for each class of shape factors ρ_i, equation (11) is obtained, which yields the variation of porosity $\Phi_i(l)$, as a function of l:

$$\Phi_i(l) = T_i \sum_{j=0}^{j=6} a_j \left(\frac{-2\gamma\cos\theta}{l} K_i \right)^j \quad (11)$$

The entire group of N classes of shape factors leads to equation (12):

$$\phi(l) = \sum_{i=1}^{N} T_i \sum_{j=0}^{j=6} a_j \left(\frac{-2\gamma\cos\theta}{l} K_i \right)^j \quad (12)$$

The porosity measured by image analysis corresponds to the ratio of the crack surfaces to the surface of the optical microscope's observation field. This surface porosity is of course not equal to the volume porosity, but the ratio between these two variables may be considered as constant. Under this assumption, the relative porosity percentages of each class of cracks are not affected. The classes are defined by dividing the range of the shape factors between 1 and ρ_{max}. This subdivision is performed in such a way that inside each class, the relative variation of the modelisation constant must be within 1%.

The results are represented in figure 3 for the R.S. model, and in figure 4 for the E.S. model.

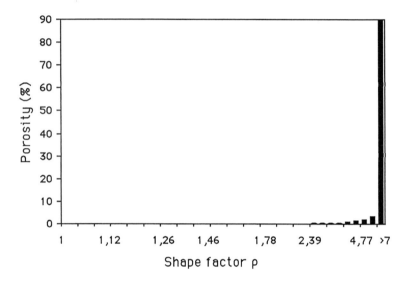

Figure 3 *Distribution of porosity as a function of the shape factor ρ (R.S. Model)*

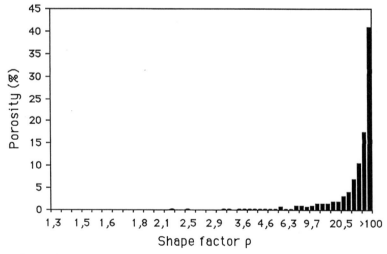

Figure 4 *Distribution of porosity as a function of the shape factor ρ (E.S. Model)*

Thus, it is possible to mathematically reconstruct a porosity curve as a function of 1, using the numerized curve $\Phi(P) = f(P)$ within each class of shape factors, the boundaries of the variation domain of l (named l_i in the "i"th class) are determined by the minimal pressure P_{min} (1 bar) and the maximal pressure $P_{max.}$ (2000 bars). According to equation (9):

$$l_{i(min.)} = \frac{-2\gamma \cos \theta}{P_{max.}} K_i \quad (13)$$

$$l_{i(max.)} = \frac{-2\gamma \cos \theta}{P_{min.}} K_i \quad (13')$$

For the group of N classes and for each modelisation, the boundaries of the variation domain of 1 are determined by equation (14):

$$\frac{-2\gamma \cos \theta}{P_{max.}} K_{min.} \leq 1 \leq \frac{-2\gamma \cos \theta}{P_{min.}} K_{max.} \quad (14)$$

4 RESULTS AND DISCUSSION

As an example, figure 5 gives the percentage of cumulated porosity as a function of l (μm) for a granite sample from Auriat, France, extracted from 984.8m beneath the earth's surface.

The corresponding curve, using the C.S. model, was obtained by the polynomial regression of degree 6 of the experimental values obtained by mercury porosimetry. The corresponding curves, using the R.S. and E.S. models, were calculated by applying equation (12) to each one.

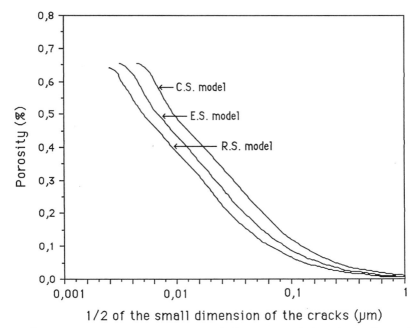

Figure 5 *Cumulated porosity as a function of l (Granite sample from Auriat, France)*

    In figure 6, the porosity distribution as a function of l (μm) is represented for each model.

    For the three models, the percentages of cumulated porosity and the distribution histograms of the porosity as a function of l, are similar. Whichever model is considered, the histograms in figure 6 present two maxima of which one is very pronounced. This maximum corresponds to the following values of 1 (nm): 6.6 (C.S. model); 4.7 (E.S. model); 3.5 (R.S. model). The maxima according to the E.S and R.S. models, have shifted near the smaller values of l, compared to the maximum of the C.S. model. The shift factors are: 4.7/6.6 = 0.71 et 3.5/8.4 = 0.53, respectively, being 28% and 45% of relative difference.

    Generally speaking, the corresponding curves to the E.S. and R.S. models are grosso modo shifted according to a factor greater than $K_{e(lim.)} = 0.707$ and $K_{r(lim.)} = 0.500$, respectively.

    These results are logical, taking into account the distribution histograms of the porosity percentages of this granite measured by image analysis (figure 3 and figure 4). Reiterating that 90% of porosity corresponds to the cracks of which the shape factor is ≥ 7 (K_e (lim.) = 0.707), for the E.S. model, whereas for the R.S. model, only 41% of porosity corresponds to the cracks in which the shape factor is ≥ 100 (K_r (lim.) = 0.500). For both models, if all the cracks had their shape factors greater than or equal to these two limits, the shifts would respectively be according to a factor 0.707 and 0.500, corresponding to the relative maximum shifts of 29% and 50%.

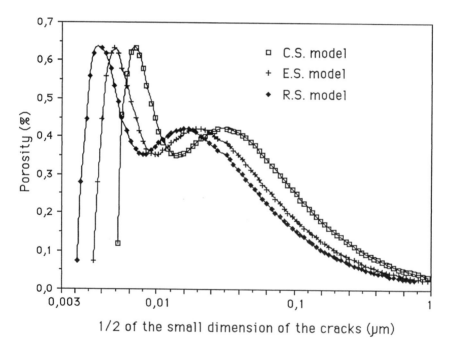

Figure 6   *Distribution histograms of porosity as a function of l*

5 CONCLUSION

In the study of crack networks (granite), by mercury porosimetry, the parameter to take into account in the quantification of penetrated volumes by mercury is the small dimension 2l of the model cracks with elliptical (E.S. model) or rectangular (R.S. model) cross-sections.

The use of these models for the analysis of the volumes penetrated by mercury and of the distribution histograms of the porosity volumes, as a function of l, leads to curves similar to those obtained from Washburn's cylindrical model (C.S. model). They are simply shifted, in relation to the cylindrical model, to the weaker values of l, using a factor, at the best at the lowest value: 0.707 (E.S. model) and 0.500 (R.S. model). Thus, although the rectangular model is the most representative of the structure of a crack in a granite, Washburn's model, applied to the a crack network, gives a first approximation of probable results, assuming the small dimension 2l of each crack as the diameter of a cylindrical pore.

However, whichever model used, the higher the constrictivity of the crack network, the more the distribution of the volume of cracks as a function of their width and the distribution of the penetrated volumes by mercury are different. Note that, the modelisation presented in this paper does not pretend to resolve this problem.

References

1. E. W. Washburn, *Proc. of the Nat. Acad. of Sciences*, 1921, **7**, 115.
2. R. G. Jenkins and M. B. Rao, *Powder technology*, 1984, **38**, 177
3. M. M. Dubinin, *Chem. Rev.*, 1960, **60**, 235.
4. C. Dubois, A. Alvarez Callera, S. Bassot and A. Chambaudet, 'Gas geochemistry', C. Dubois Ed. ISBN 0-905927-79-6. Special issue of Environmental Geochemistry and Health, 1995, 357.
5. C. Vuilleumier, PhD Thesis, University of Franche-Comté France, 1990.
6. E. Boeglin, D.E.A., University of Franche-Comté France, 1989.

CHARACTERIZATION OF SINGLE PORES AND POROUS SOLIDS

Z. SOBOTKA

Mathematical Institute of Czech Academy of Sciences
150 00 Prague 1, Žitná 25, Czech Republic

1 INTRODUCTION

The author introduces dimensionless parameters characterizing the shape properties of single pores and configurations with volume ratios of pores in porous bodies. The two-dimensional structure of plane sections through porous solids is characterized by two-dimensional structural models corresponding to the square unit representative areas. These models include the resultant void rectangle which represents the configuration and volume ratios of dispersed pores.

The orientated three-dimensional porous structures are approximately characterized by three-dimensional orthogonal models represented by the unit cube including the resultant characteristic void prism corresponding to the configuration, shapes and volumes of pores.

A particular attention is paid to seeking for the characteristic directions of porous bodies with indistinct structure. The dominant directions represented by the orientation of the orthogonal structural models are determined on the basis of configuration, shape and size of dispersed pores in the unit representative cube. The presented structural models lead to the union of phenomenological and structural points of view.

2 SINGLE PORE CHARACTERIZATION

2.1 Volume characterization of pores

In the general case, the single pores and cavities are characterized by the radii of minimum circumscribed sphere r_c, maximum inscribed sphere r_i and distance c between the centres of both spheres. Such a characterization does not depend on the orientation of pores. For the shape analysis of single pores, three-dimensionless parameters can be defined[4]:

$$\delta = \frac{r_i}{r_c}, \ \vartheta = \frac{c}{r_c}, \ \varphi_s = \frac{3V_p}{4\pi r_c^3},$$

(1)

where V_p is the volume of the pore.

The parameter φ_s representing the ratio of actual volume of the pore and that of the minimum circumscribed sphere expresses the spherical fullness and can be regarded as a measure of the regularity of pores.

For the orientated structure, the measure of prismatic fullness seems to be more suitable[4]:

$$\varphi_p = \frac{V_p}{a_c b_c c_c}, \qquad (2)$$

where a_c, b_c and c_c are the edge lengths of minimum circumscribed prism. From the considerations concerned with prismatic measures, two other dimensionless shape characteristics can be derived:

$$\lambda_c = \frac{b_c}{a_c}, \ \mu_c = \frac{c_c}{a_c}. \qquad (3)$$

These parameters can define the elongation or flatness of individual pores. If they are equal to one, they represent a spherical or cubic pore.

2.2 Characterization of pore profiles

The measurements for the shape analysis in space are relatively complex. Therefore, the shape analysis of pore profiles is advantageous for its greater simplicity. By analogy with Eq. (1), we obtain the parameters δ and ϑ of the particle profiles where r_c denotes the minimum circumscribed circle, r_i that of the maximum inscribed circle and c the distance between their centres as shown in Fig. 1. The circular two-dimensional pore fullness is expressed by

$$\varphi_p = \frac{A_p}{\pi r_c^2}. \qquad (4)$$

For the analysis of pore profiles of the orthotropic structures, the rectangles situated in the principal directions of orthotropy seems to be suitable.

2.3 Equivalent rectangles characterizing the pore profiles

For characterization of pore profiles, the stereological methods can be applied. Drawing a series of parallel cress-sections through a body with orientated structure, we obtain various images of individual pores. The plane section through an orientated pore k in the section s is shown in Fig. 2.

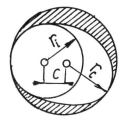

Figure 1 *A pore prifile with inscribed and circumscribed spheres*

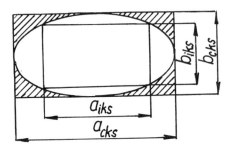

Figure 2 *A pore profile with inscribed and circumscribed rectangles*

The side lengths of the maximum inscribed rectangle are a_{iks} and b_{iks} and the side lengths of the corresponding circumscribed rectangle situated in the principal directions of orthotropy are a_{cks} and b_{cks}.

The cross-sections of individual orientated pores can be replaced by the equivalent rectangles having the same areas A_{ks} as the actual pores and the ratio of side lengths given by

$$\lambda_{ks} = \lambda_{iks} + \frac{A_{ks} - A_{iks}}{A_{cks} - A_{iks}}(\lambda_{cks} - \lambda_{iks}), \tag{5}$$

where A_{csk} is the area of the orientated circumscribed rectangle and A_{iks} is that of the maximum inscribed rectangle and

$$\lambda_{cks} = \frac{b_{cks}}{a_{cks}}, \quad \lambda_{iks} = \frac{b_{iks}}{a_{iks}}. \tag{6}$$

The parameter λ_{ks} given by Eq. (5) represents an important characteristic of shape of pore profiles.

2.4 Equivalent prisms

An analogous procedure can be applied for determining the equivalent prisms having the same volumes V_k as that of pores. The side lengths of the circumscribed prism with volume V_{ck} situated in the principal directions of orthotropy are a_{ck}, b_{ck} and c_{ck} and those of the corresponding maximum inscribed prism are a_{ik}, b_{ik} and c_{ik}. The ratios of side lengths of equivalent prisms are

$$\lambda_k = \lambda_{ik} + \frac{V_k - V_{ik}}{V_{ck} - V_{ik}}(\lambda_{ck} - \lambda_{ik}), \quad \mu_k = \mu_{ik} + \frac{V_k - V_{ik}}{V_{ck} - V_{ik}}(\mu_{ck} - \mu_{ik}), \tag{7}$$

where

$$\lambda_{ck} = \frac{b_{ck}}{a_{ck}}, \ \mu_{ck} = \frac{c_{ck}}{a_{ck}}, \ \lambda_{ik} = \frac{b_{ik}}{a_{ik}}, \ \mu_{ik} = \frac{c_{ik}}{a_{ik}} \tag{8}$$

and the ratios of side lengths of circumscribed and inscribed prisms.

3 STRUCTURAL MODELS OF POROUS MEDIA

3.1 Two-dimensional models

The author has introduced the two-dimensional structural models corresponding to the unit representative areas of plane cross-sections of porous materials. The profiles of individual dispersed pores are replaced by the equivalent rectangles which are then composed into the resultant rectangle in the unit representative area 2 as shown in Fig. 3. The area of this resultant rectangle and the ratio of their side lengths are expressed by

$$A_s = \sum_{k=1}^{n} A_{ks}, \ \lambda_s = \frac{\beta_{2s}}{\alpha_{2s}} = \frac{1}{A_s} \sum_{k=1}^{n} A_{ks}\lambda_{ks}. \tag{9}$$

Drawing p parallel plane sections through a porous body, the average value and ratio of side lengths of the resultant rectangle are given by

$$A = \frac{1}{p} \sum_{s=1}^{p} A_s, \ \lambda = \frac{\beta_2}{\alpha_2} = \frac{1}{A} \sum_{s=1}^{p} A_s\lambda_s. \tag{10}$$

Since the area of the weighted resultant rectangle is

$$A = \alpha_2\beta_2 = \lambda\alpha_2^2 = \frac{\beta_2^2}{\lambda}, \tag{11}$$

we obtain the weighted lengths characterizing the orthotropic configuration of porous structure

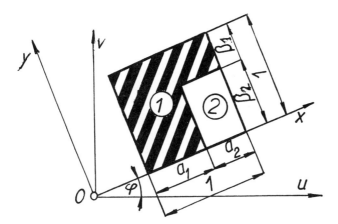

Figure 3 *The orientated orthogonal plane model of porous material*

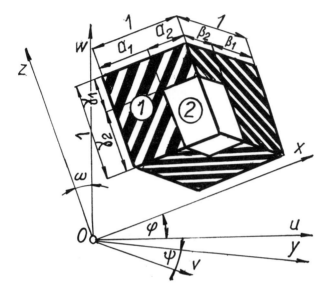

Figure 4 *The orientated three-dimensional model of a porous body*

Many porous materials have very irregular structure so that is is difficult to find the dominant directions of its orientation and to choose the corresponding coordinate system. For these disarranged porous structures, the orientation of the two-dimensional orthogonal models is determined from the configuration, shape and size of individual disordered pore profiles.

$$\alpha_2 = \sqrt{\frac{A}{\lambda}}, \ \beta_2 = \sqrt{\lambda A}. \tag{12}$$

In this case, we can choose an arbitrary system with the coordinate axes u and v. The minimum circumscribed and maximum inscribed rectangles defining the particle profiles form different angles φ_{cks} and φ_{iks} with the u-axis. The angle φ_{ks}, defining the orientation of an equivalent rectangle with the area A_{ks} can be determined by interpolation in an analogous manner as the ratio of the side lengths given by Eq. (5) according to the relation:

$$\cos \varphi_{ks} = \cos \varphi_{iks} + \frac{A_{ks} - A_{iks}}{A_{cks} - A_{iks}} (\cos \varphi_{cks} - \cos \varphi_{iks}). \tag{13}$$

Composing n equivalent rectangles in the unit representative area of one plane sections into the resultant rectangle with the area A_s given by Eq. (9), we obtain the angle defining its orientation from the expression

$$\cos \varphi_s = \frac{1}{A_s} \sum_{k=1}^{n} A_{ks} \cos \varphi_{ks}. \tag{14}$$

The average value of the directional angle determining the orientation of two-dimensional model is determined on the basis of angles φ_s in p parallel sections through a porous body:

$$\cos \varphi = \frac{1}{A} \sum_{s=1}^{p} A_s \cos \varphi_s. \tag{15}$$

3.2 Three-dimensional models

In an analogous manner as the equivalent rectangles, the n individual equivalent prisms are composed into the resultant prism in the unit representative cube with characteristic side lengths α_2, β_2 and γ_2 as shown in Fig. 4. Its volume is

$$V = \sum_{k=1}^{n} V_k. \tag{16}$$

The ratios of edge lengths of th resultant prism can be determined by the method of moments as follows

$$\lambda = \frac{1}{V} \sum_{k=1}^{n} V_k \lambda_k, \ \mu = \frac{1}{V} \sum_{k=1}^{n} V_k \mu_k. \tag{17}$$

Since the volume of the resultant representative prism is expressed by

$$V = \alpha_2 \beta_2 \gamma_2 = \lambda \mu \alpha_2^3 = \frac{\mu \beta_2^2}{\lambda^2} = \frac{\lambda \gamma_2^3}{\mu^2}, \tag{18}$$

the characteristic lengths are given by

$$\alpha_2 = \sqrt[3]{\frac{V}{\lambda \mu}}, \ \beta_2 = \lambda \sqrt{\frac{V}{\mu}}, \ \gamma_2 = \sqrt[3]{\frac{\mu^2 V}{\lambda}}. \tag{19}$$

In the case of irregular porous structures, the characteristic directions of porous bodies we determined by the directional cosines with respect to the choosen arbitrary coordinate axes u, v and w. The orientation of the maximum prism inscribed in a pore is defined by the angles φ_{ik}, ψ_{ik} and ω_{ik} and that of the minimum circumscribed prism by different angles $\varphi_{ck}, \varphi_{ck}$ and ω_{ck}. The directional cosines of equivalent prism with respect of the coordinate axed u, v and w are then given by

$$\cos \varphi_k = \cos \varphi_{ik} + \frac{V_k - V_{ik}}{V_{ck} - V_{ik}} (\cos \varphi_{ck} - \cos \varphi_{ik}), \tag{20}$$

$$\cos \psi_k = \cos \psi_{ik} + \frac{V_k - V_{ik}}{V_{ck} - V_{ik}} (\cos \psi_{ck} - \cos \psi_{ik}), \tag{21}$$

$$\cos \omega_k = \sqrt{1 - \cos^2 \varphi_k - \cos^2 \varphi_k}. \tag{22}$$

The volume of the resultant characteristic prism in the unit representative cube is again given by Eq. (16) and the ratios of its edge lengths follow from

$$\lambda = \frac{\beta_2}{\alpha_2} = \frac{\cos \varphi}{V \cos \psi} \sum_{k=1}^{n} \frac{V_k \lambda_k \cos \psi_k}{\cos \varphi_k}, \tag{23}$$

$$\mu = \frac{\gamma_2}{\alpha_2} = \frac{\cos \varphi}{V \cos \omega} \sum_{k=1}^{n} \frac{V_k \mu_k \cos \omega_k}{\cos \varphi_k}. \tag{24}$$

The directional cosines of the three-dimensional orthogonal model and simultaneously of the characteristic prism are define by

$$\cos \varphi = \frac{1}{V} \sum_{k=1}^{n} V_k \cos \varphi_k, \quad \cos \psi = \frac{1}{V} \sum_{k=1}^{n} V_k \cos \psi_k, \tag{25}$$

$$\cos \omega = \sqrt{1 - \cos^2 \varphi - \cos^2 \psi}. \tag{26}$$

4 CONCLUSION

The single pores are characterized by dimensionless parameters following from circumscribed and inscribed spheres or prisms. The pore profiles are defined by circles and rectangles. The equivalent prisms and rectangles correspond to the size, shape and orientation of individual pores.

In order to characterize the porosity of materials, the author has introduced the two-dimensional and three-dimensional structural models consisting of unit representative square or cube, respectively, which include the rectangular or prismatic regions representing the relative amount and configuration of dispersed pores. These models can represent two kinds of orthotropy arising either from orthotropy of the material or from the configuration of dispersed pores.

The presented characterization can be applied for schematism of materials with voids such as concrete, ceramics, rocks, soils and also for that of human bones in biomechanics. The structure of spongy bones consisting of trabeculae and marrow space can approximately be also represented by orthogonal structural models. The orientated marrow cavities can be replaced by the equivalent prisms with the edges situated in the principal directions of bone structure, namely in the directions of the principal stresses involved by the most frequent loading of skeleton.

The research was supported by the Grant No. 106/96/0938 of the Grant Agency of Czech Republic.

References

1. Z. Sobotka, *Proc. of the Fifth International Congress on Rheology*, Tokyo, 1969, **1**, 105.

2. Z. Sobotka, *J. of Macromolecular Science-Physics*, 1971, B5, 393.

3. Z. Sobotka, 'Rhology of Materials and Engineering Structures', Elsevier, Amsterdam, Oxford, New York, Tokyo, 1984, Chapter 3, p. 123.

4. Z. Sobotka, *4. European Symposium on the Particle Characterization*, Nürnberg, 1989, **2**, 599.

POLYMERIC MATERIALS WITH A NOVEL TYPE OF POROSITY

M.P.Tsyurupa, A.S.Shabaeva, L.A.Pavlova, T.A.Mrachkovskaya and V.A.Davankov

Institute of Organo-Element Compounds
Russian Academy of Sciences
Moscow, 117813, Russia

1 INTRODUCTION

Adsorption of significant amounts of organic substances from a gaseous phase or an aqueous solution onto the surface of a solid material is well known to imply that the adsorbent used exposes a highly developed surface area. There exist two major types of organic porous solids which exhibit reasonable values of inner surface area. These are activated carbons and polymeric adsorbents.

Porous activated carbons are usually obtained by a two-step process. Natural or synthetic organic materials are first subjected to pyrolysis under appropriate conditions, which is then followed by an activation through a high-temperature treatment in a mixture of steam, air and carbon dioxide. Owing to the removal of volatiles and partial oxidation, relatively small voids are created in the final product. The diameter of pores depends upon the conditions of the activation process, the inner surface area amounting to values as high as 500 to 2000 m^2/g.

The porous structure of polymeric adsorbents, in particular, macroporous styrene-divinylbenzene copolymers, results from the micro phase separation during the crosslinking copolymerization of the initial mixture containing the co-monomers and an inert diluent. The thermodynamic incompatibility of the copolymer formed with the diluent results in the formation of two separate phases in each copolymer bead. Insofar as the polymeric phase represents a densely crosslinked network, it cannot collapse when the diluent is removed, so that the volume which was occupied by the diluent becomes that of pores. The inner surface area of macroporous styrene-divinylbenzene adsorbents, as a rule, ranges from 300 to 500 m^2/g; less frequently it can achieve a value of 900 m^2/g (Amberlite XAD-4).

Both activated carbons and macroporous copolymers are heterogeneous two-phase materials exhibiting true internal interface. No other types of porous organic solids have been known, thus far. This paper represents a fundamentally new polymeric material. Having a homogeneous, one-phase structure, it nevertheless possesses low density thus displaying porosity of a new type. These materials have been referred to as hypercrosslinked polystyrene sorbents Styrosorb.

2 SYNTHESIS AND STRUCTURE OF HYPERCROSSLINKED SORBENTS

The synthesis of the hypercrosslinked polystyrene sorbents consists in that the preformed beads of a copolymer of styrene with divinylbenzene are subjected to an ultimate swelling with ethylene dichloride and an extensive additional crosslinking by means of a large amount of a bifunctional reagent capable of forming rigid bridges. Such a kind of bridge with very restricted conformational mobility arises when the styrene copolymer reacts with monochlorodimethyl ether in the presence of a Friedel-Crafts catalyst:

$$-CH_2-CH- \qquad -CH_2-CH- \qquad -CH_2-CH-$$

$$\xrightarrow[-CH_3OH]{CH_3OCH_2Cl} \qquad CH_2Cl \qquad \xrightarrow{-HCl} \qquad CH_2$$

$$-CH_2-CH- \qquad -CH_2-CH- \qquad -CH_2-CH-$$

Theoretically, if one mole of styrene repeating units reacts with 0.5 mole of the cross-agent, phenyl rings are connected pair-wise by methylene group, thus resulting in the formation of the polymeric network with a formal crosslinking degree of 100%. However, in the real product, a certain portion of phenyl rings remains unsubstituted whereas an equivalent part of rings converts into bridges of a more complex, three-functional structure. The residual phenyl groups can be further involved into reaction and incorporated into the network by using more than 0.5 mole of the monochlorodimethyl ether. Obviously, the portion of the trifunctional bridges increases, as well.

The hypercrosslinked polymers thus obtained are glassy transparent beads when both in dry or swollen state. They practically do not scatter X-rays under small angles, and no signs of inhomogeneity can be revealed by means of the electronic microscopy techniques. This indicates that the hypercrosslinked polystyrene exhibits a homogeneous, one phase structure. Nevertheless, the polymers have low density and absorb at low temperatures high amounts of inert gases, as if the sorbents would exhibit an inner surface area of as much as 1000-1800 m^2/g.

Another impressive property of these polymers is their ability to swell. From a traditional point of view, one could expect the hypercrosslinked polymer, all phenyl rings of which are involved in the formation of rigid cross-bridges, to maintain constant its volume when contacting a solvent, even a thermodynamically good one, such as toluene. However, the hypercrosslinked polystyrenes are capable of increasing their volume in toluene by a factor of 2 to 3, which is characteristic of conventional copolymers of styrene with just 1 to 0.3 per cent of divinylbenzene. More importantly, the same high extent of swelling is observed with typical non-solvents for polystyrene, hexane or methanol, which do not cause any swelling of styrene-DVB copolymers.

The above distinguishing characteristics of the hypercrosslinked polystyrene can be easily understood by taking into account the conditions of its preparation. The rapid crosslinking of polystyrene chains in many points of the swollen bead does not create any prerequisites for a phase separation. The bead continues to incorporate the major part of the solvent. Methylene groups, by combining phenyl groups of two neighboring polymeric chains, provide formation of a network, an ensemble of mutually condensed non-planar cycles. These cycles consist of at least three pairs of phenyl rings. The rings are large and can change noticeably their conformation. A cooperative conformational rearrangement of a large number of cycles explains the ability of the hypercrosslinked network to change significantly its volume. This occurs at drying and repeated swelling of the bead.

When ethylene dichloride is removed from the swollen homogeneous polymer after accomplishing of the crosslinking reaction, many rigid spacers prevent the polymeric chains from approaching each other closely and forming a dense polymeric phase in the dry material. The reduced density of chain packing in the homogeneous polymer (0.7 g/cm³, in contrast to 1.04 g/cm³ for the starting copolymer beads) is responsible for the "apparent porosity" of the hypercrosslinked polystyrene, its permeability to gasses and solvents. On the other hand, contraction of the network during the removal of the synthesis medium is accompanied inevitably by appearing of strong inner strains in the dried polymer. Both the absence of densely packed impermeable domains and a strongly expressed tendency to relaxation of the above inner strains through swelling and expanding the volume account for exceptionally high adsorption activity of polymers with respect to numerous organic substances. Structure and peculiar properties of the hypercrosslinked polystyrene sorbents Styrosorb have been described in detail elsewhere[1].

3 SORPTION OF VAPOURS OF HYDROCARBONS

The process of sorption of hydrocarbon vapours was examined using Styrosorb 2. The sorbent displayed an apparent inner surface area of 1000 m²/g; its pore volume was 0.22 cm³/g (in dry state) and pore diameters amounted to about 2-3 nm. Figure 1 shows adsorption isotherms for n-hexane and n-pentane vapours onto Styrosorb 2 and activated carbon AR-3. As can be seen, the adsorption capacity of Styrosorb 2, which corresponds to the initial pore volume of the dry sorbent, is achieved already at the relative pressure of p/p_s=0.04. This small concentration of the sorbate molecules is sufficient to saturate completely the activated carbon and zeolite. Contrary to this, the capacity of Styrosorb increases dramatically with a further increase of the relative pressure, the ultimate adsorption volume amounting to 1,2 cm³/g. This signifies a strong swelling of Styrosorb in the vapours of the sorbate. Direct experiments showed [2] that swelling starts almost simultaneously with the sorption, long before the pentane molecules fill in the initial pores of the dry material. Therefore, the sorbate causes expansion of the hypercrosslinked sorbent

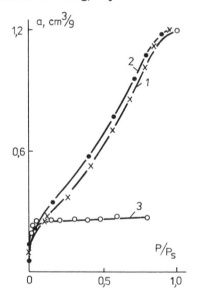

Figure 1 *Sorption isotherms for pentane (1) and hexane (2,3) vapours at 20 °C on Styrosorb 2 (1,2) and activated carbon AR-3 (3(,)*

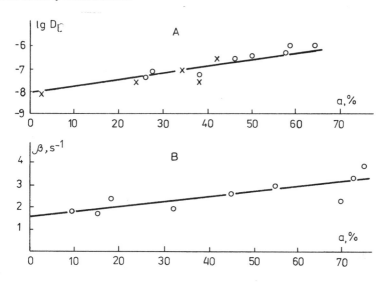

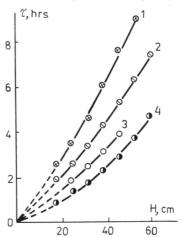

Figure 2 A: *Effective diffusion coefficients of hexane (o) and pentane (+) as a function of the degree of bead filling;* B: *Coefficients of pentane mass transfer as a function of sorption*

resulting in the formation of new sorption space accessible in the following to additional portions of pentane.

The expansion of sorbent volume in the course of adsorption strongly affects the rate of diffusion of the organic sorbates. The process begins slowly, however, as sorption proceeds, the effective diffusion coefficients, D_e, gradually raise and soon exceed the starting values of D_e by two order of magnitude (Figure 2). Contrary to this, desorption occurs quickly: almost the whole amount of alkanes can be removed within 100 seconds. The last portions of sorbates, only, are removed slower.

The increase in the coefficients of diffusion with adsorption tells unusually upon the regularities of the hydrocarbon extraction from air under dynamic conditions. A column of 60 cm x 2.0 cm ID packed with beaded Styrosorb 2 was used to investigate the dynamics of sorption in detail. The intensity of the process has been characterized by the coefficient of mass transfer, $1/\beta = 1/\beta_1 + 1/\beta_2$ where β_1 and β_2 are the outer and internal coefficients of mass transfer, respectively. The calculated values of Bio criterion, ranging from 1183000 to 1120 for the adsorption interval of 0.6 to 9.8 mmol/ml, shows unambiguously that the limiting

Figure 3 *The dependence of bed protecting time on the bed depth at the pentane concentration of 14.1 (1), 23.9 (2), 34.2 (3) and 59.2% (4)*

stage of n-pentane vapours sorption is the internal mass transfer, $\beta_1 >> \beta_2$ and $\beta \approx \beta_2$. Since β_2 is determined entirely by the value of D_e, the coefficient of mass transfer proves to be the higher, the larger is the depth of the bed (i.e., the value of adsorption). It explains the non-linear dependence of column protecting time on the bed depth, which is especially pronounced at high concentrations of pentane (Figure 3).

When percolating mixtures of n-pentane vapours with air through a sorbent bed, a significant generation of heat the moment of breakthrough, was observed to accompany the sorption. To obtain more information about the process, 7 sensors were mounted along the column. Their response to the sorbate concentration and the temperature of sorbent bed in each section are presented in Figure 4. The self-heating of sorbent beads affects the process in two ways: on the one hand, the heat generated accelerates the diffusion of the sorbate, but, on the other hand, it decreases the adsorption capacity.

With the concentration of n-pentane of 59.2 %, there is a sharp temperature wa e, about 40 °C high, moving along the column. Since the heat evolves at the sorption front, and

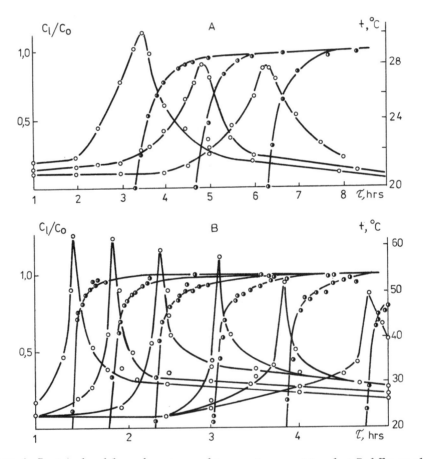

Figure 4 *Pentane breakthrough curves and temperatures registered at 7 different check points along the column as a function of time. The sorbate concentration, 11.5 %wt. (A) and 59.2 (B); the flow rate, 0.009 (A) and 0.0113 m/sec (B).*

there is not much of the carrier gas present in the mixture, the sorption zone and the heat wave move almost simultaneously. About 7 to 10 cm behind the sorption zone, the bed temperature decreases to 4-6 °C above the initial level. Nevertheless, a complete cooling of the sorbent bed requires several hours. Therefore, only 80 % of the total sorbent capacity prove to be used up rapidly at the moment of breakthrough at the column outlet, whereas the remaining 20 % represent the long "tails" of the breakthrough curve.

When the concentration of n-pentane is not very high, 11.5 %, longer time is required to saturate the sorbent column. The heat evolving through sorption will partially be consumed by the carrier gas, so that the height of the heat wave is smaller as compared to the previous experiment. Moreover, the carrier gas transports the heat ahead of the sorption zone. A definite increase in the temperature of the bed is always observed to occur long before the arrival of the sorption zone (Figure 4). In the total, heat evolution should be less expressed in the case of diluted pentane vapours, and the portion of the column capacity used up at the moment of breakthrough.

Regeneration of Styrosorb in the above column was performed by steam of 105 °C at a rate of 0.02 m/sec. Almost 90 % of the hydrocarbon adsorbed was removed within 20-30 minutes. The regenerated sorbent contained no more than 5 % water which did not affect negatively the subsequent sorption cycle. 100 sorption-desorption cycles were performed without any decrease in the sorption capacity or mechanical destruction of sorbent beads.

4 SORPTION OF ORGANIC SUBSTANCES FROM AQUEOUS SOLUTIONS

Hypercrosslinked sorbents Styrosorb exhibit high adsorption activity with respect to various organic substances dissolved in water, such as phenols, chlorophenols, aliphatic and aromatic hydrocarbons and acids, synthetic organic dyes, natural colour bodies of sugar syrups or cultural liquids in microbiological synthesis, pesticides, lipids, antibiotics, etc. In all cases the capacity of Styrosorbs exceeded substantially that of conventional macroporous styrene-divinylbenzene adsorbents. Many aspects of potential employing Styrosorbs in adsorption technologies were reviewed previously [3]. Here, we would like to consider only some examples demonstrating the inapplicability of standard approaches to describe adsorption properties to the Styrosorb type adsorbents.

Figure 5 presents sorption isotherms for phenol, benzene, toluene and L-tryptophan on Styrosorb 2 and a hypercrosslinked sorbent "Macronet Hypersol" MN-200 produced by

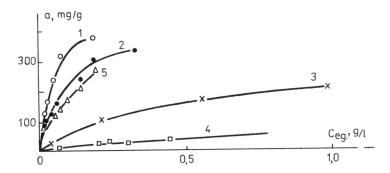

Figure 5 *Sorption isotherms for toluene (1) ,benzene (2,5), phenol (3) and L-tryptophan (4) on MN-100 (1-4) and Styrosorb 2 (5)*

Purolite International Ltd. (UK) on industrial scale. The latter sorbent possesses heterogeneous structure. Beside the micropores which are characteristic of Styrosorb 2, MN-200 has large transport pores of 100 nm in diameter. Quite evident is the fact that at low sorbate concentration, the neutral hydrophobic sorbents absorb non-polar benzene and toluene to a significantly larger extent than the polar phenol or amino acid. In no case the isotherms reach the value of ultimate saturation of the sorbent. Attempts have been made to evaluate these limiting adsorption capacities for each of the above compounds by applying commonly used equations of Freundlich, Langmuir and Dubinin-Radushkevich.

Theoretically, the upper limiting value of sorption could be expected to equal the sum of the volume of micropores and the swelling of the polymeric phase with benzene or toluene. Styrosorb 2 is a microporous material with a toluene uptake of 2.34 ml/g, due to swelling and filling up the micro pores. MN-200 takes up 1.50 ml/g of the solvents. The total pore volume of dry MN-200 amounts to 1.1 ml/g, with at least 0.9 ml/g representing macropores. The latter cannot be expected to be filled with the adsorbate. Therefore, the limiting value of adsorption for this material should be lower, about 0.5-0.6 ml/g. Consider now the results of calculations according to known isotherm equations (Table).

First of all it should be noted that despite the different shapes of isotherms obtained for the different sorbates, they all produce straight lines in linear coordinates of equations presented in the Table and acceptable correlation coefficients (marked in italic), with the exception of the isotherm of benzene sorption on MN-200 plotted in the coordinates of Freundlich. For other sorbates, however, the Freundlich equation predicts unreasonably high values of the limiting sorption capacity which exceeds many times the above given estimations. Neither Langmuir, nor Dubinin equations describe the sorption of benzene on the one-phase Styrosorb 2: they produce a value of A_0 which is too low, even in comparison with the capacity obtained experimentally. According to approaches by Langmuir and Dubinin, the limiting sorption of benzene and toluene on MN-200 amounts to a range between 0.36 and 0.56 ml/g, which can be regarded as acceptable. For L-tryptophan, however, these equations predict very low limiting values of sorption, lower than

Table Mathematical description of the sorption isotherms

Type of isotherm	MN-200				Styrosorb
	benzene	toluene	trypto-phan	phenol	benzene
Freundlich: $lga=lgK+1/nlgC_{eq}$ K, ml/g	6.14 *0.9972*	non-linear	1.83 *0.9848*	2.15 *0.9923*	6.03 *0.9888*
Langmuir: $1/a=1/A_0+1/bA_0C_{eq}$ A_0, ml/g	0.36 *0.9825*	0.56 *0.9716*	0.054 *0.9967*	0.272 *0.9961*	0.179 *0.9457*
Dubinin: $lga=lgA_0+(RT/E)^2(lgC_s/C_{eq})^2$ A_0, ml/g	0.39 *0.9869*	0.49 *0.9563*	0.075 *0.9869*	0.59 *0.9987*	0.29 *0.9543*
Capacity reached experimentally a_{ex}, ml/g	0.38	0.41	0.085	0.35	0.31

C_s = the limiting solubility of the sorbate at 20 °C; C_{eq} = the equilibrium concentration of solution; a = current capacity; K and A_0 = the ultimate capacity; R = gas constant; T = temperature; E = adsorption energy; n and b = constants.

the experimentally obtained result. Finally, by using the approach of Langmuir one arrives at a low limiting sorption value for phenol (smaller that that obtained in the experiment), and only the approach of Dubinin predicts an expected value of A_0. However, in the latter case A_0 for the polar phenol proves to be higher than the corresponding values for non-polar toluene and hexane. This sequence is unreasonable, again.

The above equations are well known to be developed for adsorption on porous solids with a constant porosity. Styrosorb 2 is capable of increasing its volume by a factor of 3 at swelling with pure solvent. Undoubtedly, swelling with the sorbate takes place when sorption occurs from aqueous solutions, as well, though, to a smaller extent. Usually, volume changes of beads of MN-200 are much smaller, of about 5 %. However, swelling of a heterogeneous structure does not need to be affine, and swelling of the hypercrosslinked polymeric phase of MN-200 may be pretty high. In any case, the swelling which accompanies the sorption is responsible for the inapplicability of traditional approaches for the description of sorption behaviour of Styrosorbs.

Acknowledgment

The authors acknowledge "Purolite International Ltd" (UK) for support and cooperation in developing new type of sorbents.

Literature

1. V.A.Davankov and M.P.Tsyurupa, *React.Polym.*, 1990, **13**, 27.
2. G.I.Rosenberg, A.S.Shabaeva, V.S.Moryakov, T.G.Musin, M.P.Tsyurupa and V.A.Davankov, *React.Polym.*, 1983, **1**, 175.
3. M.P.Tsyurupa, L.A.Maslova, A.I.Andreeva, T.A.Mrachkovskaya and V.A.Davankov, *React.Polym.*, 1995, **25**, 69.

PROCESSING OF MICROPOROUS/MESOPOROUS SUBMICRON-SIZE SILICA
SPHERES BY MEANS OF A TEMPLATE-SUPPORTED SYNTHESIS

Ch. Kaiser, G. Büchel, S. Lüdtke, I. Lauer, K.K. Unger

Institut für Anorganische Chemie und Analytische Chemie
Johannes Gutenberg-Universität, J.J. Becherweg 24, 55099 Mainz, Germany

1 INTRODUCTION

Since the introduction of the synthesis of monodisperse nonporous spherical silica beads by Stöber et al.[1] the subject has attracted ongoing attention by scientists in colloid chemistry. Three modifications of the synthesis have been reported[2-4] and the formation mechanism of the beads was investigated[5-7]. The material has found wide-spread application as fillers for polymers, additive to improve the slipperiness of polyester films, additive to cosmetic preparations and as packing in High Performance Liquid Chromatoraphy (HPLC) to perform ultrafast separations of biopolymers[8,9].

This work reports for the first time on the synthesis of porous Stöber beads by using an organotrimethoxysilane as an additive in the reaction mixture. The silane is chemically bonded in the bulk of the beads. Removal of the organic residue by calcination results in porous beads. The main objective of the study was to adjust and to control the pore structure parameters of the final products by the reaction conditions.

2 EXPERIMENTAL

2.1 Synthesis of porous beads

2.1.1 Synthesis of composite materials. Aqueous ammonia (32% w/w, reagent grade, Merck, Darmstadt, Germany) was mixed with deionised water and absolute ethanol (Merck, Darmstadt, Germany) at room temperature (the certain amounts of the used components are referred in table 1). The freshly destilled tetraethoxysilane (TES, Wacker Chemie, Burghausen, Germany) was combined with n-octadecyltrimethoxysilane (n-OTMS, Aldrich-Chemie, Steinheim, Germany) and added to the solution under stirring. The solution was stirred for 1 to 4 hours, depending on the reaction conditions. An opalescence appeared after two minutes changing to a white turbidity which is indicative for the formation of particles.

2.1.2 Drying of the composites. Ammonia was removed from the composite dispersion by distillation. Then the product was subjected to an azeotropic distillation with toluene to remove water. The ethanol/toluene mixture was removed under vacuum from the dispersed silica composites. The remaining white, fine powder was dried at 150°C and consisted of discrete, organo-modified silica composite beads.

    2.1.3 Pyrolysis. The powder was then subjected to calcination at 550 °C with a heating rate of 1 °C/min for five hours. A white and fine powder of pure porous silica beads was obtained.

    The weight loss and heat changes during the pyrolysis were monitored by thermal analysis. Fig.1 shows a typical pattern.

    A major weight loss is observed between 200 and 500 °C indicating the burn-off of the organic residue. The mass loss becomes insignificant above 600 °C.

Table 1   *Synthesis conditions of organo-silica composites with 500 nm average particle size using n-octadecyl trimethoxysilane.*

Sample No.	C(TES) / mol l^{-1}	C(n-OTMS) / mmol l^{-1}	C(NH$_3$) / mol l^{-1}	C(H$_2$O) / mol l^{-1}	Volume of EtOH / ml
1	0.22	7.1	1.88	9.3	74.4
2	0.22	8.3	1.88	9.3	74.6
3	0.22	11.8	1.88	9.3	74.5
4	0.22	23.6	1.88	9.3	74.0
5	0.22	47.2	1.88	9.3	73.0

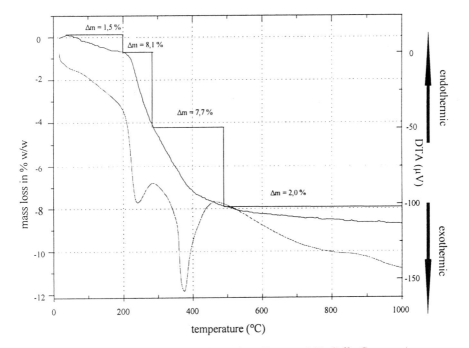

Figure 1   *Thermal analysis data of sample 4 (Linseis, L81, Selb, Germany)*

3 RESULTS AND DISCUSSION

3.1 Influence of the concentration of n-OTMS on the pore structure parameters

The concentration of the silane was varied between 7 to 47 mmol/l in the starting reacting solution (see table 1). The resulting nitrogen isotherms are shown in Fig.2.

Sample 1 indicates a reversible isotherm typical for a microporous solid, while sample 4 indicates a slight appearance of a hysteresis between the desorption and adsorption branches about $p/p_0 \sim 0.4$. Sample 5 has a type IV isotherm typical for mesoporous solids. Table 2 lists the pore structure parameters of sample 1-5 calculated from the nitrogen isotherms.

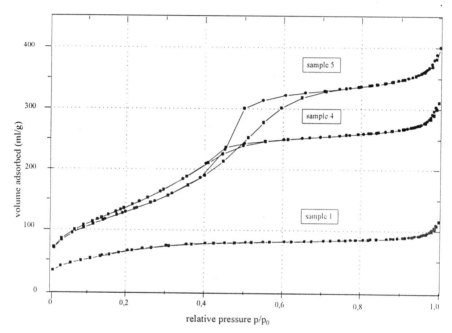

Figure 2   *Nitrogen sorption isotherms of samples 1,4 and 5 at 77K (Micromeritics, ASAP 2000, Neuss, Germany)*

Table 2   *Specific surface area, a_s, specific pore volume, v_p, according to the Gurwitch-rule and the average pore diameter, p_d, calculated from the desorption branch of the nitrogen isotherm by the BJH method[10].*

Sample No.	a_s /m^2 g^{-1}	v_p /cc g^{-1}	p_d(BJH) /nm
1	164	0.16	3.1
2	193	0.17	3.3
3	289	0.23	3.2
4	460	0.45	3.9
5	443	0.59	5.3

The pores of the beads span the micropore and low mesopore size range. Surprisingly, the average pore diameter of sample 2 to 5 range only between 2 to 5 nm. The pore size distribution of these samples (not shown) is rather narrow.

The specific surface area and the specific pore volume increase with the amount of n-OTMS used in the starting solution. There is a direct correlation between a_s(BET) from nitrogen sorption measurements and the weight loss determined by thermogravimetry in the range between 200 and 500 °C (Fig.3).

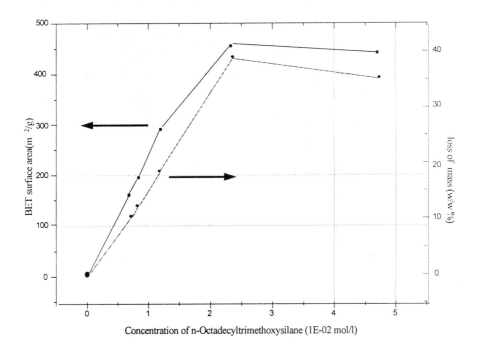

Figure 3   *Specific surface area, a_s, and the weight loss during ignition, Δm, as a function of the amount of n-OTMS in the starting solution.*

This is a direct evidence that the n-OTMS acts as a porogen as in the synthesis of macroporous polymers.

3.2 Particle size and particle morphology

The particle size is controlled by the same parameters (concentration of water and ammonia) as in the synthesis of nonporous Stöber particles. The beads are spherical in morphology and possess a narrow particle size distribution.

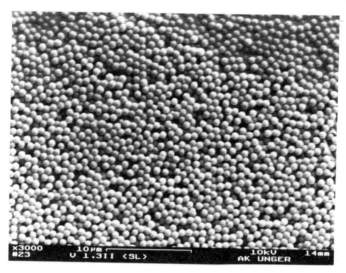

Figure 4   *Electron scanning micrograph of sample 4 (DSM-962, Zeiss, Germany)*

3.3 Pore formation process

The porosity of the beads is formed by the removal of organic residue (n-octadecyl groups). The specific pore volume of the products is proportional to the amount of n-OTMS in the starting solution. Deviations from the proportionality are observed at sample 5. Sample 5 also falls out off the linear relationship between a_s(BET) and Δm as a function of the amount of n-OTMS (s. Fig.2). This might be explained by the fact that at this concentration a part of the silane is not implemented into the organo-silica beads. The key question in the understanding of the pore formation mechanism is: How is the n-OTMS bonded in the bulk of organo-silica composite and how are the products distributed over the bulk of the beads? Furthermore, one might expect that the silane acts as a template similar to tetraalkylammonium salts in the synthesis of MFI types of zeolites[11].

Hydrolysis and condensation of the two species (TES and n-OTMS) is markedly different. n-OTMS hydrolyses much slower and the reaction products also follow a slower condensation process than those of TES. This was simply monitored by detecting the weight loss of beads as a function of the reaction time (Fig 5).

The weight loss can be assumed to be proportional to the mass of n-OTMS into the bulk phase.

The structure of organo-silica composite and the way of burning of the n-octadecyl groups can be investigated by ^{29}Si solid state NMR spectroscopy. Another way is to identify reaction products as the function of reaction time.

It is somewhat surprising that the average pore diameter increases only slightly with increase of the n-OTMS concentration in the starting solution from 3 to 5 nm. One might draw an analogy to the synthesis of MCM-41 using octadecyltrimethylammonium

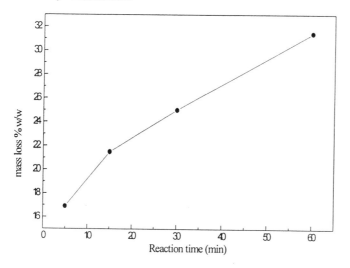

Figure 5   *Weight loss determined by thermogravimetry as a function of the reaction time. Sample similar to sample 4*

bromide as surfactant template. The average pore diameter is about 4 nm using n-hexadecyltrimethylammonium bromide. Pore formation of MCM-41 is discussed on the basis of micelles composed organic-inorganic assemblies between the surfactant and the silicate where the hydrophobic tails of the surfactant are directed to the center core of the micelles, the positively charged head groups are directed towards the outer surface. Their charge is compensated by the oppositely charged silica species. This charge matching is not the case in the formation process of the Stöber particles, because positively charged groups are lacking. Nevertheless there could be a kind of ordering of hydrophobic groups leading to the average pore diameter being about twice of the thickness of an extended n-octadecyl chain. Transmission electron micrographs of the porous silica beads show no ordered pore structure as observed for MCM-41 samples. The pores are randomly distributed. However, more detailed electron microscopic studies have to be performed to elucidate the pore structure and to find out whether n-OTMS acts as a real template or simply as a porogen.

4 CONCLUSION

N-octadecyltrimethoxysilane added to the staring solution at the Stöber process results in organo-silica beads of controlled particle size with a narrow particle size distribution. After drying the organic residue is removed by ignition at 550 °C leaving porous beads with a pronounced micropore and mesopore structure. The specific surface area and the specific pore volume are proportional to the amount of n-OTMS in the starting solution indicating that the silane is responsible for the pore space formation. The average pore diameter of the beads vary between 3 and 5 nm. This might be an indication that the n-OTMS forms a kind of micelle structure. However, to evidence this hypothesis further work is needed.

Acknowledgements

The financal support of the DFG (project number Un 26/40-1) is gratefully acknowledged.

References

1. W. Stöber, A. Fink and E. Bohn, *J. Colloid Interface Sci.*, 1968, **26**, 62.
2. K.K. Unger and H. Giesche, US patent 4,775,520.
3. E. sz. Kovats, L. Jelinek, C. Erbacher, Europian patent EP0.574.642 A1.
4. T.J. Barder, P.D. Dubois, US patent 4.983.369.
5. A.K. Van Helden, J.W. Jansen and A. Vrij, *J. Colloid Interface Sci.*, 1981, **81**.354
6. T. Matsoukas and E. Gulari, *J. Colloid Interface Sci.*, 1988, **124**, 252.
7. A.P. Philips and A. Vrij, *J.Chem.Phys.*, 1987, 87, 5634.
8. Ch. Kaiser, M. Hanson, K.K. Unger, Synthesis, Charcterization and Modification of Nonporous Silica Beads in Microspheres, Microcapsules and Liposomes, R. Ashady (ed.), in print (1996)
9. C. Kaiser, M. Hanson, H. Giesche, J. Kinkel, and K.K. Unger in E. Pelizzetti (ed.), Fine Particles Science and Technology, Kluver Academic Publisher, Boston 1996, pp. 71-84.
10. E.P. Barrett, L.G. Joyner, P.H. Halenda, J.Am.Chem.Soc., 1951, **73**, 373
11. J.C. Jansen, The Preparation of Molekular Sieves, in H. Van Bekkum, E.M. Flanigen, J.C. Jansen (eds.), Introduction to Zeolithe and Practice, *Stud. Surf. Sci. Catal.*, Vol. 58, Elesevier, Amsterdam, 1991, pp. 77-130.

Observation of Interface Curvature of Capillary Condensed Phase in Slit-shaped Nano Pore with a New MD Simulation Method

Minoru Miyahara, Tomohisa Yoshioka, and Morio Okazaki

Department of Chemical Engineering,
Kyoto University,
Kyoto 606-01 Japan

ABSTRACT

We developed a new molecular dynamics (MD) scheme to observe condensed phase with its interface to gas phase, introducing "potential buffering field" through which adsorbed phase could interact with an imaginary gas phase. Besides its ability to hold an interface, the simulation cell enabled us to conduct a MD simulation that allowed change in the number of molecules to attain equilibrium with desired vapor-phase pressure, like GCMC simulation. By taking another choice for setting of the cell, the number of the molecules stayed constant but the equilibrium pressure could be obtained easily by a new technique of "particle counting method". Either of the choices gave equilibrium relation with simple algorithm. Some adsorption simulations within slit-like pores ranging from 2nm to 4nm were carried out to obtain the isotherms for the pores. Adsorption phenomena could be observed from monolayer adsorption on a pore wall under a low relative pressure to the capillary condensation under a high relative pressure, and the adsorption equilibrium relation could be found. The interface of the condensation phase was observed to have somewhat skewed shape rather than to be a complete semicircle. Considering the non-uniformity of interface curvature, which should result from the effect of pore-wall potential, we established a new condensation model for nano-pores with a simple expression relating the equilibrium pressure and pore width. Critical condensation pressures and the shape of gas-condensate interface calculated from the model agreed well with the results of the MD simulations to show the validity of our model.

1 INTRODUCTION

Quantitative evaluation of micro structure of porous solids, especially in the order of nanometers, is of crucial importance since many of industrially important porous media have pores of the order of nanometers, or nano-pores. Further, recent advance in production technology for porous media has been yielding such materials of great interest as carbon nanotube[1] and MCM-41[2], which have quite unique structures in the order of nanometers. The demand for accurate and convenient characterization method in this range of pore size will never stop increasing. It has been pointed out that the Kelvin equation will give a faulty estimation of pore size in nano-pores[3-5]. Nevertheless this model is still often being employed today, possibly because of its handiness to relate the condensation pressure and pore diameter. Thus more information of adsorbed phase in pore, from different angles, should be sought out to establish an expression that is accurate enough for nano-pores as well as handy enough for spread use.

The condensation even at smaller pressure than the saturated pressure should be possible firstly because of the presence of the curved gas-condensate interface. Therefore, if the interface of the condensed phase in a pore can be directly observed, useful information for the condensation mechanism may be obtained. Since it is impossible to make such gas-condensate interface in a usual grand canonical Monte Carlo method, we employed a molecular dynamics method basically.

In this study, we developed a new molecular dynamics (MD) simulation scheme for the observation of adsorption/condensation phenomena in nano-pores. The effort has been made to establish a MD scheme for obtaining information of the interface of condensed phase with gas phase. Another important demand for the simulation scheme was that it should give equilibrium pressure of an adsorbed phase relatively easily. For satisfaction of the above two demands, we introduced a field that connected the pore space and an imaginary gas phase. This setting of the simulation cell allowed us to conduct a MD simulation that allowed change in the number of molecules in the system to attain equilibrium state with given equilibrium pressure, like GCMC simulation. In such sense this scheme could be called as a GC-like MD. By taking another choice for the setting of the cell, the number of the molecules stays constant but the equilibrium pressure can be obtained easily by a scheme that we call "particle counting method". By using MD simulation as an ideal experimental system, we proposed and verified a new model for condensation that can give a relation between condensation pressure and pore size for possible use in the characterization of nano-porous media. In the model, we took into account of the pore wall potential and the curvature dependency of surface tension. The feature of the model is non-uniformity in curvature of gas-condensate interface caused by external potential field. We examined the model on the agreement in the critical condensation pressures and from the aspects of the shape of gas-condensate interface.

2 SIMULATION SCHEME

We introduce a new MD method as an idealized experimental system for adsorption/condensation phenomena.

2.1 Unit cell and "Potential Buffering Field"

Figure 1 shows the unit cell developed in this study. This unit cell signifies a part of a slit-like pore. In the middle of the cell we have a potential field with a 'full' potential exerted by the pore walls. On the other hand, since the absolute value of external potential in the vapor phase is zero, there should exist slope of potential between the vapor phase and the pore space. The slope produces force that pushes adsorbate particles into the pore. To model this relation between the vapor phase and the adsorption phase, we settle a *border* with vapor phase at a location distant l_B from the edge of the *full* potential field as shown in Figure 1. Between the two location we set a potential field to connect the pore space and the vapor phase. This space plays a role of *buffering* the potential difference between pore and gas phase. Then we call it "Potential Buffering Field (PBF)". We approximated the potential in PBF simply to vanish linearly, at each location over pore width, from that within the pore to be zero at the border plane. As for the effect of length of PBF, l_B, we have confirmed that it should be longer than cut-off distance of force.

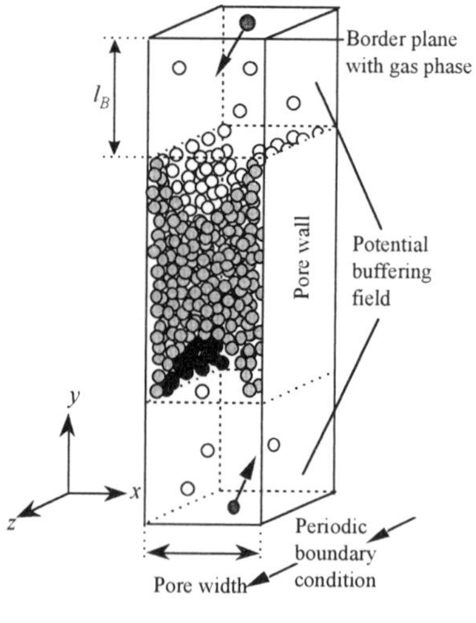

The periodic boundary condition is applied in z-direction. The temperature of the simulation system was controlled by the velocity scaling in a usual manner.[6]

Figure 1   *Schematic figure of the unit cell.*

2.2 Potential functions

We used argon-like LJ particles as adsorbate. The Lennard-Jones(12-6) potential Eq. (1), was used for the interactions between adsorbate particles.

$$E_{ij} = 4\varepsilon_{gg}\left[\left(\frac{\sigma_{gg}}{r_{ij}}\right)^{12} - \left(\frac{\sigma_{gg}}{r_{ij}}\right)^{6}\right] \qquad (1)$$

Where ε and σ are the energy and size parameter of LJ particle, r_{ij} interparticle distance, and subscript g refers to adsorbate gas. The cut-off distance of $3.5\sigma_{gg}$ was used. No long-range correction was made, however, the critical condition for condensation can be evaluated in terms of a reduced pressure relative to a saturated vapor pressure for a given cut-off distance.

As for the potential in the pore, we employed LJ (9-3) potential, as an example, to represent a pore between two semi-infinite solids consisted of LJ carbon-like particles. The interaction of one adsorbate particle and one of the two solids is shown by Eq. (2).

$$E_{is} = \frac{4\varepsilon_{gs}N_s\pi\sigma_{gs}^{3}}{3}\left[\frac{1}{15}\left(\frac{\sigma_{gs}}{R_i}\right)^{9} - \frac{1}{2}\left(\frac{\sigma_{gs}}{R_i}\right)^{3}\right] \qquad (2)$$

Where N_s is the number density of particle in the solid wall and R_i the distance from the pore wall to adsorbate particle i. Subscript s refers to solid wall. The overall potential distribution in the pore is described in the form of summation of the two potential field.

The values of the parameters are as follows. $\varepsilon_{gg}/k=120K$, $\varepsilon_{gs}/k=58K$, $\sigma_{gg}=0.34nm$, $\sigma_{gs}=0.34nm$ [7], and $N_s=114nm^{-3}$ [8]. The temperature T was set to be 87K, the normal boiling point of argon. The time mesh of the computation was 10fs, and the Verlet method[9] was used to integrate the equations of motion numerically.

2.3 Setting at the border and equilibrium pressure

2.3.1 Permeable border: GC-like MD simulation By making the border plane open and admitting the change of the number of particles, our MD method make it possible to follow an actual progress in formation of adsorption phase, in addition to obtain an equilibrium state for desired equilibrium pressure. In the GC-like MD simulation, particles are allowed to come into and go out from the system through the border plane. A particle which has succeeded in climbing up the slope of the potential in PBF is simply taken out from the system as a desorbed particle. On the other hand, particles are produced on the border plane at the frequency corresponding to the desired vapor-phase pressure, which is an input of the simulation. The vapor-phase pressure is decided assuming ideal behavior in gas phase. The production of particles can be made to have the frequency in proportion to the pressure in average. Some results are shown in the section **4.1**.

2.3.2 Reflective border: Particle counting method We can establish and observe equilibrium states in our unit cell by setting the border plane as an elastic wall with complete reflection. This is a kind of *NVT* simulations, but the greatest feature is that the adsorbed phase could have an interaction with an imaginary gas phase at the border plane. Just by counting the number of collisions with time after an equilibration, one can directly calculate the equilibrium pressure, provided ideal-gas law holds with the pressure and temperature. The determination of equilibrium vapor pressure by the Widom's particle insertion method[10,11] would be of further difficulty in this study because of the heterogeneity and hi density of the fluid within pore. Thus, our technique, the "particle counting method", should work quite effectively. In order to verify the "particle counting method", we have compared the equilibrium relative vapor pressure with the chemical potentials obtained with particle insertion method. Consequently these two values were in good agreement to demonstrate the validity of the particle counting method for the determination of the equilibrium vapor pressure of adsorbed phases.

3 CONDENSATION MODEL

3.1 Basic concept

What is aimed at here is to give a simple concept and model to describe the condensation phenomena in nano-pores. In order not to lose simplicity, the model treats fluid in pore as a continuum throughout and the derivation is based on an idealized interface of tension.   The case of slit-like pore was studied.

An equilibrium state between fluid in pore and gas phase of low pressure is considered below.   Taking bulk liquid as the standard state, chemical potential of vapor phase is given by Eq. (3), with the ideal gas approximation.

$$\text{vapor phase :} \qquad \mu_g = \mu_0 + kT \ln\left(P_g / P_{sat}\right) \qquad\qquad (3)$$

Where P is pressure, μ is chemical potential, k is Boltzmann constant. Subscript 0 and *sat* refer *standard state* and *saturation*, respectively.   Considering capillary effect to bring reduced pressure in condensed phase, and the effect of pore wall potential, chemical potential of the adsorption phase is given by Eq. (4), with the assumption of constant molar volume in the condensation phase against pressure.

$$\text{condensation phase :} \qquad \begin{aligned} \mu_{ads} &= \mu_0 + \mu_c + \mu_s \\ &= \mu_0 + \left(P_{ads}(x) - P_g\right)V_p + \Psi(x) \end{aligned} \qquad (4)$$

Where V_p is volume of one particle. Subscript *ads* and *c* refer *adsorption phase* and *curvature*, respectively. $\Psi(x)$ is the difference between the potential by solid pore walls and the potential that a molecule would feel if the pore wall consisted of liquid of same molecules.   The value of μ_{ads} should not depend on the location in the adsorption phase.   Nevertheless, the third term of the right hand side of Eq. (4) is a function of the location in the pore.   This means that the second term is also the function of the location.   Since the pressure difference across the interface is directly connected with the curvature of the interface by the Young-Laplace equation, the curvature of the interface should also be a function of the location.   The equilibrium condition $\mu_g=\mu_{ads}$, combined with Eqs. (3), (4) and Young-Laplace equation, yields:

$$kT \ln\left(P_g / P_{sat}\right) = -\gamma V_p / \rho(x) + \Psi(x) \qquad\qquad (5)$$

Where $\rho(x)$ is radius of interface curvature at x and γ is surface tension.   This equation gives local curvature of the condensed phase that varies with the location in pore corresponding to the variation in $\Psi(x)$.   We also take into account of the curvature dependency of surface tension[12] in the proposed model.

Based on Eq. (5), the equation to describe the relation between critical pore size and equilibrium vapor pressure can be derived.   Surface adsorption film of thickness t exists in the vicinity of the pore wall, while the condensation phase fills the inner portion of the pore having the interface with gas phase that touches to the adsorption film with zero contact angle.   Taking the origin of x at the center of pore, $\rho(x)$ and x should satisfy the relation of Eq. (6) due to the geometrical relation, where ϕ represents the angle formed between the tangential line on the interface and a line parallel to y-axis.

$$-\rho(x)\, d\phi \sin\phi = dx, \quad \text{or} \quad \frac{1}{\rho(x)} = \frac{d\cos\phi}{dx} \qquad\qquad (6)$$

The following differential equation is obtained from Eqs. (5) and (6), writing $\Delta\mu = -kT\ln(P_g/P_{sat})$.

$$\frac{d\cos\phi}{dx} = \frac{\Delta\mu + \Psi(x)}{\gamma V_p} \qquad\qquad (7)$$

Boundary conditions for forming the gas-condensate interface with contact angle being zero are $\phi=0$ at $x=x_0$ and $\phi=\pi/2$ at $x=0$.   We use the LJ(9-3) potential as $\Psi(x)$ because of its simplicity as well as its good performance in describing multilayer adsorption.   The potential function in Eq. (7) should be obtained from superposition of potentials from both sides of pore walls, and from subtraction of corresponding potential for liquid state. $\Psi(x)$ thus should be given as Eq. (8).

$$\Psi(x) = -A\left(x + d_h\right)^{-3} + B\left(x + d_h\right)^{-9} - A\left(d_h - x\right)^{-3} + B\left(d_h - x\right)^{-9} \tag{8}$$

Integrating Eq. (5) with this potential, and applying the boundary conditions, we obtain Eq. (9), which must be satisfied at the critical condition for condensation.

$$1 = \frac{1}{\gamma V_p}\left[\Delta\mu x_0 + \frac{A}{2}\left\{\left(d_h + x_0\right)^{-2} - \left(d_h - x_0\right)^{-2}\right\} - \frac{B}{8}\left\{\left(d_h + x_0\right)^{-8} - \left(d_h - x_0\right)^{-8}\right\}\right] \tag{9}$$

Where $A = \dfrac{2\pi}{3}\left(N_s \varepsilon_{gs}\sigma_{gs}^{6} - N_g \varepsilon_{gg}\sigma_{gg}^{6}\right)$, $B = \dfrac{4\pi}{45}\left(N_s \varepsilon_{gs}\sigma_{gs}^{12} - N_g \varepsilon_{gg}\sigma_{gg}^{12}\right)$, d_h is one half of pore width. x_0 should satisfy $-\Delta\mu = \Psi(x_0)$ according to its definition. The repulsive term in Eq. (9) can be negligible because the effect of repulsive potential decreases rapidly with increase of distance from a pore wall, and it hardly influences condensation phenomenon around the inner portion of the pore.

For a given $\Delta\mu$, corresponding critical pore width below which condensation occurs can be easily obtained with Eq. (9) using a simple iteration technique to determine d_h and x_0 simultaneously. This formula would be helpful in use for the characterization of meso-porous media. In this study, we used LJ(9-3) potential as an example of $\Psi(x)$, but one may choose any potential function that may be suitable to express the interaction between adsorbate and pore wall for a given adsorption system.

3.2 Shape of the interface

The Kelvin model implies a constant curvature of the interface. On the other hand, the curvature is a function of coordinates x in the pore as shown in Eq. (5). Therefore, the shape of the interface is not semi-cylindrical in the proposed model. Integration of both sides of Eq. (7) from surface of adsorption film on pore wall to arbitrary point yields Eq. (10), in which surface tension γ is treated strictly as a variable.

$$\cos\phi - 1 = \frac{1}{V_p}\int_{x_0}^{x}\frac{1}{\gamma(x)}\left[\Delta\mu + \Psi(x)\right]\mathrm{d}x \tag{10}$$

Since the relation of $dy/dx = 1/\tan\phi$ holds, denoting $y = y_c$ at the center of the pore, integration of the relation under the boundary condition of $y = y_c$ at $x = 0$ gives Eq. (11).

$$y = y_c + \int_0^x \frac{\mathrm{d}x}{\tan\phi} \tag{11}$$

Because ϕ can be related to x by Eq. (10), the integration of Eq. (11) can be executed numerically to obtain the shape of gas-condensate interface. The results are discussed later in section **4.3**.

4 RESULTS AND DISCUSSION

4.1 Example of GC-like MD Simulation

A simulation with permeable border was carried out in a slit-like pore of 2nm in width. The value of relative pressure was 0.4. Figure 2 shows the development of the adsorption phase. From Figure 2, we can see the change of the adsorption phase from the state of adsorption layers on solid surfaces to that of condensation. Around 20ns, the first layer of the adsorption film was formed. Then after 34ns, the density of adsorbed phase increased rapidly. This point corresponded to the formation of the condensation phase as seen in Figures 2(d) and (e). Once the condensation phase is formed, the rate of the phase growth was very high. After 85ns, almost whole space within the pore was filled with adsorbate particles, and the system reached an equilibrium state. We did not analyze the result in detail this time. We hope that this simulation method could give certain contribution to the understanding of dynamic feature of adsorption/desorption processes.

4.2 Adsorption Isotherm

With the reflective border, several runs were conducted with various numbers of particles within pores of various widths ranging from 2nm to 4nm to obtain the adsorption equilibrium relation. The density of the adsorption phase within the pore in the equilibrium state against relative pressure is shown in Figure 3. The formation of the condensation phase with curved gas-condensate interface was seen at the point corresponding to the step rise in the adsorption isotherm. Thus, the critical pressure for capillary condensation can be evaluated by this method. The value for each pore width is summarized in Table 1.

4.3 Verification of condensation pressure

The Kelvin equation describes the effect of curved surface for condensation in the following form for slit pores with zero contact angle.

$$d_s = \frac{2\gamma V_p}{kT \ln\left(P_{sat} / P_g\right)} + 2t \qquad (12)$$

Where d_s is pore width and t is the thickness of adsorption film on a pore wall. In the calculation of P_{sat}/P_g which satisfied Eq. (12), for a given d_s, we used γ which was calculated from potential parameters of the adsorbate particle used in this simulation referring to the study of Chapela *et al*[13]. V_p was decided from the density of a simulated normal bulk liquid system. The condensation pressures calculated from Eq. (12) for each size of pores are also shown in Table 1. The relation between critical relative pressure for condensation and pore width calculated from Eq. (12) (Kelvin model) and that obtained by solving Eq. (9) numerically (proposed model) are shown in Figure 4, with the MD simulation result. They are also summarized in Table 1.

From Figure 4, it is seen that the Kelvin model underestimate the pore width comparing to the proposed model. When these two condensation models are compared with the MD simulation result, the condensation pressures by the Kelvin model are obviously larger than those of the simulation result. On the other hand, the

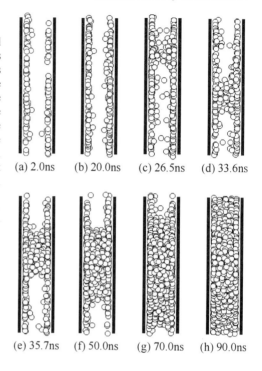

(a) 2.0ns   (b) 20.0ns   (c) 26.5ns   (d) 33.6ns

(e) 35.7ns   (f) 50.0ns   (g) 70.0ns   (h) 90.0ns

Figure 2 *Development of the adsorption phase observed in the GC-like MD simulation for a slit-like pore of 2nm in width at $P_g/P_{sat}=0.4$.*

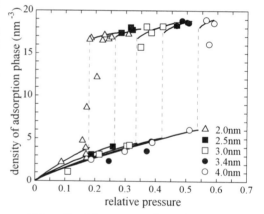

Figure 3 *Adsorption isotherm for several size of slit-like pores calculated by MD simulation. The locations of phase transition are indicated by dashed lines.*

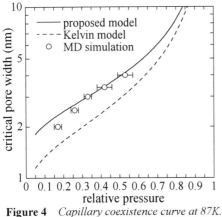

Figure 4 *Capillary coexistence curve at 87K.*

Table 1 *Comparison of condensation pressures for Kelvin model, proposed model and MD simulation results.*

Pore width (nm)	Kelvin model	Proposed model	MD simulation
2.0	0.30	0.088	0.17±0.02
2.5	0.42	0.21	0.26±0.02
3.0	0.51	0.33	0.33±0.02
3.4	0.57	0.41	0.42±0.04
4.0	0.64	0.51	0.53±0.04

relation between pore width and condensation pressure calculated by the proposed model showed good agreement with MD simulation result in the range of 3.0nm to 4.0nm. Therefore, in evaluating such nano-scale pores, the proposed model succeeds to give more accurate pore size. As for the pore of 2nm in width, the proposed model shows deviation from that of MD simulation result. The difference might have been caused by the failure of the continuous fluid approximation in such small adsorption space as 2nm in width.

4.4 Verification of shape of gas-condensate interface

The shape of the interface in simulations were determined as follows. Lines in Figure 5 indicate some density profiles in the direction of y at various x in a pore of 3nm under the relative pressure at which condensation phase has formed in our MD simulation. These lines were obtained by fitting the distribution in the direction of y to a hyperbolic tangent function. In the figure, we regarded y coordinates where the density equals to mean value of that at $y=0$ and $y=4$nm as the location of gas-condensate interface at given x. We plotted the interface locations for several x in Figure 6. The shape of gas-condensate interface under the critical relative pressure calculated from Eq. (11) is also shown in Figure 6 as a solid line. The dotted line in Figure 6 indicates the shape of semi-cylindrical interface calculated from the modified Kelvin model under the same relative pressure. Comparing the shape of interface by the proposed model with that by

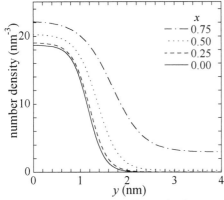

Figure 5 *Density profiles in the direction of y at x=0, 0.25, 0.5, 0.75 (nm).*

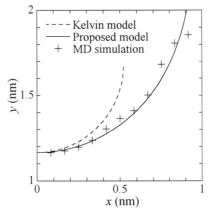

Figure 6 *Shape of gas-condensate interface.*

the Kelvin one, it is clear that the proposed model gives a smaller interface curvature than the Kelvin model. This result is reasonable considering that the potential compensation by the pore wall works on the forming of condensation phase in the proposed model. Moreover, the shape of interface which was observed in the MD simulation shows good agreement with that of the proposed model. This result supports the validity of the proposed model as for the shape of gas-condensate interface.

5 CONCLUSIONS

In this study, we introduced the concept of the *border plane* with an imaginary vapor phase and *potential buffering field* to establish a new molecular dynamics scheme for adsorption/condensation in pores. By adopting different setting of the border plane, we could performed 2 types of simulation of GC-like and NVT scheme. The benefit of our MD method would be, first, the vapor phase pressure of the system is easily obtained with simple algorithm. Secondly, the gas-condensate interface can directly be observed. By using the MD simulation as an experimental system, we proposed and verified a new condensation model which described the condensation phenomenon in nano-scale pore. The proposed model was verified from the viewpoint of the critical condensation pressure and the shape of gas-condensate interface by using the MD simulation. As a result, the critical condensation pressure predicted by this model showed good agreement with the MD simulation result in the range of 3.0nm to 4.0nm. As for the shape of gas-condensate interface, the calculated shape from the proposed model for a pore of 3nm showed good agreement with the MD simulation result.

The feature of the adsorption phase such as the shape of the gas-condensate interface could be examined in detail by the present simulation method, and was utilized for studies on adsorption phenomena in nano-scale pores in this study. Consequently, it was suggested that the proposed condensation model must be useful in characterization of nano porous media.

Acknowledgment

Computation time was provided by the Supercomputer Laboratory, Institute for Chemical Research, Kyoto University.

References

1. S. Iijima, *Nature*, 1991, **354**, 56.
2. J. S. Beck et al., *J. Am. Chem. Soc.*, 1992, **114**, 10834.
3. R. Evans, U. M. B. Marconi, and P. Tarazona, *J. Chem. Soc. Faraday Trans. 2*, 1986, **82**, 1763.
4. B. K. Peterson and K. E. Gubbins, *Mol. Phys.*, 1987, **62**, 215.
5. O. Kadlec, *Carbon*, 1989, **27**, 141.
6. M. P. Allen and D. J. Tildesley, 'Computer Simulation of Liquids', Oxford University Press, New York, 1990, Chapter 7, p.230.
7. S. Sokolowski and W. Steele, *Mol. Phys.*, 1985, **54**, 1453.
8. W. A. Steele, 'The Interaction of Gases with Solid Surface', Pergamon Press, New York, 1974, Chapter 2, p.51.
9. L. Verlet, *Phys. Rev.*, 1967, **159**, 98.
10. B. Widom, *J. Chem. Phys.*, 1963, **39**, 2808.
11. B. Widom, *J. Stat. Phys.*, 1978, **19**, 563.
12. J. C. Melrose, *Ind. and Eng. Chem.*, 1968, **60**, 53.
13. G. A. Chapela and G. Saville; *J. Chem. Soc. Faraday Trans. 2*, 1977, **73**, 1133.

TEXTURAL ADSORPTION CHARACTERIZATION ACCORDING TO THE TWOFOLD DESCRIPTION OF POROUS MEDIA

V. Mayagoitia[1], F. Rojas[1], I. Kornhauser[1] and J. Salmones[1,2].

[1]Departamento de Química. Universidad Autónoma Metropolitana-Iztapalapa.
Apartado Postal 55-534, México 13, D.F. 09340 México.
[2]STI. Instituto Mexicano del Petróleo.
Apartado Postal 14-805, México 9, D.F. 07300 México.

1 INTRODUCTION

Textural characterization of porous solids from adsorption and capillary measurements is a task pursued for a long time. Prior to any attempt of this kind, a serious effort should be made to take into account the most significant characteristics of the porous structure, specially those concerned with the processes undergone by fluids within the pores during a textural determination. Many authors have conceived the porous structure as an interconnected network of voids, but the advances and difficulties of these conceptions have never been fully imagined. A recent approach called the 'twofold description' of such media allows a natural and direct consideration of phenomena such as adsorption, imbibition or intrusion in porous solids, that promise a more precise and critical evaluation of structural parameters.

Textural adsorption characterization according to the twofold description[1] highly regards the peculiarities of real porous structures: (i) first of all, since two different kinds of entities, sites (cavities, antrae) and bonds (capillaries, passages) alternate to form the network, a twofold size distribution is the relevant information to obtain, rather than a mere single 'pore-size distribution' as it has been traditionally considered, (ii) most porous networks are not merely fully-random media: neighbouring elements are appreciably correlated in size, (iii) the state (filled or emptied with fluid) of entities as function of their size is considered, and (iv) all kind of possible interactions, assisted or hindered, are envisaged, specially the cooperative phenomena occurring when the wetting phase is displacing the non-wetting one.

Texture characterization consists in the determination of the twofold size distribution by deconvolution of the expression for the (experimentally observable) degree of filling by liquid of the whole structure in a volume basis, in terms of the relative pressure of the adsorptive. This is not an easy matter, since therein lie two important sets of quantities: the degrees of filling of the two kinds of elements in terms of their sizes, referred to the specific process involved (adsorption, desorption), and always considering size correlations among entities.

Previously, a general development has been outlined[2] describing some of the unavoidable limitations concomitant to textural determinations by adsorption methods. Methods such as that of Barrett, Joyner and Halenda and the percolation one, reveal just as the limits of the twofold model. Although up till now the objective of determining unequivocally the structure of porous materials is far from being attained, we hope that the examples presented here will provide an idea about some of the actual problems and possibilities provided by adsorption measurements for elucidating the texture of porous materials.

In this work a sequential procedure will be followed to deal with the textural properties of porous solids. First the principal aspects of the twofold description of porous structures will be presented, and under these terms a classification of porous media in five types will be recalled. Next, interpretation and treatment of experimental data obtained from adsorption hysteresis loops will be made with the aim of determining the twofold (site and bond) pore size distribution of some special materials.

2 TWOFOLD DESCRIPTION OF THE POROUS MEDIUM

Morphology of porous media has been conceived on the basis of the so-called twofold description[1]. This approach has proved useful for the study of different capillary processes such as capillary condensation and evaporation, imbibition and immiscible displacement[3,4]. In this treatment the porous network is conceived as constituted by two kinds of alternated and intercommunicated void elements: sites and bonds. The number of bonds meeting at a site is the connectivity, C; every bond connecting two sites. The size distribution functions for sites and bonds, $F_S(R)$ and $F_B(R)$, are each one normalized and expressed on a number of elements basis, so that the probabilities to find a site or a bond having a size **R** or smaller are, respectively:

$$S(\mathbf{R}) = \int_0^R F_s(R)\, dR \quad ; \quad B(\mathbf{R}) = \int_0^R F_B(R)\, dR \tag{1}$$

The following 'Construction Principle' emerges from the very definitions of 'site' and 'bond': the size of any bond is always smaller than or at most equal to the size of any of the sites to which it leads. The principle is of the uppermost importance when overlap between site and bond size distributions is considerable. In such a case, perhaps the most frequent one, some bonds have larger sizes than those of some sites, but then these elements can never be first neighbours, this constraint being precisely the cause of size correlations. Two self-consistency laws are needed to guarantee the fulfilment of the above principle. The first law states that a sufficient number of bonds (smaller than or at most equal to the sites to which they will be connected) must be provided by the bond size-distribution in order that the whole set can be fitted with the corresponding set of sites:

First Law $B(R) \geq S(R)$ for every R   (2)

The joint probability density for the event of finding a site of size R_S and one of its bonds of size R_B, can be expressed as:

$$\rho(R_S \cap R_B) = F_S(R_S)\, F_B(R_B)\, \phi(R_S, R_B) \tag{3}$$

where ϕ is a correlation function to be written afterwards. The second law has a local character and prevents the union of neighbouring sites and bonds that would violate the construction principle:

$$\text{Second Law} \qquad \phi(R_S, R_B) = 0 \qquad \text{for} \quad R_B > R_S \qquad (4)$$

If during the topological assignation of sizes the randomness is kept at a maximum, but fulfilling the Construction Principle, the most verisimilar (likely) form of ϕ for the correct case, $R_B \leq R_S$, is obtained[1]:

$$\phi(R_S, R_B) = \frac{\exp\left(-\int_{S(R_B)}^{S(R_s)} \frac{dS}{B-S}\right)}{B(R_S) - S(R_S)} = \frac{\exp\left(-\int_{B(R_B)}^{B(R_s)} \frac{dB}{B-S}\right)}{B(R_B) - S(R_B)} \qquad (5)$$

ϕ implies topological size correlations that promote a size segregation effect, i.e., there appear regions of big sites and bonds grouped together, alternated with other regions of smaller reunited elements. Size segregation intensifies with the overlap between the size distributions. When overlap is almost complete, segregation is extreme. However when the overlap is zero, elements are free to accommodate in a fully random way ($\phi = 1$, since in this case the construction principle could never be violated). The consequences of size correlations on the development of capillary processes in porous solids are of the utmost importance.

Porous networks have been classified in five types[3], according to the relative positions of their site and bond size distributions. Type I corresponds to the case of zero overlap, with distributions widely separated from each other; the smallest site is at least twice the size of the biggest bond. Type II is a more general case of zero overlap, with distributions somewhat apart, the smallest site is smaller than twice the biggest bond. Type III, is an extreme case of zero overlap; the distributions lie very close to each other the size of the biggest site is smaller or at most equal to twice the size of the smallest bond. Type IV is the most general case of intermediate overlap, there are sites smaller than some bonds; which of course are not interconnected to each other. Type V arises when the overlap between distributions is near completeness, the segregation effect induced by ϕ is a maximum and the network structuralizes into homogeneous domains in which elements possess almost the same size.

Every one of these types presents a very particular behaviour during the course of capillary processes. Each pattern of the confined phases or fluids is astonishingly characteristic for each type of structure, specially when the wetting phase is the advancing one (e.g. capillary condensation or imbibition). The fingerprint of porous morphology, controlled by ϕ, is always present in all kinds of capillary processes[4].

3 ANALYSIS OF ADSORPTION HYSTERESIS LOOPS

Classical treatments to determine the pore size distribution, have ignored two important phenomena: pore blocking effects[5] during the evaporation process along the descending boundary curve (DBC) and cooperative irreversible phenomena during capillary condensation[6] along the ascending boundary curve (ABC). Such phenomena seriously mislead the results of the pore size distributions obtained from both the ABC and the DBC. Cooperative phenomena must be inevitably taken into account to estimate the twofold pore size distribution.

Experimental data provide θ_v, the global degree of filling of the network by liquid, in volume as function of the relative pressure of vapour, p/p^0. θ_v is related to $\theta_S(R)$ and $\theta_B(R)$, the degrees of filling for sites and bonds of size R, both given in a number of elements basis:

$$\theta_V = \frac{\int\limits_0^\infty \theta_S(R)\, V_S(R)\, F_S(R)\, dR + \frac{1}{2}\, C \int\limits_0^\infty \theta_B(R)\, V_B(R)\, F_B(R)\, dR}{\int\limits_0^\infty V_S(R)\, F_S(R)\, dR + \frac{1}{2}\, C \int\limits_0^\infty V_B(R)\, F_B(R)\, dR} \tag{6}$$

where V(R) is the volume of each element of size R, either site or bond as indicated by the subscripts S and B respectively, and considering that there are C/2 bonds per site in the network.

The objective of a textural treatment consists in determining F_S and F_B as functions of R, from eqn. (6). The general solution is complex, but still accessible for certain types of porous structures, such as Types I, III and V.

Characteristics of the capillary condensation and evaporation phenomena and appearance of the hysteresis loops for the different types of porous structures can be resumed as follows. For Type I, sites and bonds behave independently during condensation. At low vapour pressures, there is only condensation in bonds. Condensation in sites starts when all bonds are filled with liquid. Because of the small size of bonds compared to sites, sudden evaporation takes place after the isotherm traces an extended plateau and at a well-defined percolation threshold an abrupt descent occurs. The hysteresis loop is extremely wide. For Type II, the hysteresis cycle is narrower but still wide, condensation in sites occurs at an earlier stage than for type I, the importance of the adsorbed layer is greater than before. For Type III the hysteresis loop shows very steep ABC and DBC, because intense cooperative phenomena arise during the condensation and evaporation processes. There is a plateau of the DBC at high relative vapour pressures because of pore blocking. Type IV shows a narrow hysteresis cycle, both ABC and DBC are sloping curves. Type V is an extreme case of porous structures, because it is constituted by alternated domains each formed by elements of the same size, condensation in bonds automatically causes the filling of sites. The evaporation process is free of any percolative effect and hysteresis is only due to the different interfacial geometries in adsorption and desorption; usually the hysteresis loop is very narrow.

The method to determine the twofold pore size distribution will be: (i) previous to any uncritical treatment of data, the hysteresis loop should be analyzed to evidence some crucial facts, specially if the porous network is a connected structure and not merely a collection of independent pores (as the interstices in an array of parallel plates), if the solid is rigid and/or if capillary condensate does not break down during evaporation (i.e. its tensile strength is not surpassed), (ii) if all the preceding conditions are fulfilled, one can safely continue by determining from the form of the hysteresis loop to which type of porous structure the sample belongs, and (iii) to apply the specific treatment[3,7] for dealing with the incumbent porous type to consider cooperative behaviour, pore blocking and size topological correlations between pores.

4 PREPARATION OF ADSORBENTS AND ADSORPTION ISOTHERMS

Two substrates were chosen to be analyzed according to the twofold description of porous media: a zirconia-silica sample and a monodisperse globular carbon specimen. Brief preparation details are given below.

The ZrO_2-SiO_2 porous support impregnated with platinum was obtained from thermal treatment of a gel precursor[8]. Tetraethoxysilane (TEOS) and zirconia (IV) butoxide were dissolved together in ethyl alcohol and left reacting for two hours at 351 K. The system was cooled to 278 K where a stoichiometric 1 N H_2SO_4 aqueous solution was added slowly to induce the co-hydrolysis of the zirconia-silica hydroxide. The acid acted as the peptizing agent and provided sulphate groups that increase the acidity of the material. The $Pt/SO_4^{2-}(ZrO_2$-$SiO_2)$ catalysts were prepared by adding platinum (0.5 wt %), using a solution of H_2PtCl_6 in ethanol. The final material was obtained by drying the hydroxides at 383 K followed by calcination at 673 K for five hours and then reduced at the same temperature in hydrogen during one hour. The zirconia-silica sample chosen for textural studies was sample ZS with a support weight composition of 10% ZrO_2-90% SiO_2.

A globular carbon sample made of colloidal monosized spheres was obtained from poly(acrilonitrile) (PAN) precursors, after oxidation in air and subsequent carbonization under an inert atmosphere. The precursors were obtained from a monodisperse PAN lattice containing spheres dispersed in n-hexane and stabilized by a block copolymer. Preparation details are given elsewhere[9]. The specific sample used for adsorption studies was labelled as carbon 3.5.3 treated at 1473 K, made of spheres with a mean diameter of 208 nm and with a specific surface area of 16.5 m^2/g. The size and the random packing of the globular carbon were determined and seen by electron microscopy.

Isotherms loops of nitrogen at 76 K were determined for the ZS sample in a volumetric-static instrument ASAP 2000 from Micromeritics Corp. Benzene adsorption at 298 K was determined for carbon 3.5.3 using a gravimetric apparatus that included a Cahn microbalance.

5 RESULTS AND DISCUSSION

Sample ZS was thought to be a type V structure; this was confirmed by the textural results. The isotherm (Figure 1) show a narrow hysteresis loop with no upper desorption plateau, i.e. no pore blocking was evident. The isotherm did not show a sudden breakage at relative pressures of about 0.4. The globular silica spheres are surrounded by zirconia crystallites; that intercalate between them preventing their sintering[10] and rendering an open porous structure. Random sphere packings can be modelled as an array of irregular tetrahedra[11]; the central voids of the tetrahedra are surrounded by four windows that communicate to neighbouring sites. Analysis of the ABC and DBC curves of a type V structure should lead to the same distribution function, since sites and bonds are about the same size and hysteresis would be only due to different menisci geometries during adsorption and desorption. Sites were assumed as spherical voids while bonds were imagined as cylindrical capillaries. The following equations were used to determine the twofold distribution[2]:

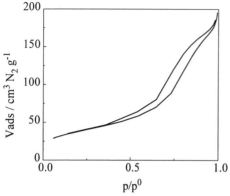

Figure 1 *Nitrogen adsorption on sample ZS at 76 K*

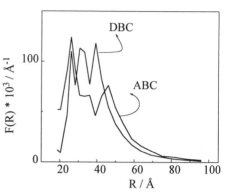

Figure 2 *Twofold pore size distribution function for sample ZS*

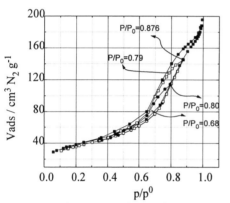

Figure 3 *Primary descending scanning curves of nitrogen at 76 K on sample ZS*

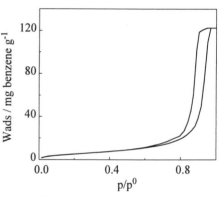

Figure 4 *Benzene adsorption on carbon 3.5.3 at 298 K*

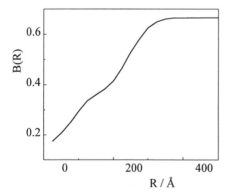

Figure 5 *Plot of B(R) vs R for carbon 3.5.3*

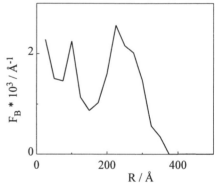

Figure 6 *Bond size distribution for carbon 3.5.3*

$$S\left(\frac{R_C}{2}\right) = B\left(\frac{R_C}{2}\right) = \frac{\int_0^{\theta_V(R_c/2)} d\theta_V / V(R)}{\int_0^1 d\theta_V / V(R)} \qquad \text{for the ABC} \qquad (7)$$

$$S(R_C) = B(R_C) = \frac{\int_0^{\theta_V(R_c)} d\theta_V / V(R)}{\int_0^1 d\theta_V / V(R)} \qquad \text{for the DBC} \qquad (8)$$

Along the ABC, $d\theta_V$ was due both to condensation in sites (induced by phase transitions in their neighbouring bonds), and to an increase in the thickness of the adsorbed layer. The Kelvin equation considering a cylindrical meniscus and a correction term due to the adsorption potential[12] was used to determine the critical radius of curvature R_C at which condensation starts. The thickness of the adsorbed layer was approximated using an equation[13] that relates this quantity with the relative pressure of the vapour. The DBC was analysed in similar terms, but in this case the meniscus shape was assumed as hemispherical. The twofold pore-size distribution for sample ZS is shown in Figure 2. In this case there is a close agreement between the size distributions calculated from the ABC and DBC curves, indicating that we have indeed a Type V structure. Additional evidence for this assertion was provided by primary descending scanning curves. Figure 3 shows that scanning curves incorporate to the DBC as soon as the largest liquid-filled pore entities (i.e. those which were just filled at the inversion point on the ABC) start evaporating the liquid at a relative pressure calculated from the corrected Kelvin equation while considering a cylindrical meniscus. The p/p^0 values shown inside Figure 3 correspond to the inversion point of the scanning curve and to the pressure at which the DBC and the scanning curve should be reunited. The agreement is satisfactory.

Carbon 3.5.3 would correspond to a Type III porous structure. The sorption isotherm (Figure 4) shows very steep ABC and DBC curves and a desorption plateau is evident. The sizes of pores are relatively large, since condensation in general occurs at high relative pressures. The packing is random and sintering and distortion of the particles ocurred[9]. The characteristic of this type of structure is the cooperative character and homogeneous filling of sites along the ABC (from the onset of condensation all sites are in a supersaturated state); while percolative vapour behaviour appears at the DBC. If bonds are assumed as volumeless windows then $\theta_V = \theta_S$ and for the ABC we have[2]:

$$\theta_S = \theta_B{}^C + C\theta_B{}^{C-1}(1-\theta_B) \qquad (9)$$

$$\theta_B = B(R_C/2) + [1 - B(R_C/2)][1-(1-\theta_B{}^{C-1})^2] \qquad (10)$$

while for the percolation process along the DBC:

$$B(R_C) = \frac{\theta_V{}^{1/C} - \theta_V{}^{(C-1)/C}}{1 - \theta_V{}^{(C-1)/C}} \qquad (11)$$

θ_V should be corrected taken into account the adsorbed layer thickness and adsorption potential[14] as above. From the ABC it is then possible to calculate $B(R_C/2)$ and $B(R_C)$ from the DBC. No information can be obtained for S(R), because in a Type III structures bonds control both the condensation and evaporation processes. Condensation in sites is possible if C or at least C-1 of the surrounding bonds are filled; desorption starts until bonds proceed to evaporate their condensate. Figure 5 shows B(R) as function of the pore radius, these results come from ABC and DBC data. The maximum value of B corresponds to (C-2)/(C-1), i.e. 0.667 for C = 4, which means that the percolation threshold will appear when one third of the bonds are susceptible of evaporating their condensate. In consequence there is no information about the size distribution of the largest pores. The pore size distribution of bonds can be obtained by derivation of B(R) (see Figure 6). Additional information about $F_B(R)$ and even $F_S(R)$ can be provided by scanning curves[7].

6 CONCLUSION

Sorption data, when properly analyzed, should lead to a complete or at least partial but confident determination of the textural properties of adsorbents.

Acknowledgement

Thanks are given to Consejo Nacional de Ciencia y Tecnología of México for support to carry out this work under Project No. 5387-E.

References

1. V. Mayagoitia, M. J. Cruz and F. Rojas. *J. Chem. Soc., Faraday Trans. 1*, 1989, **85**, 2071.
2. V. Mayagoitia, *Catalysis Lett.*, 1993, **22**, 93.
3. V. Mayagoitia, F. Rojas and I. Kornhauser, *J. Chem. Soc. Faraday Trans. 1*, 1988, **84**, 785.
4. V. Mayagoitia, F. Rojas, I. Kornhauser, G. Zgrablich and J. Riccardo, in 'Characterization of Porous Solids III', eds. J. Rouquerol, F. Rodríguez-Reinoso, K. S. W. Sing and K. K. Unger, Elsevier, Amsterdam, 1994, p. 141.
5. H. W. Quinn and R. McIntosh, in 'Surface Activity', Vol. 2, ed. J. H. Schulman, Butterworths, London, 1957, p. 122.
6. V. Mayagoitia, F. Rojas and I. Kornhauser, *J. Chem. Soc., Faraday Trans. 1*, 1985, **81** 2931.
7. V. Mayagoitia, B. Gilot, F. Rojas and I. Kornhauser, *J. Chem. Soc. Faraday Trans. 1*, 1988, **84**, 801.
8. J. Salmones, R. Licona, J. Navarrete and P. Salas, *Catalysis Lett.*, 1996, **36**, 135.
9. D.H. Everett and F. Rojas, *J. Chem. Soc., Faraday Trans. 1*, 1988, **84**, 1455.
10. M. Deeba, R. J. Farrauto and L. K. Lui, *Appl. Catal. A-Gen*, 1995, **124**, 339.
11. G. Mason, *J. Colloid Interface Sci.*, 1971, **35**, 279.
12. J. R. Philip, *J. Chem. Phys.*, 1977, **66**, 5069.
13. D. Dollimore and G. R. Heal, *J. Colloid Interface Sci.*, 1970, **33**, 508.
14. F. Rojas. PhD Thesis, University of Bristol, 1982.

DETERMINATION OF THE SURFACE AREAS AND PORESIZES OF ORGANIC MATTER IN SOILS

Marjo C. Mittelmeijer-Hazeleger, Hubert de Jonge[*] and Alfred Bliek
Depart. of Chemical Engineering   [*]Depart. of Physical Geography and Soil Science
University of Amsterdam   University of Amsterdam
Nieuwe Achtergracht 166   Nieuwe Prinsengracht 130
1018 WV Amsterdam   1018 VZ Amsterdam

1 INTRODUCTION

The wider scope of our work relates to the understanding of the dispersion and sorption behavior of pollutants in soils. For this reason the last decades environmental scientists are interested in measuring the surface areas of soils.

Soils are very complex materials, consisting of minerals, clay and organic material. For this reason normally the surface area of soils is measured and corrected afterwards for the mineral content. In the environmental literature there is consensus that Soil Organic Matter is the primary adsorbent for hydrophobic organic compounds, even in wet soils and sediments[1,2].

Until recently[3] nitrogen adsorption measurements have been recommended as the standard method to determine surface areas[4] despite the fact that large discrepancies exist between the surface areas measured by nitrogen adsorption and by retention of polar adsorbates, as ethylene glycol monoethyl ether (EGME), ethylene glycol and suchlike. For instance Chiou et al.[5] reported values for the surface areas determined by nitrogen adsorption as low as .7 m^2/g for natural soils up to 18 m^2/g for a freeze dried humic acid soil. These surface areas are in sharp contrast to the values of 560-800 m^2/g reported by Chiou et al.[6] determined by EGME retention. This difference is subject to debate in the environmental literature[6-11]. The presented density measurements of the soils, with EGME as displacement medium, together with adsorption measurements of some small hydrocarbons and carbon dioxide are helpful in resolving the above mentioned contradicting results.

2 EXPERIMENTAL

2.1 Techniques and experimental conditions

Adsorption studies using N_2 at 77K and carbon dioxide, ethyne, ethene and ethane at 273 K, were performed on a Carlo Erba Sorptomatic 1800 and 1990. Equilibrium conditions were set to a smaller change in pressure than .132 mbar in 15 min. Some samples needed more than 22 hours to reach sorption equilibrium.

In table 1 the critical diameter, the specific surface area of an adsorbed molecule (Am) and the density of the adsorbed phase (ρ_{liq}) is given. The values of ρ_{liq}, except for the one of nitrogen, are corrected according to Dubinin[12]. This is necessary because the liquid in an adsorbed phase is more dense than in a normal liquid, especially in the temperature region near the critical temperature. Specific surface areas for the adsorbates used (Am) are calculated by assuming a spherical shape and a hexagonal close packing. *Viz.* Am=$1.091(M/N. \rho_{liq})^{2/3}$ in which M is the molecular mass and N is the Avogadro number.

Table 1 *Constants of the gases used.*

gas	*Am* (nm²)	ρ_{liq} (g/cm³)	*Critical diameter* (nm)
Nitrogen	.164	.808	.30
Carbon dioxide	.179	1.100	.28
Ethane	.239	.491	.44
Ethene	.225	.497	.42
Ethyne	.204	.539	.24

Helium density measurements of all the samples were performed in a Micromeretics pycnometer. These measurements needed equilibrium times of about 5 min. For the density measurements with EGME a normal glass pycnometer was used. During density measurements with EGME a true equilibrium was not reached and we arbitrarily took the values obtained after ten days.

2.2 Samples

We selected soils with a low or negligible clay content and no carbonates. The ash content of the samples was either low or consisted of coarse silica with negligible surface area.

Different types of soils are collected from the most humified horizons of the organic topsoil at the surface under aerobic conditions (indicated as O_h) or under anaerobic conditions (peat). As humic acids are often used as reference materials for soils, we investigated three samples prepared from a commercially available humic acid (Aldrich H1,675-2). The humic acid, as received, was in the sodium salt form. The origin of the samples is given in Table 2.

Table 2 *List of the codes of the samples studied and description of their origin.*

Sample code	Origin
O_hS	Taken in a Spruce forest at the Veluwe, The Netherlands
O_hF	Taken in a Douglas fir forest at the Veluwe, The Netherlands
O_hP	Taken in a Scotch pine forest at the Veluwe, The Netherlands
Peat	Taken from a peat bog in Drenthe, The Netherlands
HA1	Prepared by dissolving in demineralized water (1:50 wt), brought to pH 2 with 0.1 M HCl, allowed to settle for 24 hrs. Centrifuged for 30 min at 2000 rpm. Washed with 0.5 M $CaCl_2$ and 0.01 M $CaCl_2$.
HA2	Same as above, except the solution was brought to pH 5.
HA3	The mother liquor of the HA2 sample was brought to pH 4, allowed to settle for 24 hrs, centrifuged. The remaining mother liquor was brought to pH 3, allowed to settle for 24 hrs, centrifuged. Washed with 0.5 M $CaCl_2$ and 0.01 M $CaCl_2$.

All soils were first air dried at 323K during 72 hours and passed over a 2 mm sieve. Washing with $CaCl_2$ of the humic acid samples was necessary in order to obtain Ca-complexed insoluble humates. After preparation these samples were freeze dried.

Before each adsorption and density measurements, the samples were dried and outgassed at 333 K under vacuum ($<10^{-3}$ mbar) for at least 24 hours. The characterization of the samples by elemental composition and ash content is given in Table 3.

Table 3 *Elemental composition, ash content and external surface area of the samples.*

| Sample code | Elemental analysis (% DAF) | | | | Ash |
	C	H	O	N	(D %)
O_hS	54.1	7.9	35.4	2.6	51.7
O_hF	55.3	5.4	36.7	2.6	24.4
O_hP	59.5	4.4	34.2	1.9	11.2
Peat	56.4	5.3	36.8	1.5	13.2
HA1	56.5	5.2	37.4	.9	19.9
HA2	61.7	7.3	30.5	.5	25.1
HA3	46.2	7.0	46.3	.6	19.6

DAF= Dry and ash free basis

3 RESULTS

In figure 1 the nitrogen adsorption isotherms of the soil and humic acid samples are shown. The isotherms look quite similar, except for the HA2 sample, that demonstrates a more or less linear isotherm.

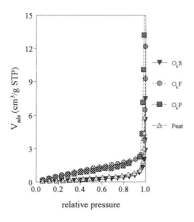
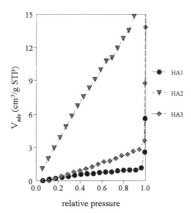

Figures 1 *Nitrogen adsorption isotherms.*

Figure 2 shows the CO_2 adsorption isotherms for the soil and humic acid samples. For all the samples at least two desorption points were measured. In none of the cases was any hysteresis found.

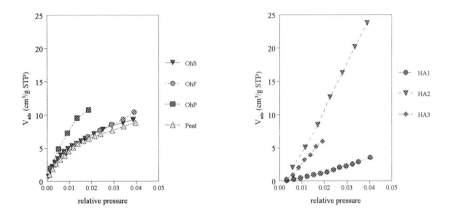

Figure 2 *Carbon dioxide isotherms*

The CO_2 isotherms of the humic acid samples are different from those of the soils. In case of the latter, the adsorption isotherms are curved, while they are more or less linear for the humic acid samples.

A plot of the adsorption isotherms of CO_2 according to Dubinin-Radushkevich[13,14] (figure 3) results in nearly identical slopes for the soil samples. Also the slopes for the HA samples are similar, but they differ from the ones for the soils.

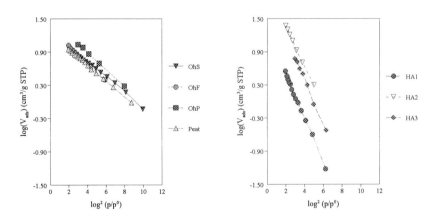

Figure 3 *Dubinin lines of the carbon dioxide isotherms*

As an example of the adsorption of hydrocarbons on soils the isotherms on the peat sample are given in figure 4.

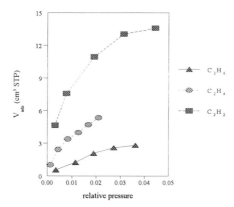

Figure 4 *The adsorption isotherms of ethane, ethene and ethyne on the peat sample*

This figure shows clearly that the amount adsorbed, for a given relative pressure, increases with decreasing critical diameter of the adsorbate. The other soil samples show a more or less similar behaviour. For the humic acid samples such an effect was only seen for the HA3 sample.

In table 4 the results of all the adsorption measurements are summarized. The monolayer equivalent surface areas for nitrogen are calculated according to the BET[15] theory. The theory used to calculate the monolayer equivalent surface area of the other adsorbates is the one of Dubinin Kaganer[16]. The pore volume, as measured with nitrogen, was taken at a relative pressure of 0.95. The micro-pore-volume of the other adsorbates was calculated according to Dubinin Radushkevich[13]. From the data presented in this table it can be seen that the value for the monolayer equivalent surface area and the porevolumes for nitrogen are always one or two orders of magnitude lower than for the other adsorbates.

Table 4 *Summarization of the adsorption results*

Sample	Monolayer equivalent surface area (m²/g)					V_p (cm³/g)				
	C_2H_2	CO_2	C_2H_4	C_2H_6	N_2	C_2H_2	CO_2	C_2H_4	C_2H_6	N_2
O_hS	73	84	47	21	.4	.0285	.0310	.0259	.0079	.0013
O_hF	95	98	53	19	3.8	.0329	.0363	.0297	.0070	.0031
O_hP	144	152	98	4	3.8	.0564	.0564	.0409	.0018	.0067
Peat	129	82	70	44	1.0	.0507	.0300	.0293	.0187	.0016
HA1	84	97	83	98	4.0	.0331	.0360	.0322	.0419	.0016
HA2	556	575	595	554	28.4	.2183	.2130	.2319	.2336	.0265
HA3	437	460	304	263	6.5	.1718	.1709	.1187	.1118	.0044

As many environmental scientists use EGME retention as a measure for the surface area of soils[8], we assessed the density of the samples not only with helium as a displacement media, but also with EGME. The results are shown in table 5.

Table 5 *Results of the density measurements*

	ρ (cm^3/g)	
Sample	Helium	EGME
O_hS	1.83	2.55
O_hF	1.68	3.14
O_hP	1.50	1.65
Peat	1.44	1.72
HA1	1.65	1.98
HA2	1.34	1.50
HA3	1.57	1.58

From the table it can be observed that the values of the density measured with helium are always lower than the values measured with EGME.

4 DISCUSSION

A linear adsorption isotherm for nitrogen (HA2 sample, figure 1) is normally only observed for samples that are non-porous or macroporous in nature.

The linearity of the adsorption isotherms of carbon dioxide on all the humic acid samples (figure 2) indicates that the micropores in these samples are much larger than the micropores in the soil samples. The same was reported by Martín-Martínez et al.[17] for a series of activated anthracites, were the adsorption isotherm became more linear with increase of the degree of activation. Increase in the degree of activation results in widening of the micropores. This was interpreted as a porefilling mechanism for small pores (up to two molecular dimensions wide) and surface coverage for larger pores.

Also the observation that the slopes in the Dubinin-plots (figure 3) are more negative for the humic acid samples than for the soil samples is an indication for the fact that humic acid samples have larger micropores than the soil samples.

For all samples clearly the monolayer equivalent surface areas as measured by CO_2 adsorption are much higher than the surface areas measured by N_2 (table 4). This phenomenon is known since long[18-21] for samples containing small pores, or pores that have constrictions of about 0.5 nm[22] and is attributed to a slow and activated diffusion process at 77K. In our case it is also possible that the structure of the samples becomes to rigid to allow nitrogen to enter the pores.

Studying the other results given in table 4, and taking into account the uncertainty in the values for the density of the gas and the specific surface area of the adsorbed molecule, we can state that the monolayer equivalent surface area and the micro-pore-volume increases with decreasing critical diameter of the adsorbate. Except for the peat sample all the other samples show no further increase in micro-pore-volume or surface area between carbon dioxide and ethyne (critical diameter 0.28 and 0.24 nm respectively).

From the combination of the critical diameters of the adsorbates and the micro-pore-volumes we can obtain a micropore size-distribution (figure 5).

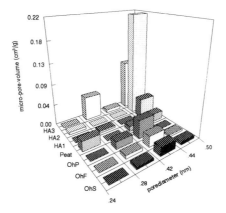

Figure 5 *Micro pore size distribution of the samples*

From figure 5 one may observe that the poresize of the humic acid samples is larger than that of the soil samples.

EGME densities were consistently found to be higher than those in helium. As it is impossible that a large molecule as EGME can enter pores that are inaccessible to helium (by far the smallest molecule), a different mechanism must be operative here. It is therefore tentatively proposed that EGME tends to alter the structure of the SOM. This could arise from dissolution of the soil in EGME, or from diffusion of EGME into the internal structure of the SOM, promoting localized swelling or separation of lamellae. A similar effect has been noticed earlier for coals with methanol[23,24].

5 CONCLUSIONS

1. Nitrogen adsorption, although the most frequently used technique to establish the surface area, is not suited for determination of the surface area of soils. This might arise from the extreme slow (activated) diffusion at the adsorption temperature of 77K.
2. EGME retention measurements are also not suited for determining the surface area of soils, as we showed that EGME can enter or create voids in the soil that are inaccessible for helium.
3. Samples prepared from humic acid do not represent a proper reference material for soils, because of their aberrant pore structure.
4. A fairly good method to determine the surface area of soils is the adsorption of carbon dioxide at 273 K.
5. The soil samples investigated are microporous. These micropores have dimensions of about 0.24 to 0.5 nm.
6. A good method to determine the micro-pore-size-distribution of soils is to measure the adsorption of carbon-dioxide, ethyne, ethene and ethane at 273 K.

REFERENCES

1. C.T. Chiou, in *Reactions and Movement of Organic Chemicals in Soils*, (Ed. D.M. Linn, SSSA Spec. Publ. 22, SSSA and ASA, Madison, Wi.) 1989, **1**.
2. K.M. Scow, in *Sorption and Degradation of Pesticides and Organic Chemical in soils*, (SSSA Spec. Publ. 32, SSSA and ASA, Madison, Wi.) 1993, **73**.
3. H. de Jonge and M.C. Mittelmeijer-Hazeleger, Environ. Sci. Technol. 1996, **30**, 408.
4. C.T. Chiou, J.-F. Lee and S.A. Boyd, Environ. Sci. Technol., 1992, **26**, 404.
5. C.T. Chiou, J.-F. Lee and S.A. Boyd, Environ. Sci. Technol., 1990, **24**, 1164.
6. C.T. Chiou, P.E. Porter and D.W. Schmedding, Environ. Sci. Technol., 1983, **17**, 227.
7. J.J. Pignatello, in *Reactions and Movement of Organic Chemicals in Soils*, (Ed. D.M. Linn, SSSA Spec. Publ. 22, SSSA and ASA, Madison, Wi.) 1989, **45**.
8. C.T. Chiou, J.J. Peters and V.H. Freed, Science 1979, **206**, 831.
9. C.T. Chiou and D.W. Rutherford, Environ. Sci. Technol., 1993, **27**, 1587.
10. W.J. Weber, P.M. McGinley and L.E. Katz, Environ. Sci. Technol., 1992, **26**, 1955.
11. P.M. McGinley, L.E. Katz and J.W. Weber Jr., Environ. Sci. Technol., 1993, **27**, 1524.
12. M.M. Dubinin, Chem. Rev. 1960, **60**, 235.
13. M.M. Dubinin and L.V. Radushkevich, Proc. Acad. Sci. USSR, 1947, **55**, 331.
14. M. Polanyi, Verb. Deutsch. Physik. Gis. 1914, **16**, 1012.
15. S. Brunauer, P.H. Emmet and E. Teller, J. Amer. Chem. Soc. 1939, **60**, 309.
16. M.G. Kaganer, Zhur. Fiz. Khim. 1959, **33**, 2202.
17. J.M. Martín-Martínez, R. Torregrosa-Marciá and M.C. Mittelmeijer-Hazeleger, Fuel 1995, **74** (1), 111.
18. H. Marsh and W.F.K. Wynne-Jones, Carbon, 1964, **1**, 269.
19. F.A.P. Maggs, Research 1953, **6**, 513.
20. P. Zwietering and D.W. van Krevelen, Fuel, 1954, **33**, 331.
21. J. Garrido, A. Linares-Solano, J.M. Martín-Martínez, M. Molina-Sabio and F. Rodríguez-Reinoso, Langmuir, 1987, **3**, 76.
22. T.G. Lamond and H. Marsh, Carbon, 1964, **1**, 281.
23. R.E. Franklin, Trans Faraday Soc., 1949, **45**, 274.
24. R.E. Franklin, Trans Faraday Soc., 1949, **45**, 668.

PORE STRUCTURE AND PHENOL ADSORPTION CHARACTERISTICS OF CARBONACEOUS ADSORBENTS PREPARED FROM COAL REJECT

F.Haghseresht, G.Q. Lu[*], H.Y. Zhu and J. Keller

Department of Chemical Engineering
The University of Queensland
St. Lucia QLD 4072 Australia

ABSTRACT

A number of carbonaceous adsorbents were prepared by oxidising coal reject under different conditions prior to pyrolysis. The extent of porosity is found to be dependent on the severity of oxidation. Their micropore volumes were determined using t-plot, the DR equation with and without correction for mesopore contribution to the adsorption process. The slope of the DR plot is shown to be a good indicator of the micropore size which is supported by the trend of phenol adsorption capacities. It is demonstrated that the values of micropore volume obtained from the DR equation with correction for mesopore adsorption is more realistic and closer that obtained by the t-plot method. Micropore size distributions of the adsorbents developed as obtained using an improved MP method also reveal that pre-oxidation creates more new micropores in the adsorbents.

1 INTRODUCTION

With ever increasing consumption of activated carbons in environmental pollution control a great attention has been focused on the development of cost-effective adsorbents as substitutes. An example of such effort is the development of suitable adsorbents from coal reject for the removal of SO_x and NO_x from flue gases[1]. In another study, eluthrites were modified for the removal of phenolic compounds from water[2].

Physical adsorption on microporous carbon is understood to be a very complex process and can not be described by any single theory. Although micropores are classified to be less than 2Å, evidences of two types of micropores are established[3]: the narrow micropores are filled at low relative pressures (<0.01) and the wider micropores are filled at high relative pressures (<0.2). The mechanism of micropore filling is dependent on the ratio of the pore width to the molecular diameter rather than the absolute value of the pore dimension[4].

The DR equation has been used extensively for gaining a better understanding of the microporous structure of activated carbons. The major drawback of the DR equation is that its linear range varies from carbon to carbon[3,5]. Besides, the contribution of the external surface area to the adsorption process in the low relative pressures can lead to large errors[6]. In this work, a series of adsorbents is prepared by oxidising coal reject in nitric acid under different conditions. Their pore structures are analysed from N_2 adsorption isotherms by DR, t-plot and an improved MP method. The results of pore structure are then related to their adsorption capacities for phenol.

2 EXPERIMENTAL

2.1 Material

The coal reject sample used in this study was obtained from New Hope Collieries, Queensland. It was dried at 100°C for two hours and the fraction between 45 and 355 μm was used.

2.2 Preparation

10 g of the dried coal reject sample was refluxed in 100 ml of nitric acid solution of desired concentration. The temperature and time length of refluxing are the main factors varied to investigate the evolution of porosity in the product. Subsequently, 4 to 5 g of the oxidised samples were then carbonised for 1.5 h at 600°C in a N_2 (99.995%) flow.

The samples were labelled as follows: the number inside the bracket represents the acid concentration and the number on the left indicates the treatment temperature and the value on the right indicates the treatment period in hour. For example, a label 80(6.22)6 represents the sample which was subjected a treatment with a 6.22 M nitric acid solution for 6 hours at 80°C while 0(0)0 means no chemical treatment. All samples were subjected to carbonisation at 600°C with or without acid treatment.

2.3 Nitrogen adsorption

Nitrogen adsorption isotherms of the dried adsorbents were determined at 77 K, using NOVA-1200 (Quantochrome, USA). Samples were degassed for 12 hours at 300°C under a vacuum of 10^{-2} torr, prior to the adsorption analysis.

2.4 Phenol Adsorption

For each experiment 0.05-0.15 g of the adsorbent equilibrated with 50 ml of a phenol solution of known concentration under continuous shaking for three days. The solutions were then separated by filtration and the phenol content remained in the liquid was determined using a Varian DMS90 UV Spectrophotometer at a wavelength of 269 nm. The adsorbed amount was calculated from the difference between the final phenol concentration after equilibration and a phenol solution of the same concentration without any adsorbent, left in the water bath for the same period of time.

3 RESULTS AND DISCUSSION

3.1 N_2 Isotherms

The isotherms of the raw coal reject and sample 0(0)0 are shown in Figures 1. The isotherm of the starting material is a typical Type II isotherm, indicating that the solid is nonporous. In contrast, a significant enhancement in adsorption at low pressures can be observed from the isotherm of 0(0)0, which is due to micropores created by carbonisation at 600°C. The isotherm has an evident character of Type I (Langmuir type) isotherms, while the plateau of the isotherm is not high, reflecting a small volume of micropores. Another important feature of the isotherm is the hysteresis extending to very low relative pressure. According to Gregg and Sing[6], such a hysteresis is associated with very narrow micropores, often of a size close to the dimension of nitrogen molecules, about 4 Å.

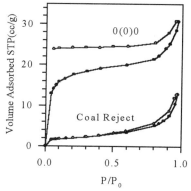

Figure 1 *Adsorption/Desorption isotherm of coal reject and the pyrolysed untreated coal reject*

Isotherms of the adsorbents treated at various conditions are illustrated in Figure 2. They all manifest the character of Type I isotherms, and have hysteresis extending to low pressures.

However, the heights of the plateau of these isotherms, are much higher than that of 0(0)0 indicating a further increase in the pore volume of the adsorbents due to acid treatment prior to carbonisation.

The height of the isotherm plateau was found to increase to a maximum at the concentration of $6.22M$ and then decreases.

It is found that the size of the hysteresis decreases with increasing acid concentration. This was also found to be true, but to a lesser extent when coal reject was treated with the acid of the same concentration but for different length of time. This pattern clearly indicates that the fraction of the larger micropores has increased with the increase in the severity of oxidation.

3.2 Quantitative Analysis of pore structure

Quantitative data of the porosity in the developed adsorbents can be derived from their N_2 isotherms. The calculated results are summarised in Table 1 and discussions on the results as well as the principle of calculations are as follows.

3.2.1 Specific surface area and BET C BET surface area (S_{BET}) and BET C values were calculated using the adsorption data in a P/P_0 range from 0.01 to 0.1 (Table 1). BET C value should be positive, and indicates the magnitude of the adsorbent-adsorbate interaction energy.

Table 1 shows that by treating coal reject with nitric acid before carbonisation, BET C increases with the acid concentration and reaction time. One possible contributing factor is that the fraction of the surface oxygen complexes increases thus resulting in a more polarised surface and stronger interaction with N_2 molecules of quadruples compared to the non-polarised carbon surface. Since the pH of the slurries of the adsorbents acid pre-treated are lower than that of 0(0)0 as shown in Table 1, it can be deduced that the surface oxides of the pre-treated samples are generally acidic. Acidic surface oxides are believed to have significant influence on the adsorption of phenol[7].

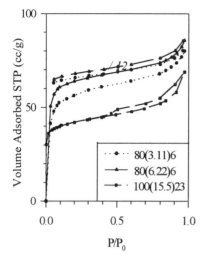

Figure 2 *Adsorption/Desorption isotherms of the adsorbents prepared : effects of pre-treatment conditions*

3.2.2 Micropore volume and external surface area from t-plot

By using the *t*-plot method micropore volume (V_{mic}), *external* surface area (S_{ext}) which include the area of external surface and the pores except for micropores are calculated. V_{mic} and the micropore surface area (S_{mic}) are calculated from the linear portion of the t-plot between relative pressures of 0.4 and 0.8 and by the BET equation. Micropore surface area is the difference between S_{BET} and S_{ext}. The reason for choosing the relative pressure range was because up to P/P_0 of 0.4, most micropores are filled by N_2 and above 0.8 the capillary condensation occurs in mesopores, which may result in underestimating the micropore volume and overestimating the external surface area.

By comparing S_{BET} of the adsorbent 0(0)0 in Table 1 with acid pre-treated samples, it can be seen that the increase in microporosity has caused a significant increase in S_{BET} of the adsorbents whose precursor materials were pre-treated.

Table 1 Surface and adsorption properties of the developed adsorbents

Sample No.	S_{BET} (m²/g)	BET C	V_{mic} (cc/g)	V_{DR} (cc/g)	E_{ads} (kJ/g)	$V_{DR}{}^c$ (cc/g)	pH	$E_{ad}{}^c$ (kJ/g)	X_p (mg/g)
Coal Reject	8.00	191	0	0.001>	-	0.001>	n/a	-	-
0(0)0	68.51	92	0.022	0.032	5.29	0.023	7.76	5.65	8.448
80(3.11)6	226	110	0.074	0.083	5.85	0.096	5.53	6.24	10.981
80(6.22)6	220	183	0.086	0.106	6.83	0.089	6.07	7.51	14.897
80(9.33)6	212	412	0.073	0.089	7.86	0.075	5.34	9.25	14.474
80(12.44)6	246	1858	0.068	0.083	8.42	0.069	6.14	10.94	15.664
80(15.5)6	211	4430	0.072	0.077	10.92	0.072	6.03	12.98	26.130
100(15.5)23	160	1759	0.051	0.064	12.94	0.051	6.72	16.06	16.761
80(6.22)15	251	654	0.083	0.106	8.08	0.085	n/a	10.92	17.153
80(6.22)23	249	8299	0.079	0.106	8.86	0.081	n/a	15.37	19.078

Figure 3 shows that the average micropore size increases as the acid concentration increases until a maximum concentration of 15.5M where the micropore size decreases significantly. In another study[7], we showed that this was attributed to the blockage of micropores by nitrates formed in acid pre-treatment. The average micropore size (width) shown in Figure 3 is obtained by assuming a slit pore shape and calculated by:

$$w = d_h / 2 = \frac{2V_{mic}}{S_{mic}} \tag{1}$$

Figure 4 shows that the micropore volume increases with an increase in acid concentration to a maximum and then decreases.

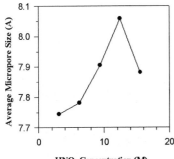

Figure 3 *Effect of HNO₃ concentration on the average micropore size*

3.2.3 Micropore volume and adsorption energy from the DR equation

Dubinin and his co-worker's theory has been widely applied for porous carbons. In general, micropore volume V^{DR}_{mic} and adsorption energy E_{ads} can be calculated from the adsorption data using the original DR equation:

$$\frac{V}{V_0} = \log V_0 - D\log^2\left(\frac{P_0}{P}\right) \qquad (2)$$

where $D = B\left(\frac{T}{\beta}\right)^2$

B is the structural constant and β the similarity coefficient. A plot of log V *vs* $\log^2(P_0/P)$ in the low and medium pressure range would yield a straight line. From the intercept and slope the micropore volume and the energy of adsorption can be obtained, respectively. It is important to note that the application of this theory is theoretically restricted to the micropore range. Therefore, contributions from mesopore and external surface should be subtracted from the overall adsorption.

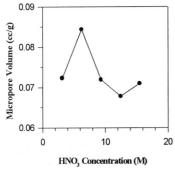

Figure 4 *Variation of micropore volume with change in acid concentration*

As shown in Figures 1 and 2, limited capillary condensation is observed for the adsorbents developed in this work. Therefore, on the mesopore and external surface, multilayer N_2 adsorption is the prevailing mechanism. The adsorption behaviour is closely similar to that on non-porous solid of the same surface nature, for example, the non-porous carbon used as the reference in *t*-plot method. If the surface area of the reference is S_{ref} and external surface area of the sample S_{ext}, then at a certain relative pressure P/P_0, the multilayer adsorption on external surface and surface of mesopores, $V_{ext, ad}$, is given by[6]:

$$V_{ext,ad} = V_{ref} \cdot \frac{S_{ext}}{S_{ref}} \qquad (3)$$

where V_{ref} is the volume adsorbed on the reference material at P/P_0. Hence, the adsorption amounts except for micropores can be estimated and are then subtracted from the original isotherm in order to obtain the contribution of micropores. The micropore adsorption thus obtained were used to plot the DR equation as shown in Figure 5.

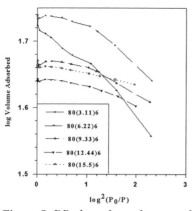

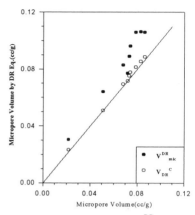

Figure 5 *DR plots of samples treated with different concentrations of acid.*

Figure 6 *Comparison of V^{DR}_{mic} and V_{DR}^{c} with V_{mic}*

The micropore volume and adsorption energy derived by this procedure are termed as V_{DR}^{c} and E_{DR}^{c}, respectively. They are compared with the corresponding data obtained from the DR equation without correction(Table 1). By plotting V_{mic} vs V^{DR}_{mic} and V_{DR}^{c} as shown in Figure 6, the effect of the correction for mesopore adsorption is observed. It shows that the V_{DR}^{c} values are less than V^{DR}_{mic} values and closer to V_{mic} values.

The $V^{c}_{DR, mic}$ is in good agreement with the V_{mic}. Therefore, the correction is necessary, because the DR equation is strictly applicable to micropore adsorption.

3.2.4 Pore Size distribution An effort to derive quantitative micropore size distribution from N_2 adsorption data has been made by using the improved MP method suggested recently by Zhu *et al*[8]. Although it is difficult to obtain information in pore size range below 7 Å using this method, some meaningful data can be obtained. Figure 7 shows plots of differential volume adsorbed versus micropore size for three samples prepared. Line "a" shows that the peak in the lower end of the distribution shifts to higher value with the increase in severity of oxidation and when coal reject was oxidised very harshly, that peak disappears. This explains the pattern of decrease in the size of the hysteresis within the adsorbents until sample 100(15.5)23 whose hysteresis loop closes. The pattern of shift of the peak marked with line "b" is not very significant although

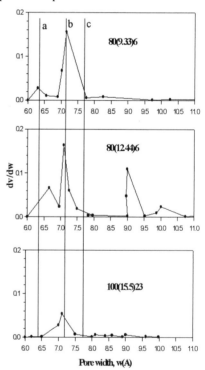

Figure 7 *Micropore size distributions of three pre-treated samples.*

indicating decreasing micropore volume with acid concentration. It is found that pre-oxidation leads to the formation of new micropores which is reflected in the increased values of the DR energy of adsorption[9], as shown in Table 1.

3.3 Phenol Adsorption

After carrying out a number of experimental runs it was discovered that the adsorption capacity of the adsorbent (X_p mg/g) reached equilibrium in a 80ppm phenol concentration. This concentration was then used to obtain the equilibrium capacity for comparison of the adsorbents and gaining further insight into their pore structure. The variation in X_p can be explained in terms of the pore structure of the adsorbents.

As shown in Figure 7, peaks after the line "c" appear with increasing acid concentration, but then disappear under sever oxidation conditions. The phenol adsorption capacity increases with an increase in the concentration of HNO_3 during pre-treatment. However, when coal reject is oxidised at the condition of sample 100(15.5)23, the adsorption capacity drops considerably. Dubinin showed that an increase in effective micropore size led to the increase in the gradient of the DR slope[10]. Marsh[11] also demonstrated that the values of the DR slope can be used as semiquantitative indicators of average pore size distribution: lower or higher values of the gradient represent narrower or wider pores respectively. The plot of the DR slope *vs* phenol adsorption (Figure 8) can be interpreted as follows: as the slope become smaller, new micropores that progressively become narrower are formed, thus leading to the increase in X_P until a maximum point after which the micropores become too narrow to be accessible to phenol molecules.

Figure 9 shows that X_p increases with increasing BET C. This is expected due to the increase in surface oxides. In other words, as the number of surface oxides increases, the amount of phenol adsorbed also increases. It is understood that phenol adsorption takes place through the formation of electron donor-acceptor complexes between the aromatic ring of phenol and surface oxides[12].

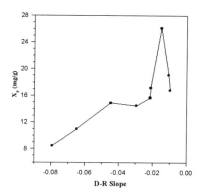

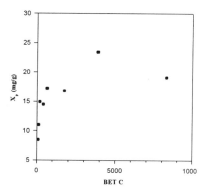

Figure 8 *Variation of X_p with DR slop*   **Figure 9** *Variation of X_p with BET C.*

4 CONCLUSION

In this work it is shown that pre-treating coal reject with nitric acid at different conditions prior to pyrolysis leads to adsorbents that have much higher surface area than the starting material and that of the untreated char. Adsorbents developed have different degree of porosity, depending on the conditions of pre-oxidation. It is also shown that the increase in the severity of pre-oxidation results in higher microporosity and more surface oxide complexes.

The slope of the DR plot is found to be a good indication of micropore size which is also supported by the trend of phenol adsorption capacities. Through the application of the DR equation, it is demonstrated that the values of micropore volume obtained from the DR equation with correction for the mesopore adsorption is more realistic and closer that obtained by the t-plot method.

REFERENCES

1. Lu, G. Q., and Do, D.D., *Carbon*, **29(2),** 1991, 207.
2. Zhonghua, Hu, and Vansant, E.F., *Catalysis Today*, **17**, 1993, 41.
3. Carrott, P.J.M., Roberts, R.A., and Sing, K.S.W., *Carbon*, **25(1),** 1987, 59.
4. Rodrigues-Reinoso, F, Garrido, M.J., Martin-Martinez, J., Molina-Sabio, M., and Torregrosa, R., *Carbon*, **27**, 1989, 23.
5. McEnaney, B., *Carbon*, **26(3)**, 1988, 267.
6. Gregg, S.J and Sing., K.S.W., Adsorption, Surface Area and Porosity, Academic Press,   1982.
7. Haghseresht, F., Lu, G.Q., and Keller, J.   Carbon, submitted, 1996
8. Zhu, H., Y., Lu, G.Q., Zhao, X.S., and Do, D.D., Langmuir, Submitted, (1996).
9. Parra, J.B., Pis, J.C., De Sousa, J.C., Pajares, J.A., and Bansal, R.C., *Carbon*, **34(6)**, 1996, 783.
10. Dubinin, M.M., *Advan. Colloid Interface Scie.* **2**, 1968, 217.
11. Marsh, H., *Carbon*, **25**,1987, 49.
12. Moreo-Castila, C., Rivera-Utrilla, J., Lopez-Raman, M.V., and Carrasco-Marin, F., *Carbon*, **33(6)**, 1995, 841.

CHARACTERIZATION OF SILICA OBTAINED FROM THE DISSOLUTION OF OLIVINE: A NITROGEN PHYSISORPTION AND THERMOPOROMETRIC STUDY

D.J.Lieftink, P.van Os, B.G.Dekker, H.Talsma[*] and J.W.Geus

Department of Inorganic Chemistry
Debye Institute, Utrecht University
P.O.Box 80083
3508 TB Utrecht
The Netherlands

[*] Dept. of Pharmaceutics, Utrecht Institute for Pharmaceutical Sciences

Abstract

The reaction of the natural occurring magnesium-silicate olivine with sulfuric acid produces a colloidal silica solution. The silica texture has been analyzed by nitrogen physisorption and thermoporometry. Both methods show comparable pore size distributions. The texture of the silica changes during the reaction from ink-bottle shape pores to cylindrical pores. The ink-bottle texture is less resistent against drying than the cylindrical pore system as shown by thermoporometry analysis.

1 INTRODUCTION

Silica is widely used in various applications *e.g.* fillers for rubber, as a support for catalysts and as starting material in the preparation of synthetic zeolites and clayminerals. Nowadays silica is produced by the melting of sand with soda and dissolving the prepared solid hydrothermally in water. The solution prepared this way is called: waterglass. The main disadvantage of this process is the high temperature necessary to melt the sand. Silica is prepared from water-glass solutions mainly by varying the concentrations of the waterglass, the temperature and the pH. Waterglass has a pH of about 14. By decreasing the pH to lower values of about 5-6 gelation and/or precipitation of silica particles can be observed.[1] Variations in the precipitation parameters cause a lot of different textures to be prepared.

This study deals with silica prepared at a totally different way: silica is precipitated after the dissolution of the natural occurring silicate olivine $(Mg,Fe)_2SiO_4$ in 3 molar sulfuric acid. The dissolution brings magnesium, iron and monomeric silica into solution. The monomeric silica polymerizes and forms silica particles. These silica particles can be separated form the magnesium-iron sulfate solution, washed and dried. The product is a white, fluffy silica. The main difference with the waterglass process is the high acidity. The sulfuric acid is neutralized from a pH of -0.5 up to about 1, and in this range the silica is prepared.

The most important feature of silica is its texture or the size of the particles and the way they are packed. The pore structure which determines if the silica is suitable for *e.g.* catalytic reactions. There are several ways to determine the texture. In this study we will compare a nitrogen physisorption, with a not widely used one: thermoporometry. Thermoporometry has the advantage of measuring both wet and dried silica. The other techniques can only be applied to the dried materials. From literature it is shown that thermoporometry and other texture analysis methods give similar results for the techniques.[2,3,4,5]

In this paper the changes of the texture of the silicas produced by the dissolution of olivine in sulfuric acid will be studied as a function of the dissolution reaction rate.

2 EXPERIMENTAL

A stoichiometric quantity of olivine is added to 1.5 liter 3M sulfuric acid, which has been heated to 70°C. The olivine has been sieved in two grainsize fractions: a very fine one (vf) with grains smaller than 90μm and a coarse one with grains between 425 and 1000μm (c). The two experiments are called 70vf and 70c hereafter. At the start the pH was about -0.5. When the pH had become 1 the reaction was terminated because at higher pH values iron hydroxides will precipitate on the silica.

The reactions have been carried out in double walled glass reaction vessels. The reaction mixture was vigorously stirred to avoid settling of the olivine grains. From each mixture two samples have been taken: the first one in the beginning of the reaction at a pH of -0.3and the second at the end at a pH of about 0.8.

The colloidal suspension was washed with demi-water. The presence of sulfate ions was checked by the addition of Ba-acetate and the subsequent precipitation of Ba-sulfate. When no Ba-sulfate precipitation was observed anymore, the silica was washed once more. We assumed that no salts were present in the silica anymore. The silica was divided in two portions, one was analyzed directly by thermoporometry, the other was dried at 120°C and subsequently analyzed by thermoporometry and nitrogen physisorption.

2.1 Texture Analysis

Nitrogen physisorption experiments have been carried out using a Micromeritics ASAP 2400 at 77K. From both the adsorption and the desorption branch the pore size distributions have been calculated using the Kelvin equation.[6] The specific surface area of the samples is calculated after the BET method.[7]

Thermoporometry has been carried out using a DSC (TA 2920 system) The temperature calibration was carried out using In and Ga, the heat capacity using In. Dried samples of silica have been saturated with water before analysis. The wet samples could be measured directly. The temperature programme of the analysis was as follows:
1. freezing the water to -60°C to solidify both bulk and pore water.
2. warming up to-1.5°C to melt the pore water. The bulk water outside the pores remains frozen.
3. cooling from-1.5 to -30°C. Starting measurement
4. warming up to -1.5°C.
After the measurement the sample is cooled again and subsequently heated to room temperature.

From the heatflow data and freezing temperature the pore volume and pore size have been calculated after Brun *et al.*[8]

3 RESULTS

3.1 Nitrogen Isotherms And BET Analysis

The nitrogen physisorption experiments showed different types of isotherms for the silica samples. Typical exemples are presented in figure 1 and 2. The both isotherms look like a type IV isotherm in the IUPAC classification. The hysteresis of the pores as shown in figure 1 can possibly be related to an ink bottle shape of the pores[9]. The hysteresis loop as shown in figure 2 may be related to a more cylindrical poresystem. The experiment 70vf produces silica with a poresystem of figure 1 both in the beginning as at the end of the reaction. The reaction 70c shows a isotherm as shown in figure 1 in the beginning and at the end one as shown in figure 2. Reaction 70vf took about three hours, reaction 70c about 20 hours.

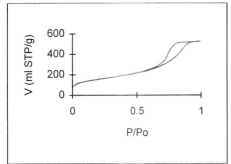
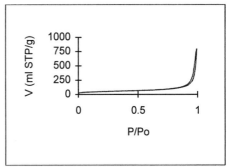

Figure 1 N_2 *physisorption isotherm of silicas produced at exp. 70vf and the first sample of exp. 70c.*

Figure 2. N_2 *physisorption isotherm of silica produced at the end of exp. 70c.*

The BET specific surface area of the samples of experiment prepared with the very fine grain size fraction (vf) remains constant during the reaction at about 380 m^2/g. The specific surface area of silicas produced with the coarse grained olivine (c) diminishes from 400 m^2/g down to about 150 m^2/g.

3.2 Thermoporometry

The pore size distributions of the first silica samples (pH =-0.3) from both reactions measured by thermoporometry are presented below. Figure 3 shows the distribution before drying and figure 4 after drying. It is clear form these figures that the porediameter decreases upon drying. The width of the distribution decreases as well.

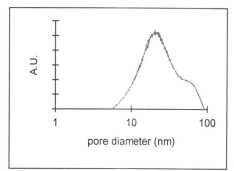

Figure 3 *Pore size distribution obtained by thermoporometry. Sample 70vf-1, before drying*

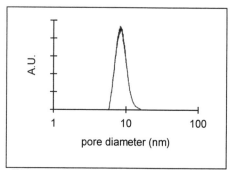

Figure 4 *Pore size distribution obtained by thermoporometry. Sample 70vf-1 after drying*

The pore size distributions of the silicas prepared at the end of the dissolution reactions (pH=0.8) are different for the two reactions. The 70vf experiment produces silica with a comparable distribution before and after drying as presented in figures 3 and 4. The silica produced at 70c conditions shows distributions as presented in figures 5 and 6, reps. before and after drying. It is clear from these figures that the pore structure does not change very much upon drying. The mean pore diameter of the dried samples has increased from about 10 nm to about 60 nm.

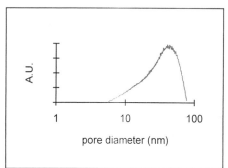

Figure 5 *Pore size distribution obtained by thermoporometry. Sample 70c-2, before drying*

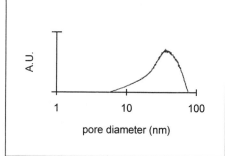

Figure 6 *Pore size distribution obtained by thermoporometry. Sample 70c-2 after drying*

3.3 Pore Size Distributions From Nitrogen Physisorption

The dried samples of the produced silicas have been analyzed by nitrogen physisorption, and the pore size distribution have been calculated using the Kelvin equation using both the adsorption and the desorption branch. Only a slight difference in mean pore diameter using the adsorption or desorption branch can be observed as can be concluded from table 1(last two columns). The pore size distribution of silica sample 70vf-1 is presented in figure 7, the 70c-2 distribution is presented in figure 8. These distributions are calculated from the desorption data.

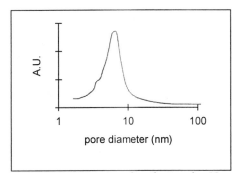

Figure 7 *Pore size distribution by N₂-physisorption of silica 70vf-1*

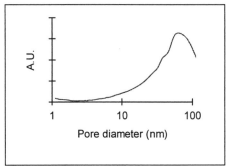

Figure 8 *Pore size distribution by N₂-physisorption of silica 70c-2*

The pore size distributions of the samples 70vf-2 and 70c-1 have the same shape and almost the same peak position as the one shown in figure 7. The distributions measured by thermoporometry have the same positions and geometry. The 70c-2 distributions show some differences in peak position. The adsorption branch yields larger porediameters than the desorption data.

Table 1 *Characteristics of the pore size distributions analyzed by thermoporometry and nitrogen physisorption.*

sample	Thermoporometry wet		Thermoporometry dry		N₂-phys	
	peak pos. (nm)	peak width (nm)	peak pos. (nm)	peak width (nm)	peak pos. ads	peak pos. des
70c-1	30	60	15	25	15	15
70c-2	40	75	40	70	105	70
70vf-1	20	80	9	8	9	7
70vf-2	35	95	14	15	15	12

4 DISCUSSION

This study shows that the porestructure of the silica analyzed with both thermoporometry and nitrogen physisorption give similar results with respect to the mesopores of the silica. Thermoporometry can give information about pores larger than about 7nm upto 80nm (diameter), while nitrogen physisorption can give additional information about micropores.

Two important features of the silica have been studied by the experiments described. The influence of drying on the texture has been analysed and the evolution of the porestructure during the dissolution reaction has been followed. In the beginning of each dissolution reaction, silicas with the same type of porestructure is produced. The (ink-bottle) porestructure is rather weak and not very resistant against the forces caused by drying. This texture is stable during the reaction using the very fine olivine. The specific surface area remains stable during the reaction also, which indicates that the size of the silicaparticles does not change. From transmission electron microscopy (TEM) we observed also a stable particle size during the reaction.

The shape of the isotherms shows a conversion of the porestructure from ink-bottle pores to more cylindrical pore systems when the dissolution reaction proceeds slowly using the coarse grained olivine. The porediameter increases and the porestructure is resistant to the drying forces. The specific surface area of the silicas produced during this experiment decreases strongly, which is partly an effect of particle size. The particles grow during the reaction and neck-growth occurs. Both features have been observed by TEM. The effect of a longer residence time in the reactionvessel is the strengthening of the porestructure.

The preparation of the silica particles can be explained by the model of nucleation and growth of LaMer.[10] During a slow dissolution reaction the silica monomers which are released by the olivine cause particle growth of the silica. Also precipitation at the concave points between two silica particles takes place.[11] This aging of the silica particles produces a strong network of cylindrical pores which does not collapse on drying.

A dissolution reaction using very fine olivine brings silica monomers in solution very fast, which results in a constant new-forming of silica particles which form loosely packed clusters of particles which have large pores with small entrances. Because of the lose packing of the particles, drying results in collapsing of the porestructure as observed by thermoporometry.

5 CONCLUSIONS

The texture of silica prepared by the dissolution of olivine in sulfuric acid can be analyzed by thermoporometry in both the wet and the dried stage. This can be an advantage when the drying behavior of silica is studied. Conventional methods can only be used when the silica is dried.

The texture of the olivine silica changes during the dissolution reaction. In the beginning loose particles are produced, which stick together and leave a ink-bottle shape pore-structure. Later this porestructure changes to a more cylindrical structure. This happens when the dissolution reaction proceeds slowly. When the reaction goes quickly, there is constant nucleation and production of the ink-bottle poresystem.

Thermoporometry shows that the ink-bottle structure is not very stable against drying forces. The cylindrical structure resists these forces due to aging of the silica particles.

Acknowledgment
P.Elberse and J.Raaymakers are thanked for the nitrogen physisorpiton experiments.

6 REFERENCES

1. R.K. Iler, 'The Chemistry of Silica', John Wiley & Sons, New York, 1979.
2. J.-F.Quinson, M. Astier and M. Brun, Applued Catalysis, 1987, **30**, 123.
3. C. Jallut, J.Lenoir, C.Bardot and C. Eyraud, J.membrane Sci., 1992, **68**, 271.
4. M.K.Titulaer, M.J den Exter, H. Talsma, J.B.H.Jansen and J.W. Geus. J.non-cryst.solids, 1994, **170**, 113.
5. K. Ishikiriyama and M. Todoki, J.coll.interf. sci., 1995, **171**, 103.
6. E.P. Barret, L.G. Joyner and P.P. Halenda, J.Am.Chem.Soc., 1951, **73**, 373.
7. S. Brunauer, P.H. Emmet and E. Teller, J.Am.Chem. Soc. 1938, **60**, 309.

8. M. Brun, A. Lallemand, J.-F. Quinson and C. Eyraud, Thermochimica acta, 1977, **21**, 59.

9. S.J. Gregg and K.S.W. Sing, Adsorption, surface area and porosity. Academic Press, London, 1967.

10 V.K. .LaMer and R.H. Dinegar, J.Am.Chem.Soc. 1950, **72**, 4847.

11. L.L. Hench and J.K. West, Chem.Rev, 1990, **90**, 33.

CHARACTERIZATION OF THE POROSITY OF M41S SILICATES BY ADSORPTION AND Xe NMR

M.A. Springuel-Huet[a], J. Fraissard[a], R Schmidt[b], M. Stöcker[b], and W. C. Conner[c]

[a]Laboratoire de Chimie des Surfaces, casier 196, Université P. et M. Curie, 4 place Jussieu, F-75252 Paris Cedex 05, France; [b]SINTEF Applied Chemistry, Postboks 124, Blindern, N-0314 Oslo, Norway; [c]Dept. Chem. Eng., Univ. Massachusetts, Amherst, MA 01003 USA

1 INTRODUCTION

[129]Xe NMR has been extensively used to characterise porous materials like zeolites and related molecular sieves, mineral clays etc... [1,2] In particular, for zeolites, an empirical relation has been found between the pore size (or rather the mean free path of a Xe atom in the pores, i. e. the average distance travelled by Xe between two successive collisions against the pore surface) and the chemical shift, δ, of adsorbed Xe, obtained by extrapolation of the δ-Xe concentration variations to zero concentration. The bigger the mean free path the smaller the chemical shift. This relation is useful to study the porosity of zeolites or to detect and characterize defects or structure intergrowths.[3] But it should be adapted to study the porosity of mesoporous silicas.[4]

The new family of regular mesoporous materials M41S[5] whose members possess uniform mesopores, between 15 and 100 Å, is interesting to test the validity, in this mesopore size range, of the empirical relation mentioned above.

In this paper we studied three purely siliceous M41S samples. Two samples of MCM-41 (denoted A and B) which present a hexagonal array of regular one-dimensional pores, and one sample of MCM-48 (denoted C) which possesses a tri-dimensional pore network were studied by [129]Xe NMR. These results are compared to the pore morphology characterized by both N_2 adsorption and desorption at 77 K. These sorption analyses were adjusted to reflect the measured relationship between the thickening of the adsorbed layer and the relative pressure, P/P_o.

2 EXPERIMENTAL

2.1 Sample Preparation

The MCM-41 samples were prepared from sodium silicate which is poured into a solution of sulphuric acid. The formed gel is then added to the template solution $[C_{14}H_{29}(CH_3)_3NBr]$. After having been heated at 100 °C for 144 h, the gels were washed and dried in air at ambient temperature. A calcination at 540 °C for 1 h in flowing nitrogen removed the template. For MCM-48 sample, the silicium source was tetraorthosilicate (TEOS), the synthesis medium was basic (NaOH) and the template was $C_{16}H_{33}(CH_3)_3NCl$. The detailed synthesis procedure has been published elsewhere[6,7].

2.2 Experimental Techniques

Nitrogen adsorption-desorption isotherms were measured at 77 K with a Carlo Erba instrument using a conventional volumetric technique.

Samples were evacuated at 500 K under vacuum overnight before xenon was adsorbed at 296K in a pressure range 10-1600 torrs. ^{129}Xe NMR spectra were obtained at 110.688 MHz on a Bruker MSL 400 spectrometer at ambient temperature. Typically 1000 to 50000 scans are recorded to have a sufficiently high signal to noise ratio. The repetition time was 1 s.

Coadsorption of water and xenon were also performed on evacuated sample A. When a desired amount of water was adsorbed, xenon was trapped by cooling the sample with liquid nitrogen.

3 RESULTS AND DISCUSSION

3.1 Xenon Experiment

Xenon adsorption isotherms show that at a given pressure samples A and B adsorb more Xe than sample C. At 1000 torrs, they adsorb 4×10^{20}, 3.5×10^{20} and 2×10^{20} Xe atoms per gram respectively that is much less than the quantity adsorbed by zeolites, due to larger pores and thus larger void fraction.

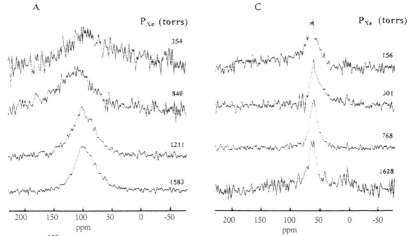

Figure 1 ^{129}Xe NMR spectra of Xe adsorbed in sample A (MCM-41) and C (MCM-48)

In contrast with zeolites, the chemical shift of adsorbed xenon is independent of the xenon pressure (Figure 1) as it has been already observed in case of compressed amorphous[8] or porous silicas.[4] This constant chemical shift has been interpreted in terms of an exchange between Xe in an adsorbed state on the silica surface (corresponding chemical shift δ_a) and gaseous Xe in the pores (corresponding chemical shift δ_g). Thus the observed chemical shift, δ_{obs}, is a weighted average of δ_a and δ_g. If Xe pressure is not too high, below a few atmospheres, δ_g is considered to be zero (it is the chemical shift reference) and δ_{obs} is proportional to δ_a.,

$$\delta_{obs} = \frac{p_a}{p_a + p_g} \delta_a,$$

where p_a and p_g are the populations of adsorbed and gazeous xenon respectively.

The linewidth is rather large between 20 and 120 ppm decreasing when the pressure increases. For sample C (MCM-48) it is smaller, between 15 and 25 ppm.

The chemical shift of Xe adsorbed in ZSM-5, ω and VPI-5 zeolites, whose structure consists of unidimensional channels, have been reported for comparison (see Table 1). As it appears clearly, if the empirical relation between δ_{obs} and the pore size obtained with zeolites is valid for those mesoporous silicas, the chemical shift of Xe adsorbed in pores with dimension larger than 12 Å should be smaller than 49 ppm. However, the observed chemical shifts are much larger. Nevertheless they decrease when the pore size increases.

Sample C (MCM-48) does not precisely fit this trend. It is not surprising since in a three dimensional network the mean free path of a Xe atom is much larger compared to a unidimensional pore network with the same pore dimensions. Therefore the ^{129}Xe chemical shift should be larger.

Table 1 *Chemical shift of adsorbed xenon*

Sample	Chemical Shift /ppm	pore size/ Å
A	100	18
B	80	29
C	60	26
ZSM-5[a]	100-110	5.4
ω[a]	75	7.4
VPI-5[b]	49	12.1

[a] from ref. 9, [b] from ref. 10

The high values of δ_{obs}, compared to 49 ppm, may reflect greater interaction, compared to zeolites, between Xe and the pore surface, i.e. a higher value of δ_a. This hypothesis is consistent with the high limiting value of δ_{obs} obtained for sample B at low temperature, about 140 ppm (Figure 2). Indeed at low temperature p_g can be considered negligible and if we assume that Xe-Xe interactions on the pore surface are not too important then $\delta_{obs} = \delta_a$. Taking into account the small amount of adsorbed xenon, this assumption seems to be reasonable. For instance the corresponding value for a ZSM-5 zeolite has been found to be about 100 ppm.[11]

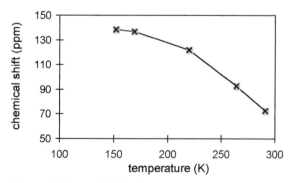

Figure 2 *Chemical shift variation against temperature*

To verify the higher Xe-surface interaction in those mesoporous solids, we adsorbed water molecules in order to mask the pore surface from xenon. In microporous solids like zeolites, co-adsorption of water molecules induces an increase in the chemical shift due to a decrease in the pore size. However, we observed a decrease in the chemical shift in the present experiment.

Indeed the spectrum shows several signals. For P/P_0 (H_2O) = 7.9 torrs (Fig. 3b), there are three major lines, one at about 95 ppm corresponding to pores without water and the others at about 70 and 15 ppm, broader, are due to pores which have adsorbed water. The presence of two broad lines indicates that the adsorption of water is inhomogeneous. The width of the highly shifted signal drastically decreases after water adsorption. This can be explained by the fact that water adsorbs on surface heterogeneities, leaving a more homogeneous surface accessible to Xe. A further addition of water molecules causes a slight decrease in the chemical shift of these two lines (Fig. 3). We can conclude that the water molecules have effectively masked the surface to xenon and therefore the xenon-surface interaction is reduced, even drastically (very low chemical shift of the third signal). Nevertheless the pore size has not been significantly changed.

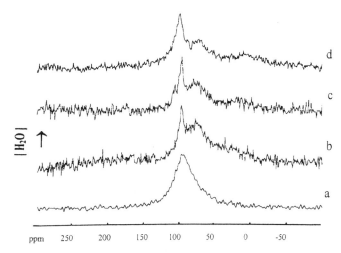

Figure 3 *Evolution of the spectrum with increasing water content, without H_2O (a), and increasing amount of adsorbed water (b) (c) (d)*

Finally these NMR results prove that the unexpected high values of the chemical shift of xenon adsorbed in these mesoporous silicas is due to a strong Xe-surface interaction, compared to zeolite. The empirical relation between the chemical shift and the pore size established with zeolites should be modified to fit the present data. The relation proposed by Terskikh and coll. for porous silica gels cannot be used for these samples certainly because of the different nature of the surface. Further studies will involve a larger set of MCM-41 with various pore size in order to obtain the new relation.

3.2 N_2 Adsorption

The adsorption and desorption isotherms for the MCM-48 sample is shown in Figure 4.

The sorption isotherms are quite similar; however, there are several notable differences. There is little hysteresis in either isotherm except for the slight hysteresis for the MCM-48 sample and an even lesser hysteresis for the MCM 41 sample. The hysteresis at P/P_0 = ~0.45 is often seen for these samples but may be due to the common tensile strength artefact.[12]

Both of the initial isotherms are quite steep to the pressure where pore filling is occurring (at P/P_0 = ~0.36 for the MCM-41 sample and P/P_0 = ~0.26 for the MCM-48 sample). Application of the BET equation gives surface area estimates of 1215 m^2/g for sample B and 1092 m^2/g for sample C. The "C" values from these analyses are 110 and 90, respectively; the BET estimates would seem reasonable.

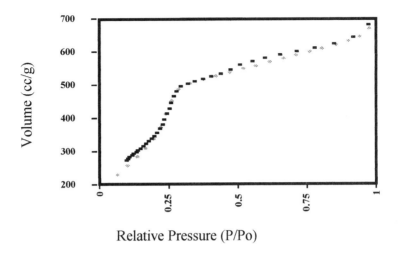

Figure 4 *Nitrogen ad-desorption isotherms at 77 K for sample C, MCM-48.*
Upper points are for desorption; lower for adsorption.

Pore structure analyses were done on both adsorption and desorption isotherms. First the FHH analysis was performed employing the conventional values of the constants in the Halsey equation for t (the thickness of the adsorbed layer) and P/Po (the relative pressure):

$$t = 6.04 \ [P/Po]^{1/3}$$

However, the steepness of the adsorption isotherms below the region of pore filling suggests that this "conventional" equation will over estimate the smaller pores.[13] A more appropriate relationship between the relative pressure and the thickness of the adsorbed layer was estimated by fitting the data in this region, below pore filling. The following relationship was found for sample B:

$$t = 5.22 \ [P/Po]^{1/1.97}$$

It was very difficult to obtain a good fit for sample C since there are few desorption data points lower in pressure than the pressure where adsorption occurs and above the region where the BET equation indicates that a monolayer is formed (V_m = 250 cm^3/g). If anything, the indications are that the Halsey exponent should have been lower. However, there were few experimental points available for sample C and so we chose to use 5.22 for the Halsey constant and 1/1.97 for the analyses of both samples B and C. Based on these analyses, the modified pore dimension estimates from adsorption and desorption for MCM 41 (sample B) is shown in Figure 5. Both adsorption and desorption give a dominant peak at ~14.5-15 Å radius. In addition, a small peak is seen in the desorption analysis at ~20 Å. If only the desorption analysis had been performed, one might have incorrectly concluded that the pore distribution was bimodal. The adsorption analysis confirms the single pore dimension and further shows that the pores are essentially cylindrical and straight. During the sorption analysis the cumulative surface area is calculated to account for the exposed surface area at any point in the sorption. The modified analysis gives ΣS = 1127 m^2/g during adsorption and $\Sigma S = 1128$ m^2/g during desorption which compares well with the BET estimate of the surface area of 1215m^2/g.

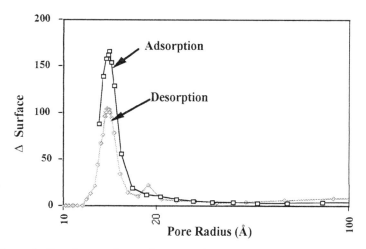

Figure 5 *Distribution of pore dimensions based on incremental surface area for MCM-41 (B) as analyzed using the modified Halsey constants.*

As for MCM 41, both adsorption and desorption show a dominant peak at the same dimension, ~11 Å for sample C. This corresponds to P/Po = ~ 0.25 which is about as low in relative pressure as the adsorption analysis can be comfortable started. Note, however, that the desorption analysis also gives evidence for a second peak at ~20 Å which is also seen for sample B. Again, this is probably the common artefact found for many samples.[12]

The difference between the modified and the conventional analyses are illustrated in Figure 6. The conventional analyses represents a shift in the measured pore dimensions to estimate larger pores than found in the modified analyses.

Table 1 summarizes the adsorption and desorption analyses by both conventional and the modified analyses. The differences in estimated pore dimensions between the samples and between conventional and modified analyses are seen. Both adsorption and desorption analyses give similar pore dimensions (as they should for cylindrical pores). The conventional analysis yields 10% larger pores than the modified analyses.

The difference between the surface areas calculated during the adsorption or desorption for P/Po > 0.25 are not evident for these samples.[13] This is because the dominant pore dimensions calculated during adsorption and desorption are the same for each of these samples. The sorption hysteresis for most solids results in higher surface areas calculated during desorption analyses where the porous network measures constrictions while it measures openings during adsorption.

Both analyses of the surface areas of MCM 41 (sample B) agree well with that estimated by application of the BET method; however, there is a substantial difference for the MCM 48 (sample C). This could be due to extreme surface roughness for this sample as might be expected for the three-dimensional porous network. As a result, there is considerable surface that is occluded at relative pressures below P/Po = 0.25.

Note that the average wall thickness (between adjacent pores) for these samples is ~8Å based on the estimated surface areas. The walls are thin.

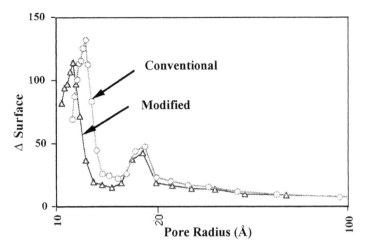

Figure 5 *Comparison of desorption analyses for MCM 48 (sample C) of the surface area distribution versus pore radius.*

Table 1 *Sorption Analyses for MCM Samples B and C*

	Surface Area m²/g (BET)	Adsorption r_p (Å)	ΣS(P/Po >0.25)	Desorption r_p (Å)	ΣS(P/Po >0.25)
B (MCM-41)	1215				
Conventional		15.3	1235	15.1 +20.4	1263
Modified		14.5	1128	14.2 +19.9	1128
C (MCM-48)	1093				
Conventional		12	558	12.4 +20	700
Modified		11.1	696	11.3 +20	642

4 SUMMARY and CONCLUSIONS

The Xe NMR studies show a large chemical shift which would suggest xenon interacting with small pores; however, as seen in Figure 1, the chemical shift does not increase with Xe pressure as found for zeolites. This is reminiscent of the studies of Xe adsorbed in samples of silica which possessed rough surfaces.[8] These large chemical shifts that are invariant with pressure suggest a facile exchange between free-like Xe in the pores and Xe interacting with the surface. The Xe on the surface is subject to strong interactions as expected at a kink in the surface layer.

The changes in the Xe NMR spectrum after the surface is exposed to water supports this picture as the chemical shift changes due to preadsorbed water. However, it is evident that there are several changes in the environment of adsorbed Xe. The exchange between Xe in the pores and that on the surface is attenuated while the band at 95 ppm shows that some surface sites remain vacant and accessible for Xe at least for the amount of adsorbed water we used. These hypotheses explain the data but are not unique in rationalizing the Xe spectra.

However it is apparent that the Xe NMR reflects a form of Xe that strongly interacts with the surface of MCM materials. The constraints on this physisorbed Xe represent a confined environment which may be ascribed to surface roughness. Thus, it is necessary to include heterogeneity of the interactions between adsorbing species and MCM materials.

The adsorption studies of nitrogen physisorption are consistent with this picture. The decrease in the Halsey exponent would be expected for a surface that is not atomically smooth. There is a slight shift between the corrected pore size distributions (based on a modified Halsey equation) but it is not significant since there is little hysteresis for these samples as measured by N_2 adsorption at 77K.

The modified analyses provide a consistent calculation for the surface area for MCM-41 but not for MCM-48, a three dimensional network of smaller pores. It is apparent that a significant fraction of estimated surface area (based on the BET model) is already occluded during the initial stages of physisorption to P/Po = 0.3. The BET model presumes that the surfaces are flat but these comparisons confirm the rough (and heterogeneous) nature of the surface for many MCM materials. We are also convinced that the comparison of adsorption and desorption analyses confirm that the apparent pores calculated during desorption at ~20 Å is an artefact as it is not seen during adsorption and is not seen for other MCM materials. Otherwise, the modified sorption analyses are entirely consistent with the XRD analyses of these novel materials.

References

1. D. Raftery and B. F. Chmelka, *NMR Basic Principles and Progress*, 1994, **30**, 111
2. M.-A. Springuel-Huet, J.-L. Bonardet and J. Fraissard, *Appl. Magn. Reson.*, 1995, **8**, 427
3. M.-A. Springuel-Huet, T. Ito and J. Fraissard, *Stud. Surf. Sc. Catal.*, 1984, **18**, 13
4. V. V. Terskikh, I. L. Mudrakovskii and V. Mastikhin, *J. Chem. Soc., Faraday Trans*, 1993, **89-23**, 4239
5. J. S. Beck, J. C. Vartuli, W. J. Roth, M. E. Leonowicz, C. T. Kresge, K. D. Schmitt, C. T. -W. Chu, D. H. Olson, E. W. Sheppard, S. B. McCullen, J. B. Higgins and J. L. Schlenker, *J. Am. Chem. Soc.*, 1992, **114**, 10834 and C. Kresge, M. Leonowisz, W. Roth, J. Vartuli and J. Beck, *Nature*, 1992, **359**, 710
6. R. Schmidt, M. Stöcker, E. Hansen, D. Akporiaye and O. H. Ellestad, *Microp. Mater.*, 1995, **3**, 443
7. R. Schmidt, M. Stöcker, D. Akporiaye, E. H. Tìrstad and A. Olsen, *Microp. Mater.*, 1995, **5**, 1
8. W. C. Conner, E. L. Weist, T. Ito and J. Fraissard, *J. Phys. Chem.*, 1989, **93**, 4138
9. Q. J. Chen, M. A. Springuel-Huet, J. Fraissard, M. L. Smith, D. R. Corbin and C. Dybowski, *J. Phys. Chem.*, **96**, 10914 (1992)
10. Q. J. Chen, J. Fraissard, H. Cauffriez and J. L. Guth, *Zeolites*, **11**, 534 (1991)
11. Q. J. Chen and J. Fraissard, *J. Phys. Chem.*, 1992, **96**, 1809
12. S. J. Gregg and K. S. W. Sing, "Adsorption, Surface Area and Porosity." Academic Press, London, 1982.
13. W. C. Conner, *Journal of Porous Materials*, 1996, **3-2**, 1

MERCURY POROSIMETRY APPLIED TO LOW DENSITY XEROGELS

R. Pirard, B. Heinrichs and J.P. Pirard,

Université de Liège, Laboratoire de Génie Chimique, Institut de Chimie (B6)

Sart-Tilman, B-4000, Liège, Belgium.

1 INTRODUCTION

Usually, the interpretation of mercury porosimetry data is made using the Washburn's equation which is based on the hypothesis that the observed volume variation is due to the progressive intrusion of mercury into the pores.[1] Most porous materials are indeed penetrated under an isostatic mercury pressure. This intrusion phenomenon can be easily evidenced by submitting non-powdered but monolithic pieces to mercury porosimetry. In that case, it can be seen that, during the pressurization and depressurization, the volume variation as a function of the pressure is reversible with, generally, a large hysteresis. Most of the time, the phenomenon is only partially reversible and some mercury remains trapped in the material. The presence of mercury can be seen by microscope observation and the residual quantity can be measured by weighing or chemical analysis.[2] This shows that the observed volume variation mechanism is really the mercury intrusion-extrusion in the pore network.

However, some materials among the most porous, such as aerogels, show no trace of trapped mercury after a mercury porosimetry experiment and, in that case, the intrusion mechanism reality is dubious. Moreover, if the volume variation is very small or non-existent during the depressurization, it is certain that the actual mechanism is a irreversible mechanical shrinking under the isostatic mercury pressure .[3]

Mineral aerogels have been extensively studied during the last decade.[4, 5] They are synthesized by drying of mineral oxides gels in temperature and pressure conditions over the critical temperature and pressure of the interstitial liquid existing in the wet gel. The obtained material is a very porous and a very low density solid. In most cases, the solid phase represents scarcely a very small fraction of the total volume. The aerogels are formed by particles aggregated into filaments which in turn form a three-dimensional cross-linked structure.[4] Those materials have obviously a very weak mechanical compressive strength. Aerogels submitted to mercury porosimetry are not intruded by mercury but are irreversibly compacted by the isostatic mercury pressure. Indeed, during the pressure increase, one observes a progressive volume variation as a function of pressure and, during the pressure release, the volume variation is practically equal to zero, showing the phenomenon irreversibility. After a mercury porosimetry experiment until 200 MPa and a return to atmospheric pressure, the apparent volume of a monolithic sample is largely reduced and it is free from trapped mercury. In this case, the variation volume mechanism is not intrusion but a mechanical compaction under the isostatic pressure of mercury.

In a previous work, [6] the pore size distribution of aerogels first compacted under various mercury pressures was measured by the nitrogen adsorption-desorption method. It has been shown that during compaction, the isostatic mercury pressure P crushes completely pores of size larger than a limit size L, and leaves the smaller size pores unaltered. The relation between L and P has been determined as $L = k / P^{0.25}$. The collapse mechanism of the pores has been modeled as the buckling of the brittle filaments of mineral oxide under an axial compressive strength. The pressure which causes the buckling of a cubic structure is given by the Euler's law

$$P = \frac{n\pi^2 EI}{L^4} \quad \text{with} \quad I = \frac{\pi d^4}{64}$$

The k value is therefore a function of E, the elastic modulus of the mineral oxide filaments, d, the diameter of filaments and n which varies with the geometry of the structure and its stiffness. Experimentally, in the case of an aerogel made of 50% SiO_2 and 50% ZrO_2, the k constant was estimated to be 47 nm.MPa$^{1/4}$.

In the case of aerogels which collapse under the isostatic mercury pressure, the knowledge of the k value allows to determine the pore volume distribution in relation with the pore size. In first approximation and if it is certain that the buckling is responsible of the volume variation in the whole pressure range until 200 MPa, the above value (47 nm.MPa$^{1/4}$) can be used for k constant. However it is preferable to determine k in each particular type of aerogel composition. This determination needs to measure the limit size L of largest pores of aerogels first submitted to mercury isostatic pressure P and it generally needs a large and hard work.

Recently, a new type of materials so-called " low density xerogels " has been synthesized. These materials have a pore volume larger than 2 cm^3/g which is similar to aerogels pore volume but they are much tougher against crushing. The particular behavior of these materials submitted to mercury porosimetry confirms the soundness of the buckling theory in the mercury porosimetry interpretation applied to aerogels and permits to determine easily the k constant characteristic of each sample.

2 EXPERIMENTAL

Low density xerogels are silica gels synthesized according to the sol-gel process by hydrolysis and condensation of tetraethylorthosilicate (TEOS) in alcoholic solution. The gel is dried by solvent evaporation at room temperature and atmospheric pressure. The addition of a small proportion of ethylenediaminetriethoxysilane (EDAS) to the TEOS alcoholic solution avoids the complete shrinkage of material during the drying. Many low density xerogels were synthesized, varying slightly the initial composition. The porous volume obtained are between 1.5 and 6 cm^3/g which are characteristic values of aerogel material type.

All this low density xerogels submitted to mercury porosimetry exhibit a similar behavior and we have selected one of them for complete description. TEOS and EDAS in ethanolic solution are hydrolyzed by an aqueous 0.2 M NH_4OH solution in ethanol at room temperature and under magnetic stirring. The mixture whose the composition in molar ratio TEOS/EDAS/ethanol/water/NH_3 equals 1/0.06/10.6/5/0.02, forms a gel after

40 minutes and is kept at 70°C in a tight flask for aging. After 72 hours, the flask is cooled, opened and the solvent is allowed to evaporate at room temperature by decreasing slowly the pressure. The last traces of organic matter are removed by burning in air at 400°C during 20 hours. After drying and calcination, the gel is shrunk at a volume ratio of 77% and its final bulk density is 0.3 g/cm^3.

Mercury porosimetry is performed on a Carlo Erba Porosimeter 2000. Specific surface area is measured by nitrogen adsorption-desorption isotherm analysis on a Carlo Erba Sorptomatic 1900. Samples, first partially compacted under isostatic mercury pressure were examined using a Zeiss Axioplan optical microscope.

3   RESULTS

The pore volume V_{Hg} measured by mercury porosimetry is equal to 2.7 cm^3/g and the specific surface area S_{BET} measured by nitrogen sorption is equal to 305 m^2/g.

Figure 1 shows the volume variation curve as a function of mercury pressure obtained for the low density xerogel monolith, the synthesis of which is given above. The curve has an unusual aspect; the slope abruptly changes at a pressure of 26 MPa. Below this pressure, the shape of the curve is usual in the case of aerogels. Above 26 MPa, the curve is characteristic of strong materials having small size pores.

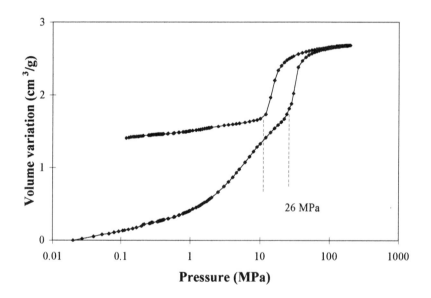

Figure 1 : *Mercury porosimetry curve (volume variation as a function of mercury pressure) of a typical low density xerogel*

The curve corresponding to the pressure decrease, can also be divided in two distinct parts separated by a sudden change of slope. The first part, at high pressure, shows a large volume variation reversibility with, however, a clear hysteresis. The second part at low pressure shows, on the contrary, a very small volume variation.

In order to understand this behavior, an other monolithic sample from the same batch has been submitted to mercury porosimetry until 20 MPa, taken out of the device and examined with microscope. Then, the same sample has been submitted to mercury porosimetry until 200 MPa and observed again with microscope. As shown on Figure 2, the behavior is the same as the one previously observed but the two parts of the curves are better separated. The visual and microscopic examination of samples is particularly interesting. The sample, submitted to 20 MPa and set back to atmospheric pressure, is free from trapped mercury as shown by microscope observation and absence of weight variation. The sample shape stays similar to the original one, but its size has clearly varied; the longest size, taken as reference, initially 8.5 mm shrinks to 7.4 mm which corresponds to a volume shrinkage of 33%. These observations and the practical irreversibility of the volume variation show that the pore collapse is the mechanism actually responsible of volume variation at pressure below 20 MPa.

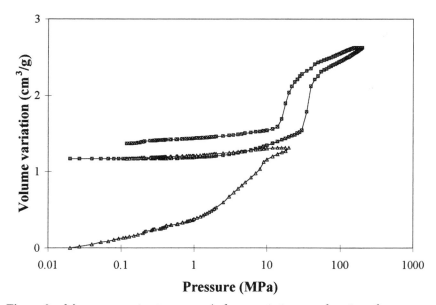

Figure 2 : *Mercury porosimetry curves (volume variation as a function of mercury pressure)*   Δ : *until 20 MPa* .   □ : *until 200 MPa*

After the experiment until 200 MPa followed by the return to the atmospheric pressure, the observation of the sample by microscope shows very numerous mercury droplets trapped in the pore network. The surface of the sample is strongly damaged and its appearance suggest that mercury created craters and cracks during the extrusion. Due to the partial fragmentation of sample, weighting and size measurement cannot be done but it is obvious that the intrusion-extrusion mechanism can be invoked to explain the volume variation at pressure above 26 MPa.

4 DISCUSSION

Similar porosimetry results are often interpreted by classical intrusion theory [7]. That brings to the conclusion that the material is composed of large and porous particles. The low pressure curve gives the distribution of cavities between the particles and the high pressure part of the curve gives the pore distribution inside the particles. This interpretation is correct if the volume variation is due to the intrusion mechanism in the whole range of pressure within the limits of mercury porosimetry.

In the present case, the change of mechanism at the pressure $P_c = 26$ MPa is obvious. The lower pressure part of the curve responds to the buckling equation

$$L = k / P^{0.25} \qquad\qquad (1)$$

and in the higher pressure part, the intrusion equation $L = -4\sigma\cos\alpha/P$ must be employed. The mercury surface tension value ($\sigma = 473 \ 10^{-5}$ N/cm) and the contact angle ($\alpha = 140°$) usually considered [8] allow to compute the pore size by the expression

$$L = 1500/P \qquad\qquad (2)$$

with P in MPa and L in nm.

At pressure P_c at which the change of mechanism occurs, the pore size can be obtained indifferently by either expressions (1) and (2). This allows to determine the k constant with

$$k = 1500/P^{0.75} \qquad\qquad (3)$$

In the present case, the computed k value is 130 nm.MPa$^{1/4}$.

The cumulative pore volume as a function of pore size (figure 3) must be computed using the buckling equation until $P_c=26$ MPa, and the intrusion equation for higher pressure. The volume distribution curve as a function of pore size can be obtained by differentiation of the cumulative volume curve.

The so obtained composite curve shows a continuous and monomodal distribution, according to what is expected for that type of materials which are often described as fractal from the size of particle to the size of largest clusters aggregates.[9] Immediately around the point of change of mechanism, the curves show a perturbation indicating that the change is not perfectly sharp but is progressive on about 5 MPa.

The fact that an intrusion mechanism succeeds a compaction one for an increasing pressure confirm that the compaction is really a buckling phenomenon. During this process, at a given pressure, the disappearance of large size pores yields a volume diminution. The small size pores stay unaltered; their size is not reduced by the isostatic pressure. Consequently, small size pores are finally intruded by mercury at a sufficiently high pressure. Indeed it has been demonstrated that penetration is impossible if the pressure reduces simultaneously all the pore sizes.[10]

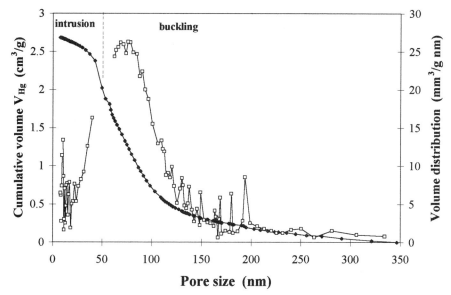

Figure 3 : *Cumulative pore volume (◆) and volume distribution (□) as a function of pore size*

5 CONCLUSION

A thorough observation of the behavior of low density xerogels submitted to mercury porosimetry shows two successive mechanisms which have to be invoked to explain quantitatively the volume variation curve as a function of pressure. At low pressure, the material behaves as an aerogel and collapses under the mercury isostatic pressure. This is due to the buckling of the brittle filaments constituting the solid skeleton of material. Above the transition pressure, which is characteristic of the material composition and microstructure, the mercury can intrude into the network of small pores not destroyed during the compression at low pressure. The succession of these two mechanisms on one material confirms that the collapse is due to a buckling mechanism and allows to determine the constant *k* which appears in the buckling equation. The successive use of buckling equation and intrusion equation allows to compute the complete pore volume distribution curve in relation with the pore size.

Acknowledgments

Authors thank "le Fonds National de la Recherche Scientifique, les Services de la Programmation de la Politique Scientifique and the Ministère de la Région Wallonne, Direction générale des Technologies et de la Recherche" for their financial support.

References

1. A.J. Lecloux, in "Catalysis Science and Technology", J.R. Anderson and M. Boudart (eds.), Springer, Berlin, 1981, Vol.2, p.171.
2. A.R. Minihan, D.R. Ward, and W. Whitby, in "The Colloid Chemistry of Silica", Proc. 200th. Nat. Meeting American Chemical Society, Washington DC, 26-31 Aug. 1990, in "Advances in Chemistry Series", H.E. Bergna (ed.), American Chemical Society, Washington DC, 1994, Vol. 234, pp 341-355.
3. L. Duffours, T. Woignier, and J. Phalippou, *J. Non-Cryst. Solids,* 1996, **194**, 283-290.
4. C.J. Brinker and G.W. Scherer, in "Sol-Gel Science : The Physics and Chemistry of Sol Gel Processing", Academic Press, San Diego, CA, 1989.
5. J. Fricke, in "Aerogels", Proc.1st. Int. Symp. Würzburg, 23-25 Sept. 1985, J. Fricke (ed.), Springer, Berlin, 1986, p.13.
6. R. Pirard, S. Blacher, F. Brouers, and J.P. Pirard, *J. Mater. Res.*, 1995, **10**, 2114-2119.
7   D.M. Smith, G.P. Johnston, and A.J. Hurd, *J. Colloid Interface Sci.*, 1990, **135**, 227-237
8. E. Steven Vittoratos and P.R. Auburn, *J. Catal.*, 1995, **152**, 415-418.
9. R. Vacher, T. Woignier, J. Phalippou, J. Pelous, and E. Courtens, *J. Non-Cryst. Solids,* 1988, **106**, 161-165.
10. G.W. Scherer, D.M. Smith, and D. Stein, *J. Non-Cryst. Solids.* 1995, **186**, 309-315.

STUDY OF KELEX 100 (METALLIC ION EXTRACTANT) IMPREGNATED IN POROUS ORGANIC POLYMERS USING NITROGEN ADSORPTION AT 77K, WATER ADSORPTION AND 129-Xe NMR SPECTROSCOPY.

S. Esteban,* J.L. Bonardet** and G. Cote*
*Laboratoire de chimie analytique, ESPCI, 10 rue Vauquelin 75005, Paris, France.
**laboratoire de chimie des surfaces, UPMC case 196, place Jussieu, Tour 55, 75252 Paris cedex 05 France.

1 INTRODUCTION

Extraction chromatography by means of a complexing substance adsorbed on high specific area porous polymers offers the advantages of extraction by solvent and those of ion exchange chromatography without having the disadvantages (1).However, the studies published to date are mainly concerned with the extractive performance of different impregnated supports and are not interested in the physicochemical behaviour of the active reagent..The aim of the present work is to understand better the physical chemistry of impregnated solids which can be used in hydrometallurgy with a view to optimizing the various properties in a rational manner. To illustrate such a study we have chosen to interest ourselves in the porosity of porous solid polymers impregnated with Kelex 100. Kelex 100 is an extractant well known in liquid-liquid extraction and for which many physicochemical data are available. Moreover, supports loaded with Kelex have an obvious economic interest, since they could be used to extract germanium or gallium ions from electrolytic baths of zinc sulfate. Such an operation would serve two objectives: recovery of an element of high added-value and also elimination of an inhibitor of zinc electrolysis. The experimental methods applied for this purpose are nitrogen adsorption-desorption measurements, water activity measurements and 129-Xe NMR spectroscopy of adsorbed xenon.

2 EXPERIMENTAL

2.1 Nature of the products.

Two porous solids were used to prepare the phase impregnated with Kelex 100. Firstly, Amberlite XAD 1180, a hydrophobic of copolymer of divinylbenzene and ethylvinylbenzene, and, secondly, Amberlite XAD7, a hydrophilic polymethacrylic ester. The polymers occur in the form of spherical particles of relatively homogeneous dimensions (0.3 <diameter <1 mm)

Kelex (Schering) is an odourless, viscous, greenish oil containing 90% of the chelating agent : 7-(4-ethyl,1-methyl,octyl)-8 hydroxyquinoline.

Impregnation was performed by the so-called "dry" method: the polymer grains, previously washed and dried, are immersed in the impregnation solution obtained by dissolving Kelex 100 in heptane; the mixture is homogenized by stirring, then the solvent is slowly eliminated under vacuum at room temperature. The impregnated support is finally left in an oven at 100°C for 12 h.

2.2 Experimental techniques

Before any adsorption, the samples are degassed at 60°C for 12 h. The nitrogen adsorption-desorption isotherms at 77K are obtained by volumetry on home-made equipment, the accuracy of the measurements being about + 5%. A complete adsorption-desorption cycle requires about 40 points and takes about 15h. Xenon is adsorbed at 27°C, the temperature of the NMR probe; spectra are obtained on a Bruker MSL400 at the xenon resonance frequency of 110.7 MHz. To obtain a single spectrum in some cases requires up to 10^5 accumulations with a time delay of 1s. The chemical shifts are expressed in ppm relative to xenon gas at zero pressure.

The activity of water is measured on a a_w-center apparatus (Novasina). After impregnation with water the samples are placed in a cell thermostated to + 0.1°C connected to a captor sensitive to the water vapour pressure. The activity is given by the ratio P/P_0, where P is the pressure read on the captor and P_0 is the saturation vapour pressure at the temperature considered.

3 RESULTS AND DISCUSSION.

3.1 Nitrogen adsorption -desorption isotherms at 77K

Figures 1 and 2 represent the nitrogen adsorption-desorption isotherms at 77K for samples XAD 7 and XAD 1180 with different Kelex. loadings. For samples without Kelex or lightly loaded (< 0.6g/g),theses isotherms show a hysteresis loop characteristic of mesopores. When the extent of impregnation is increasing (> 0.6 g/g) this loop disappears and the isotherms go from type IV to type II in the Brunauer classification. Analysis of the t-plots which can be deduced from these isotherms leads to the conclusion that there are both micro and mesoporous zones in the clean and lightly loaded samples (<0.2 g/g) On the other hand, when the Kelex content is over 0.2 g/g the micropores are completely filled or blocked by the Kelex. The values of the constant C, which can be calculated from the BET isotherm in the range. $0.03 < P/Po < 0.3$, decrease when the degree of impregnation increases. They are very different for the clean polymers, 237 and 77 for XAD 7 and XAD 1180, respectively, but fall to 35 and 32 when the Kelex loading is over 1 g/g. This result shows that superficial fluid-solid interactions are much stronger for the hydrophilic support and also that Kelex impregnation reduces these interactions.

Figures 3 and 4 show the evolution of the BET specific areas of the impregnated polymers (curves 1). As expected, the specific area decreases monotically when the degree of impregnation increases; above a loading of 1.5 g/g the impregnated solids have virtually zero area. Since the shape of the graphs is related to the evolution of the micropore (obtained from the t-plots, curve 2) and mesopore areas (obtained from the difference: $S_{BET} - S_{micro} = S_{méso}$ curve, 3) it is more instructive. Thus figure 3 for the hydrophilic support, XAD 7 indicates a maximum in the variation of the mesopore area for a Kelex content of 0.1 g/g, whereas there is no such maximum for the

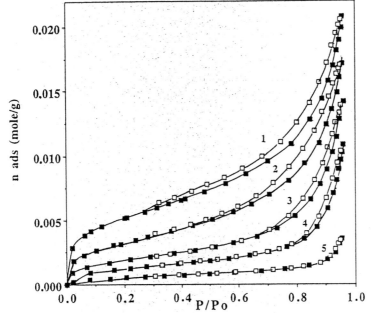

Figure 1 : *Adsorption (full symbols)-desorption (open symbols) isotherms of N2*
at 77K on XAD7 samples :
1: 0.07 g/g; 2: 0.18 g/g; 3: 0.40 g/g; 4: 0.58 g/g; 5: 1.03 g/g.

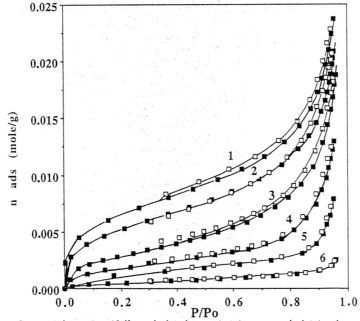

Figure 2 : *Adsorption (full symbols)-desorption (open symbols) isotherms of N2*
at 77K on XAD1180 samples :
1: 0,0 g/g; 2: 0,05 g/g; 3: 0,21g/g; 4: 0,40 g/g; 5: 0,64 g/g; 6: 1,15 g/g.

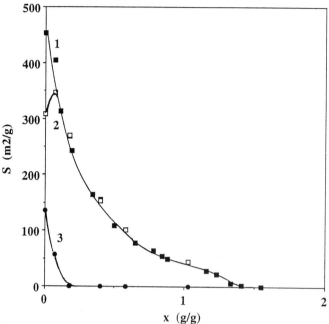

Figure 3:   *BET surface area (1), mesoporous surface (2) and microporous surface (3) versus kelex loading for XAD7 samples.*

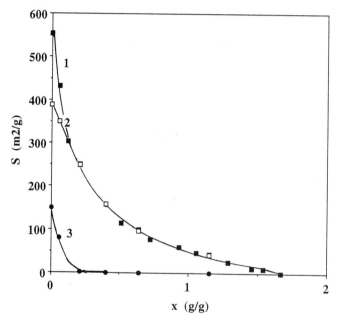

Figure 4:   *BET surface area (1), mesoporous surface (2),and microporous surface (3) versus kelex loading for XAD1180 samples.*

hydrophobic polymer XAD 1180. This result can only be explained in terms of two opposing phenomena: Kelex impregnation of the polymer leads to blocking of the micropores and mesopores (reduction of the area); this same impregnation causes also swelling of the XAD 7 polymer lattice, which is more reticulated than the polymer XAD 1180, and consequently increases the area. Up to an impregnation of 0.1 g/g swelling is more important than blocking; beyond this, it is the opposite. This interpretation is confirmed by measurements on water adsorption by exposing the polymers to a saturated vapour phase at ambient temperature; this shows that XAD 7 swells by about 70% whereas XAD 1180 is not modified. Another confirmation of this interpretation is provided by analysis of the pore distribution curves obtained by the BJH method (2). While the general form of the isotherms is similar for the two supports the pore distribution curves $dV/dr = f(r)$ (figure 5 et 6) show that they behave very differently as regards Kelex fixation. For the hydrophobic support (XAD 1180) the pore distribution shows that when the amount of extractant on the support is increased there is a progressive reduction of the specific area and the pore volume with however, a certain discrimination as regards the pore size : Kelex 100 appears to be distributed randomly over all the micro and mesopores but, naturally, the smaller pores are filled or blocked more quickly and disappear first. In the case of the hydrophilic polymer (XAD 7) much the same conclusion can be drawn as regards the Kelex distribution; however, comparing the $dV/dr = f(r)$ curves of the neat solid and the impregnated solids we can observe an increase of the number of micropores in the range 8-20 nm that reveals expansion of the acrylic lattice upon penetration by Kelex. The internal structure of the grains is seriously disturbed, no doubt because of the flexibility of the macromolecular chains. This swelling of the lattice is thus confirmed.

3.2 Measurement of the activity of water

Figure 7 shows the variation of the activity of water with the water content of polymer XAD 7 more or less impregnated with Kelex 100. We can see that for Kelex-free polymethacrylate the activity of water is very different from 1 up to a value of m_{H2O} of the order of 0.5 g/g and does not approach 1 until beyond 1 g/g (figure 7-4). The following interpretation can be advanced: at low content the water molecules interact rather strongly with the clean XAD 7 polymer and cause swelling, no doubt by dipolar interactions between the surface of the solid and water, the dipole moment of the surface of this polymer being 1.8 D, which is very close to that of water (1.84 D). Then, as the amount of water increases, the sites of the clean polymer likely to interact are consumed to such an extent that beyond a certain level there is excess water whose activity is close to 1. In the presence of Kelex the same phenomenon arises but since initially the active sites of the support are already occupied (at least in part) by the extractant, a lower amount of water (relative to the clean XAD 7 polymer) is required to reach an excess of water or a high activity of water (Figures 7-2 to 4).

In the case of Amberlite XAD 1180 we have observed that, as soon as the first traces of water are introduced into the solid, the activity of water becomes close to 1, and this whatever the Kelex content. These observations show that again the XAD 1180 polymer whether clean or impregnated has no affinity for water and that even a small amount of water is as free in the pores of the support.

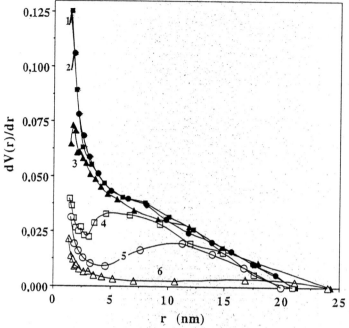

Figure 5: *pore size distribution for XAD1180 samples:*
1: 0.0 g/g; 2: 0.05 g/g; 3: 0.21 g/g; 4: 0.40 g/g; 5: 0.64 g/g; 6: 1.15 g/g

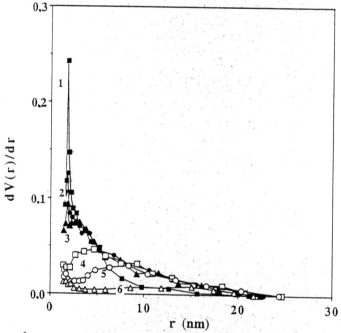

Figure 6 : *pore size distribution for XAD7 samples :*
1: 0 g/g, 2:0.07 g/g, 3:0.18 g/g, 4:040 g/g, 5:0.58 g/g, 6:1.03 g/g

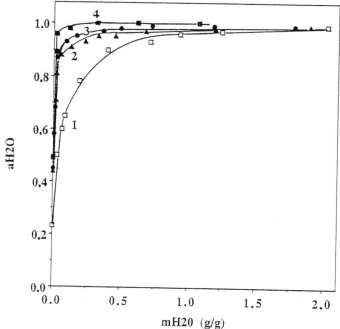

Figure 7: *Water activity versus water content for kelex loaded XAD7 samples: 1: 0,0 g/g; 2: 0,33 g/g, 3: 0,50 g/g; 4: 1,03 g/g*

3.3 129-Xe NMR results

The NMR spectroscopy of adsorbed xénon (129-Xe) is a very useful technique for studying certain physical properties of porous solids. Developed at the beginning of the 80s by Fraissard et Ito (3) it was first applied to zeolites, then extended to other porous materials, such as clays, amorphous gels and polymers. This method is the subject of several reviews(4-6).

In our case, whatever the support studied, we observe that the chemical shift is practically independent of the xenon concentration for a given Kelex content (table 1).However, the value of the chemical shift depends on the degree of impregnation and decreases when this increases. Beyond a loading of 0.21 g/g it is no longer possible to detect a signal, even at high xenon pressure (1000 torrs). These results show two things: firstly, the signal observed corresponds to xenon adsorbed in the micropores of the solid, since it is no longer detected when the microporosity has disappeared (loading> 0.2 g/g) and morever, this microporosity is open and the adsorbed xenon exchanges rapidly with the xenon gas in the rest of the solid (chemical shift indépendent of the pressure). This result is similar to that found by Conner et al (7) for compacted silica gels.

Table 1 : *Chemical shift in ppm, of adsorbed xenon versus xenon concentration.*

[Xe]10^{20}atom/g	1	1.5	1.7	2	2.3	2.5	2.8	3.2
XAD7 0g/g		132			134			135
XAD7 0.07g/g		119				116		
XAD7 0.18g/g		106						
XAD1180 0:g/g	130	129					129	
XAD1180 0.05g/g	125		123			125		
XAD1180 0.21g/g				107				

4 CONCLUSION

These results indicate that the two polymeric supports impregnated with Kelex. 100 behave very differently. Impregnation of the hydrophilic polymer leads to swelling of the lattice, while the hydrophobic compound is unaffected. For both polymers Kelex is randomly impregnated and the microporous zone disappears when the content is greater than a 0.2 g/g.

References

1. A. Warshawsky, "Ion exchanged and solvent extraction", J.A. Marinski and Y. Marcus M.Dekker, New York, 1981,**8**, 229.
2. E.P. Barrett, L.G. Joyner, P.P. Halenda. *J. Am; Chem. Soc.*, 1951, **73**, 373.
3. T. Ito , J. Fraissard. "Proceedings of the fifth international. Conference on zeolites", L.V.C. Rees, 1980, Heyden, London, p 510.
4. P.J. Barrie, J. Klinowski, *Prog.. NMR Spectr.* 1992, **24**, 91.
5. D. Raftery, B.F. Chemkla. *NMR Basic Principles and Progress*, 1994, **30**, 11.
6. M.A. Springuel-Huet, J.L. Bonardet, J. Fraissard. *Appl. Magn. Res.*, 1995, **8**, 427.
7. W.C. Conner, E.L. Weist, T. Ito, J. Fraissard. *J. Phys. Chem.* 1989, **93**, 4138.

AN ANOMALOUS ADSORPTION OF ZSM-5 ZEOLITES

M.E. Eleftheriou and C. R. Theocharis,

Department of Natural Sciences,
University of Cyprus,
P.O. Box 537, Nicosia, Cyprus.

ABSTRACT

The surface properties of synthetic template-free Na-ZSM-5 microporous zeolites were studied by isothermal N_2 adsorption measurements and FTIR / DRIFTS spectroscopy. An anomalous adsorption type I isotherm obtained for one of the samples of high $NaOH/SiO_2$ degassed at 300K was attributed to the change in Al distribution and formation of defects effected by the increase in -OH and thus, of pH.

1 INTRODUCTION

The last twenty years have witnessed a particular interest in zeolites, regarding their use as cartalysts in the chemical industry, as ion exchangers, as well as their environmental applications. These microporous aluminosilicate solids have the potentials of being used in the protection of ecosystems, from waste water and gas treatment to further use as water softeners in detergent builders replacing the undesired polyphosphates.

ZSM-5 zeolites belong to the class of synthetic zeolites of the MFI-type structure, which is of the form $Na_n [Si_{96-n} Al_n O_{192}] .16H_2O$ (n≤8).[1-5] Their three-dimensional pore structure presents a particular interest as its framework consists of two different 10-membered oxygen rings.[2] The resulting channel distribution has important effects upon their adsorptive and catalytic properties. The synthesis of such solid structures can be monitored so that crystals suitable of specific catalytic applications may be formed. Recently there have been developed certain systems , known as "template-free" systems,[6-8] which require no organic material to be present during synthesis, thus necessitating no calcination prior to use. This present work studies the surface properties of three ZSM-5 samples that had been synthesized without the presence of organic base under varying conditions.

2 EXPERIMENTAL SECTION

The samples B1 - B3, in the Na^+ form, were prepared[8] at Mainz University according to the classical method of hydrothermal crystallization of an aluminosilicate gel in the presence of sodium hydroxide solution without the addition of an organic template. Table 1 lists the data of their synthetic parameters.

Table 1 *Synthetic parameters of samples B1 - B3*

Sample	H_2O / SiO_2	SiO_2 / Al_2O_3	$NaOH/SiO_2$	Time / hrs	Temp. / K
B1	80	50	0.25	72	453
B2	80	50	0.30	113	453
B3	80	50	0.40	66	453

Nitrogen adsorption isotherms were carried out at 77K using an automated apparatus (Micromeritics ASAP 2000), with a starting pressure of 0.13 Pa. Prior to isotherm determination, the samples were heated in an oven at 473K for 72 hours and, then, were outgassed for 24 hours at 473K (sample B1), 300K and 623K (samples B2, B3). Fourier transform infrared (FTIR) spectra and diffuse reflectance FTIR spectra (DRIFTS) were measured on a Shimadzu 8000 Series spectrophotometer.

3 RESULTS AND DISCUSSION

The adsorption isotherm (Figure 1) obtained for the adsorption of nitrogen on sample B1, degassed at 473K is of type I character, usually associated with micropore filling.[9] Similar adsorption isotherms were also given by the sample B2. The hysteresis observed all along the pressure axis of the isotherms for both zeolitic samples is due to the aggregation of the relatively small crystallites of the sample (12 - 20 μm).

Sample B3, however, gave an interesting form of adsorption isotherm after outgassing at 300K. The expected type I shaped isotherm exhibits an anomalous descending curvature towards higher relative pressures, as depicted in figure 2. Repeated measurements were carried out for the same sample of the zeolitic solid B3, both at 300K and after degassing the sample at 623K prior to analysis. The isothermal experimental results are tabulated in table 2 whilst figures 3 and 4 show the nitrogen adsorption isotherms of the repeated measurements obtained at 300K and 623K respectively.

Table 2 *Nitrogen adsorption isotherm results given by a sample of B3*

Sample of B3 degassed at x K	BET Surface Area $/ m^2 g^{-1}$	Micropore Surface Area $/ m^2 g^{-1}$	Total Pore Volume $/ cm^3 g^{-1}$	Weight Loss %
300 - Initial	46	39	0.021	4.3
300 - Repeat 1	32	23	0.018	
300 - Repeat 2	11	2	0.010	
623 - Repeat 3	303	278	0.126	7.7

The fact that microporous solids generally possess high surface areas[9] makes the values of the BET and the micropore surface areas (at $p/p^0 = 0.995$) given by the samples degassed at room temperature (300K) seem relatively low. All the repeated measurements obtained by the same sample of template-free B3 show a uniform and gradual fall for samples outgassed at 300K (Figure 3). It is worth noting the appreciable increase in all surface results of the sample degassed at 623K. This is accompanied by a change in the shape of the isotherm (Figure 4). The values of the BET surface area were greater than those of the micropore surface area in all the measurements, thus indicating the microporosity of the sample. The internal surface area in microporous solids is relatively higher than the external one that includes surface defects, such as cracks and prominences.

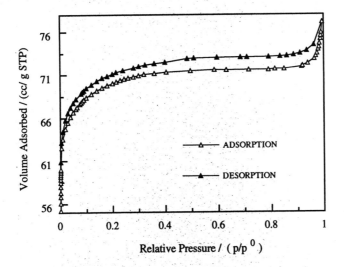

Figure 1 *Adsorption isotherm of B1 - 473K*

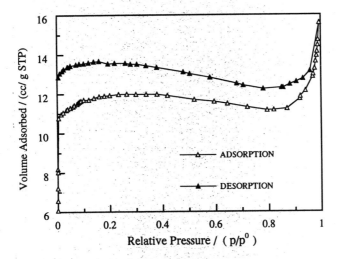

Figure 2 *Adsorption isotherm of B3 - 300K*

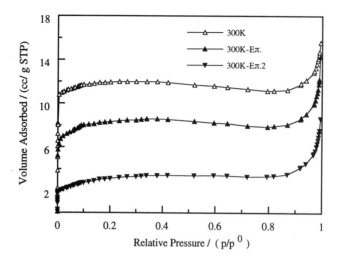

Figure 3 *Adsorption isotherms of B3 / 300K - Repeats*

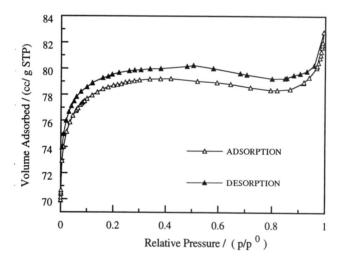

Figure 4 *Adsorption isotherm of B3 / 623K - Repeats*

This anomalous behaviour of the zeolitic sample B3, regarding the sigmoidal shape of the adsorption isotherm, seems to be associated with particular values of NaOH/SiO$_2$ ratio whereas reaction time also seems important. Various studies [10] have shown that the hydroxide ion aids the transport of silicate and aluminate species in the reaction mixture, thus increasing the rate and size of crystal growth (from 12 μm - B1 to 35 μm).[8] An increase in the -OH$^-$ concentration usually decreases the induction time, resulting in the time of synthesis being influenced by the NaOH/SiO$_2$ ratio.

Directly related to the hydroxide ion concentration also seems to be the type of silicate species present in the solution. Aluminium ions affect the crystallization process in the system by controlling the degree of hydration it possesses. Increasing the hydroxyl concentration, and thus the pH of the system, step-wise hydration of Al^{3+} to $Al\ (OH)^{2+}$ ....$Al\ (OH)_4$ is performed, thus altering the distribution of aluminium in the sample B3.[10] This is directly related to the role of the pH in the mixture, in which along with its increase, there is an increasing formation of internal structure defects during synthesis. This is because in high pH, due to the presence of isolated $Al\ (OH)_4^-$, the complete polycondensation of the silicate species does not occur which give rise to $\equiv$Si-O$^-$ groups and also to Si-atom vacancies. The presence of sodium also has an effect on the amount of defects present in the crystal structure since the Al-O-Na bonds that are being formed prevent the complete formation of the framework.[11]

Sodium ion has another important role in the synthesis of such zeolites. It can be regarded as a "template"[8] and a structure-directing agent during the reaction as it interacts with water molecules to form ordered clathrate structures.[10] These cation-water templates subsequently act as void fillers in the growing zeolite framework and stabilize the double ring silicates.

Therefore, in accordance to the above, the apperance of the anomalous shape of the isotherm regarding zeolite B3 is due to the reorganization of clathrate structures within the zeolitic voids and areas of defects. The increase in the -OH$^-$ concentration resulted to the altering of aluminium distribution in the sample and to an increase of internal structure defects during synthesis. The presence of Na$^+$ also had an effect on the amount of defects present in the crystal structure. Thus, along with the increase of structural defects, there exists a greater number of clathrate molecules moving around within the voids that are formed during N$_2$ adsorption.

The presence of hysteresis, the adsorption and desorption curves that do not meet at low relative pressures and the gradual decrease of the values of the surface properties at 300K are all related to this reorganization of clathrate water molecules within the zeolitic voids and the simultaneous defect formation. The removal of molecular water from the pores of the zeolite by an increase in temperature creates more space for N$_2$ adsorption. This is the cause for the increase in the experimental values of the surface characteristics. A further reorganization within the pores seems to lead to the closure of the adsorption / desorption paths at very low relative pressures. The persistent presence of clathrate water is evidenced by FTIR and DRIFTS spectra, in which the characteristic hydrogen-bonded -OH peaks at $3435\ cm^{-1}$ and $1640\ cm^{-1}$ are present at all working temperatures.

Acknowledgements

The authors thank Dr. M. P. O. Keung for the provision of the samples and the University of Cyprus for a studentship to M.E.

References

1. J. Jansen, in *Catalysis & Adsorption of Zeolites*, Eds. G. Ohlman et al, Elsevier, Amsterdam, 1991, **77**.
2. D. H Olson et al., J. Phys. Chem., 1981, **85**, 2238.
3. H. van Koningsveld, J. C. Jansen and H. van Bekkum, Zeolites, 1990, **10**, 235.

4. E. M. Flanigen, R. W. Grose, J. M. Bennet, J. P. Cohen, J. V. Smith, R. M. Kirchner and R. L. Patton, Nature, 1978, **271**, 512.

5. D. H. Olson, W. O. Haag & R. M. Lago, J. Catalysis, 1980, **61**, 390.

6. A. Tißler, Ph.D Thesis, Mainz University, Germany, 1989.

7. R. W. Grose and E. M. Flanigen, US Patent No.4 257 **885**, 1985.

8. M. P. O. Keung, Ph.D Thesis, Brunel University, U.K., 1993.

9. K. S. W. Sing & S. J. Gregg, 'Adsorption, Surface Area and Porosity', Academic Press, New York, 1982.

10. E. G. Derouane and Z. Gabeica, J. Solid State Chemistry, 1986, **64**, 296.

11. P. L. Llewellyn, Ph.D Thesis, Brunel University, U.K., 1992.

EFFECT OF THERMAL TREATMENT ON THE POROSITY OF SILICA GELS: STUDIES BY THERMAL DESORPTION OF LIQUIDS

J. Goworek, W. Stefaniak and T. Goworek[*]

Maria Curie Skłodowska University
Faculty of Chemistry, 20-031 Lublin, Poland

[*]Maria Curie Skłodowska University
Institute of Physics, 20-031 Lublin, Poland

1 INTRODUCTION

Porous solids have found use in many fields as water purification, oil recovery, gas and liquid adsorption and catalysis. Examples of these materials include adsorbents, chromatography packings, porous polymers, catalysts supports. Nitrogen adsorption measurements are routinely used for determining the pore size distribution (PSD) of solids[1]. In our previous papers, we reported results of an investigation of the porosity by using thermogravimetric technique. A new approach was proposed for the analysis of desorption curves of liquid wetting the porous solids perfectly. High resolution thermogravimetry, which utilises the heating procedure developed by Paulik and Paulik, has been used in the desorption experiments[2-4]. Full details of this method and its applications have been described elsewhere[5-10]. The desorption curves representing the weight-loss against temperature can be converted to a plots volume loss vs. pore radii $\Delta V = f(R_k)$ using the Kelvin equation. Differentiating of these dependencies the pore size distribution can be calculated. The PSD curves for silica gels, aluminum oxides, carbonaceous materials and porous organic membranes calculated in the manner described in Refs.[5-10] are very similar to those calculated from the low temperature adsorption/desorption data of nitrogen. The differences between PSDs obtained by different techniques were explained by the presence of the liquid film of thickness τ which remains on the wall of pores after evacuation of the core of pore. For cylindrical pores pore radius = core radius + τ. The TG method according to our suggestion leads to a core size distribution. To convert it into a pore size distribution requires knowledge of the thickness τ as a function of temperature which is unavailable at present. The nitrogen adsorption method, however, leads to pore size distributions after corrections have been made with respect to the surface film thickness. The thickness of the adsorbed film is commonly calculated from Halsey equation[11-12]. Alternatively, a standard isotherm is used to calculated the thickness of the adsorbed layer[1].

Recently we reported the effect of heat–treatment on the textural and adsorption characteristic of silica gels usually used in gas chromatography[13]. Silica gels after pretreatment at higher temperature appears as model porous interfaces for studying the influence of surface irregularities on the desorption process in thermogravimetric experiments. Comparison of the desorption curves from TG experiments and PSDs obtained by using two different techniques for thermally modified silica gels is the aim of the present paper.

2 EXPERIMENTAL

Five samples of silica gel Si–100 from Merck, Germany after pretreatment at 200, 400, 600, 800 and 1000°C were used in the experiment. The physical characterization data for these silica samples have been previously reported[13].

The adsorption/desorption isotherms of nitrogen at –195°C were measured with an automated apparatus Sorptomatic 1800 (Carlo Erba, Italy). Surface areas were calculated from the linear form of the BET equation over the linear range of relative pressure between about 0.05 to 0.4. A value of 16.2Å/molecule was used for the cross sectional area of the nitrogen molecule. The pore size distribution curves were calculated by the BJH method[14]. The results of the nitrogen adsorption method are collected in Table 1. Column 5 of Table 1 contains the concentrations of surface silanols α_{OH} taken from Ref.[13].

Thermogravimetric analysis was performed by using thermal analyser Derivatograph C (MOM, Hungary). Acetone and carbon tetrachloride (Merck, Germany), puriss grade were used as a wetting liquids. Prior to experiment liquid adsorbates were dried and stored over 3Å and 4Å molecular sieves.

Derivatograph of Paulik, Paulik and Erdey system is equipped in quasi-isothermal program which enables to maintain the equilibrium conditions as close as possible. The character of quasi-isothermal program lies in the condition that the temperature is raised at the fixed rate until transformation begins in the sample. Next, apparatus keeps constant temperature. When the transformation is completed, the process runs again in non–isothermal conditions. The desorption of liquid from pores of different dimensions may be treated as the process which is a sum of several isothermic processes. In other words, when temperature increases, the liquid/vapour equilibria for succeeding groups of pores are attained.

The samples for TG experiment, in the form of paste, were prepared by adding an excess of liquid adsorbate to the dry adsorbent and placed in the platinum crucible of the conical type[4]. This type of crucible is mostly appropriate because makes it possible to keep over the sample self-generated atmosphere of liquid vapours. The samples were outgassed before experiments to facilitate the penetration of pores by the adsorbate.

3 RESULTS AND DISCUSSION

Figure 1 illustrates the influence of the concentration of surface silanols on the adsorption of acetone from solutions in benzene. As is seen, the excees adsorption $n^{\sigma(n)}$ of acetone substantially decreases when the concentration of OH groups on the silica surface decreases. However, for both silicas the adsorption of acetone is preferenial and positive. Observed effect suggests that the surface silanols play the main role in the adsorption of acetone from binary solutions.

Let us consider the desorption of pure liquids from silica surface. Thermodesorption curves of carbon tetrachloride and acetone for investigated silica samples are shown in Figures 2 and 3, respectively. Segment 1 of these curves represents the bulk liquid outside the pores. Intensive evaporation at this stage of the desorption takes place at the boiling point of the liquid (perpendicular segment). Segment 2 corresponds to the capillary-

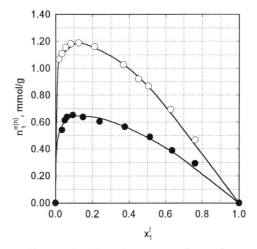

Figure 1 *Specific surface excees isotherms of acetone from benzene solution on silica gel Si-100(200) – open points, and Si-100(800) – filled points.*

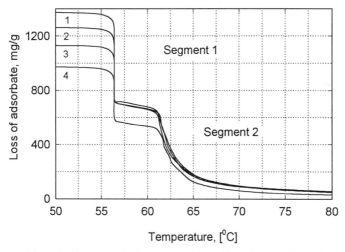

Figure 2 *Thermogravimetric curves of desorption of acetone for various silica gels; 1 – Si-100(200), 2 – Si-100(400), 3 – Si-100(600) and 4 – Si-100(800).*

condensed liquid within the pores together with the adsorbed film on the walls of pores and is, therefore, a measure of the total pore volume. The lower parts of desorption curves for silicas with different concentration of surface silanols are almost identical, which may indicate that the chemistry of the surface does not influence the thickness of the surface film. This effect is observed for all the silicas studied, irrespective of the type of the adsorbate. Identity of the ends of desorption curves is especially surprising in the case of acetone for which surface silanols are the main adsorption centers.

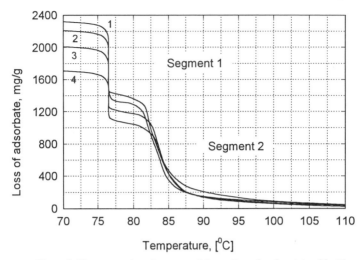

Figure 3 *Thermogravimetric curves of desorption of carbon tetrachloride for various silica gels; 1 – Si-100(200), 2 – Si-100(400), 3 – Si-100(600) and 4 – Si-100(800).*

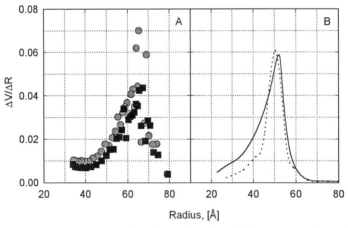

Figure 4 *Pore–core size distributions for silica gels Si-100(200) (circles and solid line) and Si-100(800) (squares and broken line) obtained by using different methods: A – nitrogen method; B – thermogravimetry.*

Figure 4 shows pore size distribution curves for Si-100(200) and Si-100(800) derived from nitrogen adsorption and desorption of acetone. It appears that for both silica samples the location of the peak of PSD is identical. Thermal treatment eliminates part of pores with smaller dimensions. This effect is observed in the case of two different techniques.

A comparison of the PSDs from nitrogen method and TG experiment for various liquid adsorbates is presented in Figures 5 and 6. In the case of carbon tetrachloride the distributions obtained by the different method are close together. For acetone systems the

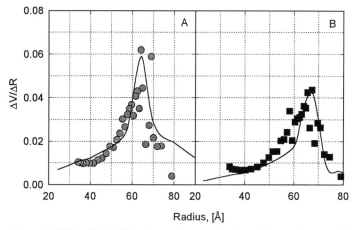

Figure 5 *PSDs obtained by using various methods; points – nitrogen method, solid lines – thermogravimetry (desorption of CCl₄); A – Si-100(200), B – Si-100(800).*

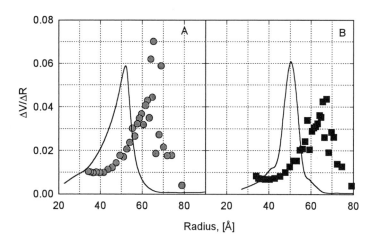

Figure 6 *PSDs obtained by using various methods; points – nitrogen method, solid lines – thermogravimetry (desorption of acetone); A – Si-100(200), B – Si-100(800).*

PSDs derived from TG data are shifted toward smaller radii in comparison to the nitrogen method. Similar effect is observed usually in the case of silica gel and strongly polar components like water or methanol[15]. The difference of the peak location of pore size distribution curves may be ascribed to the presence of strongly bonded liquid film remaining on the walls of pores after their emptying. It is noteworthy that the difference of R_{peak} (TG) - R_{peak} (N₂) is similar for both silica samples. In other words, the thickness of the surface film is independent of the concentration of surface silanols. This effect correlates reasonably with the similarity of the shape of TG curves for various silica gel samples (see Figures 2 and 3).

A comparison of the total pore volumes estimated by the different methods can be seen in Table 1. It appears that in the case of TG experiment the total pore volumes V_p are similar for acetone and carbon tetrachloride. These values correlate very well with the values of V_p from the nitrogen method.

Table 1 *Parameters characterizing the porous structure of investigated silica gels*

Silica	α_{OH}	Nitrogen method			TG method		
	[OH/nm²]	S_{BET} [m²/g]	V_p [cm³/g]	R_{Peak} [Å]	Adsorbate	V_p [cm³/g]	R_{Peak} [Å]
Si-100 (200°C)	5.83	315	1.10	66 ±3	acetone	0.91	52 ±10
					CCl₄	0.98	64 ±3
Si-100 (400°C)	3.86	319	1.08	66 ±3	acetone	0.90	51 ±10
					CCl₄	0.92	65 ±3
Si-100 (600°C)	2.21	301	1.05	66 ±3	acetone	0.92	52 ±10
					CCl₄	0.84	64 ±3
Si-100 (800°C)	1.01	230	0.88	65 ±3	acetone	0.76	50 ±10
					CCl₄	0.75	67 ±3

3.1 Correction of Desorption Curves for Instrumental Response

There is to note that the instrumental response in the case of definite pore size is not a delta function. The evaporation rate is highest at the boiling point, but this rate is non–zero also at lower temperatures. For definite boiling temperature we obtain the instrumental reaction $P(T)$, and the mass loss vs. temperature observed experimentally $M(T)$ is thus a convolution of instrumental reaction with the mass loss distribution expected in an ideal case $R(T)$ (i.e. not distorted by the $P(T)$). Real pore size distribution can be derived from $R(T)$, not from $M(T)$.

$$M(T) = \int P(T - T')R(T')dT \tag{1}$$

where $P(T - T')$ is normalized to unit area. The function $M(T)$ is done experimentally, as well as $P(T)$, which can be taken from the experiment with non–porous material. The functions $M(T)$ and $P(T)$ were approximated by respective histograms, thus instead of integration we have the summing:

$$M(T_i) = \sum_j P(T_i - T_j)R(T_j) \tag{2}$$

The formula (2) represents a set of equations which allow to determine the $R(T_j)$ values, i.e. to reconstruct the real desorption curve.

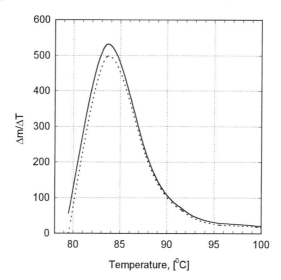

Figure 7 $\Delta m / \Delta T$ vs. T for desorption of CCl_4 from silica gel Si-100(200); solid line – initial curve, broken line – corrected curve.

Figure 7 shows for illustrative purposes the initial and corrected curves $\Delta m / \Delta T$ vs. T for carbon tetrachloride/Si-100(200) system. As is seen, the difference between these curves is very small and occurs within the lower temperatures which correspond to desorption from large mesopores.

4 SUMMARY

Summing up, one can state that the chemical character of silica surface does not influence of the shape of thermal desorption curves. Thus, PSDs derived from these curves and the surface film effects for various samples are very silmilar. The TG technique is non-destructive and can provide details of the structure of wet gels containing organic solvent. Thermogravimetry may be used for rigid materials but also for those whose texture may swell in a liquid medium. Its range of application cover a pore size range corresponding to distribution of mesopores.

5 REFERENCES

1. S.J. Gregg and K.S.W. Sing, "Adsorption, Surface Area and Porosity", Academic Press Inc., London 1982.
2. F. Paulik and J. Paulik, *Journal Thermal Analysis*, 1973, **5**, 253.
3. J. Paulik, F. Paulik and M. Arnold, *Journal Thermal Analysis*, 1987, **32**, 301.
4. F. Paulik, "Special trends in thermal analysis", John Willey & Sons, Chichester, New York, Brisbane, Toronto, Singapore 1995.
5. J. Goworek and W. Stefaniak, *Colloids Surfaces*, 1991, **62**, 135.
6. J. Goworek and W. Stefaniak, *Mat. Chem. Phys.*, 1992, **32**, 244.
7. J. Goworek and W. Stefaniak, *Colloids Surfaces*, 1994, **82**, 71.

8. J. Goworek and W. Stefaniak, in "Characterization of Porous Solids III", J. Rouquerol, F. Rodriguez-Reinoso, K.S.W. Sing and K.K. Unger, Eds., Elsevier, Amsterdam 1994, p.401.
9. J. Goworek and W. Stefaniak, *J. Thermal Anal.*, 1995, **45**, 999.
10. J. Goworek and W. Stefaniak, *Colloids Surfaces*, 1992, **69**, 23.
11. G.D. Halsey, *J. Chem. Phys.*, 1948, **16**, 931.
12. B.C. Lippens and J.H. de Boer, *J. Catalysis*, 1965, **4**, 319.
13. J. Goworek and A. Nieradka, *J. Colloid Interf. Sci.*, 1996, **180**, 371.
14. E.P. Barrett, L.G. Joyner and P.H. Halenda, *J. Amer. Chem. Soc.*, 1951, **73**, 373.
15. J. Goworek and W. Stefaniak, J. Porous Materials, *in press*.

GENERATION OF RADIALLY NONUNIFORM POROUS PARTICLES - the influence of nonuniformity and of correlation properties on fluid-solid noncatalytic reactions

A. Adrover[a] and M. Giona[b]

[a]Dipartimento di Ingegneria Chimica
Università di Roma "La Sapienza"
Via Eudossiana 18, 00184, Roma, Italy
[b]Dipartimento di Ingegneria Chimica,
Università di Cagliari
Piazza d'Armi 09123, Cagliari, Italy

1 INTRODUCTION

There has been a great deal of work on fluid-solid noncatalytic reactions and the role of non-uniform distribution of the solid reactant within initially nonporous or porous pellets[1-3]. All of these studies overlook the effects of correlations in the reactant solid distribution. Even in a porous particle of pure solid reactant, however, the solid distribution function alone cannot characterize the geometrical and topological properties of the pore network, and many other geometrical parameters, such as the spatial correlation length, should be introduced in the study of transport and reaction phenomena. In this article we present a general method to generate d-dimensional radially symmetric porous particles with an assigned void fraction distribution $\varepsilon(r)$ and a given pore-pore correlation function $C_{2\chi}(r)$. The method proposed is an extension of the procedure developed by Giona and Adrover[4] to solve the inverse problem of reconstructing porous media[5]. The method is based on the generation of a stochastic Gaussian process with specified spatial correlation properties and the subsequent application of a pointwise nonlinear filter, specifically designed for obtaining a binary (pore/matrix, 0/1) process representing the reconstructed porous structure with the assigned void-fraction distribution.

The structures generated in this way have been used to analyze the influence on fluid-solid noncatalytic reactions of spatial correlations in the solid reactant distribution, in the case of initially nonporous particles composed of a solid reactant dispersed in an inert matrix.

2 THE GENERATION METHOD

Let us consider an isotropic, radially symmetric, non-homogeneous porous particle. Under the assumption of isotropy, a generic axial cross-section of the particle is representative of the entire structure. Let us analyze a two-dimensional experimental section of the particle by means of its digital image (black and white picture). This leads to identification of the pore space $\mathcal{P}$ in terms of its characteristic function $\chi_{\mathcal{P}}(\mathbf{r})$, such that $\chi_{\mathcal{P}}(\mathbf{r}) = 1$ if $\mathbf{r} \in \mathcal{P}$ and $\chi_{\mathcal{P}}(\mathbf{r}) = 0$ elsewhere.

Knowledge of the characteristic function $\chi_P(\mathbf{r})$ makes it possible to evaluate the void fraction (porosity) $\varepsilon(r)$, $\varepsilon(r) = \frac{1}{2\pi}\int_0^{2\pi} \chi_P(r,\theta)rd\theta$, which depends exclusively on the radial coordinate r and the normalized pore-pore correlation function[a] $C_{2\chi}^{(d)}(x)$

$$C_{2\chi}^{(d)}(x) = \frac{\langle(\chi_P(\mathbf{r}) - \varepsilon(r))(\chi_P(\mathbf{r} + \mathbf{x}) - \varepsilon(|\mathbf{r} + \mathbf{x}|))\rangle}{\langle\varepsilon(r)\rangle - \langle\varepsilon^2(r)\rangle} , \tag{1}$$

where the symbol $\langle\cdot\rangle$ stands for spatial average

$$\langle f(r,\theta)\rangle = \frac{1}{2\pi}\int_0^{2\pi} d\theta \int_0^R drf(r,\theta)p^{(d)}(r) , \quad p^{(d)}(r) = d\frac{r^{d-1}}{R^d} , \tag{2}$$

R being the particle radius and $p^{(d)}(r)$ the d-dimensional radial density function ($d = 2$ for two-dimensional structures, $d = 3$ for three-dimensional).

The aim of the reconstruction is to generate two and/or three-dimensional lattice models possessing the same porosity function $\varepsilon(r)$ and the same pore-pore correlation function $C_{2\chi}^{(d)}(x)$ as the original porous structure.

The solution is presented in the form of a linear filter acting on Gaussian processes by means of a superposition of elementary correlated processes with prescribed correlation properties, as developed by Giona and Adrover[4].

Let us define the set of basis processes $\{\mathcal{Y}(\mathbf{r}, \lambda)\}$ by convoluting a system of Gaussian uncorrelated processes $\xi_\lambda(\mathbf{r})$ (with zero mean and unit variance) with a Gaussian kernel,

$$\mathcal{Y}(\mathbf{r}, \lambda) = \int_{E^d} a(\mathbf{u}, \lambda)\xi_\lambda(\mathbf{u} + \mathbf{r})d\mathbf{u} = \left(\frac{4\lambda}{\pi}\right)^{d/4} \int_{E^d} e^{-2\lambda u^2}\xi_\lambda(\mathbf{u} + \mathbf{r})d\mathbf{u} , \tag{3}$$

where E^d is the Euclidean d-dimensional space, $E^d = \{\mathbf{r} \mid -\infty < r_i < \infty , (i = 1, .., d)\}$.

Each basis process $\mathcal{Y}(\mathbf{r}, \lambda)$ remains Gaussian with zero mean and unit variance but exhibits a Gaussian decay of the correlation function[b] $C_{2\mathcal{Y}}(x, \lambda) = e^{-\lambda x^2}$.

A generic correlated Gaussian process $\mathcal{Y}(\mathbf{r})$ with correlation function $C_{2\mathcal{Y}}(x)$ can be obtained from the superposition of the basis processes $\{\mathcal{Y}(\mathbf{r}, \lambda)\}$ such as

$$\mathcal{Y}(\mathbf{r}) = \int_0^\infty p(\lambda)\mathcal{Y}(\mathbf{r}, \lambda)d\lambda , \tag{4}$$

$$C_{2\mathcal{Y}}(x) = \int_0^\infty \pi(\lambda)C_{2\mathcal{Y}}(x, \lambda)d\lambda = \int_0^\infty \pi(\lambda)e^{-\lambda x^2}d\lambda , \tag{5}$$

where $\pi(\lambda) = p^2(\lambda)$ is the weight function.

Transformation from the $\mathcal{Y}$-process to the binary process, representing the characteristic function of the reconstructed porous structure $\chi_P^R(\mathbf{r})$, is effected by means of a nonlinear filter $\mathcal{G}$ depending on the Gaussian distribution function $F_{\mathcal{Y}}$ of $\mathcal{Y}(\mathbf{r})$ and on the porosity function $\varepsilon(r)$ of the original porous medium:

$$\chi_P^R(\mathbf{r}) = \mathcal{G}(\mathcal{Y}(\mathbf{r}), \varepsilon(r)) = \begin{cases} 1 & F_{\mathcal{Y}}(\mathcal{Y}(\mathbf{r})) < \varepsilon(r) \\ 0 & F_{\mathcal{Y}}(\mathcal{Y}(\mathbf{r})) > \varepsilon(r) . \end{cases} \tag{6}$$

---

[a] It should be observed that, under the assumption of isotropy and radial symmetry, the pore-pore correlation function is a function solely of the distance $x = |\mathbf{x}|$.

[b] The basis processes $\{\mathcal{Y}(\mathbf{r}, \lambda)\}$ for different λ are uncorrelated with each other, i.e. $< \mathcal{Y}(\mathbf{r}, \lambda_1)\mathcal{Y}(\mathbf{r}', \lambda_2) >= 0$ for $\lambda_1 \neq \lambda_2$.

The definition, eq. (6), of the nonlinear filter $\mathcal{G}$ ensures statistically that the reconstructed porous structure admits the porosity function $\varepsilon(r)$. The only further condition to be imposed is that the correlation function of the reconstructed lattice $C_{2\chi R}^{(d)}(x)$ coincides with the pore-pore correlation function of the original porous structure $C_{2\chi}^{(d)}(x)$.

The correlation function $C_{2\chi R}^{(d)}(x)$ is related to the corresponding correlation function $C_{2y}(x)$ through the integral relation

$$C_{2\chi R}^{(d)}(x) = \int_0^R dr_1 \int_0^R dr_2 \int_{-\infty}^\infty dy_1 \int_{-\infty}^\infty [\mathcal{H}(y_1, y_2, r_1, r_2)]$$
$$\cdot \; g(y_1, y_2, x) p^{(d)}(r2/r_1; x) p^{(d)}(r_1) dy_2 , \tag{7}$$

with

$$\mathcal{H}(y_1, y_2, r_1, r_2) = \frac{(\mathcal{G}(y_1, \varepsilon(r_1)) - \varepsilon(r_1))(\mathcal{G}(y_2, \varepsilon(r_2)) - \varepsilon(r_2))}{\langle \varepsilon \rangle - \langle \varepsilon^2 \rangle} \tag{8}$$

where $g(y_1, y_2, x)$ is a bivariate Gaussian density function

$$g(y_1, y_2, x) = \frac{1}{2\pi(1 - C_{2y}^2(x))^{1/2}} \exp\left[-\frac{y_1^2 + y_2^2 - 2C_{2y}(x)y_1 y_2}{2(1 - C_{2y}^2(x))} \right] , \tag{9}$$

and $F^{(d)}(r_2/r_1; x) = \int_0^{r_2} p^{(d)}(r/r_1; x) dr$ is the probability distribution function of finding a point possessing the radial coordinate r_2 at a relative distance x from a point at a radial coordinate r_1 in a d-dimensional radial symmetric structure.

By defining

$$\gamma(r, r_1, x) = \frac{r^2 - (r_1^2 + x^2)}{2xr_1} , \tag{10}$$

the probability distribution function $F^{(d)}(r_2/r_1; x)$ attains the form[c]

$$F^{(d)}(r_2/r_1; x) = \begin{cases} 1 - \frac{1}{\pi}\arccos(\gamma(r_2, r_1, x)) & \text{if } (x + r_1) \leq R \text{ and } |\gamma(r_2, r_1, x)| < 1 \\ \frac{\pi - \arccos(\gamma(r_2, r_1, x))}{\pi - \arccos(\gamma(R, r_1, x))} & \text{if } (x + r_1) > R \text{ and } |\gamma(r_2, r_1, x)| < 1 \\ 1 & \text{if } \gamma(r_2, r_1, x) \geq 1 \\ 0 & \text{if } \gamma(r_2, r_1, x) \leq -1 . \end{cases}$$

$$(11)$$

Eq. (7), expressing the functional relation between $C_{2\chi R}^{(d)}(x)$ and $C_{2y}(x)$, can be integrated numerically for a given porosity function $\varepsilon(r)$. In this way, for each x, starting from a set of values $\{C_{2y_i}\} \in [-1, 1]$, it is possible to evaluate the corresponding set of values $\{C_{2\chi R_i}^{(d)}\}$. which can be used as a calibration curve[d].

By imposing $C_{2\chi R}^{(d)}(x) = C_{2\chi}^{(d)}(x)$ and by making use of the relations eqs. (7)-(11) (or equivalently of the calibration curves) it is possible to obtain the correlation function $C_{2y}(x)$. From the knowledge of $C_{2y}(x)$, the weight function $\pi(\lambda)$ can then be evaluated from eq. (5) either numerically or, where possible, analytically, following the same procedure developed by Giona and Adrover in connection with homogeneous

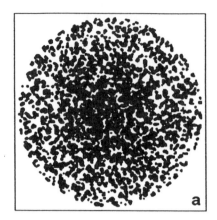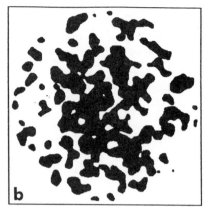

Figure 1: 2-d lattice models generated from two different basis processes $\mathcal{Y}(\mathbf{r}, \lambda)$ (for two different values of λ) by applying a nonlinear filter $\mathcal{G}$ with a linear porosity function $\varepsilon(r)$, eq.(13) ($\varepsilon_0 = 0.2$, $\varepsilon_R = 0.8$, $R = 150$ lattice units). a) $\lambda a^2 = 0.1$; b) $\lambda a^2 = 0.01$. a is the unit lattice site size

porous media. A fundamental element of the method is that, on the condition that $C_{2y}(x) \geq 0$, putting $z = x^2$, $\tilde{C}_{2y}(z) = C_{2y}(x)$, it follows that

$$\tilde{C}_{2y}(z) = \int_0^\infty \pi(\lambda) e^{-\lambda z} d\lambda \ , \tag{12}$$

and therefore $\pi(\lambda)$ is the inverse Laplace transform of the analytic continuation on the complex plane $\tilde{C}_{2y}^{(p)}(z)$ of the correlation function of the $\mathcal{Y}(\mathbf{x})$ process $\tilde{C}_{2y}(z)$. The analytic continuation $\tilde{C}_{2y}^{(p)}(z)$ of $\tilde{C}_{2y}(z)$, valid for all complex z (whose restriction to real z coincides with $\tilde{C}_{2y}(z)$), can be achieved by considering rational approximations (e.g. Padé approximants) or by means of other methods such as orthogonal polynomial expansion. In all the cases in which closed-form solutions for the inverse Laplace transform of $\tilde{C}_{2y}^{(p)}(z)$ cannot be obtained, the weight function $\pi(\lambda)$ can be evaluated numerically.

Figure 1 a)-b) shows two two-dimensional lattice models generated from the basis processes $\mathcal{Y}(\mathbf{r}, \lambda)$ with two different values of λ, by applying the nonlinear filter $\mathcal{G}$ with a linear porosity function $\varepsilon(r)$,

$$\varepsilon(r) = \varepsilon_0 + (\varepsilon_R - \varepsilon_0) r / R \ . \tag{13}$$

Figure 2 a) shows the good level of agreement attained between the porosity distribution function of the two-dimensional lattice structure in figure 1 a) (points) and the assigned linear behaviour, eq. (13) (continuous line). A good level of agreement is also attained by the pore-pore correlation function of the lattice structure of figure 1 a) (points) and the theoretical behaviour (continuous line) obtained from the numerical integration of eqs. (7)-(11), with the linear porosity function, eq. (13), and with $C_{2y}(x) = C_{2y}(x, \lambda) = \exp(-\lambda x^2)$ (figure 2b).

---

[c]It should be observed that $F^{(2)}(r_2/r_1; x) = F^{(3)}(r_2/r_1; x)$.

[d]For homogeneous porous media, the integral equation relating $C_{2\chi^R}^{(d)}(x)$ and $C_{2y}(x)$ does not depend explicitly on x and consequently there is only one calibration curve for each value of the average (and uniform) porosity $< \varepsilon > = \varepsilon$.

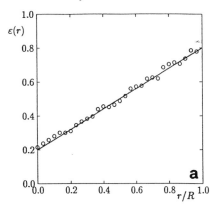
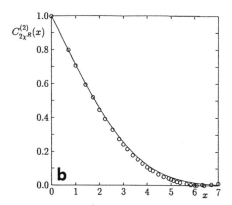

Figure 2: a) Comparison of the porosity function $\varepsilon(r)$ of the two-dimensional lattice structure of figure 1 a) (points) and the assigned linear behaviour, eq. (13) (continuous line). b) Comparison of the pore-pore correlation function $C_{2\chi_R}^{(2)}(x)$ of the lattice structure of figure 1 a) (points) and the assigned behaviour (continuous line) obtained from the numerical integration of eqs. (7)-(11), with the linear porosity function, eq. (13), and with $C_{2y}(x) = C_{2y}(x, \lambda) = \exp(-\lambda x^2)$ ($\lambda a^2 = 0.1$). x is expressed in lattice units.

3 FLUID-SOLID NONCATALYTIC REACTIONS

The method discussed above can be applied to generate generic binary structures: pore space/solid matrix, in the case of porous pellets or reactant/inert nonporous solid for initially nonporous particles.

Here we study the influence of the spatial correlation of solid reactant on fluid-solid noncatalytic reactions, focusing on initially non-porous particles with an assigned reactant solid fraction distribution.

To this end, a lattice simulator[6] was developed combining the finite difference algorithm developed by Giona et al.[7] for diffusion in the product layer with a reacting step on the reactant solid boundary, and suitable for the simulation of both linear and nonlinear kinetics in disordered lattice matrices. Further details on the lattice simulator can be found in ref. 6.

The generic fluid-solid reaction under consideration is $A(f) + bB(s) \rightarrow C(f) + dD(s)$. The pellet consists of reactant and inert solid, so that the overall pellet size R and shape do not change with reaction. Let $\varepsilon(r)$ be the reactant solid fraction distribution in the radially symmetric particle. The reacted layer is porous. The fluid species A diffuses through it and chemical reaction occurs simultaneously on the unreacted solid boundary with intrinsic isothermal kinetics $R_k(c_A)$ [mol s^{-1} m^{-2}]

$$-R_k(c_A) = k(c_A^0)c_A\psi(c_A/c_A^0), \tag{14}$$

where c_A is the molar concentration of A, $c_A^0 = c_A(r = R)$, and $\psi(c_A/c_A^0)$ is a dimensionless functional accounting for the nonlinearities in the intrinsic kinetics.

By introducing the dimensionless variables $x = r/R$, $\xi = c_A/c_A^0$ and $\tau = tk(c_A^0)/R$, the dimensionless characteristic parameters controlling the dynamics are

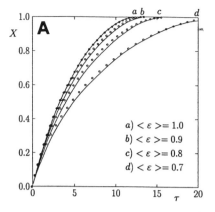
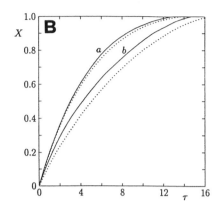

Figure 3: X vs τ obtained from lattice simulation on initially nonporous cylindrical particles ($R = 75$ l.u.). The simulation parameters are $\phi^2 = 1.0$, $\alpha = 0.1$, $\psi(\xi) = 1$ (linear intrinsic kinetics). A) Uncorrelated structures with uniform solid reactant distribution $\varepsilon(x) =< \varepsilon >$. Dots represent the solution of the corresponding continuous model, eq. (15)-(16) with $\bar{\phi}^2$ evaluated from the values of D_Σ for uncorrelated structures in Figure 4 A, curve a). B) Gaussian correlated (solid lines, $\lambda a^2 = 0.05$) and uncorrelated (dotted lines) structures with linear solid fraction distribution. a) $\varepsilon_0 = 0.7$, $\varepsilon_R = 1.0$; b) $\varepsilon_0 = 1.0$, $\varepsilon_R = 0.7$.

$\phi^2 = k(c_A^0)R/\mathcal{D}_{AC}$ and $\alpha = bc_A^0/\rho_B$, where $\mathcal{D}_{AC}$ is the diffusivity of A in C in the product layer, and ρ_B the molar density of the pure reactant B.

Figure 3 A) shows the simulation results for the conversion X vs τ in the case of a linear intrinsic kinetics $\psi(\xi) = 1$ for an initially nonporous cylindrical particle with a uniform solid reactant distribution $\varepsilon(r) =< \varepsilon >$, for different values of $< \varepsilon >$, in the completely uncorrelated case. Dots represents the solution of the corresponding shrinking-core model

$$\frac{\partial \xi}{\partial \tau} = \frac{1}{x\bar{\phi}^2}\frac{\partial}{\partial x}\left(x\frac{\partial \xi}{\partial x}\right) \quad x_c \leq x \leq 1 , \tag{15}$$

$$\xi = 1 \quad \text{at} \quad x = 1 , \qquad \frac{\partial \xi}{\partial \eta} = \bar{\phi}^2\xi \quad \text{at} \quad x = x_c ,$$

$$\frac{dx_c}{d\tau} = -\alpha\xi\,|_{x=x_c} \tag{16}$$

where $\bar{\phi}^2$ is the effective square Thiele modulus $\bar{\phi}^2 = \frac{k(c_A^0)R<\varepsilon>}{\mathcal{D}_{AC}\mathcal{D}_\Sigma(<\varepsilon>)} = \frac{\phi^2<\varepsilon>}{\mathcal{D}_\Sigma(<\varepsilon>)}$ and D_Σ the effective diffusivity in the pore space surrounded by the inert matrix. For uncorrelated structures, D_Σ can be independently obtained from EMA theories (see curve a) in figure 4 A).

Figure 3 A) shows the quantitative agreement between the simulation results and the continuous model, eqs. (15)-(16), and highlights the possibility of taking into account the disordered solid distribution by simply introducing an effective diffusivity in the space surrounded by the inert matrix and progressively transformed into the product layer.

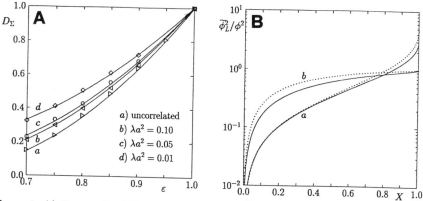

Figure 4: A) D_Σ vs ε for different values of λ. B) $\bar{\phi}_L^2(x)/\phi^2$ vs X for Gaussian correlated (solid lines, $\lambda a^2 = 0.05$) and uncorrelated (dotted lines) structures with linear solid fraction distribution. a) $\varepsilon_0 = 0.7$, $\varepsilon_R = 1.0$; b) $\varepsilon_0 = 1.0$, $\varepsilon_R = 0.7$.

With the support of this result, we went on to study the influence of correlations on fluid-solid noncatalytic reactions for initially nonporous particles characterized by a linear reactant solid distribution $\varepsilon(x) = \varepsilon_0 + (\varepsilon_R - \varepsilon_0)x$ and Gaussian decay of the correlation function $C_{2E}(x) = \exp(-\lambda x^2)$. For these structures, the behaviour of D_Σ vs ε for different values of the correlation length $L_c = 2/\sqrt{\lambda}$ has been independently evaluated from the solution of the exit-time equation and is shown in figure 4 A. Figure 4 A clearly shows that the effects of correlation on the effective diffusivity are significant only for high values of the correlation length and low values of the porosity (approaching the critical threshold).

Figure 3 B shows simulation results for the behaviour of X vs τ for two correlated initially nonporous particles characterized by the same value of λ and linear solid distribution function, and the comparison with the curves X vs τ for the corresponding uncorrelated structures (dotted curves). As can be expected, since the effective diffusivity increases with the correlation length, the total conversion time is less for correlated structures than uncorrelated. Curve a) refers to the particle with increasing solid fraction with dimensionless radius x ($\varepsilon_0 = 0.7$, $\varepsilon_R = 1.0$). In this case the effects of correlations are weak and appear only for high conversion values. Curve b) refers to the particle with decreasing solid fraction with x ($\varepsilon_0 = 1.0$, $\varepsilon_R = 0.7$). The effects of correlations are significant also for low values of the conversion. Its clear that the influence of correlation is significant only for the structure presenting lower solid fractions in contact with higher values of fluid concentration.

This result can be easily explained by analyzing the behaviour of the local effective Thiele modulus $\bar{\phi}_L^2(x) = \frac{\phi^2(1-x)\varepsilon(x)}{D_\Sigma(\varepsilon(x))}$ as a function of the conversion X for the two correlated structures (curve a) and b) in figure 4B) and for the corresponding uncorrelated ones (dotted lines).

In the case of increasing solid fraction with x, the curves $\bar{\phi}_L^2(x)/\phi^2$ vs X for the correlated and uncorrelated structures practically coincide up to higher values of the conversion (Figure 4 B, curve a)). In the case of decreasing solid fraction with x, the curves $\tilde{\phi}_L^2(x)/\phi^2$ vs X for the correlated and uncorrelated structures deviate significantly, especially for intermediate values of the conversion $0.1 \leq X \leq 0.6$. This effect shows up in the X vs τ curves, which initially deviate appreciably and then proceed practically parallel to each other.

To sum up, the influence of porosity and spatial correlations on fluid-solid noncatalytic reactions involving initially non-porous particles can be completely described by means of the effective medium diffusivity depending on both ε and λ. The effect of correlations on fluid-solid reaction evolution is significant only for low solid fractions and for particles presenting low solid fractions in contact with higher values of the fluid concentration.

4 CONCLUDING REMARKS

This article proposes a method to generate nonuniform porous media with assigned correlation function and radial porosity distribution function. The method generalizes the analysis developed by Adler et al. and by Giona and Adrover on the reconstruction of porous media.

We also study the influence of correlations in the reactant solid distribution on fluid-solid reactions starting from initially non-porous particles.

Depending on the behaviour of the radial distribution function, correlation effects may be fairly significant. In general, greater deviations from the uncorrelated case are observed for particles presenting low solid fractions in contact with higher fluid concentrations.

References

1. J. Szekely, J. W. Evans, H. Y. Sohn, *Gas-Solid Reactions*, Academic Press, New York, 1976.
2. M. P. Dudukovic, *Ind. Eng. Chem. Fundam.*, 1984, **23**, 49.
3. H. Y. Sohn and Y. N. Xia, *Ind. Eng. Chem. Res.*, 1986, **25**, 386; ibidem, 1987, **26**, 246.
4. M. Giona and A. Adrover, *AIChE J.*, 1996, **42** 1407.
5. P. M. Adler, C.G. Jacquin and J. A. Quiblier, *Int. J. Multiphase Flow*, 1990, **16**, 691.
6. A. Adrover and A. Galassini, presented at the International Conference on Chaos and Fractals in Chemical Engineering, Rome September 2-5 1996, (submitted).
7. M. Giona, A. Adrover and A.R. Giona, *Chem. Engng. Sci*, 1995, **50**, 1001.

EFFECTS OF CARBON SURFACE CHEMISTRY ON ADSORPTION OF POLAR MOLECULES

R. H. Bradley and A. Cuesta

Institute of Surface Science & Technology
Loughborough University
Loughborough
Leics. LE11 3TU

1 INTRODUCTION

The adsorption of non-polar vapours by solids occurs by induced dipole-induced dipole (dispersion) interactions as described by London[1,2]. For non-porous solids, Type II adsorption isotherms result and equilibrium data can be analysed using the BET[3] or α_s[4] methods to derive specific external surface area data. The BET model assumes an energetically heterogeneous distribution of adsorption sites implying a simple geometric mean energy of interaction across the surface however, calorimetric heat of adsorption data, measured at low surface coverages clearly indicate differential energy distributions[5,6]. For microporous solids the dominant feature of the adsorption process is the volume filling mechanism which occurs by enhanced energy effects[7,8] resulting in a Type I isotherm. In this instance the adsorption energy is a function of the micropore size distribution. For active or microporous carbons, adsorption data can be effectively analysed using the equations developed by Dubinin. Equation [1], the Dubinin-Astakhov[9] equation, can be used to obtain useful structural information from active carbons. In this equation the parameter n is related to the pore size distribution of the carbon under investigation and takes a value of between 1.5 and 7 with the lower value being characteristic of a relatively wide pore size distribution and the higher values corresponding to a narrow distribution as observed for molecular sieve materials. When $n=2$ equation [1] becomes the more general Dubinin-Radushkevich equation[10].

$$W = W_0 \, exp[-(A/\beta E_0)]^n \qquad\qquad [1]$$

In this expression W is the volume of adsorbate within the pore structure at relative pressure p/p^0 where p is the equilibrium pressure and p^0 the saturated vapour pressure at the adsorption temperature T. W is given by NV_m where N is the molar adsorption value per gram of adsorbent and V_m is the molar volume of the liquid adsorbate. W_0 is the volume of liquid adsorbed in micropores and $A = RT \, ln \, p^0/p$ (R being the gas constant). β, the so called affinity coefficient, is derived by comparing physical parameters such as molar volume with a reference vapour, traditionally benzene, such that $\beta(C_6H_6)=1$. E_0, the characteristic adsorption energy, varies from 15 to 30 kJ mol^{-1} for active carbons and is an inverse function of the average micropore width $L/nm = 10.8/(E_0 - 11.4)$. This approach therefore not only allows calculation of micropore volumes but also gives an estimate of the average micropore width.

In contrast to the above, the adsorption of polar vapours occurs at specific sites and is characterised by relatively weak interactions. For non-porous solids a Type III isotherm results whereas porous materials give a Type V curve. In both instances the initial process is effectively one of heterogeneous nucleation at specific sites followed, at higher relative pressures, by molecular clustering. At even higher pressures multilayer film formation occurs on open surfaces whereas pore filling gives rise to the limiting adsorption value associated with the Type V curve. For pure active carbons the polar sites are chemisorbed oxygen molecules whereas for impure materials other heteroatoms such as sulphur, silicon and aluminium will also act as adsorption sites.

Water vapour is probably the polar adsorptive most frequently investigated in studies involving active carbons and in this situation the interaction is primarily by hydrogen bonding giving a heat of adsorption values in the range x-y kJ.mol^{-1}. Dubinin and Serpinski[11] have proposed that analysis of the initial region of water isotherms, using the following expression, can be used to assess the concentration of primary adsorption sites via the characteristic specific amount of adsorbed water a_0:

$$a = a_0 ch(1 - ch) \qquad\qquad [2]$$

In this equation a is the equilibrium water adsorption uptake at any given value of relative pressure denoted here as h with c being the ratio between the overall rates of adsorption and desorption. a_0 is consequently a measure of the polarity of the surface or micropore structure under investigation. In the past it has been shown that the steep regions of type III isotherms, for water adsorption by carbon blacks, can be displaced to higher values of relative pressure by thermal desorption of chemisorbed oxygen[12]. Similar isotherm shifts have also been observed for active carbons and qualitative relative assessments made of polarity using equation [2][13,14]. In particular Stoeckli has demonstrated a direct relationship between a_0 enthalpy of immersion (ΔH_i) in water[15] such that

$$\Delta H_i = -25a_0 - 0.6(a_s - a_0) \qquad\qquad [3]$$

where a_s represents the limiting amount of water adsorbed as p/p_0 approaches unity.

Type III and V isotherms have a point of inflection at low values of p/p^0 which results in Type I character for this part of the isotherm. As an extension to this approach Stoeckli has recently applied equation [1] to isotherms resulting from relatively weak vapour-solid interactions in order to derive molar adsorption energies[16].

In this paper we present data substantiating the relationship between surface polarity, heat of immersion in water and polar adsorptive uptake. We propose a general relationship between polar site parameters from the Dubinin-Stoeckli approach to adsorption and surface chemical heterogeneities which can be measured spectroscopically. We also suggest future work to refine this approach.

2 EXPERIMENTAL

Water adsorption isotherms have been measured gravimetrically in a glass apparatus using calibrated silica springs or CI microbalances. The temperature of 293 K (±0.5 K) was maintained by cooling coils and thermostirers. Samples were outgassed at 523 K to constant weight and a residual vacuum of 10^{-5} torr. Surface area data for carbon blacks were measured on a Micromeritics ASAP2000 apparatus.

Surface oxygen concentrations were measured on a VG Escalab Mk1 using the $Al_k\alpha$ line, which has energy of 1486.6eV, at a power of 200 W in a vacuum of 10^{-8} torr and with the analyser in fixed transmission mode. Carbons were analysed in 15mm diameter aluminium dishes with the analysed area less than that of the sample. Chemical shift data was referenced to the main C-C/C-H peak at 284.6 eV. An analyser pass energy of 85 eV was used to obtain broad scan data which was used for surface composition calculation after subtraction of a Shirley background.

Carbon blacks were supplied by Cabot. Sample N330 was oxidised to varying levels in an O_3/O_2 fluidised bed apparatus. O_3 was generated at 0-3% using a Wallace and Tieman BA023 ozonator. No change in BET/α_s nitrogen area was detectable after treatment. PVDC chars were produced in Ar and either activated in CO_2 or heat treated in Ar.

3 RESULTS AND DISCUSSION

It is well established that the shapes of water adsorption isotherms for carbons are sensitive to surface chemistry and that either oxidation or thermal desorption of oxides can be used to change the hydrophobicity of carbons. Figure 1 shows examples of the changes in position of water isotherms for activated and heat treated PVDC carbons which result from their differing surface polarities and Table 1 contains the characteristic parameters for these materials calculated from equations [1],[2] and [3][14].

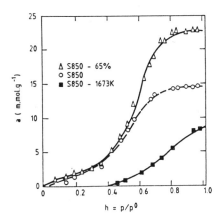

Figure 1, *Isotherms for water adsorption at 293 K by PVDC carbons (char, 65% burn off and heat treatment at 1673 K).*

Table 1, Characteristic parameters from equations 1,2 and 3. Saturation water (a_s) and nitrogen (v)   volumes. Nitrogen micropore volumes (W_0) from equation 1 (n=2). Water to nitrogen ratios (a_s/v), Characteristic adsorbed water a_0 and ΔH_i (water).

	Adsorbed Volumes				a_0 (m mol g^{-1})	
	a_s	V (cm^3 g^{-1})	W_0	a_s/V		ΔH_i (J g^{-1})
Activated S850 65%	0.41	0.54	0.48	0.76	2.99	−86.6
Activated S850 30%	0.36	0.49	0.46	0.73	1.94	−59.6
Untreated S850 char	0.26	0.41	0.38	0.63	2.23	−55.7
Heat-treated S850 1273	0.18	0.37	0.35	0.49	1.41	−40.3
Heat-treated S850 1473	0.24	0.38	0.33	0.63	0.37	−16.8
Heat-treated S850 1673	0.15	0.29	0.24	0.52	0.15	−11.8

A general decrease in the Dubinin-Serpinski parameter a_0 from the most polar (65% burn off) material to the least polar (1673 K heat treated) sample is observed which results in a decrease in the calculated ΔH_i from -86.6 Jg^{-1} to -11.8 Jg^{-1}. The overall relationship between the two parameters is shown in Figure 2.

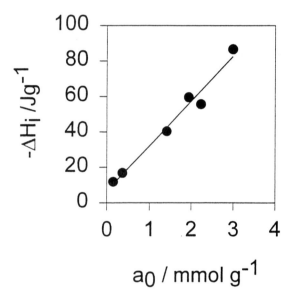

Figure 2, *Relationship between a_0 and heat of immersion from equation 3 for PVDC carbons.*

Equation [2] is empirical and results from water adsorption isotherm analysis. A significant limitation of the approach is that the parameter a_0 contains no detailed information about the specific chemical nature of the polar site(s). In recent years we have been interested in obtaining detailed physical and chemical characterisations of non-porous carbons as an aid to understanding molecular interactions within micropores. As part of this work we have been studying the relationship between carbon black surface chemistry and polarity.

Table 2, Heat of immersion data from equation 3 for PVDC carbons.

	heat of immersion			
	n-heptane		water	
carbon black	/J g^{-1}	/mJ m^{-2}	/J g^{-1}	/mJ m^{-2}
N110	−14.9	−107.5	−6.4	−45.9
N326	−8.2	−105.7	−3.6	−46.6
N330	−8.5	−108.4	−4.4	−55.8
N550	−4.5	−115.1	−2.7	−69.5
N762	−3.1	−96.2	−1.7	−53.3
K354	−13.4	−121.9	−8.3	−76.3
Monarch 1300	−61.3	−146.0	−47.2	−112.4
N330 oxidised	−8.8	−100.8	−6.7	−141.7

Figure 3 shows the first order relationship between ΔH_i (water) and total surface oxygen concentration for a series of carbon blacks as described by equation 4 which appears to be in agreement with the form of equation 2 when only the parameter reflecting surface polarity (a_0) is varied and the limiting value (a_s) is held constant. Surface composition and energy data for the carbon black systems are given in Table 2.

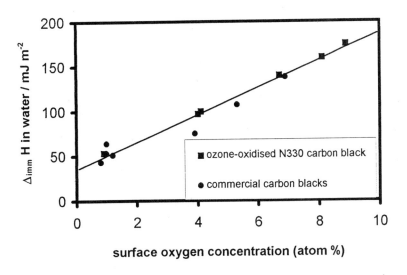

Figure 3, *Relationship between calorimetric heat of immersion in water and [O]$_T$ for non-porous carbon blacks[17].*

More recently we have demonstrated that for a wide range of non-porous carbon black surfaces, a direct relationship exists between ΔH_i (water) and the concentration of polar oxygen functionalities measured by X-ray photoelectron spectroscopy (XPS)[17].

This is described, to a first order approximation, by the following expression:

$$\Delta H_i \,/\text{J g}^{-1} = -15\text{x}10^{-3}([O]_T \text{ atom}^{-1}\text{m}^{-2})^{-1}A_S - 40(A_s) \qquad [4]$$

where $[O]_T$ is the measured total surface oxygen concentration expressed as an atomic % and A_S is the specific external surface area of the carbon in m^{2}g^{-1} measured using the BET or α_S method[15]. The heat of immersion for the unoxidised surfaces has a value of -0.04Jm[-2] for the carbon blacks studied and therefore 3-4 Jg[-1] for typical carbon blacks.

a_0 is a general term describing all of the available polar energetic heterogeneities either at a surface or within a pore whereas $[O]_T$, as used in equation 4, is a spectroscopic measurement of total external surface oxygen. The latter has the attraction of being a more exact parameter since it can be factored into individual chemical functional group components such that:

$$[O]_T = [C\text{-}OH] + [C\text{=}O] + [COOR] \text{ etc..} \qquad [5]$$

By detailed studies it may therefore be possible to obtain the energetic contribution of individual surface groups to the total energy of interaction for a surface for example enthalpy of immersion or adsorption. This approach also has the attraction that all other surface functionalities, such as nitrogen or sulphur containing groups, can also be identified and quantified. In that instance the total surface concentration of polar functional groups can be characterised.

The effects of varying $[O]_T$ for carbon blacks on the adsorption of a cationic surfactant are shown in Figure 4 in relation to data for an hydroxylated silica.

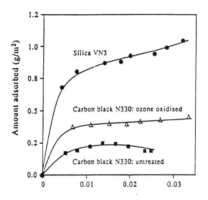

Equilibrium weight fraction of polar surfactant

Figure 4, Effects of increasing $[O]_T$ of carbon black surface from 0.9 to 6.7 on adsorption of polar surfactant in relation to hydroxylated silica.

With a view to extending our methods to porous materials we have recently been using XPS to study the chemical transformations which occur as a result of the carbonization and subsequent carbon dioxide activation of poly(*p*-phenylene terepthalamide) i.e. Kevlar, fibres. We have characterised the functional groups which result and have also observed differing chemistries which can be correlated with different activation regimes[19].

CONCLUSIONS

A direct relationship is observed between heat of immersion in water and the total surface oxygen concentration of a wide range carbon blacks. This spectroscopically measured parameter has the advantage that it incorporates all surface oxygen groups and can also be used to characterise other heteroatoms which cause polar energetic heterogeneity.

REFERENCES

1. F. London, *Z. Physik.*, 1930, **63**, 245.
2. F. London, *Z. Physik. Chem.*, 1930, **11**, 222.
3. S. Brunauer, P. Emmett and E. Teller, *J. Am. Chem. Soc.*, 1938, **60**, 309.
4. K.S.W. Sing, in 'Surface Area Determination', D. H. Everett and R. H. Ottewill eds., Butterworth, London, 1970, p.25.
5. A.V. Kiselev, *in* 'Proceedings of the Second International Congress on Surface Activity' Butterworths, London, 1957, p.168.
6. L.G. Joyner and P.H. Emmett, *J. Amer. Chem. Soc.,* 1948, **70,** 2353
7. D. H. Everett and J.C. Powl, *J. Chem. Soc. Faraday Trans.*, 1976, **72**, 619.
8. F. Stoeckli, *Helvetica Chimica Acta*, 1974, **57**, 2195.
9. M.M. Dubinin and V.A. Astakhov, *Izv. Akad. Nauk SSSR, Ser. Khim. 5*, 1971.
10. M.M. Dubinin, E.D. Zaverina and L.V. Radushkevich, *Zh. Fiz. Khim.*, 1947, **21**, 1351.
11. M.M. Dubinin and V.V. Serpinski, *Carbon*, 1981, **19**, 402.
12. A. V. Kiselev and N.V. Kovalera, *Izvest. Akad. Nauk. SSSR, Otd. Khim. Nauk 995* (transl.), 1959.
13. H. F. Stoeckli and F. Kraehenbuehl, *Carbon*, 1981, **19**, 353.
14. R.H. Bradley and B. Rand, *Carbon*, 1993, **31**, 2.
15. H.F. Stoeckli, F. Kraehenbuehl and D. Morel, *Carbon*, 1983, **21**, 589.
16. F. Stoeckli, T. Jakubov and A. Lavanchy, *J. Chem. Soc. Faraday Trans.*, 1994, **90**, 783.
17. R. H. Bradley, I. Sutherland and E. Sheng, *J. Chem. Soc. Faraday Trans.*, 1995, **91**, 3201.
18. E. Sheng, I. Sutherland, R.H. Bradley and P.K. Freakley, *Eur. Polym. J.*, 1996, **32**, 35.
19. A. Cuesta and R.H.Bradley submitted to Carbon.

EFFECT OF ASSUMED PORE SHAPE AND INTERACTION PARAMETERS ON OBTAINING PORE SIZE DISTRIBUTIONS FROM ADSORPTION MEASUREMENTS

K.L. Boulton, M.V. López-Ramón, G.M. Davies and N.A. Seaton

Department of Chemical Engineering
University of Cambridge
Pembroke Street
Cambridge CB2 3RA

1 INTRODUCTION

One of the most powerful quantifiers of the structure of porous solids is the pore size distribution (PSD). There are a variety of methods currently available for determining PSDs, but each one has a limited range of applicability. The choice of the appropriate analysis technique is dictated by the size of pores to be studied. In this work we will be focusing on microporous solids (pore widths < 2 nm).

The determination of the PSD of a solid from adsorption measurements is based on a model for adsorption in individual pores, a "single-pore model". The model consists of an assumed pore shape and a specification of the solid-fluid and fluid-fluid potential parameters. For example, PSDs for carbons are usually determined using a model with slit-shaped pores bounded by an infinite number of graphite layers. However, this is only one possible idealisation of the real structure; an alternative model pore shape could be used, or the pore wall might comprise a finite number of carbon layers. Similarly, for a given adsorbate, several sets of potential parameters are typically available, based on different types of experimental data. Thus there is considerable uncertainty involved in the construction of the single-pore model.

In this paper, we investigate the effect of this uncertainty on the PSD, using molecular simulation to solve the single-pore adsorption model.

1.1 Molecular simulation method

The grand canonical Monte Carlo (GCMC) technique, first used by Adams[1] in 1974, was employed in this work. The temperature, volume and chemical potential are the fixed macroscopic parameters (the chemical potential being calculated from the pressure and temperature using the Peng-Robinson equation of state in our case). The number of molecules is a calculated macroscopic variable. Successive microstates are produced by attempting to move, create or destroy fluid molecules. After equilibrium has been established, the simulation samples configurations or microstates characteristic of the equilibrium, and the number of molecules (n) in the system is calculated as an ensemble average over many realisations of the system. The value of n represents one point on the isotherm, at a given pressure. The whole simulation is then repeated at many different pressures to generate an entire isotherm. Many isotherms are generated to cover the expected range of pore sizes in the solid to be characterised. The PSD is then obtained by comparing the experimental adsorption isotherm of a given solid with the simulated single pore isotherms. The weighting which must be applied to each single pore isotherm such that their overall sum matches the experimental isotherm, is then the PSD.

For this investigation, the simulations were performed for a model carbon adsorbent. In all cases except where indicated otherwise, methane was used as the adsorbate.

The fluid-fluid interactions were modelled using a truncated (*i.e.* a finite range was imposed) Lennard-Jones potential. The values used for the LJ parameters were:

$\sigma_{ff} = 0.3817$ nm and $\varepsilon_{ff} / k = 148.2$ K, where k is the Boltzmann's constant.

In order to model the pore walls as being composed of a finite number of carbon layers, the solid-fluid interaction potentials of the individual layers comprising the pore wall were calculated independently, using equation (1) (which represents the potential for a single plane).

$$v_{sf}, j(z) = 2\pi\varepsilon_{sf}\rho_s\sigma_{sf}{}^2\Delta\left[\frac{2}{5}\left(\frac{\sigma_{sf}}{z+j\Delta}\right)^{10} - \left(\frac{\sigma_{sf}}{z+j\Delta}\right)^4\right], \quad (1)$$

where $v_{sf}, j(z)$ = the potential between a fluid molecule and a given carbon layer, where z = the distance between the fluid molecule and the pore wall, j is the index of the layer and Δ = the spacing between carbon layers, and ρ_s = the number density of carbon atoms per unit area of pore wall layer.

The total solid-fluid potential, v_{sf}, is then given by

$$v_{sf} = \sum v_{sf}, j(z), \quad (2)$$

The solid-fluid interaction potentials for pore walls composed of an infinite number of carbon layers were calculated using the Steele "10-4-3" potential[2], given by

$$v_{sf}(z) = 2\pi\varepsilon_{sf}\rho_s\sigma_{sf}{}^2\Delta\left[\frac{2}{5}\left(\frac{\sigma_{sf}}{z}\right)^{10} - \left(\frac{\sigma_{sf}}{z}\right)^4 - \frac{\sigma_{sf}{}^4}{3\Delta(z+0.61\Delta)^3}\right]. \quad (3)$$

The values of σ_{ss}, ε_{ss}/k and Δ used for carbon were 0.340 nm, 28.0 K and 0.335 nm respectively. σ_{sf} and ε_{sf} were calculated from the Lorentz-Berthelot rules.

The slit-shaped pore model consists of two parallel planes of infinite length and breadth, separated by the pore width. A single pore was modelled as an infinitely periodic system, with the simulation being performed in one small cell of this system. The dimensions of the simulation box were defined by the box length in the x- and y-coordinate directions, and by the pore width in the z-direction. For the square- and rectangular- shaped pore models a truncated Steele potential was used to model the interactions at the pore walls.

1.2 PSD Determination

The GCMC method is used to generate several model single-pore isotherms to cover the range of pore sizes of interest, at a given temperature. The single pore isotherms are compared with experimental data, obtained at the same temperature as used for the simulations, to extract the PSD. In mathematical terms, the PSD is obtained by solving

$$N(P) = \int_0^\infty \rho(P,w)f(w)dw, \quad (4)$$

where $N(P)$ = experimental number of moles adsorbed at pressure P, per unit mass of adsorbent; $\rho(P,w)$ = simulated single-pore isotherm for model pore of size w at pressure P; $f(w)$ = PSD.

The integral equation (4) is ill-posed and is therefore subject to unstable solutions. To alleviate this problem, the solution is stabilised using the regularisation method of Jagiełło[3], which employs a smoothing parameter. By optimising this smoothing parameter, it is possible to obtain a stable and accurate solution.

2 RESULTS AND DISCUSSION

2.1 The range of applicability of the molecular simulation method

Molecular simulation probes adsorption on a molecular scale, thus in principle it allows an accurate and detailed PSD to be extracted over the whole micropore size range. However, in practice there exists an effective upper limit to the PSD, beyond which the single-pore isotherms are essentially indistinguishable and the PSD is unreliable[5]. (This limit arises from the physics of adsorption, and is not specific to molecular simulation.) The indistinguishability of the single-pore isotherms arises in two ways: (i) if the maximum experimental pressure is sufficiently low that adsorption in the larger pores is in the Henry's law region (this is discussed by López-Ramón et al.[4] elsewhere in these proceedings); (ii) if a pore is sufficiently large that the adsorbate molecules on opposing pore walls cease to interact significantly, so that the simulated adsorption is essentially that for a non-porous solid.

Figure 1 shows isotherms obtained for a range of pore sizes, using the methane simulation program for a slit-shaped pore at 308 K.

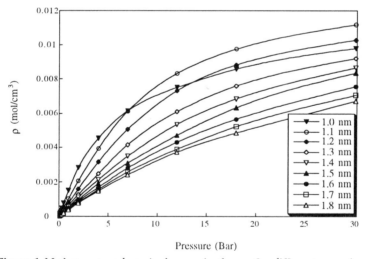

Figure 1 *Methane on carbon single-pore isotherms for different pore sizes, at 308 K*

From figure 1, the ratios of the isotherms for pore sizes 1.6, 1.7 and 1.8 nm are approximately constant (*i.e.* the adsorption per unit area is very similar) and contain little information in terms of the PSD analysis. Therefore, for the PSD determinations using methane and the slit pore model, a pore size range of 0.75 nm (the smallest pore accessible to methane) to 1.6 nm was used.

2.2 The effect of temperature

Simulations were run at 308, 333 and 373 K, and compared with experimental data obtained by Gusev et al. with BPL microporous carbon (at the same temperatures), to extract the PSD. The results are shown in figure 2.

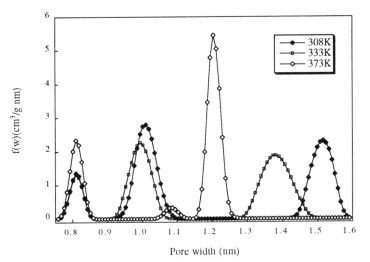

Figure 2 *Effect of temperature on PSDs of carbon, using a methane adsorbate*

From figure 2, a change in temperature has a marked effect on the PSD results. As the temperature increases the PSDs shift to smaller pore sizes, though three peaks are still evident in each case. This discrepancy can be understood when the plot in figure 1, at 308 K, is repeated at 333 and 373 K. Adsorption in larger pores becomes weaker as the temperature increases, and at 333 and 373 K the single-pore isotherms become similar by 1.5 nm (as opposed to 1.6 nm at 308 K). This restricts the range over which the PSD analysis can operate. The quality of the fit to the experimental isotherm also becomes markedly poorer with temperature, indicating an increasing uncertainty in the exact peak location. Nevertheless, there is good consistency between the PSD results at the two lower temperatures up to a pore width of about 1.3 nm.

Due to the reduction in the upper limit of reliability of the PSD as the temperature increases, the remaining PSD investigations with methane were conducted at 308 K.

2.3 The choice of inter-molecular potential parameters

The calculation of the Lennard-Jones and Steele interaction potentials requires values for the solid and fluid molecular diameters, and their potential well depths. These parameters are documented in the literature for both methane and hydrogen (used in this work). However, depending on the method used to obtain the parameters, a range of values are found. The effect on the calculated PSD of using four different pairs of Lennard-Jones parameter values in the simulation program was examined, for both methane and hydrogen. In both cases carbon was the adsorbent, BPL carbon was used for the methane analysis and Carbosieve G for hydrogen. The experimental adsorption data for hydrogen were taken from Jagiello et al[7], for methane from Gusev et al. The values for the intermolecular potential parameters were taken from Hirschfelder et al[6], see table 1.

Table 1 *Lennard-Jones potential parameters used for methane and hydrogen*

Species and code	σ (nm)	ε / k_B (K)
Methane: es0 a	0.3817	148.2
es1 c	0.3697	156.7
es2 b	0.3808	140.0
es3 b	0.3822	137.0
Hydrogen: es0 a	0.2870	29.2
es1 b	0.2915	38.0
es2 a	0.2928	37.0
es3 b	0.2968	33.3

The letters in the species and code column refer to the property used to determine the parameters: where a = virial coefficients; b = viscosity; c = thermal diffusion.

A comparison of the calculated PSDs obtained using the sets of values es0, es1, es2 and es3 for methane are shown in figure 3. The corresponding results for hydrogen are shown in figure 4. The Lennard-Jones parameters used for the solid were those given in the simulation section.

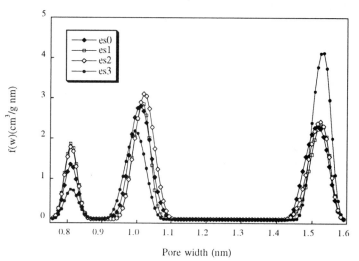

Pore width (nm)

Figure 3 *Effect of Lennard-Jones parameters on the PSD at 308 K, methane adsorbate*

From figure 3, the choice of Lennard-Jones parameters for methane has little effect on the PSD. For each set of parameters the PSD contains three peaks in almost identical locations and of comparable areas. The integrated total pore volumes are very similar in each case, as is the quality of the fit to the experimental isotherm. Thus the consistency of the PSD results is excellent.

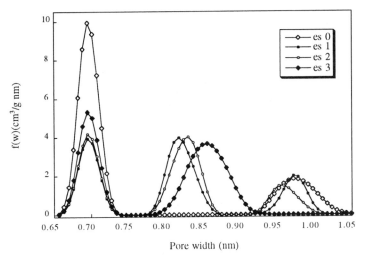

Figure 4 *Effect of Lennard-Jones parameters on the PSD at 77 K, hydrogen adsorbate*

From figure 4, using a hydrogen adsorbate some effect on the PSD is seen by changing the Lennard-Jones parameters, though there is still significant agreement. es1 and es2, for which the parameter values are very close, give comparable PSDs with three similarly-sized peaks in approximately the same locations. The profiles for es0 and es3 show only two peaks, and the distribution is shifted towards smaller pore widths. However, the total pore volumes calculated from the PSDs all agree to within 18%, and the location of the first peak (looking from left to right) is almost identical in each case.

2.4 The choice of pore wall thickness

Figure 5 shows the effect of the pore wall thickness on the PSD obtained. A simulation with infinitely thick pore walls gives a PSD which is almost identical to one obtained using three layer thick walls, and very similar to one using two layers. There is a very marked difference between the PSD obtained using one carbon layer, and those for thicker walls. As the pore walls become thinner, the solid-fluid interaction potential decreases and the equilibrium number of molecules adsorbed in the pore becomes smaller. Thus the peaks in the PSD for one layer are shifted to smaller pore sizes, to compensate for the reduction in the adsorption potential. The quality of the fit to the experimental isotherm is much poorer for one layer than for thicker walls, indicating that this pore model is not realistic. (Such a sparse structure is also at odds with what is known of the physical structure of real carbon adsorbents.) Thus we conclude that the PSD is insensitive to the assumed thickness of the pore wall.

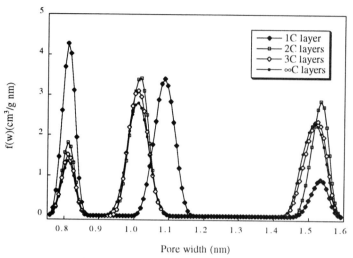

Figure 5 *Effect of pore wall thickness on carbon PSDs at 308 K, methane adsorbate*

2.5 The choice of model pore shape

Three different model pore shapes were used in the simulation, namely slits, rectangles and squares. The corresponding PSDs obtained are shown in Figure 6.

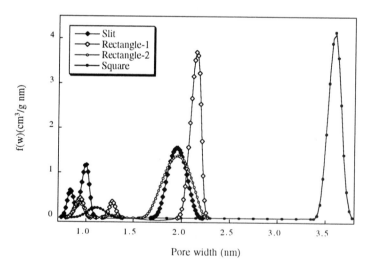

Figure 6 *Effect of pore shape on the PSD at 308 K, methane adsorbate*

Rectangle-1 and -2 in the key for figure 6 refer to the use of different PSD analysis parameters, due to uncertainty in the optimum. For both rectangle PSDs, the maximum pore-size range of the single pore isotherms used in the PSD analysis was restricted to that for the slit. For the square, the maximum pore width had to be extended considerably, due to the higher adsorption in the single pore isotherms obtained in this case.

Figure 6 demonstrates that good agreement between PSDs obtained using slit-shaped and rectangular pores is possible when the same maximum pore width is used. For both rectangle PSDs, the total pore volumes agree with that for a slit pore to within 4%, and the peak locations are similar. However, if the maximum pore width used in the rectangle PSD analysis is increased, the results begin to diverge. The PSD profile for square pores, using a much bigger maximum pore width, is significantly different. Both rectangles and squares contain corners, which represent very high energy sites due to the very close proximity of the walls. In the latter shape all four corners are close together, which enhances the effect. The discrepancies in the PSD results between slits and the other two shapes reflect the fact that regions of high energy occurring in the corners of large pores make a similar contribution to the overall isotherm as do narrow pores (wherein the walls are very close together). This issue is dealt with in detail by Jensen et al[8].

3 CONCLUSIONS

Molecular simulation is a valuable and flexible tool for estimating pore size distributions in microporous solids. The reliable range of the PSD depends on the maximum experimental pressure and temperature. As the pressure increases and the temperature decreases, the upper limit of the PSD increases. Moreover, the uncertainty in peak location increases with the average pore width of the peak. Thus peaks small pore sizes can be assigned with a much greater degree of confidence than large ones.

Deviations of a few percent in the Lennard-Jones parameter values (ε and σ) appear to have little effect on the PSD, but larger deviations can cause significant differences. Uncertainty over the pore wall thickness has little effect on the PSD (except for the unphysical case of a wall one atom thick). Model pore shapes containing corners result in a variety of different PSDs which can give comparable fits to the experimental data.

Acknowledgements

Acknowledgement is made to the donors of the Petroleum Research Fund, administered by the ACS, for partial support of this research, to Merck and Co. for sponsorship of K.L. Boulton, to the Spanish Ministry of Education and Science for a grant to M.V. López-Ramón and to the Bradlow Foundation and the Overseas Research Studentship for sponsorship of G.M. Davies.

References

1. D.J. Adams, *Molec. Phys.*, 1974, **28**, 1241.
2. W.A. Steele, 'The interaction of gases with solid surfaces', Pergamon, 1974.
3. J. Jagiełło, *Langmuir*, 1994, **10**, 2778.
4. M.V. López-Ramón, J. Jagiełło, T. Bandosz and N.A. Seaton, these proceedings.
5. V. Yu. Gusev, J.A. O'Brien and N.A. Seaton, *Langmuir*, in press.
6. J.O. Hirschfelder, C.F. Curtiss and R.B. Bird, 'Molecular theory of gases and liquids', Wiley and sons, 1964.
7. J. Jagiełło, T. Bandosz, K. Putyera and J. Schwarz, *J. Chem. Eng. Data*, 1995, **40**, 1288.
8. C.R.C. Jensen, G.M. Davies, N.A. Seaton, 'Pore models for the Monte Carlo simulation of adsorption in microporous carbons', in prepn. for submission to Carbon.

ADSORBENT APPLICATION FOR MESOPOROUS SILICATE

A. Yasutake*, N. Tomonaga*, J. Izumi*, U.M. Setoguchi** and M. Iwamoto***

*   Nagasaki R & D Center, Mitsubishi Heavy Industries, Ltd.
      5-717-1, Fukahori-machi, Nagasaki 851-03, Japan
**   Department of Applied Chemistry Nagasaki University
      1-14, Bunkyo-machi, Nagasaki 852, Japan
***   Catalysis Research Center, Hokkaido University
      Kita 11, Nishi 10, Sapporo 060, Japan

1 INTRODUCTION

Mesoporous silicate is a kind of alumino-silicate compound of which the pore diameter is 10-100 Å.

Kuroda et al. used a layer structured clay compound (kanemite) as a silica source; they intercalated cationic surfactant to kanemite to rearrange the silica bridge structure and first prepared the hexagonal structured mesoporous silicate[1]. Mobil then also prepared the same structured mesoporous silicate by means of a silicic acid dehydrated-condensation reaction at high temperature in the presence of micell with cationic surfactant of mono-alkyl ammonium compound[2].   Recently Stucky et al. reported that the same structured mesoporous silicate can be prepared by the dehydrated-condensation reaction of mono-silicic acid, which is formed by alcoxy silane hydration in the presence of micell with cationic surfactant at low temperature near room temperature[3].   As these mesoporous silicates prepared by the above-mentioned methods have more than twice the specific surface area than that of conventional zeolite, it is expected to be used as adsorbents and catalyst carriers. However, knowledge about the physical and chemical properties and their formation mechanism is not sufficient in respect to the preparation of the mesoporous silicate. Also the high temperature synthesis and the use of expensive organic silicate such as silicic acid alkoxide as a silica source impact the cost significantly. There remains also the cost reduction of mesoporous silicate from the perspective of industrial application. For cost reduction study of its preparation method, re-use of the organic template using supercritical pressure extraction and other solvent extraction was also attempted.

2 MESOPOROUS SILICATE PREPARATION

2.1 Preparation Procedure

Table 1 shows the preparation procedure of mesoporous silicate under low temperature in an acidic medium.   Sodium silicate dissolved in hydrochloric acid. (This solution is designated Solution-S.)   Cationic surfactant, which is used as an organic template, is dissolved with water. (This solution is designated Solution-T.)

When Solution-S is added to Solution-T, a white precipitate quickly appears. As the stirring is continued for several hours, the mesoporous silicate occluding the organic template is formed. This precipitate is then filtrated, dried and calcined at more than 400℃. As the organic template is removed, the mesoporous silicate, which is active for adsorption can be prepared.

Table 1 *Preparation Procedure of Mesoporous Silicates*

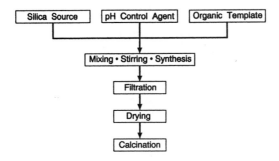

2.2 Preparation Conditions

Concerning our preparation, a wide range of temperature and pH ranges were experimented. As Figure 1 shows, it was confirmed that this mesoporous silicate could be prepared under temperatures of 25-100°C and pH values of -1 to 14. In comparison with MCM-41 and FSM-16, which have recently generated significant interest as the mesoporous silicate, this silicate synthesis reaction proceeds at relatively lower temperature while the condition is almost the same as the acidic hexagonal silicate, which was prepared with tetra ethoxy silane (TEOS) by Stucky. In the Stucky's preparation, it followed that TEOS was decomposed with an acid solution to donate mono-silicic acid, there seemed to be two possible formation processes as follows;

a) silicate network formation by the polymerized condensation on the surface of the cationic surfactant micell (Mobil)

b) association with mono-silicic acid and cationic surfactant, its micell formation and the silicate network formation by the polymerized condensation (Davis)[4]

In our study, as we used the mono-silicic acid and its oligomer, which corresponded with the decomposed compounds of TEOS, our synthesis started following the next step from TEOS hydration decomposition. Since we cannot confirm the intermediate stages, the bridge formation of silicate proceeds in the presence of mono silicate and cationic surfactant. As it was suggested by Stucky's preparation, when the silicate containing Q_4 structure

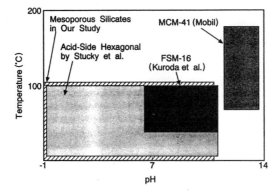

Figure 1 *Preparation conditions of mesoporous silicates*

measured by MAS-NMR, the mesoporous silicate in our study could not be formed. For confirmation of the mesoporous silicate formation in our study, the crystal structure and the degree of polymerization of the silicate was traced at each step. Figure 2 a) is the X-ray diffraction pattern of the sample just after the filtration and Figure 2 b) is the pattern after drying and Figure 2 c) is the pattern when the template is removed by calcination. As a reference, the X-ray diffraction pattern of MCM-41 is shown in Figure 2 c). As a result of X-ray diffraction patterns, the pattern of the hexagonal structure appeared just after the filtration and it became clearer after drying. There was no difference between the dried samples and the calcined samples. It indicated that the silica bridge formation seems almost completed by the drying stage. In comparison with the width of the peak (d-spacing of 100) of MCM-41 shown as a reference, the mesoporous silicate in our study had a larger FWHM value. It is evaluated that the mesoporous silicate in our study had less uniformity than MCM-41. Figure 3 a) is the MAS-NMR ^{29}Si spectrum of the silica source (Na$_4$SiO$_4$), Figure 3 b) is that after filtration, Figure 3 c) is that after drying and Figure 3 d) is that after calcination, respectively. The MAS-NMR ^{29}Si spectrum of the silica source consists of Q_0-Q_2 peaks only and Q_3, Q_4 peak cannot be observed. After filtration, the Q_0 peak almost disappeared and the Q_4 peak appeared. At the spectrum of the calcined sample, there remained only the broad peak assigned to the Q_3-Q_4 peak. With the results of these state analyses, it was confirmed that the dehydrated polymerization reaction proceeded during the stages of mother solution stirring, filtration and drying. Figure 4 shows the relationship between the pH value, the specific area, the concentration of silica and the organic template. Na$_4$SiO$_4$ was used as the silica source, cetylphridinium chloride was used as the organic template and the silica organic template molar ratio was 10. The concentration of silica gave minimum values at pH values of -1 to 0. It means that silicic acid added to Solution-T, which reacted with organic template, was converted to the organo-silicate compounds of which the solubility was very small and precipitated as a white substance. This mesoporous silicate keeps the very large value of specific surface area occurring in the pH region between -1 and 3. The precipitate seemed to maintain a stable structure in this region. In the higher pH region, the specific area decreased. When the pH value increased, the silica solubility decreased. It is considered that the silica's precipitation became comparable with the mesoporous silicate formation and the silica bridge formation may thus be hindered.

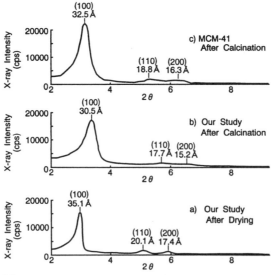

Figure 2 *X-ray diffraction patterns of mesoporous silicates*

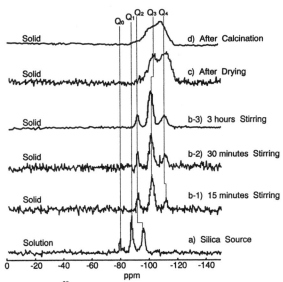

Figure 3 *^{29}Si-NMR spectra of mesoporous silicates*

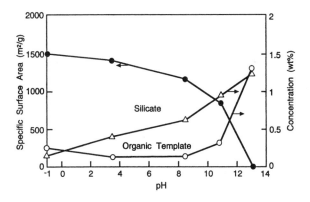

Figure 4 *pH value vs specific surface area and concentration of silicates and organic template*

Concerning the mesoporous silicate formation mechanism, as mentioned above, there are two significantly different interpretations. However, based on the state analyses used in our study, we could not confirm now the formation mechanism of this mesoporous silicate. Subsequently, we studied the recovery and recycle of mesoporous silicate using an extraction of a template. For this recovery, the CO_2 supercritical pressure extraction and the alcohol extraction were used. As shown in Figure 5, it was found that after the extraction of the template, the X-ray diffraction pattern could remain unchanged, showing that mesoporous silicate could be product without being destroyed, and thus, the feasibility of mesoporous silicate recovery using the template could be found.

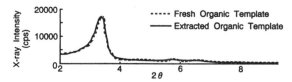

Figure 5 *X-ray diffraction patterns of mesoporous silicates*

3 ADSORPTION OF THE MESOPOROUS SILICATE

3.1 Adsorption Measurement

The adsorption measurement procedure for methyl-ethyl ketone (MEK) with the small chromatography column is shown in Figure 6. The 0.03 gram   samples of mesoporous silicate prepared in our study were put into the column. MEK concentration was adjusted to 1,000ppm with nitrogen and it was supplied to the column. As the breakthrough curve was confirmed with the outlet concentration measured by FID, the adsorption amount was determined by the breakthrough curve.

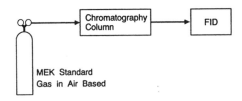

Figure 6 *Adsorption measurement apparatus*

3.2 Evaluation Result

Figure 7 shows relationship between the specific surface area and the MEK adsorption

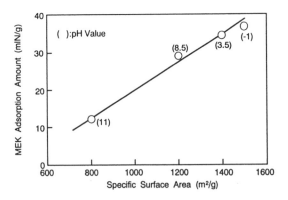

Figure 7 *MEK adsorption amount*

amount. Evaluation shown here was made, using the sample given in Figure 4. This result shows that the MEK adsorption amount increased with increasing specific surface area and that properties of mesoporous silicate synthesized in our research work could remain approximately unchanged except for its specific surface area.

Meanwhile, Figure 8 shows the adsorption property of the mesoporous silicate using the template extraction. Because the recycle of the extracted organic template showed no large difference in the adsorption property of MEK, a prospective outlook to the recycle of the organic template could be obtained.

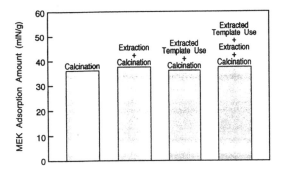

Figure 8 *Results of extracted template use*

4 FUTURE PROSPECT

As the mesoporous silicate has a surface area of more than twice that of zeolite and its surface area is close to that of activated carbon and it shows thermally more stability than activated carbon. Therefore, it shows great potential as an adsorbent and a catalyst carrier. For the volatile organic compound adsorption, the mesoporous silicate seems suitable and for the polar adsorbates such as water vapor, SO_2, H_2S, the mesoporous alumino-silicate of which the SiO_2/Al_2O_3 is small, seems suitable. The SiO_2/Al_2O_3 adjustment of this material will be easy and the optimum condition for the adsorption will be adjustable in the over recent study. For the future prospect of this material to the industrial field, a cost feasibility study was undertaken. The result is shown in Figure 9. The cost of MCM-41 is three times higher than that of high silica zeolite. It is because of the high temperature hydro-

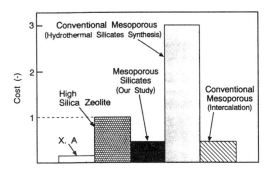

Figure 9 *Cost estimation of mesoporous silicates*

thermal synthesis and the consumption of the organic template. On the other hand, the mesoporous silicate in our study can be prepared with the inorganic silica source under low temperature, acid side condition and in this study, extracted template re-use is done. It is expected that the cost of this material is lower than that of high silica zeolite. When the functional performance of the mesoporous silicate is systematized in the near future, its application to the adsorption and the catalyst will expand significantly.

References

1. Yanagisawa, T., Shimizu, T., Kuroda, K., Kato, C., Bull. Chem. Soc. Jpn., 1990, **63**, 988
2. Beck, J.S., Vartulo, J.C., Roth, W.J., Leonowicz, M.E., Kresge, C.T., Schmitt, K.T., Chu, C. T-W., Olson, D.H., Sheppard, E.W., McCullen, S.B., Higgins, J.B., Schlenker, J.L., J. Am. Chem. Soc., 1992, **114**, 10834.
3. Huo, Q., Margolese, D.I., Ciesla, U., Demetth, D.G., Feng, P., Gier, T.E., Sieger, P., Firouzi, A., Chmelka, B.F., Schiith, F., Stucky, G.D., Chem. Moter. 1994, **6**, 1176.
4. Davis, M.E., Lobo, R.F. Chem. Moter. 1992, **4**, 756.

CHARACTERISATION BY THERMOPOROMETRY OF THE INFLUENCE OF ADDITIVES (TRITON X AND PtCl₄) ON THE POROUS STRUCTURE OF GELS DERIVED FROM TETRAETHOXYSILANE

A. Julbe*, J.F. Quinson°, A. El Mansouri* and C. Guizard*

* Laboratoire des Matériaux et Procédés Membranaires (UMR 5635 CNRS-ENSCM-UM II)- Ecole Nationale Supérieure de Chimie- 8 Rue de l'Ecole Normale- 34053 MONTPELLIER cedex 1, France.
° Laboratoire G.E.M.P.P.M. (UMR 5510 CNRS)
INSA de Lyon- 20 Avenue Albert Einstein- 69621 VILLEURBANNE cedex, France.

1 INTRODUCTION

In order to improve the characteristics of materials resulting from sol-gel techniques, additives are frequently used but their effects are not always clear and easy to evidence. In the case of sols derived from tetraethoxysilane (TEOS), drying control chemical additives (1,2) like formamide, dimethylformamide and oxalic acid have been extensively used to limit gel cracking upon drying and then prepare monoliths. The effect of surfactants, added in the pore liquid to reduce the interfacial energy and thereby decrease the capillary stresses, has been also reported to limit gel cracking (1). For sol casting, the use of organic polymers containing hydroxyl groups like polyvinylic alcohol or polyethylene glycol is also of particular interest in order to control the sol rheological properties and improve the film plasticity (3). The comparison of pore structure in silica gels with addition of several kinds of water soluble polymers (uncharged, charged and proteins) were also pointed out (4).

Surface active agents represent a specific variety of additive which has been recently largely investigated in sol-gel processing and not only to limit gel cracking. Indeed, one of the other interests of such additives concerns the control of the hydrolysis step of highly reactive alkoxides, such as titanium alkoxides, by use reverse micelles sol-gel systems (5). An other more recent application of surfactants concerns the preparation of organised lamellar, hexagonal or cubic phases ; the process is based on the concept of 'liquid crystal templating' with molecular aggregates of surfactant (6). The related studies concern mainly silicate mesostructure with cationic surfactants in hydrothermal conditions, but similar microporous silica materials can also be obtained with atmospheric conditions in the presence of an oil phase (7). The use of non-ionic surfactant to generate organised silica mesophase is reported in (8,9). Many works are currently in progress to extend this basic principle to the synthesis of a larger number of oxides (6). In all these organised phases, the inorganic framework is polymerised around surfactant aggregates.

We have recently studied in our group the effect of non-ionic surface active agents (Triton X) and of PtCl₄ on classical TEOS derived sols, gels and materials in order to prepare microporous SiO₂/Pt materials with a high microporous volume. This type of TEOS-Triton X system is not comparable to previously mentioned ones because non ionic surfactants are concerned which are added after the hydrolysis step and gelation occurs in atmospheric conditions. Surfactant molecules are then around the preformed clusters and their role is in this case the limitation of clusters condensation rate. We already demonstrated in (10) that when commercial Triton X are introduced in classical TEOS sols, they interact with silica clusters (probably by Van der Waals forces and/or hydrogen bonds) and modify the sol physico-chemical characteristics. On the basis of sols and gels characterisations (gelation time, ^{29}Si NMR, QELS and SAXS), we can assume an organic shell of surfactant to form around the clusters. The resulting steric hindrance is supposed to limit further condensation of clusters during sol ageing and explains the formation of more stable sols. The method leads to SiO_2 materials whose porous characteristics are directly related to the surfactant chain length and quantity. During the heat treatment the elimination of the organic shell around the particles produces homogeneous materials consisting of nanometric distinct silica particles with single terminal -OH groups (10).

On the basis of these original preliminary results in which the textural characterisation of gels were lacking, a complementary work has been performed, based on thermoporometry, in order to better characterise the effect of Triton X on the mesoporous structure of the corresponding gels. The effect of $PtCl_4$ on the porous structure of TEOS derived gels with and without Triton X has been also investigated.

2 EXPERIMENTAL

2.1 Sols, gels and materials preparations

Sols were prepared from commercial tetraethylorthosilicate (Aldrich). Hydrolysis of the alkoxide diluted in ethanol has been performed in acidic medium (HCl) and under reflux at 80°C for one hour. The standard sol molar composition was $TEOS:C_2H_5OH:H_2O:HCl = 1:4.5:4:0.02$. Non-ionic surface active agents (Triton X : $(CH_3)_3CCH_2C(CH_3)_2C_6H_4-(OCH_2-CH_2)_X-OH$ from Rohm and Haas company) were added in the refluxed sols cooled at room temperature. Two polyoxyethylene chain lengths have been used in this study : x=1 (Triton TX 15) and x=3 (Triton TX 35).

$PtCl_4$ is soluble in ethanol and was introduced in the cold sols after the hydrolysis step. The effect of $PtCl_4$ on the gel structure has been studied on gels with a molar ratio TX35/TEOS=0.5, by varying the molar ratio $PtCl_4$/TEOS between 0 and 0.5.

SiO_2 and SiO_2/Pt materials were prepared by firing the gel at 450°C under flowing air. A second treatment under H_2 was necessary for SiO_2/Pt materials.

The porous characteristics of gels and of the corresponding fired materials were determined respectively by water thermoporometry and N_2 adsorption-desorption.

2.2 Characterisation of gels by thermoporometry

Thermoporometry is a textural characterisation method which is well adapted to study the mesoporous structure of gels (11). The technique is based on the thermal analysis of the liquid-solid phase transformation of a capillary condensate held inside the studied porous body. The thermoporometer records the energy evolved by the solidification of the condensate when the temperature is decreased. From the solidification thermogram, it is possible to determine :
- the pore radius distribution (between 1.5 and 150 nm), from the depression of the solidification temperature, $\Delta T = T - T_o$ where T_o is the normal phase transition temperature of the liquid and T is the temperature where the phase transition is observed when this liquid is contained in the pores.
- the mesoporous volume, from the energy evolved in the phase transformation.

Before analysis, gels were washed with ethanol in order to eliminate the organic residues which are not involved in or linked to the gel network or pore surface. It is indeed necessary to make the thermoporometry measurements with a pure condensate free of any dissolved foreign molecule. After ethanol washing the porous network of gels was filled with water and thermoporometry measurements were performed with 20-40 mg of wet gels with a cooling rate of 60K/h. The effect of the ethanol washing has been studied by performing water thermoporometry after one day and three days ageing in ethanol.

3 RESULTS AND DISCUSSIONS

3.1 Effect of Triton X on the gelation time and gel porous structure

Acid catalysed hydrolysis of TEOS is known to lead to small polymers with a low crosslinking degree (12) and is then the most interesting way to prepare microporous materials. In these conditions and with the standard sol composition (without surfactant), the gelation time is t_{go}= 94 days. As shown in figure 1, the addition of surface active agent (x=3 in this case) increases the gelation time: t_g/t_{go} increases from 1 to 3.2 when TX35/TEOS increases from 0 to 1.4. This result clearly demonstrates an important effect of the surfactant molecules on the clusters condensation rates

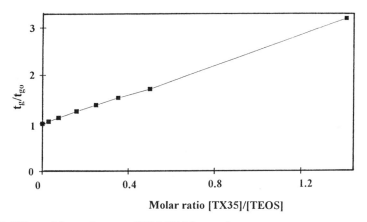

Figure 1 *Effect of the molar ratio TX35/TEOS on gelation time.*

The interactions between clusters and surfactant molecules can be evidenced with thermoporometry studies of gel porous textures. Results obtained on gels without and with Triton X (TX15, TX35) after one day and three days ageing in ethanol, are reported in table 1. They are compared to the characteristics of a gel prepared with formamide. The corresponding pore size distributions, $dV/dR = f(R)$, are reported in figure 2.

It appears that gel washing can have an important effect on its porous structure. When the gel does not contain any surfactant the washing process can lead to an increase of the mean pore radius (2.2 to 2.6 nm) probably due to the progressive departure of small silica oligomers, without any important variation of the pore volume. For gels prepared with surfactant, a one day storage in ethanol leads to a macroscopic dislocation of gels and to relatively low porous volumes (V= 259 mm³/g for TX15 and 51 mm³/g for TX35). The phenomenon of gel dislocation is enhanced for gels containing surfactant with longer chains. After three days in ethanol, the gels consolidate again (macroscopically observed) and a more uniform pore size distribution is observed by thermoporometry with an increase of the pore volume (multiplied by ~2 for TX15 and by ~3 for TX35). This unusual behaviour of gels in ethanol can be explained by a desorption of surfactant molecules which liberates Si-OH reactive groups able to condense and then to reinforce irreversibly the gel structure. The surfactant desorption from the clusters in ethanol is confirmed by elemental analysis (C, H) of dried gels before and after ethanol washing (Table 2).

Table 1 *Effect of additives and of ageing time in ethanol on the porous characteristics of TEOS derived gels studied by thermoporometry*

	Mean pore radius R (nm) measured at V/2		Porous volume V (mm³/g)	
Ageing in EtOH	*1 day*	*3 days*	*1 day*	*3 days*
Standard gel (TEOS)	2.2	2.6	425	487
[TX15]/[TEOS]= 0.5	2.3	2.2	259	401
[TX35]/[TEOS]= 0.5	2.2	2.2	51	165
[NH₂CHO]/[TEOS]= 4.5	2.1	2.1	380	406

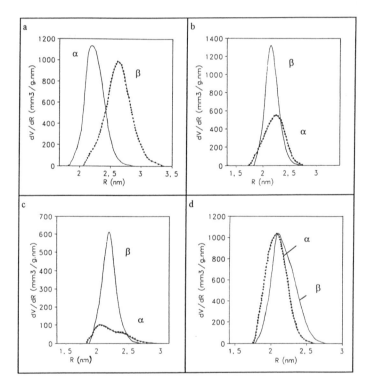

Figure 2 *Pore radius distributions measured by thermoporometry after (α) one day storage in ethanol, (β) three days storage in ethanol, on a -a) standard TEOS derived gel -b) TEOS derived gel with [TX15]/[TEOS]=0.5 -c) TEOS derived gel with [TX35]/[TEOS]=0.5- d) TEOS derived gel with [NH₂CHO]/[TEOS]=4.5.*

Table 2 *Effect of ethanol washing on C and H concentrations in dried gels with TX35/TEOS=0.5*

Sample	C (g/100g)	H (g/100g)
As prepared gel	54.3	7.8
Gel after 3 days in ethanol	14.8	3.3

Obviously the surface active agent molecules participate to the gel integrity, probably by hydrogen bonding with silica clusters whereas the number of chemical bonds Si-O-Si between clusters is limited by the steric hindrance of surfactant chains. This steric effect, more effective with longer surfactant chains, leads to gels with weak intercluster connections. A reaorganisation of the gel matrix occurs during desorption of surfactant molecules in ethanol and condensation of free Si-OH groups leads to a more rigid structure.

3.2 Effect of PtCl₄ on the gelation time and gel porous structure

In order to prepare silica microporous SiO_2/Pt with platinum, we have studied the effect of $PtCl_4$ addition in TEOS derived sols. It appears from the gelation times (Figure 3) that the effect of $PtCl_4$ is opposite to the effect of Triton X as far as it accelerates the condensation reactions like in the case of formamide addition.

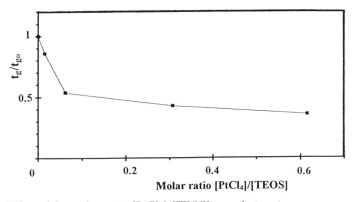

Figure 3 *Effect of the molar ratio [PtCl₄]/[TEOS] on gelation time.*

Thermoporometry results concerning the mesoporous structure of gels prepared with various quantity of $PtCl_4$ and in the presence of Triton TX35 are reported in table 3. The corresponding pore size distributions are relatively narrow and gaussian after 1 day or 3 days in ethanol. Figure 4 shows the typical pore radius distribution for this type of sample. The first observation is that the mean pore size of gels prepared in the presence of $PtCl_4$ is slightly reduced compared to gels prepared only with TX35. The second observation is that the porous structure (pore size distribution and porous volume) of the gels prepared in the presence of $PtCl_4$ and TX35 remains almost unchanged after 3 days ageing in ethanol. This relative textural stability means that these gels, as those prepared with formamide, are well reticulated. Furthermore gels are not desagregated upon ethanol washing in spite of the fact that they contain also TX35. The intercluster connexions are then stronger in the presence of $PtCl_4$; surfactant molecules, trapped in a highly reticulated structure, are less sensitive to a continuous desorption in ethanol. Finally the pore volume of the gels prepared with PtCl₄ and TX35 is very low compared to those of gels prepared without any additive and is comparable to (but slightly higher than) the pore volume of gels with only TX35. The surfactant molecules occupy a great part of the volume between the silica clusters. We can then consider that a similar gel porous structure is obtained with TX35 alone or with PtCl₄ and TX35 except that there is a greater number of intercluster connexions when PtCl₄ is used.

Table 3 *Effect of the molar ratio [PtCl₄]/[TEOS] and of ageing time in ethanol on the porous characteristics of gels prepared with various molar ratios [TX35]/[TEOS]*

Sample		Mean pore radius (nm) measured at V/2		Porous volume V (mm³/g)	
	Ageing in EtOH	*1 day*	*3 days*	*1 day*	*3 days*
[TX35] = 0.5 **[TEOS]**	[PtCl₄]/[TEOS] = 0	2.1	2.2	51	165
	[PtCl₄]/[TEOS] = 0.015	2.0	1.9	66	77
	[PtCl₄]/[TEOS] = 0.15	2.0	2.0	76	60
	[PtCl₄]/[TEOS] = 0.5	2.1	2.0	74	80
[TX35] = 0.15 **[TEOS]**	[PtCl₄]/[TEOS] = 0.015		2.0		36
[TX35] = 0 **[TEOS]**	[PtCl₄]/[TEOS] = 0.15		2.0		56

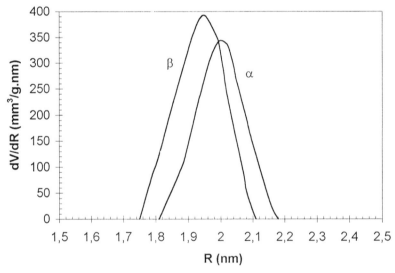

Figure 4 *Pore radius distribution of a gel with molar ratios [TX35]/[TEOS]=0.5 and [PtCl₄]/[TEOS]= 0.015, after (α) one day in ethanol, (β) three days in ethanol.*

When only $PtCl_4$ is used (TX35/TEOS=0, $PtCl_4$/TEOS=0.15) the mean pore size is unchanged and the gel mesoporous volume is slightly decreased. This result shows that the effect of TX35 in such gels is to limit the gel crosslinking degree and consequently to increase the porous volume. For a constant ratio $[PtCl_4]/[TEOS]$=0.015, if the quantity of TX35 decreases from 0.5 to 0.15, the gel pore volume is decreased by a factor 2.

3.3 Relation between the porous structure of gels and those of the final materials (SiO_2 and SiO_2/Pt)

All the materials derived from the studied gels present typically type I N_2 adsorption isotherms (figure 5) and can then be considered as microporous. As shown in table 4, the porous volume of the materials varies as a function of the nature and quantity of additives.

It is not obvious to find a direct relation between the porous volumes of gels and those of the corresponding materials obtained after the firing treatment because :
- the surfactant molecules linked to the clusters occupy a part of the gel pore volume which is liberated during the firing treatment.
- the collapse of the gels structure during drying is modified by the presence of surfactant which may modify the capillary stresses. This effect has not been studied in this work.
- thermoporometry is only able to detect mesopores in the wet gels whereas N_2 adsorption is able to detect the presence of mesopores and micropores in the heat treated gels.

Table 4 *Porous volume of materials obtained after gel firing at 450°C*

$\frac{[TX]}{[TEOS]}$	$\frac{[PtCl_4]}{[TEOS]}$	Total porous vol. (mm^3/g)
0	0	No porous vol. detected Probably ultramicro- porous material
TX15		
0.5	0	217
TX35		
	0	310
0.5	0.015	330
	0.15	293
0.15	0.015	264
0	0.15	19

Figure 5 *Typical N_2 adsorption isotherms for SiO_2/Pt samples fired at 450°C*
a)TX35/TEOS=0.5; $PtCl_4$/TEOS=0.015
b) TX35/TEOS=0.5; $PtCl_4$/TEOS=0.15
c) TX35/TEOS=0 ; $PtCl_4$/TEOS=0.15

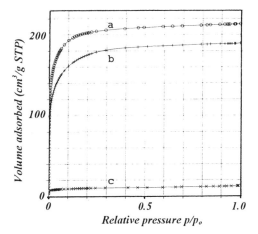

It is nevertheless possible to explain the general trend of the results on the basis of the ability of gels to collapse or not during the drying and firing treatments. It appears from thermoporometry studies that surface active agents are effective additives to limit cluster condensation but also gel collapse by steric hindrance effect. $PtCl_4$, leading to more rigid gels with a high crosslinking degree, can also be supposed to act in the same way.

These considerations are in good keeping with the data of table 4. Standard TEOS derived gels lead to materials with smaller pore sizes (probably ultramicropores not detected by N_2 adsorption) than when surfactants are used. Furthermore as already shown in (10), the pore volume is increasing with the surfactant chain length. Addition of only $PtCl_4$ in TEOS derived sols leads to microporous materials with a very low porous volume. At last, the conjunction of steric hindrance and gel rigidity effects (TX35 + $PtCl_4$) leads to the aimed highly porous microporous SiO_2/Pt materials. TEM observations reveal a good dispersion of nanometric (2-4nm) Pt particles in the silica matrix.

4 CONCLUSION

Thermoporometry revealed useful to characterize the effects of Triton X and of $PtCl_4$ on the porous structure of gels derived from tetraethoxisilane. These additives drastically modify the gelation process by two different mechanisms whose conjunction leads to highly porous microporous SiO_2/Pt materials. Surface active agents absorbed around TEOS derived clusters limit condensation reactions. A brittle gel structure is generated which leads to highly porous microporous silica after the elimination of organic chains by thermal treatment at 450°C. The effect of $PtCl_4$ is opposite to the effect of Triton X and leads to gels with a higher crosslinking degree. The conjunction of the steric hindrance and rigidity effects of respectively Triton X and $PtCl_4$ produces rigid gels in which the surfactant molecules are trapped. This is an original method to prepare highly porous microporous SiO_2/Pt supported membranes which are relatively thick ($\sim 1\mu m$).

References

1.   C.J. Brinker and G.W. Scherer, 'Sol-Gel Science', Academic Press, San Diego, 1990.
2.   G. Kordas and S.P. Mukherjee, *J. Non-Cryst. Solids*, 1986, **82**, 160.
3.   A. Larbot, A. Julbe, C. Guizard and L. Cot, *J. Memb. Sci.*, 1989, **44**, 289.
4.   S. Sato, T. Murakata, T. Suzuki and T. Ohgawara, *J. Mater. Sci.*, 1990, **25**, 4880.
5.   C. Guizard, A. Larbot, L. Cot, S. Peres and J. Rouvière, *J. de Chimie Phys.*, 1990, **87**, 1901.
6.   J.S. Beck and James C. Vartulli, *Current Opinion in Solid State & Materials Science*, 1996, **1**[1], 76.
7.   T. Dabadie, A. Ayral, C. Guizard, L. Cot, J.C. Robert and O. Poncelet, *Mat. Res. Soc. Symp. Proc.*, 1994, **346**, 849.
8.   T. Dabadie, A. Ayral, C. Guizard, L. Cot, C. Lurin, W. Nie and D. Rioult, *J. Sol-Gel Sci. and Technol.*, 1995, **4**, 107.
9.   S.A. Bagshaw, E. Prouzet and T.J. Pinnavaia, *Science*, 1995, **269**, 1242
10. A. Julbe, C. Balzer, J.M. Barthez, C. Guizard, A. Larbot and L. Cot, *J. Sol-Gel Sci. and Technol.*, 1995, **4**, 89.
11. J.F. Quinson and M. Brun, in : K.K. Unger, J. Rouquerol, K.S.W. Sing and H. Kral (Eds.), Studies in Surface Science and Catalysis Series vol. **39**, Proc. of the IUPAC Symp. Characterization of Porous Solids I, Elsevier, Amsterdam, 1988, p.307.
12. C.J. Brinker, K.D. Keefer, D.W. Schaeffer, T.A. Askink, B.D. Kay and C.S. Ashley, *J. Non-Cryst. Solids*, 1984, **63**, 45.

REDUCTION OF OCTAHEDRAL Fe(III) IN SAPONITE. INFLUENCE ON THE PILLARING ABILITY OF THE CLAY.

M.A. Vicente[1], M. A. Bañares-Muñoz[1], M. Suárez[2], J.M.M. Pozas[2], R.M. Martín[3] and M.L. Rojas[3].

[1] Departamento de Química Inorgánica, Facultad de Ciencias Químicas, Universidad de Salamanca, Plaza de la Merced S/N, 37008-Salamanca, Spain.
[2] Area de Mineralogía y Cristalografía, Departamento de Geología, Facultad de Ciencias, Universidad de Salamanca, Plaza de la Merced S/N, 37008-Salamanca, Spain.
[3] Departamento de Química Inorgánica, Facultad de Ciencias, Universidad Nacional de Educación a Distancia, Senda del Rey S/N, 28040-Madrid, Spain.

1 INTRODUCTION

The insertion of bulk metallic polycations in the interlayer region of layered clays produces intercalated compounds with high basal spacings. The calcination of these intercalated compounds yields pillared solids, in which the metallic oxide pillars "support" the separation of the clay sheets. The solids thus obtained have high values of surface area and porosity, and their thermal stability, improved with respect to the parent clay minerals, permits them to be used as catalysts, especially for the cracking of heavy oil fractions.[1 and references therein]

Saponite is a tetrahedrally charged trioctahedral smectite. Mg(II) cations usually occupy its octahedral sites, but in its ferrous variety a quantity of Fe(II, III) substitutes Mg(II); this substitution being important when the Mg/Fe ratio is lower than 5:1.[2] Saponite has been observed to have an excellent pillaring ability,[3-10] and pillared saponites have been recently tested as catalysts. [3,9]

Griffithite is a high iron content saponite from Griffith Park (California, USA), with a Mg/Fe ratio of 1.85.[11,12] The aim of the present study is to investigate the reductibility of octahedral Fe(III) cations when treating griffithite with different reducing agents, and to investigate the pillaring ability of the reduced solids thus obtained. Different concentration solutions of sodium dithionite ($Na_2S_2O_4$) and hydrazonium sulphate ($N_2H_6SO_4$) were used as reducing agents.

2 EXPERIMENTAL SECTION

2.1 Reduction of octahedral Fe(III)

Two reducing agents were used for the reduction of griffithite Fe(III) octahedral cations: sodium dithionite ($Na_2S_2O_4$) and hydrazonium sulphate ($N_2H_6SO_4$), following the method described by Rozenson & Heller-Kallai for the reduction of Fe(III) in nontronite, montmorillonite and beidellite.[13] 6 grams of griffithite were treated with 250 mL of solutions 0.05 and 0.2 N of each one of the reducing agents, and maintained in contact for 6 hours. The solids were then separated by filtration and submitted to pillaring process.

2.2 Pillaring of griffithite

The pillaring ability of the reduced solids has been investigated by intercalation with $[Al_{13}O_4(OH)_{24}(H_2O)_{12}]^{7+}$ solutions, prepared by addition of aqueous NaOH to solutions of $AlCl_3 \cdot 6H_2O$, with a ratio $OH^-/Al^{3+}=2.2$. The Al/clay relation was of 5.0 mmol Al/gram of clay. Then, the samples were centrifuged, introduced into dialysis bags and washed by dialysis until absence of chloride anions. At this moment, the samples were centrifuged and the resulting solids dried at 60°C, giving the intercalated compounds. Their calcination at 500°C for four hours yielded the corresponding pillared compounds.

2.3 Techniques

Elemental analyses were carried out by plasma emission spectroscopy, using a Perkin-Elmer emission spectrometer, model Plasma II. Previously, the solids were digested under pressure, with a nitric-hydrofluoric acid mixture, in a PTFE autoclave. X-ray diffractograms were obtained on a Siemens D-500 diffractometer at 40 kV and 30 mA (1200 W) with filtered Cu K_α line. The equipment is connected to a DACO-MP microprocessor and uses Diffract-AT software. Specific surface areas were determined from the corresponding nitrogen isotherms at 77K, obtained from a Micromeritics ASAP 2000 analyzer, after outgassing the samples at 110°C for 8 hours, with a residual pressure of 10^{-5} mm Hg. The isotherm data were processed using a computer program. [14]

3 RESULTS AND DISCUSSION

3.1 Parent saponite

The <2µm fraction of griffithite, (Griffith Park deposit, California, USA), was used as parent material. Its structural formula is: $[Si_{6.92} Al_{1.08}] O_{20} (OH)_4 [Mg_{2.92} Fe^{3+}_{1.58} Al_{0.28}] [Ca_{0.62} Na_{0.20} K_{0.04}]$ and its CEC was 86 mequiv/100g.[11] As observed, Fe^{3+} occupies about a third of octahedral positions, hence griffithite can be classified as a high iron content saponite, having a Mg/Fe ratio of 1.85.

3.2 Reduced samples

Chemical analyses show that between 10 and 25% of Fe(III) is reduced to Fe(II) during the reduction treatments. When using sodium dithionite solutions, Na^+ cations compensate the loss of positive charge produced by the reduction of iron; the content in sodium in the reduced solids being higher than in the natural sample. When using hydrazonium sulphate, no significant differences in chemical composition are observed, the charge in the layer may be compensated by the H^+ cations originating from the oxidation of hydrazonium species. These protons may also carry out an acid activation of the clay.

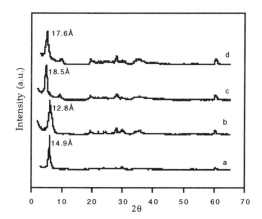

Figure 1 *XRD patterns of natural griffithite (a), natural griffithite treated with 0.05N sodium dithionite for 6 hours (b), this sample intercalated with Al_{13} polycations (c) and the latter calcined at 500°C (d).*

The reduced solids have diffraction patterns that show ordered structures similar to that of the parent saponite (Figure 1 and Table 1). Anyway, two significant changes can be observed: a decrease in the basal spacing and a widening in the 001 reflection peak. The decrease in the basal spacing can be attributed to the change in the layer exchangeable cations. Thus, parent saponite has Ca^{2+} as the only exchangeable cation, which is substituted by Na^+ in samples treated with $Na_2S_2O_4$ and by H^+ in samples treated with $N_2H_6SO_4$ (the oxidation reaction of this last compound produces nitrogen and protons). Parent saponite has a basal spacing of 14.9Å, that decreases to 12.8Å in hydrazonium-treated samples. When treating with 0.05N dithionite solution, the basal spacing is not affected, decreasing only when the more concentrated solution (0.2N) is used.

The widening in the 001 reflection peak indicates a loss of crystallinity during the reduction process. Natural griffithite is a well-ordered, very crystalline sample, having a f.w.h.m. index of 0.466°. The reduced samples are much less crystalline, their f.w.h.m. indexes varying between 0.972 and 1.550°.

SAMPLE	basal spacing (Å)	f.w.h.m. (2Θ degrees)
Natural Griffithite	14.9	0.466
0.05N Hydrazonium-treated Griffithite	12.8	1.174
0.05N Hydrazonium-treated Griffithite Al-intercalated	18.5	0.523
0.05N Hydrazonium-treated Griffithite Al-pillared	17.6	0.590
0.2N Hydrazonium-treated Griffithite	12.8	1.550
0.2N Hydrazonium-treated Griffithite Al-intercalated	18.5	0.575
0.2N Hydrazonium-treated Griffithite Al-pillared	17.3	0.706
0.05N dithionite-treated Griffithite	14.6	0.972
0.05N dithionite-treated Griffithite Al-intercalated	18.3	0.662
0.05N dithionite-treated Griffithite Al-pillared	17.4	0.725
0.2N dithionite-treated Griffithite	12.7	1.379
0.2N dithionite-treated Griffithite Al-intercalated	18.3	0.669
0.2N dithionite-treated Griffithite Al-pillared	17.5	0.830

Table 1 *Basal spacing and f.w.h.m. index obtained from X-ray diffraction patterns of natural griffithite and reduced, intercalated and pillared samples.*

The surface area of the solids obtained after reduction with sodium dithionite is similar to that of natural saponite. When treating with 0.05 and with 0.2N $Na_2S_2O_4$, the surface area of the reduced solids is about 40 m^2/g (35 m^2/g in parent griffithite). The reduction with hydrazonium sulphate increases the surface area of the solids obtained, probably because of the acid activation of the clay by the H^+ released during the oxidation of hydrazonium species. The acid activation causes the disaggregation and delamination of the clay sheets, increasing the surface area of the solids obtained, up to 63 m^2/g when treating with 0.05N $N_2H_6SO_4$ solution and up to 137 m^2/g when 0.2N $N_2H_6SO_4$ solution is used.

3.3 Intercalated and pillared samples.

Chemical analyses of intercalated and pillared solids show that values of up to 9% of Al_2O_3 are fixed during the intercalation process, values similar to those reported for the pillaring of non-ferrous saponites.

The XRD patterns of intercalated and pillared samples show that they are crystalline solids, having better ordered structures than those of the reduced solids. In the intercalated solids, the f.w.h.m. index varies from 0.523 to 0.669 degrees, increasing slightly when pillaring by calcination at 500°C.

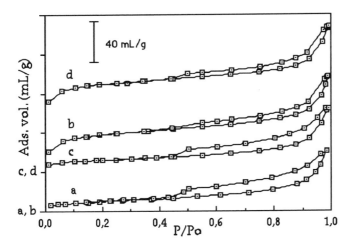

Figure 2 *Nitrogen adsorption-desorption isotherms of natural griffithite (a), natural griffithite treated with 0.05N sodium dithionite for 6 hours (b), this sample intercalated with Al$_{13}$ polycations (c) and the latter calcined at 500°C (d). (In the vertical axis, the "zero" of each representation is indicated).*

The nitrogen adsorption-desorption isotherms of natural griffithite, reduced samples and the latter intercalated and pillared with Al-oligomers are all of type I of IUPAC classification. The high adsorption of intercalated and pillared samples at low relative pressure indicates the creation of microporosity produced by these treatments. The hysteresis loops are of type H4 of IUPAC classification, associated with slit-shaped pores. The changes observed in the form of the hysteresis loops indicates a decrease of mesoporosity in intercalated and pillared solids.

When intercalating the reduced solids with Al$_{13}$ solutions, the surface area increases to about 280 m^2/g, decreasing to about 250 m^2/g for pillared compounds, calcined at 500°C. The surface area of intercalated and pillared compounds is similar for the hydrazonium and dithionite-treated solids, having values of about 270 m^2/g in intercalated samples and 230 m^2/g for the pillared solids.

t-plots of natural and reduced griffithite show an absence of microporosity, while in intercalated and pillared solids a microporous system has been generated (Figure 3). f-plots confirm that adsorption in the region of low relative pressure is about ten times higher in intercalated and pillared samples than in natural griffithite (Figure 4).

The micropore area and micropore volume of the dithionite-treated reduced solids is similar to that of natural saponite. In the hydrazonium-reduced solids, there is an increase in the values of these two magnitudes, in agreement with the creation of microporosity observed during the acid activation of griffithite. [12]

Values of micropore area and micropore volume (Table 2) confirm that most of the surface area of intercalated and pillared solids corresponds to the creation of microporosity. The micropore volume reaches values of about 0.100 cm³/g in the intercalated samples, slightly decreasing to values of about 0.080 cm³/g in the pillared solids.

The differential distribution of pores (Figure 5) shows that the intercalation and pillaring processes lead to the creation of pores at about 18Å, not observed in natural and reduced griffithite, and which appear and increase in number in the intercalated and pillared solids.

SAMPLE	S_{BET} (m^2/g)	Pore vol. (cm^3/g)	$Vol_{\mu p}$ (cm^3/g)	$S_{\mu p}$ (m^2/g)
Natural Griffithite	35	0.085	0.004	6
0.05N Hydraz.-treated Griffithite	63	0.105	0.006	12
0.05N Hydraz.-treated Grif. Al-intercalated	288	0.208	0.100	206
0.05N Hydraz.-treated Grif. Al-pillared	241	0.186	0.085	174
0.2N Hydraz.-treated Griffithite	137	0.141	0.027	56
0.2N Hydraz.-treated Grif. Al-intercalated	263	0.191	0.081	166
0.2N Hydraz.-treated Grif. Al-pillared	212	0.174	0.063	129
0.05N dithion.-treated Griffithite	42	0.076	0.005	11
0.05N dithion.-treated Grif. Al-intercalated	291	0.203	0.105	215
0.05N dithion.-treated Grif. Al-pillared	255	0.181	0.086	175
0.2N dithion.-treated Griffithite	34	0.064	0.004	8
0.2N dithion.-treated Grif. Al-intercalated	266	0.179	0.096	198
0.2N dithion.-treated Grif. Al-pillared	241	0.175	0.083	170

Table 2 *BET surface area, pore volume, micropore volume and micropore area of the different solids. ($Vol_{\mu p}$ and $S_{\mu p}$ obtained from t plot).*

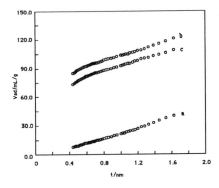

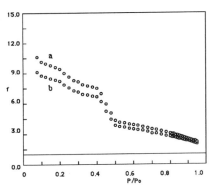

Figure 3 *t-plot of natural griffithite (a), and samples treated with 0.05N sodium dithionite and intercalated (b) and pillared (c) with Al₁₃ polycations.*

Figure 4 *f-plot of sample treated with 0.05N sodium dithionite and intercalated (a) and pillared (b) with Al₁₃ polycations. Natural griffithite is used as reference.*

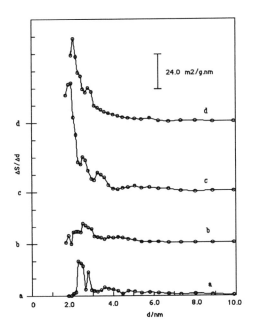

Figure 5 *Differential distribution of pores of natural griffithite (a), natural griffithite treated with 0.05N sodium dithionite for 6 hours (b), this sample intercalated with Al₁₃ polycations (c) and the latter calcined at 500°C (d). (In the vertical axis, the "zero" of each representation is indicated).*

Acknowledgements. Financial support from the Comisión Interministerial de Ciencia y Tecnología, CICYT (MAT96-0643 Project) is gratefully acknowledged.

References

1. M.L. Occelli. In: Keynotes in Energy-Related Catalysis. *Stud. Surf. Sci. Catal.,* **1988**, *35*, 101. Edited by S. Kaliaguine. Elsevier, Amsterdam.

2. C. de la Calle and H. Suquet. Vermiculite. In: Hydrous phyllosilicates. *Rev. Miner.,* **1988**, *19*, 455. Edited by S.W. Bailey. Mineralogical Society of America, Washington.

3. H. Usami, K. Tagaki and Y. Sawaki. *Chem. Lett.,* **1992**, *1992*, 1405.

4. S. Chevalier, R. Franck, H. Suquet, J.F. Lambert and D. Barthomeuf. *J. Chem. Soc. Faraday Trans.,* **1994**, *90*, 667. *Ibd.,* **1994**, *90*, 675. *Ibd.,* **1995**, *91*, 2229.

5. R.A. Schoonheydt, J. van den Eynde, H. Tubbax, H. Leeman, M. Stuyckens, I. Lenotte and W.E.E. Stone. *Clays & Clay Miner.,* **1993**, *41*, 598. *Ibd.,* **1994**, *42*, 518.

6. L. Li, X. Liu, Y. Ge, R. Xu, J. Rocha and J. Klinowski. *J. Phys. Chem.,* **1993**, *97*, 10389.

7. P.B. Malla and S. Kormaneni. *Clays & Clay Miner.,* **1993**, *41*, 472.

8. L. Bergaoui, J.F. Lambert, H. Suquet and M. Che. *J. Phys. Chem.* **1995**, *99*, 2155.

9. S. Chevalier, R. Franck, J.F. Lambert, D. Barthomeuf and H. Suquet. *Appl. Catal. A: General,* **1994**, *110*, 153.

10. L. Bergaoui, J.F. Lambert, M.A. Vicente-Rodríguez, L.J. Michot and F. Villiéras. *Langmuir,* **1995**, *11*, 2849.

11. M.A. Vicente Rodríguez, M. Suárez Barrios, J.D. López González and M.A. Bañares Muñoz. *Clays & Clay Miner.,* **1994**, *42*, 724.

12. M.A. Vicente Rodríguez, J.D. López González and M.A. Bañares Muñoz. *Micropor. Mater.,* **1995**, *4*, 251.

13. I. Rozenson and L. Heller-Kallai. *Clays & Clay Miner.,* **1976**, *24*, 271.

TEXTURE OF IMPREGNATED ACTIVE CARBON. EFFECT OF BURN-OFF

B. BUCZEK[1], A. ŚWIĄTKOWSKI[2] and S. ZIĘTEK[3]

University of Mining and Metallurgy, 30 - 059 Cracow, Poland[1]
Military Technical Academy, 00 - 908 Warsaw, Poland[2]
Military Institute of Chemistry and Radiometry, 00 - 910 Warsaw, Poland[3]

1 INTRODUCTION

Active carbons impregnated with salts of chromium, copper and silver are used for removal of HCN, ClCN, AsH_3, $CoCl_2$ from streams of polluted air. They are usually obtained by impregnation with solutions of the above mentioned salts. The kind of active carbon and its properties (specific surface area, porous structure, chemical nature of its surface) influence the activity, selectivity and stability of the impregnated product. On the other hand, a higher amount of the impregnant raises the catalytic action but reduces the physical adsorption. The knowledge of the porous structures of both the active carbon and the impregnated one may be useful when predetermining their availability for application. From this point of view the knowledge of location of the catalyst in the porous structure is also important. These problems were studied by many authors[1-3]. In our earlier investigations of the effect of texture on the properties of impregnated active carbons we have studied the preparations obtained by:

    i. separation of the active carbon (before impregnation) in the stream of air onto the fractions of various density and the intergrain burn-off[4],

    ii. abrasion of the successive layers of the active carbon grains (before and after impregnation) of the burn-off changing within the grains[5,6].

    The aim of this work was to evaluate the changes in texture of active carbons (of different overall burn-off) caused by loading with salts in the process of impregnation.

2 EXPERIMENTAL AND RESULTS

The active carbon of type A, obtained from hard coal of two industrial batches, of different duration of activation, was used in investigations. The carbon sample of a lower burn-off was denoted as AC1, while that of a higher burn-off - by AC2. Both active carbons with different burn-off were impregnated with ammonia solutions of Cr, Cu and Ag salts giving the impregnated carbons denoted as IAC1 and IAC2. The amount of loaded catalysts reached 9.1% and 10.8% for IAC1 and IAC2, respectively. The porous structure of original and impregnated active carbons was analysed using adsorption techniques. Adsorption/desorption isotherms of benzene vapours at 293 K were determined gravimetrically (Figure 1) while adsorption/desorption isotherms of argon at 77 K were determined volumetrically (Figure 2).

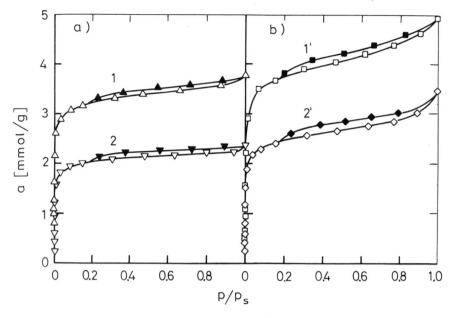

Figure 1 *Adsorption-desorption isotherms of benzene vapour at 293 K on active carbons:*
1-AC1, 1'-AC2 and impregnated active carbons: 2-IAC1, 2'-IAC2

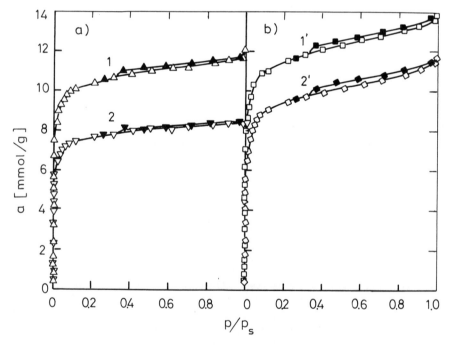

Figure 2 *Adsorption-desorption isotherms of argon vapour at 77.5 K on active carbons:*
1-AC1, 1'-AC2 and impregnated active carbons: 2-IAC1, 2'-IAC2

the Dubinin-Radushkevich equation (the limiting volume of micropores - W_o and the characteristic adsorption energy - E_o)[7,8]. The obtained structural parameters based on benzene sorption are listed (Tables 1 and 2), while those based on argon sorption (Tables 3 and 4). Moreover, the values of structural parameters presented for impregnated active carbons per 1 g of the impregnated active carbon were recalculated to the values related to 1 g of the active carbon before impregnation. They are given in the columns IAC1* and IAC2* of the tables 1-4. Such recalculated values of the parameters of porous structure may be then compared with the analogical values for non-impregnated active carbons.

3 RESULTS AND DISCUSSION

The analysis of changes in porosity caused by loading with catalysts may be carried out for both active carbons, before and after impregnation, basing on the determined sorption isotherms and the computed parameters of the structure.

From the isotherms of benzene vapours adsorption it results that the volume of micropores of the more activated carbon AC2 is about 13% greater than that of AC1, while the volume of mesopores is about 2.2 times greater. Similar differences in the values of parameters computed basing on the argon adsorption isotherms were observed, although the absolute values were slightly different from those for benzene. This fact confirms the previous observation of greater increase in the volume of mesopores than that of micropores at higher burn-off.

A similar amount of catalyst (roughly 10%) was loaded into each of the studied carbons. The comparison of the texture parameters for the active carbons before and after impregnation should inform about location of the catalysts in various kinds of pores. The differences between the values of the respective parameters are given in the column denoted with the symbol delta, (Tables 1-4). Such comparison makes possible evaluation of the part of the volumes of micro- and mesopores occupied (or blocked) by the catalyst. In the case of computations based on the benzene adsorption isotherms it was found that the volume of micropores was reduced for about 28% due to the impregnation, for both active carbons AC1 and AC2. On the other hand the volume of mesopores was reduced for 38% in case of AC1 and for about 16% in case of AC2. The computations carried out basing on the argon sorption show that the ratio of the reduction of the pores volume was somewhat lower: reaching about 19% for AC1 and about 13% for AC2. The respective values for mesopores were equal to 37% and 6%.

The differences between the values of parameters computed basing on benzene and argon adsorption isotherms may result from the different size of benzene and argon molecules (the molecular sieve effect). Another factor which may be responsible for the above mentioned differences is temperature of the adsorption measurement (77 K for argon). At this low temperature some shrinkage of pores due to shrinkage of adsorbent may occur.

In order to improve the evaluation of the changes in the porous structure of micropores the size distribution of micropores volume was calculated using the Horvath-Kawazoe equation[9] (the slit model). The HK equation was adapted for the flat arrangement of benzene molecules in the adsorbent pores[10]. The obtained size distribution curves of the volume of pores for both active carbons, before and after impregnation, are presented (Figure 3).

Table 1 *Texture Parameters of Active Carbon AC1 and Impregnated one IAC1 (Benzene at 293 K)*

Parameter	AC1	IAC1	IAC1*	Δ
W_o [cm³/g]	0.281	0.183	0.205	0.076
E_o [kJ/mol]	23.2	20.6	---	---
v_{mi} [cm³/g]	0.268	0.172	0.193	0.075
v_{me} [cm³/g]	0.066	0.036	0.040	0.025

Table 2 *Texture Parameters of Active Carbon AC2 and Impregnated one IAC2 (Benzene at 293 K)*

Parameter	AC2	IAC2	IAC2*	Δ
W_o [cm³/g]	0.315	0.211	0.232	0.083
E_o [kJ/mol]	21.9	22.0	---	---
v_{mi} [cm³/g]	0.305	0.200	0.220	0.085
v_{me} [cm³/g]	0.144	0.110	0.121	0.023

Table 3 *Texture Parameters of Active Carbon AC1 and Impregnated one IAC1 (Argon at 77.5 K)*

Parameter	AC1	IAC1	IAC1*	Δ
W_o [cm³/g]	0.297	0.214	0.240	0.057
E_o [kJ/mol]	18.5	19.8	---	---
v_{mi} [cm³/g]	0.295	0.212	0.238	0.057
v_{me} [cm³/g]	0.057	0.032	0.036	0.021

Table 4 *Texture Parameters of Active Carbon AC2 and Impregnated one IAC2 (Argon at 77.5 K)*

Parameter	AC2	IAC2	IAC2*	Δ
W_o [cm³/g]	0.322	0.253	0.278	0.044
E_o [kJ/mol]	17.4	16.7	---	---
v_{mi} [cm³/g]	0.318	0.252	0.277	0.041
v_{me} [cm³/g]	0.084	0.072	0.079	0.005

The following parameters characterizing texture of all carbon samples were computed: volume of mesopores - v_{me}, volume of micropores - v_{mi} and the parameters of

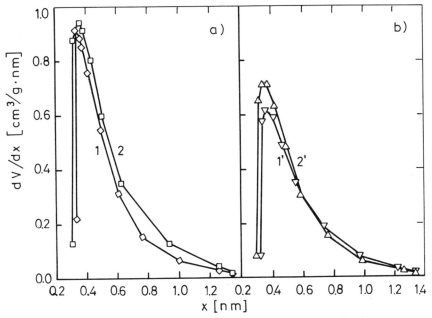

Figure 3 *Micropore size distributions a) active carbons 1-AC1, 2-AC2, b) impregnated active carbons 1'-IAC1, 2'-IAC2, using the Horvath-Kawazoe equation (benzene at 293 K)*

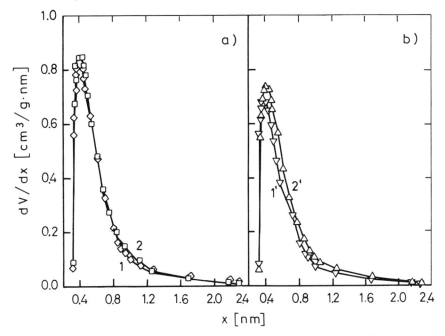

Figure 4 *Micropore size distributions a) active carbons 1-AC1, 2-AC2, b) impregnated active carbons 1'-IAC1, 2'-IAC2, using Kawazoe equation (argon at 77.5 K)*

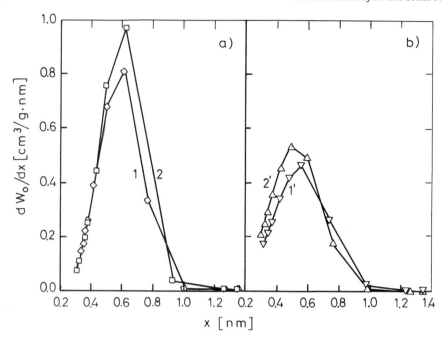

Figure 5 *Micropore size distributions a) active carbons 1-AC1, 2-AC2, b) impregnated active carbons 1'-IAC1, 2'-IAC2, using Dubinin-Stoeckli equation (benzene at 293 K)*

Data on argon adsorption were also used for determination of the distribution of the volume of micropores, modifying the HK equation to this adsorbate[11]. These results are presented (Figure 4). In the case of impregnated active carbons in computations the recalculated adsorption values, related to 1g of non-impregnated carbon were used for both adsorbates.

For the benzene the distribution of pore volume in function of their linear size (the half of the slit width at the maximum of the distribution curve for the slit model) was also computed. In this case the Dubinin-Stoeckli equation was used[11]. The distribution curves (recalculated for 1 g of non-impregnated active carbon) are presented (Figure 5), while the parameters of the DS equation - total volume of pores - W_o, the half of the slit width - x_o and dispersion - δ are listed (Table 5).

The comparison of the micropores distribution curves leads to the conclusion that in all cases (the active carbons before and after impregnation) the curves resulting from the HK and DS eqs. for the active carbons AC1 or IAC1 are located below the respective curves for the carbons AC2 or IAC2. This means that irrespective of the type of adsorbate and the applied method of characterization of the carbon texture the active carbon of the lower degree of burn-off and the impregnated active carbon obtained from this carbon have a lower contribution of microporous structures. Another interesting observation is close similarity of values of size of micropores obtained using the HK equation, both for benzene and argon as adsorbates.

Table 5 *The Dubinin-Stoeckli Equation Parameters for Carbons Before and After Impregnation*

Sample	W_o^o [cm³/g]	x_o [nm]	δ [nm]
AC1	0.281	0.586	0.135
IAC1	0.197	0.554	0.171
AC2	0.323	0.596	0.130
IAC2	0.219	0.521	0.161

The results of computations of the parameters of the DS equation for benzene as adsorbate inform about the changes of the width of micropores. The value of x_o is greater for AC2 active carbon, of a higher burn-off, than for AC1 carbon. Impregnation reduces this parameter for both active carbons (simultaneously increasing the dispersion parameter, δ), but for AC2 carbon this reduction is much greater. These observations may indicate that impregnants locating in micropores change their dimensions. It is not the only possible explanation however. Kloubek[1] showed that the layers of impregnating salts on the walls of greater pores (of the radii greater than 150 μm) posses numerous cracks forming a secondary microporous structure. Other authors[2,3] indicate blocking some part of pores of various dimensions.

4 CONCLUSION

Summing up, non-impregnated active carbons are characterized by higher width and volume of micro- and mesopores than the impregnated ones. The amount of the active phase is slightly higher for active carbons with a more developed texture. The salts are deposited mainly inside the micro- and mesoporous structure. Narrow micro- and mesopores (low burn-off) cause that salts are supported inside the mesoporous structure. On the contrary, in case of active carbon with high burn-off the active material is distributed mainly inside micropores.

Acknowledgements: The authors are grateful for partial financial support of this work from projects No 0 T00A 002 10 (MIChR) and No 10. 210. 67 (UMM).

References

1. J. Kloubek, Carbon, 1981, **19**, 303.
2. V. S. Tripathi, P.K. Ramachandran, Carbon, 1982, **20**, 25.
3. M. Molina-Sabio, V. Perez, F. Rodriguez-Reinoso, Carbon, 1994, **32**, 1259.
4. S. Ziętek, M. Mioduska, T. Żmijewski, A. Świątkowski, Mat. Chem. Phys. in print.
5. B. Buczek, S. Ziętek, A. Świątkowski, 'Fundamentals of Adsorption', ed. by M. Douglas Le Van, Kluwer Academic Publishers, Boston, 1996.
6. B. Buczek, S. Ziętek, A. Świątkowski, Langmuir, in print.

7.   M.M. Dubinin in Chemistry and Physics of Carbon (Ed. by P.L. Walker, Jr.), **vol. 2**,
      M. Dekker, New York, 1966, 51.
8.   M.M. Dubinin, Carbon, 1985, **23**, 373.
9.   G. Horváth, K. Kawazoe, J. Chem. Eng. Japan., 1983, **16**, 470.
10.   A. Świątkowski, B.J. Trznadel, S. Ziętek, Ads. Sci. Tech., in print.
11.   G. Rychlicki, A.P. Terzyk, G.S.Szymański, Polish J. Chem., 1993, **67**, 2029.

PORE VOLUME AND APPARENT DENSITY OF FINE POROUS PARTICLES

B. BUCZEK[1], E. VOGT[1], W. STEFANIAK[2]

University of Mining and Metallurgy, 30 - 059 Cracow, Poland[1]
Maria Curie - Skłodowska University, 20 - 031 Lublin, Poland[2]

1 INTRODUCTION

The knowledge of the character of porous materials is extremely important during manufacturing and checking of the quality and as well as in using. The sorptive and catalytic properties of solids are determined by the chemical character of surface and their porosity.

There is a great number of experimental methods which allows one to describe the structure of porous materials i.e. total pore volume (the volume of macro-, meso- and micropores), the average radius of pores, the apparent density, the porosity and the specific surface area. In order to determine the above mentioned properties, the commonly used adsorptive methods, the mercury porosimetry, microscopy, X-ray methods and densimetric measurement[1] are used. The properties of the investigated material, the time and price of a measurement are the most important factors when an analytical method is chosen. The interpretation of the final results requires great experience and the application of advanced mathematics. The simplest way to estimate the porous structure is the densimetric measurement, but its form cannot be applied for small particles. The literature[2] describes plenty of methods determining apparent density, but most of them are destined for particles greater than 0.5 mm.

The aim of the present paper is the determination of total pore volume and the apparent density of fine particles by thermogravimetric measurement[3] and by the comparative[4] and titration[5] methods.

2 EXPERIMENTAL AND RESULTS

2.1 Materials

The following fine porous particles were studied: cracking catalyst, carbonizate, hard coal, aluminium oxide and coke. The parameters of the porous structure such as: specific surface area using BET method (S_{BET}), volume of micropores (V_{mi}) and volume of mesopores (V_{me}) of investigated materials obtained from adsorption measurements are presented (Table 1).

Table 1 *Characterisation of the Investigated Solids by Adsorption Measurements*

Particles	Adsorbate	S_{BET} [m^2/g]	V_{mi} [cm^3/g]	V_{me} [cm^3/g]
Cracking catalyst	Nitrogen	231	--------	0.240
Carbonizate	Nitrogen	61.1	0.0077	0.380
Hard coal	Methanol	185	0.0859	0.0153
Aluminium oxide	Nitrogen	298	0.0000	0.1673
Coke	Argon	15.8	0.0075	0.0037

2.2 Description of Experimental Methods Used

2.2.1 Determination of Apparent Density by the Titration Method. In petroleum industry the particle density of free-flowing cracking catalyst[5] is estimated indirectly by measuring the pore volume. When a liquid with low viscosity and volatility, for example water, is added, the powder should remain free-flowing until the liquid has filled all the open pores. Any additional liquid coats the external surface of each particle causing immediate caking by surface tension. The apparent density was calculated from the equation 1:

$$\rho_{ap} = \frac{\rho_t}{1 + V \cdot \rho_t} \qquad (1)$$

where: ρ_{ap} - apparent density,
  ρ_t - true density,
  V   - volume of liquid used for titration per gram of solids.

The values of the apparent density obtained by the titration method are presented (Table 4).

2.2.2 Determination of Apparent Density by the Comparative Method. The comparative method based on the assumption that the minimum packed bed voidage is the same for similarly shaped particles of a narrow size range. The authors[6] worked out a procedure of prepare of samples of reference and investigated materials. The calculation of the apparent density based on bulk density measurements of these using automatic apparatus[7], securing of high accuracy and reproducibility of the results and standardisation of the conditions of measurement. Knowing the apparent density of the reference material (ρ_{apr}), the apparent density of porous particles (ρ_{ap}) was calculated from the equation 2:

$$\rho_{ap} = k \cdot \frac{\rho_{bp}}{\rho_{br}} \cdot \rho_{apr} \qquad (2)$$

where: ρ_{bp} - bulk density of porous particles,
  ρ_{br} - bulk density of the reference material,
  k   - factor defining the correlation between the shapes of the porous particles and the reference material.

Table 2 *Reference Materials of Porous Particles Investigated*

Particles	Diameter range [mm]	Reference materials
Cracking catalyst	0.04 - 0.125	glass or bronze beads
Carbonizate	0.250 - 0.560	glass or bronze beads
Hard coal	0.160 - 0.315	carborundum
Aluminium oxide	0.040 - 0.250	ceramics material
Coke	0.100 - 0.200	ceramics material

The reference materials used for the studied fine particles are shown (Table 2). The reference materials were composed according to size range of investigated samples. In this case the factor k equals to one (1.00).

The values of the apparent density obtained by the comparative method are summarised (Table 4).

2.2.3 Determination of Total Pore Volume by the Thermogravimetric Measurement and Calculation of Apparent Density. Thermogravimetric measurements of desorption of liquids[4] were carried out with a Derivatograph 1500 C (MOM, Hungary) using the quasi-isothermal program heating of the sample. The samples were prepared by immersing dry material in liquid adsorbate. The wet samples were degassed in order to remove air from the pores and complete to fill the whole pores by adsorbate. The portion (about 40-80 mg) of wet materials prepared in this way was put into a platinum crucible. The figure 1 shows the typical thermodesorption curve for a liquid which wets the adsorbent perfectly at quasi-isothermal heating program. The curve showed the dependence of sample weight loss versus temperature. The segment I of TGA curve represents the bulk liquid outside the pore structure of the adsorbent. When the first stage of desorption is completed, the temperature increases and the desorption from pores starts. The segment II corresponds to evaporation of adsorbate both from the pores and the adsorbed film on the walls of pores. The height of the segment II is a measure of the total pore volume. In order to determine it, the corresponding changes of the adsorbate mass during desorption are transformed into changes of value using the functional change of density dependently on the temperature of the adsorbate. The apparent density of the investigated solids was calculated from the equation 3:

$$\rho_{ap} = \frac{\rho_e}{1+V \cdot \rho_e}$$ (3)

where: ρ_e - effective density,
$\quad$ V - total volume from the thermogravimetric measurement.

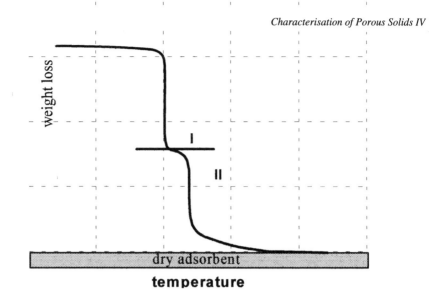

Figure 1 *TGA curve, weight loss versus temperature*

The TGA thermodesorption curves were measured for water and benzene for all studied materials. The same substances were used as the pycnometric displacements. The results obtained from the thermogravimetric measurements and effective densities are presented (Table 3).

Table 3 *The Effective Density and the Total Volume by the Thermogravimetric Measurement*

Particles	Effective density $[g/cm^3]$	Total volume $[cm^3/g]$	Effective density $[g/cm^3]$	Total volume $[cm^3/g]$
	using benzene		*using water*	
Cracking catalyst	2.395	0.27	2.413	0.31
Carbonizate	1.264	0.31	1.335	0.35
Hard coal	1.425	< 0.10	1.534	*
Aluminium oxide	2.581	0.34	2.616	0.34
Coke	1.841	< 0.08	1.859	*

  * not determined

2.3 Results

The apparent densities calculated from volumes obtained by the thermogravimetric measurements and investigated by comparative or titration methods are presented (Table 4).

Table 4 *The Apparent Density by the Comparative and by the Titration Method, and Thermogravimetric Measurement*

Particles	Apparent density [g/cm³]			
	Comparative method	Thermogravimetric measurement		Titration method
		Benzene	Water	Water
Cracking catalyst	1.431	1.459	1.380	1.601
Carbonizate	0.898	0.908	0.910	0.971
Hard coal	1.485	1.247	*	1.624
Aluminium oxide	1.368	1.375	1.384	1.429
Coke	1.411	1.605	*	1.593

There is a good agreement between results obtained from both methods (the comparative and thermogravimetric one) for the following materials: the cracking catalyst, carbonizate and aluminium oxide. The comparative method has given slightly smaller value of the apparent density than the thermogravimetric measurements for most materials. Much lower apparent density of coke derived from the comparative method can be explained by the fact that the method considered the total porosity (micro-, meso- and macropores). The thermogravimetric method does not recognise the evaporation of liquid from outside pores and macropores.

The total pore volumes were not successfully evaluated by thermodesorption of water for hard coal and coke. Due to low porosity of the above mentioned materials the thermogravimetric method is not useful in this case. The apparent density of hard coal determined by the comparative method is higher than the value calculated on the basis of evaporation of benzene. The real volume of evaporated benzene was lower than that one which was used to calculate ($V = 0.1$ cm³/g) the apparent density. However that volume was not able to determine more accurately.

All values of apparent density obtained by the titration method were higher than those derived from the thermogravimetric and comparative methods. These differences are caused by the difficulties in estimation of the caking end point.

Probably the concave meniscus is formed in the pores of the studied substance. Therefore, the smaller volume of water was used, what explains the higher values of the apparent density obtained by the titration method.

3 CONCLUSION

The titration method is not reliable for all porous materials and the end-point of titration (caking end point) is very difficult to estimate. The error of measurements depends on personal experiences.

The thermogravimetric measurements give good results of total pore volumes of micro and mesoporous materials. In the case of macroporous materials (such hard coal or coke) the determination of their volumes are not precise enough or almost impossible

The comparative method gives the possibility of determination of the apparent density of fine particle porous materials which is difficult to estimate by other methods. In order to determine the bulk density, it is suggested to apply the automatic apparatus which works under fully standarised conditions. In the case of macroporous materials (hard coal, coke) where the thermogravimetric method is useless, the comparative method turns out to be irreplaceable.

Acknowledgements: The authors are grateful for partial financial support of this work from projects No 10. 210. 49 (UMM) and No 3 T09A 110 11 (KBN).

References

1. S. Lowell, J, E, Shields, Powder Surface Area and Porosity, Chapman & Hall, London 1991.
2. B. Buczek, E. Vogt, Przemysł Chemiczny, 1993, **72**, 397 (in Polish).
3. J. Goworek, W. Stefaniak, Mat. Chem. Phys., 1992, **32**, 344.
4. A. R. Abrahamsen, D. Geldart, Power Technol., 1991, **3**, 315.
5. W. B. Innes, Oil and Gas J., 1956, **12**, 162.
6. B. Buczek, E. Vogt, Charecterisation of Porous Solids III, Studies in Surface Sience and Catalysis, Editors: J. Rouqerol, F. Rodriguez-Reinoso, K.S.W. Sing and K.K.Unger, Elsevier, Amsterdam, 1994, **87**, 339.
7. B. Buczek, E. Vogt, Inż. i Ap. Chem., 1996, **35**, 2, 17 (in Polish).

A SIMULATION STUDY OF NITROGEN ADSORPTION IN MAGNESIUM OXIDE PORES

David Nicholson and Neal A. Freeman

Department of Chemistry,
Imperial College,
LONDON SW7 2AY

M. A. Day,

ICI Materials, PO Box 90, Wilton,
Middlesborough,
Cleveland, TS6 8JE, UK.

1 INTRODUCTION

Nitrogen adsorption at 77K is one of the most extensively used tools for the characterisation of adsorbents. The majority of industrially important adsorbents are porous; in the standard classification of isotherms they are expected to produce isotherms of type I (pore sizes in the micropore range) type IV (when mesopore sizes dominate the pore size distribution or less commonly type V; associated with weak adsorbate surface interactions and strong adsorbate adsorbate interactions. Simulation is playing an ever increasing role in augmenting the interpretation of experimental data since, in contrast to many experimental systems, the adsorbent geometry is clearly defined. Several simulation studies of nitrogen adsorbates have used a one centre spherical model for the nitrogen molecule. In the early work of Walton and Quirke[2] this type of model was used to study adsorption in graphitic slit pores with continuum surfaces, these studies have been extended to cover wider pore size ranges and to include density functional theory calculations[3] . Simulations of spherical nitrogen in pores with corrugated potentials have shown that although the corrugations are quite small they can significantly modify adsorption behaviour[4]. Simulations using two centre nitrogen models go back further than the above work but have been largely confined to planar surfaces. In the pioneering work of Steele, Tildesley et al., patches of N2 on graphite were studied using molecular dynamics at low temperatures[5]. The importance of quadrupoles at these low temperatures was emphasised. Later Kuchta and Etters made a series of studies of the N2 graphite system[6]. Even at ambient temperatures, it appears that two centre models produce different isotherms for adsorption in graphitic pores from one centre models[7], although quadrupoles are not found to be important at these temperatures. Since both adsorbate structure and adsorbent surface corrugation have been found to be important, it is of interest to carry out simulations in porous where both these features are present. Here we report simulation studies of adsorption of a two centre nitrogen adsorbate at 77K into slit pores formed by the 100 planes of MgO. These systems are free of the complexities of zeolitic oxides, and have no heterogeneity resulting from disordered structures[8]. They therefore represent an idealised material which may be compared with less regular models. The pore surfaces contrast with graphitic models in being more corrugated and having different site symmetry. Experimentally microporous MgO can be prepared by controlled dehydration of Mg(OH)$_2$. The reaction proceeds via a topotactic transformation and results in well defined slit shaped pores[9].

2 SIMULATION MODEL

Nitrogen was modelled as two Lennard-Jones centres with four partial charges to represent the quadrupole. Charges of +0.373e were placed on the molecular axis at ±0.0847nm from the molecule center and -0.373e at ±0.1094nm. The potential model was based on the more elaborate potential of Kuchta and Etters[6]. This model was originally optimised to reproduce the second virial coefficient of gaseous N$_2$ and the sublimation properties of α-N$_2$. The 12-6 functions were fitted to this potential over a range of separations of the molecule centres between 0.3nm and 0.5nm for several orientations. Deviations from the original potential were small and the resulting parameters are in excellent agreement with those derived in an earlier model[10], similar to the one used here. Atom-atom Lennard Jones constants for Mg and for oxygen were taken from the work of Alavi and Macdonald[11] on the simulation of argon on planar MgO; cross parameters were calculated with the Lorentz-Berthelot combining rules. The potential parameters are summarised in Table 1.

The simulation box was constructed from two walls, 4.47nm or 15 magnesium atoms square, sandwiching the pore space. Periodic boundary conditions were applied in the *xy*-plane parallel to the surfaces. The pore width H is defined as the distance separating centres of atoms in he surfaces of opposite walls. Five different pore widths were used in the simulations (H=5.0(1),0.844(2),1.055(3),1.266(4),1.477(5)). The first is sufficiently large to be regarded as the equivalent of two isolated surfaces. The remaining four correspond to removing an integer number of (100) planes from a standard magnesium oxide structure. The (*x,y*) position of magnesium ions on one surface exactly corresponds to those on the opposing surface. In real crystals the surface atoms relax away from regular sites, but this is not reflected in these simulations.

Table 1 *Potential parameters used in the calculations*

	(ϵ/k) / K	σ / nm
N (This work)	37.8	0.3318
N (Ref 10)	36.4	0.3318
O	140.9	0.2932
Mg	6.09	0.2712

The molecule surface potential was obtained from a truncated Fourier series expansion[12,13]. For the Lennard Jones interactions, with planes separated by a distance d

$$u(l,z) = \sum_{\beta} \left[\sum_{n \geq 0} u_{0,\beta} (z + nd) + \sum_{k \neq 0} \cos(k \cdot r)(-1)^m u_{k,\beta} \right]$$

where $l=(x,y)$, k is the sum of the reciprocal lattice vectors in this cell, $m=2(k_1 + k_2)/\beta$, and $\beta=1$ and $\beta=2$ refer to the Mg and O sublattices respectively. The continuum part of the potential, $u_{o\beta}(z+nd)$ may be approximated by

$$u_{o,\beta}(z + nd) = \frac{\pi}{2a^2} \left(\frac{2C_{12}}{5} \left[z^{-10} + \frac{1}{9a(z + 0.72a)^9} \right] - \left[z^{-4} + \frac{2z^2 + 7a(z + a)}{6a(z + a)^5} \right] \right)$$

and for $l \neq 0$, the attenuation of the periodic "corrugation" contribution is

$$u_{k,\beta} = \frac{2\pi}{a^2} \left(\frac{C_{12}}{5!} \left[\frac{k}{2z} \right]^5 K_5(kz) - \frac{C_6}{2!} \left[\frac{k}{2z} \right]^2 K_2(kz) \right)$$

where $k=|k|$, K_n is the nth modified Bessel function of the second kind and $C_n=4\epsilon\sigma^n$. Values for k up to $k^2=2$ were used[11].

The electrostatic part of the potential is given by,

$$u(l,z) = \sum_{\beta} \left[\sum_{k \neq 0} \cos(k \cdot r)(-1)^m \frac{\exp(-kz) q_a q_t}{2a_s \epsilon k} \right]$$

where a_s is the area of the unit cell in the surface. These terms vanish for even values of k. It was found that better than 1% accuracy could be achieved by including terms up to $k^2=1$. Induction contributions are extremely small[14] and were not included.

Figure 1 shows the variation of the overall nitrogen surface potential across the pore for two pore widths and for three favoured adsorption sites. The total potential surface is quite complex but some general features can be distinguished from these graphs and from views of the corrugation maps taken for molecules in various orientations[15].

(i) The usual pore overlap effects are seen but the orientation effects are enhanced as the pore becomes narrower.

(ii) The strongest site found on the bare surface occurred when Nitrogen is adsorbed normal to the wall over an Mg atom. However the adsorption energy rises steeply in nearby locations which are therefore weakly adsorbing. Nitrogen vertical over O is in a repulsive field. Thus, nitrogen oriented normal to the surface is likely to be highly localised, whereas parallel oriented molecules will be mobile.

(iii) Molecules with axes oriented parallel to the surface can probe more deeply into the pore walls. Again this effect increases in smaller pores.

Simulations were carried out in the grand ensemble at 77.5K and averages taken from converged runs. The potential was cut off at 1.5nm and a long range correction was applied

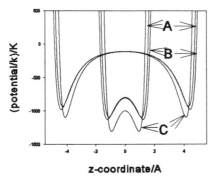

z-coordinate/A

Figure 1 *Potential energy functions for nitrogen in MgO slit pores. A: N_2 axis parallel to the surface with the molecule centre over the midpoint of O-O and Mg-Mg. B: N_2 parallel to the surface with the molecule centre over O. C: N_2 vertical to the surface with the molecule axis over Mg.*

to the attractive parts of the potential using a coarse grained singlet density. The gas phase pressure was calculated from the (corrected) chemical potential[13] and saturation vapour pressure was taken as 1bar. Isosteric heats were found from the usual fluctuation expressions[13]. Molecular orientations in the simulation were collected as ensemble averages of the functions $\cos^2\theta$ and $\cos\phi$. The angles θ,ϕ are between the molecular axis and the plane of the surface, and between the molecular axis and the x-axis respectively; the x-axis being along a line of Mg atoms.

3 RESULTS

Isotherms are displayed in two different ways in figure 2. As a fraction of monolayer coverage (main figure) and as molecular density (inset) both plotted against relative pressure on a logarithmic scale.

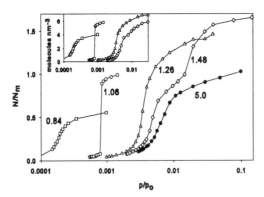

Figure 2. *Adsorption isotherms for nitrogen on MgO at 77.5K. N/Nm gives the filling in monolayer units. The inset graphs show pore density. Pore sizes in nm are shown on the graphs.*

From the plane surface isotherm (pore width 5.0nm) shown in figure 2 a monolayer of nitrogen with axes parallel to the surface, covering one face of the adsorption cell of area 19.98nm², would contain $N_m=123$ molecules, assuming a molecular area[1] of 0.162nm². If the molecular axis is normal to the surface 183 molecules can be packed into the same area. An estimate of monolayer coverage from singlet density functions taken from this simulation close to the beginning of second layer formation gives a figure of 122 molecules which agrees remarkably well with the conventional value for the area of the nitrogen molecule. Isosteric heat curves show a sharp decline at a coverage of ~123molecules (see Fig. 3) which also supports this estimate. From point B of the isotherm N_m is 110 molecules.

The most striking feature in Figure 2 is that at one particular pore width (H=1.055nm) there is a sharp transition in the isotherms which does not occur at other pore widths. This will be discussed in more detail below. Figure 2 also highlights several interesting aspects of micropore adsorption in this model system: (i) It is apparent, as has now been demonstrated in several studies, that very low relative pressures are needed to get into the Henry law region. Evidence of linearity, extrapolating through zero coverage, is found for the wider pores, but without high resolution adsorption techniques it would be easy to miss this region of the isotherm or even to arrive at a spurious value of the Henry law constant.

(ii) The approach to monolayer coverage is different in different pore sizes, but a sigmoid form in the submonolayer region is found in all cases, extending from the smallest pore width through to the open surface. The internal (chemical) width available to the molecules is approximately[7] $H'=H$-0.2nm. Thus in the smallest pore it is not possible to form a completed monolayer on both surfaces. In the three wider pores, with H'=2.6σ_N, 3.2σ_N and 3.8σ_N, the nature of the filling process beyond the monolayer depends on pore width. Clear evidence of two stage filling does not appear until the largest pore size (physical width H=1.477nm) is reached.

The specific effects of packing are also seen in the different values of maximum density. Since density continues to increase gradually with relative pressure, it is not possible to give a precise value, however it may be noted that a high density is reached more rapidly in the pore with H=1.266nm (H'=3.2) than for the other pore widths where H' is not so close to an integer.

Heats of adsorption Q are shown in figure 3, with wall (Q_w) and molecule (Q_m) components displayed separately. The interpretation of the full isosteric heats can be made in terms of these two contributions. In the two smallest pores the adsorbate part rises rapidly initially. Snapshot pictures from these simulations suggest that this can be attributed to a tendency to clustering at low coverages. The intermolecular interacting are seen to be always higher in the smallest pore for a given coverage reflecting the severe effects of confinement. In the next

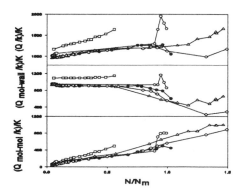

Figure 3. *Heats of adsorption. See Fig. 2 for key to symbols for different pore widths.*

largest pore there is a sharp rise in Q_m at the transition and it is seen that the interactions between adsorbate molecules then becomes much greater than that in any of the other pores below a monolayer coverage. It is interesting to note that for the two largest pore widths, Q_m continues to rise after the statistical monolayer point in contrast to the "open surface" where the conventional decrease in Q_m and in Q signals the weaker interactions associated with the beginnings of second layer adsorption. The initial heat of adsorption originates from Q_w. It has been noted in previous simulation studies[16] that significant enhancement of Q only occurs at very small pore widths and is not sufficient to account for the strong decrease in Q observed experimentally on heterogeneous surfaces. For most pore widths Q_w then remains constant up to a coverage where the attraction of other adsorbate molecules begins to compete with the walls interactions. Q_w then declines with coverage until repulsion from other adsorbate species begins to take effect. The pattern in the pore with $H=1.055$nm is different. Here there is a sharp maximum in Q close to the transition, similar to those seen in some experiments. It is seen that part of this signal comes from Q_m and part appears to come from Q_w. It is difficult to understand how Q_w can increase above its initial value to the rather high level seen here, and we are inclined to attribute this feature to the fairly large error bars associated with Q.

DISCUSSION

The results of this study demonstrate that, micropore adsorption does not always follow a regular pattern. The most striking feature is the sharp transition shown in pore size 2. This transition clearly does not correspond to capillary condensation in the conventional sense, since the internal pore width cannot accommodate more than two monolayers of nitrogen and the coverage is below a tenth of a monolayer prior to the transition.

Single particle distribution functions[15] show that after the transition the molecule centres are at an average distance of ~0.2nm from the pore centre. A simple calculation, based on the internal width of the pore given above (0.855nm) and the van der Waals length of the nitrogen molecule (0.495nm) shows that the molecular axis would make an angle of about $20°$ to the normal for the centres to be in this position. Clearly there are few molecules which lie flat on the surface. Figure 1 shows that molecules normal to the surface can have favourable energies provided that they are not directly over an O-atom. This orientation brings the quadrupole charges on one of the nitrogen atoms to within about 0.04nm of the centre of the pore on the same side as that of the adsorbate. A similarly oriented nitrogen on the opposite wall of the pore is well placed to form a highly favourable quadrupole interaction with the first molecule. There is thus a combination of 3 circumstances which together bring about a cooperative transition:

(i) The tendency of the potential corrugation to favour molecules oriented normal to the pore wall.

(ii) A critical confinement due the pore width that constrains the molecules to overlap the pore centre.

(iii) A favourable mutual attraction between molecules on opposite pore walls which is possibly augmented by the quadrupole interaction.

It seems probable that the first two conditions are essential for the transition to occur. The role of quadrupoles is less clear without further investigation. Other studies[7,17,18] suggest that quadrupoles - even as large as those on carbon dioxide - do not play a major part in determining large scale adsorption effects. On the other hand Talbot et al[19] found that herringbone ordering for nitrogen on graphite at low temperature did not occur unless the quadrupole was included in the molecular model.

Figure 4 shows a snapshot of the adsorbate on one face of pore 2 after the transition taken from a side view and from over the plane. It can be seen that most molecules are indeed in

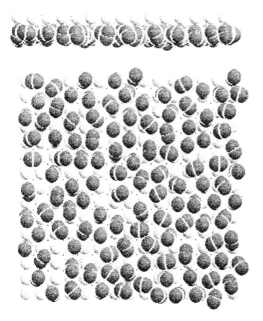

Figure 4. *Snapshot of a configuration at the top of the transition in the pore with H=1.055nm. The figure shows N2 adsorbed on one surface of the pore from the side view and from over the surface.*

orientations close to the normal to the surface. On the other hand it is surprising to see that some of the vertical molecules have a N-atom directly over a surface O which is actually a repulsive site on the bare surface (figure 1). This unexpected observation emphasises the fact that the nitrogen overlayer is not commensurate and that the minimisation of the free energy is crucially dependent on the existence and close proximity of the layer on the opposite wall. A similar deduction may be drawn from the existence of vacancies in the surface layer which accommodate the nitrogens protruding from the opposite wall. The vacancies in each layer also account for the fact that the number of molecules per unit area corresponds to a complete monolayer of flat rather than vertical molecules. This remarkable compensation effect implies that the molecular area of $0.162nm^2$ remains the best choice for surface area measurement, even though the molecular orientation does not correspond to this molecular area. It should be noted of course that the relative pressures over which adsorptin occurs are well below the accepted range of BET validity.

CONCLUSIONS

The adsorption of nitrogen at 77K in slit pores modelled as regular MgO surfaces has been simulated for a number of slit widths. A two centre model with quadrupoles was used to represent nitrogen, and dispersion and electrostatic interactions with the MgO surface were included. The isotherms and heat curves display a variety of behaviour dependent on slit width. Both packing and orientation have significant impact on the final density reached in the pores and on the form of the isotherms and heat curves. A striking feature is the appearance of a sharp transition in pores of width 1.055nm. This occurs when the nitrogen can establish a low energy structure with molecules inclined at ~20° to the surface normal, in such a way as to maximise quadrupole interacting between nitrogen on opposite walls. We conjecture that such structures would only be seen in simulations that use two centre models and corrugated surface potentials. It is possible that similar transitions could occur over a range of pore widths close to 1nm.

It is expected that first order transitions would be observed experimentally if an adsorbent with precisely defined pore widths in a very narrow range were available. Otherwise isotherms with a sharp transition would only occur as part of the total adsorption isotherm, averaged over the pore size distribution. It should be noted that high resolution is necessary to detect the significant contributions to the global isotherm in pores of this width.

The slit pore model appears to be a reasonable idealisation of MgO pores in view of the crystal structure of MgO and the likely origin of the material. The possibility of similar micropore transitions at critical pore widths in different geometries cannot be ruled out. Indeed the first transition in the nitrogen isotherm in silicalite[19]is an example of a similar, though less dramatic phenomenon.

It seems clear that methods for the characterisation of pore size distribution, using nitrogen adsorption at 77K in microporous materials, need to take account of the specific phenomena that can occur in very small pores.

Acknowledgement We thank EPSRC and ICI Ltd for their support for this work.

References

1. S. J. Gregg and K. S. W. Sing, 'Adsorption Surface Area and Porosity', Academic Press, London, 1986.

2. J. P. R. B. Walton, and N. Quirke, Molecular Simulation, 1989, **2**, 361.

3. C. Lastoskie, K. E. Gubbins, and N. Quirke, *J.Phys.Chem.* 1993, **97**, 4786.

4 D. Nicholson, *J.Chem.Soc.Faraday Trans.*, 1994, **90**, 181; T. Suzuki, K. Kaneko, and K. E. Gubbins, *Carbon Conference Newcastle*, 1996, preprint.

5 J. Talbot, D. J. Tildesley, and W. A. Steele, *Mol.Phys.*, 1984, **51**, 1331.

6. B. Kuchta, and R. D. Etters, *Phys.Rev.B*, 1987, **36**, 3400.

7. K. Kaneko, R. F. Cracknell, and D. Nicholson, *Langmuir*, 1994, **10**, 4606.

8. V. A. Bakaev, and W. A. Steele, *Langmuir*, 1992, **8**, 148.

9. M. M. L. Ribeiro-Carrot, P. J. M. Carrott, and M. M. Brotas de Carvalho, *J.Chem.Soc.Faraday Trans.* 1991, **87**, 185.

10 W. A. Steele, *J. de Phys.*, 1977, C4, 61.

11 A. Alavi, and I. R. McDonald, Mol.Phys. , 1990, **69**, 703.

12. W. A. Steele, ' Interaction of gases with solid surfaces', Pergamon, Oxford, 1974.

13. D. Nicholson and N. G. Parsonage, ' Computer Simulation and Statistical Mechanics of Adsorption', Academic Press, London, 1982.

14. R. J-M. Pellenq and D. Nicholson, *J.Phys.Chem.*, 1994, **98**, 13339.

15. N. A. Freeman and D. Nicholson, in preparation. N. A. Freeman, thesis, London University, 1993.

16 D. Nicholson, *Studies in Surface Sci. and Catalysis*, 1991, **62**, 11.

17 D. Nicholson and K. E. Gubbins, *J. Chem. Phys*, 1996, **104**, 8126.

18. A. Stubos, R. F. Cracknell, N. K. Kanellopoulos, D. Nicholson, and G. K. Papadopoulos, in preparation.

19 D. Douget, R. J-M. Pellenq, A. Boutin, A. Fuchs, D. Nicholson, *Mol. Simulation*, 1996, in press.

ADSORPTION AND THERMOGRAVIMETRIC CHARACTERIZATION OF SELECTED M41S SAMPLES

A. Sayari, C. Danumah, P. Liu and M. Jaroniec*

Department of Chemical Engineering and CERPIC
Universite Laval, Ste-Foy, QC, Canada G1K 7P4
*Department of Chemistry, Kent State University
Kent, Ohio 44242, U.S.A.

1 INTRODUCTION

Periodic mesoporous M41S silicates and related solids have a promising future as catalysts, adsorbents and hosts for a variety of advanced materials.[1,2] Their key advantages rest on their high flexibility in terms of synthesis conditions, textural and structural characteristics as well as framework compositions. Indeed, they are easy to synthesize in different structures, and their pore size may be adjusted within a wide range via different synthesis strategies.

In addition, the framework composition may be pure silica, metal cation modified silica, nonsiliceous oxides or sulfides. Since the discovery of these materials,[3] adsorption methods played a key role in their characterization.[4,5] We recently carried out adsorption studies on a series of MCM-41 samples with different pore sizes[6] and on a series of samples with different titanium contents.[7] Both low and high pressure adsorption behaviors were investigated. The present work supplements work published by Schmidt et al.[8] and deals with a comparative adsorption studies of a hexagonal MCM-41 and a cubic MCM-48 silicate samples.

2 EXPERIMENTAL

2.1 Synthesis

The MCM-41 sample was prepared as reported earlier.[9] A mixture of 5 g of Cab-O-Sil M5 silica and 32.2 g of water was stirred vigorously for 10 min, before adding a solution of 16.2 g of cetyltrimethyl ammonium bromide (CTAB) in 108.8 g of water. After an additional 10 min of stirring a solution consisting of 14.36 g of tetramethylammonium silicate (10 wt % SiO_2) and 6.82 g of sodium silicate (28 wt % SiO_2, 10 wt % Na_2O). This mixture was aged for 30 min under stirring, then transferred into a Teflon-lined autoclave and heated under autogenous pressure at 373 K for 24 h with no stirring.

The MCM-48 sample was prepared as described by Lujano et al.[10] Solution A (7.84 g of tetraethyl orthosilicate and 1.4 g of water) was added slowly to a solution B (0.76 g of NaOH and 14 g of water). After stirring for two minutes, a third solution C containing 9 g of CTAB and 26.6 g of water was added. After stirring for 5 min, a clear solution was obtained. It was transferred into an autoclave and heated at 373 K for 3 days.

The obtained solids were filtered, washed with deionized water, and dried in static air at 373 K. MCM-41 was calcined as follows. The sample was first heated under flowing dry nitrogen to 773 K at a rate of 1 K min^{-1} and kept for 2h. Then, N_2 was slowly switched to dry air for an additional period of 4 h at 773 K. A similar procedure was used for the calcination of MCM-48, except that the temperature was 813K.

2.2 Measurements

X-ray powder diffraction spectra were recorded on a Philips PW1010 and Siemens D-5000 X-ray diffractometers using nickel CuKα radiation (see Figure 1). The weight loss as a function of temperature was measured in a nitrogen atmosphere using a TA Instruments, Inc. (New Castle, DE,USA) model TGA 2950 high resolution thermogravimetric analyzer. Nitrogen adsorption-desorption isotherms were measured at 77.5 K for relative pressures from about10^{-6} to 0.995 using an ASAP 2010 volumetric adsorption apparatus from Micromeritics (Norcross, GA, USA). Prior to making adsorption measurements the samples were degassed under vacuum (about 10^{-4} Torr) at 423 K for two hours in the degas port of the adsorption analyzer.

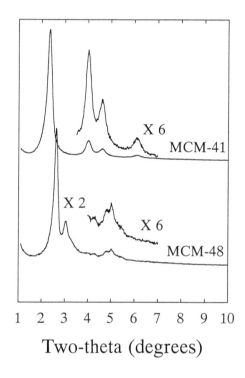

Figure 1 *A comparison of the X-ray diffraction spectra for the MCM-41 and MCM-48 samples*

3 RESULTS AND DISCUSSION

Shown in Figure 1 are X-ray diffraction patterns for both synthesized M41S samples. These patterns are analogous to those previously obtained for well ordered hexagonal and cubic structures of M41S silicates.[2,8] A comparison of nitrogen adsorption isotherms measured on both samples at 77.5 K over the entire relative pressure range is shown in Figures 2 (normal pressure scale) and 3 (logarithmic pressure scale). Both M41S samples exhibit a well pronounced step on their isotherms arising from the condensation of nitrogen inside the primary mesopores. As reported previously[4,11,12] this step is reversible for both samples and there is no evidence of hysteresis. There is a noticeable difference in the relative pressure corresponding to the condensation in primary mesopores of both samples. For the MCM-41 sample this relative condensation pressure is equal to 0.34, and is higher than the corresponding value for MCM-48, which is about 0.25. Almost the same values of the relative condensation pressure for the M41S silicates of hexagonal and cubic structures were reported by Schmidt et al.[8] This difference in the relative condensation pressures of both samples indicates that the average width of primary mesopores in MCM-48 is about 15% smaller than the corresponding value for the MCM-41 sample.[8] Another difference between the M41S samples studied can be noted at the higher range of relative pressures. In contrast to the MCM-48 sample, which does not possess a well developed secondary mesoporous structure (as evidenced by the lack of hysteresis loop in the range of high relative pressures), the adsorption isotherm for MCM-41 exhibits a small hysteresis loop in this pressure range indicating on the existence of noticeable amount of secondary mesopores. However, in the range of low pressures both isotherms are similar (see Figure 3).

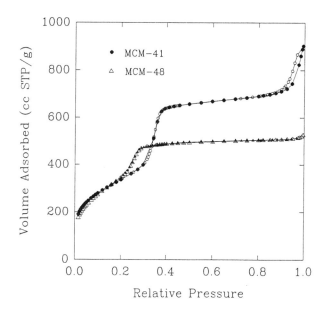

Figure 2 *A comparison of nitrogen adsorption isotherms measured at 77.5 K on the MCM-41 and MCM-48 samples*

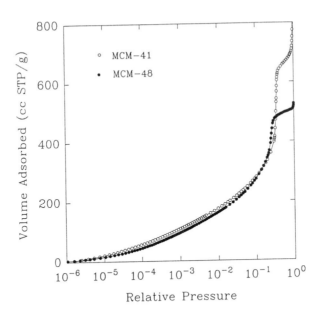

Figure 3 *Nitrogen adsorption isotherms for the MCM-41 and MCM-48 samples presented in a semi-logarithmic scale*

A comparison of the specific surface areas and pore volumes for the samples studied is given in Table 1. The total specific surface area was evaluated from the linear segment of the BET plot, whereas the total pore volume was estimated by converting the amount adsorbed at the relative pressure = 0.995 to the volume of liquid adsorbate. Microporosity of the M41S samples studied was evaluated using the t-plot[13] for pressures below the condensation pressure in primary mesopores. As can be seen from Figure 4 there is no evidence for existence of micropores. The observed deviation at the low values of the adsorbed layer thickness is caused by inadequate description of the low pressure standard adsorption using an empirical Harkins-Jura equation. The linear segment of the t-plot following the condensation in primary mesopores was used for an approximate estimation of the volume of primary mesopores and the external surface area of secondary mesopores. Table 1 contains the total quantities as well as the corresponding quantities for primary and secondary mesopores. The quantities for secondary mesopores were estimated as the difference between the total quantity and the quantity for primary mesopores.

As can be seen from Table 1 the MCM-48 sample studied had mostly primary mesopores. The mesopore size distribution calculated from the desorption isotherm for this sample using the BJH method[14] does not show second peak (see Figure 5). However, a second peak is visible for the MCM-41 sample. Both distribution curves show a sharp peak reflecting the existence of primary mesopores, which corresponds to the condensation step visible on the isotherm curves (see Figure 2).

Table 1 *Adsorption parameters for the M41S samples studied*

Parameter	MCM-41	MCM-48
BET specific surface area (m²/g)	1240	1300
Surface area of primary mesopores (m²/g)	1070	1230
Surface area of secondary mesopores (m²/g)	170	70
Total pore volume (cc/g)	1.15	0.79
Volume of primary mesopores (cc/g)	0.91	0.72
Volume of secondary mesopores (cc/g)	0.24	0.07

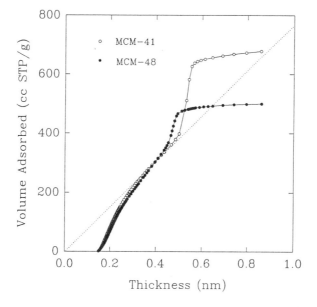

Figure 4 *A comparison of the t-plots for nitrogen adsorption on the MCM-41 and MCM-48 samples at 77.5 K.*

The low pressure parts of the adsorption isotherms shown in Figure 3 were utilized to obtain the adsorption energy distributions (see Figure 6) by inverting the adsorption integral equation.[15] The local adsorption isotherm was represented by the Fowler-Guggenheim equation, which describes localized monolayer adsorption with lateral interactions. Numerical calculations were performed for the patchwise topography of adsorption sites on the solid surface by assuming the lateral interaction energy parameter $z\omega/k_B T$ = 190 K and the regularization parameter γ = 0.01. The INTEG program was used to calculate the adsorption energy distributions for both samples.[16] It should be noted that the resulting distribution functions are numerically stable over a wide range of the regularization parameter. As can be seen from Figure 6 the adsorption energy distributions for the MCM-41 and MCM-48 samples are similar. Both distributions possess a sharp peak located at about 7 kJ/mol. Small changes in the shape of the energy distributions are visible for the adsorption energies greater than 9

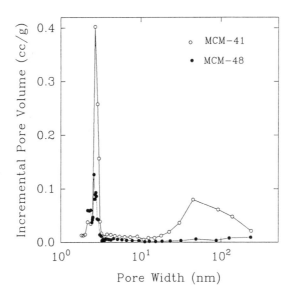

Figure 5 *A comparison of the BJH pore size distributions obtained from nitrogen desorption isotherms for the MCM-41 and MCM-48 samples*

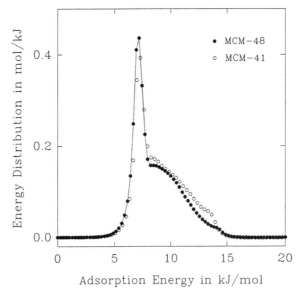

Figure 6 *A comparison of the adsorption energy distributions calculated from submonolayer nitrogen adsorption data for the MCM-41 and MCM-48 samples*

kJ/mol. In the energy range between 9 and 14 kJ/mol the distribution curve for MCM-41 lies slightly above the curve for the MCM-48 sample, whereas for the adsorption energies greater than 14 kJ/mol the curve for MCM-48 is on the top. However, nitrogen is rather a poor molecule to probe the difference in the energetic heterogeneity of the M41S silicates studied since both samples possess the same chemical nature. It appears that thermogravimetric curves shown in Figure 7 provide much more information about surface properties of both M41S samples. As can be seen from this figure the weight losses of these samples at 373K are different. The initial weight loss of MCM-48, which is associated with thermodesorption of physically adsorbed water,[17] is almost twice greater that the corresponding value for MCM-41. In addition, the MCM-48 sample shows a noticeable weight loss at 573K, which can be related to the decomposition of surface silanols.[17,18] The observed differences in the thermogravimetric curves for both samples indicate that MCM-48 is more hydrophilic and possesses a greater amount of surface silanols than MCM-41. A similar conclusion was reported by Schmidt et al.[8] on the basis of ^{29}Si MAS NMR studies.

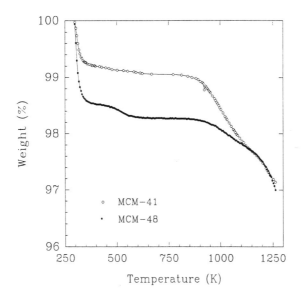

Figure 7 *A comparison of the thermogravimetric curves for the MCM-41 and MCM-48 samples*

4 CONCLUSIONS

Comparative adsorption studies of well ordered mesoporous silicates of hexagonal (MCM-41) and cubic (MCM-48) structures revealed that their energetic heterogeneity evaluated from low pressure nitrogen adsorption isotherms is rather similar. However, their structural properties are different. The average width of primary mesopores for the MCM-41 sample is about 15%

greater than that for MCM-48. In addition, the MCM-41 sample studied possessed about 20% of secondary mesopores, whereas their presence in the corresponding MCM-48 material was much lower. In addition, thermodesorption of physically adsorbed water studied by high resolution thermogravimetry revealed that MCM-48 is more hydrophilic material and possessed a greater amount of surface silanols than the corresponding MCM-41 sample.

References

1. A. Sayari, Stud. Surf. Sci. Catal., 1996, **102**, 1.
2. A. Sayari, Chem. Mater., 1996, **8**, 1840.
3. C.T. Kresge, M.E. Leonowicz, W.J. Roth, J.C. Vartuli and J.C. Beck, Nature, 1992, **359**, 710.
4. P.J. Branton, P.G. Hall and K.S.W. Sing, J. Chem. Soc., Chem. Commun., 1993, **1993**, 1257.
5. P.I. Ravikovitch, S.C.O. Domhnaill, A.V. Neimark, F. Schuth and K.K. Unger, Langmuir, 1995, **11**, 4765 and references therein.
6. M. Kruk, M. Jaroniec and A. Sayari, J. Phys. Chem., 1996, submitted.
7. M. Kruk, M. Jaroniec and A. Sayari, Micrporous Mater., 1996, submitted.
8. R. Schmidt, M. Stocker and O.H. Ellestad, Stud. Surf. Sci. Catal., 1995, **97**, 149.
9. A. Sayari, C. Danumah and I.L. Moudrakovski, Chem. Mater., 1995, **7**, 813.
10. J. Lujano, Y. Romero and J. Carrazza, Stud. Surf. Sci. Catal., 1995, **97**, 489.
11. P.J. Barnton, P.G. Hall, K.S.W. Sing, H. Reichert, F. Schuth and K. Unger, J. Chem. Soc. Faraday Trans. 1994, **90**, 2965.
12. P.L. Llewellyn, Y. Grillet, F. Schuth, H. Reichert and K.K. Unger, Microporous Mater. 1994, **3**, 345.
13. S.J. Gregg and K.S.W. Sing, 'Adsorption, Surface Area and Porosity', Academic Press, London, 1982.
14. E.P. Barrett, L.G. Joyner and P.P. Halenda, J. Am. Chem. Soc. 1951, **73**, 373.
15. M. Jaroniec and R. Madey, 'Physical Adsorption on Heterogeneous Solids', Elsevier, Amsterdam, 1988.
16. M. v.Szombathely, P. Brauer and M. Jaroniec, J. Comput. Chem. 1992, **13**, 17.
17. R.P.W. Scott, 'Silica Gel and Bonded Phases', Wiley, Chichester, 1993.
18. E.F. Vansant, P. van der Voort and K.C. Vrancken, 'Characterization and Chemical Modification of the Silica Surface', Elsevier, Amsterdam, 1995.

ANALYSIS OF PORE STRUCTURE OF ISOTROPIC GRAPHITE USING IMAGE PROCESSING

K. Oshida and T. Nakazawa

Department of Electronics
and Computer Science
Nagano National College
of Technology
Nagano 381 Japan

N. Ekinaga

Tokai Carbon
Co. Ltd.
Hofu Yamaguchi
747 Japan

M. Endo

Faculty of
Engineering
Shinshu
University
Nagano 380
Japan

M. Inagaki

Faculty of
Engineering
Hokkaido
University
Sapporo 060
Japan

1 INTRODUCTION

High density isotropic graphite, which is made from coke powder using cold isostatic press method, is used as electrodes for electric discharge processing, and in structural materials of nuclear reactors because of its arbitrary shaping, heatproof, lightweight, and durable nuclear characteristics. Recently, this material is used in various tools for semiconductor crystal growth. With its increasing applicability, this type of graphite needs to have higher purity and higher strength.

Assessment and analysis of the fracture dynamics parameters are important especially in fracture toughness, because destruction of carbon materials happen generally in the case of propagation of cracks which are latent in the carbon. Up to the present, the fracture toughness of the high density isotropic graphite has been studied using experimental methods,[1, 2] and it is suggested that the parameters of strength of the graphite are strongly correlated to pore structure of the graphite. The pore structure, however, has not been analyzed quantitatively, and the relation between the parameters of strength and the pore structure has not been explained clearly.

In this paper, the pore structure of the graphite has been analyzed by means of a polarizing microscope combined with a digitized image analyzer. Distribution of the area, the size, the number of pores, circularity, and fractal dimensions of cross-sections of pores in each samples are measured and analyzed quantitatively. The relation between these parameters of pore shape, and the parameters of strength, such as elastic modulus, bending strength, plane strain fracture toughness (K_{IC}), and critical crack opening displacement (COD_f), has been investigated.

2 MATERIALS FOR THE STUDY

Six kinds of commercially available isotropic graphite materials, which are G320, G330, G347, G520, G530 and G540 made by Tokai Carbon, have been used for this study. All the samples are made from the same kind of coke powder but of different particle sizes. Sample numbers correspond to the particle size. Figure 1 shows polarizing micrographs of cross-sections of the samples, which were observed on highly polished faces of the samples with the size of $20 \times 20 \times 20 mm^3$ under a magnification value of 200. The pores of each sample look dark in the picture and are different in size and shape.

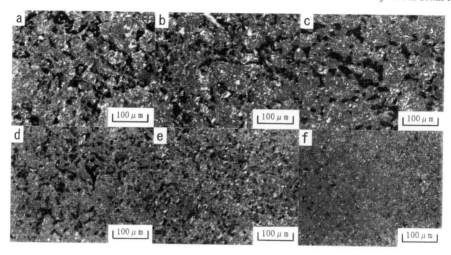

Figure 1 *Polarizing micrographs of cross-sections of the isotropic high-density graphite samples (a) G320, (b) G330, (c) G347, (d) G520, (e) G530 and (f) G540*

Characteristics of these parameters of the samples are summarized in Table 1. The sample G320 has the least bulk density value of 1.735g/cm³. All the samples except G320 are considered to be high density isotropic graphite. Elastic modulus and bending strength were measured using rectangular pieces of the sample cut from the bulk. In order to measure K_{IC} and COD_f, the notched rectangular pieces were used as samples. Elastic modulus and bending strength show a tendency to increase according to the order of sample numbers from G320 to G540. K_{IC} and COD_f values show different tendencies from elastic modulus and bending strength, that is, K_{IC} of G530 and COD_f of G347 display the highest values. In this paper, we

Table 1 *Properties of the isotropic high-density graphite samples*

Sample	Bulk density (g/cm³)	Elastic modulus (kg/mm²)	Bending strength (kg/cm²)	K_{IC} (MPa√m)	COD_f (mm)	Pore diameter* (μm)
G320	1.735	882	253	0.63	0.064	2.7
G330	1.788	1026	397	0.80	0.067	2.3
G347	1.842	1123	522	0.99	0.072	3.0
G520	1.842	1122	480	0.96	0.071	2.0
G530	1.848	1232	709	1.04	0.068	1.6
G540	1.802	1302	914	0.87	0.046	0.6

* Peak value measured by mercury porosimetry.

study the destruction mechanism of the material by using image analysis of the micrographs.

3 APPLICATION OF IMAGE ANALYSIS TO THE GRAPHITE SAMPLES

3.1 Digitization of the Micrographs

An image analysis system was used to analyze the isotropic graphite. The hardware system consists of a workstation, a personal computer, a color display, an image scanner and a color image printer. The micrograph is digitized using the image scanner driven by the personal computer, and the digital image is transferred from the personal computer to the workstation for analysis. The data consist of an array of 512×512 pixels. Each pixel has 8 bits depth (256 gray scale). One pixel size is $3.18 \times 3.18 \mu m^2$ when the magnification of the microscope is 50, and $0.802 \times 0.802 \mu m^2$ when the magnification is 200. In this paper, the digitized micrographs are called original images. The original image of G330 is shown in Figure 2(a). The pores are look bright in Figure 2(b), which is the negative image of (a). Since shapes of the pores appear clearly and brightness of the pores is different from the other solid parts, it is easy to obtain a binary image of the pore from Figure 2(b) by determining the adequate threshold level of brightness. Shapes and distributions of the pores were measured from the binary image as shown in Figure 2(c).

3.2 Frequency Analysis

A frequency analysis of the original image provides an efficient means to analyze the orientation of structure of the isotropic graphite. A two-dimensional fast Fourier transform (FFT) was carried out on the original images. Power spectra were calculated, and the results are shown in Figure 3, which indicates the spatial frequency distribution.[3, 4] The central area of the power spectrum corresponds to the low frequency and the central point indicates the direct current component, which is the brightness of the original image. The spatial frequencies increase in the outer area from the center. Since the power spectra were obtained from the original images, they show the changing contrast of the image corresponding to the cross-section structure of the sample. All of the power spectra in Figure 3 indicate concentric configuration. The micrographs show random cross-section images of the samples, due to the fact that the samples were cut in arbitrary direction. The observation of Figure 3 suggests

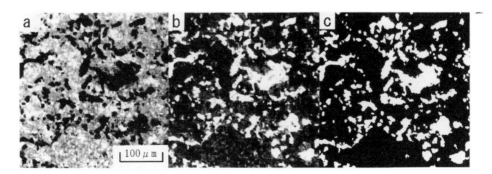

Figure 2 *(a) Digitized image of the sample G330, (b) inverse image of (a), and (c) binary image obtained from (b)*

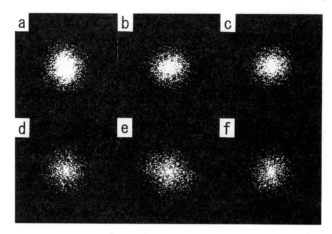

Figure 3 *Power spectra of digitized image of the sample (a) G320, (b) G330, (c) G347, (d) G520, (e) G530 and (f) G540*

that the structure of the samples are isotropic in three-dimensional space and consist of many frequency components.

3.3 Measurement of the Pore Structure

Distribution of the area size and the number of the cross-section of the pore of each sample were measured from the binary images in order to obtain parameters of the pore distribution. In addition, porosity which is the ratio of pore area to the total cross-section area was measured. The micrographs with magnification values of 50 were used in the measurement in order to get average data of a large area. Before measuring the distribution of the area size and the number of the cross-section of the pore, the pores which lie on the frame of the image were removed to avoid unexpected effect on the measurement.

Circularity (C_r) and fractal dimension (D) were used as parameters of the shape of the pores.[5, 6] The micrographs of magnification value of 200 were used for calculation of C_r and D in order to obtain detailed information of the pore shape. C_r is given by

$$C_r = \frac{4\pi S}{L^2} \tag{1}$$

where S is the area of the cross-section of the pore, and L is the perimeter of the cross-section of the pore. C_r equals 1 for a circle-shaped pore. When the cross-section shape of the pore becomes distorted, C_r becomes lower than 1.

D is calculated from S and L using the equation of

$$D = 2\frac{\Delta \log L}{\Delta \log S}. \tag{2}$$

D shows the complexity of the periphery of the cross-section of the pore.[3] If the periphery is straight, D is 1. When the periphery becomes more complex, D increases from 1 to 2.

4 RESULTS AND DISCUSSION

4.1 Characteristics of the Pore Structure

Figures 4(a) and (b) show the distribution of pore area in G330 and G530, respectively. There are a large number of pores with sizes over $800\mu m^2$ in G330, while on the other hand, 80% of the pores in G530 are under $200\mu m^2$.

Circularity (C_r) for G330 which is obtained by equation (1) is shown in Figure 5 (a). G in Figure (a) is the gradient of a line obtained by the least square method. R is the correlation coefficient which indicates linearity of the plots. C_r decreases with the increase of area size

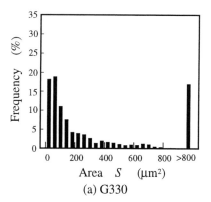

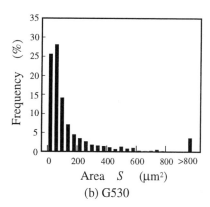

(a) G330   (b) G530

Figure 4 Histograms of pore frequency vs. pore area S

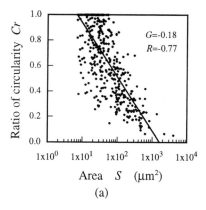

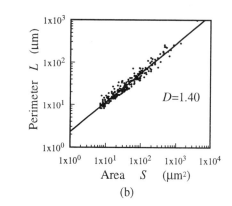

(a)   (b)

Figure 5 Plot of (a) the ratio of circularity vs. the area of the pore, and (b) the perimeter vs. the area of the pore for G330

Table 2 Characteristics values of the pore structure obtained from the image analysis

Sample	Number	Porosity	Average area	Average ratio of circularity	Fractal dimension
	N (1/mm^2)	(%)	$\overline{S}$ (μm^2)	$\overline{C_r}$	D
G320	414	21.9	545	0.65	1.40
G330	446	25.3	506	0.69	1.40
G347	480	21.1	395	0.66	1.33
G520	822	21.7	244	0.68	1.46
G530	1275	20.9	155	0.71	1.47
G540	2731	12.0	31	0.79	1.56

of cross-section of the pore (S). As the pore size increases, it is likely that the pore shape will become more distorted.

A log-log plot of the perimeter of the cross-section of the pore (L) versus S for G330 is shown in Figure 5 (b). The line in the figure is obtained by the least square method. There is a high correlation between L and S. D in the figure is the fractal dimension which is calculated from the gradient of the line by equation (2).

All the measured parameters of the pore structure are summarized in Table 2. N, $\overline{S}$ and $\overline{C_r}$ are the number of the cross-sections of the pores, the average area and the average circularity of these cross-sections, respectively. N increases and $\overline{S}$ decreases in the order of sample numbers from G320 to G540. It is thought that the differences of the sizes of the coke powder affected these parameters.

4.2 Relationship between the structure of the pores and the parameters of strength

The relationships between N and (a) elastic modulus, (b) bending strength, (c) K$_{IC}$, and (d) COD$_f$ are shown in Figures 6 (a)-6 (d), respectively. Elastic modulus and bending strength increase with the increase of N and the decrease of $\overline{S}$, and the material becomes hard and solid. There are peaks of K$_{IC}$ and COD$_f$ with the increase of N. The relation between mechanical properties and characteristics of the pore structure obtained from the analysis are summarized in Table 3.

In the sample with a large $\overline{S}$ and a small N, there is a large space in the pore and the pore shape is distorted. Since this sample has a fragile structure, K$_{IC}$ and COD$_f$ have low values. We propose a destruction process model of the sample. In the sample whose N is large and $\overline{S}$ is small the hitting probability of a crack by external stress is high and the average extended length of the crack is short. If the crack hits a pore, the extension of the crack is stopped temporarily at the pore. A new crack occurs at the weakest part of the internal face of the pore and extends again by continuous external stress. Therefore, the values of K$_{IC}$ and COD$_f$ are high. If $\overline{S}$ becomes smaller, the pores are so small that they can not inhibit the crack expansion, and K$_{IC}$ and COD$_f$ decrease. The shape of the pore with a high C_r is not easy to destroy.[7] Since there are two shape factors which determine whether the crack extends or not in the same sample, the correlation between K$_{IC}$ and $\overline{C_r}$ and the correlation between K$_{IC}$ and D are not clear at this time. In order to advance the analysis of pore shape, detailed

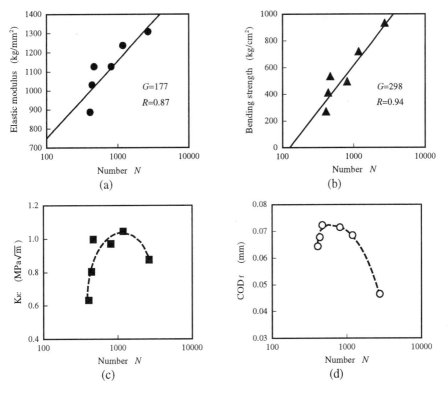

Figure 6 Relation between number N of the pore and (a) elastic modulus, (b) bending strength, (c) K_{IC}, (d) COD_f

observation using a electron microscope is currently underway.

5 CONCLUSION

The polarizing microscopic images of the isotropic graphite are analyzed using image processing. The power spectra of the digitized images are obtained by the two-dimensional FFT. The structure of the pores and the other solid parts of the graphite are almost isotropic. Distribution of the area size, the number of the pores, fractal dimension, and circularity of cross-section of the pores are measured and are analyzed quantitatively. Elastic modulus and bending strength increase with the increase of the number of small pores. A structure which includes many pores increases in K_{IC} and COD_f, because crack extension can be inhibited by the pores. This work shows that image processing is useful for analysis of micrographs to study pore structure.

ACKNOWLEDGEMENTS

The authors are indebted to Mr. H. Shin of Tokai Carbon for useful discussions. The authors

Table 3 Relations between mechanical properties and characteristics of the pore structure obtained by the image analysis

Mechanical properties	Elastic modulus (kg/mm^2)	Bending strength (kg/cm^2)	K_{IC} (MPa$\sqrt{m}$)	COD$_f$ (mm)
Number	△	△	Peak	Peak
N (1/mm^2)	(R=0.87)	(R=0.94)	(N=1000)	(N=1000)
Average area	▼	▼	Peak	Peak
$\overline{S}$ (μm^2)	($R = -0.94$)	($R = -0.94$)	(S=250μm^2)	(S=350μm^2)
Average ratio of	△	△	×	▼
circularity $\overline{C_r}$	(R=0.81)	(R=0.91)	(R=0.24)	($R = -0.83$)
Fractal dimension	△	△	×	▼
D	(R=0.65)	(R=0.73)	(R=0.11)	($R = -0.77$)

△: Plus correlation, ▼: minus correlation, and ×: no correlation.
Peak: There is a peak of the correlation. R: Efficient of correlation.

acknowledge Prof. T. Oku of Ibaraki University for useful discussions and suggestions. The authors (KO) and (TN) wish to thank Mr. K. Ikeda of Nagano National College of Technology (NNCT) for helpful discussions and suggestions. Part of the work (KO) was supported by the special research and education program fund of NNCT.

References

1. P. Marshall and E. K. Priddle, *Carbon*, 1973, **11**, 541.
2. S. Sato, H. Awaji and H. Akuzawa, *Carbon*, 1978, **16**, 95.
3. K. Oshida, K. Kogiso, K. Matsubayashi, K. Takeuchi, S. Kobayashi, M . Endo, M. S. Dresselhaus and G. Dresselhaus, *J. Mater. Res.*, 1995, **10**, 2507.
4. M. Endo, K. Oshida, K. Kobori, K. Takeuchi, K. Takahashi and M. S. Dresselhaus, *J. Mater. Res.*, 1995, **10**, 1461.
5. K. Oshida, N. Ekinaga, M. Endo and M. Inagaki, *TANSO*, 1996, [No. 173], 142 (in Japanese).
6. B. B. Mandelbrot, 'The Fractal Geometry of Nature', W. H. Freeman and Company, New York, 1982, p. 14.
7. N. Warren, *J. Geophys. Res.*, 1973, **78**, 352.

ADSORPTION STUDIES OF CARBON DIOXIDE ON KF1500, X2MH6/8 AND PX21 BY MEANS OF A HIGH PRESSURE VOLUMETRIC DEVICE.

A. Guillot, S. Follin, L. Poujardieu

CNRS-IMP, Institut de science et génie des Matériaux et Procédés
Université
66860 Perpignan Cedex France

1 INTRODUCTION

The aim of our Lab is to develop heat or refrigerating process mainly based on chemical reactions. In order to investigate adsorption refrigerating process possibilities we needed adsorption isotherms in a pressure and temperature ranges close that of the process; with this aim we set up a high pressure volumetric device. Using the Dubinin methods it has been possible to calculate the performance of a basic adsorption cycle, taking the characteristics of the adsorbate and the adsorbent into account (1). We present in this study the results of high pressure measurements on 3 activated carbons and their exploitation by the DR methods.

2 EXPERIMENTAL

Three samples of activated carbons, representative of various degrees of activation have been studied :
The X2MH6/8 provided by Takeda Chem. Ind. Ltd is a molecular sieve carbon
The KF1500 provided by Toyobo Co is a cellulose-based ACF.
The Maxsorb 30-SP provided by Kansai Coke & Chemical Co Ltd which is named usually PX21 results from a chemical activation..

3 APPARATUS AND PROCEDURE

CO_2 adsorption isotherms up to 20 Bar have been managed from 253K to 353K using a high volumetric device described in an article to be published. With this apparatus we can reach the relative pressure of one at 253K. The data accuracy remains smaller than 4% all over the experimental range. Complementary measurements up to 1 Bar from 224K to 323K have been done using a thermobalance coupled to a calorimeter TG-DSC111 Setaram.
Prior to their use all sample were outgassed at 200°C for 12 hour under secondary vaccum set by a thermomolecular pump.

4 RESULTS AND DISCUSSION

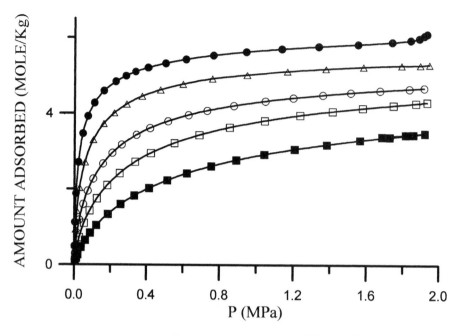

Figure 1 *X2MH 6/8 ISOTHERMS* ●253K △273K ○298K □323K ■353K

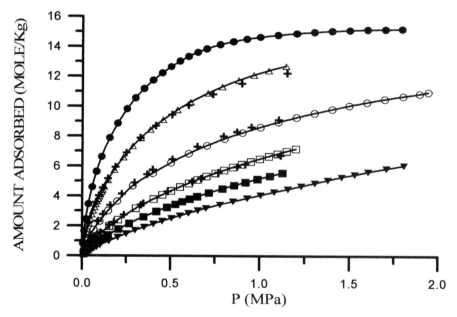

Figure 2 *KF 1500 ISOTHERMS* ● 253K △273K ○298K □323K ■343K ▼363K
  + TOKUNAGA

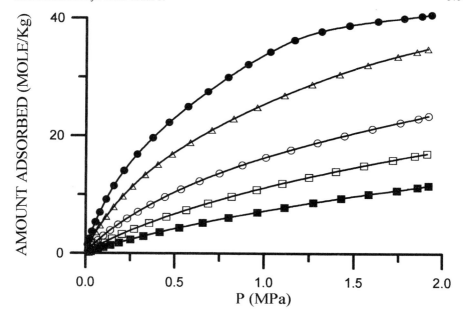

Figure 3 *MAXSORB ISOTHERMS* ● 258K △ 273K ○ 298K □ 323K ■ 353K

The experimental results are presented on figures 1, 2 and 3. Those of Tokunaga *et al.* (2) obtained on KF1500 (at 273K, 298K and 323 K up to 12 Bar). are represented on figure 2 for comparison.

The classical relation of Dubinin-Radushkevich (3) has been applied to our results, following the expression :

$$W = W_0 \exp \left[(-A / \beta) \right]^2$$

W represents the volume filled at temperature T and pressure P, $A = RTLn (P_S / P)$ is the Polanyi potential, P_S being the saturated vapour pressure, β is the affinity coefficient of the adsorbate (for the CO_2 we assume $\beta=0,35$).

A plot of Ln (W) vs. A^2 leads to parameters W_0 and E_0. It has been shown (4) that E_0 (given in Kj/mole), called characteristic energy of the micropore system is related to the average micropore size L by the relation :

$$L (nm) = 10.8 / (E_0-11.4) \quad (2).$$

This relation is valid for L< 1.8-2 nm. In order to estimate the microporous adsorbable volume W one has to estimate the density of the adsorbed phase for each temperature. Among the different expressions proposed in the litterature Ozawa's one (5) was choosen.

$$\rho = \frac{\rho_e}{\exp[\alpha(T - T_e)]}$$

$\alpha = 2,5 \ 10^{-3}$
T_e Triple point Temperature

As the saturated vapor is not defined above the critical temperature T_C we have used above 304K the Dubinin approximation :

$$P_S = P_C \left(\frac{T}{T_C} \right)^2$$

Considering the non ideality of CO_2, fugacities were used instead of pressures. . The figures 4,5,6 show the D.R. plots for the three adsorbants.

4.1 X2MH6/8 MSC

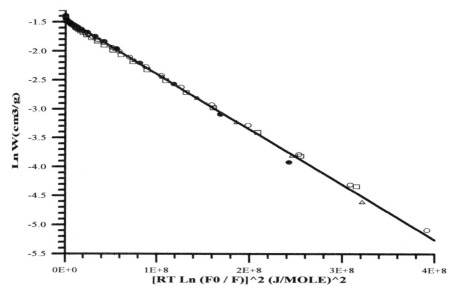

Figure 4 *X2MH 6/8 DR Plot* ● 253K △273K ◯298K ☐323K

In the case of X2MH6/8 the DR plot is linear, the same was observed on a CMS carbon by F.Stoeckli *et al.* (6) from low and high pressure experiments with CO_2. Our results leads to a micropore volume of 0,236 cm^3 / g and to a high value of characteristic energy E_o of 26.7 Kj/mole, from wich the calculated width is 0.6 nm.

4.2 KF1500

For the KF1500 we delimit 4 linear regions on the DR plot. The inflection points appear near 0.005, 0.033, 0.18 of P/P_S at 273K.

The deduced values of E_O and L are reported on Table 1. K. Kakei *et al.*(7) have characterised the same material using N_2 adsorption measurements at 77K, They delimit, on the DR plot, 3 linear domains from which the deduced values of E_o and W_o reported in Table I are very close to our results for the domains II, III and IV; they cannot observe the domain I due to the restricted diffusion of the N_2 molecules at 77K.

Taking account of the X2MH6/8 results the domain I could correspond to narrower micropores.

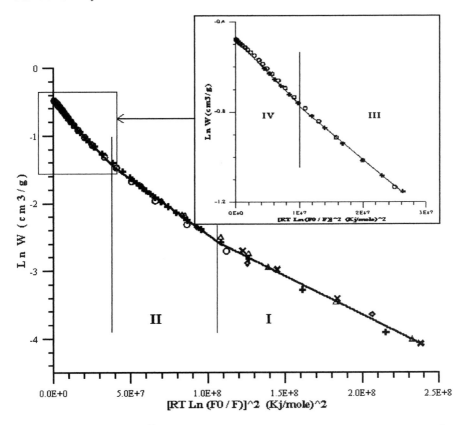

Figure 5 *KF 1500 DR Plot* ⭕253K ➕273K : high pressure
◆274K △299K ✖326K : thermogravimetry

Table 1 *Linear regions of KF1500 DR plot*

Domain	Range of Pr at 273K	W_o (cm^3/g)	E_o $(Kj/mole)$	L (nm)
I	0.005	0.26	26.7	0.7
II *Kakei*	0.005-0.033	0.44 *0.48*	22.3 *21.4*	1.0
III *Kakei*	0.033-0.180	0.58 *0.55*	18.6 *17.4*	1.5
IV *Kakei*	0.180-0.92	0.62 *0.62*	16.8 *16.1*	

Since the critical dimensions of molecules of N_2 and CO_2 are very similar, the multistage micropore filling mechanism proposed by K. Kakei *et al.*(7) accounts for our results. ACF should arise from defaults of graphitic layers upon activation; the micropore sizes should be integral multiples of *ca.* 0.35nm of the thickness of the graphite layer. The micropore sizes calculated for domains I, II and III could account for the filling of respectively the bilayer-,trilayer-and four-layer-sized micropores. For the high relative pressures range, the last domain IV, in view of the very low value of E_0 could correspond to wider micropore filling or surface adsorption on external surface.

4.3 Maxsorb-30

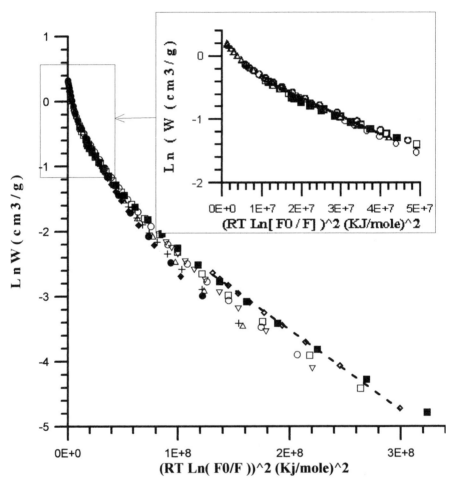

Figure 6 *MAXSORB DR Plot* ● 258K △273K ○298K □323K ■353K : high pressure
◆249K +273K ▽298K ◇323K : Thermogravimetry

Unlike the others the DR plot of the Maxsorb-30 differs completely. We can note a lack of linearity as it is observed for strongly activated carbons. Morever we can observe a dispersion of the data for the great potential range, they seem to be temperature dependant. For this domain the high temperature data leads to a high value of characteristic energy (26 Kj/mole) associated to narrower micropores but these ones normally disappear upon strong activation. As suggested by F. Stoeckli *et al.* (8), in this case, the initial part of the DR plot corresponds to the adsorption on the walls of relatively large micropores before the filling in volume. As the Maxsorb-30 is chemically activated, the number of functional surface groups could be high and this reinforces this interpretation. On the low potential domain the fit of the DR plot leads to a low value of E_o of 16.8 Kj/mole close to that find on the domain IV of the KF1500 and so related to wider micropore filling or surface coating.

So for the Maxsorb, we can't conclude anything about microporosity from the DR treatment, since the major part of adsorption could involve surface mechanisms.

5 CONCLUSIONS

This study shows that the DR analysis of experimental data above critical temperature is not problematic provided that using correct relations for estimation of the vapour saturated pressures and the density of the adsorbed phase.

Due to a lack of high pressure measurements, CO_2 has been usually used to complete N_2 characterisation; using our experimental device the CO_2 and N_2 microporosity characterisations lead to comparable results for the same range of potential. However, CO_2 characterisation gives more complete informations since it covers a larger domain of relative pressures.

We show that the DR analysis of experimental data above critical temperature is not problematic provided that using correct relations for estimation of the vapour saturated pressures and the density of the adsorbed phase.

Lastly the problems encountered with the characterisation of strongly activated carbons is not specific to CO_2 measurements; as mentionned by F. Stoeckli *et al.* (8) the use of additional calorimetry measurements should provide better pictures of the microporous structure. At least CO_2 measurements give much information with regard to adsorption surface than N_2 in view of the low relative pressure range investigated.

Morever if surface adsorption prevails over micropore filling the use of Polanyi potential function should not be the good answer.

Acknowledgment
This work was part of a CNRS-ECOTECH contract N° 94 N80/0086 with the support of the ADEME.

References

1.   S. Follin, V. Goetz and A. Guillot, Ind. and Eng. Chem. Res., 1996, **35**, 2632.
2.   N. Tokunaga, M. Abe, T. Nitta, T. Katamaya, J. Chem. Eng. Jap., 1988,**21**, 316.
3.   D.C. Bansal, J.B. Donnet, H.F. Stoeckli, 'Active Carbon', M. Dekker, New York, 1988.
4.   F. Stoeckli, P.Rebstein, L. Ballerini, Carbon, 1990, **28**, 907.
5.   S. Ozawa, S. Kusimi, Y. Ogino, J. of Coll. Interface Sci., 1976, **56**, 83.
6.   F. Stoeckli, D. Huguenin, A. Greppi, T. Jabukov, A. Pribylov, S. Kalashnikov, A. Fomkin, A. Pulin, N. Regent, V. Serpinski, 1993, Chimia, **47**, 213.
7.   K. Kakei, S. Ozeki, T. Suzuki, K. Kaneko, J. Chem. Soc. Faraday Trans. 1990, **86**, 371.

8.   F. Stoeckli, D. Huguenin and A. Greppi, 1993, J. Chem. Soc. Faraday Trans. 1993, **89**, 2055.

Comparison between different techniques to determine porosity of Aerogels

J.F. QUINSON[1], B. CHEVALIER[2], E. ELALOUI[3], R. BEGAG[3]

[1] INSA de Lyon, Laboratoire GEMPPM, (UMR CNRS 5510)
20, Avenue A. Einstein, F-69621 Villeurbanne Cedex

[2] Centre Scientifique et Technique du Bâtiment, Service Materiaux
24, rue J. Fourier, F-38400 Saint Martin D'Heres

[3] Université Claude Bernard LYON 1, Laboratoire LACE
43, Boulevard du 11 Novembre 1918, F-69622 Villeurbanne Cedex

1 INTRODUCTION

Transparent monolithic silica aerogels produced by sol-gel process and dried under supercritical conditions are easy to manufacture and have found increasing interest as a transparent double window insulation on the laboratory scale[1-3]. For building applications, one of the most important characteristic is the vision through the window. Generally some aerogels are more or less transparent due to diffusion in the material. This diffusion is the Rayleigh scattering because the size of the particles is similar with the wavelength of the light. So it was interesting to measure the size of the constitutive elements and the porosity of aerogel as function of the type and nature of the precursor used.

Silica monoliths are obtained by high temperature supercritical drying[4].The samples were produced at UCB Lyon by the team of Professor G.M. Pajonk. The following gives a short review of the fabrication of our samples. We have used polyethoxydisiloxane (PEDS), as precursor synthesised by PCAS (Produits Chimiques et Auxiliare de Synthèse), and described by the following formula[5]:

$$OR \quad OR$$
$$| \quad |$$
$$--- (--- Si --- Si ---)_{\overline{n}} ---$$
$$| \quad |$$
$$OR \quad OR$$

where R can be C_2H_5 or H.

The precursors used for fabrication are P600 and P750. These new prepolymerized precursors have the advantage of partial hydrolysis and therefore they are very prone to condensation reactions provided they are acid- or base- catalysed as description by Pajonk and al. [5].

Precursors P600 and P750 have the same silica weight ratio as tetraethoxysilane (TEOS) (28.8%). Those PEDS-P_x are obtained following the reaction:

$$Si(OC_2H_5)_4 + nH_2O \rightarrow PEDS\text{-}P_x + C_2H_5OH$$

where n is the number of moles of water added (n<2) per mole of TEOS.

The index x, is defined as $x = n/2 \times 1000$. For example, if n=1.5, then P900 is obtained. All the samples tested in porosity measurement are obtained with 40% of precursor in solvent ethanol.

A paper from R. Pirard[6] shows that aerogels submitted to a pressure of mercury are compacted and not intruded by mercury. So for this reason, we use only thermoporometry and nitrogen adsorption techniques to determine the textural characteristics of our aerogels.

2 EXPERIMENTS

2.1 Thermoporometry

Thermoporometry is a textural characterisation method which is well adapted to study the mesoporous structure of gels[7]. For thermoporometry, the filling of aerogels pores by the condensate liquid induces a capillary strain proportional to the surface tension between the liquid and the skeleton of the aerogel, and apt to destroy the fragile texture of the sample. In our case the condensate used for thermoporometry is water which is a liquid with a strong surface tension (73 dynes/cm). To avoid the texture of aerogel to collapse the following procedure has been retained :

a- The aerogel sample is kept in vaporous ethanol during 20 hours, and then filled with liquid ethanol. Ethanol has a low surface tension (24 dynes/cm).

b- The ethanol is slowly exchanged with water, by using ethanol-water mixture of increasing water concentration in order to raise slowly the surface tension until the sample is completely fill with pure water.

2.2 Nitrogen adsorption

Specific surface area and pore size distribution measurements were carried out by nitrogen adsorption/desorption on an ASAP 2000 from Micrometrics. Samples were outgased at 300 °C for a least 6 hours prior the measurements. A five-point BET analysis ($0.01 \leq P/P_0 \leq 0.29$), N_2 molecular cross sectional area of 0.162 nm²) was conducted to obtain surface areas. Pore size distributions were determined from desorption isotherm over a pore radii range from 1.5nm to 100 nm.

3 RESULTS

The variations of textural parameters measured by thermoporometry, as a function of the nature of precursor and catalyst, are recorded in table 1 for all aerogels prepared with 40% in volume of precursor.

Pore size distribution curves and cumulative porous volume curves of these samples are respectively shown in figures 1 and 2.

Sample	Vp (cm³/g)	R_{max} (nm)	S (m²/g)	Catalyse
P600 17	1.65	11.9	287	HF
P600 19	1.86	8.8	465	NH₄OH
P750 27	1.05	5.6	426	HF
P750 31	0.82	4.8	531	NH₄OH

Table 1 *Variation of the textural characteristics in function of the type of precursor and the nature of catalyse.*

Whatever the catalyst is, a shift of pore radii toward smaller values and a decrease of the mesoporous volume are observed for the precursor P750.

Moreover, it is worth noting that mesopores of the basic catalysed aerogels are smaller than those of the acid ones. This observation is not in agreement with literature data on silica Xerogels[8,9] and Aerogels[10,11] : actually under (NH₄OH) basic conditions, the gelation of silicon alkoxides is faster than (HCl, HNO₃) acid conditions, and the pore size distribution of silica gels shifts toward larger mesopores. The gelation time of aerogels prepared using HF as catalyst is very short (1 minute), which can explain this unusual textural evolution.

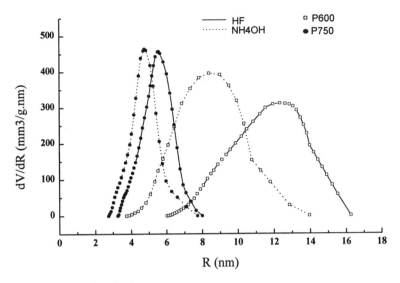

Figure 1 *Pore size distribution curves*

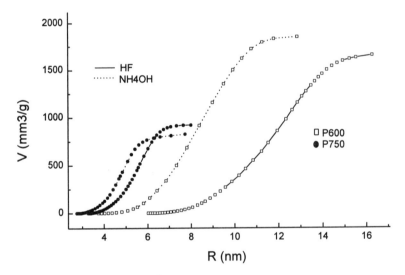

Figure 2 *Cumulative porous volume curves*

The nitrogen adsorption - desorption isotherms (figure 3) obtained for aerogels are of type IV according to the BDDT classification[12] with hysteresis loops of type E from the Boer classification.

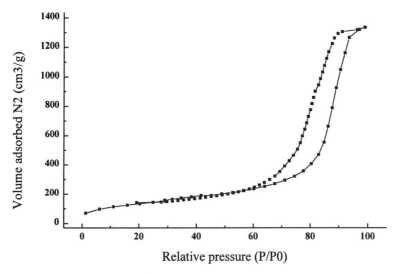

Figure 3 *Typical Nitrogen adsorption isotherm at 77 °K.*

At high relative pressures close to $P/P_0 = 1$, these isotherms reach a plateau. These different observations indicate that the aerogels contain no macropore and are mainly mesoporous.

The surface areas deduced from thermoporometry are compared with BET surface areas in table 2.

Sample	S BET (m²/g)	S Th (m²/g)	Catalyst
P600 17	291	287	HF
P600 19	491	465	NH₄OH
P750 27	529	426	HF
P750 31	727	531	NH₄OH

Table 2 *Comparison of surface areas deduced from water thermoporometry (S BET) with BET surface areas (S Th).*

For all aerogels, the BET surface areas are larger than the values determined by thermoporometry. Thermoporometry allows only to calculate the surface of mesopores. So, the differences noted between BET and thermoporometry values can be explained, in part, assuming that the aerogels contain some micropores.

Table 3 compares the pore radii R_{max}, taken at the maximum of mesopore distribution curves obtained by water thermoporometry and nitrogen desorption.

Sample	R_{max} N_2 (nm)	R_{max} H_2O (nm)	Catalyst
P600 17	10.2	11.9	HF
P600 19	6.4	8.8	NH_4OH
P750 27	4.6	5.6	HF
P750 31	4.5	4.8	NH_4OH

Table 3 *Comparison of the pore radii, R_{max}, measured by water thermoporometry and N_2 desorption*

A shift of the maximum radius, R_{max}, toward larger values of pores radii is noted for the pore size distributions deduced from thermoporometry.

These results are consistent with the fact that thermoporometry gives the actual size of the cavities while N_2 desorption measures the inlet size. They prove that the shape of pores in studied aerogels cannot be described in terms of cylinders.

4 CONCLUSION

In this study of the texture of silica aerogels, thermoporometry and N_2 adsorption - desorption appear to be complementary methods. The comparison of results deduced from these two techniques leads of better knowledge of the texture of aerogels : the aerogels appear as mainly mesoporous material. No macroporosity is detected and the existence of some micropores is assumed.

References

1.   M. Rubin and C.M. Lampert, *Solar Energy Material* 7 (1983), 393

2.   K.I.Jensen, J. *Non - Cryst. Solids* 145 (1992), 237

3.   V. Vittver, J. *Non - Cryst. Solids* 145 (1992), 233

4.   G.M.Pajonk, Sol-Gel Processind and Applications, ed.T.A. Attia (Plenum, New York, 1994) p. 209

5.   G.M.Pajonk, E. Elaloui,,M. Durant, J.L. Chevalier, B. Chevalier and P. Achard, in Sol-Gel Processing and Applications, ed. Y.A. Attia (Plenum, New York, 1994) p. 275

6.   R. Pirard, S. Blacher, F. Brouers, and J.P. Pirard, *J. Mater. Res.*, V. 10, N°. 8, Aug 1995, p. 2114

7   J.F. Quinson, M. Brun, in K.C. Unger, J. Rouquerol K.S.W. Sing and H. Kral (Eds), Studies in surface science and catakysis Series vol. 39,Proc. of the IUPAC Symp characterization of Porous Solids, Elsevier, Amsterdam, 1987, p 307

8.   L.C. Klein and G.J. Garvery, in : Better Ceramic Through Chemistry, MRS Meeting, eds. C.J. Brinker, D.E. Clark, and D.R. Ulrich Vol 32 (1984) p.33

9.   P.Yu, H.Tiu and Y. Wang, *J. Non - Cryst.* Solids 52, (1982) 511

10. D.W. Schaefer, C.S. Brinker, J.P. Wilcoxom, D.Q. Wu, J.C. Philips and B. Chu in : Better Ceramic Through Chemistry III, MRS Meeting, eds. C.J. Brinker, D.E. Clark, and D.R. Ulrich Vol 121 (1988) p. 691

11. M.Pauthe, J.F. Quinson, H. Hdach, T. Woignier, J. Phalippou, G.W. Scherer, J. *Non - Cryst. Solids*, 130 (1991) 1

12. A.J. Lecloux, in Catalysis Science and Technology, edited by J.R. Anderson and M. Boudart (Springer-Berlin, 1981), Vol. 2, p. 171

MULTISTEP POROSITY DEVELOPMENT IN CLINOCHLORE UPON HEATING

F. Villiéras, J. Yvon, M. François, G. Gérard and J. M. Cases

Laboratoire Environnement et Minéralurgie,
INPL and URA 235 CNRS,
Ecole Nationale Supérieure de Géologie,
BP 40, 54 501 Vandœuvre les Nancy, France.

ABSTRACT

Clinochlore is a magnesian trioactahedral chlorite. It is refered to a 2:2 clay mineral due to the regular stacking of 2:1 talc-like layers and Al-Mg hydrotalcite-like interlayer octahedral sheets. The structural hydroxyls of the octahedral sheets dehydroxylate around 500°C. This reaction generates a microporous maze between the 2:1 sheets. Accessibility of micropores to argon and water vapor is low due to the bad connectivity of the microporous network and the small width between 2:1 layers. The microporous network collapses around 750°C due to the dehydroxylation of 2:1 units. The subsequent recrystallization is observed from 800°C to 1100°C. This process generates mesopores due to ionic diffusion from some regions of the particles to other regions in which high temperature phases are growing.

1 INTRODUCTION

The dehydroxylation, and subsequent transformations, of hydrous minerals lead often to micropores and/or mesopores development. Such phenomena have been observed in the case of magnesium and aluminum hydroxides[1-3]. In the case of heated clay minerals, porosity development has never been described except in the case of clinochlore[4-6]. Clinochlore is a magnesium chlorite. It contains 2:1 layers which can be compared to talc. The charge defect due to Si by Al substitutions is compensated by a hydroxide interlayer sheet which can be compared to a Mg-Al hydrotalcite. Their dehydroxylation occurs around 750 and 500°C, respectively. Structural data and water vapor adsorption studies have suggested that the dehydroxylation of the octahedral sheet generates micropores due to the formation of Al_2O_3 and MgO clusters between 2:1 layers.

The aim of this article is to study more precisely the porosity of clinochlore heated between 500 and 1100°C. The microporous network generated between 500 and 700°C will be characterized using low pressure argon adsorption which is compared to water vapor adsorption results. Then, the products obtained after dehydroxylation of the 2:1 layers and the subsequent recrystallization will be studied using water vapor and nitrogen adsorption-desorption.

2 MATERIALS AND METHODS

The chlinochlore used in this study comes from the orebody of Trimouns (Pyrénées, France) and was provided by Talc de Luzenac S.A. This chlorite was formed during the hydrothermal transformation of micaschists and was referred to chlorite M in ref. 6. It contains about 80% clinochlore, 18% talc and traces of accessory minerals such as dolomite, apatite, zircon, illmenite, graphite[4,6].

30g of powder were calcined during 2 hours, in a furnace equilibrated at the desired temperature before sample introduction. Heating temperatures varied between 400 and 1100°C. Talc dehydroxylation, observed above 800°C, does not generate any textural changes[4] and its contribution to adsorption isotherms will remain roughly constant for all the studied temperatures.

Nitrogen adsorption-desorption isotherms were carried out at 77K on classical step by step all-glass volumetric equipment. Outgassing conditions were 200°C during 15 hours under a residual pressure of 0.1Pa. Surface areas were calculated using BET and t-plot treatments. The mesopore distribution was determined using the parallel pore model proposed by Dellon and Dellyes[7].

Low pressure argon adsorption isotherms were carried out at 77K on quasi-equilibrium high resolution volumetric equipment[4,8,9]. Outgassing conditions were 470°C during 15 hours under a residual pressure of 10^{-3}Pa. More than 2000 experimental data points are recorded up to P/Po=0.15. This allows to plot the derivatives of adsorption isotherms as a function of ln(P/Po) which were simulated using the Derivative Isotherm Summation (DIS) method[9,10].

Water vapor adsorption-desorption isotherms were carried out using the experimental apparatus described in ref. 5. Samples calcined up to 800°C were outgassed at 470°C during 15 hours and samples calcined from 850 to 1100°C were outgassed at 200°C during 15 hours under a residual pressure of 0.1Pa.

3 RESULTS AND DISCUSSION

2.1 The microporous network generated at 500°C

Structural studies of clinochlore heated between 500 and 700°C show that the basal spacing of the mineral decreases from 14.1 to 13.8Å[4-5]. Infrared spectroscopy shows the presence of molecular water in the heated chlorite particles which is confirmed by thermal analysis. Water vapor adsorption experiments have suggested that these water molecules are trapped into narrow micropores[5].

The presence of micropores is confirmed by argon adsorption experiments as the derivatives of adsorption vs ln(P/Po) feature drastic changes when dehydroxylation occurs (Figure 1). The argon derivative isotherms on the sample heated at 400°C is typical of adsorption on external faces of clay minerals[11] as it features two main peaks at ln(P/Po) around -4 and -7kT corresponding to the adsorption on basal and lateral faces, respectively. An additional peak (-11 kT) corresponds to adsorption on high energy sites[11]. External surfaces are also observed for samples calcined at 450 and 750°C. Between these two calcination temperatures, external surfaces are not observed anymore due to overlaps with peaks corresponding to adsorption in micropores. Adsorption isohterms on external surfaces and in micropores have been modelled using BET-Hill equation[9] and the generalized Dubinin-Asthakov equation[10] respectively. Using the DIS method, at least 4 microporous domains can be distinguished:

- The first one is a high energy domain centered around -12 kT. Adsorbed quantities are almost constant and negligible compared to the adsorbed quantities in the other adsorption domains.

- The second and third domains are centered around -10 kT and -8.5 kT, respectively. Adsorbed quantities depend on calcination temperatures (Figure 2).

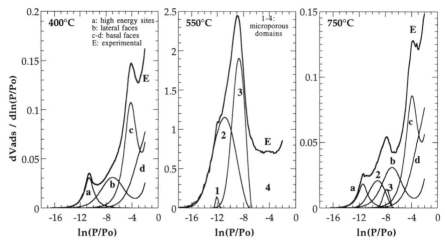

Figure 1 *Argon derivative adsorption isotherms an DIS decomposition*

- The gas adsorbed in domain 4, corresponds to a residual quantity adsorbed in the lowest energy part of the adsorption isotherm up to the BET monolayer pressure. It has been calculated by substracting to the BET monolayer capacity the adsorbed quantities in domain 1, 2 and 3 and on external surfaces.

The presence of different microporous domains indicates that the micropores are heterogeneously distributed. The position of the two main peaks (domain 2 an 3) remains roughly constant with calcination temperature, meaning that their sizes are not changing significantly. However, the geometric volumes (converted by assuming that argon is condensed in micropores in solid state, density 1.624[12]) decrease with temperature increase (Figure 2). The maximum volume is observed at 500°C, i.e. at dehydroxylation temperature. It decreases sharply between 550 and 650°C and is almost nil at 750°C, i.e. when dehydroxylation of 2:1 layers starts.

In the case of $Mg(OH)_2$[1], it has been established that all the microporous volume is not probed by nitrogen as brucite dehydroxylation generates a badly connected network between MgO clusters. It was also observed that the probed volume decreases with calcination temperature due to MgO rearrangement and connectivity decreases. The same

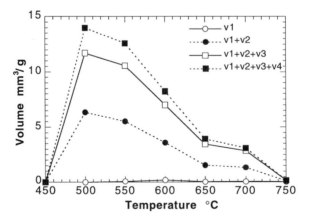

Figure 2 *Evolution of the geometric volumes calculated from solid argon*

situation is observed in the case of dehydroxylated chlorite as the measured microporous volumes decrease with temperature increases. Furthermore, the expected microporous volume of heated chlorite, derived from the theoretical contraction of brucite into MgO, would be around 50 mm³/g and is definitely higher than the maximal measured volume, i.e. 14 mm³/g. The microporous network then appears be badly connected.

Water vapor adsorption-desorption experiments[5] at 303K confirm these results. First, the results obtained by thermogravimetric analysis, infrared spectroscopy and water vapor adsorption have shown that 1) rehydroxylation of MgO into $Mg(OH)_2$ does not occur as observed by Riebeiro-Carrott et al.[2] as oxide clusters are stabilized by the 2:1 layer walls and 2) water molecules have to pass through very narrow micropores to diffuse into or out of the solid particles. Then, high outgassing temperatures (>300°C) must be applied to release the trapped water and high pressure is required to fill the micropores. The comparison of argon microporous volumes with ice volumes derived from water vapor adsorption experiments shows different behaviors depending on the calcination temperature (Figure 3).

The argon microporous volumes of domains 3 and 4 correspond roughly to the volume filled by water vapor at BET monolayer (after external surfaces correction) (Figure 3a). Then, it can be concluded that water vapor first adsorb in the widest micropores. Diffusion of water into the smallest micropores (domain 2) is obtained at higher pressure, up to saturation pressure. All the micropores remain filled when desorption at 303K is achieved. This final volume corresponds to the total argon microporous volume for the samples calcined at 500 and 550°C (Figure 3b).

At higher temperatures, the total volume probed by argon is lower than the total volume probed by water. Furthermore, for samples calcined at 650 and 700°C, Ar volume is close to water BET volume (Figure 3b). It can then be concluded that water can diffuse toward narrow micropores which can not be probed by argon. This phenomenon can be assigned to the fact that water at 303K can diffuse more easily than argon at 77K by exchange of molecules adsorbed around constrictions.

The generated network can then be considered as a microporous maze sandwiched between 2:1 layers. The theoretical width between 2:1 layers derived from basal spacing should be around 5Å. This size is low enough to prevent easy adsorption of argon or water. The observed micropores are probably the most external ones. The apparent size distribution suggested by low pressure argon adsorption is not fully explained at present

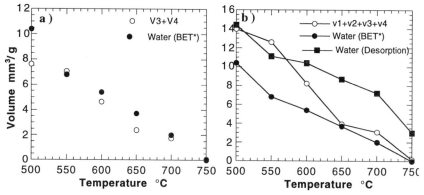

Figure 3 *Comparison between argon and water microporous volumes. a) Comparison between argon V3 + V4 volume and water vapor BET volume (* minus external surface contribution). b) Comparison between Argon total volume, BET and undesorbed water volumes*

time. The connectivity between these external micropores decreases above dehydroxylation temperature due to differences in MgO-AlO cluster arrangement.

2.2 The mesoporous network generated above 800°C

The dehydroxylation of magnesium hydroxide generates micropores which are converted to mesopores at higher temperature due to rearrangement and crystallization processes[1-2]. In the case of clinochlore, the microporous network disappears when dehydroxylation of the the 2:1 layers occurs. The obtained product is first structurally disordered and crystallized in high temperatures oxides, enstatite, forsterite and spinelle, above 800°C[4]. The crystallinity of these phases increases with calcination temperature.

Nitrogen adsorption-desorption isotherms obtained on the fired products show that the specific surface area increases from 2.2 to 5.0 m²/g when calcination temperature increases from 800 to 950°C and decreases to 3.5 m²/g for the sample heated at 1100°C (Table 1). t-plot curves reveal a constant microporosity around 0.12 cm³/g STP (0.5 m²/g). The increase in specific surface area can then be assigned to the formation of mesopores inside the reacting particles. This is confirmed by the development of desorption hysteresis (Figure 4). The shape of the desorption curves depends on calcination temperature. However, in all cases, a vertical branch is observed around P/Po = 0.52 meaning that the final capillary desorption of mesopores is controlled by constrictions. The constriction size is independent of calcination temperature. The height of the vertical branch increases from 800 to 850°C and remains constant up to 1100°C.

Water vapor adsorption-desorption isotherms confirm the mesoporous character of the fired products (Figure 5). As in the case of nitrogen, the microporous volume derived from t-plot[13] is independent of calcination temperature (0.18 cm³/g STP). The undesorbed quantity at zero relative pressure decreases between samples heated at 750 and 800°C and outgassed at 470°C. This indicates that the previously formed microporous network disappears. In the case of water vapor, the vertical branch is observed at P/Po = 0.37 and

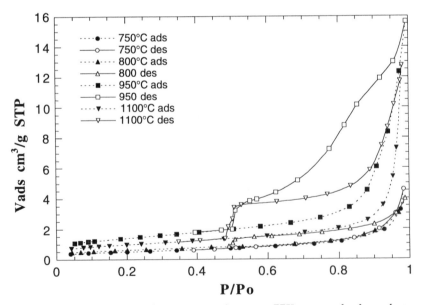

Figure 4 *Nitrogen adsorption-desorption isotherms at 77K on samples heated at and above 2:1 layers dehydroxylation temperature*

Table 1 *Results derived from BET treatment of water vapor and nitrogen adsorption isotherms on clinochlore calcined between 750 and 1100°C*

Calcination temperature °C	N_2 specific surface area m^2/g	water vapor BET monolayer volume cm^3/g STP	H_2O cross sectional area σ_w Å^2
750	2.2	0.6	16
800	2.2	0.6	16
850	3.8	1.3	11
900	4.4	1.3	13
950	5	1.4	14
1000	4.7	0.9	22
1100	3.5	0.6	28

its position remains constant. On the contrary to nitrogen the height of the vertical branch decreases when calcination temperature increases. Furthermore, the shape of water vapor desorption curves are different from nitrogen ones particularly for samples fired above 900°C.

To explain this phenomenon, water vapor and nitrogen quantities adsorbed at monolayer are compared after microporous volumes substraction. It then possible to derive water molecule cross sectional area, σ_w (Table 1). The first decreases of σ_w is not surprising as talc hydrophobicity should decrease due to its dehydroxylation which starts at 850°C[4]. The observed increase of σ_w above 850°C from 11 to 28Å^2 shows that the fired products become less hydrophilic due to high temperature phases crystallization.

The surface tension decrease of the fired products explains the differences between nitrogen and water vapor desorption isotherms. For a constant mesopore width, capillary

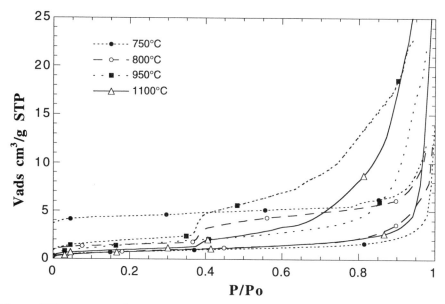

Figure 5 *Water vapor adsorption-desorption isotherms at 303K on samples heated at and above 2:1 layers dehydroxylation temperature*

desorption of water should occur at higher relative pressure for the samples calcined at the highest temperatures, which is consistent with experimental results.

Due to uncertainity on solid surface tension, water contact angle at solid-liquid-vapor interface and their evolution with temperature, mesopore size distribution can not be determined accurately from water vapor desorption isotherms. In the case of nitrogen, this effect should be negligible due to its low liquid-vapor surface tension ($8.85 mJ/m^2$ at 77K compared to $71.2 mJ/m^2$ for water at 303K). Mesopore size distribution were then derived from nitrogen desorption isotherms assuming the parallel model[7]. The calculated constriction width is 28 Å. Looking at the initial chlorite struture, this value corresponds roughly to two 2:2 units.

The variations in shape of the desorption branch with calcination temperature also indicate pore size variations. Mesopore distributions (Figure 6) confirm that additional pore widths can be observed. Their size increases with calcination temperature and the obtained values are again multiples of 13.5-14 Å: 42, 54 and 67 Å for the sample calcined at 900, 950 and 1000°C respectively. For the sample heated at 1100°C a broad peak is centered around 140 Å.

These observations can be interpreted in terms of atomic diffusion during crystallization processes. Indeed, the growth of enstatite, forsterite and spinel crystallites in selected parts of the particles is achieved by silicon, magnesium, aluminum and oxygen supply from other parts where the solid disapears, creating mesopores. The fact that pore sizes can be derived from structural properties of the chlorite itself indicates that the diffusion mechanism is still controlled by these initial crystallographic properties.

CONCLUSIONS

The two steps dehydroxylation of clinochlore generate specific porous networks. A schematic representation of the obtained textures are displayed Figure 7.

Due to the dehydroxylation of the interlayer octahedral sheets, a microporous maze is formed between 2:1 layers. Argon and water vapor adsorption is only observed in the

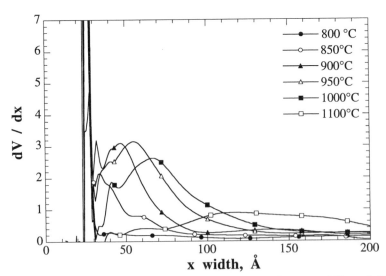

Figure 6 *Mesopore distribution of clinochlore calcined between 800 and 1100°C calculated assuming a parallel geometry of pores[7]*

external shell of the particles and in the case of water vapor, the filling of the narrowest micropores is controlled by diffusion.

The dehydroxylation of the 2:1 layers and the subseqent recrystallization in enstatite, forsterite and spinel generates a mesoporous network do to crystallites growth at the expense of vanishing zones. The surface of the crystallized products become hydrophobic at high temperature.

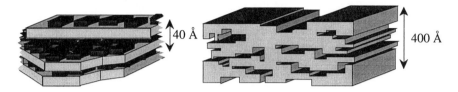

Figure 7 *Schematic representation of the microporous network formed in samples heated between 500 and 700°C (left) and the mesoporous network observed in samples fired above 750°C*

Acknowledgements
This research was supported by Talc de Luzenac S.A. and the European Economic Community: Program Raw materials, grant Ma 2M CT90 0036 DTEE.

References

1. H. Naono, *Colloids and Surfaces*, 1989, **37**, 55.
2. M.M.L. Riebeiro-Carrott, P.J.M. Carrott, M. M. Brotas de Carvalho and K. S. W. Sing, 'Characterization of porous solids II', F. Rodriguez-Reinoso, J. Rouquerol, K. S. W. Sing and K. K. Unger Eds., Elsevier, Amsterdam, 1991,p 635.
3. J. Rouquerol, F. Rouquerol and M. Ganteaume, *J. Catal.*, 1979, **57**, 220.
4. F. Villiéras, PhD Thesis, INPL, Nancy, 1993.
5. F. Villiéras, J. Yvon, M. François, J. M. Cases, F. Lhote and J.-P. Uriot, *Applied Clay Science*, 1993, **8**, 147.
6. F. Villiéras, J. Yvon, J. M. Cases, P. De Donato, F. Lhote and R. Baeza, *Clays and Clay Minerals*, 1994, **42**, 679.
7. J.F. Delon and R. Dellyes, *C.R. Acad. Sci.*,1967, **265**, 1661.
8. Y. Grillet, F. Rouquerol and J. Rouquerol, *J. Chim. Phys.*, 1977, **1974**, 179.
9. F. Villiéras, J. M.Cases, M. François, L. Michot and F. Thomas, *Langmuir*, 1992, **8**, 1789.
10. F. Villiéras, L. J. Michot, J. M. Cases, M. François and W. Rudzinski, *Langmuir*, 1996, in press.
11. F. Villiéras, L. J. Michot, J. M. Cases, I. Berend, F. Bardot, M. François, G. Gérard and J. Yvon, 'Equilibria and Dynamics of Gas Adsorption on Heterogeneous Solid Surfaces'. W. Rudzinski, W.A. Steele and G. Zgrablich Eds., Elsevier, Amsterdam, 1996, Chapter 11 in press.
12. Anonymous, 'Gas Encyclopedia', L'air Liquide, Amsterdam, Elsevier, 1976, p. 89.
13. J. Hagymassy, S. Brunauer and R. SH. Mikhail, *J. Colloid Interface sci.*, 1969, **29**, 485.

COHESIVE AND SWELLING BEHAVIOUR OF CHARGED INTERFACES: A (NVT) MONTE-CARLO STUDY.

Roland. J.-M. Pellenq, Alfred Delville and Henri van Damme

Centre de Recherche sur la Matière Divisée
CNRS and Université d'Orléans
45071 Orléans, cedex 02, FRANCE.

1 INTRODUCTION

This work is a first step towards the understanding of the cohesive behaviour of hydrated cement particles CSH (in the mineralogist notation C, S and H stand for "calcium oxide, "silicon oxide" and "hydrated" respectively). Such a material, although disordered, is made of calcium oxide layers covered with silicate chains (1,2). Therefore, it presents a structural resemblance with clay minerals such as the well-known Montmorillonite. The inter-layer distance in CSH is *ca* 14 Å and contains water molecules and calcium ions (2). CSH and Portlandite are the major products of the hydration reaction of the C3S particles (anhydrous cement particles) (3). The cohesive behaviour of CSH layers is of crucial importance in the understanding of the mechanical properties of cement gels since Portlandite (CaOH$_2$) is known to contribute very little to the hardness of cement materials. Some years ago, it has been stated that CSH are zeolite-type materials; two adjacent layers being chemically bound through Si-O-Ca-O-Si bridges across the inter-layer void space (4). However, there are also some experimental evidences showing that the inter-layer distance decreases upon heating from 14 Å to 11 Å at temperatures higher than 120°C (this corresponds to the loss of water molecules) (5). Interestingly, the 14 Å distance can be fully recovered when exposing an heated sample to water vapour at room temperature (6). The magnitude of the variation of the inter-layer distance (about 3 Å) and its reversibility are not compatible with the idea of chemically bound layers. Furthermore, a recent ^{29}Si NMR study of CSH (1) has shown that the number of SiO$_4$ units at the surface of the calcium oxide layers decreases as the calcium to silicon ratio (C/S) increases (the SiO$_4$ units being replaced by calcium ions). This further indicates that the cohesion of two adjacent layers in the real CSH material (high C/S ratio) cannot be attributed to chemical bonds since there are less silicon oxide tetrahedra available to build hypothetic bonds between layers. Therefore, it is more reasonable to consider the cohesion between two adjacent CSH layers as the result of electrostatic interactions which will only exist if those layers have a non-zero electric charge. This is consistent with Wulf's law for crystal growth which states that a layered structure is the result of an in-plane bonding process different in nature (chemical bonding) than that in the normal direction to the planes. Recent high quality *ab initio* calculations on a single Tobermorite sheet (7), a pseudo-crystalline model for pure CSH (2,8), (see Figure 1), have shown that bonding in the calcium oxide layer and in the silicate chains, is iono-covalent: the (partial) charges on the different species obtained from a Mulliken analysis are -0.9e, 1.8e, 1.5e, 0.2e for oxygen, silicon, <u>structural</u> calcium and hydrogen atoms respectively (*e* is the elementary charge). The amount of <u>inter-layer</u> calcium ions can vary and dictates the mechanical properties of the whole material. The calcium to silicon ratio ranges from 0.66 to 1.5. According to the structure of Tobermorite (8), a C/S ratio of 0.66 corresponds to a fully <u>neutral</u> protonated structure in which all oxygen atoms involved in dangling bonds of silicate chains are saturated with protons; in this case, there is no calcium ion in the inter-layer void space.

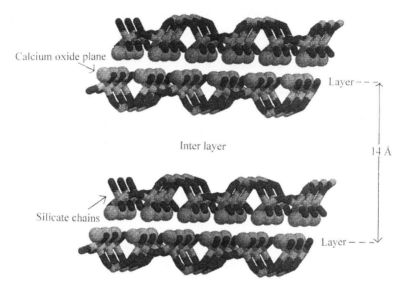

Calcium oxide plane

Layer - - -→

Inter layer

14 Å

Silicate chains

Layer - - -↓

Figure 1: *The layered structure of Tobermorite.*

Interestingly enough, such a material is known to have none of the mechanical properties that one expects for a cement gel (especially as far as hardness is concerned). Conversely, a C/S ratio ranging from 0.83 to unity corresponds to a deprotonated structure in Hamid's model for Tobermorite. In this case, the number of calcium ions in the inter-layer void space compensates for the charge on the layers. One can determine the surface density of charge in Tobermorite by dividing (minus) the number of vacant proton sites by the area of an elementary layer (found from the unit cell dimensions): we found σ_w in the range $\{-2.4, -4.8\}\ 10^{-2}\ e/\text{Å}^2$. It is interesting to note that the value of σ_w for Tobermorite is 3 to 7 times larger than that for a typical clay mineral such as Montmorillonite (9) for which $\sigma_w = -7.3\ 10^{-3}\ e/\text{Å}^2$. In this study, we consider $\sigma_w = -3.0\ 10^{-2}\ e/\text{Å}^2$ as an average value for Tobermorite.

2 THE MODEL SYSTEM

2.1 Calculation of the potential energy

The fact that both Montmorillonite and Tobermorite are layered materials leads us to consider these materials within the frame of the (anisotropic) Primitive Model (10,11). In this model, we assume a uniform electric charge density on planes σ_w; therefore the layers are considered as structure-less charged walls. Thus a given material is only characterized by its surface density of charge. It is important to mention that, to our knowledge, no theoretical or simulation investigations have been undertaken for such a high value of the surface charge density as that found for Tobermorite: the upper value found in the literature for such a model is $\sigma_w = -1.7 \; 10^{-2} \; e/\text{Å}^2$ (9). The solvent (in this case the water molecules) is taken as a dielectric continuum characterized by its dielectric constant at 298 K, $\varepsilon_r = 78.5$. The ions in the inter-layer void space are considered as charged hard spheres; their size being determined from their solvated radius (R_{sol}) *ie* 2.39 Å and 2.125 Å for calcium and sodium respectively. The simulations were run within the frame of the Canonical ensemble *ie* at fixed number of ions, volume and temperature (298 K). The generation of the Markov chain of configurations in Monte-Carlo simulations depends on the total interaction energy which is the sum of the ion-ion and the ion-wall interactions. Periodic boundary conditions and minimum image convention were applied in the x and y directions but not in the z direction perpendicular to the walls. The ion-ion potential for two kl ions separated by a distance r_{kl} is therefore written as:

$$U^{ii}(r_{kl}) \;=\; \frac{q^2}{4\pi\epsilon_0\epsilon_r r_{kl}} \qquad r_{kl} > d \tag{1}$$

$$U^{ii}(r_{kl}) \;=\; \infty \qquad r \leq d$$

where q is the ionic charge and d the (solvated) ionic diameter ($d=2R_{sol}$). The ion-wall potential is given by:

$$U^{iw}(\Delta z) \;=\; \frac{q\sigma_w W}{4\pi\epsilon_0\epsilon_r} \; F\!\left(\frac{\Delta z}{W}\right), \qquad z = \frac{\Delta z}{W} \tag{2}$$

with $F(Z) = G(Z) - 2\pi Z$ and

$$G(Z) \;=\; 4\log\!\left\{\frac{\sqrt{\frac{1}{2}+Z^2}+\frac{1}{2}}{\sqrt{\frac{1}{4}+Z^2}}\right\} - Z\left\{-\pi + 2\arcsin\frac{-(Z^2+\frac{1}{4})^2-\frac{1}{8}}{(Z^2+\frac{1}{4})^2}\right\} \tag{3}$$

Δz is the ion-wall distance. Relations (1) and (2) hold for all ions in the simulation box of volume $V=W^2L$ where W is the box size and L the separation between the walls (see Figure 2). Note that there is no cut-off distance in the calculation of the ion-ion interactions. W is directly related to the total number of the counter-ions inclosed in the inter-layer space N, their electric charge q, and the surface charge density σ_w through the following equation:

$$W \;=\; \sqrt{\frac{-Nq}{2\,\sigma_w}} \tag{4}$$

This equation states the electroneutrality of the system: the total ionic charge compensates exactly for the charge onto the walls. Due to the relatively small size of the simulation box compared to the characteristic length of coulombic interactions, it is necessary to include some corrections in order to calculate accurately the energy of the system. One way to

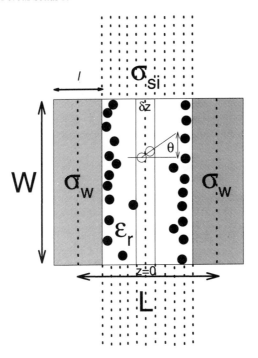

Figure 2: *Schematic picture of the model system with uniformly charged walls enclosing the neutralyzing counter-ions, the layer thickness l is 13Å (see text).*

achieve this, is to consider virtual charged sheets in between the two walls (see Figure 2). The surface charge density σ_{si} of those sheets is obtained in a self consistent manner from the ionic density distribution calculated in the simulation as a block average (12). The interaction between an ion in the simulation box and one of these virtual sheets is given by:

$$U^{is} = -\frac{q\,\sigma_{si}\,W}{4\pi\epsilon_0\epsilon_r}\,F(Z) \qquad (5)$$

The interaction of the walls with a virtual sheet is:

$$U^{ws} = -\frac{\sigma_w\sigma_{si}\,W^3}{4\pi\epsilon_0\epsilon_r}\,\{F(Z)\} \qquad (6)$$

In order to have the total energy of the system, two other terms need to be considered. The first is the self energy of the ions which can be written as:

$$U^{self-ion} = \frac{N\,q^2}{8\pi\epsilon_0\epsilon_r R_{solv}} \qquad (7)$$

The second is the self energy of the two walls given by:

$$U^{self-wall} = \frac{\sigma_W^2 \, W^3}{4 \pi \epsilon_0 \epsilon_r} \left\{ F\left(\frac{L}{W}\right) + F(0) \right\} \tag{8}$$

The two last terms are constant for a given number of ions and a given separation between the walls. They are not required in the Monte-Carlo procedure itself but are important in the study of the system stability when approaching the thermodynamic limit *ie* as W (or N) increases (see equation 4).

2.2 Calculation of the force on the walls

The force between the walls is directly related to the z-component of the pressure in the system. In Canonical Monte Carlo simulations, this force per unit area is the osmotic pressure. The z-component of the pressure can be written as the sum of three terms (12):

$$P = P_{elec} + P_{ideal} + P_{contact} \tag{9}$$

where P_{elec} is the electrostatic force per unit area acting across a fictitious plane at z, P_{ideal} $=\rho(z)kT$ is the kinetic contribution with $\rho(z)$ being the ionic density at z and $P_{contact}$ is proportional to the number of ions in contact (collisions) through the plane located at z. It has been shown that the pressure can be obtained most accurately by calculating each of its components at the midplane of the system (10). The electrostatic pressure can be obtained from the derivative with respect to the z variable of the interaction energy previously defined as the sum of the ion-ion, ion-wall and the wall-wall interactions including all long-range corrections. The contact pressure at the midplane is extrapolated from the ensemble average of $cos\theta$ of each pair of ions pertaining to different sides of the simulation box (with respect to the midplane) (10). Finally, P_{ideal} is obtained from the ionic density in a volume $W^2\delta z$ spanning the midplane (see Figure 2). Most of the simulations were run with 200 ions. Simulation with 1600 ions give essentially identical results as far as the total energy per ion, the pressure and the ionic profile are concerned. Figure 3 shows an excellent agreement between our result and those reported by Valleau *et al* (12) for $q=+2e$, $R_{sol}=3$ Å, $\epsilon_r=78.5$ and $\sigma_w=-1.4 \ 10^{-2} \ e/Å^2$. It is interesting to note that (i) the electrostatic contribution to the total pressure is always negative (ii) both the ideal gas and the contact terms are positive (iii) all these three terms are strongly distance dependent. The total pressure is therefore the result of a fine balance between these three contributions.

3 DISCUSSION AND CONCLUSION

Here after, we report some pressure calculations for different value of the surface charge density (characterizing a given material) as a function of the inter-layer distance. Since the simulations were run at constant volume, a positive pressure would correspond to a swelling system *ie* in a real experiment, the inter-layer distance would increase until the mechanical equilibrium is recovered. Conversely, a negative pressure indicates a cohesive system *ie* in a real experiment, the inter-layer separation would decrease to compensate for this negative pressure. Figure 4 shows the variation of the osmotic pressure with the wall

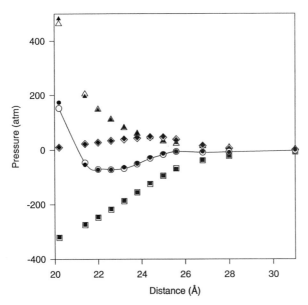

Figure 3: *Comparison between the results given in Figure 3 of reference (12) (black symbols) and our simulation data (white symbols); circle: total pressure, square: electrostatic pressure, triangle: ideal gas contribution and diamonds: contact term (see text).*

separation for the Ca^{2+} and Na^+/Montmorillonite systems respectively. It is clear that the Ca^{2+}/Montmorillonite system exhibits a cohesive behaviour for distances larger than 21.5 Å. This attraction is at relatively long range (at 35 Å, the pressure is still negative) and the minimum (-8.9 atm) is reached at 24 Å. The pressure becomes positive at 35 Å and goes through a maximum at around 40 Å before a continuous decrease. The overall behaviour is in agreement with experiment although pressure oscillations at short range (solvation forces), experimentally observed (13), are not reproduced in the simulation because of the continuum approach used to describe the solvent (11). It is also interesting to note the good agreement between the simulation results and the predictions of HNC theory (9,14). In the case of sodium ions in Montmorillonite, the pressure remains always positive indicating a swelling behaviour (9,15). Figure 5 shows pressure/separation curves for the Ca^{2+} and Na^+ in Tobermorite. Again the Na^+/Tobermorite system is continuously swelling while the Ca^{2+}/Tobermorite system exhibits a strong cohesive behaviour with a deep minimum (-580 atm) at 19.2 Å. One can estimate at -0.03 J/m^2 the cohesive energy between the two plates. This value falls in between that for a metal (-0.5 J/m^2) and that for a molecular cluster (-0.005 J/m^2). Thus this model, although simplistic is able to reproduce the observed cohesive behaviour of cement materials and the swelling behaviour of sodium/clays systems. Figure 6 presents a two dimensional plot of the pressure as a function of the surface charge density and of the plate separation for calcium ions. It is interesting to note the existence of a pressure minimum for a critical value of the surface charge density $\sigma_w^{crit}=-3 \ 10^{-2} \ e \ /Å^2$ at a distance of 19.5 Å. The analysis of the different contributions to the pressure indicates that (i) the electrostatic pressure (negative) becomes more attractive with increasing value of σ (ii) the ideal gas term (positive) gives the largest contribution to the repulsive part of the total pressure for $\sigma < \sigma_w^{crit}$; the contact pressure

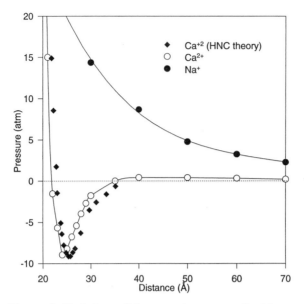

Figure 4: *Variations of the osmotic pressure for Montmorillonite.*

being negligible (iii) conversely for $\sigma > \sigma_w^{crit}$, the contact pressure starts to increase dramatically and diverges; it becomes at $\sigma_w = -4 \ 10^{-2} \ e/\text{Å}^2$, the main component of the repulsive part of the total pressure. The deep pressure minimum obtained at $\sigma_w = \sigma_w^{crit}$, progressively vanishes and becomes positive for $\sigma_w = -3.9 \ 10^{-2} \ e/\text{Å}^2$ for an inter-layer separation of 19.5 Å. We also note the existence of a secondary minimum for values of σ close to σ_w^{crit} (see also Figures 3 and 5).

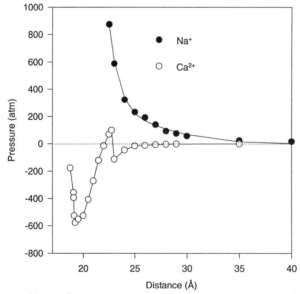

Figure 5: *Variations of the osmotic pressure for Tobermorite.*

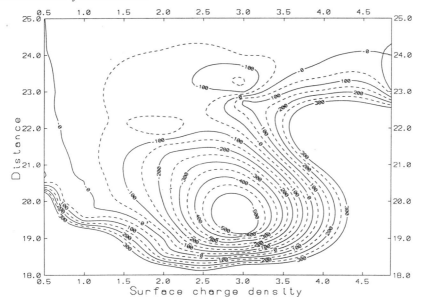

Figure 6: *The total pressure (atm) as a function of L (Å) and $10^2\sigma_w(e/Å^2)$.*

This is the consequence of the brutal vanishing of the contact pressure when the layer separation allows the formation of two distinguishable counter-ion layers. The absolute distance between the layers is then equal to two ionic diameters. The existence of a deep minimum at short distance thus corresponds to a situation in which it is not possible to define two electric double layers. A more detailed picture of the interaction between charged layers should include both the atomic roughness of the surfaces and a molecular description of the solvent in order to describe phenomena such as solvation forces (16,17).

References

1. Y. Klur, PhD Thesis, Université Pierre et Marie Curie, Paris, 1996.
2. H. F. W. Taylor, *Z. fur Kristall.*, 1992, **202**, 41.
3. D. Damidot, A. Nonat, P. Barret, *J. Am. Ceram. Soc.*, 1990, **73**, 3319.
4. S. Komarneni, S. Guggenheim, *Mineral. Mag.*, 1988, **52**, 371.
5. X. Cong, R. J. Kirkpatrick, *Cement and Concrete Research*, 1995, **25**, 1237.
6. C. Ménager, M. Crespin, R. J.-M. Pellenq, A. Delville, to be published.
7. F. Marinelli, R. J.-M. Pellenq, A. Delville, to be published.
8. S. A. Hamid, *Z. fur Kristall.*, 1981, **154**, 189.
9. R. Kjellander, S. Marcelja, J. P. Quirk, *J. Coll. Int. Sci.*, 1988, **126**, 194.98.
10. L. Gulbrand, B. Jönsson, H. Wennerström, P Linse, *J. Chem. Phys.*, 1984, **80**, 2221.
11. A. Delville, 'Hydration and Setting of Cements', A. Nonat, J. C. Mutin eds., Spon, 1992.
12. J. P. Valleau, R. Ivkov, G. M. Torrie, *J. Chem. Phys.*, 1991, **95**, 520.
13. R. Kjellander, S. Marcelja, R. M. Pashley, J. P. Quirk, *J. Phys. Chem.*, 1988, **92**, 6489.
14. R. Kjellander, S. Marcelja, *J. Phys. Chem.*, 1986, **90**, 1230.
15. A. Delville, P. Laszlo, *New J. Chem.*, 1989, **13**, 491.
16. A. Delville, *J. Phys. Chem.*, 1993, **97**, 9703.
17. J. N. Israelachvili, R. M. Pashley, *Nature*, 1983, **306**, 249.

CALCULATION OF ADSORPTION ISOTHERMS AND THERMODYNAMIC QUANTITES FOR ARGON ADSORBED IN CARBON SLIT-LIKE MICROPORES WITH DIFFERENT PORE WIDTHS USING THE DENSITY FUNCTIONAL THEORY.

N. Neugebauer, M. v. Szombathely, and P. Bræuer

Institute of Physical and Theoretical Chemistry
The University Leipzig, Germany

1 INTRODUCTION

There is a big demand for theoretical methods to describe the gas- and liquid-phase adsorption on solid surfaces. The density functional theory (DFT) approach for inhomogeneous systems is capable in an outstanding way to predict the adsorption both on planar surfaces and in micro-, meso- und macropores in good agreement with experimental results. In contrast to the classic computer simulation methods like e.g. the Molecular Dynamics (MD) oder Monte-Carlo-Simulation (MC) the DFT represents a statistic thermodynamic model with the applicability to interfacial systems of macroscopic range.

The basic rules for modelling in density functional theory consist of a description of the fluid structure influenced by an external field as good as possible, whereby attaching greatest importance on observation of the thermodynamic relations valid in inhomogeneous systems. Also, macroscopic characteristics should be possible to be calculated correctly as e.g. adsorption isotherms or phase transitions.

In this paper the DFT was used to calculate isotherms of argon adsorbed in slit-like micropores with different pore widths at several temperatures. So, only overlap and dispersion interactions appear and particle-particle as well as particle-wall interactions are independent on any orientation. For this situation the atomic DFT approaches are applicable in their weighted-density approximation (WDA) variants. The approaches by Tarazona [4], Meister & Kroll [5], Groot & Eerden [6], and Curtin & Ashcroft [7] differ in formalism and treatment of terms for the non-ideal (excess) contribution of the hard sphere reference to the free energy and in their consistency with exact statistic thermodynamic relations from the liquid theories. The very accurate and efficient approach by Curtin & Ashcroft was used for the calculations.

In this paper the adsorption isotherm has been calculated at five temperatures and six pore widths. For each point of the calculated isotherms an equilibrium density profile is yield giving supplemental informations to an better comprehension of the adsorption process and thermodynamics. In a characteristic way the adsorption isotherms support the separate stages of the adsorption process in dependence on the pore width and pressure as adsorption on the walls and then condensation inside the pores. The thermodynamic quantities, to be investigated, are accessible from the dependence of the isotherms on the temperature. In this paper the differential quantities were calculated from the adsorption isotherms and isosteres using common thermodynamic relations.

2 THEORY

On solid-fluid interfaces permanently site dependent potentials appear as the cause for an inhomogeneous structure of the fluids nearby the phase boundary. The external potential $v_{ext}(\mathbf{r})$ of the solid causes a spatial structure of the fluid density nearby the wall. A common basic idea of all the DFT models is the approach and mathematic treatment of the free energy as a functional of density. There exist several good proven approximations for the free energy functional $F[\rho]$, especially its excess part, which allows the calculation for a number of inhomogeneous systems. Within the iteration equations the excess free energy term is responsible for the inhomogenity of the fluid inherited by the external field, but the cause is the external field itself [1].

The slit-like pore model consists of two walls, facing parallel to each other with the distance $z = H$ and having an infinite extension in x- and y- direction.

With regard to conditions relevant for experiments, the volume V, the temperature T and the chemical potential μ are hold constant. The statistic of such a system is described by the grand canonical ensemble $\Xi(V, T, \mu)$. The corresponding grand potential functional Ω for an inhomogeneous fluid influenced by an external potential is

$$(2.1) \qquad \Omega_V[\rho] = F[\rho] + \int d\mathbf{r}\, \rho(\mathbf{r}) \left\{ v_{ext}(\mathbf{r}) - \mu \right\} .$$

μ is the chemical potential, the one particle density ρ is a function of site, $\rho(\mathbf{r})$ is called density profile. At thermodynamic equilibrium the grand potential is found in its minimum $(d\Omega(T, V, \mu) = 0)$, and the chemical potentials of adsorbed and bulk phase are equal.

$$(2.2) \qquad \left.\frac{\delta\Omega_V[\rho]}{\delta\rho(\mathbf{r})}\right|_{\rho=\rho_{eq.}} = 0 \quad ; \quad \mu^{bulk} = \mu .$$

$$(2.3) \qquad (2.1) \text{ and } (2.2) \text{ give} \qquad \mu^{bulk} = v_{ext}(\mathbf{r}) + \left.\frac{\delta F[\rho]}{\delta\rho(\mathbf{r})}\right|_{\rho(\mathbf{r})=\rho_{eq.}(\mathbf{r})}$$

as basis for functional differentiation. (2.3) expresses the overall constancy of the chemical potential through the adsorbed inhomogeneous fluid.

The pertubative theoretical decomposition of the pair interaction potential $\phi(r)$ into reference and pertubative part enables the decomposition of the free energy term $F[\rho]$ into a reference, as the rule the hard sphere, $F[\rho]_{ref} = F[\rho]_{hs}$, and into a pertubative, $F_A[\rho]$, contribution. The hard-sphere (reference) part is decomposed into an ideal, $F_{id}(\rho)$, and a nonideal (excess) part $F_{ex}[\rho]$,

$$(2.4) \qquad F[\rho] = \underbrace{F_{id}[\rho] + F_{ex}[\rho]}_{F_{hs}[\rho]} + F_A[\rho] \quad .$$

$$(2.5) \qquad F_{id}[\rho] = \int d\mathbf{r}\, \rho(\mathbf{r})\, \psi_{id}\big(\rho(\mathbf{r})\big) ,$$

$$(2.6) \qquad F_{ex}[\rho] = \begin{cases} \int d\mathbf{r}\, \rho(\mathbf{r})\, \psi_{ex}\big(\rho(\mathbf{r})\big) & \rightarrow \quad LDA \\ \int d\mathbf{r}\, \rho(\mathbf{r})\, \psi_{ex}\big(\bar{\rho}(\mathbf{r})\big) & \rightarrow \quad WDA \end{cases} ,$$

$$(2.7) \qquad F_A[\rho] = \frac{1}{2} \iint d\mathbf{r}_1\, d\mathbf{r}_2\, \rho(\mathbf{r}_1)\, \rho(\mathbf{r}_2)\, \phi_A(|\mathbf{r}_1 - \mathbf{r}_2|) .$$

In (2.7) the pair correlation function $g(r)[\rho]$ is approximatively set to 1. A DFT approach belongs to the local density approximation (LDA) methods, when the hard-sphere free excess energy per particle ψ_{ex} is modelized as exclusively dependent on the density at regarded site $(\psi_{ex}(\rho(\mathbf{r})))$. If ψ_{ex} goes into the modelization as a function of all densities found in a fixed range around the site $\mathbf{r}$, $\psi_{ex}[\rho](\mathbf{r})$ can be described as a function of a so-called weighted ($\bar{\rho}(\mathbf{r})$, Tarazona, Curtin & Ashcroft) or reference ($\rho_0(\mathbf{r})$, Meister & Kroll, Groot & Eerden) density on site $\mathbf{r}$, and it refers to a weighted density approximation (WDA). The nonlocality of $\psi_{ex}[\rho](\mathbf{r})$ by itself follows by the appropriate volume of a real particle.

The exact functional differentiation of the free energy term (2.4) gives the iteration equation (2.8), in which no model specific information is embodied yet,

$$(2.8) \qquad \rho_{eq.}(\mathbf{r}_1) = \frac{1}{\Lambda^3} \exp \left\{ \frac{1}{k_B T} \left[\mu^{bulk} - \frac{\delta F_{ex}[\rho]}{\delta \rho(\mathbf{r}_1)} \bigg|_{\rho = \rho_{eq.}} \right. \right.$$

$$\left. \left. - \int d\mathbf{r}_2 \, \rho_{eq.}(\mathbf{r}_2) \, \phi_A(r_{1,2}) + v_{ext}(\mathbf{r}_1) \right] \right\} .$$

Exclusively the formulation of the $\frac{\delta F_{ex}[\rho]}{\delta \rho(\mathbf{r}_1)}$ - term, which contains the variation of the excess part of the hard sphere free energy, is modified differently within the mentioned atomar DFT approaches. $\frac{\delta F_{ex}[\rho]}{\delta \rho(\mathbf{r})}$ is to be regard as the excess chemical potential.

2.1 The excess free energy functional $F_{ex}[\rho]$ (2.6)

The weighted density $\bar{\rho}(\mathbf{r})$ is determined as an average of the local density $\rho(\mathbf{r})$ over a fixed volume range with the use of a weight function w,

$$(2.9) \qquad \psi_{ex}[\rho](\mathbf{r}) = \psi_{ex}(\bar{\rho}(\mathbf{r})) \; ; \quad \bar{\rho}[\rho](\mathbf{r}) = \int d\mathbf{r}_2 \, \rho(\mathbf{r}_2) \, w(r_{1,2}, \bar{\rho}[\rho](\mathbf{r}_1)) ,$$

under the normalization $\int d\mathbf{r} \, w(r, \rho) = 1$. With the directive of best approximation the approach by Tarazona model 3 [4] and that by Curtin & Ashcroft [7], respectively, apply a weight function, which depends on the weighted density as functional of density itself. With (2.9) the general iteration equation (2.8) is able to be specified for WDA approaches to

$$\frac{\delta F_{ex}[\rho]}{\delta \rho(\mathbf{r}_1)} = \psi_{ex}(\bar{\rho}(\mathbf{r}_1)) + \int d\mathbf{r}_2 \, \rho(\mathbf{r}_2) \, \psi'_{ex}(\bar{\rho}(\mathbf{r}_2)) \frac{\delta \bar{\rho}(\mathbf{r}_2)}{\delta \rho(\mathbf{r}_1)} ,$$

and for the approach by Curtin & Ashcroft, using a density dependent weight function to

$$(2.10) \qquad \frac{\delta \bar{\rho}(\mathbf{r}_2)}{\delta \rho(\mathbf{r}_1)} = \frac{w(r_{2,1}, \bar{\rho}(\mathbf{r}_2))}{1 - \int d\mathbf{r}_3 \, \rho(\mathbf{r}_3) \, w(r_{3,2}, \bar{\rho}(\mathbf{r}_2))} .$$

2.2 The weight function by Curtin & Ashcroft [7]

From liquid theories an exact relation is known between the n-point direct correlation function $c_n(\mathbf{r}_1, ..., \mathbf{r}_n)[\rho]$ and the n-th variation of free excess energy with respect to the density, for $n = 2$ the renormalization condition is

$$(2.11) \qquad \frac{\delta^2 \beta F_{ex}[\rho]}{\delta \rho(\mathbf{r}_2) \, \delta \rho(\mathbf{r}_1)} = -\frac{\delta c_1[\rho](\mathbf{r}_1)}{\delta \rho(\mathbf{r}_2)} = -c_2[\rho](\mathbf{r}_1, \mathbf{r}_2) .$$

$w(r, \rho)$ can be calculated exactly in the scope of 2-point correlation. In the case of homogeneous fluid the renormalization equation by Curtin & Ashcroft [7],

$$-k_B T \, c(r_{1,2}, \rho) = 2 \, \psi'_{ex}(\rho) \, w(r_{1,2}, \rho)$$

$$+ \rho \, \psi''_{ex}(\rho) \int d\mathbf{r}_3 \, w(r_{3,1}, \rho) \, w(r_{3,2}, \rho)$$

$$(2.12) \qquad\qquad + 2 \, \rho \, \psi'_{ex}(\rho) \int d\mathbf{r}_3 \, w'(r_{3,1}, \rho) \, w(r_{3,2}, \rho) \, ,$$

is valid. Its solution can be calculated using *Fourier* transformation which leads to the inhomogeneous differential equation (2.13), see Fig. 1.

$$(2.13) \qquad \begin{aligned} c_k(\rho) = &- 2 \, \Delta\psi'(\rho) \, w_k(\rho) \\ &- \rho \, \Delta\psi''(\rho) \, w_k^2(\rho) \\ &- 2 \, \rho \, \Delta\psi'(\rho) \, w_k(\rho) \, w'_k(\rho) \, . \end{aligned}$$

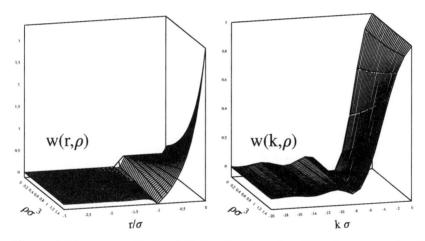

Figure 1 *The renormalized weight function in real and Fourier space using the analytic accessible expression $c(r_{12}, \rho)$ by Percus & Yevick from the Percus Yevick equation via Laplace transformation [2, 3].*

3 THERMODYNAMICS

3.1 Results

Fig. 2 shows the calculated equilibrium density profiles for some pressures choosen in each pore seperately. These profiles show clearly the formation of a first monolayer at small increasing pressure on the pore walls. If this layer is nearly filled, then the formation of the second layer starts, which is less strong structurized. After about 2 - 3 layers are formed, the pore is filled suddenly. This happens when condensation takes place, and the profiles convergate to the liquid density in direction of pore centre.

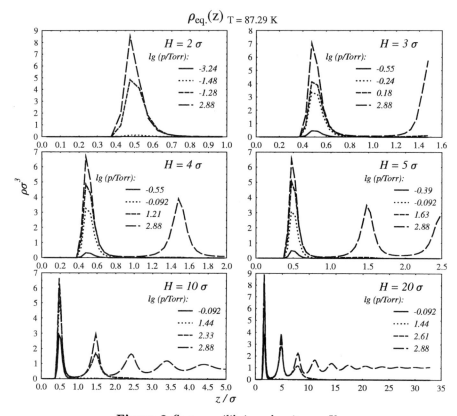

Figure 2 *Some equilibrium density profiles*

The bulk liquid density of argon $\rho^{liq} = 0.7513\,\sigma^{-3}$ at $T = 87.29\,K$, $p = 760\,Torr$. Because the bulk gas density under this conditions $\rho^{bulk} = 0.00308\,\sigma^{-3}$, the adsorption excess $\Gamma^{\sigma} \approx \Gamma^{s}$, where Γ^{s} is the absolute adsorption amount.

In Fig. 3 the process of the adsorption is observable excellently on the example of the $10\,\sigma$ pore. There three layers are formed with increasing pressure before the adsorbed phase condenses at the relative pressure $p/p_{sat} = 0.403$ or $\lg(p/Torr) = 2.49$, respectively.

The excess isotherms can be calculated by integration of the equilibrium density profiles

$$(3.14) \qquad \Gamma^{\sigma}(p) = \left.\frac{n^{\sigma}}{A}\right|_{p} = \left.\int_{0}^{\frac{H}{2}} dz\left(\rho(z) - \rho^{bulk}\right)\right|_{p},$$

$$(3.15) \qquad \hat{\Gamma}^{\sigma} = \frac{2\,\Gamma^{\sigma}}{H}, \qquad \left[\hat{\Gamma}^{\sigma}\right] = \frac{particle}{\text{Å}^{3}\,(pore\ volume)}.$$

The terracelike appearance of the isotherms at low pressures (Fig. 4) shows the layerwise adsorption. The step in the graph means the condensation of adsorbed phase inside the pore. The values of $\hat{\Gamma}^{\sigma}$ for $p/p_{sat} \to 1$ are a little greater than these of the

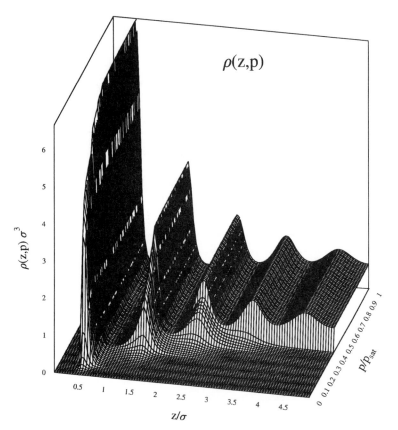

Figure 3 *The density as function of the distance z to the pore wall and of relative pressure as a suitable way to visualize the adsorption process inside the 10 σ pore.*

bulk liquid density caused by the wall potential within the pore. It is doubtful to interprete the formation of the second layer inside the $3\,\sigma$ and $4\,\sigma$ pore as *condensation* because only four argon atoms can be placed side by side into the $4\,\sigma$ pore, e.g.

3.2 The differential molar excess enthalphy $\Delta H_{diff}^{\sigma} = -Q_{st}^{\sigma}$

This quantity (Fig. 5), in general called the isosteric heat of adsorption, is calculated from the isosteres $\ln p = f(1/T)_{\hat{\Gamma}^{\sigma}}$ using (3.16)

$$(3.16) \qquad -Q_{st}^{\sigma} = \Delta H_{diff}^{\sigma} \equiv H_{diff}^{\sigma} - H^{bulk} = -Q_{st}^{\sigma} = RZ \left(\frac{\partial \ln p}{\partial 1/T}\right)_{\hat{\Gamma}^{\sigma}} .$$

The vaporisation enthalpy of Ar $\Delta_v H(87,29\,K) = 6.51\,kJ/mole$. In Fig. 5, first of all, the trend can be seen, that the value of Q_{st}^{σ} attains to the range of $\Delta_v H$ with inreasing pore width H, and the condensation of the adsorbed phase becomes more and more similar to that of the free bulk, because the contribution of the wall potential v_{ext} becomes smaller. In all six pores the first particles are adsorbed onto the walls directly,

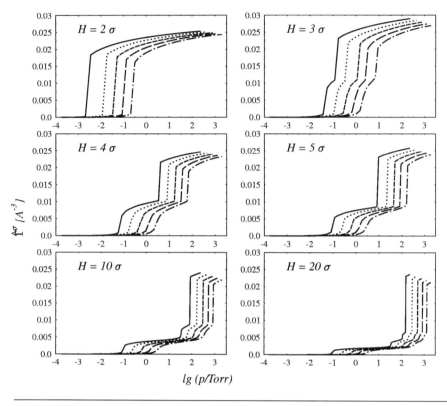

$$\widehat{f}^{\sigma} \; [A^{-3}]$$

lg (p/Torr)

—— $T = 77.29\ K$, ······ $T = 82.29\ K$, ––– $T = 87.29\ K$, ––· $T = 92.29\ K$, ·— $T = 97.29\ K$

Figure 4 *Adsorption isotherms for argon in slit-like graphite pores*

for what reason Q_{st}^{σ} is about 10 $kJ/mole$ for the lowest pressure independent on the pore width.

The $2\,\sigma$ pore represents a special case, because, conditioned by geometries, there only two layers are between the walls (Fig. 2) and therefore all adsorbed particles feel the same wall potential. Hence the suddenly increase up to the liquid density causes the step of Q_{st}^{σ} in Fig. 5, whose value is an additive effect of the particle-wall interaction $(v_{ext}(z \approx \sigma/2) = \text{const.})$ and the particle-particle interaction. The increase is about $\Delta_v H$. Inside the $3\,\sigma$ pore two different layers are formed, two layers on the walls and one central layer. The layers on the walls are exposed to an another wall potential than the layer in the pore middle.

From the $4\,\sigma$ pore upwards, the tendence is visible that Q_{st}^{σ} attains to $\Delta_v H$ at such pressures, where the pore centre is filled suddenly by the adsorptive, respectively. Looking at the values of Q_{st}^{σ} on this pressures, the answer can be given to the still open problem, when the pore is great enough, to speak about condensation in it. Fig. 5 allows the conlusion to speak about a bulk-like condensation in pores of the widths 10σ and greater. The $20\,\sigma$ pore gives a value for Q_{st}^{σ} at the condensation point, that agrees with the vaporisation enthalphy of argon. Within the $10\,\sigma$ and $20\,\sigma$ pores the small local maximum after the condensation comes from the fact, that the next particles tendentially will be placed first into the energetic prefered layer on the wall and than

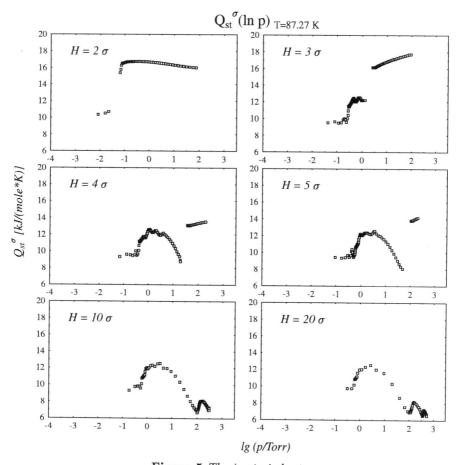

Figure 5 *The isosteric heat*

more and more into the layers in the pore middle, where the wall potential is smaller. In principal, a further increase of pressure leads to an increasing structuring of the adsorbed phase inside the condensed pore.

REFERENCES

[1] D.Henderson. *Fundamentals of Inhomogeneous Fluids.* Marcel Dekker, Inc., 1992.

[2] J.K.Percus and G.J.Yevick. *Phys.Rev.*, 110(1), 1958.

[3] J.P.Hansen and I.R.McDonald. *theory of simple liquids.* academic press, second edition, 1986.

[4] P.Tarazona. Free-energy density functional for hard spheres. *Phys.Rev.A*, 31(4):2672, 1984.

[5] R.D.Groot. Density functional models for inhomogeneous hard sphere fluids. *Molec.Phys.*, 60(1):45, 1987.

[6] R.D.Groot and J.P.van der Eerden. Renormalized density-functional theory for inhomogeneous fluids. *Phys.Rev.A.*, 36(9):4356, 1987.

[7] W.A.Curtin and N.W.Ashcroft. 'weighted-density functional theory of inhomogeneous liquids and the freezing transition'. *Phys.Rev.A*, 32(5):2909, 1985.

CHARACTERIZATION OF HETEROGENEOUS ADSORBENTS USING DATA FROM ISOTHERMS AT TWO DIFFERENT TEMPERATURES

K. Koch, M. v. Szombathely, N. Neugebauer

University of Leipzig, Institute of Physical and Theoretical Chemistry
Linnéstraße 2, D-04103 Leipzig, Germany

1 INTRODUCTION

Adsorption on heterogeneous solid surfaces can be described by means of an adsorption integral equation [1]. The total adsorption amount is the integral over the product of an local adsorption isotherm model and a distribution function of an adsorbent property. Inversion of this integral equation allows the calculation of the distribution functions for pore size or adsorption energy. For this calculation it is necessary to make certain assumptions about the adsorption model, i.e., for the estimation of pore size distributions the adsorption energy parameter has to be fixed, and the energy distribution function can only be calculated assuming a certain pore geometry and pore size. Consequently, the choice of an appropriate local adsorption model is of great importance for an accurate characterization of a given adsorbent. A sensible test for the consistency of an adsorption model is its capability to describe temperature dependency of adsorption. Furthermore the introduction of multi temperature data in the evalution of adsorption data can improve stability and accuracy. In this work we present studies based on simulated DFT isotherms and measured Argon gas isotherms at two different temperatures. We investigate the influence of introduction of data from different temperatures for the calculation of energy and pore size distributions. To make use of the extended amount of data from different temperatures we apply an extended adsorption integral equation [2]. The stability and accuracy of the solution of this equation is compared with results from the single isotherm approach.

2 THEORY

2.1 Adsorption integral equation

The total fractional surface coverage for a heterogenous solid is expressed by the following adsorption integral equation:

$$\Theta_{total}(p) = \int_{x_{min}}^{x_{max}} \Theta_{local}(p, x) F(x) dx \quad (1)$$

Θ_{total} ... measured fractional surface coverage

Θ_{local} ... local isotherm model

$F(x)$ ... distribution function

x ... adsorbent property

If we formulate the adsorption model Θ_{local} as a function of pore width or adsorption energy, the inversion of this equation provides the pore size distribution (PSD) or adsorption energy distribution function (AED) for the solid. For numerical calculation discretization of the integral equation is required. This leads to a system of m linear equations, where m is the number of measurement points. There are two points that need to take into special consideration: First, from the mathematical point of view eq.1 is a Fredholm integral equation of the first kind. Equations of this type are known as numerically ill posed problems. This means that the usual least squares solution method is not utilizable to calculate the distribution function, because experimental errors in the measured data are transfered to the resulting function $F(x)$ in such a way that the solution can be completely disturbed by oscillations. It is then not possible to extract any useful information from the experimental data. Special mathematical methods are necessary to overcome this problem. We apply the *regularisation* method [3, 4] which uses additional stabilizing constraints for the solution.

The second important point is the choice of the local isotherm function Θ_{local} for the integral kernel. The model must be capable to describe the special characteristics of the investigated adsorbent. The introduction of data from different temperatures into the solution of the integral equation provides a way to verify adsorption models as well as to improve numerical behaviour.

2.2 Extended adsorption integral equation

If we state, that the distribution function is temperature independent we can introduce the temperature as an additional variable as follows:

$$\Theta_{tot}(\mathbf{p}, \mathbf{T_i}) = \int_{x_{min}}^{x_{max}} \Theta_{loc}(\mathbf{p}, \mathbf{T_i}, \mathbf{x})\mathbf{F}(\mathbf{x})d\mathbf{x} \tag{2}$$

With a number of n measured isotherms, i ranges from 1 to n. The number of equations for the resulting discretized linear problem now equals the sum of data points of all used n isotherms. This extends the dimension of the data basis leading to more stability and accuracy. The kernel function has to be calculated for the appropriate isotherm temperatures. In the following text we will only investigate the effect of two different temperatures.

3 SIMULATED DATA

3.1 Adsorption model

The applicability of the extended adsorption integral equation was first tested with simulated isotherm data. We calculated synthetic isotherms using a DFT local isotherm model [5, 6] for a slit pore with a width of 10 Å(see Fig. 1/right). For the AED a function with two Gaussian peaks was postulated (Fig. 1/left). To simulate experimental uncertainty a random error of 1 percent was added on the data.

The regularisation method was applied for recalculation of the distribution function with the DFT model using two single isotherms and with combined data from both isotherms for the relative pressure range from $1 \cdot 10^{-6}$ to 0.05. The recalculated AEDs (Fig. 2) and isotherms (Fig. 3) are in good agreement with the postulated ones for single and combined calculation. The shoulder in the peak of the 77 K AED that is caused by

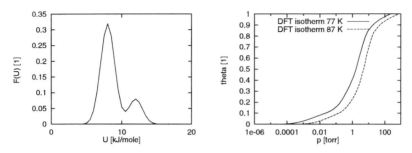

Figure 1: Postulated adsorption energy distribution function and adsorption isotherms for a slit pore of 10 Å (DFT model)

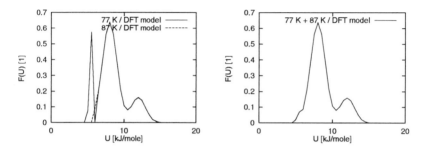

Figure 2: Recalculated AED with DFT local isotherm model

the simulated error disappears in the combined computation. To show the importance of the choice of an appropriate model in the next step the Langmuir model was used as local isotherm model for the recalculation of the AED from the simulated data. Although we can estimate an AED with the Langmuir model (Fig. 4) that describes the adsorption behaviour for the single isotherm result (Fig. 5/left) the Langmuir model fails to reproduce the temperature dependency of the isotherms (Fig. 5/right). The negative parts in the single temperature Langmuir AED already give an indication that the model can not describe adsorption behaviour correctly. The combined calculation with two temperatures shows clearly the non-adequacy of the model. The high peak at low energy is caused by the lateral interaction that is not contained in the Langmuir

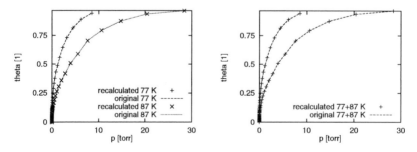

Figure 3: Recalculated isotherms with DFT local isotherm model

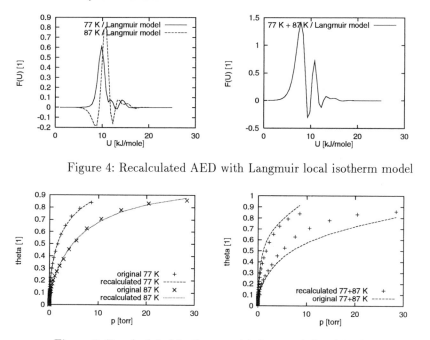

Figure 4: Recalculated AED with Langmuir local isotherm model

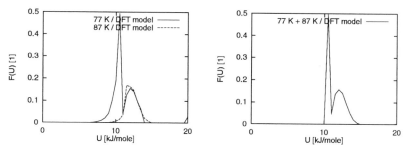

Figure 5: Recalculated isotherms with Langmuir local isotherm model

model and thus mapped as solid-fluid interaction energy. The peak at 10 kJ/mole arises from the energetic heterogeneity of the second kind caused by the 10 Å slit pore model. Furthermore, the stronger interaction caused by the micropore potential shifts the whole distribution function towards higher energies.

3.2 Enhanced data basis

To investigate the effect of the enhanced data basis on the numerical behaviour of the problem the recalculation was carried out with isotherm data in the range from $p_{rel} = 1 \cdot 10^{-6}$ to $1 \cdot 10^{-4}$. Fig. 6 shows that the enhanced data basis leads to more

Figure 6: Recalculated AED ($p_{rel} = 1 \cdot 10^{-6}$ to $1 \cdot 10^{-4}$), single and combined calculation

detailed information in the solution for the AED. Note, that this is not only due to the enlarged number of data points. There is also more information inherent in temperature dependency of adsorption model. This is especially useful if only a few or less accurate measurement points are available in the low pressure region of one isotherm. The combination of two isotherms can than improve the results.

4 EXPERIMENTAL DATA

The experimental isotherms are determined using an ASAP 2010 automatic gas adsorption device from MICROMERITICS. The measurements where carried out for 77.2 K and 87.3 K (T_b for N_2 and Ar, respectively) using Argon as adsorptive.

4.1 Adsorption energy distribution

Fig. 7 shows the energy distribution function of a graphitized carbon (lab sample) calculated with Langmuir model, for two single isotherms and combination of both 77 K and 87 K.

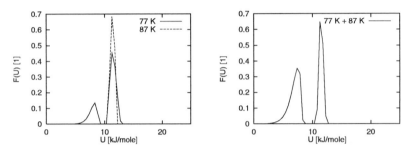

Figure 7: AED for graphitized carbon, Langmuir model

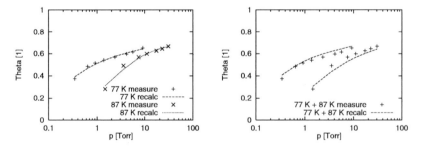

Figure 8: Measured and recalculated adsorption isotherms for graphitized carbon, Langmuir model

The recalculated isotherms shown together with the measurement data in Fig. 8 are in good agreement with the experimental data. However, the calculation with a Fowler-Guggenheim local isotherm model that takes into consideration lateral interaction (random distribution approximation) shows a slightly better description for this sample (Fig. 9 and Fig. 10).

Next the AED for a carbon fibre sample was calculated using the Langmuir model. Fig. 12 and Fig. 11 illustrate that the Langmuir model seems to work well in this case. The recalculated isotherms reproduce the experimental data almost perfectly. The distribution function consists of three peaks. Since this sample is supposed to have an

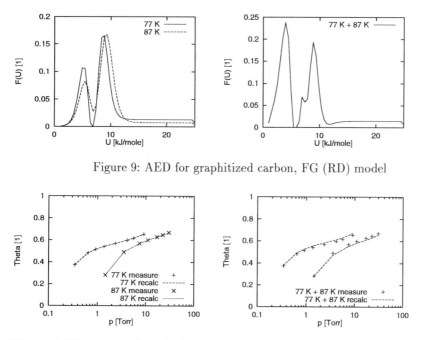

Figure 9: AED for graphitized carbon, FG (RD) model

Figure 10: Measured and recalculated adsorption isotherms for graphitized carbon, FG (RD) model

energetically homogeneous surface because it consists of nearly one hundred percent of carbon, the peaks are not realistic. The calculated AED shows the energetic heterogeneity of the second kind caused by pore geometry effects. The Langmuir model cannot describe this kind of heterogeneity. Because of the less complex isotherm shape the Langmuir model , however, is capable to describe the experimental data numerically correct. This leads to the following conclusion: If an adsorption model does not describe the experimental data with its temperature dependency this gives a suggestion concerning non-adequacy of the model. But even if a model seems to describe the data numerically correct it is important to ask for the physicochemical sense of the results.

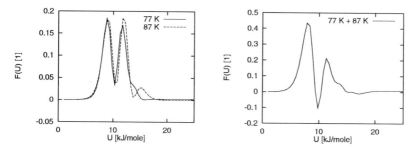

Figure 11: AED for carbon fibre

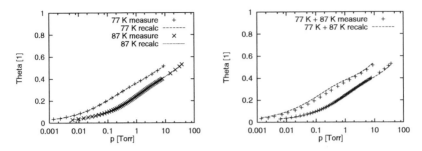

Figure 12: Measured and recalculated adsorption isotherms for carbon fibre

4.2 Pore size distribution

Fig. 13 shows the isotherms from Ambersorb XEN 563 and XEN 572 active carbons

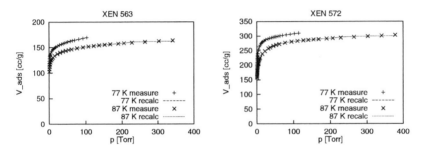

Figure 13: Measured and recalculated adsorption isotherms for active carbons XEN 563 and XEN 572 (single isotherm calculation)

at 77 K and 87 K in the relative pressure range $1 \cdot 10^{-5}$ to 0.8. In Fig. 14 the pore size distribution (DFT model) for the range of 2 Å to 30 Å is shown. Comparison with

Table 1: elementary analysis data for active carbons XEN 563 and XEN 572

sample	C[%]	H[%]	N[%]	S[%]	O[%]	Cl[%]
XEN563	82,20	1,95	0,00	10,49	1,62	0,83
XEN572	93,19	0,46	0,00	1,19	1,81	0,64

the data from the elementary analysis (Tab. 1) gives an explanation for the shift of the PSD. Since the DFT model, which was used as integral kernel, has to be formulated for energetic homogenous surface, the different adsorption sites in the XEN active carbons are mapped as different pore sizes. Since the PSDs are rather complex, they can describe temperature dependency and the combined calculation doesn't provide more information than the single isotherm computation. However it is not possible to calculate PSD and AED at the same time from only two isotherms, the comparison of the PSD for two temperatures can help to evaluate the energetic homogeneity of an experimental sample. The PSD for an ideal energetic homogeneous substance should be the same within the narrow temperature range from 77 to 87 K.

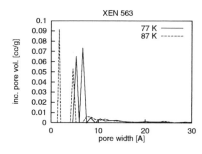
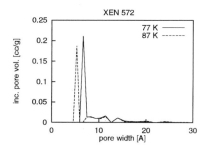

Figure 14: Pore size distribution for active carbons XEN 563 and XEN 572 for two different temperatures (single isotherm calculation)

5 CONCLUSION

Isothermic data from several temperatures can improve reliability and accuracy of the results for the PSD and AED. The application of an extended adsorption integral equation can be used to verify adequacy of adsorption models. More stable solutions can be achieved even if only few measurement points in single isotherms are available. The effort is not remarkable higher because, data for 77 K and 87 K are often already available or additional measurements can be carried out with established automated techniques. Care must be taken to use accurate data especially in the low pressure range to avoid introduction of additional errors resulting from measurement uncertainties.

Financial support of this project by Deutsche Forschungsgemeinschaft (Graduierten-kolleg "Physikalische Chemie der Grenzflächen") is gratefully acknowledged.

References

[1] W. RUDZINSKI AND D. H. EVERETT. "Adsorption of Gases on Heterogeneous Surfaces". Academic Press (1992).

[2] M. V. SZOMBATHELY, K. KOCH, AND N. NEUGEBAUER. Simultaneous solution of the adsorption integral equation from a set of isotherms at different temperatures: a way for more detailed characterization of heterogeneous adsorbents. In "Proceedings of Fifth International Conference on Fundamentals of Adsorption" (1995).

[3] M. V. SZOMBATHELY AND P. BRÄUER. *Journal of Computational Chemistry* **13**, 17–32 (1992).

[4] P. C. HANSEN. *Inverse proplems* **8**, 849–872 (1992).

[5] P. TARAZONA. *Phys. Rev. A* **31**, 2672–2679 (1993).

[6] C. LASTOSKIE, K. E. GUBBINS, AND N. QUIRKE. *J. Phys. Chem.* **97**, 4786–4796 (1993).

K. Murata and K. Kaneko

Department of Chemistry, Faculty of Science, Chiba University

1 INTRODUCTION

The adsorption of supercritical gases, such as O_2, N_2, NO, CO_2, CO, H_2, and CH_4, in micropores has gathered much attention from both fundamental and practical aspects. We, however, have no established approach to describe the molecular ensemble of the supercritical gas confined in a micropore. We propose to describe the molecular ensemble of a supercritical gas in a micropore using the concept of Quasi-vapor (Confined fluid) and Organized fluid. This states of quasi-vapor and organized fluid are evidenced by experimental results on methane by carbon micropores.

2 TWO PHASE MODEL OF MICROPORE FLUIDS

Bulk fluids are classified into vapor, liquid and supercritical fluid. How can we understand the state of the supercritical gas in the micropore? It is presumed that the supercritical gas is transformed into confined and organized fluids in the micropore, which depends on the pore width, since adsorbed molecules of a supercritical gas are strongly interacted with each other in the micropore. Recent related studies on supercritical gas adsorption by microporous carbons showed the formation of molecular clusters or organized molecular assembly for supercritical gases. The clusters or organized molecular assemblies for the supercritical gas should behave like vapors. The quasi-vaporized supercritical gas can be adsorbed in the micropore to from the adsorbed layer. We designate the quasi-vapor and the adsorbed layer by confined fluid and organized fluid respectively. This two phase model is plausible, if we admit it, the upper shift of the critical temperature of the confined fluid with the decrease of the pore width in addition to the Fisher's downward shift[1,2], as shown in Figure 1. Figure 1 shows this model.

3 EXPERIMENTAL

Microporous carbon AX21 and $Mg(OH)_2$-AX21 were used. The preparation of $Mg(OH)_2$-AX21 is shown in Figure 2. The microporosity of these samples was determined by N_2 adsorption at 77 K. The adsorption isotherms of supercritical methane on carbon samples were determined over the pressure range up to 10 MPa at 273, 293, and 303 K. The buoyancy effect was corrected by use of the true density determined by the He high pressure measurement at 303 K.

The total pore volume W_0 was obtained by α_s analysis of N_2 adsorption isotherm. The inherent pore volume W_L for supercritical methane, which is governed by the molecule-pore potential, was estimated by Langmuir equation.

4 RESULTS AND DISCUSSION

As all of the quasi-vaporized molecules of supercritical gas in the micropore

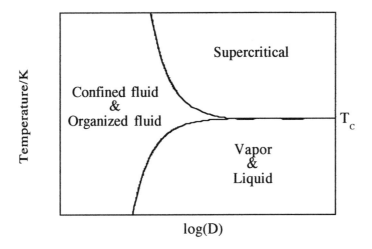

Figure 1   *Critical temperature shift as a function of pore width D and 3 phase region. T_C is bulk critical temperature.*

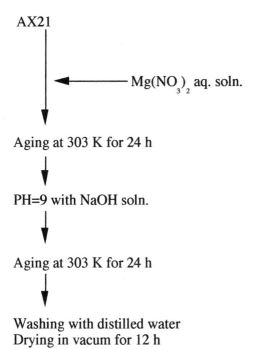

Figure 2   *Preparation of Mg(OH)$_2$-AX21*

are not perfectly adsorbed in the micropore, there should be two states in the micropore. The presence of two states of a supercritical gas can be evidenced by a thermodynamic approach. The supercritical DR analysis[3] provides the P_{0q} of the quasi-vaporized fluid, and thereby the temperature dependence of P_{0q} should suggest the macroscopic state of the quasi-vaporized fluid. The experiment on supercritical methane adsorption on microporous carbon can give a good example for the above model.

We measured the high pressure adsorption isotherms of CH_4 on AX21 and $Mg(OH)_2$-AX21 at different temperatures. $Mg(OH)_2$-AX21 denotes the fine $Mg(OH)_2$ particles-dispersed AX21; 3-4 wt % $Mg(OH)_2$ particles are deposited on AX21. The microporosity of $Mg(OH)_2$-AX21 is shown in Table 1. Figure 3 shows the temperature dependence of CH_4 adsorption isotherms of AX21 and $Mg(OH)_2$-AX21. The dispersion of $Mg(OH)_2$ enhances the CH_4 adsorption noticeably. The marked enhancement effect of $Mg(OH)_2$ and MgO dispersion on CH_4 adsorption was also observed in case of ACF (Kaneko et al, 1993).[4] Although the exact mechanism on the enhancement effect is not clear, a weak chemisorptive interaction of the dispersed $Mg(OH)_2$ with supercritical CH_4 should be associated with it. The lower the adsorption temperature, the greater the amount of CH_4 adsorption. This adsorption behavior is indicative of a typical physical adsorption in Figure 3.

Table 1 *Micropore parameters of AX21 and $Mg(OH)_2$-AX21 for adsorption measurement of supercritical CH_4*

Sample	Micropore volume $ml\ g^{-1}$	Surface area $m^2 g^{-1}$	Pore width nm
AX21	1.49	2400	1.30
$Mg(OH)_2$-AX21	1.01	1330	1.56

The supercritical DR analysis was applied to these adsorption isotherms, leading to the P_{0q} as a function of adsorption temperature. Figure 4 shows the van't Hoff plots of P_{0q} for AX21 and $Mg(OH)_2$-AX21. The good linearity indicates the presence of the thermodynamic equilibrium between two phases. The slope of the van't Hoff plot provides the enthalpy change, ΔH_t between two states. Table 2 shows ΔH_t values for AX21 and $Mg(OH)_2$-AX21 and the ΔH_V of bulk CH_4 is shown for comparison.

Table 2   *Enthalpy change ΔH_t for transition of organized fluid to confined fluid for supercritical methane on AX21 and $Mg(OH)_2$-AX21*

sample	ΔH_t $kJ\ mol^{-1}$
Enthalpy change of vaporization for bulk methane	8.18
AX21	13.4
$Mg(OH)_2$-AX21	13.0

The ΔH_t for $Mg(OH)_2$-AX21 is greater than that for AX21, which indicates the stabilization of the adsorbed state of CH_4 in the micropore of $Mg(OH)_2$-AX21. The ΔH_t, values are much greater than ΔH_V of bulk CH_4. This difference suggests that the adsorbed state of CH_4 is more stabilized than the bulk liquid state of CH_4. Therefore we must distinguish the adsorbed phase in the micropore from the bulk liquid phase. We propose to introduce organized and confined fluid fluids states two phases. The organized fluid corresponds to the adsorbed molecular assembly thorough quasi-vaporization, whereas the confined fluid is used for the quasi-vapor in the micropore. As the molecular density of the confined fluid is much less than that of the organized fluid, the amount of the adsorption is mainly determined by the amount of the organized fluid. The ΔH_t is the enthalpy change of the transition from the organized fluid to confined fluid.

This model should be helpful to understand the adsorption behavior of supercritical gases in micropores.

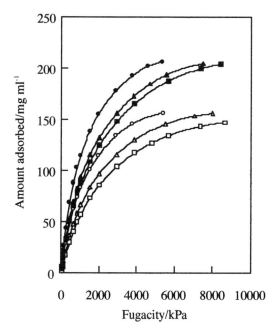

Figure 3   *The temperature dependence of the adsorption isotherms of supercritical methane on AX21 and Mg(OH)₂-AX21*

AX21	○:273 K,	△:293 K,	□:303 K
Mg(OH)₂-AX21	●:273 K,	▲:293 K,	■:303 K

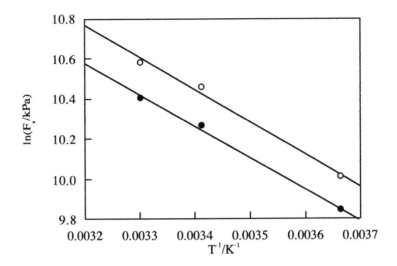

Figure 4   *The van't Hoff plots of the F_S (fugacity of quasi-saturated vapor pressure) of supercritical methane adsorption on AX21 and Mg(OH)$_2$-AX21*

    AX21   ○

    Mg(OH)$_2$-AX21   ●

Reference
1   M. E. Fisher and H. Nakanishi, J. Chem. Phys., 1981, **75**, 5857
2   H. Nakanishi and M. E. Fisher, J. Chem. Phys., 1983, **78**, 3279
3   K. Kaneko, K. Shimizu and T. Suzuki, J. Chem. Phys., 1992, **97**, 8705
4   K. Kaneko, K. Murata, K. Shimizu, S. Camara, and T. Suzuki, Langmuir, 1993, **9**, 1165

SURFACE CHARACTERISTICS AND PORE STRUCTURE OF ACTIVATED CHAR PREPARED FROM MIXTURE OF COAL TAR PITCH AND A GRAPHITE-FeCl₃ INTERCALATION COMPOUND

A. Albiniak[a], E. Broniek[a], J. Kaczmarczyk[a], A. Jankowska[a], T. Siemieniewska[a], B. McEnaney[b], X. S. Chen[b], G. Furdin[c] and D. Bégin[c]

[a] Laboratory for Lignites and Carbon Adsorbents, Institute of Chemistry and Technology of Coal, Technical University of Wrocław, Poland
[b] School of Materials Science, University of Bath, Bath, BA2 7AY, UK
[c] Laboratoire de Chimie du Solide Minéral, Université H. Poincaré, Nancy, France

1 INTRODUCTION

Carbonisation of unmodified coal-derived liquids usually yields unreactive cokes with low porosity that cannot be easily developed into activated carbons by conventional methods. Addition of Lewis acids to coal-derived liquids can substantially modify the course of carbonisation leading to reactive carbon structures in high yield. G. Furdin[1-2] proposed a novel approach to the introduction of the Lewis acid, FeCl₃, by using a graphite-FeCl₃ intercalation compound. This approach circumvents the problems associated with the volatility and hygroscopic character of pure FeCl₃.

Our preliminary research,[3-4] concerning some aspects of porosity development in chars from mixtures of coal tar pitch with graphite-FeCl₃, varying the concentration of the intercalation compound and degree of activation, gave encouraging results as to the possibility of microporosity and mesoporosity development in these materials.

The aim of the present work was to survey some of the specific characteristics of the pore structure of the steam activated chars, as revealed by adsorption of polar adsorptives - water vapour and ammonia, contrasted with adsorption of much less specific adsorptives like nitrogen, benzene and carbon dioxide. In this paper we have focused on one material - a steam activated char from mixture of coal tar pitch with a graphite-FeCl₃ intercalation compound.

2 EXPERIMENTAL

The initial char was prepared, by G. Furdin and co-workers from a mixture of coal tar pitch with 8 wt.% graphite-FeCl₃ intercalation compound at the final heat treatment temperature (HTT) of 750°C. This char has been additionally heat treated in a thermogravimetric apparatus at 800°C in a stream of argon, followed by steam activation at 800°C to achieve a burn-off of 75 %.

For the activated sample, adsorption results for water vapour and ammonia at 298 K were obtained, in addition to the previously presented[3] adsorption data for nitrogen at 77 K, and benzene and carbon dioxide at 298 K. Sorption measurements were carried out by a static technique, in a gravimetric apparatus (McBain balances). For the computation of parameters characterising the pore structure of the chars, benzene and carbon dioxide adsorption data were used applying a procedure described before.[5-6]

3 RESULTS AND DISCUSSION

3.1   Adsorption of Benzene, Nitrogen and Water

The influence of the chemical nature of adsorbed benzene, nitrogen, and water molecules on the position and shape of the resulting adsorption isotherms is illustrated in Figure 1 and Table 1.

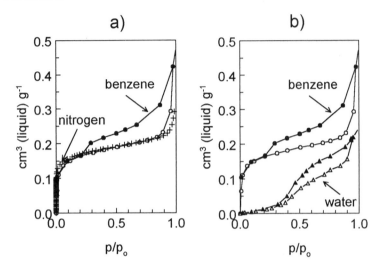

Figure 1   *Adsorption isotherms on the steam activated char of: a) benzene at 298 K and nitrogen at 77 K, and b) benzene and water at 298 K. Open points adsorption, closed points desorption.*

The contrast between the almost superposed benzene and nitrogen adsorption isotherms in Figure 1a (this and similar results have been reported before[3,5-6]) and the pronounced difference in the isotherms of benzene and water (Figure 1b), is clearly visible.

For benzene and nitrogen, the values of V_m (and so those of S_{BET}), as well as the ranges of relative pressures for rectilinearity in the BET co-ordinates, are almost identical; the difference between the adsorption mechanisms of these two kinds of molecules is reflected only by differing values of the BET parameter C (Table 1), which is related to their heats of adsorption.

Table 1   *Values of BET parameters from nitrogen (77 K) and benzene (298 K) adsorption isotherms*

Adsorptive	V_m cm^3/g	S_{BET} m^2/g	C	Rectilinearity range for p/p_o
Benzene	0.138	382	180	0.01-0.15
Nitrogen	0.142	399	950	0.01-0.13

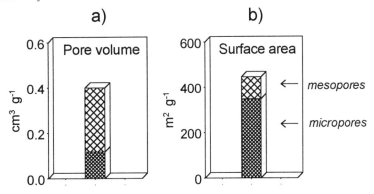

Figure 2 *Pore volumes (a) and effective surface areas (b) corresponding to micropores and mesopores of the activated char, calculated from benzene and carbon dioxide isotherms at 298 K.*

For the activated char only about 25 % of the total adsorption volume is due to micropores with widths in the range 0.4 - 2 nm,[*] while 75 % of the volume corresponds to mesopores, Figure 2a. However, when effective surface areas are considered (i.e., the sum of the volume of micropores expressed as a surface area and the surface area of mesopores), this ratio is reversed, Figure 2b. It is clear that it is the presence of micropores which has the main impact on the value of the effective surface area of this activated char.

The shape of the water adsorption isotherm in Figure 1b is typical for adsorption on activated carbons, whose surfaces are usually hydrophobic. There is negligible adsorption in the low relative pressure region of the isotherm [7] followed by a steep rise in the isotherm at higher relative pressures, probably caused by adsorbate-adsorbate interactions involving hydrogen bonds, resulting in formation of clusters of water molecules.

When Gurvitsch volumes[†] for benzene and water are compared, the volume of adsorbed water: 0.220 cm^3 (liquid)/g is distinctly lower than that of benzene - 0.410 cm^3 (liquid)/g. This suggests a lower adsorbate density for water than that for bulk liquid water. In a study of unactivated and activated PVDC-based carbons Bradley and Rand[8] found lower saturation adsorption volumes for water (water Gurvitsch volumes), when compared with nitrogen.

These authors stated that water saturation adsorption volumes are lower than the respective nitrogen micropore volumes. This statement is not in accord with our results concerning the activated GIC char. Assuming that the values of V_m in Table 1 (0.138 cm^3 (liquid)/g for benzene and 0.142 cm^3 (liquid)/g for nitrogen) are close to the benzene and

---

[*] The absence of very narrow micropores (widths below 0.4 nm) was deduced[5-6] from comparison of benzene and carbon dioxide adsorption data at 298 K.

[†] It was not possible to read from the isotherms in Figure 1 the exact limiting adsorption volumes (saturation volumes) at $p/p_0 \approx 1$, due to the steep rise of these isotherms in the high relative pressure region. Therefore, the Gurvitsch volumes were taken as volumes to $p/p_0 = 0.96$. This value of relative pressure, in case of benzene corresponds to the pore width of 50 nm, if slit-shaped mesopores are assumed and if the thickness of the adsorbed film, with Elftex 120 as standard adsorbent, is taken into account.

nitrogen micropore volumes, it is clear that the water Gurvitsch volume of 0.220 cm³ (liquid)/g is higher than these volumes. A possible explanation for the difference between the results of Bradley and Rand [8] and ours is that the water Gurvitsch volume for the GIC activated char includes a significant contribution from mesopore filling. In the case of the highly microporous PVDC-based carbons studied by Bradley and Rand mesopore filling would not be so significant. We were not able to separate the contributions from micropore and mesopore filling by water using the equations of Dubinin[9] and Dubinin and Serpinsky[10], since these equations were unreliable when applied to adsorption of water vapour on the steam activated GIC char. Water Gurvitsch volumes that were higher than nitrogen micropore volumes were also obtained by other authors for activated chars from aramid fibres[11-12].

3.2. Adsorption of Carbon Dioxide and Ammonia

The isotherms of carbon dioxide and ammonia adsorption at 298 K are presented in Figure 3. For both adsorptives the shapes of the isotherms are similar, although smaller amounts of carbon dioxide than of ammonia are adsorbed at the same temperature and in the same range of pressures.

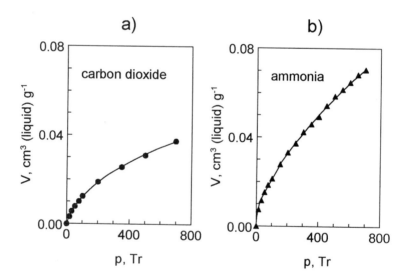

Figure 3 *Adsorption isotherms on the steam activated char of:*
a) carbon dioxide at 298 K, and b) ammonia at 298 K.

To better compare adsorption of carbon dioxide and ammonia at 298 K, the adsorption data are plotted according to the Dubinin-Radushkevich (DR) equation[13] (Figure 4). Corrections resulting from adsorption on the surface of mesopores (the mesopore surface area of the activated char was close to 100 m²/g) were taken into account[14], using standard adsorption isotherms of carbon dioxide and ammonia at 298 K on the nonporous carbon Elftex 120. These corrections were subtracted from the experimental data only for ammonia, because in case of carbon dioxide their values were negligible.

The DR equation, in the co-ordinates ln V *versus* [RT ln $(p_o/p)]^2 / \beta^2$ is expected to give a rectilinear relationship. Further, as the affinity coefficient β is taken into account (the values for β for carbon dioxide and ammonia were taken as 0.37 and 0.28, respectively), the isotherms for carbon dioxide and ammonia should be positioned on the same characteristic curve.

It is clear from Figure 4, that the carbon dioxide and ammonia data are not superposed and that a linear DR relationship is only found for carbon dioxide, (Figure 4a). For both the corrected and uncorrected ammonia isotherms (Figure 4b), it is difficult to define a rectilinear range in the DR plot.

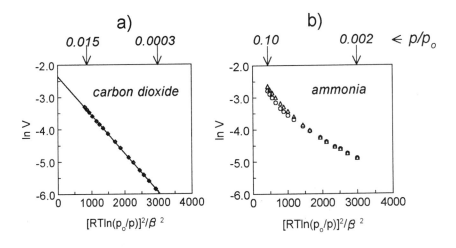

Figure 4 *DR plots for adsorption on the steam activated char of: a) carbon dioxide at 298 K, b) ammonia at 298 K: circles -data corrected for adsorption on the surface of mesopores, triangles - data without corrections.*

The ammonia isotherms, plotted in DR co-ordinates, were further analysed using the carbon dioxide results as reference data, since, compared to ammonia, carbon dioxide can be considered as a non-specific adsorptive.[15] Therefore, in Figure 5 both carbon dioxide and ammonia isotherms have been plotted. For clarity, the data points for carbon dioxide are omitted in Figure 5 and only corrected ammonia adsorption data are shown; similar plots for the uncorrected ammonia data are not presented. The values of DR micropore volume: $V_{o,DR}$ and characteristic energy of adsorption $E_{o,DR}$, are presented for carbon dioxide and the corrected and uncorrected ammonia data in Table 2.

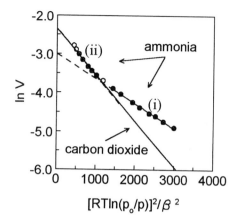

Figure 5 *DR plots for adsorption on the steam activated char of carbon dioxide at 298 K and ammonia at 298 K (data corrected for adsorption on the surface of mesopores); full circles indicate the regions of rectilinearity, open circles are those which were not included in the straight lines.*

$[RTln(p_o/p)]^2/\beta^2$

Table 2 *Values of DR parameters from carbon dioxide (298 K) and ammonia (298 K) adsorption isotherms*

Adsorptive	$V_{o, DR}$ cm^3/g	$E_{o, DR}$ kJ/mol	Rectilinearity range for p/p$_o$:
Carbon dioxide	0.095	29.0	0.0003-0.015
Ammonia *(corrected for adsorption in the mesopores)*	0.050 0.097	39.5 28.5	0.002-0.015 [a] 0.03-0.07 [b]
Ammonia *(not corrected)*	0.049 0.105	39.6 29.7	0.002-0.01 [a] 0.015-0.07 [b]

[a] *Section (i) in Figure 5;* [b] *section (ii) in Figure 5*

It follows from Figure 5 and Table 2, that it is for ammonia adsorption in the region of somewhat higher relative pressures, (ii) Figure 5, that a good accord with the values of the carbon dioxide DR parameters was achieved; this accord was particularly satisfactory, for the corrected ammonia results. In this case the values of $V_{o, DR}$ and of $E_{o, DR}$ for carbon dioxide and ammonia are almost identical (0.095 cm^3/g and 0.097 cm^3/g, respectively, and 29.0 kJ/mol and 28.5 kJ/mol, respectively).

The observed upward deviations of ammonia adsorption on the characteristic curve in the region of lower relative pressures, (i) Figure 5, point to specific interactions between the polar ammonia molecules and active centres on the surface forming the walls of the pores of the activated char. The active sites may include relics of the Lewis acids inserted into the char by means of the graphite-FeCl$_3$ intercalation compounds. The particularly high value of $E_{o, DR}$ (close to 40 kJ/mol) seems to confirm this. This high $E_{o, DR}$ value cannot result from adsorption in very narrow micropores (utramicropores), because - as was mentioned in Section 3.1 - this category of pores is not present in the steam-activated GIC char.

4 CONCLUSIONS

The char produced from a mixture of coal tar pitch with addition of a graphite-FeCl$_3$ intercalation compound was steam activated to a 75 % burn-off. The BET surface area of this char, determined using either nitrogen or benzene, is close to 400 m^2/g and results mainly from the presence of micropores in the char. The presence of micropores suggests that the activated chars may be suitable for gas adsorption applications. A well developed mesoporosity (75% of total adsorption volume) should facilitate the kinetics of gas adsorption and also may be useful for liquid phase adsorption processes. Very high values of the characteristic energy of adsorption (E$_{0, DR}$ = 40 kJ/mol), calculated from enhanced ammonia adsorption at low relative pressures, indicate the presence of specific adsorption centres on the internal surface of the activated char. The presence of such specific adsorption centres may be useful in applications of these adsorbents in catalysis. Thus the activated char is characterised by a pore system and adsorptive properties that suggest that the intercalated pitches are promising precursors for preparation of specific carbon adsorbents.

5 REFERENCES

1. G.Furdin, D.Bégin, J.F.Mareché, D.Petitjean and E.Alain, *Carbon*, 1994, **32**, 599.
2. D.Bégin, E.Alain, G.Furdin and J.F.Mareché, *Fuel*, 1995, **74**, 139.
3. A.Albiniak, E.Broniek, J.Kaczmarczyk, A.Jankowska, T.Siemieniewska, B.McEnaney, X.S.Chen, G.Furdin, D.Bégin, E.Alain and J.F.Mereché, 'Proceedings of the 8th International Conference on Coal Science, Oviedo, Spain, September 10-15, 1995, Coal Science', Ed. J.A.Pajares and J.M.D.Tascón, Elsevier Science B.V., Amsterdam 1995, pp.1141-1144.
4. A. Albiniak, E.Broniek, J.Kaczmarczyk, A.Jankowska, T.Siemieniewska, B.McEnaney, G.Furdin, D.Bégin, J. F. Mareché, E.Alain, *'Matériaux Carbonés Adsorbants, Carbochimie - Environnement, PICS 119, France-Pologne, Réunions Annuelles, Zvenigorod (pres de Moscou), Russie'*, 25-28 September, 1995, Proceedings, pp. 113-117.
5. Siemieniewska, T., Tomków, K., Kaczmarczyk, J., Albiniak, A., Grillet, Y. and François, M., *Energy and Fuels*, 1990, **4**, 61-70.
6. Siemieniewska, T., Tomków, K., Kaczmarczyk, J., Albiniak, A., Broniek, E., Jankowska, A., Grillet, Y. and François, M., 'Characterisation of Porous Solids III, Studies in Surface Science and Catalysis 87', Ed. J.Rouquerol, F.Rodriguez-Reinoso, K.S.W.Sing and K.K.Unger, Elsevier Science Publishers, Amsterdam 1994, pp. 695-704 .
7. S.J.Gregg, K.S.W.Sing, 'Adsorption, Surface Area and Porosity', 2nd Edition, Academic Press, London, 1982, p. 268.
8. R.H.Bradley and B.Rand, *Carbon*, 1993, **31**, 269.
9. M.M.Dubinin, *Carbon*, 1980, **18**, 355.
10. M.M.Dubinin and V.V.Serpinsky, *Carbon*, 1981, **19**, 402.
11. J.J.Freeman, J.B.Tomlinson, K.S.W.Sing, C.R.Theocharis, *Carbon*, 1993, **31**, 865.

12. J.J.Freeman, J.B.Tomlinson, K.S.W.Sing, C.R.Theocharis, *Carbon,* 1995, **33**, 795.
13. M.M.Dubinin, *Carbon,* 1989, **27**, 457.
14. M.M.Dubinin, ' Characterisation of Porous Solids, Studies in Surface Science and Catalysis 39', Ed. K.Unger, J.Rouquerol, K.S.W.Sing and H. Kral, Elsevier Science Publishers, Amsterdam 1988, pp. 127-137.
15. H.Marsh, *Fuel,* 1965, **44**, 253.

6 ACKNOWLEDGEMENTS

The authors are grateful for financial support provided by the Technical University of Wrocław, Poland (from the Polish Committee for Scientific Research). Part of the research was sponsored by the University of Bath and the CVCP ORS Award scheme of the United Kingdom.

MODELLING OF GAS SORPTION PROCESSES IN SHALLOW FIXED BED ADSORBERS

E.P.J. Mallens, M.P.M. Cnoops and J.J.G.M. van Bokhoven

TNO-Prins Maurits Laboratory
Research Group Respiratory Protection and Air Purification
P.O. Box 45
2280 AA Rijswijk
The Netherlands

SUMMARY

A mathematical model was developed to simulate the adsorption process of organic components in a filter bed of activated carbon, mainly focused on the application of gasmask canisters. In this area an adequate description of the initial breakthrough of components is of most interest and special attention has been given to this point. The adsorption process of toluene in a packed bed of activated carbon was considered in this study. The applied model included plug flow and axial dispersion to describe the mass transport in the axial direction. Mass transfer limitation between the bulk of the gas phase and the outer surface of the particle was accounted for. Mass transport inside the particle was described by a diffusive contribution assuming Fick's law with an effective molecular diffusion coefficient as transport parameter. Equilibrium between the gas phase and solid phase concentration was assumed throughout the particle. A computer program was written to solve the set of partial differential equations. The values of the model parameters were determined experimentally or estimated from literature correlations. The model was validated by comparison of simulated and experimentally determined toluene outlet concentrations in a packed bed of activated carbon, applying a continuous feed of a mixture of toluene and nitrogen. Parameters varied were the bed length of 0.01 to 0.035 m_r, the superficial gas velocity of 0.127 to 0.191 $m^3_f\,m^{-2}_r\,s^{-1}$, and the toluene inlet concentration of $2.25\ 10^{-3}$ to $4.83\ 10^{-3}$ $kg\,m^{-3}_f$. The agreement between measured and simulated outlet concentrations is good. However, the agreement becomes slightly less at low normalised toluene concentrations, the more so when the bed length increases. The effects of variations in the superficial gas velocity and inlet concentrations are described well.

1 INTRODUCTION

A modelling effort of the breakthrough behaviour of organic components in gasmask filter beds of activated carbon has been ongoing within the research group Respiratory Protection and Air Purification of TNO-Prins Maurits Laboratory. Usually, the filter is a shallow packed bed consisting of activated carbon granules, sometimes of spherical or cylindrical shape.

Developing an adequate model to predict breakthrough curves presents several advantages. First, insight is gained into the physical processes that determine the breakthrough behaviour of the filter bed. Secondly, mathematical models are helpful in the introduction of design improvements to the filter, for example with respect to particle and bed dimensions as well as geometry. Finally, an experimentally validated model reduces future experimental efforts significantly.

The aim of the present study was to develop a mathematical model to simulate the adsorption process of organic components on activated carbon at almost dry conditions *i.e.* a relative humidity of 50 % or lower. The model predictions are compared with measured toluene outlet concentrations for shallow packed beds at ambient temperatures.

2 MATHEMATICAL MODEL

2.1 Introduction

A mathematical model of the filter bed was developed, consisting of two continuity equations for the adsorptive, *i.e.* one over the gas phase and one over the solid phase. Both phases were assumed to be isothermal, which is valid since the heat effects of the adsorption process are negligible due to the low concentrations of the adsorptive. The transport mechanism in the gas phase was described by a convective contribution assuming plug flow and a contribution accounting for mixing in the axial direction. The resistance for mass transfer between the bulk of the gas phase and the outer surface of the particle was located in a fictitious stagnant film around the particle. Intraparticle gas transport was described by a diffusive contribution assuming Fick's law. Equilibrium between the gas phase and solid phase concentration was assumed at each position throughout the particle, since the elementary adsorption step is much faster compared to the mass transport inside the pores of the particle.

2.2 Model Equations

The continuity equation for the adsorptive over the gas phase is given by equation 1, with the initial condition and boundary conditions given by equations 2 to 4.

$$\varepsilon_b \frac{\partial C_b}{\partial t} = D_{ea} \frac{\partial^2 C_b}{\partial z^2} - u_s \frac{\partial C_b}{\partial z} - (1 - \varepsilon_b) a_v k_f (C_b - C_i) \tag{1}$$

$$C_b \big|_{t=0} = 0 \text{ for } 0 \leq z \leq L_b \tag{2}$$

$$C_b \big|_{z=0} = C_{in} \text{ for } t > 0 \tag{3}$$

$$\frac{\partial C}{\partial z} \bigg|_{z=L} = 0 \text{ for } t > 0 \tag{4}$$

Equation 1 describes the bulk gas concentration as a function of time, at every axial position in the bed. The first term on the right hand side describes the mixing in the axial direction with the effective axial diffusion coefficient as transport parameter. The second term describes convective transport in axial direction, while the last term accounts for the mass transfer from the bulk gas phase to the outer surface of the particle. The driving force for the latter is the concentration gradient over the fictitious stagnant gas film around the carbon particle.

The continuity equation for the adsorptive over the solid phase is given by equation 5, with the initial and boundary conditions given by equations 6 to 8, respectively.

$$\left(\frac{\partial q}{\partial C}+\varepsilon_p\right)\frac{\partial C}{\partial t}=\frac{D_{eff}}{r^2}\frac{\partial}{\partial r}\left(r^2\frac{\partial C}{\partial r}\right)$$ (5)

$$C|_{t=0}=0 \text{ for } 0\le r\le R_p$$ (6)

$$D_{eff}\frac{\partial C}{\partial r}\bigg|_{r=r_p}=k_f\,(C_b-C_i) \text{ for } t>0$$ (7)

$$\frac{\partial C}{\partial r}\bigg|_{r=0}=0 \text{ for } t>0, \text{ leading to } \left(\frac{\partial q}{\partial C}+\varepsilon_p\right)\frac{\partial C}{\partial t}=3D_{eff}\frac{\partial^2 C}{\partial r^2}\bigg|_{r=0}$$ (8)

Equation 5 describes the diffusion and adsorption of the adsorptive inside the pores of the adsorbent particle, which are considered to be spherical. The first term between brackets on the left hand side is the derivative of the adsorptive loading as a function of the gas phase concentration and accounts for the adsorption equilibrium inside the particle. The diffusion process inside the particle is described with an effective molecular diffusion coefficient as transport parameter. This description can be interpreted as diffusion in the macropores only.

2.3 Adsorption Isotherm Equation

The adsorption isotherms of the adsorptives of primary concern are described well using the Dubinin-Radushkevich adsorption isotherm, given by equation 9.

$$\frac{q}{q_{max}}=e^{\frac{-b_0 T^2}{\beta^2}\log^2\left(\frac{c}{c_s(T)}\right)}$$ (9)

The values of b_o and q_{max} were estimated by the regression of equation 9 to the experimentally determined sorption equilibrium data for the corresponding set of adsorptive and adsorbent material.

2.4 Numerical Method

The model describing the filter bed consists of two partial differential equations and corresponding initial and boundary conditions. The approach used to solve this set of equations is the so-called *numerical method of lines*[1]. The equations are solved by placing a grid of discretisation points in the bed in axial direction. From each of the axial grid points, a number of grid points extends in the radial direction of the adsorbent particle. The derivatives of concentration to axial and radial spatial co-ordinates, z and r, in equations 1 and 5 respectively, were calculated using a simple three-point or five-point discretisation[1] (routines DSS002 and DSS012). As a result, the partial differential equations were transformed into a set of ordinary differential equations in time, for each grid point. A computer program was written in FORTRAN to solve this set of equations, with the use of a standard numerical integration routine, D02NCF, from the NAG-library[2]. The amount of time required to solve the complete set of equations typically amounted to two hours on a Pentium[©] 100 MHz PC when 21 radial and 41 axial grid points were used, for a clock time of about four hours. The required time was approximately proportional to the third power of the number of grid points, due to the integrator routine.

2.5 Model Parameters

The values of the parameters used in the simulations are shown in Table 1. The gas mixture consisted of nitrogen and toluene, the latter present in very low concentrations. Therefore, the physical parameters of the gas mixture were based on the values for pure nitrogen. The values of q_{max} and b_o followed from regression of the Dubinin-Radushkevich isotherm equation, see section 2.3, to experiments, while the value of ß was calculated[3]. The activated carbon applied consists of cylindrical particles with a diameter of 1 mm and an average length of 3 mm. Since the continuity equation over the solid phase is based on a spherical particle shape, an equivalent diameter was calculated. The latter is based on an identical surface to volume ratio for a spherically shaped particle compared to the cylindrical particles applied in the experiments. This equivalent diameter amounts to 1.3 mm. The experiments were carried out at 298 K and the activated carbon was placed in a quartz tube with a diameter of $5 \ 10^{-2}$ m. The values for the bed porosity and the particle porosity were determined in separate experiments.

Table 1 *Parameter values used in the simulation of the adsorption process of toluene in a packed bed of activated carbon*

Parameter	Value	Unit	Parameter	Value	Unit
q_{max}	324	$kg \ m^{-3}_{pt}$	D_{eff}	$8.8 \ 10^{-6}$	$m^3_f \ m^{-1}_{pore} \ s^{-1}$
b_0	$1.42 \ 10^{-6}$	K^{-2}	ε_b	0.372	$m^3_f \ m^{-3}_r$
β	1.194	--	ε_{pt}	0.55	$m^3_f \ m^{-3}_{pt}$
d_{pt}	$1.3 \ 10^{-3}$	m_{pt}	Pe_{ax}	2	-
C_{sat}	0.125	$kg \ m^{-3}_f$	ρ	1.18	$kg \ m^{-3}$
T	298	K	μ	$18.3 \ 10^{-6}$	$kg \ m^{-1} \ s^{-1}$

The effective axial diffusion coefficient was calculated from the axial Peclet number. Typical Reynolds numbers are between 10 and 20 and in this range the axial Peclet number is approximately equal to 2 for cylindrical particles[4]. The mass transfer coefficient was determined on the basis of the Chilton-Colburn factor, j_D, see equation 10.

$$k_f = j_D \ Re \ Sc^{1/3} \frac{D}{d_p} \tag{10}$$

An empirical correlation[5], valid for Reynolds numbers between 0.01 and 1500, and Schmidt numbers between 0.6 and 7000 was used for the calculation of the j_D factor, see equation 11.

$$\varepsilon_b j_D = 0.765 \, \text{Re}^{-0.82} + 0.365 \, \text{Re}^{-0.386} \tag{11}$$

The effective diffusion coefficient in the particle has been taken equal to the molecular diffusion coefficient. Variables are the bed length, 0.01 to 0.035 m_r, the superficial gas velocity, 0.127 to 0.191 $m^3_f \, m^{-2}_r \, s^{-1}$, and the toluene inlet concentration, 2.25 10^{-3} kg m$^{-3}_f$ to 4.83 10^{-3} kg m$^{-3}_f$.

3 RESULTS AND DISCUSSION

Simulated and experimentally determined breakthrough curves of toluene are compared to determine the model characteristics and to validate the mathematical model presented in section 2.2.

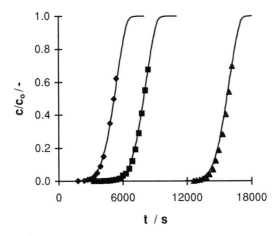

Figure 1 *Normalised outlet concentrations of toluene as a function of time when continuously feeding a mixture of toluene and nitrogen to a packed bed of activated carbon with $C_0 = 2.25 \, 10^{-3}$ kg m$^{-3}_f$ and $u_s = 0.127 \, m^3_f \, m^{-2}_r \, s^{-1}$. Bed length: ◆ = 0.01 m_r,* ■ *= 0.015 m_r and* ▲ *= 0.03 m_r.*

Figure 1 shows the simulated and experimentally determined outlet concentrations of toluene as a function of time for three different bed lenghts, a superficial gas velocity of 0.127 $m^3_f \, m^{-2}_r \, s^{-1}$ and a toluene inlet concentration of 2.25 10^{-3} kg m$^{-3}_f$. The same results are presented in Figure 2 on a semilogarithmic scale, to allow a better comparison at low normalised toluene concentrations. The agreement between simulated and experimentally determined values is good, see Figure 1 and 2, especially given the fact that the values of the mass transport coefficients have been calculated from literature correlations for fixed beds. The model only slightly overpredicts the outlet concentrations of toluene at a bed length of 0.03 m_r, especially at low toluene outlet concentrations, as shown by Figure 2.

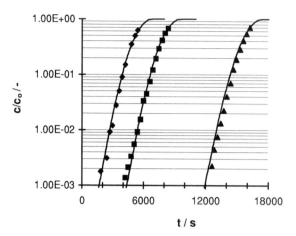

Figure 2 *Normalised outlet concentrations of toluene as a function of time when continuously feeding a mixture of toluene and nitrogen to a packed bed of activated carbon with, $C_0 = 2.25 \ 10^{-3} \ kg \ m^{-3}{}_f$ and $u_s = 0.127 \ m^3{}_f \ m^{-2}{}_r \ s^{-1}$. Bed length: ◆ = 0.01 m_r, ■ = 0.015 m_r and ▲ = 0.03 m_r.*

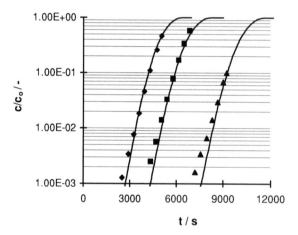

Figure 3 *Normalised outlet concentrations of toluene as a function of time when continuously feeding a mixture of toluene and nitrogen to a packed bed of activated carbon bed with $C_0 = 2.35 \ 10^{-3} \ kg \ m^{-3}{}_f$ and $u_s = 0.191 \ m^3{}_f \ m^{-2}{}_r \ s^{-1}$. Bed length: ◆ = 0.015 m_r, ■ = 0.02 m_r and ▲ = 0.03 m_r.*

The simulated and measured toluene outlet concentrations as a function of time for three different bed lengths, a superficial gas velocity of $0.191 \ m^3{}_f \ m^{-2}{}_r \ s^{-1}$ and an inlet concentration of $2.35 \ 10^{-3} \ kg \ m^{-3}{}_f$ are shown in Figure 3. A good agreement is observed, which means that the model is able to account well for changes in the mass transport rates due to a change in the superficial velocity. The agreement is less for low normalised toluene outlet concentrations and the more so when the bed length increases.

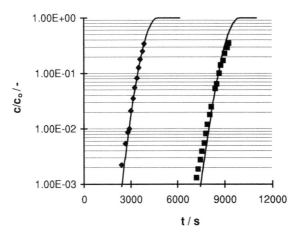

Figure 4 *Normalised outlet concentrations of toluene as a function of time when continuously feeding a mixture of toluene and nitrogen to a packed bed of activated carbon, with $C_0 = 4.83 \ 10^{-3}$ kg m^{-3}_f and $u_s = 0.127 \, m^3_f \, m^{-2}_r s^{-1}$. Bed length:* ◆ = *0.015 m_r,* ■ = *0.035 m_r.*

Figure 4 shows the simulated and measured outlet concentration profiles for a toluene inlet concentration of $4.83 \ 10^{-3}$ kg m^{-3}_f and a superficial velocity of 0.127 $m^{-3}_f m^{-2}_r s^{-1}$. The effects of an increase in the inlet concentration are well simulated by the model, resulting in a good agreement with the measured outlet concentrations, though for low concentrations again a slight difference occurs between simulated and measured values. The relative deviation between the calculated and measured time interval during which a given toluene outlet concentration is reached, is always less than 4 %.

4 CONCLUSIONS

A mathematical model was developed to simulate the adsorption process of toluene in shallow packed beds of activated carbon. The model includes plug flow and axial dispersion to describe the mass transport in the axial direction. Mass transfer limitation between the bulk of the gas phase and the outer surface of the particle is accounted for. Mass transport inside the particle is described by a diffusive contribution assuming Fick's law with an effective molecular diffusion coefficient as transport parameter. The agreement between simulated and measured toluene outlet concentrations is good. However, the agreement becomes slightly less at low normalised toluene concentrations, the more so when the bed length increases.

List of symbols

Roman

a_v	external surface area of the particles	$m^2_i \, m^{-3}_{pt}$
b_0	microporosity coefficient	K^{-2}

C	gas phase concentration inside the particle	kg m^{-3}$_f$
C_0	inlet concentration	kg m^{-3}$_f$
C_b	bulk gas phase concentration	kg m^{-3}$_f$
C_i	concentration at the outer surface of the particle	kg m^{-3}$_f$
C_s	vapour concentration adsorptive	kg m^{-3}$_f$
D_{ea}	effective axial diffusion coefficient	m^3$_f$ m^{-1}$_r$ s^{-1}
D_{eff}	effective diffusion coefficient inside the particle	m^3$_f$ m^{-2}$_r$ s^{-1}
j_D	Chilton-Colburn factor	--
k_f	mass transfer coefficient	m^3$_f$ m^{-2}$_i$ s^{-1}
L_b	bed length	m$_r$
q	adsorptive loading	kg m^{-3}$_{pt}$
q_{max}	maximum adsorptive loading	kg m^{-3}$_{pt}$
r	radial coordinate inside the particle	m$_{pt}$
T	temperature	K
t	clock time	s
u_s	superficial gas velocity	m^3$_f$ m^{-2}$_r$ s
z	axial coordinate inside the filter bed	m$_r$

Greek

β	affinity coefficient	--
ε_b	bed porosity	m^3$_f$ m^{-3}$_r$
ε_p	particle porosity	m^3$_f$ m^{-3}$_{pt}$
μ	dynamic viscosity	kg m^{-1} s^{-1}
ρ	density fluidum	kg m^{-3}$_f$

Dimensionless groups

Pe_{ax}	Peclet number $= u_s d_p / D_{ea}$
Re	Reynolds number $= \rho u_s d_p / \mu$
Sc	Schmidt number $= \mu / (\rho D_m)$

Subscripts

f	fluidum
i	interface
pt	particle
r	reactor

References

1. C. A. Silebi and W. E. Schiesser, 'Dynamic Modeling of Transport Process Systems', Academic Press Inc., San Diego, 1992.
2. NAG, 'Fortran Library Manual', NAG Ltd., Wilkinson House, Oxford, 1991.
3. A. E. Duisterwinkel, *Carbon*, 1993, **31**, 8, 1354.
4. D. J. Gunn, *Chem. Eng. Sci.*, 1987, **42**, 2, 363.
5. P. N. Dwivedi and S. N. Upadhyay, *Ind. Eng. Chem., Process Des. Dev.*, 1977, **16**, 2, 157.

THE CHANGES OF SURFACE AREA AND POROSITY OF CARBON FIBRES DURING THEIR ACTIVATION

S. F. Grebennikov[1], A. V. Koulitchenko[1] and J. T. Williams[2]

[1] St. Petersburg State University of Technology and Design,
18 B.Morskaya str.,
St. Petersburg 191186,
RUSSIA.

[2] De Montfort University,
The Gateway,
Leicester,
LE1 9BH,
ENGLAND.

1 INTRODUCTION

The fine structure of activated carbon fibres (ACF) is not well known. Methods of X-ray scattering can be used only for the determination of the average size of crystallites and pores but do not give information about interrelation of pores and their shape. This information is necessary for a range of practical and theoretical aims, for example prediction of sorption of gas or liquid molecules with different shape, interpretation of kinetic data, and for the calculation of diffusion coefficients where an effective size of the area with uniform structure is required. Because of the uncertainty of this effective size, the diffusion coefficients differ by several orders of magnitude for the same type of active carbon granules and ACF.

Interrelations between structural elements of solids can be determined by methods of structural modelling. For analysis of the fine structure of ACF, a method of mathematical modelling present in [1] was used. In accordance with this method, contacted carbon crystallites create microporous zones which have micropores of two types: -

innercrystallite micropores, that are flat slits, created by basic facets of crystallites, and intercrystallite micropores, which have prismatic facets as walls, the shape of which is considered approximately as slit-like with width 2x.

The elementary microporous zone (EMZ) is a group of jointed micropores, with determinational size 1. In some cases EMZs are jointed, while in others, flat slit like mesopores exist between outer facets of EMZs. Mesopores are of width 2q, volume V_{ME} and specific surface area S_{ME}. In this case structure parameters can be calculated from:

$$V_m = W_0 + l/d \tag{1}$$

where V_m = total volume of microporous zone;
         W_0 = volume of micropores;
         d   = apparent density

$$l = bV_m / S_{ME} \tag{2}$$

where l = EMZ size:
         b = average number of sides of EMZ, which participate in mesopore creation
                                                                 b = 4

$$N = S_{ME}^3 / b^3 V_m^2 \tag{3}$$

where N = total number of EMZ, in mass unit of adsorbent

$$q = V_{ME} / S_{ME} \tag{4}$$

where q = half-width of mesopores

$$Z = W_0 / N \chi^3 \pi \, 5.54 \tag{5}$$

where Z = number of micropores in EMZ

$$\eta = W_0 d, \tag{6}$$
with $\chi = \eta/(1+\eta)$

where η = ratio of volume of intercrystallite pores to volume of innercrystallite
         pores
and χ = ratio of volume of intercrystallite pores to total volume of micropores

$$n = (1-\eta)W_0 /(\varpi N) \tag{7}$$

where n = number of crystallites in EMZ;
and ϖ = volume of a micropore

$$m = [\varpi(1+V_m / W_0)]^{1/3} \tag{8}$$

where m = crystallite size

$$N_0 = (1 + \eta)nN, \tag{9}$$

where N_0 = total number of micropores in mass unit of adsorbent
For calculation of the above parameters, W_0, S_{ME}, V_{ME}, and d were evaluated experimentally.

2 EXPERIMENTAL

ACF were produced from carbonised viscose fibres (CF) by their oxidation in H_2O (series I and II), $H_2O + CO_2$ (series III) and CO_2 (series IV) at temperatures of 1000 - 1200 K for different times.

The burn off degree α was used as an integral characteristic of activation

$$\alpha = (m_0 - m_t) / m_0 \tag{10}$$

where m_t, m_0 = fibres mass at time t and t = 0

Initial carbonised fibres did not contain open pores.

Methods of gas absorption (adsorptives: benzene, acetone, water at T = 298 K), mercury porosimetry, helium pycnometry and X-ray scattering were used to assess characteristics of ACF porous structure.

Initial data of microporous structure were obtained based on the theory of volume filling of micropores [2], the basic equation of which is:

$$W = W_0 \exp\left[-\left[\, RT\ln\left(P_0 / P\right) / \beta E_0\right]^2\right] \tag{11}$$

where   W = volume of micropores filled at relative pressure P / P_0 and temperature T;

     W_0 = total volume of micropores;

     β = affinity coefficient;

and E_0 = characteristical energy of benzene absorption

E_0 depends on micropore halfwidth

$$x = (13.29 - 153 \cdot 10^{-5} E_0) / E_0 \tag{12}$$

Volume of mezopores, V_{ME}, was determined as the difference between sorption volume at $P/P_0 = 1$ and volume of micropores W_0. Surface square of mesopores was calculated by the thermodynamic method of Kiselev [3].

The volume of macropores, V_{MA}, was determined by mercury porosimetry. Density, d, was determined using picnometric helium gas. Specific surface area was determined by a thermodynamic method. The consequence of the thermic equation of sorption $F(a, T, \Delta G) = 0$ is that during water vapour sorption by ACF, the statistical monomolecular layer is created at minimal value of sorption entropy. This corresponds to the point of inflection of the sorption isotherm in coordinates ΔG - a, i.e. plotted in a vs $\Delta G = RT\ln P/P_0$ axes.

The surface area, S_{MI}^T, of micropores was determined experimentally from the difference of sorption at the point of inflection and the sorption on the mezopores surface. The micropores surface area, S_{MI}^D, was determined independently from the hypothesis of slit-shape pores of Dubinin's method [2]. Both S_{MI}^T and S_{MI}^D values are in agreement and lead to the preliminary conclusion about slit-shape pore creation in ACF.

3 RESULTS AND DISCUSSION

As an example, the parameters of porous structure of ACF of Series I and II are given in Table1. All parameters of microstructure are increasing with increasing degree of burn off and the apparent density is tending to the density of graphite, showing that it is mostly the amorphous carbon phases burning off.

Table 1 *Characteristic values of porous structure of the ACF*

α (%)	Pore Volume (cm³g⁻¹)			x	S_{ME}	d
	V_{MI}	V_{ME}	V_{MA}	(nm)	(m²g⁻¹)	(g cm⁻³)
Series I						
17	0.28	0.04	0.00	0.58	41	1.78
20	0.31	0.07	0.00	0.62	48	1.79
25	0.34	0.09	0.01	0.71	43	1.81
30	0.39	0.10	0.02	0.73	83	1.84
39	0.44	0.16	0.04	0.76	109	1.92
46	0.62	0.28	0.06	0.95	189	1.98
Series II						
22	0.13	0.09	0.02	0.57	34	1.96
38	0.25	0.15	0.04	0.62	73	2.01
56	0.37	0.23	0.07	0.64	90	2.06
64	0.51	0.25	0.09	0.69	105	2.09

Calculated parameters of ACF microstructure are shown on Figure 1. Adequacy of the model is confirmed by the following:
1) Halfwidth of mesopores is between 1 and 2.6 nm, in accordance with measured sizes of equivalent radii of mesopores on the Kelvin equation.
2) Crystallite sizes r and m correspond to X-ray structure analysis data which give L_a and L_c between 1 - 2 nm.
3) m/r = 2.3 - 2.5, is in good correlation with L_a / L_c, which did not change during the activation process.

The model presented above gives the opportunity to calculate the micropores surface area S_{MI}^M as a product of unit micropore surface area and the number of micropores. The micropores surface area could be compared with surface areas S_{MI}^T and S_{MI}^D.

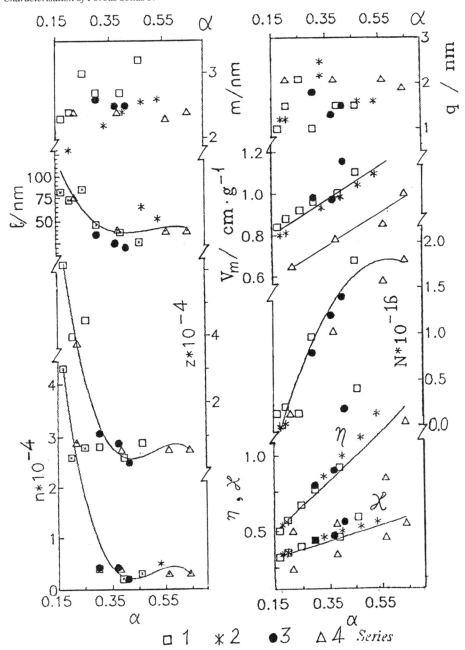

Figure 1 *The generalised curves of the development of the ACF microstructure elements during the activation process*

Table 2 *Specific surface micropore areas of ACF (Series II)*

α (%)	S_{MI}^{T} (m²g⁻¹)	S_{MI}^{D} (m²g⁻¹)	S_{MI}^{M} (m²g⁻¹)
18	544	608	624
33	702	693	710
47	790	755	772
53	843	802	822

Data in Table 2 show agreement of results of three independent determinations. The size of EMZ could be interpreted as a microfibril length, that satisfactorily corresponds to X-ray structural analysis data. This value is greater than in activated carbons [1]. This is in accordance with an idea of anisotropy structure of carbon fibres (CF) and anisotropy structure of activated carbons (AC). Thus, the model yields data that are in agreement with the real structure of ACF.

Development of CF microstructure in the process of oxidation is shown in Figure 1. The volume of micropore zones increases as the burning off increases, N, η, χ are gradually increasing. The size of EMZ strongly decreases until 30% burn off and then with further activation the size does not change. Probably this is because of burning off of amorphous carbon between fibrils leads to the creation of EMZ. Further activation associated with pore creation and increasing micropore sizes in fibrils, as number of crystallites, n, in EMZ and number of pores, Z, at $\alpha > 30$ - 40 % is not changing substantially.

Mesopore sizes correspond to the lower limit of capillary phenomena and they could be attributed to the transition type between supermicropores and mezopores.

Crystallite size, m, changes little during the activation process. Thus, intercrystallite pore creation probably could be normal for CF only at low degrees of burning off (as parameters η and χ are increasing gradually). This is confirmed by stability of parameters L_a and L_c, during the activation process.

4 CONCLUSION

Based on model interpretations, it is possible to affirm that slit-shape pores are created in ACF as in AC. These pores together with crystallites form microporous zones. Anisotropic structure of CF leads to the creation of only micro- and mesopores during the activation process. The volume of macropores is not substantial. The length of microporous zones sustantially increases in comparison with AC.

Independently of the activation process conditions, porous structure elements are developing in accordance with the same laws, i. e. the generalised curves could be used for prediction of burning degree to obtain the CF porosity that is needed.

The model presentation of micropore zone size helps to interpret data of kinetic measurements.

During the activation process two porous systems are created - the system of transport pores and the system of micropores (microporous zone). Diffusion coefficients in the two systems differ substantially because of sterical limitations for molecules transfering in micropores. For the correct calculation of effective diffusion coefficient the bidispersed model of porous solid [5] should be used, and not the homogeneous model. The information required to calculate the size of microporous zones for the bidispersed model is presented in this paper.

References

1.   M. M. Dubinin, D.V. Fedoseev, Proceedings of the USSR Academy of Sciences, Chemistry, 1982,2,246.
2.   M. M. Dubinin, "Characterisation of Porous Solids 1", Elsevier, Sci. Publishers B.V., Amsterdam, 1988,127.
3.   A. V. Kiselev, Proceedings. of the Second Int. Congress of Surface Activity, London, 1957, No.2., 189.
4.   S. F. Grebennikov et al., Proceedings of the USSR Academy of Sciences, Chemistry, 1983, 3, 498.
5.   A. M. Voloshchuk, PhD Thesis, Inst. of Phys. Chem., Academy of Sci. of Russia, 1989.

MESOPOROUS SILICAS FROM MIXED TEMPLATES

Sumio Ozeki and Mitsuaki Yamamoto
Department of Chemistry, Faculty of Science, Chiba University
1-33 Yayoi-cho, Inage-ku, Chiba 263, Japan

Kazunori Nobuhara
Fuji Silysia Chemical LTD.,
1846, 2-chome, Kozoji-cho, Kasugai-shi, Aichi 487, Japan

1 INTRODUCTION

The mesoporous silicas, such as MCM-[1,2] and FSM-type silicas,[3] have a pore with a narrow size distribution and non-interconnecting networks, as shown in a striking, beautiful, sharp step on N_2 adsorption isotherm at 77 K. Therefore, the potential materials have been paid attention to from the point of view of both fundamentals (a standard adsorbent) and application (a host substance for, e.g., catalysts).

The size and shape of pores of the silicas can be controlled by template quarterly alkylammonium molecules; the length and number of hydrocarbon chain dominate the unit size (or pore size) and structure (hexagonal and lamellar structure) of organosilicates, respectively. Also, the solubilization of small organic molecules, such as mesitylene,[1b,4] into organosilicates may produce large, controlled pores. It seems that the structure of the organosilicates should be limited by a geometry of a template unit (supramolecules or suprastructure). Thus, we might obtain any type of mesoporous materials by the control of the geometry of a template unit, using methods such as mixing a few templates and localization of solubilizates.

We prepared organosilicates from kanemite dispersions containing two kinds of quarterly alkylammonium chloride ($C_nH_{2n+1}N(CH_3)_3Cl$: n=8, 12, 16; $(C_{12}H_{25})_2N(CH_3)_2Cl$; $C_{12}H_{25}NH_3Cl$) at 338 K. Three type of mixed surfactant systems are selected: a short chain/long chain mixture, a small head/large head mixture and a single chain/double chain mixture. Kanemite ($NaHSi_2O_5 \cdot 3H_2O$) is known as a layered material having ion exchange ability. We also used various mixtures of hexadecyltrimethyl-ammonium bromide and sodium salicylate as a template unit, which showed viscoelasticity due to thread-like micelles. These results showed utility for control of the size and shape of the pores of silicas.

2 EXPERIMENTAL SECTION

2.1 Materials

2.1.1 Mixed Quarterly Alkylammonium Systems   Organosilicates were prepared from aqueous kanemite dispersions containing two kinds of quarterly alkylammonium chloride DCl ($C_nH_{2n+1}N(CH_3)_3Cl$: n=8, 12, 16; $(C_{12}H_{25})_2N(CH_3)_2Cl$; $C_{12}H_{25}NH_3Cl$) at pH

7 - 8 and 338 K for 1 week. Each surfactant will be referred to as CnMt, C12^{2}, and C12, respectively. Total concentration of surfactant solutions mixed at various molar fraction (X = 0, 0.25, 0.5, 0.75, and 1) was 0.1 mol/dm^3. Kanemite was obtained by vacuum-drying a water glass (SiO$_2$/Na$_2$0 = 2.03 (molar ratio); Fuji Chemical Co., Ltd.) at room temperature, subsequent calcination of the dried disodium silicate at 973 K for 6 h, and dispersion in distilled water (2g/L).

Prepared organosilicates, referred to like CnMt using the name of surfactants used, were calcined at 973 K in air for 6 h.

2.1.2 Viscoelastic Systems   Another type of organosilicate was synthesized from aqueous water-glass solutions containing hexadecyltrimethylammonium bromide (DBr) and sodium salicylate (NaSal) as a template unit at pH 8.5 and 338 K for 3 days. The composition (molar ratio) of the reaction solutions was SiO$_2$/D = 1 and Sal/D = 0 - 2 ($\equiv$x$_{sal}$). The concentration of D (and thus SiO$_2$) was 0.05 mol/dm^3, irrespective of the composition. The reaction solutions in the x$_{sal}$ range 0.5 - 1.5 showed viscoelasticity due to thread-like micellar formation.

Prepared organosilicates were separated by centrifugation at 10,000 rpm for 15 min, dried at room temperature, and subsequently calcined at 973 K in air for 6 h.

2.2 Characterization

The samples were examined by X-ray (CuKα) diffraction (Rigaku Denki Geiger Flex 2028). The d-spacing from the (100) peak, d$_{100}$, was calculated. Amounts of nitrogen and carbon atoms in the organosilicates prepared were determined by elemental analyses. The intercalated amount of surfactants (x$_R$^{Int}) was estimated from nitrogen contents and expressed in the unit of mmoles/(g silicate). N$_2$ adsorption on calcined samples, dried at 523 K and 10^{-5} Torr for 1 h, was measured by gravimetric method at 77 K. Pore size distributions and specific surface areas of calcined samples were calculated from the Cranston-Inkley method (the CI method) and the B.E.T. method, using the N$_2$ adsorption isotherms, respectively.

3 RESULTS AND DISCUSSION

3.1 Mixed Quarterly Alkylammonium/Kanemite Systems

3.1.1 C8Mt/C16Mt Systems   Figure 1 shows amounts of intercalated alkylammonium ion (= amount of N) as a function of the molar fraction of C16Mt, X$_{16}${$\equiv$C16Mt/(C16Mt + C8Mt)}. The intercalated amount increased stepwise from ca. 1.0 to 2.2 mmoles/(g silicate) or from D/Si (molar ratio)=0.058 to 0.15 beyond X$_{16}$=0.25, and became almost constant at X$_{16}$ $\geq$ 0.5. Thus, alkylammonium ions should be intercalated by ion exchange at definite sites, and C16Mt ion would be selectively intercalated at least at X$_{16}$ $\geq$ 0.5.

The calcined C16Mt showed a typical hexagonal diffraction pattern. The addition of C8Mt to C16Mt made the hexagonal pattern vague, and no appreciable diffraction peaks appeared at X$_{16}$ $\leq$ 0.25. The d$_{100}$ values of the hexagonal samples were constant (4.0 ± 0.2 nm) at X$_{16}$ $\geq$ 0.5. In the same region (X$_{16}$ $\geq$ 0.5), N$_2$ adsorption isotherms showed a clear step in the relative pressure region of *P*/*P$_0$* = 0.4. The isotherm of X$_{16}$=0.25 had a faint step at around *P*/*P$_0$* = 0.4, but there was no step on the isotherm of X$_{16}$=0 (C8Mt). The pore size for X$_{16}$ = 0.5, 2.7 nm in diameter, was smaller than that of a calcined C16Mt sample (X$_{16}$=1), suggesting that a short chain ammonium can regulate a pore size. Figure 1 includes pore size from the CI method and B.E.T. surface area.

These also demonstrate a step between X_{16}=0.25 and 0.5.

These results suggest that ion exchanged alkylammonium ions (C8Mt and C16Mt) between silicate layers should form rodlike, mixed micelles, which would lead to silicate-layer-bending. The hexagonal pores formed in such a process must have a definite size which is limited by the structure of the silicate layer. If $HSi_2O_5^-$ is a structural unit, the pore size must be discrete, irrespective of the composition of the mixed micelles.

3.1.2 C12Mt/C12 Systems   The amount of intercalated C12 was 7.1 mmoles/(g silicate) or D/Si=0.49 and three times larger than that of C12Mt. This intercalated amount is almost equal to the ion-exchange limit (7.3 mmol/ g silicate) or $D/HSi_2O_5^-$ =1. With the addition of C12Mt to C12, total intercalated amount monotonically decreased (Figure 2). This suggests mixed intercalation of the two surfactants. Also, the slightly convex curve indicates weak selectivity of C12 in the intercalation.

All intercalated silicates showed (100) peak, and (110) and (200) peaks became manifest by mixing the two surfactants at the molar fraction $\dot{X}_{12Mt}\{\equiv C12Mt/(C12Mt + C12)\}$=0.5 and 0.75. Probably, they all have a hexagonal structure. The d_{100} value decreased with increasing the fraction of C12Mt from 5.2 nm to 3.7 nm, and became unchanged at $X_{12Mt} \geq 0.5$. These trends seem to arise from the geometrical packing efficiency due to head/tail size balance or the curvature of a micellar surface.

The calcination of the organosilicates from the mixed systems made even (100) peak very small, but brought about two small (100) peaks corresponding to the (100) peaks of the calcined C12Mt and C12 silicates; 3.6 and 4.9 nm, respectively. At X_{12Mt}=0.25 no clear XRD peaks appeared. The XRD pattern of the calcined silicas from the miced systems seems to change systematically with increasing X_{12Mt}, whose tendency was quite different from that of the corresponding organosilicates.

The N_2 adsorption isotherm of the calcined C12Mt sample corresponds to a reversible type IV isotherm with a vague step at around $P/P_0 = 0.25$, as shown in Figure 3. The addition of small amount ($X_{12Mt} = 0.75$) of C12 reduced N_2 adsorption and lead to an irreversible type IV isotherm with a faint step at around $P/P_0 = 0.45$. Further

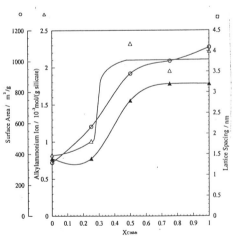

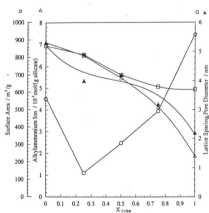

Figure 1 *Total amount of intercalated alkylammonium ion (Δ), pore size from the Cl method (▲) and the B.E.T. surface area (o) as a function of the molar fraction of C16Mt, $X_{16}\{\equiv C16Mt/(C16Mt + C8Mt)\}$, in the C16MtCl/C8MtCl sys-tems.*

Figure 2 *Total amount of intercalated alkylammonium ion (Δ), d_{100} of inter-calated silicates (□), pore size from the Cl method (▲) and B.E.T. surface area (o) as a function of the molar fraction of C12Mt, X_{12Mt} $\{C12Mt/(C12Mt + C12)\}$, in the C12MtCl/C12Cl systems.*

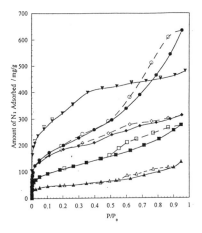

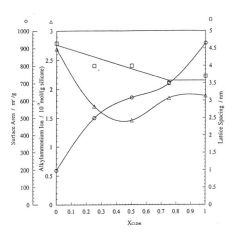

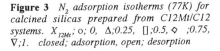

Figure 3 *N_2 adsorption isotherms (77K) for calcined silicas prepared from C12Mt/C12 systems. X_{12Mt}; o; 0, Δ;0.25, $\square$;0.5, $\diamond$;0.75, ∇;1. closed; adsorption, open; desorption*

Figure 4 *Total amount of intercalated alkylammonium ion (Δ), d_{100} ($\square$) and B.E.T. surface area (o) as a function of the molar fraction of C12Mt, $X^2_{12Mt}\{\equiv C12Mt/ (C12Mt + C12^2)\}$, in the C12MtCl/C12^2 systems.*

addition of C12 made their isotherms a typical type IV, and gradually diminished amount of N_2 adsorbed to lead to a minimum at $X_{12Mt} = 0.25$. The N_2 adsorption isotherm of the calcined C12 sample ($X_{12Mt} = 0$), which showed a typical type IV isotherm with a large hysteresis at high pressure region, was markedly different from that of $X_{12Mt} = 0.25$. This large effect of small amount of C12Mt on the C12 structure cannot be deduced from changes in the intercalated amounts and the XRD patterns of the intercalated samples. Thus, during calcination, some structural change should be induced.

3.1.3 C12Mt/C12^2 Systems   The intercalation amount decreased with the addition of C12Mt from 2.7 mmoles/(g silicate) for a C12^2 sample ($X^2_{12Mt}\{\equiv C12Mt/(C12Mt + C12^2)\}=0$) to 1.9 mmoles/(g silicate) for a C12Mt sample ($X^2_{12Mt} = 1$), through a minimum at $X^2_{12Mt} =0.5$ (Figure 4), suggesting a repulsive interaction between C12Mt and C12^2 molecules. The intercalated C12^2 sample had a lamellar structure having $d_{100} = 4.7$ nm. Adding C12Mt to the C12^2 system, presumably the structure was converted into a hexagonal one. At $X^2_{12Mt} \leq 0.5$, the intercalated samples seems to be similar structure among them, considering the X^2_{12Mt} dependence of the XRD results and the intercalated amount. The intercalated samples of $X^2_{12Mt} = 0.75$ and 1 were hexagonal and similar each other in d_{100} values (3.7 nm) and intercalated amounts, but after calcination these samples were distinguished from each other in d-spacing and pore size distribution (and pore size; see Figure 6). This again demonstrates that the pore structure should change during calcination, as shown in the C12Mt/C12 systems.

Figure 5 shows N_2 adsorption isotherms of the calcined samples. The addition of small amount of C12^2 to C12Mt ($X^2_{12Mt} = 0.75$ and 0.5) made the step sharper than in the C12Mt system, because of increase in the curvature of micellar surfaces. The sample from the two chain surfactant ($X^2_{12Mt} = 1$) had a little pore (Figure 6) and a small specific surface area (200 m^2/g), as suggested by no XRD peaks. The B.E.T. surface area monotonically increased with X^2_{12M} up to 820 $\pm$ 100 m^2/g (Figure 4).

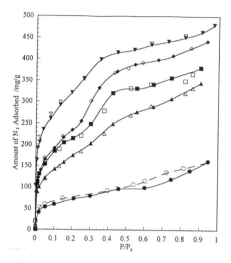

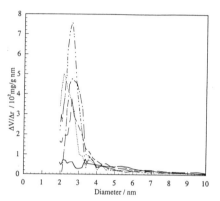

Figure 5 N_2 *adsorption isotherms (77K) for calcined silicas prepared from the C12Mt/C12^2 systems.*
X^2_{12Mt}; o;0, Δ;0.25, $\square$;0.5, $\diamond$;0.75, ∇;1.
closed; adsorption, open; desorption

Figure 6 *Pore size distribution (the Cl method) from N$_2$ adsorption isotherms for calcined silicas prepared from the C12Mt/C12^2 systems.*
X^2_{12Mt}; ———— ;0, — — — — ;0.25, —·—·—;0.5, —··—··;0.75, ··········· ;1

3.2 Viscoelastic Systems

Aqueous hexadecyltrimethylammonium bromide (DBr) solutions containing sodium salicylate (NaSal) has viscoelastic properties in certain conditions. For example, it is well known that the relaxation time in the viscoelastic system containing 0.1 mol/dm^3 DBr at 298 K comprises three regions of Sal/D ratio (x_{sal}); $x_{sal} \leq 1$, $1 < x_{sal} \leq 2$ and $x_{sal} \geq 2$.[4] At $x_{sal} \leq 1$, non-entangled, short rodlike micelles at $x_{sal} \ll 1$ changes into a fully-entangled thread-like micelles beyond about $x_{sal} = 0.3$. At $1 < x_{sal} \leq 2$, it is believed that pseudo-networks comprising thread-like micelles may be formed.

Adding sodium silicate (SiO$_2$/D = 1) to the above systems (0.05 mol/dm^3 DBr), viscoelastic behavior of the solution was observed by eye at 298 K in the region of $x_{sal} \approx$ 0.75-2. The viscoelasticity was diminished with elevating temperature up to 338 K.

The intercalated amounts of D (or N) and Sal as a function of x_{sal} are shown in Figure 7. Intercalated D increased linearly with x_{sal} and became constant (2.7 mmoles/(g silicate)) in the x_{sal} range 1 to 2. Intercalated Sal also increased in parallel with the intercalated amount of D at $x_{sal} < 1$. On the other hand, intercalated Sal increased stepwise at around $x_{sal} = 1$ and reached 2.9 ± 0.3 mmoles/(g silicate) at $x_{sal} = 2$. This means that Sal was intercalated or solubilized into D-micelles as an ion-pair with D (1:1

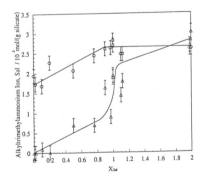

Figure 7 *Intercalated amounts of hexadecyltrimethylammonium ion (D) and salicylate ion (Sal) as a function of Sal/D ratio (x_{sal}) in the DBr/NaSal systems.* o; D, Δ; Sal

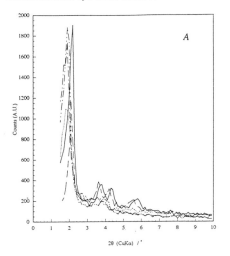

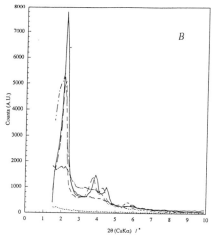

Figure 8 *X-ray diffraction patterns of the intercalated silicates (A) and the calcined silicas (B) from the hexa-decyltrimethylammonium bromide and sodium salicylate systems.*
A. x_{sal}: ——— ;0, — — — — ;0.2, —·—·— ;0.76, —··——;1.09, ------ ;2
B. x_{sal}: ——— ;0, — — — — ;0.2, —·—·— ;0.5, —··——;1.09, ------ ;2

complex) at x_{sal} < 1 and that the intercalated Sal/D ratio (x_{sal}^{Int}) is ca. 0.35 at x_{sal} = 1 (an extrapolated value) and ca. 1 at x_{sal} = 2. Since D content remaind unchanged and Sal content increased with x_{sal} at 1<x_{sal}<2,; D paired only with Sal by ion-exchange of $HSi_2O_5^-$ with Sal⁻. Then, the replacement may induce the structural change from hexagonal to lamellar (see below), presumably, because of the reduction of the interactions silicate ions with D which should be responsible for forming the hexagonal structure due to hydrocarbon chain interaction.

From the above composition analysis, a mechanism for the organosilicate formation may be deduced:

$$nD^+ + nSiO^- \rightarrow (D^+ \cdot SiO^-)_n \quad \text{for } x_{sal} = 0$$

$$nD^+ + nSiO^- + m(D^+ \cdot Sal^-) \rightarrow (D^+ \cdot SiO^-)_n(D^+ \cdot Sal^-)_m \quad \text{for } 0 < x_{sal} < 1$$

$$nD^+ + nSiO^- + n/2(D^+ \cdot Sal^-) \rightarrow (D^+ \cdot SiO^-)_n(D^+ \cdot Sal^-)_{n/2} \quad \text{for } x_{sal} = 1$$

$$(D^+ \cdot SiO^-)_n(D^+ \cdot Sal^-)_{n/2} + k(Na^+ \cdot Sal^-)$$
$$\rightarrow (D^+ \cdot SiO^-)_{n-k}(D^+ \cdot Sal^-)_{n/2+k}(Na^+ \cdot SiO^-)_k \quad \text{for } 1 < x_{sal} < 2$$

$$(D^+ \cdot SiO^-)_n(D^+ \cdot Sal^-)_{n/2} + n(Na^+ \cdot Sal^-)$$
$$\rightarrow (D^+ \cdot Sal^-)_{3n/2}(Na^+ \cdot SiO^-)_n \quad \text{for } x_{sal} = 2$$

The Sal-cointercalated orgnosilicates at x_{sal} < 0.2 showed a typical hexagonal diffraction pattern (Figure 8). Probably, the intercalated organosilicates prepared at x_{sal} < 1 are hexagonal, because calcination of them gave a clear (100) peak, although their XRD peaks, besides (100) peak, are obscure. (Usually, a lamellar structure is destroyed by calcination). On the other hand, the organosilicates prepared at 1≤x_{sal}≤2 may be lamellar, as suggested by comparison of their XRD patterns between before and after calcination. The d-spacing increased with the ion-pair solubilization from 4.0 nm to 4.7 nm for before calcination and from 3.7 nm to 4.4 nm for after calcination. Beyond x_{sal} = 2, d-spacing is almost constant, 4.7 nm, for the organosilicates and 4.4 nm for the calcined one, as long as a (100) peak appeared.

The N_2 adsorption isotherms of the calcined silicates for x_{sal} < 1 showed a clear step in the medium pressure region, whose onset shifted with Sal-intercalation to a higher pressure, as shown in Figure 9. The step was the sharpest at x_{sal} = 0.2. The isotherms for x_{sal} ≥ 0.5 all had appreciable external surface area, as suggested by the adsorption-increase in the high pressure region. As Sal-content increased at $x_{sal} > 1$, adsorption amount decreased and a step on the isotherm disappeared. The samples prepared at around $x_{sal} = 1$ ($1 \le x_{sal} < 1.1$) tend to be different from run to run, and gave a reversible type IV isotherm with a small step or a type II isotherm (no hysteresis), as seen in Figure 9 ($x_{sal} = 1$ and 1.09). At $x_{sal} = 2$, the isotherm seems to be type II. The pore size distributions became somewhat broader with Sal-addition, but still sharp, and the pore size changed from 3.2 nm ($x_{sa} = 0$) to 4.2 nm ($x_{sal} = 1$).

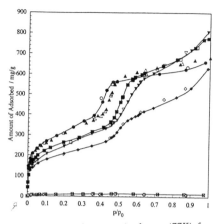

Figure 9 *N_2 adsorption isotherms (77K) for calcined silicas prepared from the hexadecyltrimethylammonium bromide and sodium salicylate systems.*
x_{sal}; 0;0, $\triangle$;0.2, $\square$;0.5, ∇;0.97, $\diamondsuit$;1.09, $\boxtimes$;2. closed; adsorption, open; desorption.

The structure of the organosilicates changed with the molar ratio (x_{sal}^{Int}) of salicylate ion to C16Mt ion in the organosilicates, according to the bulk composition (x_{sal}) at which the structural changes of bulk solutions containing thread-like micelles should occur. Thus, thread-like micelles must work as a template for the organosilicate structure.

4 CONCLUSION

Mesoporous silicas were prepared from mixed template solutions of quarterly ammonium surfactants by using a kanemite as a silicate source. Three different mixed template systems were examined: a short chain/long chain mixture, a small head/large head mixture and a single chain/double chain mixture. The results showed that the mixed template method is a potential procedure to regulate the pore structure of mesoporous silicas, such as size and shape, by geometrical factors (a head/tail size balance or micellar surface curvature), silicate/alkylammonium interactions, a bending energy of silicate layer/chain interaction balance   However, there is no accurate structure of kanemite available for understanding a mechanism of intercalation of alkylammonium ion and the following silicate-layer-folding. Also, the samples obtained here were not necessarily sufficient to extract some conclusions, probably because the intercalation of long chain ammonium ions into kanemite is not so easy. Therefore, another experiment using sodium silicate as a silica source is proceeding. Anyway, the results suggest that even when inhomogeneous surfactants are used as a template, it is possible to prepare homogeneous mesoporous silicas, as long as mixed micelles are formed or selective intercalation may occur. When the micro-phase separation occurs in micellar phases, two kinds of pore may coexist. This means that practically homogeneous mesoporous silicas may be prepared even if a industrial quarterly alkylammonium is used

The mesoporous silicas were obtained by the organosilicates prepared from viscoelastic silicate solutions containing sodium salicylate (NaSal) and hexadecytrimethylammonium bromide (DBr). The structure of the organosilicates changed with the molar ratio (x_{sal}^{Int}) of salicylate ion (Sal) to hexadecytrimethylammonium ion (D) in the organosilicates (DSi), corresponding to the composition (x_{sal}) at which the structural changes of bulk solutions containing thread-like micelles should occur. Thus, the thread-like micelles may work as a template for the organosilicate structure. In low content of Sal ($x_{sal}<1$), the ion pair (DSal) may be solubilized by the organosilicates (DSi), but the hexagonal structure was maintained. At around $x_{sal}=1$, x_{sal}^{Int} increased from 0.35 to 0.7 and the lamellar structure appeared, indicating that the hexagonal-to-lamellar transition may occur at around $x_{sal}=1$. At $1<x_{sal}<2$, the D content remaind unchanged, but the Sal content increased; x_{sal}^{Int} approached 1 as X_{sal} approached 2. This means that D paired only with Sal by ion-exchange of Si with Sal: the replacement may induce the structural change from hexagonal to lamellar, presumably, because of the reduction of the interactions silicate ions with D which should be responsible for forming the hexagonal structure due to hydrocarbon chain interaction. The further addition of NaSal seemed to form the lamellar structure having the repeating layer of 1.3 nm. By calcination of the organosilicates, the mesoporous silicas were obtained from the hexagonal organosilicates, but not from the lamellar organosilicates, except for those at around $x_{sal}=1$.

5 ACKNOWLEDGEMENT

We thank Mr. Junichi Miyamoto and Mr. Tomonari Hayashi for preliminary synthetic experiments.

References

1. a. C. T. Kresge, M. E. Leonowicz, W. J. Roth, J. C. Vartuli and J. S. Beck, *Nature*, 1992, **359**, 710. b. J. S. Beck, J. C. Vartuli, W. J. Roth, M. E. Leonowicz, C. T. Kresge, K. D. Schmitt, C. T.-U. Chu, D. H. Olsen, E. W. Sheppard, S. B. McCullen, J. B. Higgins and J. L. Schlenker, *J. Am. Chem. Soc.*, 1992, **114**, 10834. c. J. S. Beck, J. C. Vartuli, G. J. Kennedy, C. T. Kresge, W. J. Roth and S. E. Schramm, *Chem. Mater.*, 1994, **6**, 1816. d. J. C. Vartuli, C. T. Kresge, M. E. Leonowicz,C. T.-U. Chu, S. B. McCullen, I. D. Jhonson and E. W. Sheppard, *Chem. Mater.*, 1994, **6**, 2070. e. J. C. Vartuli, K. D. Schmitt, C. T. Kresge, W. J. Roth, M. E. Leonowicz, S. B. McCullen, S. D. Hellring, J. S. Beck, J. L. Schlenker, D. H. Olson and E. W. Sheppard, *Chem. Mater.*, 1994, **6**, 2070.
2. a. A. Monnier, F. Schüth, Q. Huo, D. Kumar, D. I. Margolese, R. S. Maxwell, G. D. Stucky, M. Krishnamurty, P. Petroff, A. Firouzi, M. Janicke and B. F. Chmelka,, *Science*, 1993, **261**, 1299. b. Q. Huo, D. I. Margolese, U. Ciesla, P. Feng, T. E. Gier, P. Sieger, R. Leon, P. M. Petroff, F. Schüth and G. D. Stucky, *Nature*, 1994, **368**, 317. c. Q. Huo, D. I. Margolese, U. Ciesla, D. G. Demuth, P. Feng, T. E. Gier, P. Sieger, A. Firouzi, B. F. Chmelka, F. Schüth and G. D. Stucky, *Chem. Mater*, 1994, **6**, 1176.
3. a. T. Yanagisawa, T. Shimizu, K. Kuroda and C. Kato, *Bull. Chem. Soc. Jpn.*, 1990, **63**, 988. b. S. Inagaki, Y. Fukushima and K. Kuroda, *J. Chem. Soc. Chem. Commun.*, **1993**, 680. c. S. Inagaki, Y. Fukushima and K. Kuroda, *Stud. Surf. Sci. Catal*, 1994, **84**, 125.
4. T. Shikata, H. Hirata, T. Kotaka, *Langmuir*, 1987, **3**, 1801.

BORON DISTRIBUTION IN MFI-TYPE ZEOLITES :
INTEREST OF ADSORPTION MICROCALORIMETRY AT 77 K

C. Sauerland, P. L. Llewellyn, Y. Grillet & J. Rouquerol

Centre de Thermodynamique et de Microcalorimétrie du C.N.R.S., 26 Rue du 141ème RIA, 13003 Marseille (FRANCE)

ABSTRACT

This study presents the influence of a post-synthesis treatment of boron-containing MFI-type zeolites (different calcination procedures and humidification) on the adsorption interactions followed by isothermal microcalorimetry at 77 K coupled with quasi-equilibrium adsorption volumetry. The adsorption of two species (Ar and CO) is compared and discussed in view of the texture and surface chemistry of the samples. The splitting of the microcalorimetric curve in the fluid-like to solid-like transition region of the adsorption of argon would seem to highlight surface textural modifications due to a partial deboronation of one of the samples prepared.

1 INTRODUCTION

The principal industrial interest in MFI-type zeolites stem from their pore structure leading to, for example, shape selectivity in the case of catalysis. The catalytic activity of such materials results from the number of Brönsted or Lewis sites present. These sites are most often the T(III) atoms (trivalent atoms such as Al, Fe, B ...) which are substituted for silicium in the crystal structure.

An estimation of the properties of potential catalysts (as well as all porous or divided solids) can be obtained from physisorption studies at 77K using simple probes. A comparison of the results obtained with non-polar atoms and polar molecules gives an insight into the textural and chemical nature of the surface. The use of large crystals permits the phenomena due to the external surface to be virtually ignored allowing the pore properties to be uniquely studied. It is possible to couple low temperature physisorption with isothermal microcalorimetry to further characterise these materials.

Fundamental physisorption studies on silicalite-I, the pure silica MFI-type analogue, have shown unusual adsorbate phase transitions [1-3]. For non-polar atoms such as argon and krypton, a single phase transition is observed of the type [1,2] :

fluid-like   ⇔   solid-like

indicating a change in adsorbate behaviour from that of dynamic disorder to that of commensurate order. For polar molecules such as nitrogen and carbon monoxide a double transition is observed [1,3] :

fluid-like   ⇔   network fluid   ⇔   solid-like

where a partially ordered mobile phase is characterised in addition to those of the initial dynamic disordered state and final commensurate ordered state. The transitions are characterised by steps in the isotherm, peaks in the adsorption enthalpy curve and significant changes in the neutron diffraction patterns [1,2].

Samples of more applied interest such as Al-MFI and Fe-MFI zeolites also induce adsorbate transitions at low temperatures. These transitions are affected by, not only the extent of T(III) substitution, but also by the T(III) distribution within the crystal (zoning) [4]. These phase transitions are also modified by partial or total pore blockage, for example on n-nonane preadsorption [5].

In the present study, B-MFI type zeolites are studied by adsorption volumetry at 77K coupled with isothermal microcalorimetry. This study has arisen due to the relative difficulty to examine the boron distribution incorporated into the network on the scale of individual crystals. X-ray emission mapping, NMR and neutron diffraction all encounter problems due to the relatively small size of the boron atom. Differential thermal analysis permits a determination of the Si/B ratios as well as the distribution in B-MFI zeolites with the template still present, however template free crystals can not be investigated as it the template decomposition which is analysed [6].

A single as-synthesised B-MFI sample is taken for this study. However four different pre-treatment protocols are used, allowing the influence of calcination conditions and humidification to be investigated. Sample Controlled Thermal Analysis (SCTA) is used to calcine the samples via a "soft chemistry" route and this is compared with a traditional calcination protocol. Part of these sample were maintained in a wet atmosphere to see the effect of this humidification on the samples.

2 EXPERIMENTAL

2.1 Adsorbents

The studied samples were prepared at Ecole Nationale Superieure de Chimie in Mulhouse (France) by an alkaline-free, fluoride route [7]. The molar ratio of Si/B=50 was obtained by atomic absorption spectroscopy. The boron distribution in the as-synthesised sample is homogeneous as shown by the method developed by Soulard *et al.* [6]. The sample was calcined up to 873 K following two different procedures : ① sample BC via Sample Controlled Thermal Analysis (SCTA)[8] (under a dynamic, residual pressure of 10^{-1} Pa) which proffers a predetermined rate of template decomposition and avoids the development of cracks in large crystals upon calcination, or, ② sample BO via a linear increase of temperature under an oxygen flow. Afterwards, part of each of these two samples are hydrated in a relative water pressure of 0.8 over a saturated NH_4Cl solution for 14 days (sample BCW and sample BOW respectively). Before each adsorption experiment the samples are outgassed to 473 K via SCTA, here in order to ensure that the samples are prepared in a reproducible manner. The final pressure above the sample after pre-treatment is inferior to 10^{-1} Pa.

Table 1. *Sample preparation of adsorbents used in this study.*

Sample	calcination procedure	humidification in $p_{(H2O)}/p^0=0.8$
BC	① SCTA.	no
BCW	① SCTA.	14 days
BO	② linear heating, oxygen flow	no
BOW	② linear heating, oxygen flow	14 days

2.2 Adsorptives

The adsorptives, argon and carbon monoxide, used in this study are of high purity grade (>99.9995 % and >99.998 % purity respectively) and obtained from Alphagaz (Air Liquide, France). The argon atom which is spherical and non-polar interacts in a non-specific manner with a surface, thus allowing a textural study of the adsorbent. The carbon monoxide molecule is non-spherical and polar, with both a dipole (0.39×10^{-3} Cm) and quadrupole moment (-12.3×10^{-40} Cm^2). It interacts in a specific manner with a surface, allowing the influence of the adsorbent field gradient to be investigated. Carbon monoxide is preferred to nitrogen (quadrupole moment of -5.0×10^{-40} Cm^2) as a "specific probe molecule" as it gives rise to stronger specific interactions due to its greater cumulative moment.

2.3 Experimental Techniques

The adsorption isotherms at 77 K are obtained via a volumetric technique employing a quasi-equilibrium (extremely slow and constant) method of adsorptive introduction [9]. This latter technique may by coupled with isothermal microcalorimetry [10] allowing a high resolution determination of the differential enthalpies of adsorption (or net differential enthalpies, if the enthalpy of vaporisation at the experimental temperature is deducted from the previous ones).

3 RESULTS & DISCUSSION

The enthalpy curve obtained for the adsorption of carbon monoxide on the B-MFI type zeolites calcined via Sample Controlled Thermal Analysis (SCTA) are shown in figure 1. These curves are similar to those obtained with Al- and Fe-MFI type zeolites with a homogeneous T(III) distribution [4].

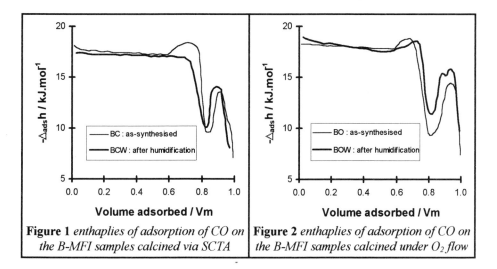

Figure 1 *enthaplies of adsorption of CO on the B-MFI samples calcined via SCTA*   **Figure 2** *enthaplies of adsorption of CO on the B-MFI samples calcined under O$_2$ flow*

The initial part of the curve up to V/Vm =0.7 is interpreted as the result of the decrease in adsorbate-adsorbent interactions and increase in adsorbate-adsorbate interactions. The peak in the curves at around V/Vm =0.9 is indicative of the fluid-like to solid-like transition [3]. The passage via the network fluid is not observed due to the diffuse nature of this transition when a significant amount of T(III) atoms are present in the zeolite. The humidification process seems not to significantly effect the enthalpy curve which would suggest that the zeolite is not modified.

Figure 2 shows the enthalpy curves obtained for the adsorption of carbon monoxide on the samples obtained via a traditional calcination protocol. The initial part of the curves are similar to those in figure 1 and the values in this region are also very similar. The interesting point in this figure is the splitting of the peak at around V/Vm =0.9 for the humidified sample. This has also been seen before for Fe-MFI samples with a significant amount of zoning [4] suggesting that the adsorbate phase transition occurs at slightly different pressures in the silica rich part and iron rich part of the crystals. This would seem surprising in the present study as the as-synthesised sample has a homogeneous boron distribution and the sample calcined via SCTA does not show such behaviour. From the results in figure 2 however, it is certain that humidification after traditional calcination does modify the samples microporous surface. Carbon monoxide adsorption suggests that this modification could be either of a chemical or textural nature. To explore this further, the adsorption of argon is compared and shown in figure 3.

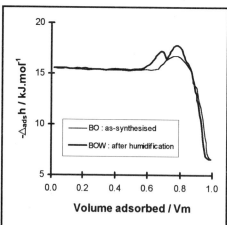

Figure 3 *enthaplies of adsorption of Ar on the B-MFI samples calcined under O₂ flow*

The enthalpy curves obtained for argon on the samples obtained via a traditional calcination protocol shown in figure 3 are rather interesting. The as-synthesised sample shows a curve which is very similar to those found in the past for Al and Fe MFI type zeolites [2]. The initial part of the curve shows a decrease in adsorbate - adsorbent interactions and increase in adsorbate - adsorbate interactions. The increase in enthalpy at around V/Vm = 0.8 is indicative of the fluid-like to solid-like transition.

The sample treated in a humid atmosphere shows a significant splitting of the peak in enthalpy curve at around V/Vm = 0.8. This is the first time to the authors' knowledge that such behaviour has been shown and is indicative of a textural modification of the pore surface. This splitting of the calorimetric curve could be explained by a "local" deboronation, for example in the rim portion of the zeolites, occurring during the humidifcation prior to outgassing which probably leads to a textural modification. This may be a local deboronation producing zones with a minute "roughness", able to slightly change the equilibrium pressure and/or coverage requested to produce the fluid-like to solid-like transition of the argon (or carbon monoxide) adsorbed on these zones.

Such a behaviour is not seen for the adsorption of argon on the samples treated via SCTA. This highlights the interest of SCTA for the preparation of high quality model zeolite samples.

To explain the unusual behaviour seen for the humidified samples calcined traditionally, it is known that at high temperatures, the framework boron in H-[B]-MFI is mainly trigonally co-ordinated and is very susceptible to water attack. In the presence of oxygen, water will be formed by oxidation of the template which leads to unstable B(OH)x species while calcination in an inert atmosphere preserves the stable tetrahedral/saturated co-ordination [11]. One may also imagine that the partial extraction of boron which leads to the formation of silanol nests is accompanied by a textural modification : the pore channel might be slightly larger within definite areas of the crystals.

CONCLUSION

Isothermal microcalorimetry at 77K coupled with the adsorption of argon and carbon monoxide is shown to be a preferment tool for the detection of anomalies in the surface of B MFI type zeolites.

A B MFI sample synthesised with a homogeneous boron distribution and calcined under oxygen atmosphere is shown to be partially deboronated. This "local" deboronation was shown by the splitting of the enthalpy peak in the fluid-like to solid-like adsorbate transition region for the adsorption of carbon monoxide. For the first time, a 'textural zoning' (geometric size of the extraction sites) could be observed by argon adsorption at 77 K.

REFERENCES

[1] U. Müller, H. Reichert, E. Robens, K. K. Unger, Y. Grillet, F. Rouquerol, J. Rouquerol, D. Pan & A. Mersmann, *Fresenius Z. Anal. Chem.* 1989, **39**, 433.

[2] P. L. Llewellyn, J.-P. Coulomb, Y. Grillet, J. Patarin, H. Reichert, G. Andre & J. Rouquerol, *Langmuir* 1993, **9**, 1846.

[3] P. L. Llewellyn, J.-P. Coulomb, Y. Grillet, J. Patarin, H. Lauter, & J. Rouquerol, *Langmuir* 1993, **9**, 1852.

[4] P. L. Llewellyn, Y. Grillet & J. Rouquerol, *Langmuir* 1994, **10**, 570.

[5] Y. Grillet, P. L. Llewellyn, M. B. Kenny, F. Rouquerol & J. Rouquerol, *Pure & Appl. Chem.* 1993, **65**, 2157.

[6] M. Soulard, S. Bilger, H. Kessler & J.-L. Guth, *Zeolites* 1987, **7**, 463.

[7] J.-L. Guth, H. Kessler & R. Wey, in *'Proc. 7th. Int. Zeolite Conf. Tokyo 1986'*, (Eds. Y. Murakami, A. Iijima & J.W. Ward) Elsevier: Amsterdam, 1986, p121.

[8] J. Rouquerol, *Thermochimica Acta*, 1989 **144**, 209.

[9] J. Rouquerol, F. Rouquerol, Y. Grillet & R. J. Ward, *In 'Characterisation of Porous Solids'*; (Eds. K.K. Unger, J. Rouquerol, K. S. W.,Sing, & H. Kral) Elsevier: Amsterdam, 1988, p67.

[10] J. Rouquerol, *Thermochimica Acta*, 1985, **96**, 377.

[11] R. de Ruiter, A. P. M. Kentgens, J. Grootendorst, J. C. Jansen & H. van Bekkum, *Zeolites* 1993, **13**, 128.

A SIMPLE ACCELERATED PROCEDURES FOR CARBON TETRACHLORIDE ACTIVITY AND RETENTIVITY IN ACTIVE CARBON.

A.M. Zakareya

Occupational Health Department
High Institute of Public Health
Alexandria University

1 INTRODUCTION

Active carbons used in gas and vapor systems are normally evaluated by tests based on the adsorption and desorption of carbon tetrachloride. The adsorption stage gives a value designated as the carbon tetrachloride activity, which is basically a measure of pore volume of active carbon. The desorption stage furnishes a value termed the carbon tetrachloride retentivity, which is a measure of irreversible adsorption.

The activity of carbon sample is measured as weight percent pickup, at equilibrium, of carbon tetrachloride from a dry air stream saturated with the vapor of carbon tetrachloride. The entire procedures are standardized and described by the American society for Testing and Materials (ASTM) D3467-76.[1] On the other hand, the retentivity indicated the weight percent of CCl_4 retained by carbon sample first saturated by CCl_4 and then subjected to air lowing for 6 hours.[2]

The present work objective is to outline a simple accelerated routine procedure to determine carbon tetrachloride activity and retentivity of active carbon samples.

2 EXPERIMENTAL

Four granular samples of Egyptian activated coals, sized averagely 2.33 mm, have been used in the present study. The tested sample is packed in a teflon tube (i.d. 11 mm) at a height of 50 mm, and a pressure drop of 0.6 ± 0.05 inch of water at 1l/min is maintained. Analytical reagent carbon tetrachloride (BDH) is injected into the sample by a microsyringe (0-100 µl).

The test assembly is shown in Figure (1). It is a closed system, through which a limited volume of air (5.7 l) is circulated. Its main constituents are a bellows pump and an infra red gas analyzer.(Miran 1A)

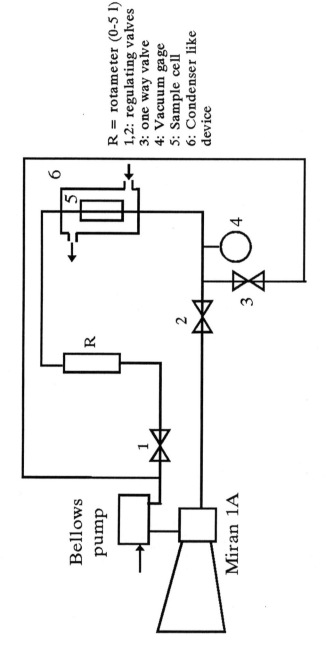

R = rotameter (0-5 l)
1,2: regulating valves
3: one way valve
4: Vacuum gage
5: Sample cell
6: Condenser like device

Figure (1): Schematic diagram of test assembly.

2.1 Activity Test - adsorption stage

The principle of the activity test is to inject sequentially 50 µl of carbon tetrachloride in circulating air at a flowrate of 1 l/m into the sample which is maintained at $25 \pm 1°C$. Sample saturation is realized when the precalibrated Miran 1A scale reading is that equivalent to 50 µl of CCl4. i.e., the injected volume is totally unadsorbed by the sample. Carbon tetrachloride activity in accordance to this method is computed from the following formula:

$$CCl_4 \ \text{Activity(A)} = \left[\frac{(B - 50) \ XC \ X \ 10^{-3}}{D} \right] X \ 100$$

where B= Total volume injected in µl
       C= Specific gravity of CCl4 at 25°C (1.594)
       D= Carbon sample weight in grams

In case, low activity carbon is saturated by less than 50 µl, the proportion injected is decreased to 10 µl with the same sequences.

2.2 Retentivity test-desorption stage

The main steps involved in the retentivity test, after Miran 1A is zeroed, are:
i.   The air flow is bypassed at a flowrate of 5l/min to reach the maximum attainable vacuum which might be 400 ml of mercury, at the same time, the water bath temperature is raised to 95°C.
ii. The air flow is switched back to pass through the sample for 30 minutes, after which the scale reading of Miran 1A is recorded. This reading represented the desorbed amount of the sample. The retentivity is computed by the following simple formula:

$$CCl_4 \ \text{Retentivity} = \left[A - \frac{E}{D} \right] X \ 100$$

where E = adsorbed amount of CCl4 in grams.

3 RESULTS AND DISCUSSION

Results of activity and retentivity tests, under the conditions mentioned earlier, for four Egyptian activated coals (S1-S4) are presented in Table 1 for each sample, 5 test runs have been made.

Table 1: CCl_4 Activity And Retentivity Results Mean % ± SD

Sample No.	Activity	Retentivity
S1	45.32±1.8	94.6 ±2.5
S2	27.06±1.19	93.16 ±1.03
S3	29.32±0.55	79.3 ±2.05
S4	44 ±2.86	88.56 ± 2.37

The results presented in table 1 are relative results, i.e., S1, having the highest activity and retentivity values, is the best among the four samples; however, there is no absolute measure to evaluate the four samples. There are three alternatives to solve this problem, first to accept the present procedures as a comparative relative test, second, to check a lot of samples having varying activities and retentivities measured by accepted standared methodes, e.g., ASTM D3467-76, and try to find a correlation, third to take samples having 100% activity by ASTM test to pass through the entire present procdures to get the corresponding value to which other values would be referred.

A brief comparison between the present activity test and that of ASTM D3467-76 is presented in Table 2.

Table 2: Comparison between the present activity test and ASTM test.

Parameter	Present test	ASTM Test
CCl4 concentration (ppm)	12,658	39,746
Flow rate (l/min)	1	1.67
Carbon bed volume (Cm3)	4.75	16.7
Handled volume of air (l)	5.7	50 -
Superficial velocity (m/min)	10.5	10

The main differences between the present test and that of ASTM are the less volume handled and the lower concentration of CCl4 encountered which makes it somewhat safer. However it needs a detector (MIRAN 1A) which is not the case with ASTM method.

CONCLUSION

The presented procedures for the determination of CCl4 activity and retentivity seem to be simple, relatively safe and accelerated. The two tests could be done by the same set, where activity test might take few minutes, while retentivity test about 35 minutes. However, it's hard to say that the present activity procedures are comparable to those of ASTM, unless a definite correlation is found between the two methods.

REFERENCES

1. Annual Book of ASTM Standards, American Society for Testing and Materials, Part 30, Test Method D3467-76, Philadelphia, PA, 1979.

2. W. Hassler, "Activated Carbon", MacGrow Hill, USA, 1978, Part VI, P. 306.

SYNTHESIS AND MECHANISM OF SWOLLEN MCM-41

Peter J. Branton, Julianne Dougherty, Gordon Lockhart and John W. White

The Research School of Chemistry, The Australian National University, Canberra ACT 0200, Australia

1 INTRODUCTION

The disclosure by Mobil scientists in 1992[1,2] of a new family of highly uniform mesoporous silicates and aluminosilicates has generated considerable interest. It extends the range of molecular sieve materials into the large pore region. One member of this family, MCM-41, consists of an hexagonal arrangement of cylindrical pores, adjustable in size from *ca.* 2 to 10 nm diameter. In general, pore size can be increased by increasing the alkyl chain length of the template surfactant molecule or by the addition of an organic auxilliary molecule such as mesitylene. In this case, the increase in pore size is due to 'swelling' through the solubilization of these molecules in the hydrophobic region of the surfactant template.

The work reported here concentrates on the synthesis and mechanism of formation of larger pore size MCM-41 through swelling effects by the addition of auxilliary organic molecules. In 1995, Edler and White[3] reported the room temperature formation of siliceous MCM-41. The synthesis gel for this system was found to be highly ordered at room temperature and able to retain its structure upon drying and calcination. Using the same start system, swelling effects were determined using mesitylene, dodecane and octadecane as the auxiliary molecules. The process was followed using Small Angle X-ray Scattering (SAXS) and for the final calcined materials, adsorption isotherm measurements. The effect of temperature and time on the as-synthesized gels were studied using *in situ* SAXS measurements.

2 EXPERIMENTAL

Samples of MCM-41 were synthesized using the method of Edler and White[3]; 8.2 g of cetyltrimethylammonium bromide (CTAB) was dissolved in 25 g of water at *ca.* 80°C. Mesitylene, dodecane or octadecane swelling agent was added directly to the CTAB solution so as to obtain a mole fraction of CTAB:Swelling agent of 1:0.83. This mixture was added with rapid stirring to 9.5 g of a sodium silicate solution (~14% NaOH, ~27% SiO_2) in 20 g of water and acidified with 0.6 g of sulphuric acid. The resulting gel was stirred for 15 min, after which time a further 10 g of water was added and stirring continued for another 30 min.

The small angle X-ray scattering machine used in the present work has been described elsewhere.[4] Samples were contained in 1.5 mm glass capillaries. For dried powder samples, measurements were performed at room temperature. Capillaries containing as-synthesized gels were sealed and SAXS measurements recorded immediately at room temperature and at 50°C, 75°C and 100°C for time periods up to 20 h. Temperatures were accurate to ±0.2°C.

Nitrogen isotherms were measured at 77 K using a conventional manual volumetric technique. Outgassing temperatures of 120°C were used prior to isotherm measurements and outgassing continued until a vacuum better than 2×10^{-4} Torr was achieved.

3 RESULTS AND DISCUSSION

All d-spacing results as determined from SAXS measurements are summarized in Table 1. The mole fraction of CTAB : Swelling agent used in the present study was equal to one used by Beck *et al*[2]. They obtained a pore diameter of 65 Å in the calcined material (as determined by argon physisorption) from an aluminosilicate whose pore size had been increased by the addition of mesitylene. Dodecane was found to give the largest d-spacing followed by mesitylene and octadecane. It has been noted elsewhere that addition of the swelling agent directly to a solution of surfactant prior to mixing with silicate solution gives a larger d-spacing than when the swelling agent is added to the gel as the final ingredient.[2]

Table 1   *SAXS d-spacing measurements*

Swelling Agent	*SAXS*	*d-spacing / Å*		
	Plain	Mesitylene	Dodecane	Octadecane
as-syn gel	43	54	66	46
Dried as-syn gel	39	42	54	-
Gel heated at 75°C and dried	42	44	66	-
Gel heated at 75°C, dried and calcined	38	39	60	-
Gel heated at 100°C and dried	-	-	52 (very broad)	59
Gel heated at 100°C, dried and calcined	-	-	62 (very broad)	53

Octadecane is a solid at room temperature with a melting point of 28°C. Following SAXS *in situ* heat treatment of the gel at 100°C, the octadecane melts and is able to dissolve in the hydrophobic region of the CTAB surfactant, the d-spacing increasing from 46 Å to 52 Å after 30 mins, 53 Å after 60 mins, 57 Å after 4 h with a maximum value of 58 Å measured after 18 h of heat treatment.

The pore size distribution was greatest for the larger pores obtained using dodecane. The effect of stirring was investigated to see whether or not some of the aggregation of surfactant rods used as template was altered to give a different distribution of hexagonal mesopores. Another gel was prepared using the same conditions as before, this time stirring more slowly. The SAXS pattern of the resulting gel again yielded a broad peak but with maxima at 52 and 64 Å, showing that efficient mixing is crucial to a good single product and that the aggregation of surfactant rods is quite sturdy. Small peaks in the SAXS patterns of some gels with a d-spacing of 26 Å were due to the lamellar crystalline form of CTAB which was present in excess.

On washing the as-syn gels with warm water (to remove excess CTAB) and drying, a decrease in d-spacing was observed in all cases, being most obvious when using mesitylene. This is evidence that the swollen materials are not fully stable at room temperature, in contrast to the same system using no swelling agent.

The effect of heat on the as-syn gels was investigated in order to try and retain the enlarged structures upon gel drying. The effect of heat was followed as a function of time using *in situ* SAXS measurements. For the MCM-41 gel swollen with mesitylene, no change in peak intensity or d-spacing was observed on heating at 50°C. The effect of heat at 75°C is shown in Fig.1. Here the peak at 54 Å decreases in intensity and very slightly in d-spacing, at the same time a new peak appears at 43 Å. Only one broad peak with d-spacing 43 Å is observed after 6 h heat which then sharpens upon longer heating times. The same result is observed for heating at 100°C, the process being accelerated. Gel washing with warm water and drying following heat treatment at 75°C for 20 h shows the retention of the peak at 44 Å with some evidence of a smaller less defined peak overlapping at *ca*. 40 Å. This is attributed to "standard" non-swollen MCM-41. Calcination at 450°C for 4 h reduces the d-spacing further and any effect due to swelling is lost. A less well-defined material than the standard MCM-41 is the final product. Calcination of all materials caused peak broadening.

Figure 1   *Effect of Heat at 100°C on MCM-41 swollen with Mesitylene*

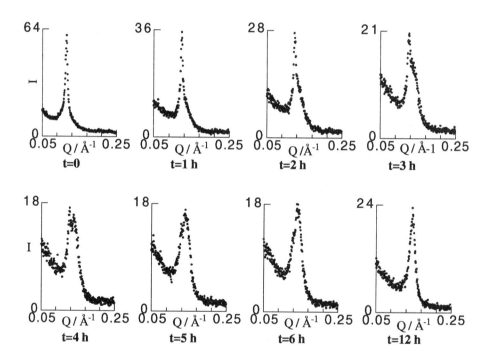

The adsorption isotherm of nitrogen at 77 K on the mesitylene swollen MCM-41 is shown in Fig.2 and is of Type IV in the IUPAC classification[5] of isotherms with reversible hysteresis. The BET surface area and all pore characteristics derived from the isotherm are shown in Table 2. Despite the questionable validity of the Kelvin equation for mesopores with diameter of *ca*. 4 nm, a pore diameter range, d_p, of 2.4-3.5 nm was calculated assuming hemispherical meniscus and zero contact angle with allowance made for multilayer adsorption on the pore walls.[6] This result is in good agreement with the SAXS measurement. The adsorption isotherm of nitrogen at 77 K on the standard MCM-41 (heating the as-syn gel at 75°C for 20 h, warm water washing and drying, and calcination at 450°C for 4 h) is also shown in Fig.2 as a comparison. Surface areas, S(BET), have been determined from a BET analysis and by assuming the molecular area of nitrogen to be 0.162 nm^2 in the completed monolayer. The range of mesopore filling has been denoted as P/P$_0$. Values of the total mesopore volume, V_p, have been obtained from the volumes adsorbed at P/P$_0$=0.95, by assuming that the pores have been filled with condensed liquid adsorbate.[6] (For materials with larger more ill-defined external surface areas, values of V_p have been taken at P/P$_0$=0.90).

Figure 2 *Adsorption isotherm for nitrogen at 77 K on plain MCM-41 and MCM-41 swollen with mesitylene*

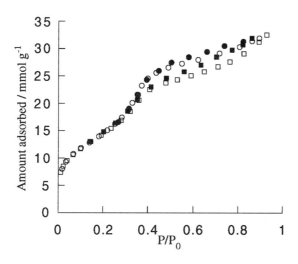

Circles denote plain MCM-41. Squares deonte MCM-41 swollen with mesitylene. Open symbols denote adsorption. Filled symbols denote desorption.

Table 2 *Adsorption of nitrogen at 77 K on calcined MCM-41 materials.*

Adsorbent	$S(BET)$/m^{2}g^{-1}	P/P_0	d_p/nm	V_p/cm^{3}g^{-1}
Standard	1215	0.27-0.41	2.4-3.6	1.12*
Mesitylene swollen MCM-41	1215	0.26-0.41	2.4-3.5	1.08*
Dodecane swollen MCM-41 (75°C heat prior to calcination)	1005	0.42-0.67	3.4-6.9	1.32
Dodecane swollen MCM-41 (100°C heat prior to calcination)	940	0.42-0.67	3.4-7.0	1.17
Octadecane swollen MCM-41	1020	0.43-0.65	3.5-6.7	1.19

* V_p calculated at P/P$_0$=0.90.

For the MCM-41 swollen with dodecane, heating at 75°C caused no change in the gel structure. Subsequent water washing and drying, following heat treatment at 75°C for 20 h, showed only a small decrease in d-spacing. On calcination at 450°C for 4 h, a product with pore diameter of 60 Å is produced. SAXS results for the heat treatment as a function of time on the as-syn gel at 100°C are shown in Fig.3, in which the peak at 66 Å broadens and an additional peak at 38 Å (presumably due to standard MCM-41) appears. After 10 h heating, another much smaller peak is evident at 42 Å. Water washing and drying, following heating at 100°C for 20 h causes a larger fall in d-spacing than with heat at 75°C for 20 h and a broader peak in the SAXS pattern with evidence again of more than one pore size distribution. The result of calcination is very similar to that following heat treatment at 75°C with a broad distribution of mesopores.

Nitrogen isotherms on the calcined materials using dodecane and gel heat treatment at 75 and 100°C for 20 h are virtually indistinguishable, as can be seen in Fig.4. A hysteresis loop is now present and the parameters obtained from isotherm analyses are shown in Table 2. The major difference is in the larger external area for the material following heat treatment at 75°C, and hence the total pore volume, V_p.

Figure 3   *Effect of Heat at 100°C on MCM-41 swollen with Dodecane*

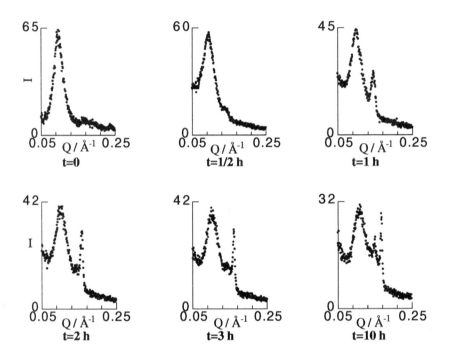

The use of octadecane as swelling agent produced the most crystalline large pore material with several visible orders of reflection. Some 'tailing' in the (100) hexagonal phase reflection peak of the dried and calcined sample occurred and was probably caused by the time delay between gel heating and washing and drying, causing some of the octadecane to solidify again.

A sharper peak in the SAXS pattern was reflected by the narrower hysteresis loop in the nitrogen adsorption isotherm shown in Fig.5a. Isotherm parameters are shown in Table 2. The α_s-plot in Fig.5b (using non-porous hydroxylated silica as the reference material)[7] provides evidence that, despite the SAXS peak at 53 Å tailing towards smaller pore sizes, there is no evidence of micropores. The sharp step in the α_s-plot demonstrates a narrow pore size distribution, and the plateau at higher relative pressures indicates a small external area.

Figure 4 *Adsorption isotherm for nitrogen at 77 K on MCM-41 swollen with dodecane*

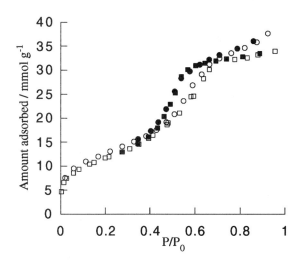

Circles denote gel heating at 75°C for 20 h prior to calcination. Squares denote heat at 100°C for 20 h prior to calcination. Open symbols denote adsorption. Filled symbols denote desorption.

Figure 5 *(a) Nitrogen adsorption isotherm at 77 K and (b) α_s plot for MCM-41 swollen with octadecane*

(a) **(b)**

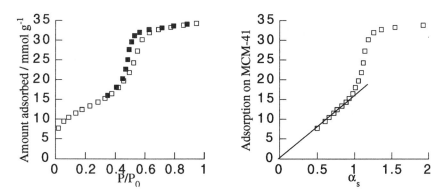

Open symbols denote adsorption, Filled symbols denote desorption.

4 CONCLUSIONS

Pore diameter increase in siliceous MCM-41 through swelling effects are highly dependent on the organic swelling molecule, the temperature and the heating time. Using mesitylene as the swelling agent, it was not possible to retain the increased pore diameter of the as-synthesized gel. Using both dodecane and octadecane, it was possible to obtain a calcined material with increased pore diameter. Dodecane produced the largest pore diameter with the broadest distribution of mesopores centred around 60 Å. The use of octadecane as swelling agent produced the most crystalline material with a pore diameter of 53 Å.

Acknowledgment

We thank Mr C. Tomkins, glass-blower in the Research School of Chemistry, for the construction of the adsorption apparatus.

References

1. C.T.Kresge, M.E.Leonowicz, W.J.Roth, J.C.Vartuli and J.S.Beck, *Nature*, 1992, **359**, 710.
2. J.S.Beck, J.C.Vartuli, W.J.Roth, M.E.Leonowicz, C.T.Kresge, K.D.Schmitt, C.T-W.Chu, D.H.Olson, E.W.Sheppard, S.B.McCullen, J.B.Higgins and J.L.Schlenker, *J.Am.Chem.Soc.*, 1992, **114**, 10834.
3. K.J.Edler and J.W.White, *J.Chem.Soc., Chem.Commun.*, 1995, 155.
4. M.Aldissi, S.Henderson, J.W.White and T.Zemb, Mat.Sci.Forum, 1988, **27/28**, 437.
5. K.S.W.Sing, D.H.Everett, R.A.W.Haul, L.Moscou, R.A.Pierotti, J.Rouquerol and T.Siemieniewska, *Pure Appl. Chem.*, 1985, **57**, 603.
6. S.J.Gregg and K.S.W.Sing, 'Adsorption, Surface Area and Porosity', Academic Press, New York, 1982, 2nd edn.
7. M.R.Bhambhani, P.A.Cutting, K.S.W.Sing and D.H.Turk, *J.Colloid Interface Sci.*, 1972, **38**, 109.

The Preparation and Analysis of Microporous Silica Membranes for Gas Separation

R.M. de Vos, A. Nijmeijer, K. Keizer, H. Verweij
Laboratory for Inorganic Materials Science, Department of Chemical Technology, University of Twente, PO Box 217, 7500 AE Enschede, The Netherlands

Abstract

Supported silica membranes are prepared by dipcoating. To control the processes from the support to the end product, pore-size measurements using different techniques have proven to be a very valuable tool. The supports have pore radii in the order of 70-80 nm and are characterized by mercury intrusion porosimetry, the γ-alumina intermediate layer has a pore diameter of 3.5 nm and is characterized by permporometry and the silica toplayer has a pore diameter of 0.5 nm and is characterized by physical adsorption and permeation measurements.

Key words: membrane, silica, mercury intrusion porosimetry, permporometry, physical adsorption, gas permeation.

1   Introduction

Microporous ceramic membranes, which have a pore radius typically smaller than 1 nm [1] are very attractive for gas separation processes due to their high temperature and chemical stability [2, 3]. The application of ceramic membranes in the so-called membrane reactors has also proven to be very promising [4-7]. The use of membrane reactors in chemical industry is mainly motivated by the equilibrium shift which is caused by selective permeation of reaction products. This leads to a higher conversion or to the same conversion at a lower temperature or pressure.

Sol-gel modification of mesoporous γ-alumina membranes (which have a pore radius between 1 nm and 25 nm [1]) with polymeric silica sols has proven to be a very succesful process for the preparation of microporous membranes [2, 8]. In this process a very thin toplayer of silica is dipcoated upon the mesoporous membrane. Due to the fact that these toplayers are so thin (typically 30-100 nm), the gas permeation rates through these membranes are very high. Gas separation factors are in the order of 300 for H_2-CH_4 and in the order of 10 for H_2-CO_2. To obtain these high separation factors, a very small pore-size distribution in the defect free toplayer is necessary, but to obtain such a homogeneous toplayer, a very well defined and homogeneous intermediate layer is necessary. Pore-size measurement techniques such as mercury intrusion porosimetry for the support, permporometry for the intermediate layer and gas adsorption techniques for the toplayer have proven to be very useful in the search to obtain more homogeneous membranes.

2   Preparation

The microporous membranes are prepared according to the following route. First an α-alumina support is prepared. There are two ways to prepare these supports. The first method is die pressing a commercial available α-alumina powder (PAI, Philips) isostatically at 0.1 MPa and afterwards sintering at 1300 °C. The other method is the so-called colloidal filtration method in which a colloidal suspension is made of pure alumina powder (AKP30, Sumitomo) in a nitric acid solution and the use of ultrasound. This suspension is filtered using a water jet pump, dried overnight at ambient temperature and fired at 1100 °C for 1 hour. After firing the supports are machined to the required dimensions and polished until a shiny surface is obtained.

The γ-alumina layer is prepared by dipcoating with a boehmite (γ-AlOOH) solution, which is prepared using the preparation method as described in refs. 8-10. In this method aluminium-tri-sec-butoxide (Janssen Chimica) is added drop-wise to water, which is, under vigorous stirring, maintained at 90 °C. After adding the alkoxide, the reaction mixture is kept at 90 °C under stirring, in order to remove the reaction product butanol. During the complete synthesis time the water/alkoxide ratio is maintained at 70:1. After this, the slurry is peptized with 1 N HNO_3 by refluxing for 20 hours. The resulting clear sol is very stable and has an opaque appearance. The pH of this sol is about 4 and the final AlOOH concentration is 1.0 M.

For the preparation of the layer, the boehmite sol is diluted with a PVA solution. The dipping time is a few seconds and the complete process is performed in a class 1000 cleanroom, to avoid inhomogeneities (defects) caused by dust particles.

The dipped supports are dried at 40 °C and 60% R.H. prior to calcination at 600 °C. After calcination the complete dipping process is repeated to repair any possible defects, which are left after the first dipping procedure. The thus obtained layer has a thickness of about 1 μm. See Figure 1.

Figure 1:   SEM picture of a support with γ-alumina intermediate layer.

On this intermediate layer a silica layer is dipcoated. The silica sol is prepared by an acid catalysed hydrolysis of tetra-ethyl-ortho-silicate (TEOS, Aldrich) in ethanol (Merck) [8, 11]. The sols are prepared by drop-wise addition of acidified water to the TEOS/ethanol solution, followed by refluxing for 3 hours at 80 °C under stirring. To obtain the dipping solution, the silica sol is diluted 19 times with ethanol. The supports with the intermediate γ-alumina layers are dipcoated with the dipping solution and a dipping time of a few seconds, after which the membranes are allowed to dry for a few seconds. The complete dipping process is carried out under class 1000 cleanroom conditions again. The membranes are calcined for 3 hours at 400 °C. This process is repeated once again to obtain a defect free silica toplayer with a thickness of about 30 nm. This silica layer is shown in Figure 2.

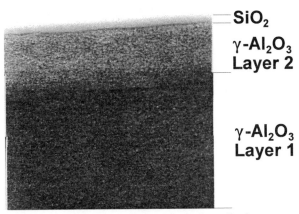

Figure 2: TEM picture of the silica toplayer with two γ-alumina intemediate layers

3   Pore-size analysis and permeation measurements

3.1   Mercury intrusion porosimetry

The pore size distribution of the used supports are measured by mercury intrusion porosimetry. A typical pore size distribution of a colloidal filtrated α-alumina support is given in Figure 3.

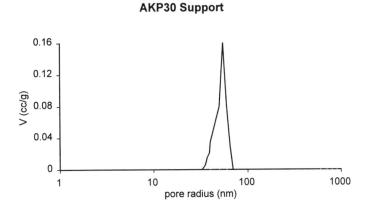

Figure 3:   Typical pore size distribution of a colloidal filtrated α-alumina support determined by mercury intrusion porosimetry

As can be seen from Figure 3, the mean pore radius of a colloidal filtrated AKP30 support is 70 nm with a rather small distribution. The porosity of these supports is typically about 35%. An isostatical pressed support has a mean pore radius of 80 nm and a porosity of 50%.

3.2   Permporometry

The pore size distribution of the γ-alumina intermediate layer is determined by permporometry. The advantage of this method is that only active pores in the materials are measured. In contrast to Mercury Intrusion Porometry and physical adsorption. It is very useful to test a few membranes of a batch to spot pinholes etc., before dipping the silica layer. A typical pore diameter of a γ-alumina intermediate layer is 3.5 nm. The permporometry measurements are performed on a home-made apparatus as described in [12]. In this method the oxygen permeation through the membrane is measured as function of the relative vapor pressure of cyclohexane. The experiment is performed in the desorption mode. At high pressures all the pores in the γ-alumina toplayer will be blocked by condensed cyclohexane. When the pressure is lowered, first the larger pores will be opened and oxygen can permeate through these pores. The amount of permeated oxygen is detected by a gas-chromatograph (Varian 3300). Since the relative vapor pressure corresponds to a certain Kelvin radius, a pore size distribution can be calculated from a plot of the oxygen permeation as a function of Kelvin radius. The Kelvin equation is given by:

$$\ln P_{rel} = -\frac{\gamma_s V_{mol}}{RT}\left(\frac{1}{r_{k,1}} + \frac{1}{r_{k,2}}\right)\cos\theta$$

P_{rel} is the relative vapour pressure of the condensable vapour, γ_s the surface tension of the liquid-vapour interface, V_{mol} the molar volume of the condensed vapour, R the gas constant, T the temperature, θ the contact angle which the liquid makes with the pore wall and $r_{k,1}$, $r_{k,2}$ the Kelvin radii, which are the radii of curvature of the vapour-liquid interface.

3.3   Physical adsorption

Pore-size measurements on supported silica membranes is not possible with the current techniques because of small amount of material which has to be determined. So the physical adsorption measurements have to be performed on unsupported silica. The unsupported silica was obtained by simply drying the dipping solution. The thus obtained dry gel was given the same heat treatment as the membranes. The physical adsorption measurements were performed on a Carlo Erba Sorptomatic 1990. The nitrogen adsorption isotherms of these microporous systems will be of type I [1]. For calculation of the pore-size from the isotherms, the modified Horváth-Kawazoe method was used. One has to keep in mind however, that nitrogen can interact with the hydroxyl-groups of the silica which can cause problems with interpreting the measurement data [13]. More information about these measurements can be found in [14]. From the measurements it follows that unsupported silica has a mean pore diameter of about 0.5 nm, which means that, when used in membranes, molecular sieving properties should be observed. These molecular sieving properties have been found in the gas permeation measurements, which will be discussed below. This indicates that, although non-supported systems were used for the measurements, the pore size determined for these systems are approaching the pore sizes which are found in the gas permeation measurements and which cause the observed molecular sieving properties.

3.4   Gas permeation

Gas permeation measurements were performed on a home built apparatus based on the dead end method under pressure control. In the dead end method no sweep gas is used at the permeation side of the membrane. The outlet is connected directly with the atmosphere or vacuum. A differential pressure tranducer (validyne) is used for determining the differential pressure over the membrane. The apparatus is described more extensively in [2, 8]. The basis of the measurement is to apply a certain pressure difference of a certain gas to the membrane and to measure the resulting flow. The permeation is now defined as the ratio of flux through the membrane and the pressure

difference accross the membrane. It is expressed in mol m^{-2} s^{-1} Pa^{-1}. More information about the mechanisms of gas transport through microporous systems can be found for example in [15].
A typical graph of the permeation of several gases at different temperatures for a state of the art silica membrane is given in Figure 4.

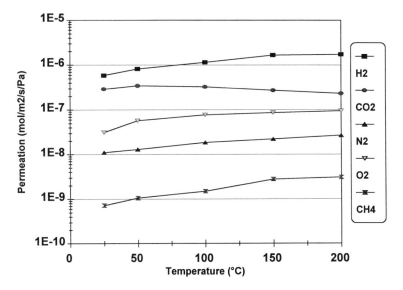

Figure 4:   Permeation of different gases at several temperatures. Measured at a standard silica membrane and a pressure difference of 1 bar.

From Figure 4 it can be seen that there is a significant difference in the permeation of CH_4 and the permeation of H_2. Recently permselectivities of more than 300 are found. The permselectivity of a material is simply defined as the flux of the first component devided by the flux of the second one. These high permselectivities indicate that the membranes exhibit molecular sieving properties. For these properties pore diameters in the range of 0.5 nm are necessary.

4   Discussion and conclusions

With the use of sol-gel techniques it is very good possible to prepare microporous asymmetric silica membranes which exhibit molecular sieving properties. To optimize the preparation method and for standard analysis of the intermediate products (i.e. support and support with intermediate layer) different porosity measurement techniques have proven to be a very valuable tool. With the use of these techniques also the degradation behaviour of the membranes under process conditions, for example steam reforming conditions (20-30 bar at 500-600 °C and in the presence of water vapor) can be followed. However because the physical adsorption techniques available for the analysis of microporous materials are not yet suitable for supported systems, there is still a rather big uncertainty in the determination of the pore sizes of these systems. A good method for measuring the pore sizes of supported microporous systems would be a huge improvement in the research on the preparation of these systems and the determination of their behaviour under severe process conditions. This knowledge is needed before the membrane separation processes will be applied more commonly in chemical industry and the use of membrane reactors can take a high flight.

5   References

[1]   K.S.W. Sing, D.H. Everett, R.A.W. Haul, L. Moscou, R.A. Pierotti, J. Rouqérol, T. Siemiewska *Pure Appl. Chem.,* **57** (1985) 603.

[2]   R.S.A. de Lange, J.H.A. Hekkink, K. Keizer, A.J. Burggraaf, *Microporous Mater.,* **4** (1995) 169.

[3]   R. Soria, *Catal. Today,* **25** (1995) 285.

[4]   J.N. Armor, *Catal. Today,* **25** (1995) 199.

[5]   S. Uemiya, N. Sato, H. Ando, T. Matsuda, E. Kikuchi, *Appl. Catal.,* **67** (1991) 223.

[6]   J. Zaman, A. Chakma, *J. Membrane Sci.,* **92** (1994) 1.

[7]   G. Saracco, G.F. Versteeg, W.P.M. van Swaaij, *J. Membrane Sci.,* **95** (1994) 105.

[8]   R.S.A. de Lange, Thesis University of Twente, 1993.

[9]   A.F.M. Leenaars, K. Keizer, A.J. Burggraaf, *J. Mater. Sci.* **19** (1984) 1077.

[10] A.F.M. Leenaars, K. Keizer, A.J. Burggraaf, *J. Colloid Interface Sci.,* **105** (1985) 27.

[11] W.J. Elferink, B.N. Nair, R.M. de Vos, K. Keizer, H. Verweij, *J. Colloid Interface Sci.,* **180** (1996) 127.

[12] G.Z. Cao, J. Meijerink, H.W. Brinkman, A.J. Burggraaf, *J. Membrane Sci.,* **83** (1993) 221.

[13] S. Kondo, "Colloidal Silicas", Chapter 1.4 in *Adsorption on New and Modified Inorganic Sorbents,* Studies in Surface Science and Catalysis vol. 99, A. Dabrowski, V.A. Tertykh eds., Elsevier Science B.V. (1996).

[14] R.S.A. de Lange, K. Keizer, A.J. Burggraaf, *J. Porous Mater.,* **1** (1995) 139-53.

[15] R.J. Uhlhorn, Thesis University of Twente, 1990.

STANDARDIZATION OF SURFACE GEOMETRY CHARACTERIZATION

E. ROBENS[1], K.-F. KREBS[2], K. MEYER[3]

[1] Institut für Anorganische und Analytische Chemie der Johannes Gutenberg-Universität, D-55099 Mainz (Germany)
[2] Merck KGaA, LAB CHROM Synthese, D-64271 Darmstadt (Germany)
[3] Bundesanstalt für Materialforschung und -prüfung (BAM), Zweiggelände Adlershof, Rudower Chaussee 5, D-12489 Berlin (Germany)

The most important parameters which characterize the surface of dispersed or porous solids are
* density
* specific surface area,
* specific pore volume,
* pore size distribution and
* particle size distribution.

Since our review[1] published in the proceedings of COPS II in Alicante; some new work on standardization of measuring methods has been done. Therefore, we present a supplement in the appended tables.

Standard procedures for adsorption measurements and for the characterization of surface area and pore size distribution are formulated in two IUPAC recommendations[2,3]. Guidelines for measurements have been published in the COPS III proceedings[4]. These documents provide a basis for pure and applied research. Besides, national and international standardization authorities issue official rules for instrumentations and measuring procedures which allow for the comparison of industrial products. A survey on British activities has been published recently[5] and German standards are collected in a paperback[6]. Future plans of British and German working groups are concerned with the determination of micropores. Different methods of particle counting and characterization are collected in a VDI manual[7]. An appreciable number of standards are concerned with sampling and division of the sample, which are not included in the appended table. We recommend to keep to these rules, because bad preparations may effect a remarkable measuring error. Basis for the choice of symbols and units are ISO 31 and ISO 1000. Correspondingly the IUPAC "Green Book"[8] may be used. In both, indices are not employed in elevated position to avoid confusion with exponents.

An important alteration in the list of standardization committees (Table 3) should be noted: The German group and the secretariat of ISO/TC 24/SC4: "Sizing methods other than sieving" are now incorported in the Normenausschuß Bau im DIN Deutsches Institut für Normung e.V., D-10772 Berlin. The ISO working group is engaged in elaborating methods for the surface characterization of dispersed materials and just issued an ISO standard on the BET method. The list of standards (Table 4) was actualized and addresses included. The list covers also standards which describe methods of other kind to characterize the surface structure. A great number of standards is concerned with ad- and absorption of water by various materials:

ASTM C 121, C 272, C 373, D 570, D 1815, D 2842, D 3285, D 3816/M, D 5455, E 946

DIN 1230-3, 1996-8, 4051, 7708-2/3, 7746-2, 16890, 16911, 16913-2, 16955, 16956, 16958, 18132, 18180, 19538, 40634-1, 52103, 52251-2, 52351, 52364, 52459, 52617, 53129, 53330, 53338/-2, 53433, 53434, 53495, 53184, 53923, 54540-4, 57291-268750, 68752,

EN 99, 121, 15 9, 176, 177, 178, 186-1/2, 187-1/2, 188, 532, 382-2, 2155-2, 2378, 20535

IEC 811-1-3

ISO 62, 535, 2417, 2508, 2896, 5635, 5637, 6783, 8361-1/2, 8787

LN (Luftfahrt-Norm) 29820, 65336

OENORM B 3006, B 3122/-2, 3234, 7013, 7017, 7018, 7735-1/2, 7873-7, 53495

SEV-ASE 3621-1-3

VDE 0291-2

With regard to reference materials and manufacturers, Table 1 and 2 of our former review[1] may be consulted. New porous/finely dispersed reference materials (CRM) have been developed in the German Federal Institute for Materials Research and Testing (BAM) in 1996. It was certified with respect to the specific surface area, specific pore volume and mean pore radius by means of the gas adsorption method using krypton and nitrogen[9]. The CRMs were produced in compliance with BCR guidelines[10], ISO guidelines[11] and are available by BAM[12]. A new project was created within the 4th framework of the BCR programme "Standards, Measurement and Testing" (SMT) 1994-1998: SMT4-CT 95-2025: Certification of pore size reference materials (1996-98). It is concerned on materials with micro- and mesopores.

AFNOR		**Association Française de Normalisation**, Tour Europe, F-92080 Paris la Défence Cedex, France
X 11-601	79	Tamisage et granulométrie - Détermination de l'aire massique ou volumique des poudres par perméabilimétrie - Méthode de Lea et Nurse
X 11-602	77	Détermination de l'aire massique des poudres par divers méthodes de perméabilimétrie à l'air
X-11-620	94	Détermination de la surface spécifique des solides par adsorption de gaz à l'aide de la méthode BET
X-11-621	75	Détermination de l'aire massique (surface spécifique) des poudres par adsorption de gaz - Méthode B.E.T.:Mésure volumétrique par adsorption d'azote à basse temperature
X 11-622	77	- Variantes de la méthode de base
X 11-630	81	Granulométre - Vocabulaire
X 11-632	83	- Expression des resultats experimentaux d'analyse granulométrique
X 11-634	88	- Caractérisation de la taille et de la forme des éléments d'une population granulaire
X 11-635	85	- Représentation des distributions granulométriques - Partie 1: Modèles de réference
X 11-636	85	- - Partie 2: Adjustement d'une courbe experimentale à un modèle de réference. Cas du tamisage
X 11-640	79	- Analyse granulométrique des poudres fines sur tamiseuse à depression d'air
X 11-642	82	- Tamisage en milieu liquide des poudres de granulométrie inférieure à 200 micromètres
X 11-660	83	- Analyse granulométrique par microscopie optique - Généralités sur le microscope
X 11-661	84	- Détermination de la taille des particules d'une poudre - Méthode par microscopie optique
X 11-666	84	- Analyse granulométrique des poudres - Méthode par diffraction

X 11-667		- Méthode optique par laser - Mesurage du temps de transition
X 11-670	89	- Analyse granulométrique de particules en suspension dans un électrolyte par utilisation du compteur à variation de résistance
X 11-680	80	- Triage par fluides - Analyse granulométrique par sédimentation par gravité dans un liquide
X 11-681	82	- Analyse granulométrique par sédimentation par gravité dans un liquide - Méthode de la pipette
X 11-682	84	- - Méthode par occultation - Photosédimentomètre
X 11-683	81	- Analyse granulométrique d'une poudre par sédimentation par gravité à hauteur variable dans un liquide - Méthode par mesure d'absorption de rayons X
X 11-684	85	- Analyse granulométrique par sédimentation cumulative dans un liquide immobile - Méthode de la balance de sédimentation
X 11-685	83	- Analyse granulométrique par sédimentation centrifuge dans un liquide immobile par rapport à l'axe de centrifugation
X 11-690	84	- Triage par gravité dans un fluide en mouvement (Lévigation - Elutration)
X 11-693	83	- Analyses granulométriques - Liquides de suspension et agents dispersants
X 11-695	87	- Caractérisation des séparations granulométriques
X 11-696	89	Granulométrie par analyse d'images

AIA		**Asbestos International Association**, 68 Gloucester Place, GB - London W1H 3HL, U.K.
RTM1	82	Reference method for the determination of the concentration of asbestos fibres
RTM2	84	Method airborne asbestos and other inorganic fibres by scanning electron microscopy

ASTM		**American Society for Testing and Materials**, 1916 Race Street, Philadelphia 3, PA, USA
C 115		Wagner turbidimeter. Dito: AASHTO T 98, ANSI A 1.7
C 204	68	Standard for the determination of the fineness of Portland cement by air flow. Blaine method. Dito: Federal Test Method Standard 158 + Method 2101, AASHTO T 153
C 373	88	Determination of apparent density, dry density, water absorption and porosity of burnt porcelain and flint ware
D 1193		Specification for Reagent Water
D 2355 T	65	Determination of isotherms of decolorization
D 3663	84	Test Method for Surface Area of Catalysts
D 3766	86	Definitions of Terms Relating to Catalysts and Catalysis
D 3906	85	Test Method for Relative Zeolite Diffraction Intensities
D 3908	88	Test Method for Hydrogen Chemisorption on Supported Platinum on Alumina Catalysts by Volumetric Vacuum Method
D 3942	85	Test Method for Determination of the Unit Cell Dimension of a Faujasite-Type Zeolite
D 4222	83	Test Method for Determination of Nitrogen Adsorption and Desorption Isotherms of Catalysts by Static Volumetric Measurements
D 4284	88	Standard Test Method for Determining Pore Volume Distribution of Catalysts by Mercury Intrusion Porosimetry
D 4365	85	Test Method for Determining Zeolite Area of a Catalyst
D 4567	86	Test Method for Single-Point Determination of the Specific Surface Area of Catalysts Using Nitrogen Adsorption by the Continuous Flow Method

| D 4641 | 88 | Practice for Calculation of Pore Size. Distributions of Catalysts from Nitrogen Desorption Isotherms |
| D 4824 | 88 | Test Method for Determination of Catalyst Acidity by Ammonia Chemisorption |

BSI   **British Standards Institution**, BS House, 2 Park Street, GB - London W1A 2BS / BSI Sales Department, Linford Wood, Milton Keynes MK14 6LE, U.K.

2955	58	Glossary of terms
3406		Methods for the determination of particle size distributions
3406-1	86	Guide to powder sampling
3406-2	84	Recommendations for gravitational liquid sedimentation methods for powders and suspensions
3406-3	83	Air elutriation methods
3406-4	85	Optical microscope methods
3406-5	83	- Part 5: Recommendations for electrical sensing zone method (the Coulter principle)
3406-5	85	- Part 6: Recommendations for centrifugal liquid sedimentation methods for powders and suspensions
3406-7	88	Recommendations for single particle light interaction methods
4359		Determination of the specific surface area of powders
4359-1	84	Recommendations for gas adsorption (BET) methods
4359-2	87	Recommended air permeability methods
4359-4	95	Recommendations for methods of determination of metal surface area using gas adsorption techniques
		Recommended methods for the evaluation of porosity and pore size distribution. - Part 1. Mercury porosimetry
		- Part 2. Gas adsorption
		- Part 3. Challenge test
		- Part 4. Liquid expulsion

DIN   **Deutsches Institut für Normung e.V.**, D-10772 Berlin, Germany

4760	82	Gestaltabweichung, Begriffe, Ordnungssystem
4761 ≡ ISO 8785	78	Oberflächencharakter; Geometrische Oberflächentextur-Merkmale, Begriffe, Kurzzeichen
4762 ≡ISO 4287-1	89	Oberflächenrauheit; Begriffe; Oberfläche und ihre Kenngrößen
4763	81	Stufung der Zahlenwerte für Rauheit
7726 ≡ ISO 1382	82	Schaumstoffe, Begriffe und Einteilung
51 005	96	Thermische Analyse (TA) - Begriffe
51 006	90	Themische Analyse (TA) - Thermogravimetrie (TG)
51 918	86	Bestimmung der Rohdichte nach der Auftriebsmethode und der offenen Porosität durch Imprägnieren mit Wasser
51 957		Scheindichte
52 102	88	Bestimmung von Dichte, Trockenrohdichte, Dichtigkeitsgrad und Gesamtporosität
53 108	75	Prüfung von Papier und Pappe: Bestimmung der Rauhigkeit nach Bendtsen
53 193		Scheindichte
66 100	78	Körnung; Korngrößen zur Kennzeichnung von Kornklassen und Korngruppen

| 725-10(pr) | | Bestimmung der Verdichtungseigenschaften |
| 993-1 | 95 | Prüfverfahren für dichte geformte feuerfeste Erzeugnisse. - Bestimmung der Rohdichte, offene Porosität und Gesamtporosität |

GOST R   **Committee of the Russian Federation for Standardization, Metrologiy and Certification,** Leninsky Prospekt 9, Moskva 117049, Russia

| | 79 | Density, specific surface area (BET) and pore size distribution (BJH, Dollimore and Heal) |

ISO   **International Organization for Standardization,** Central Secretariat, 1 rue de Varembé, BP 56, CH- 1211 Genève 20

1382	82	Cellular materials; definitions of terms and classification
1953	72	Hard coal - Size analysis
4287-1		Surface roughness; terminology; surface and its parameters
8785	94	Surface character; geometrical characteristics of surface texture terms, definitions, symbols
9276	90	Representation of results of particel size analysis. - Part 1: Graphical representation
9277	96	Determination of specific surface area by BET gas adsorption method
10070	89	Metallic powders. - Determination of envelope-specific surface area from measurements of the permeability to air of a powder bed under steady-state flow conditions
10076	89	Determination of particle size distribution by gravitational sedimentation in a liquid and attenuation measurement

IUPAC   **International Union of Pure and Applied Chemistry,** Bank, 2, Pound Way, GB - Oxford, U.K.

Appendix II	72	Manual of symbols and terminology for physico-chemical quantities and units
Part I		Terminology and symbols in colloid and surface chemistry
Part II	76	Terminology in heterogeneous catalysis
	85	Reporting physisorption data for gas/solid systems with special reference to the determination of surface area and porosity. *Pure and Applied Chemistry* 57 (1985) 4, 603-619
	86	Reporting data on adsorption from solution at the solid/solution interface. *Pure and Applied Chemistry* 58 (1986) 968-984
	94	Recommendations for the characterization of porous solids. *Pure and Applied Chemistry* (8.1994)

SEV-ASE   **Schweizerischer Elektrotechnischer Verein,** Seefeldstrasse 301, CH - 8034 Zürich, Switzerland

| 3621-1-3 | 91 | Allgemeine Prüfung für Isolier- und Mantelwerkstoffee für Kabel und isoliere Leitungen; Teil 1. Allgemeine Prüfverfahren; Hauptabschnitt 3: Prüfverfahren zur Dichtebestimmung; Wasseraufnahmeprüfung; Schrumpfungsprüfng |

VDI   **Verein Deutscher Ingenieure e.V.,** Postfach 101045, D - 40001 Düsseldorf, Germany

| 2031 | 62 | Feinheitsbestimmungen an technischen Stäuben |

References

1. E. Robens, K.-F. Krebs: "Standardisation, reference materials and comparative measurements for surface area and pore characterisation". In: F. Rodriguez-Reinoso, J. Rouquerol, K.S.W. Sing, K.K. Unger (eds.): *Characterization of Porous Solids II*. Elsevier, Amsterdam 1991, p. 133-140.
2. IUPAC Recommendations 1984: Sing, K.S.W., Everett, D.H., Haul, R.A.W., Moscou, L., Pierotti, R.A., Rouquérol, J., Siemieniewska, T., "Reporting Physisorption Data for Gas/Solid Systems with Special Reference to the Determination of Surface Area and Porosity", *Pure & Appl. Chem.* 57 (1985) 4, 603-619.
3. IUPAC Recommendations 1994: Rouquérol, J., Avnir, D.; Fairbridge, C.W.; Everett, D.H.; Haynes, J.H.; Pernicone, N.; Ramsay, J.D.F.; Sing, K.S.W., Unger, K.K.: "Recommendations for the Characterization of Porous Solids", *Pure & Appl. Chem.* 66 (1994) 8, 1739-1785.
4. J. Rouquerol, D. Avnir, D.H. Everett, C. Fairbridge, M. Haynes, N. Pernicone, J.D.F. Ramsay, K.S.W. Sing, K.K. Unger: Guidelines for the characterizatiioon of porous solids. In: J. Rouquerol, F. Rodriguez-Reinoso, K.S.W. Sing, K.K. Unger (eds.): *Characterization of Porous Solids III*. Elsevier, Amsterdam 1994, p. 1-9.
5. R.W. Lines: British standards for particle, surface and pore characterisation. In: N.G. Stanley-Wood, R.W. Lines (eds.): *Particle Size Analysis*. Royal Society of Chemistry, Cambridge 1992, p. 40-47.
6. DIN-Taschenbuch 133: *Partikelmeßtechnik*, 3. ed., Beuth, Berlin 1990.
7. *VDI-Handbuch Reinhaltung der Luft*, VDI Verlag Düsseldorf 1984.
8. I. Mills, T. Cvitas, K. Homann, N,. Kallay, K. Kuchitsu (eds.): Quantities, Units and Symbols in Physical Chemistry. Blackwell, Oxford 1989.
9. K. Meyer, P. Klobes, B. Röhl-Kuhn: Certification of reference material with special emphasis on porous solids. *Cryst. Res. Technol.* 32 (1997) 1, 173-183.
10. BCR/48/93: *Guidelines for the Production and Certification of BCR Reference Materials*, Brussels, 1994.
11. *ISO Guide 30*: Terms & definitions used in connection with reference materials, 1992.
    ISO Guide 31: Contents of certificates of reference materials, 1981.
    ISO Guide 32 (draft): Calibration of chemical analysis and use of certified reference materials, 1994.
    ISO Guide 33: Uses of certified reference materials, 1989.
    ISO Guide 34: Quality system guidelines for the production of reference materials, 1994.
    ISO Guide 35: Certification of reference materials - General and statistical principles, 1989.
12. Bundesanstalt für Materialforschung und -prüfung (BAM), Unter den Eichen 87, D - 12205 Berlin.

CHARACTERISATION OF THE POROSITY OF 18th CENTURY PORTUGUESE DECORATIVE TILES

M.M.L. Ribeiro Carrott, J.L. Farinha Antunes and P.J.M. Carrott

Departamento de Química, Universidade de Évora,
Colégio Luís António Verney, Rua Romão Romalho, 39,
7000 Évora, PORTUGAL.

Abstract

In order to obtain more quantitative information on the role of porosity in the degradation and conservation of tiles, nitrogen adsorption isotherms at 77K and mercury intrusion-extrusion porosimetry curves have been determined on samples of 18th century tiles in different states of deterioration. The effect of subjecting the tiles to various modification procedures was also investigated.

1 INTRODUCTION

Decorative tiles, characteristically blue and, in earlier examples, yellow, have been used extensively in Portugal since the 15th century and are considered an important part of the country's artistic and architectural heritage. Unfortunately, many valuable decorative panels are now in very bad condition, having been subject to various physical and chemical degradation processes over the centuries, and there is now an urgent need to establish reliable methods for their characterisation and consolidation.

The most important causes of degradation of tiles, and also of other porous materials used in art, are those associated with the capillary transport of water into and out of the internal pore structure[1] which, amongst other things, results in a gradual attrition of the microstructure. More importantly, however, the transport of water introduces soluble salts, such as sulphates, nitrates and alkaline chlorides, into the pore network of the tiles. As a result of cyclic variations in humidity and dryness, the salts dissolve and recrystalize, thereby creating internal pressures high enough to break down the microstructure and, in this way, it might be expected that the overall porosity of the tile would increase. Therefore, one of the factors to take into account in the protective and conservative treatments of tiles should be the modification of the pore structure.

Consequently, in order to understand the mechanisms of deterioration it is fundamental to correlate the changes in pore structure with the degree of degradation. Also, the evaluation and prediction of the efficiency of the conservation treatment requires a knowledge of the resulting alteration of the pore structure. Therefore, one of the most important studies to be carried out in the field of conservation of tiles is the characterisation of pore structure.

In order to obtain information on the role of porosity, we present in this work results of the characterisation, by means of nitrogen adsorption at 77K and mercury porosimetry, of a number of 18th century tiles in different states of preservation, and investigate the effect of subjecting the tiles to various modification procedures, including treatment with a polymeric consolidant.

In comparison to studies carried out on other ceramic materials or on stone monuments, not much work has been done on tiles. Some characterisation studies of the chemical and mineralogical composition of tiles have already been carried out using X-ray diffraction, ion chromatography and scanning electron microscopy.[2-4] As far as we are aware, the work to be reported here is the first study to be published on the application of low temperature nitrogen adsorption and mercury porosimetry to decorative tiles. It is worthwhile noting that, although mercury porosimetry is a widely used technique in the field of conservation, nitrogen adsorption is practically absent, mainly due to the considerably small quantities of materials which can be analysed when samples are taken from works of art, thereby reducing the significance of the results.[1] As will be seen below, this is a very unfortunate state of affairs, as nitrogen adsorption appears to be a very promising technique in this field.

2 EXPERIMENTAL

The sample Friso is from an 18th century tile in very good condition, belonging to the collection of the Museu Nacional do Azulejo. Samples C13, G1, I32 and T32 correspond to very deteriorated tiles, from the same period as Friso, and taken from different walls of the inside of the church of S. Salvador, in Coimbra. The samples G1(250) and G1(350) were obtained by heating G1 under vacuum at 250 and 350 °C, respectively. Part of sample T32 was impregnated with a polyurethane and another part was immersed in distilled water for 24h at room temperature in order to remove soluble salts. The resulting samples are designated by T32(imp) e T32(des).

Nitrogen adsorption isotherms at 77K were determined on a conventional manual volumetric apparatus with a Datametrics Barocel 572 for pressure measurement and mercury porosimetry measurements were carried out using a Quantachrome Autoscan-500 (maximum pressure = 34 atm). In both techniques only the ceramic body (without the glaze) of the tile was used and samples were outgassed at 150°C to remove physisorbed water.

3 RESULTS AND DISCUSSION

3.1 Mercury Porosimetry

Examples of mercury intrusion-extrusion curves are given in Figure 1. In all cases, the overall shape of the curves is the same, there being extensive hysteresis and entrapment of mercury. For this reason, the extrusion curve is only shown in some cases. As has been previously pointed out,[5] the type of curve obtained is associated with a complex pore structure, in which percolation effects have a significant influence on the filling and emptying of the pores by mercury. With this type of system it is not possible to derive a pore size distribution from the pressure-volume curves.

We can, however, compare the effective pore volumes obtained from the volume of mercury intruded at 3.5 MPa. The values are given in the Table and it can be seen that there is little difference between the untreated tiles, with the exception of G1, which gives a

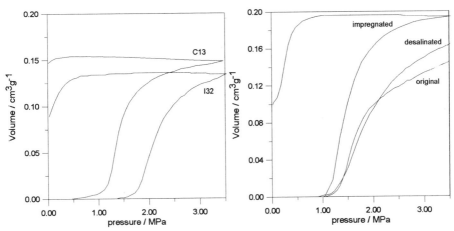

Figure 1 *Mercury intrusion extrusion curves determined on tiles I32 and C13 (left) and on T32 before and after modification (right)*

pore volume about twice that of the others. Desalination of T32 increases the pore volume although, as can be seen in Figure 1, little change in the shape of the curve is observed. Impregnation also increases the pore volume. However, it should be borne in mind that the presence of polymer will reduce the mean density of the tile and hence the increased value may be, at least partially, apparent, rather than real.

3.2 Nitrogen Adsorption on Non-modified Samples

In Figure 2 are shown the nitrogen isotherm and corresponding α_s plot (using silica as standard[6]) obtained on the sample Friso, apparently in very good condition, and the values obtained from the BET and α_s analysis of the isotherm are given in the Table. The

Table *Results of mercury porosimetry and of BET and α_S Analysis of Nitrogen Isotherms on 18th century Portuguese Tiles*

Sample	Nitrogen adsorption			Hg por.
	A_S (BET) m^2g^{-1}	C_{BET}	A_S (α_s) m^2g^{-1}	v_p cm^3g^{-1}
Friso	5.8	313	5.8	0.13
C13	4.7	44	4.7	0.14
G1	6.4	149	6.5	0.30
G1(250)	7.1	163	7.3	nd
G1(350)	10.2	144	10.3	nd
I32	10.6	192	11.0	0.13
T32	14.5	126	14.7	0.13
T32(imp)	1.2	54	1.1	0.20
T32(des)	21.0	208	(19.0)*	0.16

*Value obtained from the linear region that doesn't go through the origin.

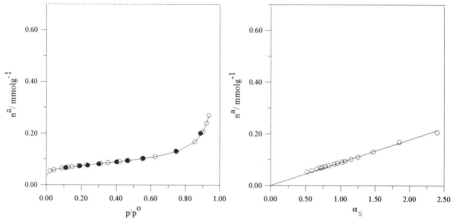

Figure 2   *Nitrogen isotherms and corresponding α_S plots on the sample Friso, in good state*

reversible type II isotherm and the linear α_s plot passing through the origin indicate the absence of microporosity and mesoporosity on this sample.

The results obtained with the deteriorated samples are clearly very different from the previous one. Nitrogen isotherms and corresponding α_s plots (using silica as standard) determined on the visibly deteriorated samples C13, G1, I32 and T32 are shown in Figure 3 and the results of the BET and α_s analyses of the isotherms are given in the Table. In all cases the shape of the isotherms and α_s plots indicate the presence of mesoporosity, associated with slits between the lamellar particles, as can be inferred from the type H3 hysteresis loop. The shape of the α_s plots and the reasonable agreement between the BET

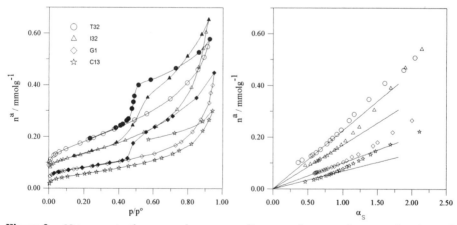

Figure 3   *Nitrogen isotherms and corresponding α_S plots on the very deteriorated samples C13, G1, I32 and T32.*

and α_s areas show the absence of microporosity, the small deviations from linearity at low pressures probably being due to differences in the chemical nature between these samples and the standard silica.

Although there are no significant differences between the 4 samples, it is seen that for C13 the C_{BET} value is considerably lower than those of the remaining samples, indicating a different chemical composition of the surface (in fact, this sample had a reddish colour while the others were creamish) but that doesn't affect the general textural characteristics in comparison with the others. It is also observed that T32 and I32 have higher specific surface areas than C13 and G1, indicating more mesoporosity, although the inverse was verified in terms of macroporosity (comparing the mercury porosimetry results of I32 and T32 with G1), which suggests a higher degree of exfoliation that can be associated with a more advanced state of degradation (in fact, these samples were taken from a wall that was definitely very damaged).

So, in general terms, these results suggest that the degradation of the tiles is associated with an increase in mesoporosity but not in macroporosity.

3.3 Nitrogen Adsorption on Modified Samples

Nitrogen isotherms and corresponding α_s plots (using silica as standard) determined on the samples G1(250) and G1(350), together with the unmodified G1, are shown in Figure 4 and those of samples T32(des) and T32(imp) are shown in Figure 5. The results of the BET and α_s analyses of the isotherms are given in the Table.

The thermal treatment of the sample G1 provokes a significant increase in the specific surface area, but the shapes of the isotherms and α_s plots remain analogous to the original, thereby indicating that there is no alteration in the type of mesoporosity nor generation of microporosity. The increase in surface area can be due to removal of chemisorbed water or decomposition of inorganic materials or even organic matter, present inside the slits between layers. In fact, we could see certain dark areas on the samples after heating, probably due to defects associated with non-stoichiometry or carbonisation of organic matter.

It can be seen from Figure 5 and Table that the process of desalination of T32 also

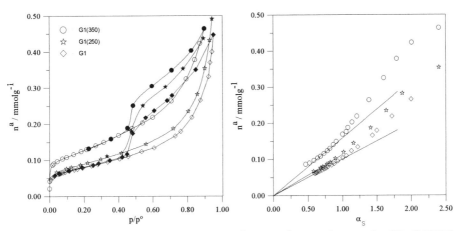

Figure 4   *Nitrogen isotherms and corresponding α_S plots on the samples G1, G1(250) and G1(350).*

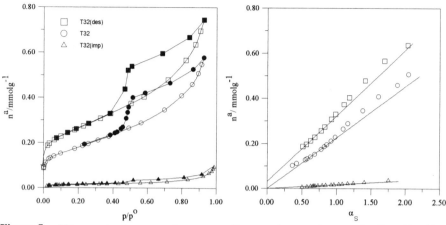

Figure 5   *Nitrogen isotherms and corresponding α_S plots on the samples T32, T32(des) and T32(imp)*

provokes an increase in surface area without alteration of the type of mesoporosity. Nevertheless, in this case some microporosity arises, as indicated by the small intercept of the α_s plot giving a value of $0,001$ cm^3(liq)g^{-1} for the micropore volume.

The fact that the thermal treatment and the immersion in water have similar consequences in the change of the texture is partially explained by the fact that both processes provoke the withdrawal of materials and consequently the opening of pores. Nevertheless, the mechanisms will be different, which could account for the appearance of microporosity only when desalinated by immersion in water. In this case, the micropores can result from the partial blocking of some of the wider pores, since the process occurs in a less controlled way than during heating under vacuum, which is done very slowly. Nevertheless, these are only possible hypotheses, as the originals were different samples.

In contrast, after impregnation with the polymer the surface area of T32 is considerably reduced, although there remains some mesoporosity as indicated by the persistence of the hysteresis loop. This could be due to an insufficient amount of polymer, or to the inaccessibility of the polymer to some narrower pores, which are accessible to nitrogen. The decrease of the C_{BET} value in relation to the non-impregnated sample, reflects the weakening of the interaction of nitrogen with the surface, expected by the surface coverage by the polymer.

4 CONCLUSIONS

The results presented have indicated that the tile in good condition is predominantly macroporous, but that degradation is associated with the development of a heterogeneous mesopore structure.

The mesoporosity is normally blocked, at least partially, by the presence of salts. However, when these salts are removed, by aqueous leaching or by thermal treatments, the pores become accessible and the surface area increases significantly.

The preliminary results also indicate that when the tile is impregnated with a polyurethane polymer the accessible mesoporosity is quite effectively reduced. Taking into account the role of porosity in the deterioration, the process looks promising for the conservation of tiles. Nevertheless, more studies need to be carried out. Particularly, it will be useful to test the effect of other consolidants and also of impregnation in different conditions, such as after desalination.

Finally, it is worthwhile remarking that, although both types of techniques should be used to properly investigate the role of pore structure, nitrogen adsorption appears to be more useful in the case of the tiles and, consequently, should be considered in the studies to be done in the field of conservation more often than so far.

Acknowledgements

The authors are grateful to the *Junta Nacional de Investigação Científica e Tecnológica* (JNICT), *Fundo Europeu para o Desenvolvimento Regional* (FEDER) and program PRAXIS XXI for financial support (contract n°. PRAXIS/2/2.1/HIS/13/94). We also thank Prof. Rui Namorado Rosa (*Physics Department, University of Évora*) for providing facilities for mercury porosimetry, Drs. João Castel-Branco Pereira and Manuela Malhoa Gomes (*Museu Nacional do Azulejo*) for provision of samples and Prof. J. Bordado (*Hoechst Portuguesa*) for the impregnation of a sample.

References

1   P. Rossi-Doria, in 'Principles and Applications of Pore Structural Characterization', ed. J.M. Haynes and P. Rossi-Doria, J.W. Arrowsmith, Bristol, 1985, p.441.

2   J.L. Farinha Antunes, MSc. Thesis, Instituto Superior Técnico, Lisbon, 1992.

3   J.L. Farinha Antunes, J. Costa Pessoa, M. Amaral Fortes and M.O. Figueiredo, *Studies in Conservation*, in press.

4   J.L. Farinha Antunes, M.O. Figueiredo, J. Costa Pessoa and M. Amaral Fortes, in 'The Ceramics Heritage', (Proc. World Ceramics Congress), ed. P. Vincenzini, Techner F.R.L., Faenza, 1995.

5   M. Day, I.B. Parker, J. Bell, R. Fletcher, J. Duffie, K.S.W. Sing, and D. Nicholson, in 'Characterization of Porous Solids III', ed. J. Rouquerol, F. Rodriguez-Reinoso, K.S.W. Sing and K.K. Unger, Elsevier, Amsterdam, 1994, p.225.

6   S.J. Gregg and K.S.W. Sing, 'Adsorption, Surface Area and Porosity', 2nd edition, Academic Press, London, 1982, p.93.

Author Index

Subject Index

Z